INSTRUCTOR'S RESOURCE GUIDE WITH COMPLETE SOLUTIONS

TO ACCOMPANY

UNDERSTANDING BASIC STATISTICS
THIRD EDITION
BRASE/BRASE

Charles Henry Brase
Regis University

Corrinne Pellillo Brase
Arapahoe Community College

Laurel Tech Integrated Publishing Services

HOUGHTON MIFFLIN COMPANY BOSTON NEW YORK

Sponsoring Editor: Lauren Schultz
Assistant Editor: Jennifer King
Editorial Assistant: Kasey McGarrigle
Senior Manufacturing Coordinator: Marie Barnes
Senior Marketing Manager: Ben Rivera

Printed in the U.S.A.

ISBN: 0-618-33363-0

123456789-BBS-07 06 05 04 03

Contents

Part IV: Complete Solutions

Part V: Test Item File Sample Questions And Answers

Part I

Teaching Hints

Suggestions for Using the Text

In writing this text, we have followed the premise that a good textbook must be more than just a repository of knowledge. A good textbook should be an agent interacting with the student to create a working knowledge of the subject. To help achieve this interaction, we have modified the traditional format, to encourage active student participation.

Each chapter begins with Preview Questions, which indicate the topics addressed in each section of the chapter. Next comes a Focus Problem that uses real-world data. The Focus Problems show the students the kinds of questions they can answer when they have mastered the material in the chapter. In fact, students are asked to solve each chapter's Focus Problem as soon as the concepts required for the solution have been introduced.

A special feature of this text are the Guided Exercises built into the reading material. These Guided Exercises, with their completely worked solutions, help the students focus on key concepts in the newly introduced material. The Section Problems reinforce student understanding and sometimes require the student to look at the concepts from a slightly different perspective than that presented in the section. Chapter Review problems are much more comprehensive. They require students to place each problem in the context of all they have learned in the chapter. Data Highlights at the end of each chapter ask students to look at data as presented in newspapers, magazines, and other media and then to apply relevant methods of interpretation. Finally, Linking Concept problems ask students to verbalize their skills and synthesize the material.

We believe that the approach from small-step Guided Exercises to Section Problems, to Chapter Review problems, to Data Highlights, to Linking Concepts will enable the instructor to use his or her class time in a very profitable way, going from specific mastery details to more comprehensive decision-making analysis.

Calculators and statistical computer software take much of the computational burden out of statistics. Many basic scientific calculators provide the mean and standard deviation. Those calculators that support two-variable statistics provide the coefficients of the least-squares line, the value of the correlation coefficient, and the predicted value of y for a given x. Graphing calculators sort the data, and many provide the least-squares line. Statistical software packages give full support for descriptive statistics and inferential statistics. Students benefit from using these technologies. In many examples and exercises in *Understandable Statistics,* we ask students to use calculators to verify answers. Illustrations in the text show TI-83 calculator screens, MINITAB outputs, and ComputerStat outputs, so that students can see the different types of information available to them through the use of technology.

However, it is not enough to enter data and punch a few buttons to get statistical results. The formulas producing the statistics contain a great deal of information about the *meaning* of the statistics. The text breaks down formulas into tabular form so that students can see the information in the formula. We find it useful to take class time to discuss formulas. For instance, an essential part of the standard deviation formula is the comparison of each data value to the mean. When we point this out to students, it gives meaning to the standard deviation. When students understand the content of the formulas, the numbers they get from their calculator or computer begin to make sense.

The third edition features Focus Points at the beginning of each section, describing that section's primary learning objectives. Also new to the seventh edition is the change from Calculator Notes to Technology Notes. The Technology Notes briefly describe relevant procedures for using the TI-83 Plus calculator, Microsoft Excel, and MINITAB. In addition, the Using Technology segments have been updated to include Excel material.

For courses in which technologies are strongly incorporated into the curriculum, we provide two separate supplements, the *Technology Guide* (for the TI-83 Plus, MINITAB, and ComputerStat) and the *Excel Guide* (for Microsoft Excel). These guides gives specific hints for using the technologies, and also give Lab Activities to help students explore various statistical concepts.

Alternate Paths Through the Text

Like previous editions, the third edition of *Understanding Basic Statistics* is designed to be flexible. The text provides many topics so you can tailor a course to fit your students' needs. The text also aims to be a *readable reference* for topics not specifically included in your course.

Table of Prerequisite Material

Chapter	Prerequisite Sections
1 Getting Started	**None**
2 Organizing Data	**1.1, 1.2**
3 Averages and Variation	**1.1, 1.2, 2.2**
4 Regression and Correlation	**1.1, 1.2, 3.1, 3.2**
5 Elementary Probability Theory	**1.1, 1.2, 2.2, 3.1, 3.2**
6 The Binomial Probability Distribution and Related Topics	**1.1, 1.2, 2.2, 3.1, 3.2, 5.1, 5.2, with 5.3 useful but not essential**
7 Normal Distributions	
with 7.4 omitted	**1.1, 1.2, 2.2, 3.1, 3.2, 5.1, 5.2, 6.1**
with 7.4 included	**add 6.2, 6.3**
8 Introduction to Sampling Distributions	**1.1, 1.2, 2.2, 3.1, 3.2, 5.1, 5.2, 6.1, 7.1, 7.2, 73**
9 Estimation	
with 9.3 and 9.4 omitted	**1.1, 1.2, 2.2, 3.1, 3.2, 5.1, 52, 6.1, 7.1, 7.2, 7.3, 8.1, 8.2**
with 9.3 and 9.4 included	**add 6.2, 6.3, 7.4**
10 Hypothesis Testing	
with 10.5 omitted	**1.1, 1.2, 2.2, 3.1, 3.2, 5.1, 5.2, 6.1, 7.1, 7.2, 7.3, 8.1, 8.2**
with 10.5 included	**add 6.2, 6.3, 7.4**
11 Inferences about Differences	
with 11.4 omitted	**1.1, 1.2, 2.2, 5.1, 5.2, 6.1, 7.1, 7.2, 7.3, 8.1, 8.2, 9.1, 10.1, 10.2, 10.3**
with 11.4 included	**add 6.2, 6.3, 7.4, 10.5**
12 Additional Topics Using Inferences	
with 12.4 and 12.5 omitted	**1.1, 1.2, 2.2, 5.1, 5.2, 6.1, 7.1, 7.2, 7.3, 8.1, 8.2, 10.1, 10.2, 10.3**
with 12.4 and 12.5 included	**add Chapter 4, 9.1, 9.2**

Teaching Tips for Each Chapter

CHAPTER 1 GETTING STARTED

Double-Blind Studies (SECTION 1.3)

The double-blind method of data collection, mentioned at the end of Section 1.3, is an important part of standard research practice. A typical use is in testing new medications. Because the researcher does not know which patients are receiving the experimental drug and which are receiving a more familiar drug (or a placebo), the researcher is prevented from subconsciously doing things that might skew the results.

If, for instance, the researcher communicates a more optimistic attitude to patients in the experimental group, this could influence how they respond to diagnostic questions or might actually influence the course of their illness. And if the researcher wants the new drug to prove effective, this could subconsciously influence how he or she handles information related to each patient's case. All such factors are eliminated in double-blind testing.

The following appears in the physician's dosing instructions package insert for the prescription drug QUIXIN™:

> In randomized, double-masked, multicenter controlled clinical trials where patients were dosed for 5 days, QUIXIN™ demonstrated clinical cures in 79% of patients treated for bacterial conjunctivitis on the final study visit day (day 6-10).

Note the phrase "double-masked." This is, apparently, a synonym for "double-blind." Since "double-blind" is widely used in the medical literature and in clinical trials, why do you suppose they chose to use "double-masked" instead?

Perhaps this will provide some insight: QUIXIN™ is a topical antibacterial solution for the treatment of conjunctivitis, i.e., it is an antibacterial eye drop solution used to treat an inflammation of the conjunctiva, the mucous membrane that lines the inner surface of the eyelid and the exposed surface of the eyeball. Perhaps, since QUIXIN™ is a treatment for eye problems, the manufacturer decided the word "blind" shouldn't appear *anywhere* in the discussion.

Source: Package insert. QUIXIN™ is manufactured by Santen Oy, P.O. Box 33, FIN-33721 Tampere, Finland, and marketed by Santen Inc., Napa, CA 94558, under license from Daiichi Pharmaceutical Co., Ltd., Tokyo, Japan.

CHAPTER 2 ORGANIZING DATA

Emphasize when to use the various graphs discussed in this chapter: bar graphs when comparing data sets; circle graphs for displaying how a total is dispersed into several categories; time plots to display how data changes over time; histograms to display relative frequencies of grouped data; stem-and-leaf displays for displaying grouped data in a way that does not lose the detail of the original raw data.

CHAPTER 3 AVERAGES AND VARIATION

Students should be instructed in the various ways that sets of numeric data can be represented by a single number. The concepts of this section can be motivated to students by emphasizing the need to represent a set of data by a single number.

The different ways this can be done that are discussed in Section 3.1: mean, median, and mode vary in appropriateness according to the situation. In many cases of numeric data, the mean is the most appropriate measure of central tendency. If the mean is larger or smaller than most of the data values, however, then the median may be the number that best represents a set of data. The median is most appropriate usually if the set of data is annual salaries, costs of houses, or any set of data which contains one or a few very large or very small values. The mode would be the most appropriate if the population was the votes in an election or Nielsen television ratings, for example. Students should get aquainted with these concepts by calculating the mean, median, and mode for sets of data, and interpreting their meanings, or which is the most appropriate.

Range, variance, and standard deviation can be represented to students as other numbers that aid in representing a set of data in that they measure how data is dispersed. Students will begin to have a better understanding of these measures of dispersion, like mean, median, and mode, by calculating these numbers for given sets of data, and interpreting their respective meanings.

Chebyshev's theorem is an important theorem to discuss with students that relates to the mean and standard deviation of *any* data set.

Finally, the mean, median, first and third quartiles, and range of a set of data can be easily viewed in a box-and-whisker plot.

CHAPTER 4 REGRESSION AND CORRELATION

Least-Squares Criteria (Section 4 .2)

With some sets of paired data, it will not be obvious which is the explanatory variable and which is the response variable. Here it may be worth mentioning that for linear regression, the choice matters. The results of a linear regression analysis will differ, depending on which variable is chosen as the explanatory variable and which one as the response variable. This is not immediately obvious. We might think that with x as the explanatory variable, we could just solve the regression equation $y = a + bx$ for x in terms of y to obtain the regression equation that we would get if we took y as the explanatory variable instead. But this would be a mistake.

The figure below shows the vertical distances from data points to the line of best fit. The line is defined so as to make the sum of the squares of these vertical distances as small as possible.

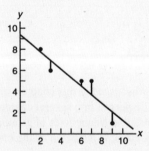

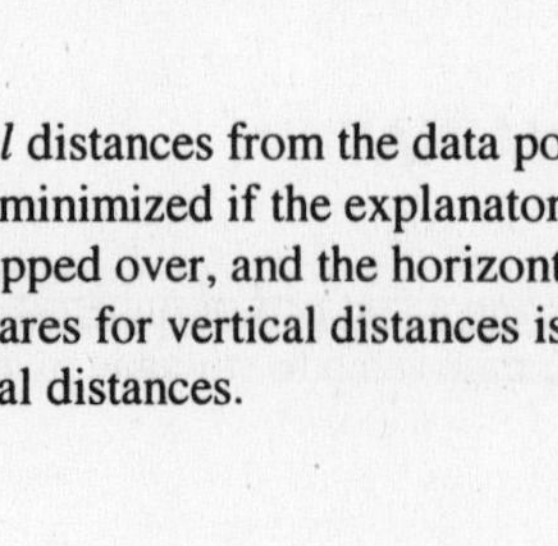

The next figure, now, shows the *horizontal* distances from the data points to the same line. These are the distances whose sum of squares would be minimized if the explanatory and response variables switched roles. With such a switch, the graph would be flipped over, and the horizontal distances would become vertical ones. But the line that minimizes the sum of squares for vertical distances is not, in general, the same line that minimizes the sum of squares for horizontal distances.

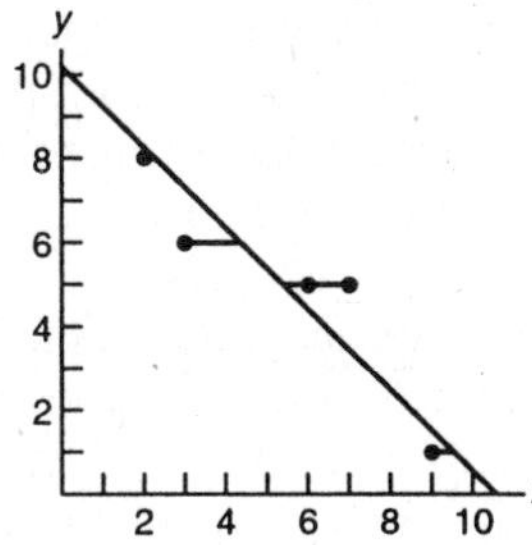

So there is more than one way, mathematically, to definite the line of best fit for a set of paired data. This raises a question: what is the *proper* way to define the line of best fit?

Let us turn this question around: under what circumstances is a best fit based on *vertical* distances the right way to go? Well, intuitively, the distance from a data point to the line of best fit represents some sort of deviation from the ideal value. We can most easily conceptualize this in terms of measurement error. If, now, we treat the error as a strictly vertical distance, then we are saying that in each data pair, the second value is possibly off but the first value is exactly correct. In other words, the least-squares method with vertical distances assumes that the first value in each data pair is measured with essentially perfect accuracy, while the second is measured only imperfectly.

An illustration shows how these assumptions can be realistic. Suppose we are measuring the explosive force generated by the ignition of varying amounts of gunpowder. The weight of the gunpowder is the explanatory variable, and the force of the resulting explosion is the response variable. It could easily happen that we were able to measure the weight of gunpowder with great exactitude—down to the thousandth-ounce—but that our means of measuring explosion force was quite crude, such as the height to which a wooden block was flung into the air by the explosion. We would then have an experiment with a good deal of error in the response variable measurement but for, practical purposes, no error in the explanatory variable measurement. This would all be perfectly in accord with the vertical-distance criterion for finding the line of best fit by the least-squares method.

But now consider a different version of the gunpowder experiment. This time we have a highly refined means of measuring explosive force (some sort of electronic device, let us say) and at the same time we have only a very crude means of measuring gunpowder mass (perhaps a rusty pan balance). In this version of the story, the error would be in the measurement of the response variable, and a horizontal least-squares criterion would be called for.

Now, the most common situation is one in which both the explanatory and the response variables contain some error. The preceding discussion suggests that the most appropriate least-squares criterion for goodness of fit for a line through the cluster of data points would be a criterion in which error was represented as a line lying at some slant, as in the figure.

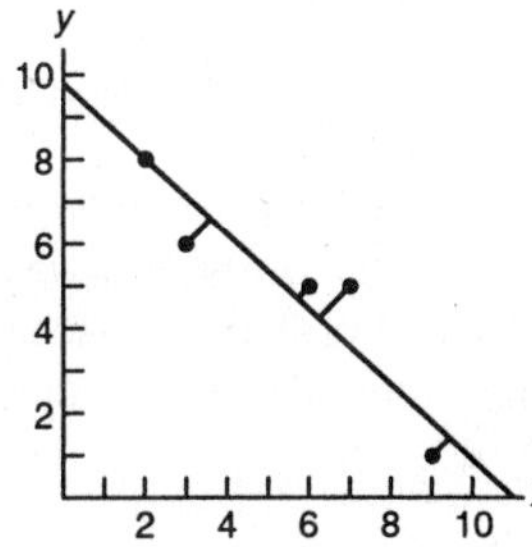

To apply such a criterion, we would have to figure out to define distance in two dimensions when the x and y axis have different units of measure. We will not try to solve that puzzle here. Instead we just summarize what we have learned: there is more than one least-squares criterion for fitting a line to a set of data points, and the choice of which criterion to use implies an assumption about which variable(s) is affected by the error (or other deviation) that moves points off the line representing ideal results.

And, finally, we now see that the standard use of vertical distances in the least-squares method *implies an assumption that the error is predominantly in the response variable.* This is often a reasonable assumption to make, since the explanatory variable is frequently a *control* variable, that is, a variable under the experimenter's control and thus generally capable of being adjusted with a fair amount of precision. The response variable, by contrast, is the one that must simply be measured and cannot be fine-tuned through an experimenter's adjustment. However, it is worth noting that this is only the typical relationship, not a necessary one (as the second gunpowder scenario shows).

Finally, it is also worth nothing that both the vertical and the horizontal least-squares criteria will produce a line that passes through the point $(\bar{x}, \bar{y})$. Thus the vertical- and horizontal-least-squares lines must either coincide (which is atypical but not impossible) or intersect at $(\bar{x}, \bar{y})$. The other thing the two lines have in common is the correlation coefficient, r. It is easy to see, looking at the formula for r, that the value of r does not depend on which variable is chosen as the explanatory one and which as the response one.

Variables and the Issue of Cause and Effect (Sections 4.2 and 4.3)

As remarked at the end of Section 4.3, the relationship between two measured variables x and y may not, in physical terms, be one of cause and effect, respectively. It often is, of course, but it may instead happen that y is the cause and x is the effect. Note that in the example where x = cricket chirps per second and y = air temperature, y is obviously the cause and x the effect. In other situations, x and y will be two effects of a common, possibly unknown, cause. For example, x might be a patient's blood sugar level and y might be the patient's body temperature. Both of these variables could be caused by an illness, which might be quantified in terms of a count of bacterial activity. The point to remember is that although the x-causes-y scenario is typical, strictly speaking the designations "explanatory variable" and "response variable" should be understood not in terms of a causal relationship but in terms of which quantity is initially known and which one is inferred.

CHAPTER 5 ELEMENTARY PROBABILITY THEORY

Ways To Think About Probability (Section 5.1)

As the text describes, there are several methods for assigning a probability to an event. Probability based on intuition is often called *subjective* probability. Thus understood, probability is a numerical measure of a person's confidence about some event. Subjective probability is assumed to be reflected in a person's decisions: the higher an event's probability, the more the person would be willing to bet on its occurring.

Probability based on relative frequency is often called *experimental* probability, because relative frequency is calculated from an observed history of experiment outcomes. But we are already using the word experiment in a way that is neutral among the different treatments of probability—namely, as the name for the activity that produces various possible outcomes. So when we are talking about probability based on relative frequency, we will call this *observed* probability.

Probability based on equally likely outcomes is often called *theoretical* probability, because it is ultimately derived from a theoretical model of the experiment's structure. The experiment may be conducted only once, or not at all, and need not be repeatable.

These three ways of treating probability are compatible and complementary. For a reasonable, well-informed person, the subjective probability of an event should match the theoretical probability, and the theoretical probability in turn predicts the observed probability as the experiment is repeated many times.

Also, it should be noted that although in statistics, probability is officially a property of *events,* it can be thought of as a property of *statements,* as well. The probability of a statement equals the probability of the event that makes the statement true.

Probability and statistics are overlapping fields of study; if they weren't, there would be no need for a chapter on probability in a book on statistics. So the general statement, in the text, that probability deals with known populations while statistics deals with unknown populations is, necessarily, a simplification. However, the statement does express an important truth: if we confront an experiment we initially know absolutely nothing about, then we can collect data, but we cannot calculate probabilities. In other words, we can only calculate probabilities after we have formed some idea of, or acquaintance with, the experiment. To find the theoretical probability of an event, we have to know how the experiment is set up. To find the observed probability, we have to have a record of previous outcomes. And as reasonable people, we need some combination of those same two kinds of information to set our subjective probability.

This may seem obvious, but it has important implications for how we understand technical concepts encountered later in the course. There will be times when we would like to make a statement, say, about the mean of a population, and then give the probability that this statement is true—that is, the probability that the event described by the statement occurs (or has occurred). What we discover when we look closely, however, is that often this can't be done. Often we have to settle for some other conclusion instead. The Teaching Tips for Sections 8.1 and 9.1 describe two instances of this problem.

CHAPTER 6 THE BINOMIAL PROBABILITY DISTRIBUTION AND RELATED TOPICS

Binomial Probabilities (Section 6.2)

Students should be able to show that $pq = p(1 - p)$ has its maximum value at $p = 0.5$. There are at least three ways to demonstrate this: graphically, algebraically, and using calculus.

Graphical method

Recall that $0 \le p \le 1$. So, for $q = 1 - p$, $0 \le q \le 1$ and $0 \le pq \le 1$. Plot $y = pq = p(1 - p)$ using MINITAB, a graphing calculator, or other technology. The graph is a parabola. Observe which value of p maximizes pq. (Many graphing calculators can find the maximum value and where it occurs.)

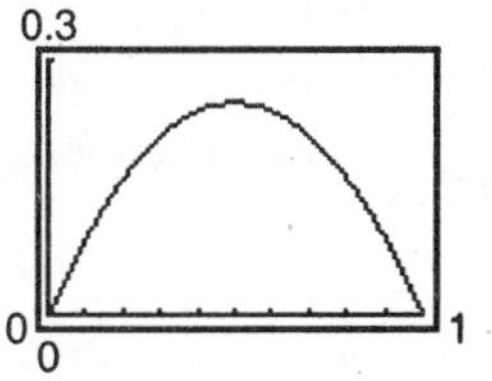

So pq has a maximum value of 0.25, when $p = 0.5$.

Algebraic method

From the definition of q, it follows that $pq = p(1 - p) = p - p^2 = -p^2 + p + 0$. Recognize that this is a quadratic function of the form $ax^2 + bx + c$, where p is used instead of x, and $a = -1$, $b = 1$, and $c = 0$.

The graph of a quadratic function is a parabola, and the general form of a parabola is $y = a(x - h)^2 + k$. The parabola opens up if $a > 0$, opens down if $a < 0$, and has a vertex at (h, k). If the parabola opens up, it has its minimum at $x = h$, and the minimum value of the function is $y = k$. Similarly, if the parabola opens down, it has its maximum value of $y = k$ when $x = h$.

Using the method of completing the square, we can rewrite $y = ax^2 + bx + c$ in the form $y = a(x - h)^2 + k$ to show that $h = -b/2a$ and $k = c - b^2/4a$. When $a = -1$, $b = 1$, and $c = 0$, it follows that $h = 1/2$ and $k = 1/4$. So the value of p that maximizes pq is $p = 1/2$, and then $pq = 1/4$. This confirms the results of the graphical solution.

Calculus-based method
Advanced Placement students have probably had (or are taking) calculus, including tests for local extrema. For a function with continuous first and second derivatives, at an extremum the first derivative equals zero and the second derivative is either positive (at a minimum) or negative (at a maximum).
The first derivative of $f(p) = pq = p(1 - p)$ is given by

$$\begin{aligned} f'(p) &= \frac{d}{dp}[p(1-p)] \\ &= \frac{d}{dp}[-p^2 + p] \\ &= -2p + 1 \end{aligned}$$

Solve $f'(p) = 0: -2p + 1 = 0$

$$p = \frac{1}{2}$$

Now find $f''\left(\frac{1}{2}\right)$: $f''(p) = \frac{d}{dp}[f'(p)]$

$$\begin{aligned} &= \frac{d}{dp}(-2p+1) \\ &= -2 \end{aligned}$$

So $f''\left(\frac{1}{2}\right) = -2$.

Since the second derivative is negative when the first derivative equals zero, $f(p)$ has a maximum at $p = 1/2$.

This result has implications for confidence intervals for p; see the Teaching Tips for Chapter 9.

CHAPTER 7 NORMAL DISTRIBUTIONS

Emphasize the differences between discrete and continuous random variables with examples of each.

Emphasize how normal curves can be used to approximate the probabilities of both continuous and discrete random variables, and in the cases when the distribution of a set of data can be approximated by a normal curve, such a curve is defined by 2 quantities: the mean and standard deviation of the data. In such a case, the normal curve is defined by the equation $y = \dfrac{e^{-\frac{1}{2}\left(\frac{x-\mu}{\sigma}\right)^2}}{\sigma\sqrt{2\pi}}$.

Review Chebyshev's Theorem from Chapter 3. Emphasize that this theorem implies that for *any* set of data

at *least* 75% of the data lie within 2 standard deviations on each side of the mean; at *least* 88.9% of the data lie within 3 standard deviations on each side of the mean, and at *least* 93.8% of the data lie within 4 standard deviations on each side of the mean.

In comparison, a set of data that has a distribution which is symmetrical and bell-shaped, in particular has a normal distribution, is more restrictive in that

approximately 68% of the data values lie within 1 standard deviation on each side of the mean;
approximately 95% of the data values lie within 2 standard deviations on each side of the mean; and
approximately 99.7% of the data values lie within 3 standard deviations on each side of the mean.

Remind students regularly that a z-value equals the number of standard deviations from the mean for data values of *any* distribution approximated by a normal curve.

Emphasize the connection between the area under a normal curve and probability values of the random variable. That is, emphasize that the area under any normal curve equals 1, and the percentage of area under the curve between given values of the random variable equals the probability that the random variable will be between these values. The values in a z-table are areas *and* probability values.

Emphasize the conditions whereby a binomial probability distribution (discussed in Chapter 6) can be approximated by a normal distribution: $np > 5$ and $n(1-p) > 5$, where n is the number of trials and p is the probability of success in a single trial.

When a normal distribution is used to approximate a discrete random variable (such as the random variable of a binomial probability experiment), the *continuity correction* is an important concept to emphasize to students. A discussion of this important adjustment can be a good opportunity to compare discrete and continuous random variables.

CHAPTER 8 INTRODUCTION TO SAMPLING DISTRIBUTIONS

Emphasize the differences between population parameters and sample statistics. Point out that when knowledge of the population is unavailable, then knowledge of a corresponding sample statistic must be used to make inferences about the population.

Emphasize the main two facts discussed from the Central Limit Theorem:

1) If x is a random variable with a normal distribution whose mean is μ and standard deviation is σ, then the means of random samples for any fixed size n taken from the x distribution is a random variable $\bar{x}$ that has a normal distribution with mean μ and standard deviation $\sigma/\sqrt{n}$.

2) If x is a random variable with *any* distribution whose mean is μ and standard deviation is σ, then the mean of random samples of a fixed size n taken from the x distribution is a random variable $\bar{x}$ that has a distribution that approaches a normal distribution with mean μ and standard deviation $\sigma/\sqrt{n}$ as n increases without limit.

Choosing sample sizes greater than 30 is an important point to emphasize in the situation mentioned in part 2 of the Central Limit Theorem above. This commonly-accepted convention insures that the $\bar{x}$ distribution of Part 2 will have a normal distribution regardless of the distribution of the population from which these samples are drawn.

Emphasize that the Central Limit Theorem allows us to infer facts about populations from sample means having normal distributions.

CHAPTER 9 ESTIMATION

Understanding Confidence Intervals (Section 9.1)

As the text says, nontrivial probability statements involve variables, not constants. And if the mean of a population is considered a constant, then the event that this mean falls in a certain range with known numerical bounds has either probability 1 or probability 0.

However, we might instead think of the population mean as itself a variable, since, after all, the value of the mean is initially unknown. In other words, we may think of the population we are sampling from as one of many populations—a population of populations, if you like. One of these populations has been randomly selected for us to work with, and we are trying to figure out which population it is, or, at least, what its mean is.

If we think of our sampling activity in this way, we can then think of the event "The mean lies between *a* and *b*" as having a non-trivial probability of being true. Can we, now, create a 0.90 confidence interval and then say that the mean has a 90% probability of being in that interval? It might seem so, but in general the answer is no. Even though a procedure might have exactly a 90% success rate at creating confidence intervals that contain the mean, a confidence interval created by such a procedure will not, in general, have exactly a 90% chance of containing the mean.

How is this possible? To understand this paradox, let us turn from mean-finding to a simpler task: guessing the color of a randomly-drawn marble. Suppose a sack contains some red marbles and some blue marbles. And suppose we have a friend who will reach in, draw out a marble, and announce its color while we have our backs turned. The friend can be counted on to announce the correct color *exactly 90% of the time* (that is, with a probability of 90%) and the wrong color the other 10% of the time. So if the marble drawn is blue, the friend will say "blue" 9 times out of 10 and "red" the remaining time. And conversely for a red marble. This is like creating an 0.90 confidence interval.

Now the friend reaches in, pulls out a marble, and announces, "blue." Does this mean that we are 90% sure the friend is holding a blue marble? *It depends on what we think about the mix of marbles in the bag*. Suppose we think that the bag contains three red marbles and two blue ones. Then we expect the friend to draw a red marble 3/5 of the time and announce "blue" 10% of those times, or 3/50 of all draws. And we expect the friend to draw a blue marble 2/5 of the time and announce "blue" 90% of those times, or 18/50 of all draws. This means that the ratio of true "blue" announcements to false ones is 18 to 3, or 6 to 1. And thus we attach a probability of 6/7 = 85.7%, not 90%, to our friend's announcement that the marble drawn is blue, even though we believe our friend to be telling the truth 90% of the time. For similar reasons, if the friend says "red," we will attach a probability of 93.1% to this claim. Simply put, our initial belief that there are more red marbles than blue ones pulls our confidence in a "blue" announcement downward, and our confidence in a "red" announcement upward, from the 90% level.

Now, if we believe that there are an *equal* number of red and blue marbles in the bag, then, as it turns out, we will attach 90% probability to "blue" announcements and to "red" announcements as well. But *this is a special case.* In general, the probabilities we attach to each of our friend's statements will be different from the frequency with which we think he is telling the truth. Furthermore: if we have *no idea* about the mix of marbles in the bag, then we will be *unable* to set probabilities for our friend's statements, because we will be unable to run the calculation for how often his "blue" statements are true and his "red" statements are true. In other words, *we cannot justify simply setting our probability equal, by default, to the test's confidence level.*

This story has two morals. (1) The probability of a statement is one thing, and the success rate of a procedure that tries to come up with true statements is another. (2) Our prior beliefs about the conditions of an experiment are an unavoidable element in our interpretation of any sample data.

Let us apply these lessons to the business of finding confidence intervals for population means. When we create a 0.90 confidence interval, we will in general *not* be 90% sure that the interval contains the mean. It could *happen* to turn out that we were 90% sure, but this will depend on what ideas we had about the population mean going in. Suppose we were fairly sure, to start with, that the population mean lay somewhere between 10 and 20, and suppose we then took a sample that led to the construction of a 0.90 confidence interval which ran from 31 to 46. We would *not* conclude, with 90% certainty, that the mean lay between 31 and 46. Instead, we would have a probability lower than that, because previously we thought the mean was outside that range. At the same time, we would be much more ready to believe that the mean lay between 31 and 46 than we were before, because, after all, a procedure with a 90% success rate produced that prediction. Our exact probability for the "between 31 and 46" statement would depend on our entire initial probability distribution for values of the population mean—something we would have a hard time coming up with, if the question were put to us. Thus, under normal circumstances, our exact level of certainty about the confidence interval could not be calculated.

So the general point made in the text holds, even if we think of a population mean as a variable. The procedure for finding a confidence interval of confidence level c does not, in general, produce a statement (about the value of a population mean) that has a probability c of being true.

Confidence Intervals for *p* (Section 9.3)

The result obtained in the Teaching Tip for Chapter 6 has implications for the confidence interval for p: the most conservative interval estimate of p, the widest possible confidence interval in a given situation, is obtained when $E = z_c\sqrt{pq/n}$ is calculated using $p = 1/2$.

CHAPTER 10 HYPOTHESIS TESTING

What a Hypothesis Test Tells Us (Sections 10.1–10.3)

The procedure for hypothesis testing with significance levels may at first confuse some students, especially since the levels are chosen somewhat arbitrarily. Why, the students may wonder, don't we just calculate the likelihood that the null hypothesis is true? Or is that really what we're doing, when we find the P value?

Once again we run the risk of confusion over the role of probability in our statistical conclusions. The P value is *not* the same thing as the probability, in light of the data, of the null hypothesis. Instead, the P value is the probability that the data would turn out the way it did, assuming the null hypothesis to be true. Just as with confidence intervals, here we have to be careful not to think we are finding the probability of a given statement when in fact we are doing something else.

To illustrate: consider two coins in a sack, one fair and one two-headed. One of these coins is pulled out at random and flipped. It comes up heads. Let us take, as our null hypothesis, the statement "The flipped coin was the fair one." The probability of the outcome, given the null hypothesis, is 1/2, because a fair coin will come up heads half the time. This probability is in fact the P value of the outcome. On the other hand, the probability that the null hypothesis is true, given the evidence, is 1/3, since out of all the heads outcomes one will see in many such trials, 1/3 are from the fair coin.

Now suppose that instead of containing two coins of known character, the sack contains an unknown mix—some fair coins, some two-headed coins, and possibly some two-tailed coins, as well. Then we can still calculate the P value of a heads outcome, because the probability of "heads" with a fair coin is still 1/2. But the probability of the coin's being fair, given that we're seeing heads, *cannot be calculated,* because we know nothing about the mix of coins in the bag. So the P value of the outcome is one thing, and the probability of the null hypothesis is another.

The lesson should be now be familiar: without some prior ideas about the character of an experiment, either based on a theoretical model or on previous outcomes, we cannot attach a definite probability to a statement about the experimental setup or its outcome.

This is the usual situation in hypothesis testing. We normally lack the information needed to calculate probabilities for the null hypothesis and its alternative. What we do instead is to take the null hypothesis as defining a well-understood scenario from which we *can* calculate the likelihoods of various outcomes—the probabilities, that is, of various kinds of sample results, given that the null hypothesis is true. By contrast, the alternative hypothesis includes all sorts of scenarios, in some of which (for instance) two population means are only slightly different, in others of which the two means are far apart, and so on. Unless we have identified the likelihoods of all these possibilities, relative to each other and to the null hypothesis, we will not have the background information needed to calculate the probability of the null hypothesis from sample data.

In fact, we will not have the data necessary to calculate what the text calls the power, $1-\beta$, of a hypothesis test. This is what the text means when it says that finding the power requires knowing the H_1 distribution. Because we cannot specify the H_1 distribution when we are concerned with things like diagnosing disease (instead of drawing coins from a sack and the like), we normally cannot determine the probability of the null hypothesis in light of the evidence. Instead, we have to content ourselves with quantifying the risk, α, of rejecting the hypothesis when it is true.

A Paradox About Hypothesis Tests

The way hypothesis tests work leads to a result that at first seems surprising. It can sometimes happen that, at a given level of significance, a one-tailed test leads to a rejection of the null hypothesis while a two-tailed test would not. Apparently, one can be justified in concluding that $\mu > k$ (or $\mu < k$ as the case may be) but not justified in concluding that $\mu \neq k$ —even though the latter conclusion follows from the former! What is going on here?

This paradox dissolves when one remembers that a one-tailed test is used only when one has appropriate information. With the null hypothesis H_0: $\mu = k$, we choose the alternative hypotheses H_1: $\mu > k$ only if *we are already sure* that μ is not less than H_1: $\mu < k$. This assumption, in effect, boosts the force of any evidence that μ does not equal k—and if it is not less than or equal to k, it must be greater.

In other words, when a right-tailed test is appropriate, rejecting the null hypothesis means concluding *both* that $\mu > k$ and that $\mu \neq k$. But when there is no justification for a one-tailed test, one must use a two-tailed test and needs somewhat stronger evidence before concluding that $\mu \neq k$.

CHAPTER 11 INFERENCES ABOUT DIFFERENCES

Emphasize that when working with paired data, it is very important to have a definite and uniform method of creating data pairs so that the data remains a dependent sample. When using Dependent Samples,

- Members in the first population are randomly selected then paired with corresponding members of a second population on a one-to-one basis.
- The value of a random variable x_1 is determined for each member in the first population and the corresponding measure, x_2, is determined for its partner in the second population.
- Then, for each pairing, the difference, $d = x_1 - x_2$, in the measured values is calculated
- Finally, perform a *t* test on the differences.

When testing the difference in means of two large independent samples, we make the hypothesis that there is no difference between the two means. Then we test to see if this hypothesis should be rejected.

When testing the difference of means for large samples, we use the normal distribution for critical values.

When studying the difference in means of two large independent samples, we can also look at the confidence interval $\mu_1 - \mu_2$, where μ_1 and μ_2 are the means of independent populations.

Be sure students know how to interpret the confidence intervals for $\mu_1 - \mu_2$ for the three cases (only negative values, only positive values, both positive and negative values);

1. If the $c\%$ confidence interval contains only negative values, we conclude that $\mu_1 - \mu_2 < 0$, and we can be $c\%$ confident that $\mu_1 < \mu_2$.
2. If the $c\%$ confidence interval contains only positive values, we conclude that $\mu_1 - \mu_2 > 0$, and we can be $c\%$ confident that $\mu_1 > \mu_2$.
3. If the $c\%$ confidence interval contains both positive and negative values, we cannot conclude that either μ_1 or μ_2 is larger.

When testing the difference of means for small samples, we use the Student's t distribution for critical values rather than the normal distribution which is used for large samples.

CHAPTER 12 ADDITIONAL TOPICS USING INFERENCE

Emphasize that both the χ^2 distribution and the F distribution are not symmetrical and have only non-negative values.

Emphasize that the applications of the χ^2 distribution include the test for independence of two factors, goodness of fit of a present distribution to a given distribution, and whether a variance (or standard deviation) has changed or varies from a known population variance (or standard deviation). The χ^2 distribution is also used to find a confidence interval for a variance (or standard deviation).

Hints for Distance Education Courses

Distance education uses various media, each of which can be used in one-way or interactive mode. Here is a representative list:

		One-way	Interactive
Medium:	Audio	Cassette tapes	Phone
	Audiovisual	Videotapes, CD-ROMs	Teleconferencing
	Data	Computer-resident tutorials, web tutorials	E-mail, chat rooms, discussion lists
	Print	Texts, workbooks	Mailed-in assignments, mailed-back instructor comments, fax exchanges

Sometimes the modes are given as asynchronous (students working on their schedules) versus synchronous (students and instructors working at the same time), but synchronous scheduling normally makes sense only when this enables some element of interactivity in the instruction.

Naturally the media and modes may be mixed and matched. A course might, for instance, use a one-way video feed with interactive audio, plus discussion lists.

THINGS TO KEEP IN MIND

Even in a very high-tech telecourse, print is a foundational part of the instruction. The textbook is *at least* as important as in a traditional course, since it is the one resource which requires no special equipment to use, and whose use is not made more difficult by the distance separating student and instructor.

Because students generally obtain all course materials at once, before instruction begins, mid-course adjustments of course content are generally not practicable. Plan the course carefully up front, so everything is in place when instruction begins.

In distance courses, students can often be assumed to have ready access to computers while working on their own. This creates the opportunity for technology-based assignments that in a traditional course might be feasible at best as optional work (for example, assignments using ComputerStat, MINITAB, or Microsoft Excel; see the corresponding guides that accompany *Understandable Statistics*). However, any time students have to spend learning how to use unfamiliar software will add to their overall workload and possibly to their frustration level. Remember this when choosing technology-based work to incorporate.

Remember that even (and perhaps especially) in distance education, students take a course because they want to interact with a human being rather than just read a book. The goal of distance instruction is to make that possible for students who cannot enroll in a traditional course. Lectures should not turn into slide shows with voice commentary, even though these may be technologically easier to transmit than, say, real-time video. Keep the human element uppermost.

All students should be self-motivated, but in real life nearly all students benefit from a little friendly supervision and encouragement. This goes double for distance education. Make an extra effort to check in with students one-on-one, ask how things are going, and remind them of things they may be forgetting or neglecting.

CHALLENGES IN DISTANCE EDUCATION

Technology malfunctions often plague distance courses. To prevent this from happening in yours:

- Don't take on too much at once. As the student sites multiply, so do the technical difficulties. Try the methodology with one or two remote sites before expanding.
- Plan all technology use well in advance and thoroughly test all equipment before the course starts.
- Have redundant and backup means for conducting class sessions. If, for instance, a two-way teleconferencing link goes down, plan for continuing the lecture by speakerphone, with students referring to predistributed printed materials as needed.
- Allow enough slack time in lectures for extra logistical tasks and occasional technical difficulties.
- If possible, do a pre-course dry run with at least some of the students, so they can get familiar with the equipment and procedures and alert you to any difficulties they run into.
- When it is feasible, have a facilitator at each student site. This person's main job is to make sure the technology at the students' end works smoothly. If the facilitator can assist with course administration and answer student questions about course material, so much the better.

In a distance course, establishing rapport with students and making them comfortable can be difficult.

- An informal lecture style, often effective in traditional classrooms, can be even more effective in a distance course. Be cheerful and use humor. (But in cross-cultural contexts, remember that what is funny to you may fall flat with your audience.)
- Remember that your voice will not reach the students with the clarity as in a traditional classroom. Speak clearly, not too fast, and avoid over-long sentences. Pause regularly.
- Early in the course, work in some concrete, real-world examples and applications to help the students relax, roll up their sleeves, and forget about the distance-learning aspect of the course.
- If the course is interactive, via teleconferencing or real-time typed communication, get students into "send" mode as soon as possible. Ask them questions. Call on individuals if you judge that this is appropriate.
- A student-site assistant with a friendly manner can also help the students quickly settle into the course.

The distance learning format will make it hard for you to gauge how well students are responding to your instruction. In a traditional course, students' incomprehension or frustration often registers in facial expression, tone of voice, muttered comments—all of which are, depending on the instructional format, either difficult or impossible to pick up on in a distance course. Have some way for students to give feedback on how well the course is going for them. Possibilities:

- Quickly written surveys ("On a scale of 1 to 5, please rate …") every few weeks.
- Periodic "How are things going?" phone calls from you.
- A student-site assistant can act as your "eyes and ears" for this aspect of the instruction, and students may be more comfortable voicing frustrations to him or her than to you.

If students are to mail in finished work, set deadlines with allowance for due to mail delivery times, and with allowance for the possibility of lost mail. This is especially important for the end of the term, when you have a deadline for turning in grades.

Cheating is a problem in any course, but especially so in distance courses. Once again, an on-site facilitator is an asset. Another means of forestalling cheating is to have open-book exams, which takes away the advantage of sneaking a peek at the text.

Good student-instructor interaction takes conscious effort and planning in a distance course. Provide students with a variety of ways to contact you:

- E-mail is the handiest way for most students to stay in touch.

- Phone; a toll-free number is ideal. When students are most likely to be free in the evenings, set the number up for your home address and schedule evening office hours when students can count on reaching you.

- When students can make occasional in-person visits to your office, provide for that as well.

ADVANTAGES IN DISTANCE EDUCATION

Compared to traditional courses, more of the information shared in a distance course is, or can be, preserved for later review. Students can review videotaped lectures, instructor-student exchanges via e-mail can be reread, class discussions are on a reviewable discussion list, and so on.

To the extent that students are on their own, working out of texts or watching prerecorded video, course material can be modularized and customized to suit the needs of individual students: where a traditional course would necessarily be offered as a 4-unit lecture series, the counterpart distance course could be broken into four 1-unit modules, with students free to take only those modules they need. This is especially beneficial when the course is aimed at students who already have some professional experience with statistics and need to fill-in-gaps rather than comprehensive instruction.

STUDENT INTERACTION

Surprisingly, some instructors have found that students interact more with one another in a well-designed distance course than in a traditional course, even when the students are physically separated from one another. Part of the reason may be a higher level of motivation among distance learners. But another reason is the same technologies which facilitate student-instructor communication—things like e-mail and discussion lists—also facilitate student-student communication. In some cases, distance learners have actually done better than traditional learners taking the very same course. Better student interaction was thought to be the main reason.

One implication is that while group projects, involving statistical evaluations of real-world data, might seem more difficult to set up in a distance course, they are actually no harder, and the students learn just as much. The web has many real-world data sources, like the U.S. Department of Commerce (home.doc.gov) which has links to the U.S. Census Bureau (www.census.gov), the Bureau of Economic Analysis (www.bea.gov), and other agencies that compile publicly-available data.

Suggested References

THE AMERICAN STATISTICAL ASSOCIATION

Contact Information
1429 Duke Street
Alexandria, VA 22314-3415
Phone: (703) 684-1221 or toll-free: (888) 231-3473
Fax: (703) 684-2037

ASA Publications
Stats: The Magazine for Students of Statistics
CHANCE magazine
The American Statistician
AmStat News

BOOKS

Huff, Darryll and Geis, Irving (1954). *How to Lie with Statistics.* Classic text on the use and misuse of statistics.

Moore, David S. (2000) *Statistics: Concepts and Controversies,* fifth edition. Does not go deeply in to computational aspects of statistical methods. Good resource for emphasizing concepts and applications.

Tufte, Edward R. (2001). *The Visual Display of Quantitative Information,* second edition. A beautiful book, the first of three by Tufte on the use of graphic images to summarize and interpret numerical data. The books are virtual works of art in their own right.

Tanur, Judith M. (1989) *Statistics: A Guide to the Unknown,* third edition. Another excellent source of illustrations.

REFERENCES FOR DISTANCE LEARNING

Bolland, Thomas W. (1994). "Successful Customers of Statistics at a Distant Learning Site." *Proceedings of the Quality and Productivity Section, American Statistical Association,* pp. 300–304.

Lawrence, Betty and Gaines, Leonard M. (1997). "An Evaluation of the Effectiveness of an Activity-Based Approach to Statistics for Distance Learners." *Proceedings of the Section on Statistical Education, American Statistical Association,* pp. 271–272.

Wegman, Edward J. and Solka, Jeffrey L. (1999). "Implications for Distance Learning: Methodologies for Statistical Education." *Proceedings of the Section on Statistical Education, the Section on Teaching Statistics in the Health Sciences, and the Section on Statistical Consulting, American Statistical Association,* pp. 13–16.

Distance Learning: Principles for Effective Design, Delivery, and Evaluation

University of Idaho website: www.uidaho.edu/evo/distglan.html

Part II

Transparency Masters

TABLE 1 Random Numbers

92630	78240	19267	95457	53497	23894	37708	79862	76471	66418
79445	78735	71549	44843	26404	67318	00701	34986	66751	99723
59654	71966	27386	50004	05358	94031	29281	18544	52429	06080
31524	49587	76612	39789	13537	48086	59483	60680	84675	53014
06348	76938	90379	51392	55887	71015	09209	79157	24440	30244
28703	51709	94456	48396	73780	06436	86641	69239	57662	80181
68108	89266	94730	95761	75023	48464	65544	96583	18911	16391
99938	90704	93621	66330	33393	95261	95349	51769	91616	33238
91543	73196	34449	63513	83834	99411	58826	40456	69268	48562
42103	02781	73920	56297	72678	12249	25270	36678	21313	75767
17138	27584	25296	28387	51350	61664	37893	05363	44143	42677
28297	14280	54524	21618	95320	38174	60579	08089	94999	78460
09331	56712	51333	06289	75345	08811	82711	57392	25252	30333
31295	04204	93712	51287	05754	79396	87399	51773	33075	97061
36146	15560	27592	42089	99281	59640	15221	96079	09961	05371
29553	18432	13630	05529	02791	81017	49027	79031	50912	09399
23501	22642	63081	08191	89420	67800	55137	54707	32945	64522
57888	85846	67967	07835	11314	01545	48535	17142	08552	67457
55336	71264	88472	04334	63919	36394	11196	92470	70543	29776
10087	10072	55980	64688	68239	20461	89381	93809	00796	95945
34101	81277	66090	88872	37818	72142	67140	50785	21380	16703
53362	44940	60430	22834	14130	96593	23298	56203	92671	15925
82975	66158	84731	19436	55790	69229	28661	13675	99318	76873
54827	84673	22898	08094	14326	87038	42892	21127	30712	48489
25464	59098	27436	89421	80754	89924	19097	67737	80368	08795

TABLE 1 *continued*

67609	60214	41475	84950	40133	02546	09570	45682	50165	15609
44921	70924	61295	51137	47596	86735	35561	76649	18217	63446
33170	30972	98130	95828	49786	13301	36081	80761	33985	68621
84687	85445	06208	17654	51333	02878	35010	67578	61574	20749
71886	56450	36567	09395	96951	35507	17555	35212	69106	01679
00475	02224	74722	14721	40215	21351	08596	45625	83981	63748
25993	38881	68361	59560	41274	69742	40703	37993	03435	18873
92882	53178	99195	93803	56985	53089	15305	50522	55900	43026
25138	26810	07093	15677	60688	04410	24505	37890	67186	62829
84631	71882	12991	83028	82484	90339	91950	74579	03539	90122
34003	92326	12793	61453	48121	74271	28363	66561	75220	35908
53775	45749	05734	86169	42762	70175	97310	73894	88606	19994
59316	97885	72807	54966	60859	11932	35265	71601	55577	67715
20479	66557	50705	26999	09854	52591	14063	30214	19890	19292
86180	84931	25455	26044	02227	52015	21820	50599	51671	65411
21451	68001	72710	40261	61281	13172	63819	48970	51732	54113
98062	68375	80089	24135	72355	95428	11808	29740	81644	86610
01788	64429	14430	94575	75153	94576	61393	96192	03227	32258
62465	04841	43272	68702	01274	05437	22953	18946	99053	41690
94324	31089	84159	92933	99989	89500	91586	02802	69471	68274
05797	43984	21575	09908	70221	19791	51578	36432	33494	79888
10395	14289	52185	09721	25789	38562	54794	04897	59012	89251
35177	56986	25549	59730	64718	52630	31100	62384	49483	11409
25633	89619	75882	98256	02126	72099	57183	55887	09320	72363
16464	48280	94254	45777	45150	68865	11382	11782	22695	41988

Source: Reprinted from *A Million Random Digits with 100,000 Normal Deviates* by the Rand Corporation (New York: The Free Press, 1955). Copyright 1955 by the Rand Corporation. Used by permission.

TABLE 2 Binomial Coefficients $C_{n,r}$

$n \backslash r$	0	1	2	3	4	5	6	7	8	9	10
1	1	1									
2	1	2	1								
3	1	3	3	1							
4	1	4	6	4	1						
5	1	5	10	10	5	1					
6	1	6	15	20	15	6	1				
7	1	7	21	35	35	21	7	1			
8	1	8	28	56	70	56	28	8	1		
9	1	9	36	84	126	126	84	36	9	1	
10	1	10	45	120	210	252	210	120	45	10	1
11	1	11	55	165	330	462	462	330	165	55	11
12	1	12	66	220	495	792	924	792	495	220	66
13	1	13	78	286	715	1,287	1,716	1,716	1,287	715	286
14	1	14	91	364	1,001	2,002	3,003	3,432	3,003	2,002	1,001
15	1	15	105	455	1,365	3,003	5,005	6,435	6,435	5,005	3,003
16	1	16	120	560	1,820	4,368	8,008	11,440	12,870	11,440	8,008
17	1	17	136	680	2,380	6,188	12,376	19,448	24,310	24,310	19,448
18	1	18	153	816	3,060	8,568	18,564	31,824	43,758	48,620	43,758
19	1	19	171	969	3,876	11,628	27,132	50,388	75,582	92,378	92,378
20	1	20	190	1,140	3,845	15,504	38,760	77,520	125,970	167,960	184,756

TABLE 3 Binomial Probability Distribution $C_{n,r}p^rq^{n-r}$

This table shows the probability of *r* successes in *n* independent trials, each with probability of success *p*.

		p																			
n	*r*	.01	.05	.10	.15	.20	.25	.30	.35	.40	.45	.50	.55	.60	.65	.70	.75	.80	.85	.90	.95
2	0	.980	.902	.810	.723	.640	.563	.490	.423	.360	.303	.250	.203	.160	.123	.090	.063	.040	.023	.010	.002
	1	.020	.095	.180	.255	.320	.375	.420	.455	.480	.495	.500	.495	.480	.455	.420	.375	.320	.255	.180	.095
	2	.000	.002	.010	.023	.040	.063	.090	.123	.160	.203	.250	.303	.360	.423	.490	.563	.640	.723	.810	.902
3	0	.970	.857	.729	.614	.512	.422	.343	.275	.216	.166	.125	.091	.064	.043	.027	.016	.008	.003	.001	.000
	1	.029	.135	.243	.325	.384	.422	.441	.444	.432	.408	.375	.334	.288	.239	.189	.141	.096	.057	.027	.007
	2	.000	.007	.028	.057	.096	.141	.189	.239	.288	.334	.375	.408	.432	.444	.441	.422	.384	.325	.243	.135
	3	.000	.000	.001	.003	.008	.016	.027	.043	.064	.091	.125	.166	.216	.275	.343	.422	.512	.614	.729	.857
4	0	.961	.815	.656	.522	.410	.316	.240	.179	.130	.092	.062	.041	.026	.015	.008	.004	.002	.001	.000	.000
	1	.039	.171	.292	.368	.410	.422	.412	.384	.346	.300	.250	.200	.154	.112	.076	.047	.026	.011	.004	.000
	2	.001	.014	.049	.098	.154	.211	.265	.311	.346	.368	.375	.368	.346	.311	.265	.211	.154	.098	.049	.014
	3	.000	.000	.004	.011	.026	.047	.076	.112	.154	.200	.250	.300	.346	.384	.412	.422	.410	.368	.292	.171
	4	.000	.000	.000	.001	.002	.004	.008	.015	.026	.041	.062	.092	.130	.179	.240	.316	.410	.522	.656	.815
5	0	.951	.774	.590	.444	.328	.237	.168	.116	.078	.050	.031	.019	.010	.005	.002	.001	.000	.000	.000	.000
	1	.048	.204	.328	.392	.410	.396	.360	.312	.259	.206	.156	.113	.077	.049	.028	.015	.006	.002	.000	.000
	2	.001	.021	.073	.138	.205	.264	.309	.336	.346	.337	.312	.276	.230	.181	.132	.088	.051	.024	.008	.001
	3	.000	.001	.008	.024	.051	.088	.132	.181	.230	.276	.312	.337	.346	.336	.309	.264	.205	.138	.073	.021
	4	.000	.000	.000	.002	.006	.015	.028	.049	.077	.113	.156	.206	.259	.312	.360	.396	.410	.392	.328	.204
	5	.000	.000	.000	.000	.000	.001	.002	.005	.010	.019	.031	.050	.078	.116	.168	.237	.328	.444	.590	.774
6	0	.941	.735	.531	.377	.262	.178	.118	.075	.047	.028	.016	.008	.004	.002	.001	.000	.000	.000	.000	.000
	1	.057	.232	.354	.399	.393	.356	.303	.244	.187	.136	.094	.061	.037	.020	.010	.004	.002	.000	.000	.000
	2	.001	.031	.098	.176	.246	.297	.324	.328	.311	.278	.234	.186	.138	.095	.060	.033	.015	.006	.001	.000
	3	.000	.002	.015	.042	.082	.132	.185	.236	.276	.303	.312	.303	.276	.236	.185	.132	.082	.042	.015	.002
	4	.000	.000	.001	.006	.015	.033	.060	.095	.138	.186	.234	.278	.311	.328	.324	.297	.246	.176	.098	.031
	5	.000	.000	.000	.000	.002	.004	.010	.020	.037	.061	.094	.136	.187	.244	.303	.356	.393	.399	.354	.232
	6	.000	.000	.000	.000	.000	.000	.001	.002	.004	.008	.016	.028	.047	.075	.118	.178	.262	.377	.531	.735
7	0	.932	.698	.478	.321	.210	.133	.082	.049	.028	.015	.008	.004	.002	.001	.000	.000	.000	.000	.000	.000
	1	.066	.257	.372	.396	.367	.311	.247	.185	.131	.087	.055	.032	.017	.008	.004	.001	.000	.000	.000	.000
	2	.002	.041	.124	.210	.275	.311	.318	.299	.261	.214	.164	.117	.077	.047	.025	.012	.004	.001	.000	.000
	3	.000	.004	.023	.062	.115	.173	.227	.268	.290	.292	.273	.239	.194	.144	.097	.058	.029	.011	.003	.000
	4	.000	.000	.003	.011	.029	.058	.097	.144	.194	.239	.273	.292	.290	.268	.227	.173	.115	.062	.023	.004
	5	.000	.000	.000	.001	.004	.012	.025	.047	.077	.117	.164	.214	.261	.299	.318	.311	.275	.210	.124	.041
	6	.000	.000	.000	.000	.000	.001	.004	.008	.017	.032	.055	.087	.131	.185	.247	.311	.367	.396	.372	.257
	7	.000	.000	.000	.000	.000	.000	.000	.001	.002	.004	.008	.015	.028	.049	.082	.133	.210	.321	.478	.698

TABLE 3 *continued*

		p																			
n	*r*	.01	.05	.10	.15	.20	.25	.30	.35	.40	.45	.50	.55	.60	.65	.70	.75	.80	.85	.90	.95
8	0	.923	.663	.430	.272	.168	.100	.058	.032	.017	.008	.004	.002	.001	.000	.000	.000	.000	.000	.000	.000
	1	.075	.279	.383	.385	.336	.267	.198	.137	.090	.055	.031	.016	.008	.003	.001	.000	.000	.000	.000	.000
	2	.003	.051	.149	.238	.294	.311	.296	.259	.209	.157	.109	.070	.041	.022	.010	.004	.001	.000	.000	.000
	3	.000	.005	.033	.084	.147	.208	.254	.279	.279	.257	.219	.172	.124	.081	.047	.023	.009	.003	.000	.000
	4	.000	.000	.005	.018	.046	.087	.136	.188	.232	.263	.273	.263	.232	.188	.136	.087	.046	.018	.005	.000
	5	.000	.000	.000	.003	.009	.023	.047	.081	.124	.172	.219	.257	.279	.279	.254	.208	.147	.084	.033	.005
	6	.000	.000	.000	.000	.001	.004	.010	.022	.041	.070	.109	.157	.209	.259	.296	.311	.294	.238	.149	.051
	7	.000	.000	.000	.000	.000	.000	.001	.003	.008	.016	.031	.055	.090	.137	.198	.267	.336	.385	.383	.279
	8	.000	.000	.000	.000	.000	.000	.000	.000	.001	.002	.004	.008	.017	.032	.058	.100	.168	.272	.430	.663
9	0	.914	.630	.387	.232	.134	.075	.040	.021	.010	.005	.002	.001	.000	.000	.000	.000	.000	.000	.000	.000
	1	.083	.299	.387	.368	.302	.225	.156	.100	.060	.034	.018	.008	.004	.001	.000	.000	.000	.000	.000	.000
	2	.003	.063	.172	.260	.302	.300	.267	.216	.161	.111	.070	.041	.021	.010	.004	.001	.000	.000	.000	.000
	3	.000	.008	.045	.107	.176	.234	.267	.272	.251	.212	.164	.116	.074	.042	.021	.009	.003	.001	.000	.000
	4	.000	.001	.007	.028	.066	.117	.172	.219	.251	.260	.246	.213	.167	.118	.074	.039	.017	.005	.001	.000
	5	.000	.000	.001	.005	.017	.039	.074	.118	.167	.213	.246	.260	.251	.219	.172	.117	.066	.028	.007	.001
	6	.000	.000	.000	.001	.003	.009	.021	.042	.074	.116	.164	.212	.251	.272	.267	.234	.176	.107	.045	.008
	7	.000	.000	.000	.000	.000	.001	.004	.010	.021	.041	.070	.111	.161	.216	.267	.300	.302	.260	.172	.063
	8	.000	.000	.000	.000	.000	.000	.000	.001	.004	.008	.018	.034	.060	.100	.156	.225	.302	.368	.387	.299
	9	.000	.000	.000	.000	.000	.000	.000	.000	.000	.001	.002	.005	.010	.021	.040	.075	.134	.232	.387	.630
10	0	.904	.599	.349	.197	.107	.056	.028	.014	.006	.003	.001	.000	.000	.000	.000	.000	.000	.000	.000	.000
	1	.091	.315	.387	.347	.268	.188	.121	.072	.040	.021	.010	.004	.002	.000	.000	.000	.000	.000	.000	.000
	2	.004	.075	.194	.276	.302	.282	.233	.176	.121	.076	.044	.023	.011	.004	.001	.000	.000	.000	.000	.000
	3	.000	.010	.057	.130	.201	.250	.267	.252	.215	.166	.117	.075	.042	.021	.009	.003	.001	.000	.000	.000
	4	.000	.001	.011	.040	.088	.146	.200	.238	.251	.238	.205	.160	.111	.069	.037	.016	.006	.001	.000	.000
	5	.000	.000	.001	.008	.026	.058	.103	.154	.201	.234	.246	.234	.201	.154	.103	.058	.026	.008	.001	.000
	6	.000	.000	.000	.001	.006	.016	.037	.069	.111	.160	.205	.238	.251	.238	.200	.146	.088	.040	.011	.001
	7	.000	.000	.000	.000	.001	.003	.009	.021	.042	.075	.117	.166	.215	.252	.267	.250	.201	.130	.057	.010
	8	.000	.000	.000	.000	.000	.000	.001	.004	.011	.023	.044	.076	.121	.176	.233	.282	.302	.276	.194	.075
	9	.000	.000	.000	.000	.000	.000	.000	.000	.002	.004	.010	.021	.040	.072	.121	.188	.268	.347	.387	.315
	10	.000	.000	.000	.000	.000	.000	.000	.000	.000	.000	.001	.003	.006	.014	.028	.056	.107	.197	.349	.599

TABLE 3 *continued*

		p																			
n	*r*	.01	.05	.10	.15	.20	.25	.30	.35	.40	.45	.50	.55	.60	.65	.70	.75	.80	.85	.90	.95
11	0	.895	.569	.314	.167	.086	.042	.020	.009	.004	.001	.000	.000	.000	.000	.000	.000	.000	.000	.000	.000
	1	.099	.329	.384	.325	.236	.155	.093	.052	.027	.013	.005	.002	.001	.000	.000	.000	.000	.000	.000	.000
	2	.005	.087	.213	.287	.295	.258	.200	.140	.089	.051	.027	.013	.005	.002	.001	.000	.000	.000	.000	.000
	3	.000	.014	.071	.152	.221	.258	.257	.225	.177	.126	.081	.046	.023	.010	.004	.001	.000	.000	.000	.000
	4	.000	.001	.016	.054	.111	.172	.220	.243	.236	.206	.161	.113	.070	.038	.017	.006	.002	.000	.000	.000
	5	.000	.000	.002	.013	.039	.080	.132	.183	.221	.236	.226	.193	.147	.099	.057	.027	.010	.002	.000	.000
	6	.000	.000	.000	.002	.010	.027	.057	.099	.147	.193	.226	.236	.221	.183	.132	.080	.039	.013	.002	.000
	7	.000	.000	.000	.000	.002	.006	.017	.038	.070	.113	.161	.206	.236	.243	.220	.172	.111	.054	.016	.001
	8	.000	.000	.000	.000	.000	.001	.004	.010	.023	.046	.081	.126	.177	.225	.257	.258	.221	.152	.071	.014
	9	.000	.000	.000	.000	.000	.000	.001	.002	.005	.013	.027	.051	.089	.140	.200	.258	.295	.287	.213	.087
	10	.000	.000	.000	.000	.000	.000	.000	.000	.001	.002	.005	.013	.027	.052	.093	.155	.236	.325	.684	.329
	11	.000	.000	.000	.000	.000	.000	.000	.000	.000	.000	.000	.001	.004	.009	.020	.042	.086	.167	.314	.569
12	0	.886	.540	.282	.142	.069	.032	.014	.006	.002	.001	.000	.000	.000	.000	.000	.000	.000	.000	.000	.000
	1	.107	.341	.377	.301	.206	.127	.071	.037	.017	.008	.003	.001	.000	.000	.000	.000	.000	.000	.000	.000
	2	.006	.099	.230	.292	.283	.232	.168	.109	.064	.034	.016	.007	.002	.001	.000	.000	.000	.000	.000	.000
	3	.000	.017	.085	.172	.236	.258	.240	.195	.142	.092	.054	.028	.012	.005	.001	.000	.000	.000	.000	.000
	4	.000	.002	.021	.068	.133	.194	.231	.237	.213	.170	.121	.076	.042	.020	.008	.002	.001	.000	.000	.000
	5	.000	.000	.004	.019	.053	.103	.158	.204	.227	.223	.193	.149	.101	.059	.029	.011	.003	.001	.000	.000
	6	.000	.000	.000	.004	.016	.040	.079	.128	.177	.212	.226	.212	.177	.128	.079	.040	.016	.004	.000	.000
	7	.000	.000	.000	.001	.003	.011	.029	.059	.101	.149	.193	.223	.227	.204	.158	.103	.053	.019	.004	.000
	8	.000	.000	.000	.000	.001	.002	.008	.020	.042	.076	.121	.170	.216	.237	.231	.194	.133	.068	.021	.002
	9	.000	.000	.000	.000	.000	.000	.001	.005	.012	.028	.054	.092	.142	.195	.240	.258	.236	.172	.085	.017
	10	.000	.000	.000	.000	.000	.000	.000	.001	.002	.007	.016	.034	.064	.109	.168	.232	.283	.292	.230	.099
	11	.000	.000	.000	.000	.000	.000	.000	.000	.000	.001	.003	.008	.017	.037	.071	.127	.206	.301	.377	.341
	12	.000	.000	.000	.000	.000	.000	.000	.000	.000	.000	.000	.001	.002	.006	.014	.032	.069	.142	.282	.540
15	0	.860	.463	.206	.087	.035	.013	.005	.002	.000	.000	.000	.000	.000	.000	.000	.000	.000	.000	.000	.000
	1	.130	.366	.343	.231	.132	.067	.031	.013	.005	.002	.000	.000	.000	.000	.000	.000	.000	.000	.000	.000
	2	.009	.135	.267	.286	.231	.156	.092	.048	.022	.009	.003	.001	.000	.000	.000	.000	.000	.000	.000	.000
	3	.000	.031	.129	.218	.250	.225	.170	.111	.063	.032	.014	.005	.002	.000	.000	.000	.000	.000	.000	.000
	4	.005	.005	.043	.116	.188	.225	.219	.179	.127	.078	.042	.019	.007	.002	.001	.000	.000	.000	.000	.000
	5	.000	.001	.010	.045	.103	.165	.206	.212	.186	.140	.092	.051	.024	.010	.003	.001	.000	.000	.000	.000
	6	.000	.000	.002	.013	.043	.092	.147	.191	.207	.191	.153	.105	.061	.030	.012	.003	.001	.000	.000	.000
	7	.000	.000	.000	.003	.014	.039	.081	.132	.177	.201	.196	.165	.118	.071	.035	.013	.003	.001	.000	.000
	8	.000	.000	.000	.001	.003	.013	.035	.071	.118	.165	.196	.201	.177	.132	.081	.039	.014	.003	.000	.000
	9	.000	.000	.000	.000	.001	.003	.012	.030	.061	.105	.153	.191	.207	.191	.147	.092	.043	.013	.002	.000
	10	.000	.000	.000	.000	.000	.001	.003	.010	.024	.051	.092	.140	.186	.212	.206	.165	.103	.045	.010	.001
	11	.000	.000	.000	.000	.000	.000	.001	.002	.007	.019	.042	.078	.127	.179	.219	.225	.188	.116	.043	.005
	12	.000	.000	.000	.000	.000	.000	.000	.000	.002	.005	.014	.032	.063	.111	.170	.225	.250	.218	.129	.031
	13	.000	.000	.000	.000	.000	.000	.000	.000	.000	.001	.003	.009	.022	.048	.092	.156	.231	.586	.267	.135
	14	.000	.000	.000	.000	.000	.000	.000	.000	.000	.000	.000	.002	.005	.013	.031	.067	.132	.231	.343	.366
	15	.000	.000	.000	.000	.000	.000	.000	.000	.000	.000	.000	.000	.000	.002	.005	.013	.035	.087	.206	.463

TABLE 3 *continued*

		p																			
n	*r*	.01	.05	.10	.15	.20	.25	.30	.35	.40	.45	.50	.55	.60	.65	.70	.75	.80	.85	.90	.95
16	0	.851	.440	.185	.074	.028	.010	.003	.001	.000	.000	.000	.000	.000	.000	.000	.000	.000	.000	.000	.000
	1	.138	.371	.329	.210	.113	.053	.023	.009	.003	.001	.000	.000	.000	.000	.000	.000	.000	.000	.000	.000
	2	.010	.146	.275	.277	.211	.134	.073	.035	.015	.006	.002	.001	.000	.000	.000	.000	.000	.000	.000	.000
	3	.000	.036	.142	.229	.246	.208	.146	.089	.047	.022	.009	.003	.001	.000	.000	.000	.000	.000	.000	.000
	4	.000	.006	.051	.131	.200	.225	.204	.155	.101	.057	.028	.011	.004	.001	.000	.000	.000	.000	.000	.000
	5	.000	.001	.014	.056	.120	.180	.210	.201	.162	.112	.067	.034	.014	.005	.001	.000	.000	.000	.000	.000
	6	.000	.000	.003	.018	.055	.110	.165	.198	.198	.168	.122	.075	.039	.017	.006	.001	.000	.000	.000	.000
	7	.000	.000	.000	.005	.020	.052	.101	.152	.189	.197	.175	.132	.084	.044	.019	.006	.001	.000	.000	.000
	8	.000	.000	.000	.001	.006	.020	.049	.092	.142	.181	.196	.181	.142	.092	.049	.020	.006	.001	.000	.000
	9	.000	.000	.000	.000	.001	.006	.019	.044	.084	.132	.175	.197	.189	.152	.101	.052	.020	.005	.000	.000
	10	.000	.000	.000	.000	.000	.001	.006	.017	.039	.075	.122	.168	.198	.198	.165	.110	.055	.018	.003	.000
	11	.000	.000	.000	.000	.000	.000	.001	.005	.014	.034	.067	.112	.162	.201	.210	.180	.120	.056	.014	.001
	12	.000	.000	.000	.000	.000	.000	.000	.001	.004	.011	.028	.057	.101	.155	.204	.225	.200	.131	.051	.006
	13	.000	.000	.000	.000	.000	.000	.000	.000	.001	.003	.009	.022	.047	.089	.146	.208	.246	.229	.142	.036
	14	.000	.000	.000	.000	.000	.000	.000	.000	.000	.001	.002	.006	.015	.035	.073	.134	.211	.277	.275	.146
	15	.000	.000	.000	.000	.000	.000	.000	.000	.000	.000	.000	.001	.003	.009	.023	.053	.113	.210	.329	.371
	16	.000	.000	.000	.000	.000	.000	.000	.000	.000	.000	.000	.000	.000	.001	.003	.010	.028	.074	.185	.440
20	0	.818	.358	.122	.039	.012	.003	.001	.000	.000	.000	.000	.000	.000	.000	.000	.000	.000	.000	.000	.000
	1	.165	.377	.270	.137	.058	.021	.007	.002	.000	.000	.000	.000	.000	.000	.000	.000	.000	.000	.000	.000
	2	.016	.189	.285	.229	.137	.067	.028	.010	.003	.001	.000	.000	.000	.000	.000	.000	.000	.000	.000	.000
	3	.001	.060	.190	.243	.205	.134	.072	.032	.012	.004	.001	.000	.000	.000	.000	.000	.000	.000	.000	.000
	4	.000	.013	.090	.182	.218	.190	.130	.074	.035	.014	.005	.001	.000	.000	.000	.000	.000	.000	.000	.000
	5	.000	.002	.032	.103	.175	.202	.179	.127	.075	.036	.015	.005	.001	.000	.000	.000	.000	.000	.000	.000
	6	.000	.000	.009	.045	.109	.169	.192	.171	.124	.075	.036	.015	.005	.001	.000	.000	.000	.000	.000	.000
	7	.000	.000	.002	.016	.055	.112	.164	.184	.166	.122	.074	.037	.015	.005	.001	.000	.000	.000	.000	.000
	8	.000	.000	.000	.005	.022	.061	.114	.161	.180	.162	.120	.073	.035	.014	.004	.001	.000	.000	.000	.000
	9	.000	.000	.000	.001	.007	.027	.065	.116	.160	.177	.160	.119	.071	.034	.012	.003	.000	.000	.000	.000
	10	.000	.000	.000	.000	.002	.010	.031	.069	.117	.159	.176	.159	.117	.069	.031	.010	.002	.000	.000	.000
	11	.000	.000	.000	.000	.000	.003	.012	.034	.071	.119	.160	.177	.160	.116	.065	.027	.007	.001	.000	.000
	12	.000	.000	.000	.000	.000	.001	.004	.014	.035	.073	.120	.162	.180	.161	.114	.061	.022	.005	.000	.000
	13	.000	.000	.000	.000	.000	.000	.001	.005	.015	.037	.074	.122	.166	.184	.164	.112	.055	.016	.002	.000
	14	.000	.000	.000	.000	.000	.000	.000	.001	.005	.015	.037	.075	.124	.171	.192	.169	.109	.045	.009	.000
	15	.000	.000	.000	.000	.000	.000	.000	.000	.001	.005	.015	.036	.075	.127	.179	.202	.175	.103	.032	.002
	16	.000	.000	.000	.000	.000	.000	.000	.000	.000	.001	.005	.014	.035	.074	.130	.190	.218	.182	.090	.013
	17	.000	.000	.000	.000	.000	.000	.000	.000	.000	.000	.001	.004	.012	.032	.072	.134	.205	.243	.190	.060
	18	.000	.000	.000	.000	.000	.000	.000	.000	.000	.000	.000	.001	.003	.010	.028	.067	.137	.229	.285	.189
	19	.000	.000	.000	.000	.000	.000	.000	.000	.000	.000	.000	.000	.000	.002	.007	.021	.058	.137	.270	.377
	20	.000	.000	.000	.000	.000	.000	.000	.000	.000	.000	.000	.000	.000	.000	.001	.003	.012	.039	.122	.358

TABLE 4 Areas of a Standard Normal Distribution

(a) Table of Areas to the Left of *z*

z	.00	.01	.02	.03	.04	.05	.06	.07	.08	.09
−3.4	.0003	.0003	.0003	.0003	.0003	.0003	.0003	.0003	.0003	.0002
−3.3	.0005	.0005	.0005	.0004	.0004	.0004	.0004	.0004	.0004	.0003
−3.2	.0007	.0007	.0006	.0006	.0006	.0006	.0006	.0005	.0005	.0005
−3.1	.0010	.0009	.0009	.0009	.0008	.0008	.0008	.0008	.0007	.0007
−3.0	.0013	.0013	.0013	.0012	.0012	.0011	.0011	.0011	.0010	.0010
−2.9	.0019	.0018	.0018	.0017	.0016	.0016	.0015	.0015	.0014	.0014
−2.8	.0026	.0025	.0024	.0023	.0023	.0022	.0021	.0021	.0020	.0019
−2.7	.0035	.0034	.0033	.0032	.0031	.0030	.0029	.0028	.0027	.0026
−2.6	.0047	.0045	.0044	.0043	.0041	.0040	.0039	.0038	.0037	.0036
−2.5	.0062	.0060	.0059	.0057	.0055	.0054	.0052	.0051	.0049	.0048
−2.4	.0082	.0080	.0078	.0075	.0073	.0071	.0069	.0068	.0066	.0064
−2.3	.0107	.0104	.0102	.0099	.0096	.0094	.0091	.0089	.0087	.0084
−2.2	.0139	.0136	.0132	.0129	.0125	.0122	.0119	.0116	.0113	.0110
−2.1	.0179	.0174	.0170	.0166	.0162	.0158	.0154	.0150	.0146	.0143
−2.0	.0228	.0222	.0217	.0212	.0207	.0202	.0197	.0192	.0188	.0183
−1.9	.0287	.0281	.0274	.0268	.0262	.0256	.0250	.0244	.0239	.0233
−1.8	.0359	.0351	.0344	.0336	.0329	.0322	.0314	.0307	.0301	.0294
−1.7	.0446	.0436	.0427	.0418	.0409	.0401	.0392	.0384	.0375	.0367
−1.6	.0548	.0537	.0526	.0516	.0505	.0495	.0485	.0475	.0465	.0455
−1.5	.0668	.0655	.0643	.0630	.0618	.0606	.0594	.0582	.0571	.0559
−1.4	.0808	.0793	.0778	.0764	.0749	.0735	.0721	.0708	.0694	.0681
−1.3	.0968	.0951	.0934	.0918	.0901	.0885	.0869	.0853	.0838	.0823
−1.2	.1151	.1131	.1112	.1093	.1075	.1056	.1038	.1020	.1003	.0985
−1.1	.1357	.1335	.1314	.1292	.1271	.1251	.1230	.1210	.1190	.1170
−1.0	.1587	.1562	.1539	.1515	.1492	.1469	.1446	.1423	.1401	.1379
−0.9	.1841	.1814	.1788	.1762	.1736	.1711	.1685	.1660	.1635	.1611
−0.8	.2119	.2090	.2061	.2033	.2005	.1977	.1949	.1922	.1894	.1867
−0.7	.2420	.2389	.2358	.2327	.2296	.2266	.2236	.2206	.2177	.2148
−0.6	.2743	.2709	.2676	.2643	.2611	.2578	.2546	.2514	.2483	.2451
−0.5	.3085	.3050	.3015	.2981	.2946	.2912	.2877	.2843	.2810	.2776
−0.4	.3446	.3409	.3372	.3336	.3300	.3264	.3228	.3192	.3156	.3121
−0.3	.3821	.3783	.3745	.3707	.3669	.3632	.3594	.3557	.3520	.3483
−0.2	.4207	.4168	.4129	.4090	.4052	.4013	.3974	.3936	.3897	.3859
−0.1	.4602	.4562	.4522	.4483	.4443	.4404	.4364	.4325	.4286	.4247
−0.0	.5000	.4960	.4920	.4880	.4840	.4801	.4761	.4721	.4681	.4641

For values of *z* less than −3.49, use 0.000 to approximate the area.

TABLE 4(a) *continued*

z	.00	.01	.02	.03	.04	.05	.06	.07	.08	.09
0.0	.5000	.5040	.5080	.5120	.5160	.5199	.5239	.5279	.5319	.5359
0.1	.5398	.5438	.5478	.5517	.5557	.5596	.5636	.5675	.5714	.5753
0.2	.5793	.5832	.5871	.5910	.5948	.5987	.6026	.6064	.6103	.6141
0.3	.6179	.6217	.6255	.6293	.6331	.6368	.6406	.6443	.6480	.6517
0.4	.6554	.6591	.6628	.6664	.6700	.6736	.6772	.6808	.6844	.6879
0.5	.6915	.6950	.6985	.7019	.7054	.7088	.7123	.7157	.7190	.7224
0.6	.7257	.7291	.7324	.7357	.7389	.7422	.7454	.7486	.7517	.7549
0.7	.7580	.7611	.7642	.7673	.7704	.7734	.7764	.7794	.7823	.7852
0.8	.7881	.7910	.7939	.7967	.7995	.8023	.8051	.8078	.8106	.8133
0.9	.8159	.8186	.8212	.8238	.8264	.8289	.8315	.8340	.8365	.8389
1.0	.8413	.8438	.8461	.8485	.8508	.8531	.8554	.8577	.8599	.8621
1.1	.8643	.8665	.8686	.8708	.8729	.8749	.8770	.8790	.8810	.8830
1.2	.8849	.8869	.8888	.8907	.8925	.8944	.8962	.8980	.8997	.9015
1.3	.9032	.9049	.9066	.9082	.9099	.9115	.9131	.9147	.9162	.9177
1.4	.9192	.9207	.9222	.9236	.9251	.9265	.9279	.9292	.9306	.9319
1.5	.9332	.9345	.9357	.9370	.9382	.9394	.9406	.9418	.9429	.9441
1.6	.9452	.9463	.9474	.9484	.9495	.9505	.9515	.9525	.9535	.9545
1.7	.9554	.9564	.9573	.9582	.9591	.9599	.9608	.9616	.9625	.9633
1.8	.9641	.9649	.9656	.9664	.9671	.9678	.9686	.9693	.9699	.9706
1.9	.9713	.9719	.9726	.9732	.9738	.9744	.9750	.9756	.9761	.9767
2.0	.9772	.9778	.9783	.9788	.9793	.9798	.9803	.9808	.9812	.9817
2.1	.9821	.9826	.9830	.9834	.9838	.9842	.9846	.9850	.9854	.9857
2.2	.9861	.9864	.9868	.9871	.9875	.9878	.9881	.9884	.9887	.9890
2.3	.9893	.9896	.9898	.9901	.9904	.9906	.9909	.9911	.9913	.9916
2.4	.9918	.9920	.9922	.9925	.9927	.9929	.9931	.9932	.9934	.9936
2.5	.9938	.9940	.9941	.9943	.9945	.9946	.9948	.9949	.9951	.9952
2.6	.9953	.9955	.9956	.9957	.9959	.9960	.9961	.9962	.9963	.9964
2.7	.9965	.9966	.9967	.9968	.9969	.9970	.9971	.9972	.9973	.9974
2.8	.9974	.9975	.9976	.9977	.9977	.9978	.9979	.9979	.9980	.9981
2.9	.9981	.9982	.9982	.9983	.9984	.9984	.9985	.9985	.9986	.9986
3.0	.9987	.9987	.9987	.9988	.9988	.9989	.9989	.9989	.9990	.9990
3.1	.9990	.9991	.9991	.9991	.9992	.9992	.9992	.9992	.9993	.9993
3.2	.9993	.9993	.9994	.9994	.9994	.9994	.9994	.9995	.9995	.9995
3.3	.9995	.9995	.9995	.9996	.9996	.9996	.9996	.9996	.9996	.9997
3.4	.9997	.9997	.9997	.9997	.9997	.9997	.9997	.9997	.9997	.9998

For z values greater than 3.49, use 1.000 to approximate the area.

TABLE 4 *continued*

(b) Confidence Interval, Critical Values z_c

Level of Confidence c	Critical Value z_c
0.75, or 75%	1.15
0.80, or 80%	1.28
0.85, or 85%	1.44
0.90, or 95%	1.645
0.95, or 95%	1.96
0.98, or 98%	2.33
0.99, or 99%	2.58

TABLE 4 *continued*

(c) Hypothesis Testing, Critical Values z_0

Level of Significance	$\alpha = 0.05$	$\alpha = 0.01$
Critical value z_0 for a left-tailed test	−1.645	−2.33
Critical value z_0 for a right-tailed test	1.645	2.33
Critical value $\pm z_0$ for a two-tailed test	±1.96	±2.58

TABLE 5 Student's *t* Distribution

Student's *t* values generated by Minitab Version 9.2

d.f. \ c / α' / α''	0.750 / 0.125 / 0.250	0.800 / 0.100 / 0.200	0.850 / 0.075 / 0.150	0.900 / 0.050 / 0.100	0.950 / 0.025 / 0.050	0.980 / 0.010 / 0.020	0.990 / 0.005 / 0.010
1	2.414	3.078	4.165	6.314	12.706	31.821	63.657
2	1.604	1.886	2.282	2.290	4.303	6.965	9.925
3	1.423	1.638	1.924	2.353	3.182	4.541	5.841
4	1.344	1.533	1.778	2.132	2.776	3.747	4.604
5	1.301	1.476	1.699	2.015	2.571	3.365	4.032
6	1.273	1.440	1.650	1.943	2.447	3.143	3.707
7	1.254	1.415	1.617	1.895	2.365	2.998	3.499
8	1.240	1.397	1.592	1.860	2.306	2.896	3.355
9	1.230	1.383	1.574	1.833	2.262	2.821	3.250
10	1.221	1.372	1.559	1.812	2.228	2.764	3.169
11	1.214	1.363	1.548	1.796	2.201	2.718	3.106
12	1.209	1.356	1.538	1.782	2.179	2.681	3.055
13	1.204	1.350	1.530	1.771	2.160	2.650	3.012
14	1.200	1.345	1.523	1.761	2.145	2.624	2.977
15	1.197	1.341	1.517	1.753	2.131	2.602	2.947
16	1.194	1.337	1.512	1.746	2.120	2.583	2.921
17	1.191	1.333	1.508	1.740	2.110	2.567	2.898
18	1.189	1.330	1.504	1.734	2.101	2.552	2.878
19	1.187	1.328	1.500	1.729	2.093	2.539	2.861
20	1.185	1.325	1.497	1.725	2.086	2.528	2.845
21	1.183	1.323	1.494	1.721	2.080	2.518	2.831
22	1.182	1.321	1.492	1.717	2.074	2.508	2.819
23	1.180	1.319	1.489	1.714	2.069	2.500	2.807
24	1.179	1.318	1.487	1.711	2.064	2.492	2.797
25	1.178	1.316	1.485	1.708	2.060	2.485	2.787
26	1.177	1.315	1.483	1.706	2.056	2.479	2.779
27	1.176	1.314	1.482	1.703	2.052	2.473	2.771
28	1.175	1.313	1.480	1.701	2.048	2.467	2.763
29	1.174	1.311	1.479	1.699	2.045	2.462	2.756
30	1.173	1.310	1.477	1.697	2.042	2.457	2.750
35	1.170	1.306	1.472	1.690	2.030	2.438	2.724
40	1.167	1.303	1.468	1.684	2.021	2.423	2.704
45	1.165	1.301	1.465	1.679	2.014	2.412	2.690
50	1.164	1.299	1.462	1.676	2.009	2.403	2.678
55	1.163	1.297	1.460	1.673	2.004	2.396	2.668
60	1.162	1.296	1.458	1.671	2.000	2.390	2.660
90	1.158	1.291	1.452	1.662	1.987	2.369	2.632
120	1.156	1.289	1.449	1.658	1.980	2.358	2.617
∞	1.150	1.282	1.440	1.645	1.960	2.326	2.576

TABLE 6 The χ^2 Distribution

d.f.\ α	.995	.990	.975	.950	.900	.100	.050	.025	.010	.005
1	0.0^4393	0.0^3157	0.0^3982	0.0^2393	0.0158	2.71	3.84	5.02	6.63	7.88
2	0.0100	0.0201	0.0506	0.103	0.211	4.61	5.99	7.38	9.21	10.60
3	0.072	0.115	0.216	0.352	0.584	6.25	7.81	9.35	11.34	12.84
4	0.207	0.297	0.484	0.711	1.064	7.78	9.49	11.14	13.28	14.86
5	0.412	0.554	0.831	1.145	1.61	9.24	11.07	12.83	15.09	16.75
6	0.676	0.872	1.24	1.64	2.20	10.64	12.59	14.45	16.81	18.55
7	0.989	1.24	1.69	2.17	2.83	12.02	14.07	16.01	18.48	20.28
8	1.34	1.65	2.18	2.73	3.49	13.36	15.51	17.53	20.09	21.96
9	1.73	2.09	2.70	3.33	4.17	14.68	16.92	19.02	21.67	23.59
10	2.16	2.56	3.25	3.94	4.87	15.99	18.31	20.48	23.21	25.19
11	2.60	3.05	3.82	4.57	5.58	17.28	19.68	21.92	24.72	26.76
12	3.07	3.57	4.40	5.23	6.30	18.55	21.03	23.34	26.22	28.30
13	3.57	4.11	5.01	5.89	7.04	19.81	22.36	24.74	27.69	29.82
14	4.07	4.66	5.63	6.57	7.79	21.06	23.68	26.12	29.14	31.32
15	4.60	5.23	6.26	7.26	8.55	22.31	25.00	27.49	30.58	32.80
16	5.14	5.81	6.91	7.96	9.31	23.54	26.30	28.85	32.00	34.27
17	5.70	6.41	7.56	8.67	10.09	24.77	27.59	30.19	33.41	35.72
18	6.26	7.01	8.23	9.39	10.86	25.99	28.87	31.53	34.81	37.16
19	6.84	7.63	8.91	10.12	11.65	27.20	30.14	32.85	36.19	38.58
20	7.43	8.26	8.59	10.85	12.44	28.41	31.41	34.17	37.57	40.00
21	8.03	8.90	10.28	11.59	13.24	29.62	32.67	35.48	38.93	41.40
22	8.64	9.54	10.98	12.34	14.04	30.81	33.92	36.78	40.29	42.80
23	9.26	10.20	11.69	13.09	14.85	32.01	35.17	38.08	41.64	44.18
24	9.89	10.86	12.40	13.85	15.66	33.20	36.42	39.36	42.98	45.56
25	10.52	11.52	13.12	14.61	16.47	34.38	37.65	40.65	44.31	46.93
26	11.16	12.20	13.84	15.38	17.29	35.56	38.89	41.92	45.64	48.29
27	11.81	12.88	14.57	16.15	18.11	36.74	40.11	43.19	46.96	49.64
28	12.46	13.56	15.31	16.93	18.94	37.92	41.34	44.46	48.28	50.99
29	13.21	14.26	16.05	17.71	19.77	39.09	42.56	45.72	49.59	52.34
30	13.79	14.95	16.79	18.49	20.60	40.26	43.77	46.98	50.89	53.67
40	20.71	22.16	24.43	26.51	29.05	51.80	55.76	59.34	63.69	66.77
50	27.99	29.71	32.36	34.76	37.69	63.17	67.50	71.42	76.15	79.49
60	35.53	37.48	40.48	43.19	46.46	74.40	79.08	83.30	88.38	91.95
70	43.28	45.44	48.76	51.74	55.33	85.53	90.53	95.02	100.4	104.2
80	51.17	53.54	57.15	60.39	64.28	96.58	101.9	106.6	112.3	116.3
90	59.20	61.75	65.65	69.13	73.29	107.6	113.1	118.1	124.1	128.3
100	67.33	70.06	74.22	77.93	82.36	118.5	124.3	129.6	135.8	140.2

Source: From H. L. Herter, *Biometrika*, June 1964. Printed by permission of Biometrika Trustees.

FREQUENTLY USED FORMULAS

n = sample size N = population size f = frequency

Chapter 2

Class Width $= \dfrac{\text{high} - \text{low}}{\text{number classes}}$ (increase to next integer)

Class Midpoint $= \dfrac{\text{upper limit} + \text{lower limit}}{2}$

Lower boundary = lower boundary of previous class + class width

Chapter 3

Sample mean $\bar{x} = \dfrac{\Sigma x}{n}$

Population mean $\mu = \dfrac{\Sigma x}{N}$

Weighted average $= \dfrac{\Sigma xw}{\Sigma w}$

Range = largest data value − smallest data value

Sample standard deviation $s = \sqrt{\dfrac{\Sigma(x-\bar{x})^2}{n-1}}$

Computation formula $s = \sqrt{\dfrac{SS_x}{n-1}}$ where $SS_x = \Sigma x^2 - \dfrac{(\Sigma x)^2}{n}$

Population standard deviation $\sigma = \sqrt{\dfrac{\Sigma(x-\mu)^2}{N}}$

Sample variance s^2

Population variance σ^2

Sample Coefficient of Variation $CV = \dfrac{s}{\bar{x}} \bullet 100$

Sample mean for grouped data $\bar{x} = \dfrac{\Sigma xf}{n}$

Sample standard deviation for grouped data $s = \sqrt{\dfrac{\Sigma(x-\bar{x})^2 f}{n-1}} = \sqrt{\dfrac{\Sigma x^2 f - (\Sigma xf)^2 / n}{n-1}}$

Chapter 4

$$SS_x = \sum x^2 - \frac{(\sum x)^2}{n}$$

$$SS_y = \sum y^2 - \frac{(\sum y)^2}{n}$$

$$SS_{xy} = \sum xy - \frac{(\sum x \sum y)}{n}$$

Least-squares line $y = a + bx$ where $b = \frac{SS_{xy}}{SS_x}$ and $a = \overline{y} - b\overline{x}$

Pearson product-moment correlation coefficient $r = \frac{SS_{xy}}{\sqrt{SS_x SS_y}}$

Coefficient of determination = r^2

Chapter 5

Probability of the complement of event A $P(not\ A) = 1 - P(A)$

Multiplication rule for independent events $P(A\ and\ B) = P(A) \bullet P(B)$

General multiplication rules $P(A\ and\ B) = P(A) \bullet P(B, given\ A)$
$P(A\ and\ B) = P(B) \bullet P(A, given\ B)$

Addition rule for mutually exclusive events $P(A\ or\ B) = P(A) + P(B)$

General addition rule $P(A\ or\ B) = P(A) + P(B) - P(A\ and\ B)$

Permutation rule $P_{n,r} = \frac{n!}{(n-r)!}$

Combination rule $C_{n,r} = \frac{n!}{r!(n-r)!}$

Chapter 6

Mean of a discrete probability distribution $\mu = \Sigma x P(x)$

Standard deviation of a discrete probability distribution $\sigma = \sqrt{\Sigma (x-\mu)^2 P(x)}$

For Binomial Distributions

r = number of successes; p = probability of success; $q = 1 - p$

Binomial probability distribution $P(r) = \frac{n!}{r!(n-r)!} p^r q^{n-r}$

Mean $\mu = np$

Standard deviation $\sigma = \sqrt{npq}$

Chapter 7

Raw score $x = z\sigma + \mu$

Standard score $z = \dfrac{x - \mu}{\sigma}$

Chapter 8

Mean of $\overline{x}$ distribution $\mu_{\overline{x}} = \mu$

Standard deviation of $\overline{x}$ distribution $\sigma_{\overline{x}} = \dfrac{\sigma}{\sqrt{n}}$

Standard score for $\overline{x}$ $z = \dfrac{\overline{x} - \mu}{\sigma / \sqrt{n}}$

Chapter 9

Confidence Interval

for $\mu\,(n \geq 30)$ $\overline{x} - z_c \dfrac{\sigma}{\sqrt{n}} < \mu < \overline{x} + z_c \dfrac{\sigma}{\sqrt{n}}$

for $\mu\,(n < 30);\, d.f. = n - 1$ $\overline{x} - t_c \dfrac{s}{\sqrt{n}} < \mu < \overline{x} + t_c \dfrac{s}{\sqrt{n}}$

for $p\,(np > 5 \text{ and } nq > 5)$ $\hat{p} - z_c \sqrt{\dfrac{\hat{p}(1-\hat{p})}{n}} < p < \hat{p} + z_c \sqrt{\dfrac{\hat{p}(1-\hat{p})}{n}}$ where $\hat{p} = r / n$

Sample Size for Estimating

means $n = \left(\dfrac{z_c \sigma}{E}\right)^2$

proportions $n = p(1-p)\left(\dfrac{z_c}{E}\right)^2$ with preliminary estimate for p

$n = \dfrac{1}{4}\left(\dfrac{z_c}{E}\right)^2$ without preliminary estimate for p

Chapter 10

Sample Test Statistics for Tests of Hypotheses

for $\mu\,(n \geq 30)$ $z = \dfrac{\overline{x} - \mu}{\sigma / \sqrt{n}}$

for $\mu\,(n < 30);\, d.f. = n - 1$ $t = \dfrac{\overline{x} - \mu}{s / \sqrt{n}}$

for p $z = \dfrac{\hat{p} - p}{\sqrt{pq / n}}$ where $q = 1 - p$

Chapter 11

Sample Test Statistics for Tests of Hypothesis

for paired differences d $\quad t = \dfrac{\overline{d} = \mu_d}{s_d / \sqrt{n}}$ with $d.f. = n - 1$

difference of means large sample $\quad z = \dfrac{(\overline{x}_1 - \overline{x}_2) - (\mu_1 - \mu_2)}{\sqrt{\dfrac{\sigma_1^2}{n_1} + \dfrac{\sigma_2^2}{n_2}}}$

difference of means small sample with $\sigma_1 \approx \sigma_2$; $d.f. = n_1 + n_2 - 2$

$$t = \frac{(\overline{x}_1 - \overline{x}_2) - (\mu_1 - \mu_2)}{s\sqrt{\dfrac{1}{n_1} + \dfrac{1}{n_2}}} \quad \text{where} \quad s = \sqrt{\frac{(n_1 - 1)s_1^2 + (n_2 - 1)s_2^2}{n_1 + n_2 - 2}}$$

difference of proportions $\quad z = \dfrac{\hat{p}_1 - \hat{p}_2}{\sqrt{\dfrac{\hat{p}\hat{q}}{n_1} + \dfrac{\hat{p}\hat{q}}{n_2}}}$ where $\hat{p} = \dfrac{r_1 + r_2}{n_1 + n_2}; \hat{q} = 1 - \hat{p};\ \hat{p}_1 = r_1 / n_1; \hat{p}_2 = r_2 / n_2$

Confidence Intervals

for difference of means (when $n_1 \geq 30$ and $n_2 \geq 30$)

$$(\overline{x}_1 - \overline{x}_2) - z_c\sqrt{\frac{\sigma_1^2}{n_1} + \frac{\sigma_2^2}{n_2}} < \mu_1 - \mu_2 < (\overline{x}_1 - \overline{x}_2) + z_c\sqrt{\frac{\sigma_1^2}{n_1} + \frac{\sigma_2^2}{n_2}}$$

for difference of means ($n_1 < 30$ and/or $n_2 < 30$ and $\sigma_1 \approx \sigma_2$) ; $d.f. = n_1 + n_2 - 2$

$$(\overline{x}_1 - \overline{x}_2) - t_c s\sqrt{\frac{1}{n_1} + \frac{1}{n_2}} < \mu_1 - \mu_2 < (\overline{x}_1 - \overline{x}_2) + t_c s\sqrt{\frac{1}{n_1} + \frac{1}{n_2}} \quad \text{where} \quad s = \sqrt{\frac{(n_1 - 1)s_1^2 + (n_2 - 1)s_2^2}{n_1 + n_2 - 2}}$$

for difference of proportions where $\hat{p}_1 = r_1 / n_1; \hat{p}_2 = r_2 / n_2;\ \hat{q}_1 = 1 - \hat{p}_1;\ \hat{q}_2 = 1 - \hat{p}_2$

$$(\hat{p}_1 - \hat{p}_2) - z_c\sqrt{\frac{\hat{p}_1\hat{q}_1}{n_1} + \frac{\hat{p}_2\hat{q}_2}{n_2}} < p_1 - p_2 < (\hat{p}_1 - \hat{p}_2) + z_c\sqrt{\frac{\hat{p}_1\hat{q}_1}{n_1} + \frac{\hat{p}_2\hat{q}_2}{n_2}}$$

Chapter 12

$x_2 = \sum \frac{(O-E)^2}{E}$ where $E = \frac{\text{(row total)(column total)}}{\text{sample size}}$

Tests of Independence $d.f. = (R-1)(C-1)$

Goodness of fit $d.f. = \text{(number of entries)} - 1$

Sample test statistic for $H_0 : \sigma^2 = k;\ d.f. = n-1$ $\chi^2 = \frac{(n-1)s^2}{\sigma^2}$

Linear Regression

Standard error of estimate $S_3^e = \sqrt{\frac{SS_y - bSS_{xy}}{n-2}}$ where $b = \frac{SS_{xy}}{SS_x}$

Confidence interval for y

$y_p - E < y < y_p + E$ where y_p is the predicted y value for x and $E = t_c S_e \sqrt{1 + \frac{1}{n} + \frac{(x-\overline{x})^2}{SS_x}}$ with $d.f. = n - 2$

Part III

Sample Chapter Tests with Answers

CHAPTER 1 TEST
FORM A

1. The Colorado State Legislature wants to estimate the length of time it takes a resident of Colorado to earn a Bachelor's degree from a state college or university. A random sample of 265 recent in-state graduates were surveyed.

   (a) Identify the variable. **1. (a)** ______________

   (b) Is the variable quantitative or qualitative? **(b)** ______________

   (c) What is the implied population? **(c)** ______________

2. For the information in parts **(a)** through **(g)** below, list the highest level of measurement as ratio, interval, ordinal, or nominal and explain your choice.

   A student advising file contains the following information.

   (a) Name of student **2. (a)** ______________

   (b) Student I.D. number **(b)** ______________

   (c) Cumulative grade point average **(c)** ______________

   (d) Dates of awards (scholarships, dean's list, ...) **(d)** ______________

   (e) Declared major or undecided if no major declared **(e)** ______________

   (f) A number code representing class standing: 1 = Freshman, 2 = Sophomore, 3 = Junior, 4 = Senior, 5 = Graduate student **(f)** ______________

   (g) Entrance exam rating for competency in English: Excellent, Satisfactory, Unsatisfactory **(g)** ______________

3. Categorize the style of gathering data (sampling, experiment, simulation, census) described in each of the following situations.

   (a) Look at all the apartments in a complex and determine the monthly rent charged for each unit. **3. (a)** ______________

   (b) Give one group of students a flu vaccination and compare the number of times these students are sick during the semester with students in a group who did not receive the vaccination. **(b)** ______________

   (c) Select a sample of students and determine the percentage who are taking mathematics this semester. **(c)** ______________

   (d) Use a computer program to show the effects on traffic flow when the timing of stop lights is changed. **(d)** ______________

CHAPTER 1, FORM A, PAGE 2

4. Write a brief essay in which you describe what is meant by an experiment. Given an example of a situation in which data is gathered by means of an experiment. How is gathering data from an experiment different from using a sample from a specified population?

4. ______________________________

5. Consider the experiment of rolling a single die. Describe how you would use a random number table to simulate the outcomes of rolling a single die. Using the following row of random numbers from the table, find the first five outcomes.

36017 98590 64180 72315 39710

5. ______________________________

6. Identify each of the following samples by naming the sampling technique used (cluster, convenience, simple random, stratified, systematic).

(a) Measure the length of time every fifth person coming into a bank waits for teller service over a period of two days.

6. (a) ______________________________

(b) Take a sample of five Zip codes from the Chicago metropolitan region and use all the elementary schools from each of the Zip code regions. Determine the number of students enrolled in first grade in each of the schools selected.

(b) ______________________________

(c) Divide the users of the computer online service Internet into different age groups and then select a random sample from each age group to survey about the amount of time they are connected to Internet each month.

(c) ______________________________

(d) Survey five friends regarding their opinion of the student cafeteria.

(d) ______________________________

(e) Pick a random sample of students enrolled at your college and determine the number of credit hours they have each accumulated toward their degree program.

(e) ______________________________

CHAPTER 1 TEST
FORM B

1. A book store wants to estimate the proportion of its customers who buy murder mysteries. A random sample of 76 customers are observed at the checkout counter and the number purchasing murder mysteries is recorded.

(a) Identify the variable. **1. (a)** __________

(b) Is the variable quantitative or qualitative? **(b)** __________

(c) What is the implied population? **(c)** __________

2. For the information in parts **(a)** through **(e)** below, list the highest level of measurement as ratio, interval, ordinal, or nominal and explain your choice.

A restaurant manager is developing a clientele profile. Some of the information for the profile follows:

(a) Gender of diners **2. (a)** __________

(b) Size of groups dining together **(b)** __________

(c) Time of day the last diner of the evening departs **(c)** __________

(d) Age grouping: young, middle age, senior **(d)** __________

(e) Length of time a diner waits for a table. **(e)** __________

3. Categorize the style of gathering data (sampling, experiment, simulation, census) for the following situations.

(a) Consider all the students enrolled at your college this semester and report the age of each student. **3. (a)** __________

(b) Select a sample of new F10 pickup trucks and count the number of manufacturer defects in each of the trucks. **(b)** __________

(c) Use computer graphics to determine the flight path of a golf ball when the position of the hand on the golf club is changed. **(c)** __________

(d) Teach one section of English composition using a specific word processing package and teach another without using any computerized word processing. Count the number of grammar errors made by students in each section on a final draft of a 20 page term paper. **(d)** __________

CHAPTER 1, FORM B, PAGE 2

4. Write a brief essay in which you discuss some of the aspects surveys. Give specific examples to illustrate your main points. **4.** ____________________

5. A business employs 736 people. Describe how you could get a random sample of size 30 to survey regarding desire for professional training opportunities. Identify the first 5 to be included in the sample using the following random number sequence.

62283 14130 55790 40133 47596 17654 **5.** ____________________

6. To determine monthly rental prices of apartment units in the San Francisco area, samples were constructed in the following ways. Categorize (cluster, convenience, simple random, stratified, systematic) each sampling technique described.

(a) Number all the units in the area and use a random number table to select the apartments to include in the sample. **6. (a)** ____________________

(b) Divide the apartment units according to number of bedrooms and then sample from each of the groups. **(b)** ____________________

(c) Select 5 Zip codes at random and include every apartment unit in the selected Zip codes. **(c)** ____________________

(d) Look in the newspaper and consider the first sample of apartment units that list rent per month. **(d)** ____________________

(e) Call every 50th apartment complex listed in the yellow pages and record the rent of the unit with unit number closest to 200. **(e)** ____________________

CHAPTER 1 TEST
FORM C

Write the letter of the response that best answers each problem.

1. A consumer research company wants to estimate the average cost of an airline ticket for a round trip within the continental United States. A random sample of 50 airfares was gathered giving an average price of $438. Identify the variable. **1.** ________

(a) Random sample of 50 airfares (b) Airline fare

(c) Consumer research company (d) Quantitative (e) $438

2. For the information in parts A. through E., choose the highest level of measurement (or cannot determine):

(a) Ratio (b) Interval

(c) Ordinal (d) Nominal (e) Cannot determine

A. Temperature of refrigerators **2. A.** ________

B. Horsepower of racecar engines **B.** ________

C. Marital status of school board members **C.** ________

D. Ratings of television programs (poor, fair, good, excellent) **D.** ________

E. Ages of children enrolled in a daycare **E.** ________

3. Categorize the style of gathering data described in each of the following situations:

(a) Sampling (b) Experiment

(c) Simulation (d) Census (e) Cannot determine

A. Give one group of people a diet supplement and another a placebo. After both groups have been on the same meal program for one month, compare the weight losses of the two groups. **3. A.** ________

B. Use a computer program to show the effects on airline traffic flow where air traffic controllers change methods. **B.** ________

C. Select a sample of consumers and determine the percentage who own cellular phones. **C.** ________

D. Determine the annual income for all employees in a company. **D.** ________

CHAPTER 1, FORM C, PAGE 2

4. Consider the following study:

Students in a limnology class took water samples from a lake to determine the temperature at different depths. Which of the following techniques for gathering data do you think was used? **4.** ________

(a) Double-blind experiment (b) Experiment

(c) Analysis of variance (d) Placebo effect (e) Observational study

5. When using a random-number table to get a list of nine random numbers from 57 to 634, you would use groups of **5.** ________

(a) 9 digits. (b) 2 digits.

(c) 1 digit. (d) 3 digits. (e) 2 digits and then 3 digits.

6. Identify each of the following samples by naming the sampling technique used.

(a) Cluster (b) Convenience

(c) Simple random (d) Stratified (e) Systematic

A. Every tenth customer entering a health club is asked to select his or her preferred method of exercise. **6. A.** ________

B. Divide the subscribers of a magazine into three different income categories and then select a random sample from each category to survey about their favorite feature. **B.** ________

C. Take a sample of six Zip codes from the Minneapolis metropolitan region and use all the car dealerships in the selected areas. Determine the number of new cars sold each month at each dealership. **C.** ________

D. Use a random number table to select a sample of books and determine the number of pages in each book. **D.** ________

E. Determine the annual salary of each of the nurses that are on duty at the time you chose to interview at the hospital. **E.** ________

CHAPTER 2 TEST
FORM A

1. The Dean's Office at Hendrix College gave the following information about numbers of majors in different academic areas: Humanities, 372; Natural Science, 415; Social Science, 511; Business Administration, 619; Philosophy, 196. Make a Pareto chart representing this information.

1.

2. Professor Hill in the Music Department kept a list of the number of students visiting his office each week for two semesters (30 weeks). The results were

15	23	17	13	3	9	7	6	8	11
16	32	27	4	20	3	28	5	6	11
20	12	8	10	25	10	8	15	11	9

(a) Make a frequency table with five classes, showing class boundaries, class midpoints, frequencies, and relative frequencies.

2. (a)

(b) Make a frequency histogram with five classes.

(b)

CHAPTER 2, FORM A, PAGE 2

(c) Make a relative frequency histogram with five classes. **(c)**

3. Jim is a taxi driver who keeps a record of his meter readings. The results for the past twenty meter readings (rounded to the nearest dollar) are given below.

15	7	9	21	19	17	8	35	22	33
46	5	24	37	51	49	57	42	12	16

Make a stem-and-leaf display of the data. **3.** ______________

4. The Air Pollution Index in Denver for each day of the second week of February is shown below. **4.**

1.7 2.4 5.3 4.1 3.2 2.0 2.5

Make a time plot for these data.

CHAPTER 2, FORM A, PAGE 3

5. A survey of 100 students was taken to see how they preferred to study. The survey showed that 38 students liked it quiet, 20 students liked the television on, 34 students liked the stereo on, and 8 students liked white noise such as in a lunch room. Make a circle graph to display this information.

5.

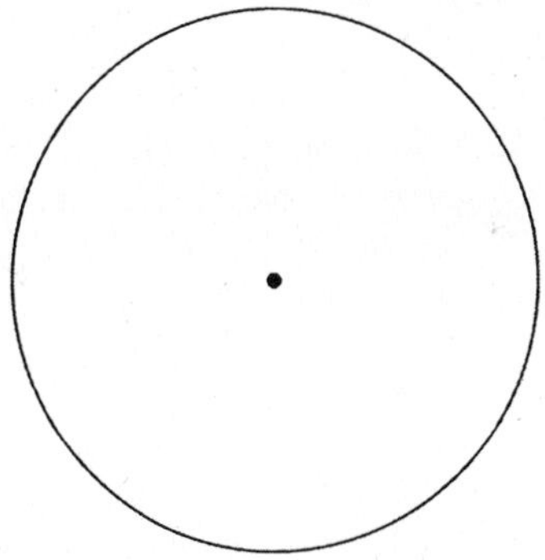

6. Of all the shoppers at a supermarket on a given day, it was determined that 71% were women under age 60, 20% were women 60 years or older, 7% were men under age 60 and 2% were men 60 years or older. Make a pie chart of this information.

6.

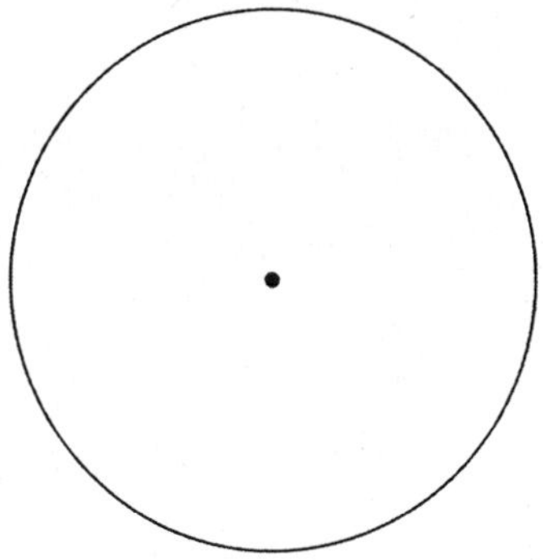

7. Make a dotplot for the data in Problem 2 regarding the number of students visiting the office. Compare the dotplot to the histogram in Problem 2.

7.

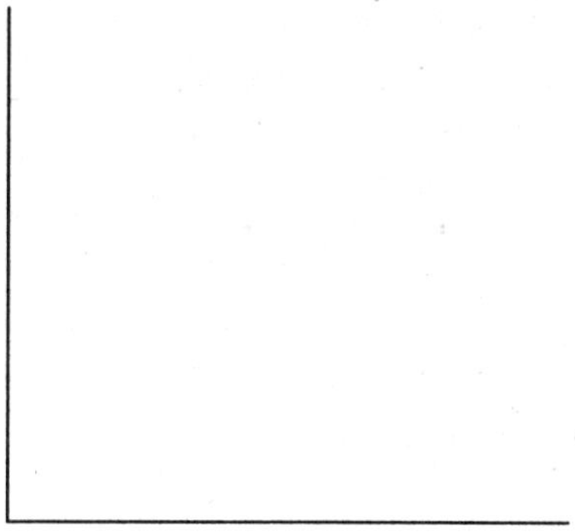

8. A sample of 20 motorists was taken from a freeway where the speed limit was 65 mph. A dotplot of their speeds is shown below. How many motorists were speeding?

8. ______________________

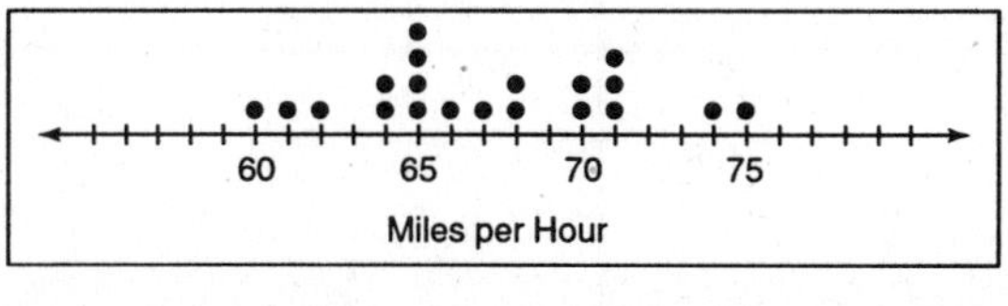

9. Following is a list of ages of participants entered in a 5K fun run. Make a stem-and-leaf display for these data and describe the distributions.

24	31	8	29	36	55	42	40	22	19	24
43	38	18	32	50	10	28	35	25	28	47

9. ______________________

CHAPTER 2 TEST
FORM B

1. A book store recorded the following sales last month by genre: Romance, 519; Murder Mystery, 732; Biography, 211; Self help, 819; Travel guide, 143; Children's books, 643. Make a Pareto chart displaying this information.

1.

2. The College Registrar's Office recorded the number of students receiving a grade of Incomplete. Results for the past 24 quarters are

28	47	19	58	63	77	53	39	93	35
42	81	62	67	71	59	48	56	75	48
63	32	46	57						

(a) Make a frequency table with five classes, showing class boundaries, class midpoints, frequencies, and relative frequencies.

2. (a)

(b) Make a frequency histogram with five classes.

(b)

CHAPTER 2, FORM B, PAGE 2

(c) Make a relative frequency histogram with five classes. (c)

3. The Humanities Division recorded the number of students signed up for the Study Abroad Program each quarter. The results are

58	26	21	29	33	47	42	38	44	56
52	64	68	59	63	36	34	45	51	50

Make a stem-and-leaf display of the data. 3. ______________

4. The Air Pollution Index in Denver for each day of the second week of February is shown below. 4.

Week	1	2	3	4	5	6
Price (S)	289	291	298	305	311	322

Week	7	8	9	10	11	12
Price (S)	316	300	290	299	291	288

Make a time plot for these data.

CHAPTER 2, FORM B, PAGE 3

5. A college senior class has 5000 students. Their graduation forms have their chosen major. There are 800 who chose social science, 400 who chose science, 1100 who chose humanities, 1400 who chose computer-related majors, 900 who chose engineering, and 400 who have yet to fill their major. Make a circle graph to display this information.

5.

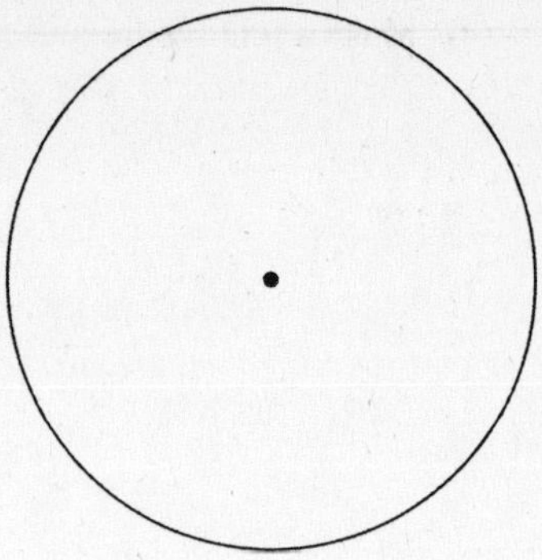

6. The school administration would like to know who is taking the city bus to school. A survey showed that the buses held 61% freshman, 25% sophomores, 12% juniors, and 2% seniors. Make a pie chart of this information.

6.

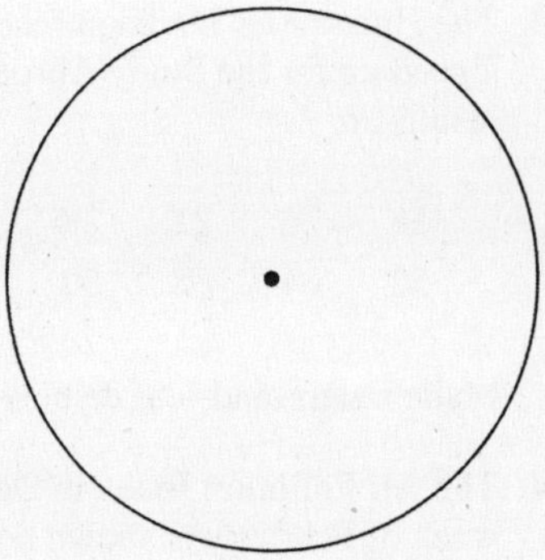

7. Make a dotplot for the data in Problem 2 regarding the number of students receiving a grade of Incomplete. Compare the dotplot to the histogram of Problem 2.

7.

8. A sample of 15 days was selected from the summer season. A dotplot of the daily high temperature is shown here. How many days were colder than 70°F?

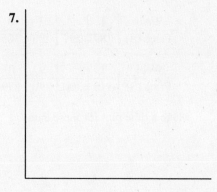

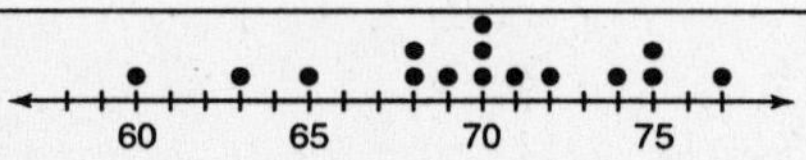

8. ______________________

9. Following is a list of diameters (in mm) of holes produced by an assembly line machine. Make a stem-and-leaf display for these data and describe the distribution.

2.3	3.7	1.2	3.6	2.4	2.6	3.7	0.9
1.8	2.5	2.5	3.0	2.8	1.7	3.1	4.1

9. ______________________

CHAPTER 2 TEST
FORM C

Write the letter of the response that best answers each problem.

1. __________ identify the frequency of events or categories in decreasing order of frequency of occurrence. **1.** __________

(a) Time plots (b) Bar graphs

(c) Pareto charts (d) Stem-and-Leaf Displays (e) Circle graphs

2. __________ are useful for quantitative or qualitative data. With qualitative date, the frequency or percentage of occurrence can be displayed. With quantitative data, the measurement itself can be displayed. Watch that the measurement scale is consistent or that a jump scale squiggle is used. **2.** __________

(a) Time plots (b) Bar graphs

(c) Pareto charts (d) Stem-and-Leaf Displays (e) Circle graphs

3. __________ display how a *total* is dispersed into several categories. This graph is very appropriate for qualitative data, or any data where percentage of occurrence makes sense. **3.** __________

(a) Time plots (b) Bar graphs

(c) Pareto charts (d) Stem-and-Leaf Displays (e) Circle graphs

4. __________ display how data change over time. It is best if the units of time are consistent in a given plot. **4.** __________

(a) Time plots (b) Bar graphs

(c) Pareto charts (d) Stem-and-Leaf Displays (e) Circle graphs

5. __________ organize and group data but also allow us to recover the original data. **5.** __________

(a) Time plots (b) Bar graphs

(c) Pareto charts (d) Stem-and-Leaf Displays (e) Circle graphs

CHAPTER 2, FORM C, PAGE 2

6. A survey of 500 teenagers was taken to see which sport was their favorite to watch on television. The pie chart below displays the results. Choose the correct data (numbers of teenagers) from which the pie chart was constructed. **6.** ________

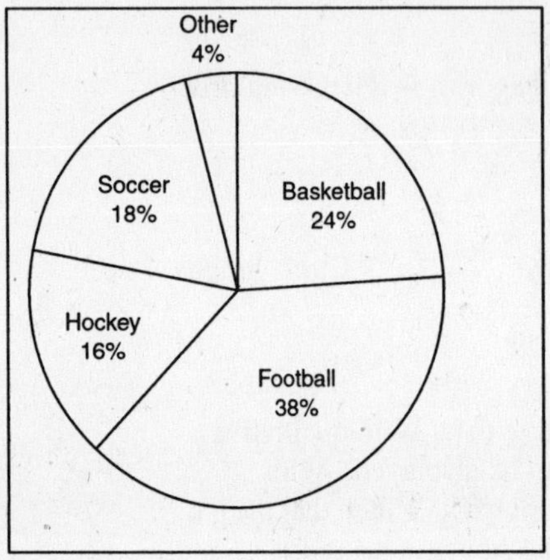

(a) Basketball, 190; football, 120; hockey, 90; soccer, 80; other, 20

(b) Basketball, 120; football, 190, hockey, 90; soccer, 80; other, 20

(c) Basketball, 20; football, 90, hockey, 80; soccer, 190; other, 120

(d) Basketball, 240; football, 380, hockey, 160; soccer, 180; other, 40

(e) Basketball, 120; football, 190, hockey, 80; soccer, 90; other, 20

7. Following is a histogram displaying the test scores for students in a statistics class.

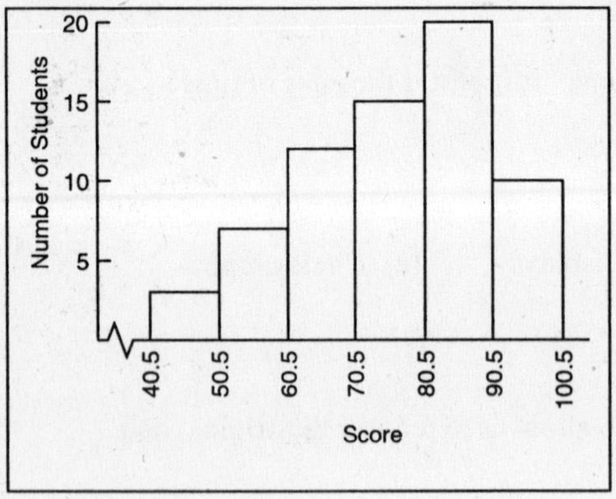

Categorize the distribution shape as **7.** ________

(a) Uniform (b) Symmetric

(c) Bimodal (d) Skewed left (e) Skewed right

CHAPTER 2, FORM C, PAGE 3

8. A sample of 12 children was taken from a daycare. A dotplot of the average number of hours of daily television viewing is shown here.

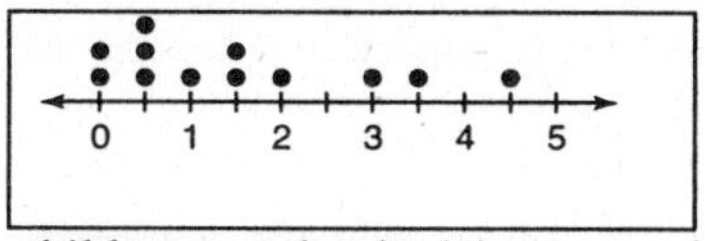

How many children watch television more than 3 hours per day? **8.** ________

(a) 2 (b) 3

(c) 9 (d) 10 (e) Cannot determine

9. Following is a list of prices (to the nearest dollar) for college textbooks. Make a stem-and-leaf display for these data. **9.** ________

2.3	3.7	1.2	3.6	2.4	2.6	3.7	0.9
1.8	2.5	2.5	3.0	2.8	1.7	3.1	4.1

(a)

0	2 2 3
1	2 7 8
2	4 4 7 9
3	1 2 5 8
4	2 4

(b)

10	2 7 8
20	0 0 4 4 7 9
30	0 1 2 5 8
40	2 4

(c)

0	9
1	2 7 8
2	3 4 5 5 6 8
3	0 1 6 7 7
4	1

(d)

1	2 7 8
2	0 4 4 7 9
3	0 1 2 5 8
4	2 4

(e)

12 17 18
20 20 24 24 27 29
30 31 32 35 38
42 44

CHAPTER 3 TEST
FORM A

1. A random sample of 18 airline carry-on luggage bags gave the following weights (rounded to the nearest pound).

12	25	10	38	12	19	8	12	17
41	7	22	10	19	12	16	5	14

Find the mean, median, and mode of these weights. **1.** ____________________

2. Find the mean and the 5% trimmed mean for the following annual salaries (in thousands) of employees in a small company. Which is most representative of the average annual salary? Why? **2.** ____________________

38.5	31.0	29.8	37.4	40.1	35.1	41.5	12.6	39.7	28.4
34.2	38.6	187.4	40.6	39.7	31.0	29.8	42.0	30.8	35.5

3. A random sample of 7 Northern Pike from Taltson Lake (Canada) gave the following lengths rounded to the nearest inch.

21 27 46 35 41 36 25

(a) Find the range. **3. (a)** ____________________

(b) Find the sample mean. **(b)** ____________________

(c) Find the sample variance. **(c)** ____________________

(d) Find the sample standard deviation. **(d)** ____________________

4. A random sample of receipts for individuals eating at the Terrace Restaurant showed the sample mean to be $\bar{x}$ = \$10.38 with sample standard deviation s = \$2.17.

(a) Compute the coefficient of variation for this data. **4. (a)** ____________________

(b) Use Chevyshev's Theorem to find the smallest interval centered on the mean in which we can expect at least 75% of the data to fall. **(b)** ____________________

CHAPTER 3, FORM A, PAGE 2

5. A random sample of 330 adults were asked the maximal amount (dollars) they would spend on a ticket to a top rated performance. The results follow where x is the cost and f is the number of people who would spend that maximal amount.

x	20	30	40	50	60
f	62	83	120	40	25

(a) Compute the sample mean.

(b) Compute the sample variance.

(c) Compute the sample standard deviation.

5. (a) ____________

(b) ____________

(c) ____________

6. A random sample of 27 skiers at Vail, Colorado gave their ages. The results were

18	25	32	16	41	52	29	58	23
62	47	56	19	22	38	15	46	33
49	52	37	26	72	44	19	24	29

(a) Give the five number summary including the low value Q_1, median, Q_3 and high value.

(b) Make a box-and-whisker plot for the given data.

(c) Find the interquartile range.

6. (a) ____________

(b)

(c) ____________

7. Sophia took a test and scored in the 79th percentile. What percentage of the scores were at or below her score? What percentage were above?

7. ____________

CHAPTER 3 TEST
FORM B

1. A veterinarian in a small animal clinic had the following record of life spans of Golden Retrievers (to the nearest year).

9	12	15	11	8	10	7	5	11	14
13	6	11	16	11	14	11	4	12	11

Find the mean, median, and mode for this data. **1.** ____________________

2. Find the mean and the 10% trimmed mean for the following annual snowfalls (in inches) for a city in northern Wisconsin. Which is most representative of the average annual snowfall? Why? **2.** ____________________

24 37 28 13 38 29 112 21 40 36

3. A random sample of 6 people, each 20 pounds overweight, volunteered to go on the same diet. After 3 months, their weight loss (in pounds) were

12 5 14 19 15 8

(a) Find the range. **3. (a)** ____________________

(b) Find the sample mean. **(b)** ____________________

(c) Find the sample variance. **(c)** ____________________

(d) Find the sample standard deviation. **(d)** ____________________

4. A large sample of Northern Pike caught at Taltson Lake (Canada) showed that the average length was $\overline{x} = 32.5$ inches with sample standard deviation $s = 8.6$ inches.

(a) Compute the coefficient of variation for this data. **4. (a)** ____________________

(b) Use Chevyshev's Theorem to find an interval centered on the mean in which we can expect at least 75% of the data to fall. **(b)** ____________________

CHAPTER 3, FORM B, PAGE 2

5. A random sample of 146 students in Chemistry 215 gave the following grade information (A = 4.0, B = 3.0, C = 2.0, D = 1.0, and F = 0). In the following table x = grade and f = number of students receiving this grade.

x	0	1	2	3	4
f	8	14	62	43	19

(a) Find the sample mean.

(b) Find the sample variance.

(c) Find the sample standard deviation.

5. (a) ____________

(b) ____________

(c) ____________

6. A random sample of 24 professors at Montana State University gave the following ages (years).

29	32	56	61	27	43	38	65
36	47	41	68	59	40	33	35
44	39	28	46	42	62	58	45

(a) Give the five number summary including the low value Q_1, median, Q_3 and high value.

6. (a) ____________

(b) Make a box-and-whisker plot for the given data.

(b)

(c) Find the interquartile range.

(c) ____________

7. Tyler took a test and scored in the 81st percentile. What percent of scores were above his score? What percent were at or below his score?

7. ____________

CHAPTER 3 TEST
FORM C

Write the letter of the response that best answers each problem.

1. The durations of a random sample of 16 commercials are below (in seconds). Find the mean, median, and mode of these durations. **1.** ________

30	12	26	7	14	35	20	30
55	8	35	18	42	15	30	10

(a) Mean = 24.19, median = 30, mode = 23

(b) Mean = 20.13, median = 21, mode = 35

(c) Mean = 24.19, median = 23, mode = 30

(d) Mean = 20, median = 24.19, mode = 30

(e) Mean = 30, median = 20, mode = 22.13

2. Find the 5% trimmed mean for the following heights of fruit trees (in feet) in Gordy's Farm Market. **2.** ________

5.2	6.5	7.4	8.0	9.1	7.6	5.4	3.1	4.7	15.0
4.1	3.7	7.4	6.9	5.5	4.0	3.9	5.2	5.8	7.2
8.2	4.9	7.3	6.1	7.2	6.5	3.4	8.6	4.7	5.8
1.3	6.0	5.8	4.3	5.4	6.5	7.6	9.0	6.7	13.4

(a) 6.05 ft (b) 6.27 ft

(c) 5.80 ft (d) 6.16 ft (e) 6.36 ft

3. A random sample of 9 Walleye Pike from Salty Lake gave the following weights rounded to the nearest pound.

3 4 11 6 5 7 6 2 4

A. Find the range. **3. A.** ________

(a) 4 (b) 5 (c) 7 (d) 8 (e) 9

B. Find the sample mean. **B.** ________

(a) 5.33 (b) 9 (c) 7 (d) 2.65 (e) 5

CHAPTER 3, FORM C, PAGE 2

C. Find the sample variance. **C.** ________

(a) 2.65 (b) 5 (c) 7 (d) 5.33 (e) 6.22

D. Find the sample standard deviation. **D.** ________

(a) 5.33 (b) 7 (c) 2.49 (d) 2.65 (e) 6.22

4. A random sample of 30 receipts for individuals shopping at the Community Drug Store showed the sample mean to be $\overline{x}$ = \$28.19 with sample standard deviation s = \$4.06.

A. Compute the coefficient of variation for this data. **4. A.** ________

(a) 6.94% (b) 14.4% (c) 694% (d) 4.32% (e) 0.48%

B. Use Chevyshev's Theorem to find the smallest interval centered on the mean in which we can expect at least 88.9% of the data to fall. **B.** ________

(a) 16.01 to 40.37 (b) 24.13 to 28.19

(c) 20.07 to 36.31 (d) 11.95 to 44.43 (e) 24.13 to 32.25

5. A random sample of 210 high school students were asked how many hours per week they spent studying at home. The results follow where x is the hours and f is the number of students who spend that amount of time.

x	3	5	7	9	11
f	30	45	70	55	10

A. Compute the sample mean. **5. A.** ________

(a) 0.025 (b) 40.29 (c) 6.71 (d) 0.149 (e) 6

B. Compute the sample variance. **B.** ________

(a) 4.871 (b) 2.212 (c) 199 (d) 210 (e) 4.894

C. Compute the sample variance. **C.** ________

(a) 4.894 (b) 4.871 (c) 2.207 (d) 2.212 (e) 14.11

CHAPTER 3, FORM C, PAGE 3

6. What are the numbers used in making a box-and-whisker plot? **6.** ________

(a) Low value, Q_1, mean, Q_3, high value

(b) Low value, Q_1, median, Q_3, high value

(c) s, mean, s^2, interquartile range

(d) Q_1, Q_2, Q_3

(e) Low value, interquartile range, high value

7. Luke took a test and scored in the 88th percentile. What percentage of the scores were above his score? **7.** ________

(a) 87% (b) 13% (c) 89% (d) 88% (e) 12%

CHAPTER 4 TEST
FORM A

For the given data, solve the following problems.

Taltson Lake is in the Canadian Northwest Territories. This lake has many Northern Pike. The following data was obtained by two fishermen visiting the lake. Let x = length of a Northern Pike in inches and let y = weight in pounds.

x (inches)	20	24	36	41	46
y (pounds)	2	4	12	15	20

1. Draw a scatter diagram. Using the scatter diagram (no calculations) would you estimate the linear correlation coefficient to be positive, close to zero, or negative? Explain your answer.

1.

2. For the given data compute each of the following.

(a) $\overline{x}$ and $\overline{y}$

(b) SS_x, SS_y, SS_{xy}

(c) The slope b and y intercept a of the least squares line; write out the equation for the least squares line.

(d) Graph the least-squares line on your scatter plot of problem 1.

2. (a) ____________________

(b) ____________________

(c) ____________________

(d) ____________________

3. Compute the sample correlation coefficient r. Compute the coefficient of determination. Give a brief explanation of the meaning of the correlation coefficient and the coefficient of determination in the context of this problem.

3. ____________________

4. If a 32 inch Northern Pike is caught, what is the weight in pounds as predicted by the least-squares line?

4. ____________________

CHAPTER 4, FORM A, PAGE 2

For the given data, solve the following problems.

A marketing analyst is studying the relationship between x = amount spent on television advertising and y = increase in sales. The following data represents a random sample from the study.

x ($ thousands)	15	28	19	47	10	92
y ($ thousands)	340	260	152	413	130	855

5. Draw a scatter diagram. Using the scatter diagram (no calculations) would you estimate the linear correlation coefficient to be positive, close to zero, or negative? Explain your answer.

5.

6. For the given data compute each of the following.

(a) $\overline{x}$ and $\overline{y}$ — **6. (a)** ____________

(b) SS_x, SS_y, SS_{xy} — **(b)** ____________

(c) The slope b and y intercept a of the least-squares line; write out the equation for the least-squares line. — **(c)** ____________

(d) Graph the least-squares line on your scatter plot of problem 8. — **(d)** ____________

7. Compute the sample correlation coefficient r. Compute the coefficient of determination. Give a brief explanation of the meaning of the correlation coefficient and the coefficient of determination in the context of this problem. — **7.** ____________

8. Suppose that the amount spent on advertising is $37,000. What does the least-squares line predict for the increase in sales? — **8.** ____________

CHAPTER 4 TEST
FORM B

For the given data, solve the following problems.

Do higher paid chief executive officers (CEO's) control bigger companies? Let us study x = annual CEO salary ($ millions) and y = annual company revenue ($ billions). The following data are based on information from *Forbes* magazine and represents a sample of top US executives.

x ($ millions)	0.8	1.0	1.1	1.7	2.3
y ($ billions)	14	11	19	20	25

1. Draw a scatter diagram. Using the scatter diagram (no calculations) would you estimate the linear correlation coefficient to be positive, close to zero, or negative? Explain your answer.

1.

2. For the given data compute each of the following.

(a) $\overline{x}$ and $\overline{y}$

(b) SS_x, SS_y, SS_{xy}

(c) The slope b and y intercept a of the least-squares line; write out the equation for the least-squares line.

(d) Graph the least-squares line on your scatter plot of problem 1.

2. (a) ________________

(b) ________________

(c) ________________

(d) ________________

3. Compute the sample correlation coefficient r. Compute the coefficient of determination. Give a brief explanation of the meaning of the correlation coefficient and the coefficient of determination in the context of this problem.

3. ________________

4. If a CEO has an annual salary of 1.5 million, what is his or her annual company revenue as predicted by the least-squares line?

4. ________________

CHAPTER 4, FORM B, PAGE 2

For the given data, solve the following problems.

An accountant for a small manufacturing plant collected the following random sample to study the relationship between x = the cost to make a particular item and y = the selling price.

x ($)	26	50	47	23	52	71
y ($)	78	132	128	70	152	198

5. Draw a scatter diagram. Using the scatter diagram (no calculations) would you estimate the linear correlation coefficient to be positive, close to zero, or negative? Explain your answer.

5.

6. For the given data compute each of the following.

(a) $\overline{x}$ and $\overline{y}$

(b) SS_x, SS_y, SS_{xy}

(c) The slope b and y intercept a of the least-squares line; write out the equation for the least-squares line.

(d) Graph the least-squares line on your scatter plot of problem 8.

6. (a) ____________

(b) ____________

(c) ____________

(d) ____________

7. Compute the sample correlation coefficient r. Compute the coefficient of determination. Give a brief explanation of the meaning of the correlation coefficient and the coefficient of determination in the context of this problem.

7. ____________

8. Suppose that the cost to make a particular item is $35. What does the least-squares line predict for the selling price?

8. ____________

CHAPTER 4 TEST
FORM C

Write the letter of the response that best answers each problem.

Does the weight of a vehicle affect the gas mileage? The following random sample was collected where x = weight of a vehicle in hundreds of pounds and y = miles per gallon.

x (lb hundreds)	26	35	29	39	20
y (mpg)	22.0	16.1	18.8	15.7	23.4

1. Based on a scatter diagram, would you estimate the linear correlation coefficient to be **1.** ________

(a) close to −1 (b) closer to 0 and negative

(c) close to 1 (d) closer to 0 and positive (e) Cannot determine

2. What is the equation for the least-squares line? **2.** ________

(a) $y = -32.55x + 0.448$ (b) $y = -32.55x - 0.448$

(c) $y = -32.55x + 0.448$ (d) $y = -0.448x + 32.55$ (e) $y = 0.448x - 32.55$

3. Compute the coefficient of determination. **3.** ________

(a) −0.941 (b) 0.941

(c) −0.970 (d) 0.970 (e) 0.965

4. If a vehicle weighs 2200 pounds, what does the least-squares line predict for the miles per gallon? **4.** ________

(a) 22.7 (b) 66.0

(c) 42.4 (d) 22.0 (e) Cannot determine

CHAPTER 4, FORM C, PAGE 2

Write the letter of the response that best answers each problem.

A graduate school committee is studying the relationship between x = an applicants' undergraduate grade point average and y = the applicants' score on the graduate entrance exam. The following random sample was collected to study this relationship.

x (GPA)	3.2	3.9	4.0	3.4	3.7	3.0
y (score)	725	788	775	647	800	672

5. Based on a scatter diagram, would you estimate the linear correlation coefficient to be **5.** __________

(a) close to −1 (b) closer to 0 and negative

(c) close to 1 (d) closer to 0 and positive (e) Cannot determine

6. What is the equation for the least-squares line? **6.** __________

(a) $y = -123.03x + 299.8$ (b) $y = 123.03x - 299.8$

(c) $y = 299.8x + 123.03$ (d) $y = -299.8x + 123.03$ (e) $y = 123.03x + 299.8$

CHAPTER 4, FORM C, PAGE 3

7. Compute the sample correlation coefficient. **7.** ________

(a) 0.587 (b) –0.766

(c) 0.875 (d) 0.766 (e) –0.586

8. If a student has a grade point average of 3.5, what does the least-squares line predict for the score on the graduate entrance exam? **8.** ________

(a) 926.3 (b) 3.5

(c) 730.4 (d) 130.8 (e) 1172.3

CHAPTER 5 TEST
FORM A

1. A random sample of 317 new Smile Bright electric toothbrushes showed 19 were defective.

(a) How would you estimate the probability that a new Smile Bright electric toothbrush is defective? What is your estimate? **1. (a)** ______________

(b) What is your estimate for the probability that a Smile Bright electric toothbrush is not defective? **(b)** ______________

(c) Either an electric toothbrush is defective or not. What is the sample space in this problem. Do the probabilities assigned to the sample space add up to one? **(c)** ______________

2. If you roll a single fair die and count the number of dots on top, what is the probability of getting a number less than 3 on a single throw? **2.** ______________

3. You roll two fair dice, a blue one and a yellow one.

(a) Find *P*(even number on the blue die and 3 on the yellow die). **3. (a)** ______________

(b) Find *P*(3 on the blue die and even number on the yellow die). **(b)** ______________

(c) Find *P*(even number on the blue die and 3 on the yellow die) or *P*(3 on the blue die and even number on the yellow die). **(c)** ______________

4. An urn contains 12 balls identical in every respect except color. There are 3 red balls, 7 green balls, and 2 blue balls.

(a) You draw two balls from the urn, but replace the first ball before drawing the second. Find the probability that the first ball is red and the second is green. **4. (a)** ______________

(b) Repeat part (a), but do not replace the first ball before drawing the second. **(b)** ______________

5. Robert is applying for a bank loan to open up a pizza franchise. He must complete a written application, and then be interviewed by bank officers. Past records for this bank show that the probability of being approved in the written part is 0.63. Then the probability of being approved by the interview committee is 0.85, given the candidate has been approved on the written application. What is the probability Robert is approved on both the written application and the interview? **5.** ______________

CHAPTER 5, FORM A, PAGE 2

6. A hair salon did a survey of 360 customers regarding satisfaction with service and type of customer. A walk-in customer is one who has seen no ads and not been referred. The other customers either saw a TV ad or were referred to the salon (but not both). The results follow.

	Walk-In	TV Ad	Referred	Total
Not Satisfied	21	9	5	35
Neutral	18	25	37	80
Satisfied	36	43	59	138
Very Satisfied	28	31	48	107
Total	103	108	149	360

Assume the sample represents the entire population of customers. Find the probability that a customer is

(a) Not satisfied

(b) Not satisfied and walk-in

(c) Not satisfied, given referred

(d) Very satisfied

(e) Very satisfied, given referred

(f) Very satisfied and TV ad

(g) Are the events satisfied and referred independent or not? Explain your answer.

6. (a) ____________

(b) ____________

(c) ____________

(d) ____________

(e) ____________

(f) ____________

(g) ____________

7. A computer package sale comes with two different choices of printers and four different choices of monitors. If a store wants to display each package combination that is for sale, how many packages must be displayed? Make a tree diagram showing the outcomes for selecting printer and monitor.

7. ____________

8. In how many ways can the 40 members of a 4H club select a president, a vice-president, a secretary, and a treasurer?

8. ____________

9. In how many different ways can a person choose three movies to see in a theater playing 11 movies.

9. ____________

CHAPTER 5 TEST
FORM B

1. A Student Council is made up of 4 women and 6 men. One of the women is president of the Council. A member of the council is selected at random to report to the Dean of Student Life.

(a) What is the probability a woman is selected? **1. (a)** ______________

(b) What is the probability a man is selected? **(b)** ______________

(c) What is the probability that the president of the Student Council is selected? **(c)** ______________

2. If you roll a single fair die and count the number of dots on top, what is the probability of getting a number greater than 2 on a single throw? **2.** ______________

3. You roll two fair dice, a white one and a red one.

(a) Find *P*(5 or 6 on the white die and odd number on the red die). **3. (a)** ______________

(b) Find *P*(odd number on the white die and 5 or 6 on the red die). **(b)** ______________

(c) Find *P*(5 or 6 on the white die and odd number on the red die)or *P*(odd number on the white die and 5 or 6 on the red die). **(c)** ______________

4. An urn contains 17 balls identical in every respect except color. There are 6 red balls, 8 green balls, and 3 blue balls.

(a) You draw two balls from the urn, but replace the first ball before drawing the second. Find the probability that the first ball is red and the second ball is green. **4. (a)** ______________

(b) Repeat part (a), but do not replace the first ball before drawing the second. **(b)** ______________

5. The Dean of Hinsdale College found that 12% of the female students are majoring in Computer Science. If 64% of the students at Hinsdale are women, what is the probability that a student chosen at random will be a woman majoring in Computer Science? **5.** ______________

CHAPTER 5, FORM B, PAGE 2

6. The Committee on Student Life did a survey of 417 students regarding satisfaction with Student Government and class standing. The results follow:

	Freshman	Sophomore	Junior	Senior	Total
Not Satisfied	17	19	23	12	71
Neutral	61	35	32	38	166
Satisfied	23	49	43	65	180
Total	101	103	98	115	417

Assume the sample represents the entire population of students. Find the probability that a student selected at random is

(a) Satisfied (with Student Government) **6. (a)** ____________

(b) Satisfied, given the student is a Senior **(b)** ____________

(c) Neutral **(c)** ____________

(d) Neutral and Freshman **(d)** ____________

(e) Neutral, given the student is a Freshman **(e)** ____________

(f) Senior, given satisfied **(f)** ____________

(g) Are the events, Freshman and neutral independent or not? Explain. **(g)** ____________

7. There are 3 different routes that Alexander can walk from home to the post office and 2 different routes that he can walk from the post office to the bank. How many different routes can Alexander walk from home to the bank? Make a tree diagram showing the outcomes for selecting the routes. **7.** ____________

8. A fishing camp has 16 clients. Each cabin at the camp will accommodate 5 fishermen. In how many different ways can the first cabin be filled with clients? **8.** ____________

9. In how many different ways can a committee of four be selected from the 24 parents attending a school board meeting? **9.** ____________

CHAPTER 5 TEST
FORM C

Write the letter of the response that best answers each problem.

1. A random sample of 420 new Ford trucks showed that 105 required repairs within the first warranty year.

A. What is the estimate for the probability that a new Ford truck will need repairs within the first warranty year? **1. A.** ________

(a) 105 (b) 0.75 (c) 315 (d) 0.25 (e) 0.20

B. What is the estimate for the probability that a new Ford truck will not need repairs within the first warranty year? **B.** ________

(a) 105 (b) 0.75 (c) 315 (d) 0.25 (e) 0.80

2. If you roll a single fair die and count the number of dots on top, what is the probability of getting an even number or a 5 on a single throw? **2.** ________

(a) $\frac{2}{3}$ (b) $\frac{1}{2}+\frac{1}{6}$ (c) $\frac{1}{12}$ (d) $\frac{5}{6}$ (e) $\frac{1}{3}$

3. You roll two fair dice, a white one and a green one.

A. Find *P*(number greater than 2 on the white die and 4 on the green die). **3. A.** ________

(a) $\frac{5}{6}$ (b) $\frac{5}{36}$ (c) $\frac{1}{9}$ (d) $\frac{1}{2}$ (e) $\frac{2}{9}$

B. Find *P*(4 on the white die and number greater than 2 on the green die). **B.** ________

(a) $\frac{2}{9}$ (b) $\frac{1}{2}$ (c) $\frac{5}{6}$ (d) $\frac{5}{36}$ (e) $\frac{1}{9}$

C. Find *P*(number greater than 2 on the white die and 4 on the green die) or *P*(4 on the white die and number greater than 2 on the green die). **C.** ________

(a) 1 (b) $\frac{2}{9}$ (c) $\frac{5}{18}$ (d) $\frac{1}{81}$ (e) $\frac{25}{36}$

CHAPTER 5, FORM C, PAGE 2

4. An urn contains 8 balls identical in every aspect except color. There is 1 red ball, 2 green balls, and 5 blue balls.

A. You draw two balls from the urn, but replace the first ball before drawing the second. Find the probability that the first ball is blue and the second is green. **4. A.** ________

(a) $\frac{5}{32}$ (b) $\frac{7}{8}$ (c) $\frac{51}{56}$ (d) $\frac{5}{28}$ (e) $\frac{3}{4}$

B. Repeat part A, but do not replace the first ball before drawing the second. **B.** ________

(a) $\frac{5}{32}$ (b) $\frac{7}{8}$ (c) $\frac{51}{56}$ (d) $\frac{5}{28}$ (e) $\frac{3}{4}$

5. The athletic coach found that 31% of the basketball players have an A average in school. If 2% of the students at the school are basketball players, what is the probability that a student chosen at random will be a basketball player with an A average? **5.** ________

(a) 62% (b) 0.62% (c) 6.45% (d) 0.0645% (e) 0.0062%

6. A hospital administration did a survey of patients regarding satisfaction with care and type of surgery. The results follow:

	Heart	Hip	Knee	Total
Not Satisfied	7	12	2	21
Neutral	15	38	10	63
Satisfied	32	16	25	73
Very Satisfied	4	22	23	49
Total	58	88	60	206

Assume the sample represents the entire population of patients. Find the probability that a patient selected at random is

A. Satisfied **6. A.** ________

(a) $\frac{32}{206}$ (b) 73 (c) $\frac{122}{206}$ (d) 122 (e) $\frac{73}{206}$

CHAPTER 5, FORM C, PAGE 3

B. Very satisfied and had knee surgery **B.** ________

(a) $\frac{109}{206}$ (b) $\frac{11}{206}$ (c) $\frac{23}{206}$ (d) $\frac{23}{60}$ (e) $\frac{23}{60}+\frac{23}{49}$

C. Neutral, given had hip surgery **C.** ________

(a) $\frac{38}{88}$ (b) $\frac{88}{206}$ (c) $\frac{38}{206}$ (d) $\frac{38}{63}$ (e) $\frac{63}{88}$

D. A knee surgery patient **D.** ________

(a) 60 (b) 206 (c) $\frac{146}{206}$ (d) $\frac{60}{206}$ (e) $\frac{2}{206}$

E. Satisfied given had heart surgery. **E.** ________

(a) $\frac{73}{206}$ (b) $\frac{58}{73}$ (c) $\frac{32}{58}$ (d) $\frac{32}{206}$ (e) $\frac{32}{73}$

F. Not satisfied and had heart surgery. **F.** ________

(a) $\frac{7}{58}$ (b) $\frac{7}{206}$ (c) $\frac{7}{21}$ (d) $\frac{21}{206}$ (e) $\frac{58}{206}$

7. George has 4 ties, 3 shirts, and 2 pairs of pants. How many different outfits can he wear if he chooses one tie, one shirt and one pair of pants for each outfit? **7.** ________

(a) 288 (b) 12 (c) 9 (d) 24 (e) 10

8. In how many ways can 12 athletes be awarded a first-place medal, a second-place medal, and a third-place medal? **8.** ________

(a) 33 (b) 220 (c) 6 (d) 1728 (e) 1320

9. In how many different ways can a student choose 3 out of 8 problems to complete on a take-home exam? **9.** ________

(a) 56 (b) 336 (c) 21 (d) 3.5 (e) 3

CHAPTER 6 TEST
FORM A

1. Sam is a representative who sells large appliances such as refrigerators, stoves, and so forth. Let x = number of appliances Sam sells on a given day. Let f = frequency (number of days) with which he sells x appliances. For a random sample of 240 days, Sam had the following sales record.

x	0	1	2	3	4	5	6	7
f	9	72	63	41	28	14	8	5

Assume the sales record is representative of the population of all sales day.

(a) Use the relative frequency to find $P(x)$ for $x = 0$ to 7. **1. (a)** ______________

(b) Use a histogram to graph the probability distribution of part (a). **(b)**

(c) Compute the probability that x is between 2 and 5 (including 2 and 5). **(c)** ______________

(d) Compute the probability that x is less than 3. **(d)** ______________

(e) Compute the expected value of the x distribution. **(e)** ______________

(f) Compute the standard deviation of the x distribution. **(f)** ______________

2. The director of a health club conducted a survey and found that 23% of members used only the pool for workouts. Based on this information, what is the probability that for a random sample of 10 members, 4 used only the pool for workouts? **2.** ______________

3. Of those mountain climbers who attempt Mt. McKinley (Denali), only 65% reach the summit. In a random sample of 16 mountain climbers who are going to attempt Mt. McKinley, what is the probability of each of the following?

(a) All 16 reach the summit. **3. (a)** ______________

(b) At least 10 reach the summit. **(b)** ______________

(c) No more than 12 reach the summit. **(c)** ______________

(d) From 9 to 12 reach the summit, including 9 and 12. **(d)** ______________

CHAPTER 6, FORM A, PAGE 2

4. A coach found that about 12% of all hockey games end in overtime. What is the expected number of games ending in overtime if a random sample of 50 hockey games are played?

4. ______________________

5. The probability that a truck will be going over the speed limit on I-80 between Cheyenne and Rock Springs, Wyoming is about 75%. Suppose a random sample of 5 trucks on this stretch of I-80 are observed.

(a) Make a histogram showing the probability that $r = 0, 1, 2, 3, 4, 5$ trucks going over the speed limit.

5. (a)

(b) Find the mean μ of this probability distribution.

(b) ______________________

(c) Find the standard deviation of the probability distribution.

(c) ______________________

6. Records show that the probability of catching a Northern Pike over 40 inches at Taltson Lake (Canada) is about 15% for each full day a person spends fishing. What is the minimal number of days a person must fish to be at least 83.3% sure of catching one or more Northern Pike over 40 inches?

6. ______________________

CHAPTER 6 TEST
FORM B

1. An aptitude test was given to a random sample of 228 people intending to become Data Entry Clerks. The results are shown below where x is the score on a 10 point scale, and f is the frequency of people with this score.

x	1	2	3	4	5	6	7	8	9	10
f	9	21	46	51	42	18	12	10	8	5

Assume the above data represents the entire population of people intending to become Data Entry Clerks.

(a) Use the relative frequencies to find $P(x)$ for $x = 1$ to 10. **1. (a)** ______________

(b) Use a histogram to graph the probability distribution of part (a). **(b)**

(c) To be accepted into a training program, students must have a score of 4 or higher. What is the probability an applicant selected at random will have this score? **(c)** ______________

(d) To receive a tuition scholarship a student needs a score of 8 or higher. What is the probability an applicant selected at random will have such a score? **(d)** ______________

(e) Compute the expected value of the x distribution. **(e)** ______________

(f) Compute the standard deviation of the x distribution. **(f)** ______________

2. The management of a restaurant conducted a survey and found that 28 of the customers preferred to sit in the smoking section. Based on this information, what is the probability that for a random sample of 12 customers, 3 preferred the smoking section? **2.** ______________

3. Of all college freshmen who try out for the track team, the coach will only accept 30%. If 15 freshmen try out for the track team, what is the probability that

(a) all 15 are accepted? **3. (a)** ______________

(b) at least 8 are accepted? **(b)** ______________

(c) no more than 4 are accepted? **(c)** ______________

CHAPTER 6, FORM B, PAGE 2

(d) between 5 and 10 are accepted (including 5 and 10)? **(d)** ______________

4. The president of a bank approves 68% of all new applications. What is the expected number of approvals if a random sample of 75 loan applications are chosen? **4.** ______________

5. The probability that a vehicle will change lanes while making a turn is 55%. Suppose a random sample of 7 vehicles are observed making turns at a busy intersection.

(a) Make a histogram showing the probability that $r = 0, 1, 2, 3, 4, 5, 6, 7$ vehicles will make a land change while turning. **5. (a)**

(b) Find the expected value μ of this probability distribution. **(b)** ______________

(c) Find the standard deviation of this probability distribution. **(c)** ______________

6. Past records show that the probability of catching a Lake Trout over 15 pounds at Talston Lake (Canada) us about 20% for each full day a person spends fishing. What is the minimal number of days a person must fish to be at least 89.3% sure of catching one or more Lake Trout over 15 pounds? **6.** ______________

CHAPTER 6 TEST
FORM C

Write the letter of the response that best answers each problem.

1. The following data are based on a survey taken by a consumer research firm. In this table x = number of televisions in household and % = percentages of U.S. households.

x	0	1	2	3	4	5 or more
%	3%	11%	28%	39%	12%	7%

A. What is the probability that a household selected at random has less than 3 televisions? **1. A.** ________

(a) 0.81 (b) 0.39 (c) 0.42 (d) 0.58 (e) 0.19

B. What is the probability that a household selected at random has more than 4 televisions? **B.** ________

(a) 0.7 (b) 0.19 (c) 0.81 (d) 0.93 (e) 0.07

C. Compute the expected value of the x distribution (round televisions of 5 or more to 5). **C.** ________

(a) 15 (b) 2.67 (c) 1.28 (d) 1.13 (e) 3.1

D. Compute the standard deviation of the x distribution (round televisions of 5 or more to 5). **D.** ________

(a) 15 (b) 2.67 (c) 1.28 (d) 1.13 (e) 3.1

2. A meteorologist found from the past year's records that it rained 17% of the days. Based on this information, what is the probability that for a random sample of 15 days, it rained 3 of those days? **2.** ________

(a) 0.17 (b) 0.67 (c) 0.28 (d) 0.13 (e) 0.22

3. Of those people who lose weight on a diet, 90% gain all the weight back. In a random sample of 12 dieters who have lost weight, what is the probability of each of the following.

A. All 12 gain the weight back? **3. A.** ________

(a) 0.90 (b) 0.282 (c) 0.540 (d) 0.142 (e) 10.8%

B. At least 9 gain the weight back? **B.** ________

(a) 0.974 (b) 0.026 (c) 0.889 (d) 1.33% (e) 0.997

C. No more than 6 gain the weight back? **C.** ________

(a) 0.004 (b) 1.000 (c) 0.999 (d) 0.000 (e) 0.531

D. From 8 to 10 gain the weight back, including 8 and 10. **D.** ________

(a) 0.085 (b) 1.17% (c) 0.336 (d) 0.387 (e) 0.118

CHAPTER 6, FORM C, PAGE 2

4. The manager of a supermarket found that 72% of the shoppers who taste a free sample of a food item will buy the item. What is the expected number of shoppers that will buy the item if a random sample of 50 shoppers taste a free sample? **4.** ____________

(a) 10 (b) 36 (c) 3 (d) 72 (e) 50

5. The probability that merchandise stolen from a store will be recovered is 15%. Suppose a random sample of 8 stores, from which merchandise has been stolen, is chosen.

A. Find the mean μ of this probability distribution. **5. A.** __________

(a) 1.02 (b) 1.07 (c) 1.01 (d) 1.14 (e) 1.2

B. Find the standard deviation of the probability distribution. **B.** __________

(a) 1.02 (b) 1.07 (c) 1.01 (d) 1.14 (e) 1.2

6. Records show that the probability of seeing a hawk migrating on a day in September is about 35%. What is the minimal number of days a person must watch to be at least 96.8% sure of seeing one or more hawks migrating? **6.** ____________

(a) 5 (b) 6 (c) 7 (d) 8 (e) 9

CHAPTER 7 TEST
FORM A

1. Each of the following curves fails to be a normal curve. Give reasons why these curves are not normal curves.

(a)

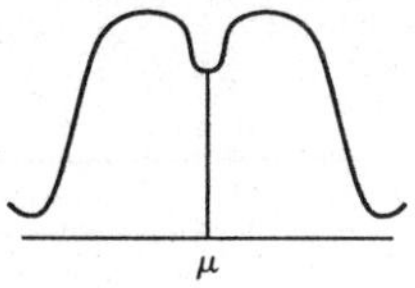

1. (a) ____________

(b)

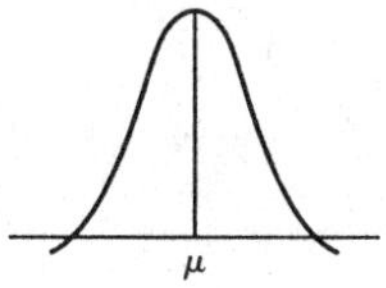

(b) ____________

2. According to the Empirical rule, for a distribution that is symmetrical and bell-shaped (in particular, for a normal distribution) approximately _______ of the data values will lie within three standard deviations on each side of the mean.

2. ____________

3. Assuming that the heights of boys in a high school basketball tournament are normally distributed, with mean 70 in. and standard deviation 2.5 in., how many boys in a group of 40 in the tournament will be taller than 75 inches?

3. ____________

4. Let x be a random variable that represents the length of time it takes a student to complete Dr. Gill's Chemistry Lab Project. From long experience, it is known that x has a normal distribution with mean $\mu = 3.6$ hours and standard deviation $\sigma = 0.5$ hour.

Convert each of the following x intervals to standard z intervals.

(a) $x \geq 4.5$ **4. (a)** ____________

(b) $3 \leq x \leq 4$ **(b)** ____________

(c) $x \leq 2.5$ **(c)** ____________

Convert each of the following z intervals to raw score x intervals.

(d) $z \leq -1$ **(d)** ____________

(e) $1 \leq z \leq 2$ **(e)** ____________

(f) $z \geq 1.5$ **(f)** ____________

CHAPTER 7, FORM A, PAGE 2

5. John and Joel are salesmen in different districts. In John's district, the long term mean sales is \$17,319 each month with standard deviation \$684. In Joel's district, the long term mean sales is \$21,971 each month with standard deviation \$495. Assume that sales in both districts follow a normal distribution.

(a) Last month John sold \$19,214 whereas Joel sold \$22,718 worth of merchandise. Relative to the buying habits of customers in each district, does this mean Joel is a better salesman? Explain.

5. (a) ____________________

(b) Convert Joel's sales last month to a standard z score, and do the same for John's sales last month. Then locate both z scores under a standard normal curve. Who do you think is the better salesman? Explain your answer.

(b) ____________________

6. The length of time to complete a door assembly on an automobile factory assembly line is normally distributed with mean $\mu = 6.7$ minutes and standard deviation $\sigma = 2.2$ minutes. For a door selected at random, what is the probability the assembly line time will be

(a) 5 minutes or less?

6. (a) ____________________

(b) 10 minutes or more?

(b) ____________________

(c) between 5 and 10 minutes?

(c) ____________________

7. From long experience, it is known that the time it takes to do an oil change and lubrication job on a vehicle has a normal distribution with mean $\mu = 17.8$ minutes and standard deviation $\sigma = 5.2$ minutes. An auto service shop will give a free lube job to any customer who must wait beyond the guaranteed time to complete the work. If the shop does not want to give more than 1% of its customers a free lube job, how long should the guarantee be (round to the nearest minute).

7. ____________________

CHAPTER 7, FORM A, PAGE 3

8. You are examining a quality control chart regarding the number of employees absent each shift from a large manufacturing plant. The plant is staffed so that operations are still efficient when the average number of employees absent each shift is $\mu = 15.7$ with standard deviation $\sigma = 3.5$. For the most recent 12 shifts, the number of absent employees were

Shift	1	2	3	4	5	6	7	8	9	10	11	12
#	6	10	7	16	19	18	17	21	22	18	16	19

(a) Make a control chart showing the number of employees absent during the 12-day period.

8. (a)

(b) Are there any periods during which the number absent is out of control? Identify the out of control periods according to Type I, Type II, Type III out of control signals.

(b) ______________________

9. Medical treatment will cure about 87% of all people who suffer from a certain eye disorder. Suppose a large medical clinic treats 57 people with this disorder. Let r be a random variable that represents the number of people that will recover. The clinic wants a probability distribution for r.

(a) Write a brief but complete description in which you explain why the normal approximation to the binomial would apply. Are the assumptions satisfied? Explain.

9. (a) ______________________

(b) Estimate $P(r \le 47)$.

(b) ______________________

(c) Estimate $P(47 \le r \le 55)$.

(c) ______________________

CHAPTER 7 TEST
FORM B

1. Each of the following curves fails to be a normal curve. Give reasons why these curves are not normal curves.

(a)

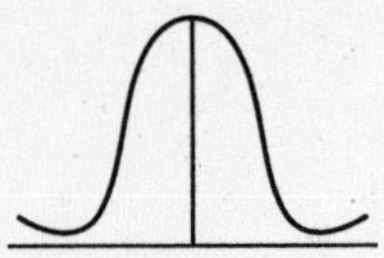

1. (a) ____________________

(b)

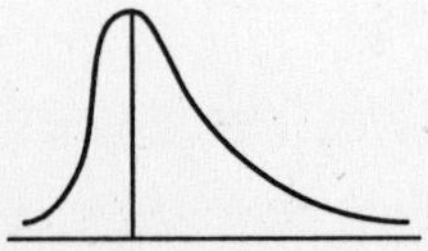

(b) ____________________

2. According to the Empirical rule, for a distribution that is symmetrical and bell-shaped (in particular, for a normal distribution) approximately _______ of the data values will lie within one standard deviation on each side of the mean.

2. ____________________

3. Assuming that the weights of newborn babies at a certain hospital are normally distributed with mean 6.5 pounds and standard deviation 1.2 pounds, how many babies in a group of 80 babies from this hospital will weigh more than 8.9 pounds?

3. ____________________

4. Let x be a random variable that represents the length of time it takes a student to write a term paper for Dr. Adam's Sociology class. After interviewing many students, it was found that x has an approximately normal distribution with mean $\mu = 6.8$ hours and standard deviation $\sigma = 2.1$ hours.

Convert each of the following x intervals to standardized z units.

(a) $x \le 7.5$ **4. (a)** ____________________

(b) $5 \le x \le 8$ **(b)** ____________________

(c) $x \ge 4$ **(c)** ____________________

Convert each of the following z intervals to raw score x intervals.

(d) $z \ge -2$ **(d)** ____________________

(e) $0 \le z \le 2$ **(e)** ____________________

(f) $z \le 3$ **(f)** ____________________

CHAPTER 7, FORM B, PAGE 2

5. Operating temperatures of two models of portable electric generators follow a normal distribution. For generator I, the mean temperature is $\mu_1 = 148°\text{F}$ with standard deviation $\sigma_1 = 25°\text{F}$. For generator II, the mean temperature is $\mu_2 = 143°\text{F}$ with standard deviation $\sigma_2 = 8°\text{F}$. At peak power demand, generator I was operating at 166°F, and generator II was operating at 165°F.

(a) At peak power output, both generators are operating at about the same temperature. Relative to the operating characteristics, is one a lot hotter than the other? Explain.

5. (a) ____________________

(b) Convert the peak power temperature for each generator to standard z units. Then locate both z scores under a standard normal curve. Could one generator be near a melt down? Which one? Explain your answer.

(b) ____________________

6. Weights of a certain model of fully loaded gravel trucks follow a normal distribution with mean $\mu = 6.4$ tons and standard deviation $\sigma = 0.3$ tons. What is the probability that a fully loaded truck of this model is

(a) less than 6 tons?

6. (a) ____________________

(b) more than 7 tons?

(b) ____________________

(c) between 6 and 7 tons?

(c) ____________________

7. Quality control studies for Speedy Jet Computer Printers show the lifetime of the printer follows a normal distribution with mean $\mu = 4$ years and standard deviation $\sigma = 0.78$ years. The company will replace any printer that fails during the guarantee period. How long should Speedy Jet printers be guaranteed if the company wishes to replace no more than 10% of the printers?

7. ____________________

CHAPTER 7, FORM B, PAGE 3

8. A toll free computer software support service for a spreadsheet program has established target length of time for each customer help phone call. The calls are targeted to have mean duration of 12 minutes with standard deviation 3 minutes. For one help technician the most recent 10 calls had the following duration.

Call #	1	2	3	4	5	6	7	8	9	10
Length	15	25	10	9	20	19	11	5	4	8

(a) Make a control chart showing the number of calls.

8. (a)

(b) Are there any periods during which the length of calls are out of control? Identify the out of control periods according to Type I, Type II, Type III out of control signals.

(b) ______________________

9. Psychology 231 can be taken as a correspondence course on a Pass/Fail basis. Long experience with this course shows that about 71% of the students pass. This semester 88 students are taking Psychology 231 by correspondence. Let r be a random variable that represents the number that will pass. The Psychology Department wants a probability distribution for r.

(a) Write a brief but complete description in which you explain why the normal approximation to the binomial would apply. Are the assumptions satisfied? Explain.

9. (a) ______________________

(b) Estimate $P(r \geq 60)$.

(b) ______________________

(c) Estimate $P(60 \leq r \leq 70)$.

(c) ______________________

CHAPTER 7 TEST
FORM C

Write the letter of the response that best answers each problem.

1. Which of the following curves is a normal curve? **1.** ________

(a)
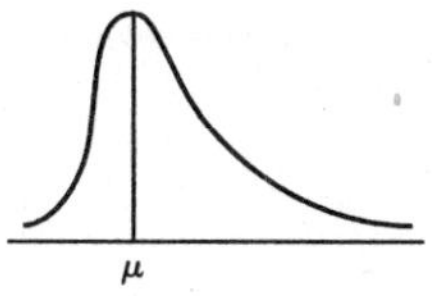

(b)
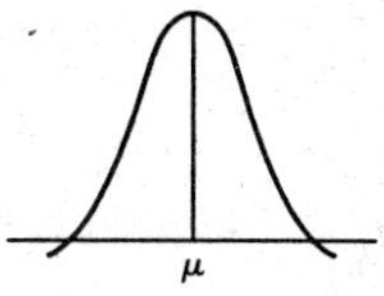

(c)
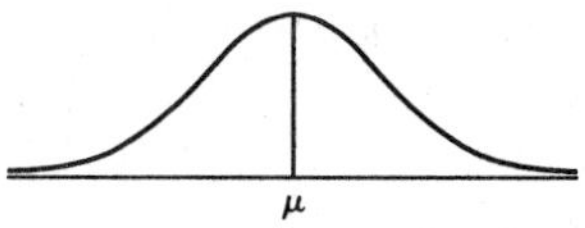

(d)
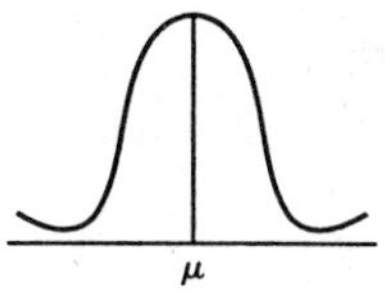

(e)
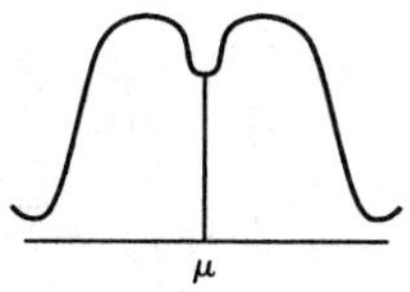

2. According to the Empirical rule, for a distribution that is symmetrical and bell-shaped (in particular, for a normal distribution) approximately _______ of the data values will lie within two standard deviations on each side of the mean. **2.** ________

(a) 75% (b) 95% (c) 68% (d) 88.9% (e) 99.7%

3. The delivery time for a package sent within the United States is normally distributed with mean of 4 days and standard deviation of approximately 1 day. If 300 packages are being sent, how many packages will arrive in less than 3 days? **3.** ________

(a) 8 (b) 96 (c) 102 (d) 198 (e) 48

4. Let x be a random variable that represents the length of time it takes a student to complete a take-home exam in Dr. Larson's psychology class. After interviewing many students, it was found that x has an approximately normal distribution with mean $\mu = 5.2$ hours and standard deviation $\sigma = 1.8$ hours.

A. Convert the x interval $x \geq 9.7$ to a standard z interval. **4. A.** ________

(a) $z \leq 2.5$ (b) $z \geq -2.5$

(c) $z \geq 4.5$ (d) $z \geq 2.5$ (e) $z \leq -2.5$

B. Convert the z interval $-1.5 \leq z \leq 1$ to a raw score x interval. **B.** ________

(a) $2.5 \leq x \leq 7$ (b) $3.44 \leq x \leq 6.66$

(c) $3.7 \leq x \leq 12.2$ (d) $-7 \leq x \leq 2.5$ (e) $-3.7 \leq x \leq -2.3$

CHAPTER 7, FORM C, PAGE 2

5. Maria and Zoe are taking Biology 105, but are in different classes. Maria's class has an average of 78% with a standard deviation of 5% on the midterm while Zoe's class has an average of 83% with a standard deviation of 12%. Assume that scores in both classes follow a normal distribution.

A. Convert Maria's midterm score of 84 to a standard z score. **5. A.** ________

(a) 0.083 (b) 0.5 (c) 0.2 (d) 1.2 (e) 6

B. Convert Zoe's midterm score of 89 to a standard z score. **B.** ________

(a) 1.2 (b) 0.5 (c) 6 (d) 0.917 (e) 2.2

C. Who do you think did better relative to their class? **C.** ________

(a) Maria (b) Zoe

(c) They performed the same (d) Neither (e) Cannot determine

6. The lifetime of a SuperTough AAA battery is normally distributed with mean $\mu = 28.5$ hours and standard deviation $\sigma = 5.3$ hours. For a battery selected at random, what is the probability that the lifetime will be

A. 25 hours or less? **6. A.** ________

(a) 0.7454 (b) 0.6604 (c) 0.2546 (d) 0.3396 (e) 0.9999

B. 34 hours or more? **B.** ________

(a) 0.8485 (b) 0.1515 (c) 1.038 (d) 0.8508 (e) 0.1492

C. between 25 hours and 34 hours? **C.** ________

(a) 0.4038 (b) 0.5962 (c) 0.1054 (d) 0.8946 (e) 2/736

7. Quality control studies for Dependable Dishwashers show the lifetime of a dishwasher follows a normal distribution with mean $\mu = 8$ years and standard deviation $\sigma = 1.2$ years. The company will replace any dishwasher that fails during the guarantee period. How long should the company's dishwashers be guaranteed if the company wishes to replace no more than 2% of the dishwashers? **7.** ________

(a) 0.16 year (b) 0.13 year

(c) 5.5 years (d) 10.5 years (e) 2.5 years

CHAPTER 7, FORM C, PAGE 3

8. When evaluating a control chart, which of the following is not a warning signal that a random variable x is out of control?

8. ________

(a) A run of nine consecutive points on one side of the center line (the line at target value μ).

(b) One point falls beyond the 3σ level.

(c) Two points fall beyond the 3σ level.

(d) At least two points lie beyond the 2σ level on the same side of the center line.

(e) At least two of three consecutive points lie beyond the 2σ level on the same side of the center line.

9. Records show that 29% of all payments to a mail-order company are submitted after the due date. Suppose 50 payments are submitted this week. Let r be a random variable that represents the number of payments that are late. Use the normal approximation to the binomial to estimate

A. $P(r \geq 20)$

9. A. ________

(a) 0.0307 (b) 0.0594 (c) 0.9406 (d) 0.9564 (e) 0.0436

B. $P(20 \leq r \leq 25)$

B. ________

(a) 0.0591 (b) 0.0585 (c) 0.0431 (d) 0.0304 (e) 0.0298

CHAPTER 8 TEST
FORM A

1. Write a brief but complete discussion of each of the following topics:

population, parameter, sample, sampling distribution,

statistical inference using sampling distributions.

In each case be sure to give a complete and accurate definition of the terms. Illustrate your discussion using examples from everyday life.

1. ______________________

2. The diameters of oranges from a Florida orchard are <u>normally distributed</u> with mean $\mu = 3.2$ inches and standard deviation $\sigma = 1.1$ inches. A packing supplier is designing special occasion presentation boxes of oranges and needs to know the average diameter for a random sample of 8 oranges. What is the probability that the mean diameter $\overline{x}$ for these oranges is

(a) smaller than 3 inches?

2. (a) ______________________

(b) longer than 4 inches?

(b) ______________________

(c) between 3 and 4 inches?

(c) ______________________

3. The manufacturer of a new compact car claims the miles per gallon (mpg) for the gasoline consumption is mound shaped and symmetric with mean $\mu = 25.9$ mph and standard deviation $\sigma = 9.5$ mph. If 30 such cars are tested, what is the probability the average mph $\overline{x}$ is

(a) less than 23 mph?

3. (a) ______________________

(b) more than 28 mph?

(b) ______________________

(c) between 23 and 28 mpg?

(c) ______________________

CHAPTER 8 TEST
FORM B

1. Write a brief but complete discussion in which you cover the following topics: What is the mean $\mu_{\bar{x}}$ and standard deviation $\sigma_{\bar{x}}$ of the $\bar{x}$ distribution based on a sample size n? Be sure to give appropriate formulas in your discussion. How do you find a standard z score corresponding to $\bar{x}$? State the Central Limit Theorem, and the general conditions under which it can be used. Illustrate your discussion using examples from everyday life.

1. ______________________________

2. Chemists use pH to measure the acidity/alkaline nature of compounds. A large vat of mixed commercial chemicals is supposed to have a mean pH $\mu = 6.3$ with a standard deviation $\sigma = 1.9$. Assume a normal distribution for pH values. If a random sample of ten readings in the vat is taken and the mean pH $\bar{x}$ is computed, find each of the following.

(a) $P(5.2 \le \bar{x})$

(b) $P(\bar{x} \le 7.1)$

(c) $P(5.2 \le \bar{x} \le 7.1)$

2. (a) ______________________________

(b) ______________________________

(c) ______________________________

3. Fire department response time is the length of time it takes a fire truck to arrive at the scene of a fire starting from the time the call was given to the truck. Response time for the Castle Wood Fire Department follows a mound shaped and symmetric distribution. The response time has mean $\mu = 8.8$ minutes with standard deviation $\sigma = 2.1$ minutes. If a random sample of 32 response times is taken and the mean response time $\bar{x}$ is computed, find each of the following.

(a) $P(8 \le \bar{x})$

(b) $P(\bar{x} \le 9)$

(c) $P(8 \le \bar{x} \le 9)$

3. (a) ______________________________

(b) ______________________________

(c) ______________________________

CHAPTER 8 TEST
FORM C

Write the letter of the response that best answers each problem.

1. Complete the following definitions

A. A __________ is a numerical descriptive measure of a sample. **1. A.** __________

(a) parameter (b) population

(c) statistic (d) statistical inference (e) probability sampling distribution

B. A __________ is a subset of measurements from the population. **B.** __________

(a) proportion (b) sample

(c) estimate (d) statistic (e) parameter

C. A __________ is a numerical descriptive measure of a population. **C.** __________

(a) parameter (b) central limit theorem

(c) statistic (d) statistical inference (e) probability sampling distribution

D. A __________ can be thought of as a set of measurements (or counts), either existing or conceptual. **D.** __________

(a) population (b) statistic

(c) estimate (d) sample (e) parameter

2. The weights of envelopes sent from an insurance office are normally distributed with mean $\mu = 12$ ounces and standard deviation $\sigma = 3.7$ ounces. The mail room clerk would like to know the average weight of 20 envelopes. What is the probability that the mean weight $\bar{x}$ is

A. lighter than 10 ounces? **2. A.** __________

(a) 0.0351 (b) 0.0078 (c) 0.2946 (d) 0.4922 (e) 0.9922

B. heavier than 13 ounces? **B.** __________

(a) 0.6064 (b) 0.3936 (c) 0.8869 (d) 0.1131 (e) 0.2743

C. between 10 and 13 ounces? **C.** __________

(a) 0.9636 (b) 0.3118 (c) 0.8791 (d) 0.1209 (e) 0.1053

CHAPTER 8, FORM C, PAGE 2

3. The manufacturer of a coffee dispensing machine claims the ounces per cup is mound shaped and symmetric with mean $\mu = 7$ ounces and standard deviation $\sigma = 0.8$ ounce. If 40 cups of coffee are measured, what is the probability that the average ounces per cup $\bar{x}$ is

A. less than 6.8 ounces? **3. A.** ________

(a) 0.0571 (b) 0.9429 (c) 0.1170 (d) 0.4013 (e) 0.5987

B. more than 7.4 ounces? **B.** ________

(a) 0.0028 (b) 0.9992 (c) 0.6915 (d) 0.0008 (e) 0.3085

C. between 6.8 and 7.4 ounces? **B.** ________

(a) 0.0579 (b) 0.2902 (c) 0.9760 (d) 0.0563 (e) 0.9421

CHAPTER 9 TEST
FORM A

1. As part of an Environmental Studies class project, students measured the circumferences of a random sample of 45 Blue Spruce trees near Brainard Lake, Colorado. The sample mean circumference was $\bar{x} = 29.8$ inches with sample deviation $s = 7.2$ inches. Find a 95% confidence interval for the population mean circumference of all Blue Spruce trees near this lake. Write a brief explanation of the meaning of the confidence interval in the context of this problem.

1. ______________________

2. Collette is self-employed, selling stamps at home parties. She wants to estimate the average amount a client spends at each party. A random sample of 35 clients' receipts gave a mean of $\bar{x} = \$34.70$ with standard deviation $s = \$4.85$.

(a) Find a 90% confidence interval for the average amount spent by all clients.

2. (a) ______________________

(b) For a party with 35 clients, use part (a) to estimate a range of dollar values for Collette's total sales at that party.

(b) ______________________

3. How long does it take to commute from home to work? It depends on several factors including route, traffic, and time of departure. The data below are results (in minutes) from a random sample of 8 trips. Use these data to create a 95% confidence interval for the population mean time of the commute.

27 38 30 42 24 37 30 39

3. ______________________

4. A random sample of 19 Rainbow Trout caught at Brainard Lake, Colorado had mean length $\bar{x} = 11.9$ inches with sample standard deviation $s = 2.8$ inches. Find a 99% confidence interval for the population mean length of all Rainbow Trout in this lake. Write a brief explanation of the meaning of the confidence interval in the context of this problem.

4. ______________________

CHAPTER 9, FORM A, PAGE 2

5. A random sample of 78 students were interviewed and 59 said they would vote for Jennifer McNamara as student body president.

   (a) Let p represent the proportion of all students at this college who will vote for Jennifer. Find a point estimate $\hat{p}$ for p. **5. (a)** ______________

   (b) Find a 90% confidence interval for p. **(b)** ______________

   (c) What assumptions are required for the calculation of part (b)? Do you think these assumptions are satisfied? Explain. **(c)** ______________

   (d) How many more students should be included in the sample to be 90% sure that a point estimate $\hat{p}$ will be within a distance of 0.05 from p. **(d)** ______________

6. A random sample of 53 students were asked for the number of semester hours they are taking this semester. The sample standard deviation was found to be $s = 4.7$ semester hours. How many <u>more</u> students should be included in the sample to be 99% sure the sample mean $\overline{x}$ is within one semester hour of the population mean μ for all students at this college? **6.** ______________

7. What percentage of college students own cellular phones? Let p be the proportion of college students who own cellular phones.

   (a) If no preliminary study is made to estimate p, how large a sample is needed to be 90% sure that a point estimate $\hat{p}$ will be within a distance of 0.08 from p. **7. (a)** ______________

   (b) A preliminary study shows that approximately 38% of college students own cellular phones. Answer part (a) using this estimate for p. **(b)** ______________

CHAPTER 9 TEST FORM B

1. A random sample of 14 evenings (6 PM to 9PM) at the O'Sullivan household showed the family received an average of $\bar{x} = 5.2$ solicitation phone calls each evening. The sample standard deviation was $s = 1.9$. Find a 96% confidence interval for the population mean number of solicitation calls this family receives each night. Write a brief explanation of the meaning of the confidence interval in the context of this problem.

1. ______________________

2. Jordan is the manager of a used book store. He wants to estimate the average amount a customer spends per visit. A random sample of 80 customers' receipts gave a mean of $\bar{x}$ = \$6.90 with standard deviation s = \$2.45.

(a) Find a 90% confidence interval for the average amount spent by all customers.

2. (a) ______________________

(b) For a day when the book store had 80 customers, use part (a) to estimate a range of dollar values for the total income on that day.

(b) ______________________

3. Mr. Crandall has assigned a term paper due at the end of the semester. He would like to know the average length of the paper. The data below are the numbers of typed pages from a random sample of 10 term papers. Use these data to create a 95% confidence interval for the population mean length of all term papers for his class.

14	20	25	10	16
8	15	12	18	9

3. ______________________

4. Computer Depot is a large store that sells and repairs computers. A random sample of 110 computer repair jobs took technicians an average of $\bar{x} = 93.2$ minutes per computer. The sample standard deviation was $s = 16.9$ minutes. Find a 99% confidence interval for the population mean time μ for computer repairs. Write a brief explanation of the meaning of the confidence interval in the context of this problem.

4. ______________________

CHAPTER 9, FORM B, PAGE 2

5. A random sample of 56 credit card holders showed that 41 regularly paid their credit card bills on time.

(a) Let p represent the proportion of all people who regularly paid their credit card bills on time. Find a point estimate $\hat{p}$ for p. **5. (a)** ____________________

(b) Find a 95% confidence interval for p. **(b)** ____________________

(c) What assumptions are required for the calculations of part (b)? Do you think these assumptions are satisfied? Explain. **(c)** ____________________

(d) How many more credit card holders should be included in the sample to be 95% sure that a point estimate $\hat{p}$ will be within a distance of 0.05 from p? **(d)** ____________________

6. Allen is an appliance salesman who works on commission. A random sample of 39 days showed that the sample standard deviation value of sales was $s = \$215$. How many more days should be included in the sample to be 95% sure the population mean μ is within \$50 of the sample mean $\overline{x}$? **6.** ____________________

7. What percentage of male athletes wear contact lenses during performances? Let p be the proportion of male athletes who wear contact lenses.

(a) If you have no preliminary estimate for p, how many male athletes should you include in a random sample to be 90% sure that the point estimate $\hat{p}$ will be within a distance of 0.05 from p. **7. (a)** ____________________

(b) Studies show that approximately 19% of male athletes wear contact lenses during performances. Answer part (a) using this estimate for p. **(b)** ____________________

CHAPTER 9 TEST
FORM C

Write the letter of the response that best answers each problem.

1. As part of a real estate company's study, the selling prices of 50 homes in a particular neighborhood were gathered. The sample mean price was $\bar{x}$ = \$234,000 with sample standard deviation s = \$28,500. Find a 95% confidence interval for the population mean selling price of all homes in this neighborhood. **1.** ________

(a) \$229,969 to \$238,031 (b) \$226,100 to \$241,900

(c) \$227,370 to \$240,630 (d) \$223,601 to \$244,399

(e) \$232,883 to \$235,117

2. Latasha is a waitress at Seventh Heaven Hamburgers. She wants to estimate the average amount each group at a table leaves for a tip. A random sample of 42 groups gave a mean of $\bar{x}$ = \$7.30 with standard deviation s = \$2.92.

A. Find a 90% confidence interval for the average amount left by all groups. **2. A.** ________

(a) \$7.19 to \$7.41 (b) \$6.14 to \$8.46

(c) \$6.56 to \$8.04 (d) \$2.50 to \$12.10 (e) \$6.42 to \$8.18

B. For a double-shift with 42 groups, use part A to estimate a range of dollar values for Latasha's total sales for that double-shift. **B.** ________

(a) \$301.98 to \$311.22 (b) \$269.64 to \$343.56

(c) \$257.88 to \$355.32 (d) \$105.00 to \$508.20 (e) \$275.52 to \$337.68

3. How healthy are the employees at Direct Marketing Industry? A random sample of 12 employees was taken and the number of days each was absent for sickness was recorded (during a one-year period). Use these data to create a 95% confidence interval for the population mean days absent for sickness.

2	5	3	7	10	0
6	8	5	11	3	1

3. ________

(a) 2.87 days to 7.29 days (b) 3.28 days to 6.89 days

(c) 3.11 days to 7.05 days (d) 2.77 days to 7.39 days

(e) 2.87 days to 7.29 days

CHAPTER 9, FORM C, PAGE 2

4. Denise is a professional swimmer who trains, in part, by running. She would like to estimate the average number of miles she runs in each week. For a random sample of 20 weeks, the mean is $\bar{x}$ = 17.5 miles with standard deviation s = 3.8 miles. Find a 995 confidence interval for the population mean number of weekly miles Denise runs. **4.** ________

(a) 15.01 miles to 19.99 miles (b) 15.07 miles to 19.93 miles

(c) 15.34 miles to 19.66 miles (d) 15.31 miles to 19.69 miles

(e) 15.08 miles to 19.92 miles

5. A random sample of 84 shoppers were interviewed and 51 said they prefer to shop alone rather than with someone such as friends or family.

A. Let p represent the proportion of all shoppers at this mall who would prefer to shop alone. Find a point estimate $\hat{p}$ for p. **5. A.** ________

(a) 0.393 (b) 51 (c) 84 (d) 0.607 (e) 0.5

B. Find a 90% confidence interval for p. **B.** ________

(a) 0.519 to 0.695 (b) 0.517 to 0.697

(c) −0.20 to 1.41 (d) 0.503 to 0.711 (e) 0.305 to 0.481

C. How many more students should be included in the sample to be 90% sure that a point estimate $\hat{p}$ will be within a distance of 0.05 from p. **C.** ________

(a) 271 (b) 187 (c) 259 (d) 283 (e) 175

6. A random sample of 61 students were asked how much they spent for classroom textbooks this semester. The sample standard deviation was found to be s = \$28.70. How many more students should be included in the sample to be 99% sure the sample mean $\bar{x}$ is within \$7 of the population mean μ for all students at this college? **6.** ________

(a) 0 (b) 65 (c) 51 (d) 4 (e) 112

CHAPTER 9, FORM C, PAGE 3

7. What percentage of college students are attending a college in the state where they grew up? Let p be the proportion of college students from the same state as that in which the college resides.

A. If no preliminary study is made to estimate p, how large a sample is needed to be 90% sure that a point estimate $\hat{p}$ will be within a distance of 0.07 from p? **7. A.** ______

(a) 139 (b) 196 (c) 340 (d) 6 (e) 138

B. A preliminary study shows that approximately 71% of college students grew up in the same state as that in which the college resides. Answer part A using this estimate for p. **B.** ______

(a) 280 (b) 113 (c) 139 (d) 114 (e) 162

CHAPTER 10 TEST
FORM A

For each of the following problems, please provide the requested information.

(a) State the null and alternate hypotheses. Will we use a left-tailed, right-tailed, or two-tailed test? What is the level of significance?

(b) Identify the sampling distribution to be used: the standard normal or the Student's t. Find the critical value(s).

(c) Sketch the critical region and show the critical value(s) on the sketch.

(d) Compute the z or t value of the sample test statistic and show it's location on the sketch of part (c).

(e) Find the P value or an interval containing the P value for the sample test statistic.

(f) Based on your answers for parts (a) to (e), shall we reject or fail to reject the null hypothesis. Explain your conclusion in the context of the problem.

1. A large furniture store has begun a new ad campaign on local television. Before the campaign, the long term average daily sales were \$24,819. A random sample of 40 days during the new ad campaign gave a sample mean daily sale of $\bar{x}$ = \$25,910 with sample standard deviation s = \$1917. Does this indicate that the population mean daily sales is now more than \$24,819? Use a 1% level of significance.

1. (a) ____________________

(b) ____________________

(c)

(d) ____________________

(e) ____________________

(f) ____________________

CHAPTER 10, FORM A, PAGE 2

2. A new bus route has been established between downtown Denver and Englewood, a suburb of Denver. Dan has taken the bus to work for many years. For the old bus route, he knows from long experience that the mean waiting time between buses at his stop was $\mu = 18.3$ minutes. However, a random sample of 5 waiting times between buses using the new route had mean $\bar{x} = 15.1$ minutes with sample standard deviation $s = 6.2$ minutes. Does this indicate that the population mean waiting time for the new route is different from what it used to be? Use $\alpha = 0.05$.

2. (a) ____________________

(b) ____________________

(c)

(d) ____________________

(e) ____________________

(f) ____________________

3. The State Fish and Game Division claims that 75% of the fish in Homestead Creek are Rainbow Trout. However, the local fishing club caught (and released) 189 fish one weekend, and found that 125 were Rainbow Trout. The other fish were Brook Trout, Brown Trout, and so on. Does this indicate that the percentage of Rainbow Trout in Homestead Creek is less than 75%? Use $\alpha = 0.01$.

3. (a) ____________________

(b) ____________________

(c)

(d) ____________________

(e) ____________________

(f) ____________________

CHAPTER 10, FORM A, PAGE 3

4. How tall are college hockey players? The average height has been 68.3 inches. A random sample of 55 hockey players gave a mean height of 69.1 inches with sample standard deviation $s = 1.6$ inches. Does this indicate that the population mean height is different from 68.3 inches? Use 5% level of significance.

4. (a) ______

(b) ______

(c)

(d) ______

(e) ______

(f) ______

5. How long does it take to have food delivered? A Chinese restaurant advertises that delivery will be no more than 30 minutes. A random sample of delivery times (in minutes) is shown below. Based on this sample, is the delivery time greater than 30 minutes? Use a 5% level of significance. Assume that the distribution of times is normal.

32	28	21	39	30	27	29
39	32	28	42	25	26	30

5. (a) ______

(b) ______

(c)

(d) ______

(e) ______

(f) ______

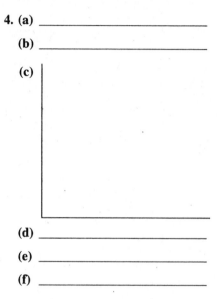

CHAPTER 10, FORM A, PAGE 4

6. A music teacher knows from past records that 60% of students taking summer lessons play the piano. The instructor believes this proportion may have dropped due to the popularity of wind and brass instruments. A random sample of 80 students yielded 43 piano players. Test the instructor's claim at $\alpha = 0.05$.

6. (a) ______________________

(b) ______________________

(c)

(d) ______________________

(e) ______________________

(f) ______________________

CHAPTER 10 TEST
FORM B

For each of the following problems, please provide the requested information.

(a) State the null and alternate hypotheses. Will we use a left-tailed, right-tailed, or two-tailed test? What is the level of significance?

(b) Identify the sampling distribution to be used: the standard normal or the Student's t. Find the critical value(s).

(c) Sketch the critical region and show the critical value(s) on the sketch.

(d) Compute the z or t value of the sample test statistic and show it's location on the sketch of part (c).

(e) Find the P value or an interval containing the P value for the sample test statistic.

(f) Based on your answers for parts (a) to (e), shall we reject or fail to reject the null hypothesis. Explain your conclusion in the context of the problem.

1. Long term experience showed that after a type of eye surgery it took a mean of $\mu = 5.3$ days recovery time in a hospital. However, a random sample of 32 patients with this type of eye surgery, were recently treated as outpatients during the recovery. The sample mean recovery time was $\bar{x} = 4.2$ days with sample standard deviation $s = 1.9$ days. Does this indicate that the mean recovery time for outpatients is less than the time for those recovering in the hospital? Use a 1% level of significance.

1. (a) ____________________

(b) ____________________

(c)

(d) ____________________

(e) ____________________

(f) ____________________

CHAPTER 10, FORM B, PAGE 2

2. Recently the national average yield on municipal bonds has been $\mu = 4.19\%$. A random sample of 16 Arizona municipal bonds gave an average yield of 5.11% with sample standard deviation $s = 1.15\%$. Does this indicate that the population mean yield for all Arizona municipal bonds is greater than the national average? Use a 5% level of significance.

2. (a) ______________________

(b) ______________________

(c)

(d) ______________________

(e) ______________________

(f) ______________________

3. At a local four-year college, 37% of the student-body are freshmen. A random sample of 42 student names taken from the Dean's Honor List over the past several semesters showed that 17 were freshmen. Does this indicate the population proportion of freshmen on the Dean's Honor List is different from 37%? Use a 1% level of significance.

3. (a) ______________________

(b) ______________________

(c)

(d) ______________________

(e) ______________________

(f) ______________________

CHAPTER 10, FORM B, PAGE 3

4. How long does it take juniors to complete a standardized exam? The long-term average is 2.8 hours. A random sample of 43 juniors gave a sample mean of $\bar{x} = 2.5$ hours with standard deviation $s = 0.8$ hour. Does this indicate that the population mean time is different from 2.8 hours? Use 5% level of significance.

4. (a) ______________________

(b) ______________________

(c)

(d) ______________________

(e) ______________________

(f) ______________________

5. How old are the customers? The owner of a comedy club has based business decisions on the average of customers historically being 30 years. A random sample of customers' ages (in years) is shown below. Based on this sample, is the average age more than 30 years? Use a 5% level of significance. Assume that the distribution of ages is normal.

37	30	26	35	45	42	51	40
43	27	39	38	46	21	28	20

5. (a) ______________________

(b) ______________________

(c)

(d) ______________________

(e) ______________________

(f) ______________________

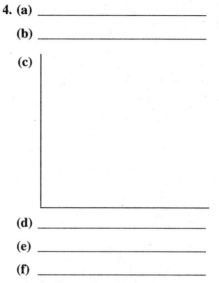

CHAPTER 10, FORM B, PAGE 4

6. The Department of Transportation in a particular city knows from past records that 27% of workers in the downtown district use the subway system each day to commute to or from work. The department suspects this proportion has increased due to decreased parking spaces in the downtown district. A random sample of 130 workers in the downtown district showed 49 used the subway daily. Test the departments' suspicion at the 5% level of significance.

6. (a) ______________________

(b) ______________________

(c)

(d) ______________________

(e) ______________________

(f) ______________________

CHAPTER 10 TEST
FORM C

Write the letter of the response that best answers each problem.

1. A small electronics store has begun to advertise in the local newspaper. Before advertising, the long term average weekly sales were \$9820. A random sample of 50 weeks while the newspaper ads were running gave a sample mean weekly sale of $\overline{x} = \$10{,}960$ with the sample standard deviation $s = \$1580$. Does this indicate that the population mean weekly sales is now more than \$9820? Test at the 5% level of significance.

A. State the null and alternate hypotheses. **1. A.** ________

(a) H_0: $\mu = 9820$; H_1: $\mu < 9820$ (b) H_0: $\mu = 9820$; H_1: $\mu > 9820$

(c) H_0: $\overline{x} = 10{,}960$; H_1: $\overline{x} > 10{,}960$ (d) H_0: $\mu = 10{,}960$; H_1: $\mu > 10{,}960$

(e) H_0: $\mu > 9820$; H_1: $\mu = 9820$

B. Find the critical value(s). **B.** ________

(a) $z_0 = 2.33$ (b) $z_0 = 1.96$

(c) $t_0 = -1.96$ (d) $z_0 = 1.645$ (e) $t_0 = -1.96$

C. Compute the z or t value of the sample test statistic. **C.** ________

(a) $z = 5.10$ (b) $z = 0.10$

(c) $t = 0.72$ (d) $z = -5.10$ (e) $t = -0.10$

D. Find the P value or an interval containing the P value for the sample test statistic. **D.** ________

(a) P value = 0.236 (b) P value = 0.460

(c) P value < 0.0001 (d) P value > 0.0001 (e) Cannot determine

E. Based on your answers for parts A–D, what is your conclusion? **E.** ________

(a) Do not reject H_0 (b) Reject H_0 (c) Cannot determine

(d) The population mean weekly sales is less than \$9820.

(e) The population mean weekly sales is more than \$10,960.

CHAPTER 10, FORM C, PAGE 2

2. The average annual salary of employees at Wintertime Sports was $28,750 last year. This year the company opened another store. Suppose a random sample of 18 employees gave an average annual salary of $\bar{x}$ = $25,810 with sample standard deviation s = $4230. Use a 1% level of significance to test the claim that the average annual salary for all employees is different from last year's average salary. Assume salaries are normally distributed.

A. State the null and alternate hypotheses. **2. A.** ________

(a) H_0: $\mu = 28{,}750$; H_1: $\mu < 28{,}750$ (b) H_0: $\mu = 25{,}810$; H_1: $\mu \neq 25{,}810$

(c) H_0: $\mu_1 = \mu_2$; H_1: $\mu_1 \neq \mu_2$ (d) H_0: $\bar{x} = 25{,}810$; H_1: $\bar{x} \neq 25{,}810$

(e) H_0: $\mu = 28{,}750$; H_1: $\mu \neq 28{,}750$

B. Find the critical value(s). **B.** ________

(a) $t_0 = 2.58$ (b) $t_0 = \pm 1.96$

(c) $t_0 = \pm 2.567$ (d) $t_0 = \pm 2.898$ (e) $z_0 = -2.567$

C. Compute the z or t value of the sample test statistic. **C.** ________

(a) $t = -2.95$ (b) $z = -2.95$

(c) $t = -2.87$ (d) $z = -12.51$ (e) $t = 2.95$

D. Find the P value or an interval containing the P value for the sample test statistic. **D.** ________

(a) Cannot determine (b) $0.01 < P$ value < 0.02

(c) P value < 0.010 (d) P value > 0.010 (e) P value < 0.005

E. Based on your answers for parts A–D, what is your conclusion? **E.** ________

(a) Do not reject H_0 (b) Reject H_0 (c) Cannot determine

(d) The average salary is less than $25,810.

(e) The average salary is different from $10,960.

CHAPTER 10, FORM C, PAGE 3

3. The owner of Prices Limited claims that 75% of all the items in the store are less than \$5. Suppose that you check a random sample of 146 items in the store and find that 105 have prices less than \$5. Does this indicate that the items in the store costing less than \$5 is different from 75%? Use $\alpha = 0.01$.

A. State the null and alternate hypotheses. **1. A.** ________

(a) H_0: $p_1 = p_2$; H_1: $p_1 \neq p_2$ (b) H_0: $p = 5$; H_1: $p \neq 5$

(c) H_0: $\hat{p} = 0.75$; H_1: $\hat{p} \neq 0.75$ (d) H_0: $p = 0.75$; H_1: $p \neq 0.75$

(e) H_0: $\hat{p} = 0.72$; H_1: $\hat{p} \neq 0.72$

B. Find the critical value(s). **B.** ________

(a) $z_0 = \pm 2.58$ (b) $z_0 = -2.33$

(c) $z_0 = \pm 1.96$ (d) $t_0 = -1.645$ (e) $t_0 = \pm 2.58$

C. Compute the z or t value of the sample test statistic. **C.** ________

(a) $t = -0.86$ (b) $t = -0.73$

(c) $z = -0.73$ (d) $z = 0.86$ (e) $z = -0.86$

D. Find the P value or an interval containing the P value for the sample test statistic. **D.** ________

(a) P value = 0.2327 (b) P value = 0.0975

(c) P value = 0.3898 (d) P value = 0.1949 (e) P value = 0.8051

E. Based on your answers for parts A–D, what is your conclusion? **E.** ________

(a) Do not reject H_0 (b) Reject H_0 (c) Cannot determine

(d) The items in the store cost less than \$5.

(e) The items in the store do not cost less than \$5.

CHAPTER 10, FORM C, PAGE 4

4. A machine in the lodge at a ski resort dispenses a hot chocolate drink. The average cup of hot hocolate is supposed to contain 7.75 ounces. A random sample of 68 cups of hot chocolate form this machine show the average content to be 7.62 ounces with a standard deviation of 0.6 ounces. Do you think the machine is out of adjustment and the average amount of hot chocolate is less than it is supped to be? Use a 5% level of significance.

A. State the null and alternate hypotheses. **4. A.** __________

(a) H_0: $\mu_1 = \mu_2$; H_1: $\mu_1 < \mu_2$ (b) H_0: $\mu = 7.75$; H_1: $\mu \neq 7.75$

(c) H_0: $\bar{x} = 7.62$; H_1: $\bar{x} < 7.62$ (d) H_0: $\mu > 7.62$; H_1: $\mu = 7.62$

(e) H_0: $\mu = 7.75$; H_1: $\mu < 7.75$

B. Find the critical value(s). **B.** __________

(a) $z_0 = -1.645$ (b) $z_0 = \pm 2.58$

(c) $z_0 = \pm 1.96$ (d) $z_0 = 1.645$ (e) $t_0 = -1.645$

C. Compute the z or t value of the sample test statistic. **C.** __________

(a) $z = -14.73$ (b) $z = -1.645$

(c) $z = -1.79$ (d) $t = -0.03$ (e) $z = -0.03$

D. Find the P value or an interval containing the P value for the sample test statistic. **D.** __________

(a) P value = 0.0734 (b) P value = 0.4880

(c) P value = 0.4880 (d) P value = 0.0367 (e) P value = 0.9633

E. Based on your answers for parts A–D, what is your conclusion? **E.** __________

(a) Do not reject H_0 (b) Reject H_0 (c) Cannot determine

(d) The cup will be over full.

(e) The machine is working fine.

CHAPTER 10, FORM C, PAGE 5

5. The average number of miles on vehicles traded in at Smith Brothers Motors is 64,000. Smith Brothers Motors has started a new deal offering lower financing charges. They are interested in whether the average mileage on trade-in vehicles has decreased. Test using $\alpha = 0.01$ and the results (in thousands) from a random sample printed below (taken after the deal started). Assume mileage is normally distributed.

39	47	62	110	58
90	50	99	41	28

A. State the null and alternate hypotheses. **5. A.** __________

(a) H_0: $\bar{x} = 62{,}400$; H_1: $\bar{x} \neq 62{,}400$ (b) H_0: $\bar{x} = 64{,}000$; H_1: $\bar{x} > 64{,}000$

(c) H_0: $\mu_d = 64{,}000$; H_1: $\mu_d < 64{,}000$ (d) H_0: $\mu = 64{,}000$; H_1: $\mu < 64{,}000$

(e) H_0: $\mu > 64{,}000$; H_1: $\mu > 64{,}000$

B. Find the critical value(s). **B.** __________

(a) $t_0 = 2.821$ (b) $z_0 = -2.33$

(c) $t_0 = -2.821$ (d) $t_0 = -3.250$ (e) $t_0 = -2.764$

C. Compute the z or t value of the sample test statistic. **C.** __________

(a) $t = -0.52$ (b) $t = -0.18$

(c) $t = -0.17$ (d) $t = -0.02$ (e) $z = -0.58$

D. Find the P value or an interval containing the P value for the sample test statistic. **D.** __________

(a) Cannot determine (b) P value = 0.4286

(c) P value = 0.4325 (d) P value > 0.250 (e) P value > 0.125

E. Based on your answers for parts A–D, what is your conclusion? **E.** __________

(a) Do not reject H_0 (b) Reject H_0 (c) Cannot determine

(d) People are trading in more vehicles.

(e) People are trading in fewer vehicles.

CHAPTER 10, FORM C, PAGE 6

6. Results from previous studies showed 79% of all high school seniors from a certain city plan to attend college after graduation. A random sample of 200 high school seniors from this city showed 162 plan to attend college. Does this indicate that the percentage has increased from that of previous studies? Test at 5% level of significance.

A. State the null and alternate hypotheses. **6. A.** ________

(a) H_0: $\mu = 0.79$; H_1: $\mu > 0.79$ (b) H_0: $p = 0.79$; H_1: $p \neq 0.79$

(c) H_0: $p = 0.79$; H_1: $p > 0.79$ (d) H_0: $\hat{p} = 0.81$; H_1: $\hat{p} \neq 0.81$

(e) H_0: $\hat{p} = 0.79$; H_1: $\hat{p} > 0.79$

B. Find the critical value(s). **B.** ________

(a) $z_0 = 1.645$ (b) $t_0 = 2.33$

(c) $z_0 = 2.33$ (d) $t_0 = \pm 2.58$ (e) $z_0 = \pm 1.96$

C. Compute the z or t value of the sample test statistic. **C.** ________

(a) $z = 0.72$ (b) $t = 1.645$

(c) $z = 0.62$ (d) $z = 1.645$ (e) $z = 0.69$

D. Find the P value or an interval containing the P value for the sample test statistic. **D.** ________

(a) P value = 0.2676 (b) P value = 0.7642

(c) P value > 0.05 (d) P value = 0.2451 (e) P value = 0.2358

E. Based on your answers for parts A–D, what is your conclusion? **E.** ________

(a) Do not reject H_0 (b) Reject H_0 (c) Cannot determine

(d) More seniors are going to college.

(e) Less seniors are going to college.

CHAPTER 11 TEST
FORM A

For each of the following problems, please provide the requested information.

(a) State the null and alternate hypotheses. Will we use a left-tailed, right-tailed, or two-tailed test? What is the level of significance?

(b) Identify the sampling distribution to be used: the standard normal or the Student's *t*. Find the critical value(s).

(c) Sketch the critical region and show the critical value(s) on the sketch.

(d) Compute the *z* or *t* value of the sample test statistic and show it's location on the sketch of part (c).

(e) Find the *P* value or an interval containing the *P* value for the sample test statistic.

(f) Based on your answers for parts (a) to (e), shall we reject or fail to reject the null hypothesis. Explain your conclusion in the context of the problem.

1. A telemarketer is trying two different sales pitches to sell a carpet cleaning service. For Sales Pitch I, 175 people were contacted by phone and 62 of these people bought the cleaning service. For Sales Pitch II, 154 people were contacted by phone and 45 of these people bought the cleaning service. Does this indicate that there is any difference in the population proportions of people who will buy the cleaning service, depending on which sales pitch is used? Use $\alpha = 0.05$

1. (a) ____________________

(b) ____________________

(c)

(d) ____________________

(e) ____________________

(f) ____________________

2. A systems specialist has studied the work-flow of clerks all doing the same inventory work. Based on this study, she designed a new work-flow layout for the inventory system. To compare average production for the old and new methods, a random sample of six clerks was used. The average production rate (number of inventory items processed per shift) for each clerk was measured both before and after the new system was introduced. The results are shown below. Test the claim that the new system increases the mean number of items process per shift. Use $\alpha = 0.05$.

Clerk	1	2	3	4	5	6
B: Old	116	108	93	88	119	111
A: New	123	114	112	82	127	122

2. (a) ____________________

(b) ____________________

(c)

(d) ____________________

(e) ____________________

(f) ____________________

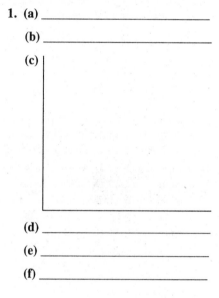

CHAPTER 11, FORM A, PAGE 2

3. How productive are employees? One way to answer this question is to study annual company profits per employee. Let x_1 represent annual profits per employee in computer stores in St. Louis. A random sample of $n_1 = 11$ computer stores gave a sample mean of $\bar{x}_1 = 25.2$ thousand dollars profit per employee with sample standard deviation $s_1 = 8.4$ thousand dollars. Another random sample of $n_2 = 9$ building supply stores in St. Louis gave a sample mean $\bar{x}_2 = 19.9$ thousand dollars per employee with sample standard deviation $s_2 = 7.6$ thousand dollars. Does this indicate that in St. Louis, computer stores tend to have higher mean profits per employee? Use $\alpha = 0.01$

3. (a) ____________________

(b) ____________________

(c)

(d) ____________________

(e) ____________________

(f) ____________________

4. How big are tomatoes? Some say that depends on the growing conditions. A random sample of $n_1 = 89$ organically grown tomatoes had sample mean weight $\bar{x}_1 = 3.8$ ounces with sample standard deviation $s_1 = 0.9$ ounces. Another random sample of $n_2 = 75$ tomatoes that were not organically grown had sample mean weight $\bar{x}_2 = 4.1$ ounces with sample standard deviation $s_2 = 1.5$ ounces. Does this indicate a difference either way between population mean weights of organically grown tomatoes compared to those not organically grown? Use a 5% level of significance.

4. (a) ____________________

(b) ____________________

(c)

(d) ____________________

(e) ____________________

(f) ____________________

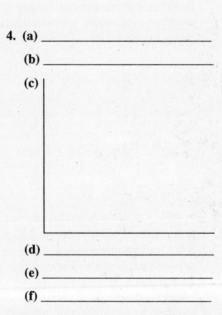

CHAPTER 11 TEST
FORM B

For each of the following problems, please provide the requested information.

(a) State the null and alternate hypotheses. Will we use a left-tailed, right-tailed, or two-tailed test? What is the level of significance?

(b) Identify the sampling distribution to be used: the standard normal or the Student's *t*. Find the critical value(s).

(c) Sketch the critical region and show the critical value(s) on the sketch.

(d) Compute the *z* or *t* value of the sample test statistic and show it's location on the sketch of part (c).

(g) Find the *P* value or an interval containing the *P* value for the sample test statistic.

(h) Based on your answers for parts (a) to (e), shall we reject or fail to reject the null hypothesis. Explain your conclusion in the context of the problem.

1. In a random sample of 62 students, 34 said they would vote for Jennifer as student body president. In another random sample of 77 students, 48 said they would vote for Kevin as student body president. Does this indicate that in the population of all students, Kevin has a higher proportion of votes? Use $\alpha = 0.05$

1. (a) ______________

(b) ______________

(c)

(d) ______________

(e) ______________

(f) ______________

2. Five members of the college track team in Denver (elevation 5200 ft) went up to Leadville (elevation 10,152 ft) for a track meet. The times in minutes for these team members to run two miles at each location are shown below.

Team Member	1	2	3	4	5
Denver	10.7	9.1	11.4	9.7	9.2
Leadville	11.5	10.6	11.0	11.2	10.3

Assume the team members constitute a random sample of track team members. Use a 5% level of significance to test the claim that the times were longer at the higher elevation.

2. (a) ______________

(b) ______________

(c)

(d) ______________

(e) ______________

(f) ______________

CHAPTER 11, FORM B, PAGE 2

3. Two models of a popular pick-up truck are tested for miles per gallon (mpg) gasoline consumption. The Pacer model was tested using a random sample of $n_1 = 9$ trucks and the sample mean was $\bar{x}_1 = 27.3$ mpg with sample standard deviation $s_1 = 6.2$ mpg. The Road Runner model was tested using a random sample of $n_2 = 14$ trucks. The sample mean was $\bar{x}_2 = 22.5$ mpg with sample standard deviation $s_2 = 6.8$ mpg. Does this indicate that the population mean gasoline consumption for the Pacer is higher than that of the Road Runner? Use $\alpha = 0.01$

3. (a) ____________

(b) ____________

(c)

(d) ____________

(e) ____________

(f) ____________

4. Students at the college agricultural research station are studying egg production of cage-free chickens compared to caged chickens. During a one week period, a random sample of $n_1 = 93$ cage-free hens produced an average of $\bar{x}_1 = 11.2$ eggs with sample standard deviation $s_1 = 4.4$ eggs. For the same period, another random sample of $n_2 = 87$ caged hens produced a sample average of $\bar{x}_2 = 8.5$ eggs per hen with sample standard deviation $s_2 = 5.7$. Does this indicate the population mean egg production for cage-free hens is higher? Use a 5% level of significance.

4. (a) ____________

(b) ____________

(c)

(d) ____________

(e) ____________

(f) ____________

CHAPTER 11 TEST

FORM C

1. A random sample of 257 dog owners was taken 10 years ago, and it was found that 146 owned more than one dog (Sample 1). Recently, a random sample of 380 dog owners showed that 200 owned more than one dog (Sample 2). Do these data indicate that the proportion of dog owners owning more than one dog has decreased? Use a 5% level of significance.

A. State the null and alternate hypotheses. **1. A.** __________

(a) H_0: $\hat{p}_1 = \hat{p}_2$; H_1: $\hat{p}_1 > \hat{p}_2$ (b) H_0: $p_1 = p_2$; H_1: $p_1 < p_2$

(c) H_0: $p_1 = \frac{146}{257} = 0.57$; H_1: $p_1 < 0.57$ (d) H_0: $p_1 = p_2$; H_1: $p_1 \neq p_2$

(e) H_0: $p_1 = p_2$; H_1: $p_1 > p_2$

B. Find the critical value(s). **B.** __________

(a) $t_0 = \pm 1.96$ (b) $z_0 = 2.33$

(c) $t_0 = -1.645$ (d) $z_0 = 1.645$ (e) $z_0 = \pm 1.96$

C. Compute the z or t value of the sample test statistic. **C.** __________

(a) $z = 0.04$ (b) $z = 1.04$

(c) $z = 0.08$ (d) $t = -0.08$ (e) $t = 1.04$

D. Find the P value or an interval containing the P value for the sample test statistic. **D.** __________

(a) P value = 0.8508 (b) P value = 0.4681

(c) P value = 0.2984 (d) P value = 0.4840 (e) P value = 0.1492

E. Based on your answers for parts A–D, what is your conclusion? **E.** __________

(a) Do not reject H_0 (b) Reject H_0 (c) Cannot determine

(d) More people own dogs.

(e) Less people own dogs.

CHAPTER 11, FORM C, PAGE 2

2. Seven manufacturing companies agreed to implement a time management program in hopes of improving productivity. The average times, in minutes, it took the companies to produce the same quantity and kind of part are listed below. Does this information indicate the program decreased production time? Assume normal population distributions. Use $\alpha = 0.05$.

Company	1	2	3	4	5	6	7
Before program (1)	75	112	89	95	80	105	110
After program (2)	70	110	88	100	80	100	99

A. State the null and alternate hypotheses. **2. A.** ________

(a) H_0: $\mu_1 = \mu_2$; H_1: $\mu_1 > \mu_2$ (b) H_0: $\mu_d = 0$; H_1: $\mu_d \neq 0$

(c) H_0: $\mu_1 = \mu_2$; H_1: $\mu_1 < \mu_2$ (d) H_0: $\mu_d = 0$; H_1: $\mu_d > 0$

(e) H_0: $\mu = 0$; H_1: $\mu < 0$

B. Find the critical value(s). **B.** ________

(a) $t_0 = 2.447$ (b) $t_0 = 1.782$

(c) $t_0 = 1.895$ (d) $z_0 = 1.645$ (e) $t_0 = 1.943$

C. Compute the z or t value of the sample test statistic. **C.** ________

(a) $t = 0$ (b) $z = 0.36$

(c) $t = 1.44$ (d) $t = 0.36$ (e) $z = -1.44$

D. Find the P value or an interval containing the P value for the sample test statistic. **D.** ________

(a) P value $= 0.1498$ (b) P value $= 0.10$

(c) P value $= 0.0749$ (d) P value > 0.250 (e) Cannot determine

E. Based on your answers for parts A–D, what is your conclusion? **E.** ________

(a) Do not reject H_0 (b) Reject H_0 (c) Cannot determine

(d) All manufacturing companies should implement the program.

(e) Some manufacturing companies should implement the program.

CHAPTER 11, FORM C, PAGE 3

3. An independent rating service is trying to determine which of two film developing ships has quicker service. Over a period of 12 randomly selected times, the average waiting period to develop a 24-exposure roll sold at Shop 1 is 58 minutes with standard deviation 3.5 minutes. The average waiting period at Shop 2 to develop a 24-exposure roll over a period of 8 randomly selected times is 53 minutes with standard deviation 4.9 minutes. Using a 1% level of significance, can we say there is a difference in the average waiting time at Shop 1 and Shop 2?

A. State the null and alternate hypotheses. **3 A.** __________

(a) H_0: $\overline{x}_1 = \overline{x}_2$; H_1: $\overline{x}_1 \neq \overline{x}_2$ (b) H_0: $\mu_1 = \mu_2$; H_1: $\mu_1 \neq \mu_2$

(c) H_0: $\mu_1 = \mu_2$; H_1: $\mu_1 > \mu_2$ (d) H_0: $\mu = 58$; H_1: $\mu \neq 58$

(e) H_0: $\mu_d > 0$; H_1: $\mu_d \neq 0$

B. Find the critical value(s). **B.** __________

(a) $t_0 = \pm 2.878$ (b) $z_0 = \pm 2.56$

(c) $z_0 = \pm 1.96$ (d) $t_0 = 2.552$ (e) $t_0 = 2.878$

C. Compute the z or t value of the sample test statistic. **C.** __________

(a) $z = 2.67$ (b) $t = 2.49$

(c) $t_0 = \pm 2.67$ (d) $z = 2.49$ (e) $t = 2.67$

D. Find the P value or an interval containing the P value for the sample test statistic. **D.** __________

(a) $0.010 < P$ value < 0.025 (b) $0.02 < P$ value < 0.05

(c) Cannot determine (d) $0.01 < P$ value < 0.02 (e) $0.005 < P$ value < 0.010

E. Based on your answers for parts A–D, what is your conclusion? **E.** __________

(a) Do not reject H_0 (b) Reject H_0 (c) Cannot determine

(d) It takes about an hour to develop film.

(e) Shop 1 takes longer than Shop 2.

CHAPTER 11, FORM C, PAGE 4

4. The personnel manager of a large retail clothing store suspects a difference in the mean amount of break time taken by workers during the weekday shifts compared to that of the weekend shifts. It is suspected that the weekday workers take longer breaks on the average. A random sample of 46 weekday workers had a mean $\bar{x}_1 = 53$ minutes of break time per 8-hour shift with standard deviation $s_1 = 7.3$ minutes. A random sample of 40 weekend workers had a mean $\bar{x}_2 = 47$ minutes with standard deviation $s_2 = 9.1$ minutes. Test the manager's suspicion at the 5% level of significance.

A. State the null and alternate hypotheses. **4 A.** __________

(a) H_0: $\mu_1 = \mu_2$; H_1: $\mu_1 < \mu_2$ (b) H_0: $\mu_1 = \mu_2$; H_1: $\mu_1 > \mu_2$

(c) H_0: $\mu_1 = \mu_2$; H_1: $\mu_1 \neq \mu_2$ (d) H_0: $\bar{x}_1 = \bar{x}_2$; H_1: $\bar{x}_1 > \bar{x}_2$

(e) H_0: $\mu_d = 0$; H_1: $\mu_d > 0$

B. Find the critical value(s). **B.** __________

(a) $t_0 = 1.645$ (b) $t_0 = \pm 1.96$

(c) $z_0 = 1.645$ (d) $z_0 = 3.34$ (e) $z_0 = 2.33$

C. Compute the z or t value of the sample test statistic. **C.** __________

(a) $t = 3.34$ (b) $t = 0.73$

(c) $z = 0.73$ (d) $z = 1.645$ (e) $z = 3.34$

D. Find the P value or an interval containing the P value for the sample test statistic. **D.** __________

(a) Cannot determine (b) P value = 0.0495

(c) P value = 0.0008 (d) P value = 0.0004 (e) P value = 0.2327

E. Based on your answers for parts A–D, what is your conclusion? **E.** __________

(a) Do not reject H_0 (b) Reject H_0 (c) Cannot determine

(d) All weekend workers take shorter breaks.

(e) Some weekend workers take shorter breaks.

CHAPTER 12 TEST
FORM A

For the given data, solve Problems 1–3.

Taltson Lake is in the Canadian Northwest Territories. This lake has many Northern Pike. The following data was obtained by two fishermen visiting the lake. Let x = length of a Northern Pike in inches and let y = weight in pounds.

x (inches)	20	24	36	41	46
y (pounds)	2	4	12	15	20

1. Compute the standard error of estimate S_e. **1.** ____________

2. The least-squares line predicts that a 32-inch Northern Pike will weigh 9.65 pounds. Find the 90% confidence interval for this prediction. **2.** ____________

3. Using the sample correlation coefficient $r = 0.996$, test whether or not the population correlation coefficient p is different from zero. Use $\alpha = 0.01$. Is r significant in this problem? Explain. **3.** ____________

For the given data, solve Problems 4–6.

A marketing analyst is studying the relationship between x = amount spent on television advertising and y = increase in sales. The following data represents a random sample from the study.

x (\$ thousands)	15	28	19	47	10	92
y (\$ thousands)	340	260	152	413	130	855

4. Compute the standard error of estimate S_e. **4.** ____________

5. The least-squares line predicts that spending \$37,000 on advertising will increase sales by \$373,500. Find the 95% confidence interval for this prediction. **5.** ____________

6. Using the sample correlation coefficient $r = 0.954$, test whether or not the population correlation coefficient p is positive. Use $\alpha = 0.05$. Is r significant in this problem? Explain. **6.** ____________

CHAPTER 12, FORM A, PAGE 2

7. Are teacher evaluations independent of grades? After midterm, a random sample of 284 students were asked to evaluate teacher performance. The students were also asked to supply their midterm grade in the course being evaluated. In this study, only students with a passing grade (A, B, or C) were included in the summary table.

	Mid Term Grade			
Teacher Evaluation	A	B	C	Row Total
Positive	53	33	18	104
Neutral	25	46	29	100
Negative	14	22	44	80
Column Total	92	101	91	284

Use a 5% level of significance to test the claim that teacher evaluations are independent of midterm grades. **7.** ____________

8. How old are college students? The national age distributions for college students is shown below.

National Age Distribution for College Students

Age	Under 26	26–35	36–45	46–55	Over 55
Clients	39%	25%	16%	12%	8%

The Western Association of Mountain Colleges took a random sample of 212 students and obtained the following sample distribution.

Sample Distribution, Western Association of Mountain Colleges

Age	Under 26	26–35	36–45	46–55	Over 55
Number of Students	65	73	41	21	12

Is the sample age distribution for the Western Association of Mountain Colleges a good fit to the national distribution? Use $\alpha = 0.05$. **8.** ____________

9. If we have a normal population with variance σ^2 and a random sample of n measurements taken from this population, what probability distribution do we use to test claims about the variance? **9.** ____________

10. A technician tested 25 motors for toy electric trains and found the sample standard deviation of electrical current to be $s = 4.9$ amperes.

If the manufacturer specifies that $\sigma = 4.1$ amperes, does the sample data indicate that σ is larger than 4.1? Use a 1% level of significance. **10.** ____________

CHAPTER 12 TEST
FORM B

For the given data, solve Problems 1–3.

Do higher paid chief executive officers (CEOs) control bigger companies? Let us study x = annual CEO salary ($ millions) and y = annual company revenue ($ billions). The following data are based on information from *Forbes* magazine and represents a sample of top US executives.

x ($ millions)	0.8	1.0	1.1	1.7	2.3
y ($ billions)	14	11	19	20	25

1. Compute the standard error of estimate S_e. **1.** ________________

2. The least-squares line predicts that a company whose CEO has an annual salary of $1.5 million will have an annual revenue of $18.74 billion. Find the 90% confidence interval for this prediction. **2.** ________________

3. Using the sample correlation coefficient $r = 0.880$, test whether or not the population correlation coefficient p is different from zero. Use $\alpha = 0.01$. Is r significant in this problem? Explain. **3.** ________________

For the given data, solve Problems 4–6.

An accountant for a small manufacturing plant collected the following random sample to study the relationship between x = the cost to make a particular item and y = the selling price.

x ($)	26	50	47	23	52	71
y ($)	78	132	128	70	152	198

4. Compute the standard error of estimate S_e. **4.** ________________

5. The least-square line predicts that an item that costs $35 will sell for $100.30. Find a 95% confidence interval for your prediction of Problem 12. **5.** ________________

6. Using the sample correlation coefficient $r = 0.994$, test whether or not the population correlation coefficient p is positive. Use $\alpha = 0.05$. Is r significant in this problem? Explain. **6.** ________________

CHAPTER 12, FORM B, PAGE 2

7. Is the choice of college major independent of grade average? A random sample of 445 students were surveyed by the Registrar's Office regarding major field of study and grade average. In this study, only students with passing grades (A, B, or C) were included in the survey. Grade averages were rounded to the nearest letter grade (e.g., 3.6 grade point average was rounded to 4.0 or A).

	Grade Average			
Major	A	B	C	Row Total
Science	38	49	63	150
Business	41	42	59	142
Humanities	32	53	68	153
Column Total	111	144	190	445

Use a 1% level of significance to test the claim that choice of major field is independent of grade average.

7. ____________________

8. The Fish and Game Department in Wisconsin stocked a new lake with the following distribution of game fish.

Initial Stocking Distribution

Fish	Pike	Trout	Perch	Bass	Bluegill
Percent	10%	15%	20%	25%	30%

After six years a random sample of 197 fish from the lake were netted, identified, and released. The sample distribution is shown next.

Sample Distribution after Six Years

Fish	Pike	Trout	Perch	Bass	Bluegill
Number	52	15	33	55	42

Is the sample distribution of fish in the lake after six years a good fit to the initial stocking distribution? Use a 5% level of significance.

8. ____________________

9. An automobile service station times the Quick Lube service for a random sample of 22 customers. The sample standard deviation of times was $s = 6.8$ minutes.

The service manager specifies that σ be 6.0 minutes. Does the sample data indicate that σ is different from 6.0? Use a 1% level of significance.

9. ____________________

CHAPTER 12 TEST
FORM C

Write the letter of the response that best answers each problem.

Does the weight of a vehicle affect the gas mileage? The following random sample was collected where x = weight of a vehicle in hundreds of pounds and y = miles per gallon.

x (lb hundreds)	26	35	29	39	20
y (mpg)	22.0	16.1	18.8	15.7	23.4

1. Compute the standard error of estimate S_e. **1.** ________

(a) 0.965 (b) −0.970
(c) 0.941 (d) 1.975 (e) 0.065

2. The least-squares line predicts that a vehicle weighing 2200 pounds will get 22.7 miles per gallon. Find a 90% confidence interval for this prediction. **2.** ________

(a) 21.1 mpg $\le y \le$ 24.3 mpg (b) 19.5 mpg $\le y \le$ 25.9 mpg
(c) 19.9 mpg $\le y \le$ 25.5 mpg (d) 19.0 mpg $\le y \le$ 26.4 mpg
(e) 20.6 mpg $\le y \le$ 24.8 mpg

3. Using the sample correlation coefficient r, test whether or not the population correlation coefficient p is different from zero. Use $\alpha = 0.01$. Is r significant in this problem? Explain. **3.** ________

(a) Do not reject H_0; r is not significant (b) Reject H_0; r is significant
(c) Do not reject H_0; r is significant (d) Reject H_0; r is not significant
(e) Cannot determine

CHAPTER 12, FORM C, PAGE 2

Write the letter of the response that best answers each problem.

A graduate school committee is studying the relationship between x = an applicant's undergraduate grade point average and y = the applicant's score on the graduate entrance exam. The following random sample was collected to study this relationship.

x (GPA)	3.2	3.9	4.0	3.4	3.7	3.0
y (score)	725	788	775	647	800	672

4. Compute the standard error of estimate S_e. **4.** ________

(a) 45.93 (b) 0.766

(c) 183.2 (d) 51.6 (e) 0.587

5. The least-squares line predicts that a student with a grade point average of 3.5 will score 730.4 on the graduate entrance exam. Find the 95% confidence interval for this prediction. **5.** ________

(a) $690.3 \le y \le 770.5$ (b) $678.1 \le y \le 782.7$

(c) $624.6 \le y \le 836.2$ (d) $629.3 \le y \le 831.5$

(e) $592.6 \le y \le 868.2$

6. Using the sample correlation coefficient r, test whether or not the population correlation coefficient p is positive. Use $\alpha = 0.05$. Is r significant in this problem? Explain. **6.** ________

(a) Do not reject H_0; r is significant (b) Do not reject H_0; r is not significant

(c) Reject H_0; r is significant (d) Reject H_0; r is significant

(e) Cannot determine

CHAPTER 12, FORM C, PAGE 3

7. A market research study was conducted to compare three different brands of antiperspirant. The results of the study are summarized below. Use a 5% level of significance to test the claim that opinion is independent of brank.

	Brand			
Opinion	A	B	C	Total
Excellent	29	37	50	116
Satisfactory	83	65	43	191
Unsatisfactory	18	9	6	33
Total	130	111	99	340

A. State the null and alternate hypotheses. **7. A.** ________

(a) H_0: Opinion and brand are dependent; H_1: Opinion and brand are independent

(b) H_0: $\mu_A = \mu_B = \mu_C$; H_1: Not all μ_1, μ_2, μ_3, are equal.

(c) H_0: Opinion and brand are independent; H_1: Opinion and brand are dependent

(d) H_0: $\sigma_1^2 = \sigma_2^2$; H_1: $\mu_1 = \mu_2$

(e) H_0: The distributions are normal; H_1: The distributions are not normal

B. What is the critical value (or critical values)? **B.** ________

(a) $\chi^2_{0.05} = 0.711$ (b) $\chi^2_{0.05} = 16.92$

(c) $z_0 = 1.645$ (d) $\chi^2_{0.05} = 9.49$ (e) $F_0 = 19.00$

C. What is the value of the sample test statistic? **C.** ________

(a) $\chi^2 = 19.00$ (b) $\chi^2 = 9.49$

(c) $t = 0.25$ (d) $F = 0.10$ (e) $\chi^2 = 21.4$

D. What is your conclusion? **D.** ________

(a) Reject H_0 (b) Do not reject H_0 (c) Cannot determine

(d) Brand A is the best.

(e) All of the brands work equally well.

CHAPTER 12, FORM C, PAGE 4

8. How much do second graders weigh? A county hospital found the weight distribution shown below.

Hospital Weight Distribution

Weight (lb)	Under 45	45–59	60–74	75–89	Over 89
Percent	7%	21%	41%	19%	12%

An elementary school within the county took a random sample of 125 second graders and obtained the following sample distribution.

Sample Distribution, Elementary School

Weight (lb)	Under 45	45–59	60–74	75–89	Over 89
Number of Second Graders	6	29	50	30	10

Is the sample weight distribution for the elementary school a good fit to the hospital distribution? Use $\alpha = 0.05$.

A. State the null and alternate hypotheses. **8. A.** __________

(a) H_0: $\mu_1 = \mu_2 = \mu_3 = \mu_4 = \mu_5$; H_1: Not all μ_1, μ_2, μ_3, are equal.

(b) H_0: Weight and percent are independent; H_1: Weight and percent are dependent

(c) H_0: $\sigma_1^2 = \sigma_2^2$; H_1: $\mu_1 = \mu_2$

(d) H_0: The distributions are the same; H_1: The distributions are different.

(e) H_0: The distributions are the same; H_1: The distributions for elementary school are higher.

B. What is the critical value (or critical values)? **B.** __________

(a) $\chi^2_{0.05} = 9.49$ (b) $\chi^2_{0.05} = 11.07$

(c) $\chi^2_{0.05} = 0.711$ (d) $\chi^2_{0.05} = 1.145$ (e) $t_0 = 2.776$

C. What is the value of the sample test statistic? **C.** __________

(a) $\chi^2 = 11.07$ (b) $\chi^2 = 5.35$

(c) $\chi^2 = 4.49$ (d) $t = 7.27$ (e) $\chi^2 = 9.49$

D. What is your conclusion? **D.** __________

(a) Reject H_0 (b) Do not reject H_0 (c) Cannot determine

(d) The elementary school has heavier second graders.

(e) The means are the same.

CHAPTER 12, FORM C, PAGE 5

9. Find the χ^2 value for each situation.

A. 1% of the area under the curve is to the right of χ^2 when *d.f.* = 18. **9. A.** __________

(a) 33.41 (b) 7.01 (c) 37.16 (d) 6.41 (e) 34.81

B. 5% of the area under the curve is to the left of χ^2 when *d.f.* = 15. **B.** __________

(a) 6.57 (b) 25.00 (c) 7.26 (d) 23.00 (e) 6.26

C. 5% of the area under the curve is to the right of χ^2 when $n = 20$. **C.** __________

(a) 10.12 (b) 31.41 (c) 32.85 (d) 30.14 (e) 10.85

10. A salesperson tested 30 sport utility vehicles for gas mileage (in miles per gallon) and found the sample standard deviation to be $s = 4.7$ mpg.

A. If the manufacturer specifies that $\sigma = 4.0$ mpg, does the sample data indicate that σ is larger than 4.0? Use $\alpha = 0.01$ and compute the critical value and the value of the sample test statistic. **10. A.** __________

(a) $\chi^2_{0.01} = 40.04, \chi^2 = 49.59$ (b) $\chi^2_{0.01} = 14.26, \chi^2 = 40.04$

(c) $\chi^2_{0.01} = 52.34, \chi^2 = 41.42$ (d) $\chi^2_{0.01} = 50.89, \chi^2 = 41.42$

(e) $\chi^2_{0.01} = 49.59, \chi^2 = 40.04$

B. What is your conclusion for the test in Part B? **B.** __________

(a) There is sufficient evidence to conclude that the standard deviation is larger than 4.0 mpg.

(b) The distributions are different.

(c) The variances are unequal.

(d) The mpg for the sample is larger than 4.0 mpg.

(e) There is insufficient evidence to conclude that the standard deviation is larger than 4.0 mpg.

Answers to Sample Chapter Tests

CHAPTER 1 FORM A

1. (a) Length of time to earn a Bachelor's degree.

(b) Quantitative

(c) Length of time it took each of the Colorado residents who earned (or will earn) a Bachelor's degree to complete the degree program.

2. Explanations will vary.

(a) Nominal

(b) Nominal

(c) Ratio

(d) Interval

(e) Nominal

(f) Ordinal

(g) Ordinal

3. (a) Census

(b) Experiment

(c) Sampling

(d) Simulation

4. Answers will vary.

5. The outcomes are the number of dots on the face, 1 through 6. Consider single digits in the random number table. Select a starting place at random. Record the first five digits you encounter that are between (and including) 1 and 6. The first five outcomes are

3 6 1 5 6

6. (a) Systematic

(b) Cluster

(c) Stratified

(d) Convenience

(e) Simple random

CHAPTER 1 FORM B

1. (a) Observed book purchase (mystery or not a mystery).

(b) Quantitative

(c) Observed book purchase (mystery or not a mystery)
of all current customers of the bookstore.

2. Explanations will vary.

(a) Nominal

(b) Ratio

(c) Interval

(d) Ordinal

(e) Ratio

3. (a) Census

(b) Sampling

(c) Simulation

(d) Experiment

4. Answers will vary.

5. Assign each of the 736 employees a distinct number between 1 and 736. Select a starting place in the random number table at random. Use groups of three digits. Use the first 30 distinct groups of three digits that correspond to employees numbers.

622 413 055 401 334

6. (a) Simple random

(b) Stratified

(c) Cluster

(d) Convenience

(e) Systematic

CHAPTER 1
FORM C

1. (b)

2. A. (b)

B. (a)

C. (d)

D. (c)

E. (a)

3. A. (b)

B. (c)

C. (a)

D. (d)

4. (e)

5. (d)

6. A. (e)

B. (d)

C. (a)

D. (c)

E. (b)

CHAPTER 2
FORM A

1.

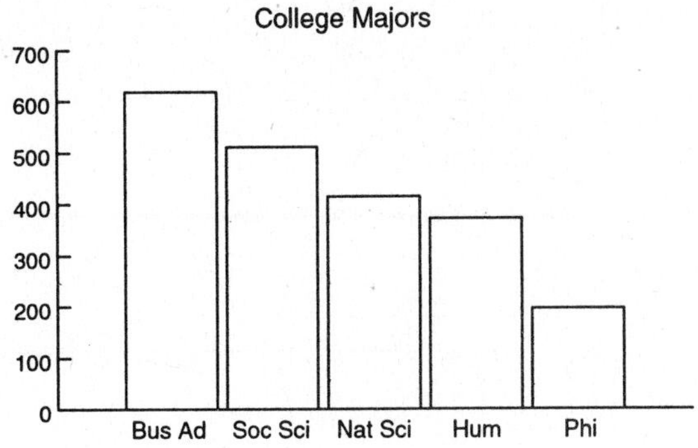

2. (a) The Class Width is 6

Frequency and Relative Frequency Table			
Class Boundaries	Freq.	Rel. Frequency	Class Midpoint
2.5 – 8.5	10	0.3333	5.5
8.5 – 14.5	9	0.3000	11.5
14.5 – 20.5	6	0.2000	17.5
20.5 – 26.5	2	0.0667	23.5
26.5 – 32.5	3	0.1000	29.5

(b)

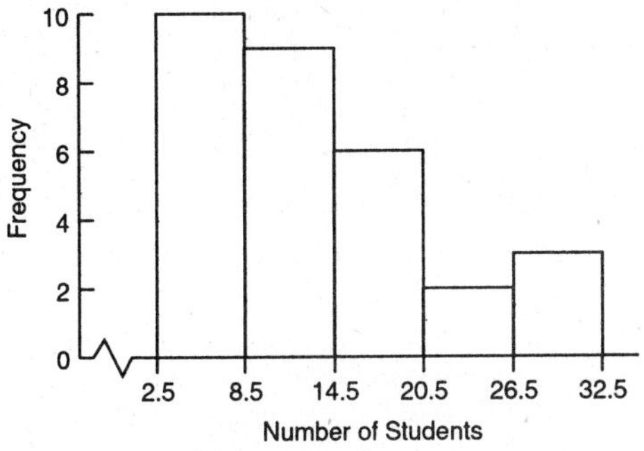

(c)

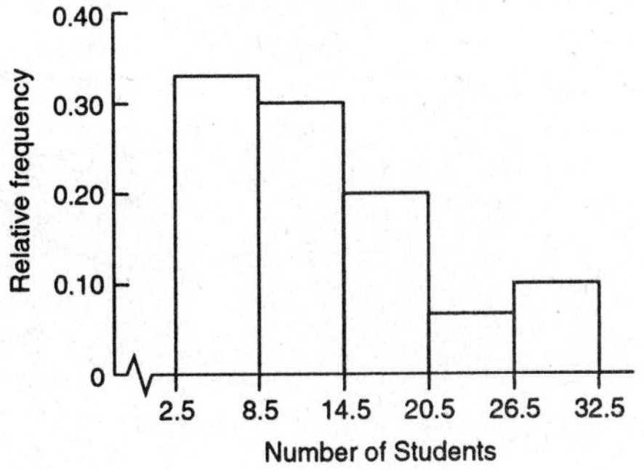

3. 0 | 5 = \$5

0	5 7 8 9
1	2 5 6 7 9
2	1 2 4
3	3 5 7
4	2 6 9
5	1 7

4.

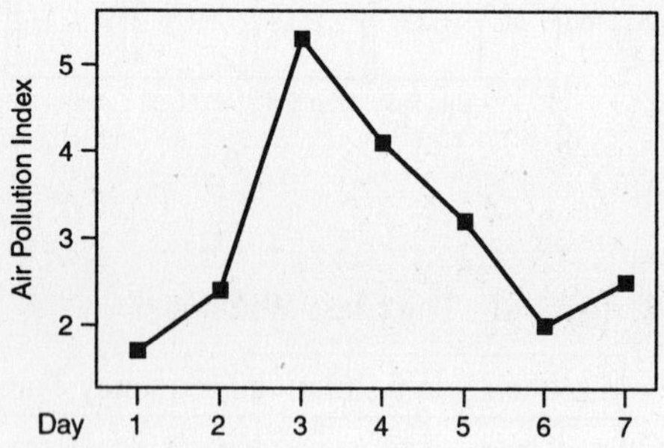

5.

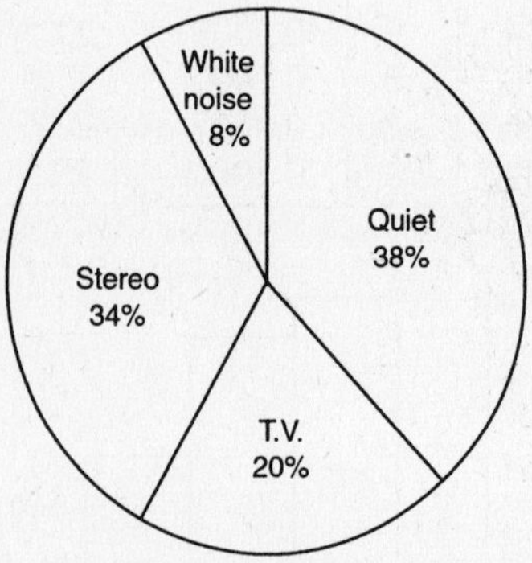

6.

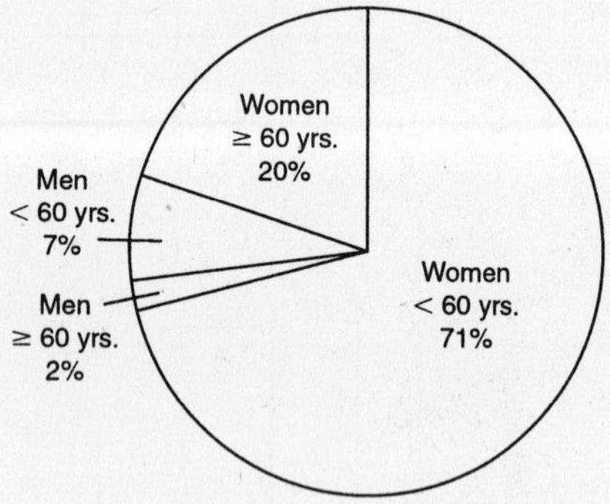

7.

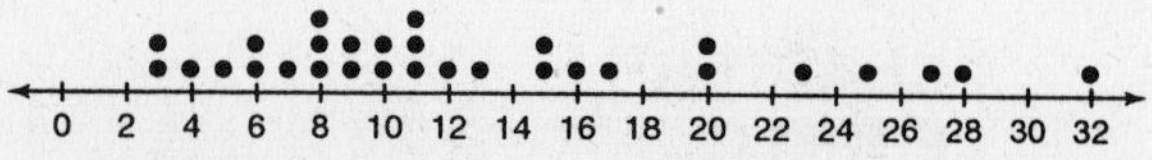

The dotplot is similar to the histogram in that most data values fall below 20.5. However, the dotplot has more detail, i.e., each data value can be seen in the dotplot.

8. 11 motorists

9. 3 | 1 = 31 years old

0	8
1	0 8 9
2	2 4 4 5 8 8 9
3	1 2 5 6 8
4	0 2 3 7
5	0 5

The distribution is fairly symmetrical.

CHAPTER 2
FORM B

1.

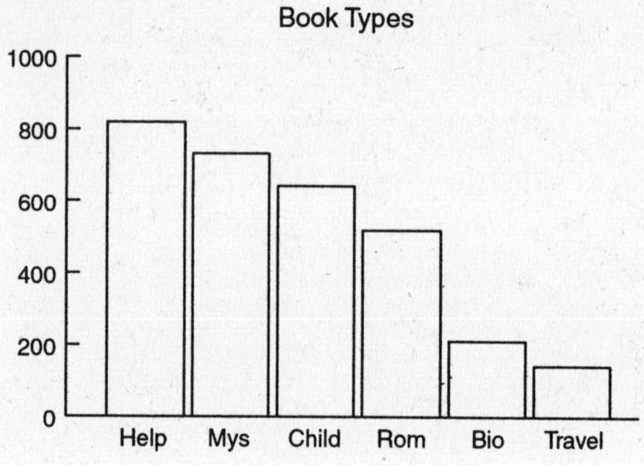

2. (a) The Class Width is 15

Frequency and Relative Frequency Table			
Class Boundaries	Freq.	Rel. Frequency	Class Midpoint
18.5 – 33.5	3	0.1250	26
33.5 – 48.5	7	0.2917	41
48.5 – 63.5	8	0.3333	56
63.5 – 78.5	4	0.1667	71
78.5 – 93.5	2	0.0833	86

(b)

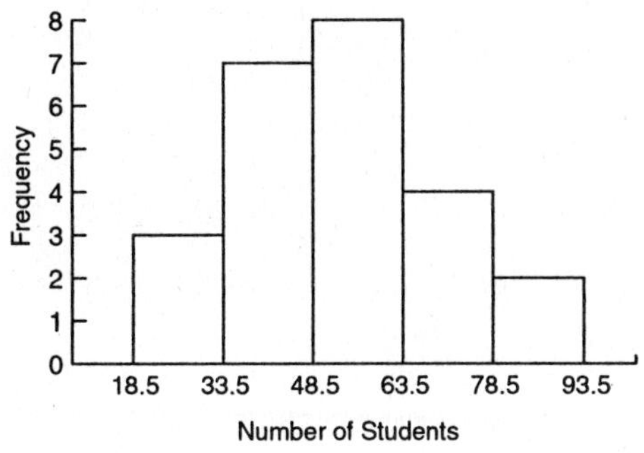

(c)

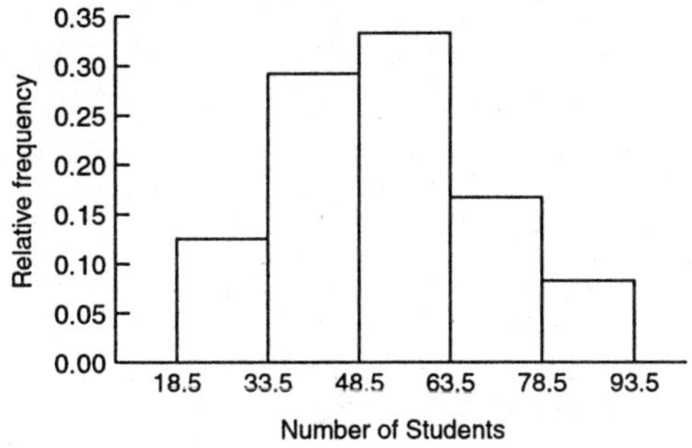

3. 2 | 1 = 21 students

2	1	6	9			
3	3	4	6	8		
4	2	4	5	7		
5	0	1	2	6	8	9
6	3	4	8			

4.

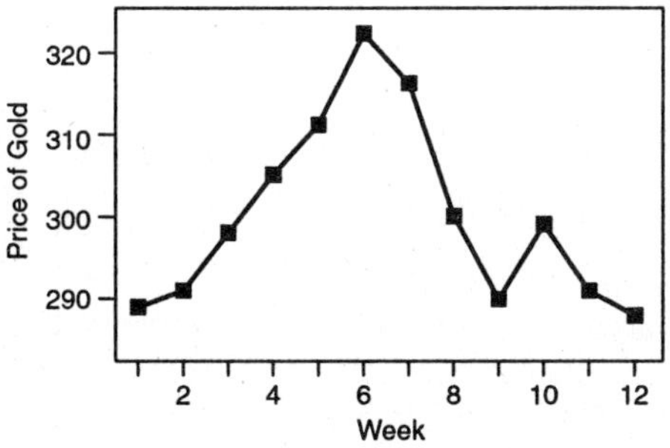

5.

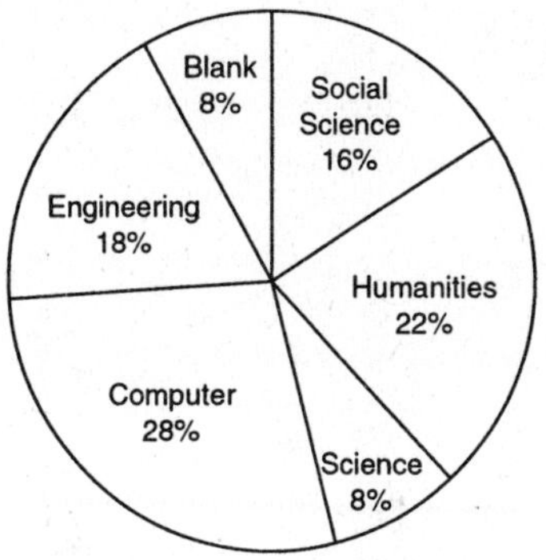

6.

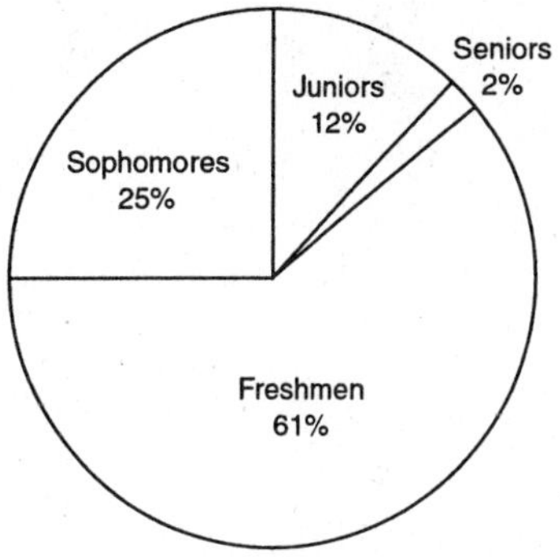

7.

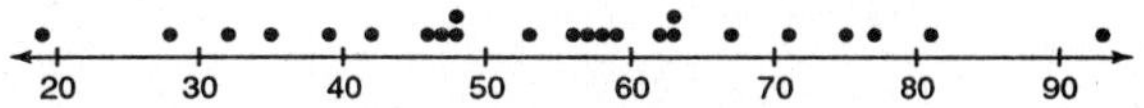

The dotplot is similar to the histogram in that most data values fall in the middle (mound-shaped). However, the dotplot has more detail, i.e., each data value can be seen in the dotplot.

8. 6 days

9. 1 | 7 = 1.7 mm

0	9					
1	2	7	8			
2	3	4	5	5	6	8
3	0	1	6	7	7	
4	1					

The distribution is fairly symmetrical.

CHAPTER 2
FORM C

1. (c)

2. (b)

3. (e)

4. (a)

5. (d)

6. (e)

7. (d)

8. (a)

9. (c)

CHAPTER 3
FORM A

1. $\bar{x} = 16.61$; median = 13;
mode = 12

2. \$42,185; \$35,761; The trimmed mean because it does not include the extreme values. Note that all the salaries are below \$42,185 except one.

3. (a) Range = 25

(b) $\bar{x} = 33$

(c) $s^2 = 81.67$

(d) $s = 9.04$

4. (a) $CV = 21\%$

(b) 6.04 to 14.72

5. (a) $\bar{x} = 36.45$

(b) $s^2 = 130.55$

(c) $s = 11.43$

6. (a) Low value = 15; $Q_1 = 23$; median = 33; $Q_3 = 49$; High value = 72

(b)

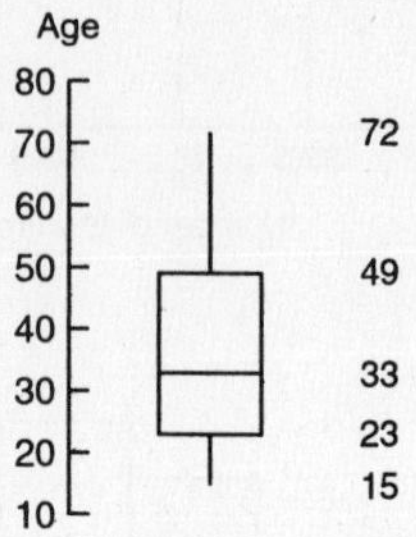

(c) Interquartile range = 26

7. 79% below; 21% above

CHAPTER 3
FORM B

1. $\bar{x} = 10.55$; median = 11;
mode = 11

2. 37.8 in.; 31.63 in.; The trimmed mean because it does not include the extreme values. Note that all but 3 values are below 37.8 in.

3. (a) Range = 14

(b) $\bar{x} = 12.17$

(c) $s^2 = 25.37$

(d) $s = 5.04$

4. (a) $CV = 26.5\%$

(b) 15.3 to 49.7

5. (a) $\bar{x} = 2.35$

(b) $s^2 = 1.02$

(c) $s = 1.007$

6. (a) Low value = 27; $Q_1 = 35.5$; median = 42.5; $Q_3 = 57$; High value = 68

(b)

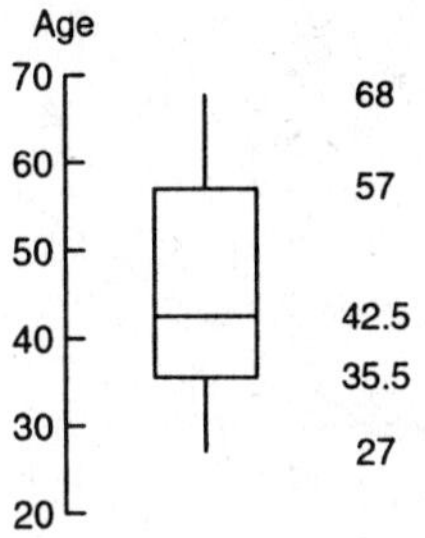

(c) Interquartile range = 21.5

7. 19% above; 81% below

CHAPTER 3
FORM C

1. (c)

2. (d)

3. A. (e)
B. (a)
C. (c)
D. (d)

4. A. (b)
B. (a)

5. A. (c)
B. (e)
C. (d)

6. (b)

7. (e)

CHAPTER 4
FORM A

1. Length of Northern Pike x versus Weight y

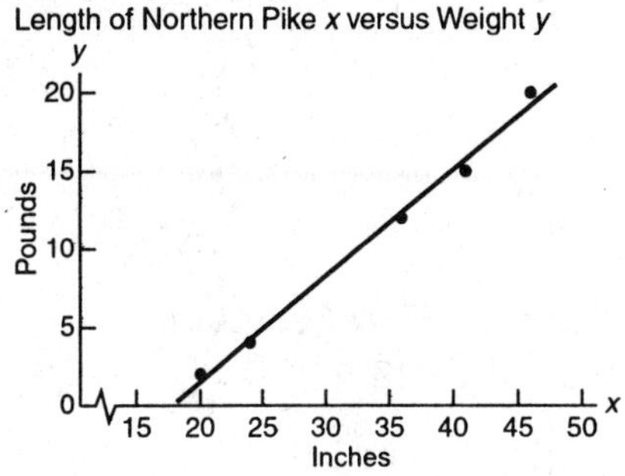

The linear correlation coefficient appears to be positive.

2. (a) $\overline{x} = 33.4$; $\overline{y} = 10.6$

(b) $SS_x = 491.2$; $SS_y = 227.2$; $SS_{xy} = 332.8$

(c) $b = 0.6775$; $a = -12.029$;
$y = 0.6775x - 12.029$

(d) See line on the graph of Problem 1.

3. $r = 0.996$; $r^2 = 0.992$; 99.2% of the variation in y can be explained by the least squares line using x as the predicting variable.

4. For $x = 32$ inches, $y = 9.65$ pounds

5.

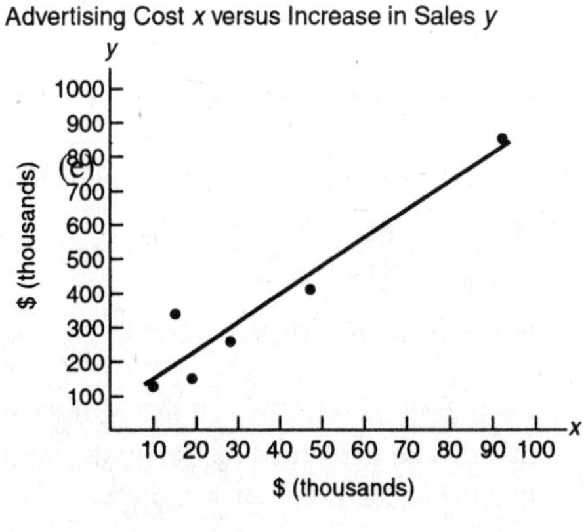

The linear correlation coefficient appears to be positive.

6. **(a)** $\bar{x} = 35.17$; $\bar{y} = 358.33$

(b) $SS_x = 4722.83$; $SS_y = 381.33$; $SS_{xy} = 39{,}030.67$

(c) $b = 8.264$; $a = 67.71$; $y = 8.26x + 67.7$

(d) See line on the graph of Problem 8.

7. $r = 0.954$; $r^2 = 0.910$; 91.0% of the variation in y can be explained by the least squares line using x as the predicting variable.

8. For $x = 37$ (or \$37,000), $y = 373.5$ (or \$373,500)

CHAPTER 4 FORM B

1. CEO salary x versus Company Revenue y

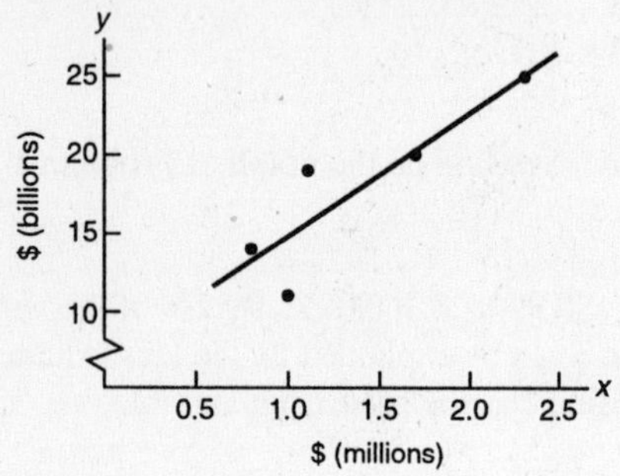

The linear correlation coefficient appears to be positive.

2. **(a)** $\bar{x} = 1.38$; $\bar{y} = 17.8$

(b) $SS_x = 1.508$; $SS_y = 118.8$; $SS_{xy} = 11.78$

(c) $b = 7.81$; $a = 7.02$; $y = 7.81x + 7.02$

(d) See the line on the graph in Problem 1.

3. $r = 0.880$; $r^2 = 0.775$; 77.5% of the variation in y can be explained by the least squares line using x as the predicting variable.

4. For a CEO salary of 1.5 million dollars, we predict an annual company revenue of 18.74 billion dollars.

5. Manufacturing Cost x versus Selling Price y

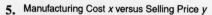

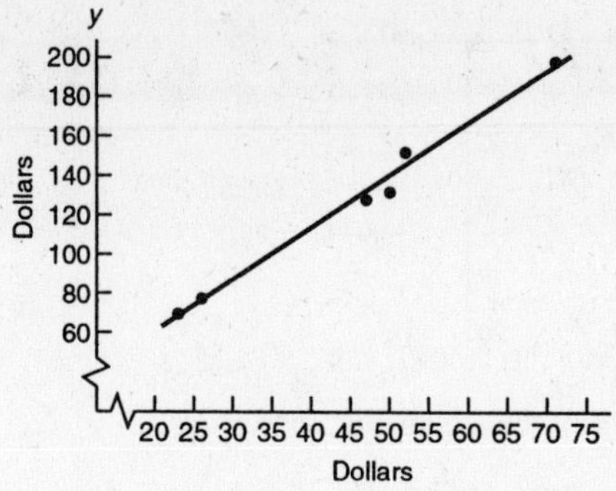

The linear correlation coefficient appears to be positive.

6. **(a)** $\bar{x} = 44.83$; $\bar{y} = 126.33$

(b) $SS_x = 1598.83$; $SS_y = 11{,}339.33$; $SS_{xy} = 4232.33$

(c) $b = 2.6471$; $a = 7.653$; $y = 2.65x + 7.65$

(d) See line on the graph of Problem 8.

7. $r = 0.994$; $r^2 = 0.988$; 98.8% of the variation in y can be explained by the least squares line using x as the predicting variable.

8. For $x = \$35$, $y = \$100.30$

CHAPTER 4 FORM C

1. (a)

2. (d)

3. (b)

4. (a)

5. (c)

6. (e)

7. (d)

8. (c)

CHAPTER 5
FORM A

1. (a) Relative frequency; 19/317 = 0.0599 or 5.99%
   (b) 1 – 0.599 = 0.9401 or about 94%
   (c) Defective, not defective; the probabilities add up to one.

2. 1/3

3. (a) 1/12 (b) 1/12 (c) 1/6

4. (a) With replacement, *P*(Red first *and* Green second) = (3/12)(7/12) = 7/48 or 0.146
   (b) Without replacement, *P*(Red first *and* Green second) = (3/12)(7/11) = 21/132 or 0.159

5. *P*(Approval on written *and* interview) = *P*(written)*P*(interview, *given* written) = (0.63)(0.85) = 0.536 or about 53.6%

6. (a) 35/360 (b) 21/360
   (c) 5/149 (d) 107/360
   (e) 48/149 (f) 31/360
   (g) No; *P*(referred) = 149/360 is not equal to *P*(referred, given satisfied) = 59/138.

7. 8;

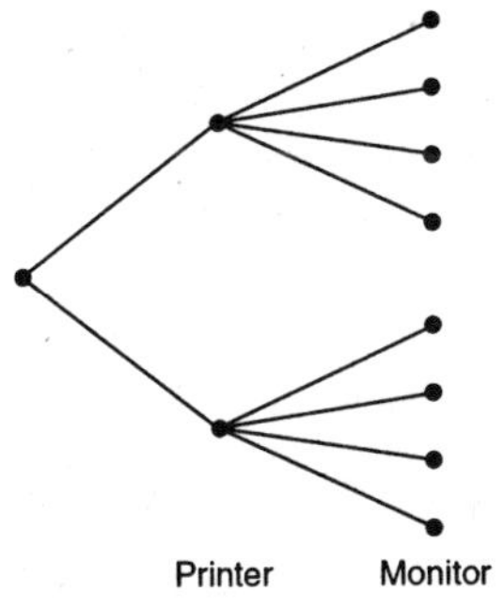

8. 40•39•38•37 = 2,193,360

9. $C_{11,3} = 165$

CHAPTER 5
FORM B

1. (a) *P*(woman) = 4/10 = 0.4 or 40%
   (b) *P*(man) = 6/10 = 0.6 or 60%
   (c) *P*(President) = 1/10 = 0.1 or 10%

2. 2/3

3. (a) 1/6 (b) 1/6 (c) 1/3

4. (a) With replacement, *P*(Red first *and* Green second) = (6/17)(8/17) = 0.166 or 16.6%
   (b) Without replacement, *P*(Red first *and* Green second) = (6/17)(8/16) = 0.176 or 17.6%

5. *P*(woman *and* computer science major) = *P*(woman)*P*(computer science major *given woman*) = (0.64)(0.12) = 0.077 or about 7.7%

6. (a) 180/417 (b) 65/115
   (c) 166/417 (d) 61/417
   (e) 61/101 (f) 65/180
   (g) No; *P*(neutral) = 166/417 is not equal to *P*(neutral, given freshman) = 61/101

7. 6;

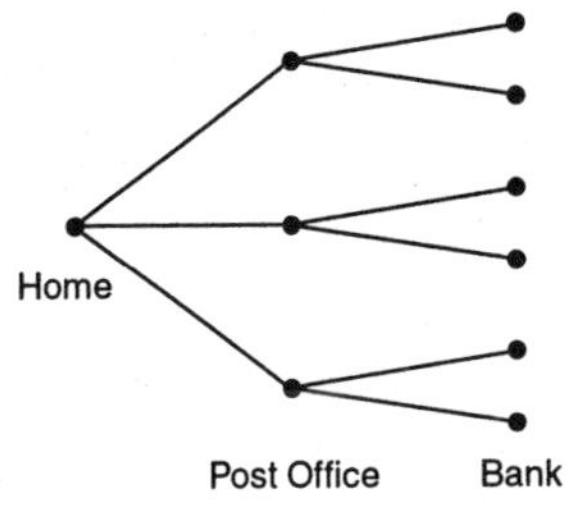

8. 16•15•14•13•12 = 524,160

9. $C_{24,4} = 10,626$

CHAPTER 5
FORM C

1. **A.** (d)
   B. (b)

2. (a)

3. **A.** (c)
   B. (e)
   C. (b)

4. **A.** (a)
   B. (d)

5. (b)

6. **A.** (e)
   B. (c)
   C. (a)
   D. (d)
   E. (c)
   F. (b)

7. (d)

8. (e)

9. (a)

CHAPTER 6
FORM A

1. **(a)** These results are rounded to three digits.
   $P(0) = 0.038$; $P(1) = 0.300$; $P(2) = 0.263$; $P(3) = 0.171$; $P(4) = 0.117$; $P(5) = 0.058$; $P(6) = 0.033$; $P(7) = 0.021$

   (b)

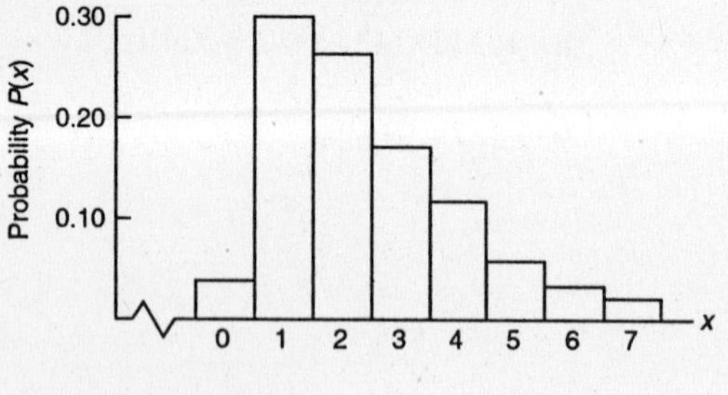

   (c) $P(2 \le x \le 5) = 0.609$

   (d) $P(x < 3) = 0.601$

   (e) $\mu = 2.44$

   (f) $\sigma = 1.57$

2. $P(4) = C_{10,4}(0.23)^4(0.77)^6 = 0.12$

3. $n = 16$; $p = 0.65$;
   Success = reach summit

   (a) $P(r = 16) = 0.001$

   (b) $P(r \ge 10) = 0.688$

   (c) $P(r \le 12) = 0.866$

   (d) $P(9 \le r \le 12) = 0.706$

4. 6

5. **(a)**

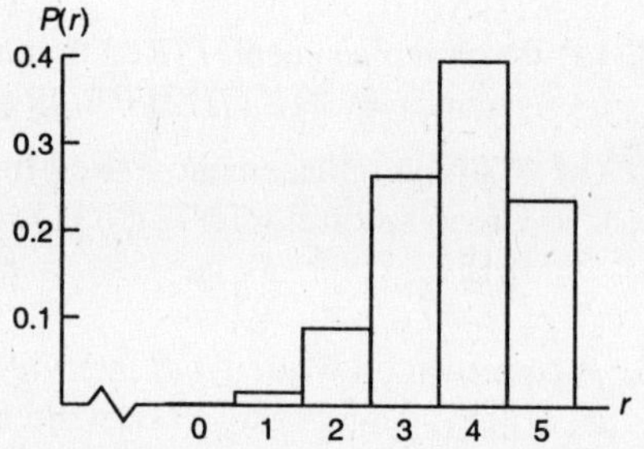

   (b) $\mu = 3.75$

   (c) $\sigma = 0.968$

6. $n = 11$ is the minimal number.

CHAPTER 6
FORM B

1. **(a)** Values are rounded to three digits.
   $P(1) = 0.066$; $P(2) = 0.092$; $P(3) = 0.202$; $P(4) = 0.224$; $P(5) = 0.184$; $P(6) = 0.079$; $P(7) = 0.053$; $P(8) = 0.044$; $P(9) = 0.035$; $P(10) = 0.022$

   (b)

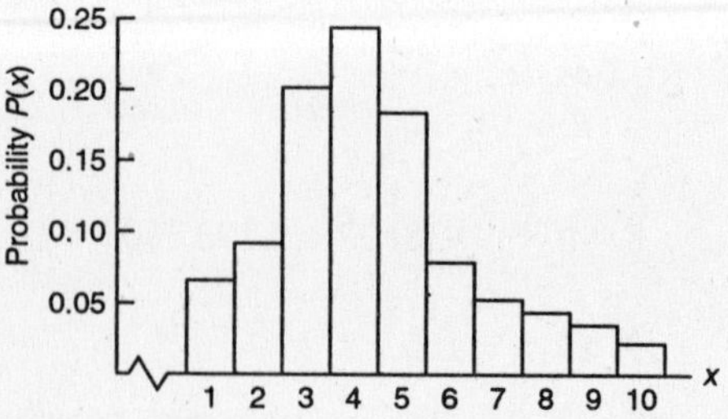

   (c) 0.641

   (d) 0.101

   (e) $\mu = 4.40$

   (f) $\sigma = 2.08$

2. $P(3) = C_{12,3}(0.28)^3(0.72)^9 = 0.25$

3. Success = accept; $p = 0.30$; $n = 55$;

(a) $P(r = 15) = 0.000$ (to three digits)

(b) $P(r \geq 8) = 0.051$

(c) $P(r \leq 4) = 0.517$

(d) $P(5 \leq r \leq 10) = 0.484$

4. 51

5. (a)

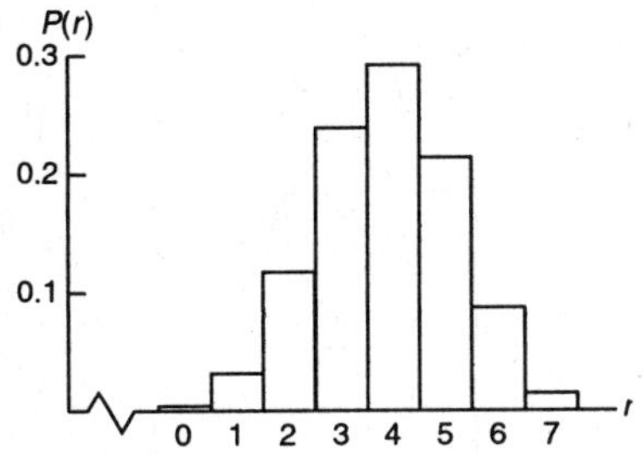

(b) $\mu = 3.85$

(c) $\sigma = 1.32$

6. $n = 10$ is the smallest number.

CHAPTER 6
FORM C

1. A. (c)
B. (e)
C. (b)
D. (d)

2. (e)

3. A. (b)
B. (a)
C. (d)
D. (c)

4. (b)

5. A. (e)
B. (c)

6. (d)

CHAPTER 7
FORM A

1. (a) A normal curve is bell-shaped with one peak. Because this curve has two peaks, it is not *normal*.

(b) A normal curve gets closer and closer to the horizontal axis, but it never touches it or crosses it.

2. 99.7%

3. 1

4. (a) $z \geq 1.8$

(b) $-1.2 \leq z \leq 0.8$

(c) $z \leq -2.2$

(d) $x \leq 3.1$

(e) $4.1 \leq x \leq 4.6$

(f) $x \geq 4.35$

5. (a) No, look at z values

(b) For Joel, $z = 1.51$. For John, $z = 2.77$. Relative to the district, John is a better salesman.

6. (a) $P(x \leq 5) = P(z \leq -0.77) = 0.2206$

(b) $P(x \geq 10) = P(z \geq 1.5) = 0.0668$

(c) $P(5 \leq x \leq 10)$
$= P(-0.77 \leq z \leq 1.5)$
$= 0.7126$

7. 29.92 minutes or 30 minutes

8. (a)

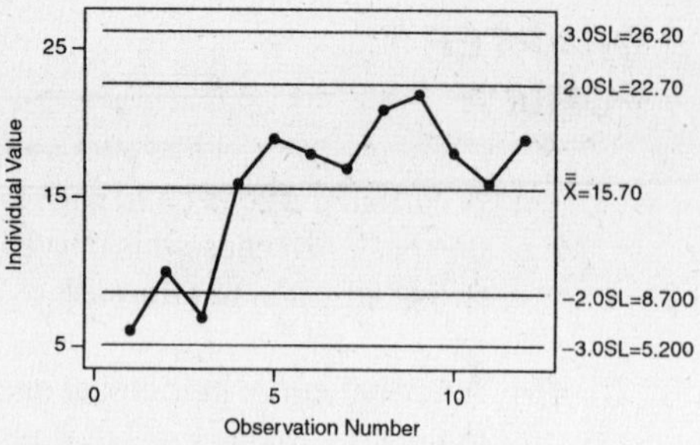

(b) Yes;
Type III for days 1 to 3;
Type II for days 4 thru 12

9. (a) Essay

(b) $P(r \le 47)$
$= P(x \le 47.5)$
$= P(z \le -0.82)$
$= 0.2061$

(c) $P(47 \le r \le 55)$
$= P(46.5 \le x \le 55.5)$
$= P(-1.22 \le z \le 2.33)$
$= 0.8789$

CHAPTER 7 FORM B

1. (a) The tails of a normal curve must get closer and closer to the x-axis. In this curve the tails are going away from the x-axis.

(b) A normal curve must be symmetrical. This curve is not.

2. 68%

3. 2

4. (a) $z \le 0.33$

(b) $-0.86 \le z \le 0.57$

(c) $z \ge -1.33$

(d) $x \ge 2.6$

(e) $6.8 \le x \le 11$

(f) $x \le 13.1$

5. (a) Generator II is hotter; see the z values

(b) $z = 0.72$ for Generator I;
$z = 2.75$ for Generator II
Generator II could be near a melt down since it is hotter.

6. (a) $P(x < 6) = P(z < -1.33) = 0.0918$

(b) $P(x > 7) = P(z > 2.00) = 0.0228$

(c) $P(6 \le x \le 7) = P(-1.33 \le z \le 2.00)$
$= 0.8854$

7. $z = 3$ years

8. (a)

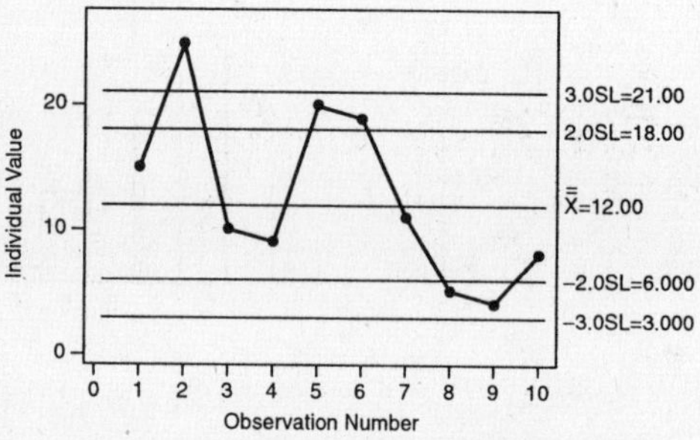

(b) Type I, day 2; Type III for days 5–6 and 8–10.

9. (a) Essay

(b) $P(r \ge 60) = P(x \ge 59.5) = P(z \ge -0.70)$
$= 0.7580$

(c) $P(60 \le r \le 70) = P(59.5 \le x \le 70.5)$
$= P(-0.70 \le z \le 1.88) = 0.7279$

CHAPTER 7 FORM C

1. (c)

2. (b)

3. (e)

4. A. (d)
B. (a)

5. A. (d)
B. (b)
C. (a)

6. A. (c)
B. (e)
C. (b)

7. (c)

8. (d)

9. **A.** (b)
B. (a)

CHAPTER 8 FORM A

1. Essay

2. **(a)** $P(\bar{x} < 3) = P(z < -0.51) = 0.3050$

(b) $P(\bar{x} > 4) = P(z > 2.06) = 0.0197$

(c) $P(3 \le \bar{x} \le 4) = P(-0.51 \le z \le 2.06)$
$= 0.6753$

3. **(a)** $P(\bar{x} < 23) = P(z < -1.67) = 0.0475$

(b) $P(\bar{x} > 28) = P(z > 1.21) = 0.1131$

(c) $P(23 \le \bar{x} \le 28) = P(-1.67 \le z \le 1.21)$
$= 0.8394$

CHAPTER 8 FORM B

1. Essay

2. **(a)** $P(5.2 \le \bar{x}) = P(-1.83 \le z) = 0.9664$

(b) $P(\bar{x} \le 7.1) = P(z \le 1.33) = 0.9082$

(c) $P(5.2 \le \bar{x} \le 7.1) = P(-1.83 \le z \le 1.33)$
$= 0.8746$

3. **(a)** $P(8 \le \bar{x}) = P(-2.15 \le z) = 0.9842$

(b) $P(\bar{x} \le 9) = P(z \le 0.54) = 0.7054$

(c) $P(8 \le \bar{x} \le 9) = P(-2.15 \le z \le 0.54)$
$= 0.6896$

CHAPTER 8 FORM C

1. **A.** (c)
B. (b)
C. (a)
D. (a)

2. **A.** (b)
B. (d)
C. (c)

3. **A.** (a)
B. (d)
C. (e)

CHAPTER 9 FORM A

1. 27.70 to 31.90 inches
We are 95% confident that the mean circumference of all Blue Spruce trees near this lake will lie between 27.70 and 31.90 inches.

2. **(a)** \$33.35 to \$36.05

(b) \$1167.25 to \$1261.75

3. $\bar{x} = 33.38$, $s = 6.46$,
$t = 2.365$,
27.98 to 38.78 minutes

4. 10.05 to 13.75 inches;
use $t = 2.878$
We are 99% confident that the mean length of Rainbow Trout is between 10.05 and 13.75 minutes.

5. **(a)** $\hat{p} = 59/78 \approx 0.756$

(b) 0.68 to 0.84

(c) $np > 5$ and $nq > 5$; Yes

(d) 122 more

6. 94 more

7. **(a)** 106

(b) 100

CHAPTER 9
FORM B

1. 4.10 to 6.30 calls;
use $t = 2.160$
We are 96% confident that the mean number of solicitation calls this family receives each night is between 4.10 and 6.30 calls.

2. (a) \$6.45 to \$7.35
(b) \$516 to \$588

3. $\bar{x} = 14.7$, $s = 5.31$, $t = 2.262$,
10.90 to 18.50 pages

4. 89.0 to 97.4 minutes
We are 99% confident that mean time for computer repairs is between 89.0 and 97.4 minutes.

5. (a) $\hat{p} = 41/56$ or 0.732
(b) 0.62 to 0.85
(c) np and nq are both greater than 5; Yes
(d) 246 more

6. 33 more

7. (a) 271
(b) 167

CHAPTER 9
FORM C

1. (b)

2. A. (c)
B. (e)

3. (a)

4. (b)

5. A. (d)
B. (a)
C. (e)

6. (c)

7. A. (a)
B. (d)

CHAPTER 10
FORM A

1. (a) H_0: $\mu = 5.3$; H_1: $\mu < 5.3$; left-tailed; $\alpha = 0.01$
(b) standard normal; $z_0 = 2.33$
(c)

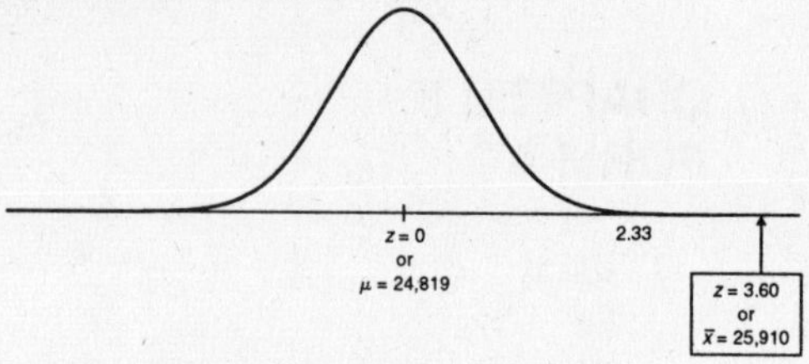

(d) $\bar{x} = 25{,}910$ corresponds to $z = 3.60$
(e) P value = 0.0002
(f) Reject H_0. There is evidence that the population mean daily sales is greater.

2. (a) H_0: $\mu = 18.3$; H_1: $\mu \neq 18.3$; two-tailed; $\alpha = 0.05$
(b) Student's t; $t_0 = \pm 2.776$; $d.f. = 4$
(c)

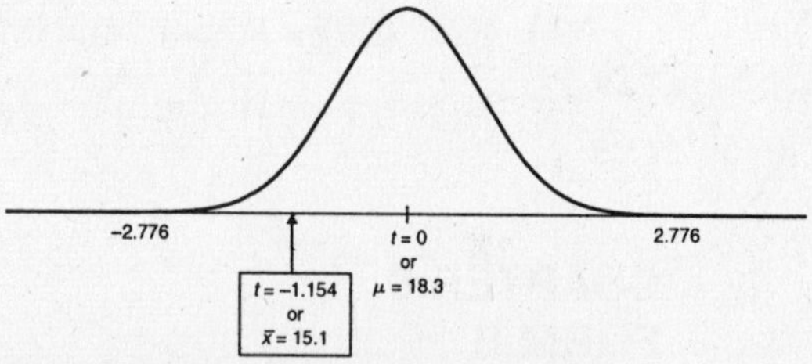

(d) $\bar{x} = 15.1$ corresponds to $t = -1.154$
(e) P value > 0.250
(f) Do not reject H_0. There is not enough evidence to conclude that the waiting times are different.

3. (a) H_0: $p = 0.75$; H_1: $p < 0.75$; left-tailed; $\alpha = 0.01$
(b) standard normal; $z_0 = -2.33$
(c)

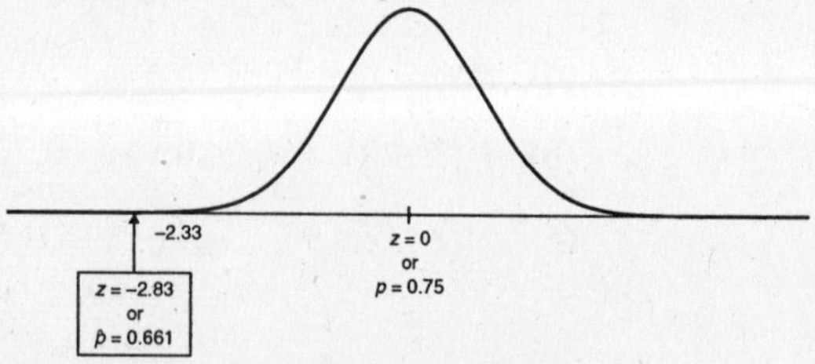

(d) $\hat{p} = 125/189 = 0.661$ corresponds to $z = -2.83$
(e) P value = 0.0023
(f) Reject H_0. There is evidence that the proportion of Rainbow Trout is less than 75%.

4. **(a)** H_0: $\mu = 68.3$; H_1: $\mu \neq 68.3$; two-tailed; $\alpha = 0.05$
(b) standard normal; $z_0 = \pm 1.96$
(c)

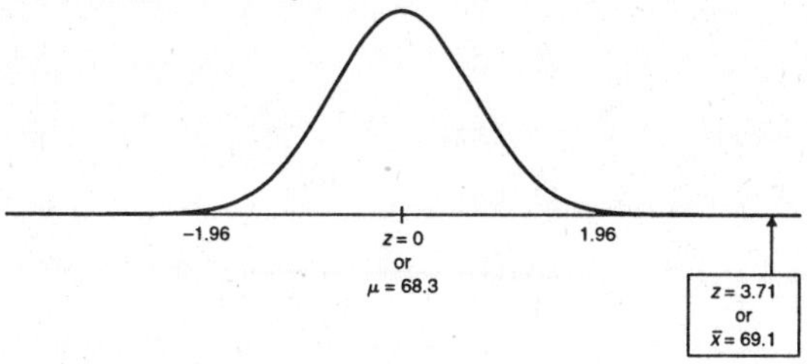

(d) $\bar{x} = 69.1$ corresponds to $z = 3.71$
(e) P value < 0.0001
(f) Reject H_0. There is evidence that the population mean height is different from 68.3 inches.

5. **(a)** H_0: $\mu = 30$; H_1: $\mu > 30$; right-tailed; $\alpha = 0.05$
(b) Student's t; $t_0 = 1.771$; $d.f. = 13$
(c)

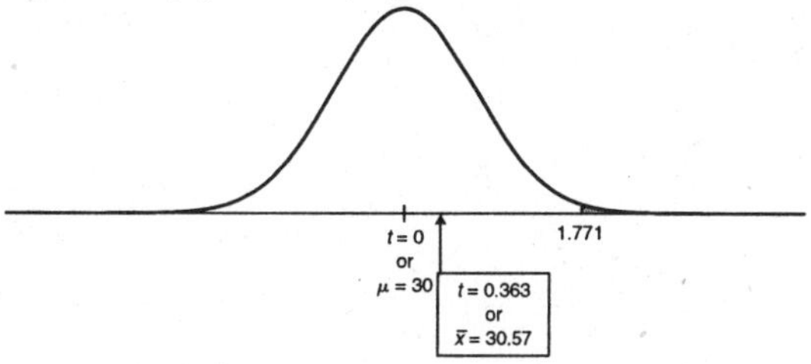

(d) $\bar{x} = 30.57$ corresponds to $t = 0.363$
(e) P value > 0.25
(f) Do not reject H_0. There is not enough evidence to conclude that the delivery time is greater than 30 min.

6. **(a)** H_0: $p = 0.6$; H_1: $p < 0.6$; left-tailed; $\alpha = 0.05$
(b) standard normal; $z_0 = -1.645$
(c)

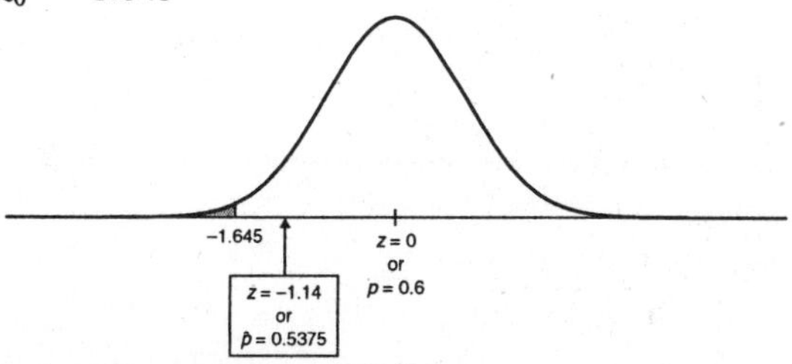

(d) $\hat{p} = 43/80 = 0.5375$ corresponds to $z = -1.14$
(e) P value $= 0.1271$
(f) Do not reject H_0. There is not enough evidence to conclude that the proportion is less than 60%.

CHAPTER 10
FORM B

1. **(a)** H_0: $\mu = 5.3$; H_1: $\mu < 5.3$; left-tailed; $\alpha = 0.01$
(b) standard normal; $z_0 = -2.33$
(c)

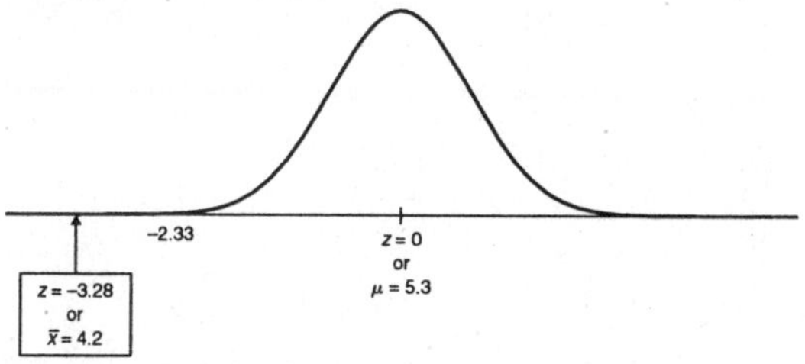

(d) $\bar{x} = 4.2$ corresponds to $z = -3.28$
(e) P value $= 0.0005$
(f) Reject H_0. There is evidence that the average recovery time is less as an outpatient.

2. **(a)** H_0: $\mu = 4.19$; H_1: $\mu > 4.19$; right-tailed; $\alpha = 0.05$
(b) Student's t; $d.f. = 15$; $t_0 = 1.753$
(c)

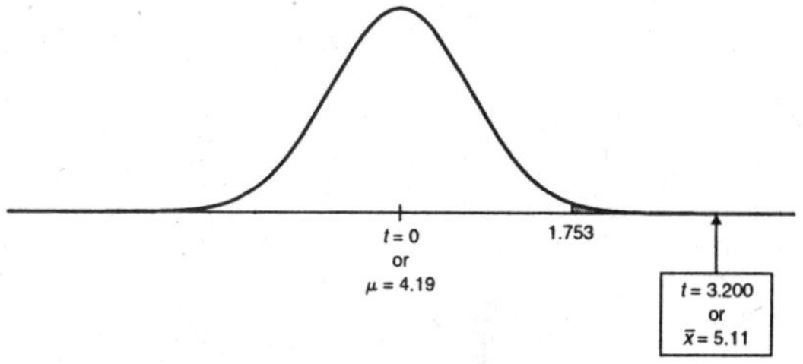

(d) $\bar{x} = 5.11$ corresponds to $t = 3.200$
(e) P value < 0.005
(f) Reject H_0. There is evidence that the average yield of Arizona municipal bonds is higher than the national average yield.

3. **(a)** H_0: $p = 0.37$; H_1: $p \neq 0.37$; two-tailed; $\alpha = 0.01$
(b) standard normal; $z_0 = \pm 2.576$
(c)

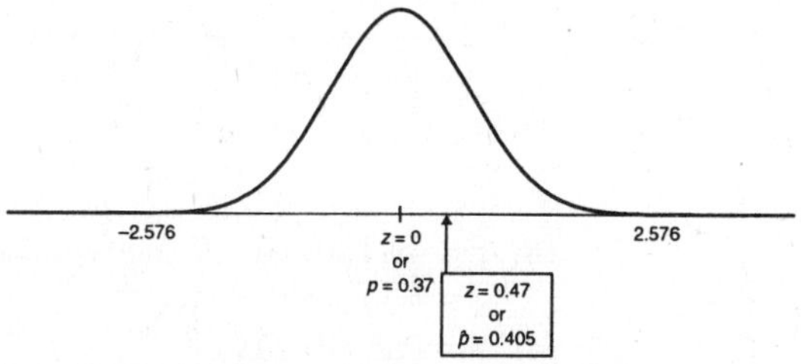

(d) $\hat{p} = 17/42 = 0.405$ corresponds to $z = 0.47$
(e) P value $= 0.6384$

(f) Do not reject H_0. There is not evidence to conclude that the proportion of freshmen on the Dean's List is different from that in the college.

4. (a) H_0: $\mu = 2.8$; H_1: $\mu \neq 2.8$; two-tailed; $\alpha = 0.05$

(b) standard normal; $z_0 = \pm 1.96$

(c)

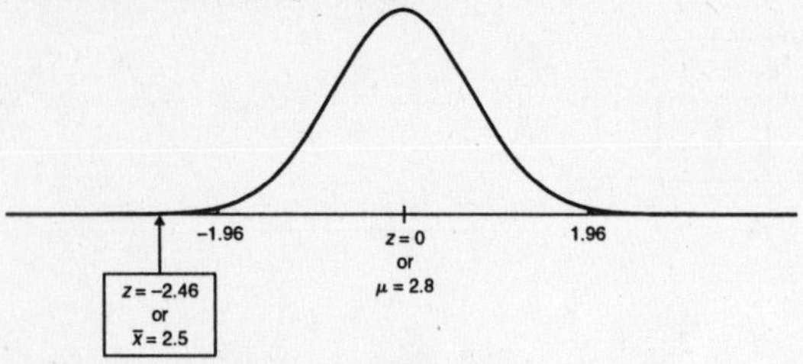

(d) $\overline{x} = 2.5$ corresponds to $z = -2.46$

(e) P value $= 0.0138$

(f) Reject H_0. There is evidence that the population mean time is different from 2.8 hours.

5. (a) H_0: $\mu = 30$; H_1: $\mu > 30$; right-tailed; $\alpha = 0.05$

(b) Student's t; $t_0 = 1.753$; $d.f. = 15$

(c)

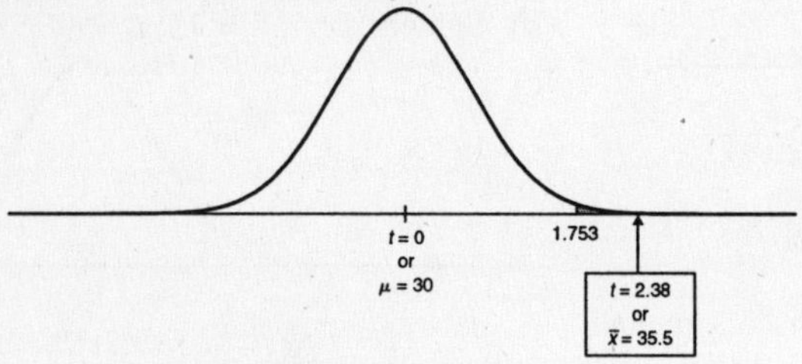

(d) $\overline{x} = 35.5$ corresponds to $t = 2.38$

(e) $0.01 < P$ value > 0.025

(f) Reject H_0. There is evidence that the average age is greater than 30 years.

6. (a) H_0: $p = 0.27$; H_1: $p > 0.27$; right-tailed; $\alpha = 0.05$

(b) standard normal; $z_0 = 1.645$

(c)

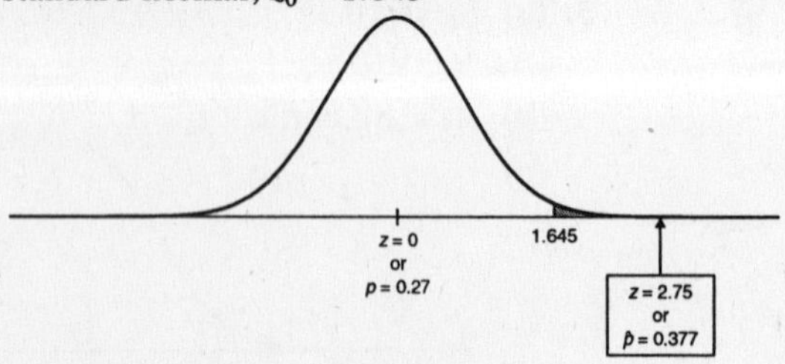

(d) $\hat{p} = 49/130 = 0.377$ corresponds to $z = 2.75$

(e) P value $= 0.003$

(f) Reject H_0. There is sufficient evidence to conclude that the proportion is greater than 27%.

CHAPTER 10 FORM C

1. A. (b)
B. (d)
C. (a)
D. (c)
E. (b)

2. A. (e)
B. (d)
C. (a)
D. (c)
E. (b)

3. A. (d)
B. (a)
C. (e)
D. (c)
E. (a)

4. A. (e)
B. (a)
C. (c)
D. (d)
E. (b)

5. A. (d)
B. (c)
C. (b)
D. (e)
E. (a)

6. A. (c)
B. (a)
C. (e)
D. (d)
E. (a)

CHAPTER 11
FORM A

1. (a) H_0: $p_1 = p_2$; H_1: $p_1 \neq p_2$; two-tailed; $\alpha = 0.05$

(b) standard normal; $z_0 = \pm 1.96$

(c)

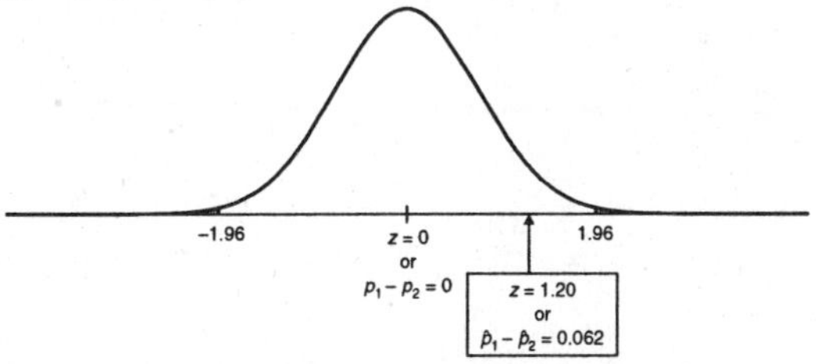

(d) $\hat{p}_1 = 62/175 = 0.354$; $\hat{p}_2 = 45/154 = 0.292$; $\hat{p}_1 - \hat{p}_2 = 0.062$ corresponds to $z = 1.20$

(e) P value = 0.2302

(f) Do not reject H_0. There is not enough evidence to conclude that there is a difference in population proportions of success between the two different sales pitches.

2. (a) H_0: $\mu_d = 0$; H_1: $\mu_d < 0$; left-tailed; $\alpha = 0.05$

(b) Student's t; $d.f. = 5$; $t_0 = -2.015$

(c)

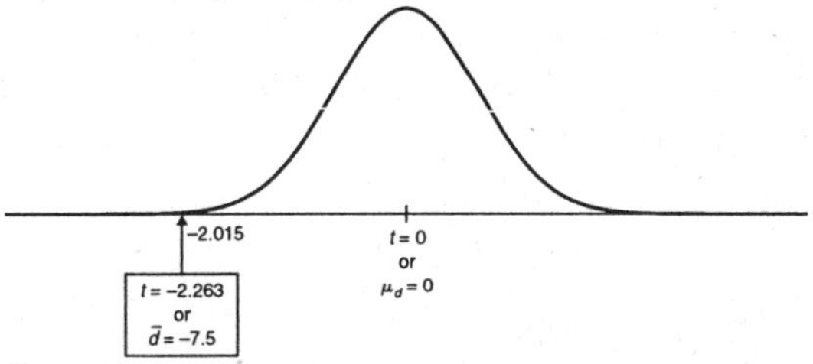

(d) $\bar{d} = -7.5$ corresponds to $t = -2.263$

(e) $0.025 < P \text{ value} < 0.050$

(f) Reject H_0. There is evidence to conclude that the new process increases the mean number of items processed per shift.

3. (a) H_0: $\mu_1 = \mu_2$; H_1: $\mu_1 > \mu_2$; right-tailed; $\alpha = 0.01$

(b) Student's t; $d.f. = 18$; $t_0 = 2.552$

(c)

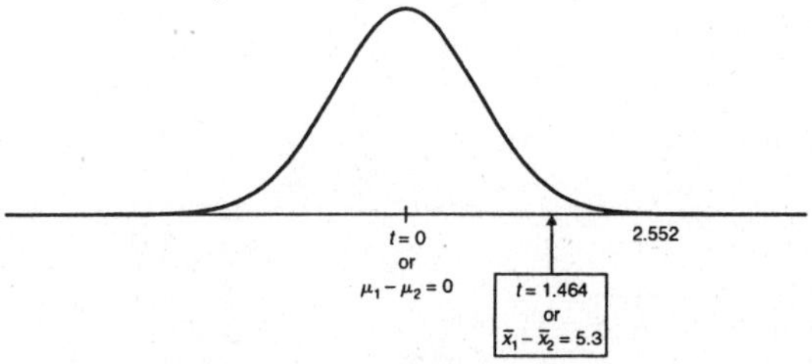

(d) $\bar{x}_1 - \bar{x}_2 = 5.3$ corresponds to $t = 1.464$

(e) $0.075 < P \text{ value} < 0.10$

(f) Do not reject H_0. There is not sufficient evidence to show that the population mean profit per employee in computer stores is higher than those for building supply stores.

4. (a) H_0: $\mu_1 = \mu_2$; H_1: $\mu_1 \neq \mu_2$; two-tailed; $\alpha = 0.05$

(b) standard normal; $z_0 = \pm 1.96$

(c)

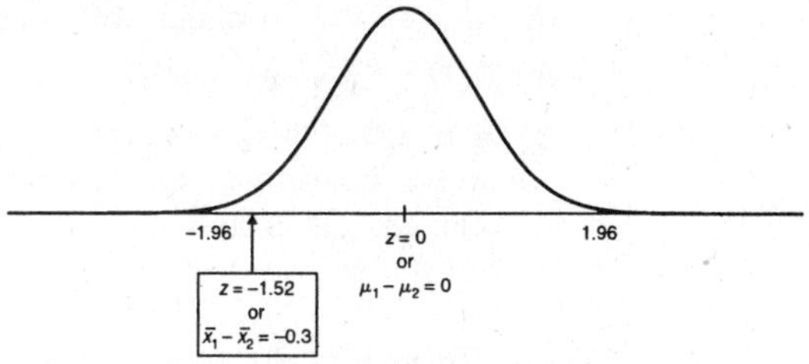

(d) $\bar{x}_1 - \bar{x}_2 = -0.3$ corresponds to $z = -1.52$

(e) P value = 0.1286

(f) Do not reject H_0. There is not enough evidence to say that the mean weight of tomatoes grown organically is different than that of other tomatoes.

CHAPTER 11
FORM B

1. (a) H_0: $p_1 = p_2$; H_1: $p_1 < p_2$; left-tailed; $\alpha = 0.05$

(b) standard normal; $z_0 = -1.645$

(c)

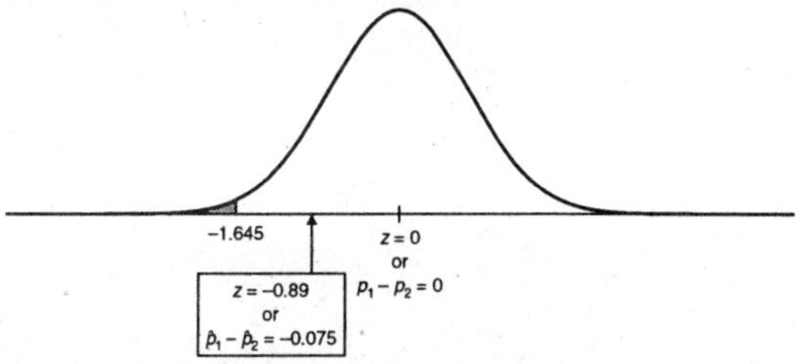

(d) $\hat{p}_1 = 0.548$ = proportion in favor of Jennifer; $\hat{p}_2 = 0.623$ = proportion in favor of Kevin; $\hat{p}_1 - \hat{p}_2 = -0.075$ corresponds to $z = -0.89$

(e) P value = 0.1867

(f) Do not reject H_0. There is not enough evidence to say that the proportion who plan to vote for Kevin is higher.

2. (a) H_0: $\mu_d = 0$; H_1: $\mu_d < 0$; left-tailed; $\alpha = 0.05$

(b) Student's t; $d.f. = 4$; $t_0 = -2.132$

(c)

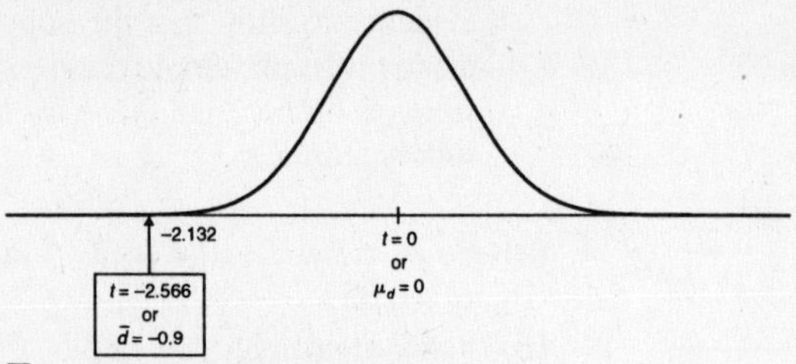

(d) $\bar{d} = -0.9$ corresponds to $t = -2.566$

(e) $0.025 < P$ value < 0.050

(f) Reject H_0. There is evidence that the average time for runners at higher elevation is longer.

3. (a) H_0: $\mu_1 = \mu_2$; H_1: $\mu_1 > \mu_2$; right-tailed; $\alpha = 0.01$

(b) Student's t; $d.f. = 21$; $t_0 = 2.518$

(c)

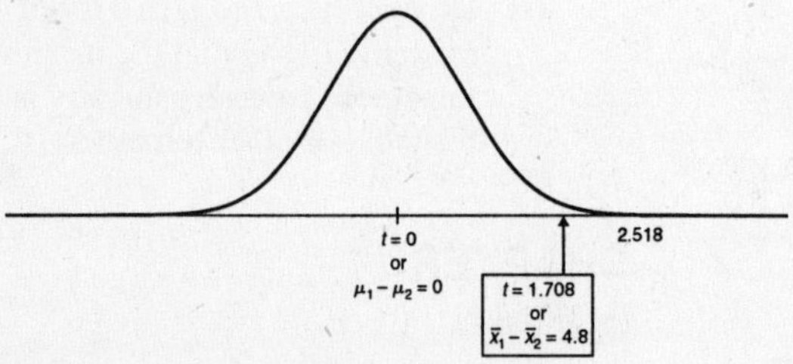

(d) $\bar{x}_1 - \bar{x}_2 = 4.8$ corresponds to $t = 1.708$

(e) $0.05 < P$ value < 0.075

(f) Do not reject H_0. There is not enough evidence to conclude that the average mileage for the Pacer is greater.

4. (a) H_0: $\mu_1 = \mu_2$; H_1: $\mu_1 > \mu_2$; right-tailed; $\alpha = 0.05$

(b) standard normal; $z_0 = 1.645$

(c)

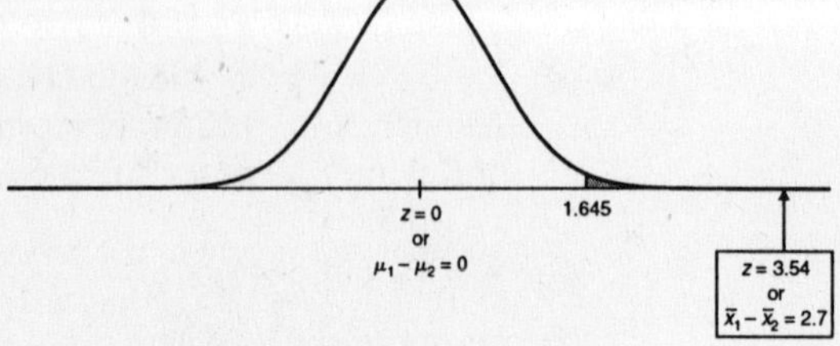

(d) $\bar{x}_1 - \bar{x}_2 = 2.7$ corresponds to $z = 3.54$

(e) P value $= 0.0002$

(f) Reject H_0. The evidence indicates that the average egg production of cage-free chickens is higher.

CHAPTER 11 FORM C

1. A. (e)
B. (d)
C. (b)
D. (e)
E. (a)

2. A. (d)
B. (e)
C. (c)
D. (b)
E. (a)

3. A. (b)
B. (a)
C. (e)
D. (d)
E. (a)

4. A. (b)
B. (c)
C. (e)
D. (d)
E. (b)

CHAPTER 12
FORM A

1. $S_e = 0.757$
2. $7.70 \le y \le 11.61$ pounds
3. H_0: $\rho = 0$; H_1: $\rho \ne 0$; $t_0 = \pm 5.841$; sample $t = 19.31$; Reject H_0; There is evidence that the population correlation coefficient is not equal to 0. r is significant.
4. $S_e = 89.19$
5. $\$105{,}900 \le y \le \$641{,}100$
6. H_0: $\rho = 0$; H_1: $\rho > 0$; $t_0 = 2.132$; sample $t = 6.37$; Reject H_0; There is sufficient evidence that the population correlation coefficient is positive. r is significant.
7. H_0: Teacher evaluations are independent of midterm grades;
   H_1: Teacher evaluations are not independent of midterm grades;
   $\chi^2 = 43.68$; $\chi^2_{0.05} = 9.49$;
   Reject H_0. There is evidence to say that at the 5% level of significance teacher evaluations are not independent of midterm grades.
8. H_0: The distribution of ages of college students is the same in the Western Association as in the nation.
   H_1: The distribution of ages of college students is different in the Western Association than it is in the nation.
   $\chi^2 = 15.03$; $\chi^2_{0.05} = 9.49$;
   Reject H_0. There is evidence that the distribution of ages is different.
9. Chi-square distribution
10. H_0: $\sigma = 4.1$; H_1: $\sigma > 4.1$; $\chi^2_{0.01} = 42.98$;
    $\chi^2 = 34.28$; Do not reject H_0. There is not enough evidence to reject $\sigma = 4.1$.

CHAPTER 12
FORM B

1. $S_e = 2.988$
2. $11.01 \le y \le 26.47$ billion dollars for annual company revenue
3. H_0: $\rho = 0$; H_1: $\rho \ne 0$; $t_0 = 5.841$; sample statistic $t = 3.21$; Do not reject H_0; The sample statistic is not significant. We do not have evidence of correlation between CEO salary and company revenue at the 1% level of significance. r is not significant.
4. $S_e = 5.826$
5. $\$82.38 \le y \le \$118.22$
6. H_0: $\rho = 0$; H_1: $\rho > 0$; $t_0 = 2.132$; sample $t = 18.17$; Reject H_0; There is evidence that the population correlation coefficient is positive. r is significant.
7. H_0: The choice of college major is independent of grade average.
   H_1: The choice of college major is not independent of college average.
   $\chi^2_{0.05} = 13.28$; $\chi^2 = 2.64$;
   Do not reject H_0. There is not enough evidence to conclude that college major is not independent of grade average.
8. H_0: The distribution of fish fits the initial stocking distribution.
   H_1: The distribution of fish after six years does not fit the initial stocking distribution.
   $\chi^2_{0.05} = 9.49$; $\chi^2 = 66.78$;
   Reject H_0. There is enough evidence to conclude that the fish distribution has changed.
9. H_0: $\sigma = 6$; H_1: $\sigma \ne 6$; critical values are 8.03 and 41.40. $\chi^2 = 26.97$;
   Do not reject H_0. There is not enough evidence to conclude that the standard deviation is different from 6.

CHAPTER 12
FORM C

1. (a)

2. (c)

3. (b)

4. (a)

5. (e)

6. (b)

7. **A.** (c)
B. (d)
C. (e)
D. (a)

8. **A.** (d)
B. (a)
C. (c)
D. (b)

9. **A.** (e)
B. (c)
C. (d)

10. **A.** (e)
B. (e)

Part IV

Complete Solutions

Chapter 1 Getting Started

Section 1.1

1. (a) The variable is the response regarding frequency of eating at fast-food restaurants.
   (b) The variable is qualitative. The categories are the number of times one eats in fast-food restaurants.
   (c) The implied population is responses for all adults in the U.S.

2. (a) The variable is the miles per gallon.
   (b) The variable is quantitative because arithmetic operations can be applied the mpg values.
   (c) The implied population is gasoline mileage for all new 2001 cars.

3. (a) The variable is student fees.
   (b) The variable is quantitative because arithmetic operations can be applied to the fee values.
   (c) The implied population is student fees at all colleges and universities in the U.S.

4. (a) The variable is the shelf life.
   (b) The variable is quantitative because arithmetic operations can be applied to the shelf life values.
   (c) The implied population is the shelf life of all Healthy Crunch granola bars.

5. (a) The variable is the time interval between check arrival and clearance.
   (b) The variable is quantitative because arithmetic operations can be applied to the time intervals.
   (c) The implied population is the time interval between check arrival and clearance for all companies in the five-state region.

6. Form B would be better. Statistical methods can be applied to the ordinal data obtained from Form B, but not to the answers obtained from Form A.

7. (a) *Length of time to complete an exam* is a ratio level of measurement. The data may be arranged in order, differences and ratios are meaningful, and a time of 0 is the starting point for all measurements.
   (b) *Time of first class* is an interval level of measurement. The data may be arranged in order and differences are meaningful.
   (c) *Class categories* is a nominal level of measurement. The data consists of names only.
   (d) *Course evaluation scale* is an ordinal level of measurement. The data may be arranged in order.
   (e) *Score on last exam* is a ratio level of measurement. The data may be arranged in order, differences and ratios are meaningful, and a score of 0 is the starting point for all measurements.
   (f) *Age of student* is a ratio level of measurement. The data may be arranged in order, differences and ratios are meaningful, and an age of 0 is the starting point for all measurements.

8. (a) *Salesperson's performance* is an ordinal level of measurement. The data may be arranged in order.
   (b) *Price of company's stock* is a ratio level of measurement. The data may be arranged in order, differences and ratios are meaningful, and a price of 0 is the starting point for all measurements.
   (c) *Names of new products* is a nominal level of measurement. The data consist of names only.
   (d) *Room temperature* is an interval level of measurement. The data may be arranged in order and differences are meaningful.

(e) *Gross income* is a ratio level of measurement. The data may be arranged in order, differences and ratios are meaningful, and an income of 0 is the starting point for all measurements.

(f) *Color of packaging* is a nominal level of measurement. The data consist of names only.

9. **(a)** *Species of fish* is a nominal level of measurement. Data consist of names only.

(b) *Cost of rod and reel* is a ratio level of measurement. The data may be arranged in order, differences and ratios are meaningful, and a cost of 0 is the starting point for all measurements.

(c) *Time of return home* is an interval level of measurement. The data may be arranged in order and differences are meaningful.

(d) *Guidebook rating* is an ordinal level of measurement. Data may be arranged in order.

(e) *Number of fish* caught is a ratio level of measurement. The data may be arranged in order, differences and ratios are meaningful, and 0 fish caught is the starting point for all measurements.

(f) *Temperature of the water* is an interval level of measurement. The data may be arranged in order and differences are meaningful.

Section 1.2

1. Essay

2. Answers vary. Use groups of 3 digits.

3. Answers vary. Use groups of 4 digits.

4. Answers vary. Use groups of 3 digits.

5. **(a)** Assign a distinct number to each subject. Then use a random number table. Group assignment methods vary.

(b) Repeat part (a) for 22 subjects.

(c) Answers vary.

6. Answers vary. Use single digits with odd corresponding to heads and even to tails.

7. **(a)** Yes, it is appropriate that the same number appears more than once because the outcome of a die roll can repeat. The outcome of the 4th roll is 2.

(b) No, we do not expect the same sequence because the process is random.

8. Answers vary. Use groups of 3 digits.

9. **(a)** Reasons may vary. For instance, the first four students may make a special effort to get to class on time.

(b) Reasons may vary. For instance, four students who come in late might all be nursing students enrolled in an anatomy and physiology class that meets the hour before in a far-away building. They may be more motivated than other students to complete a degree requirement.

(c) Reasons may vary. For instance, four students sitting in the back row might be less inclined to participate in class discussions.

(d) Reasons may vary. For instance, the tallest students might all be male.

10. In all cases, assign distinct numbers to the items, and use a random-number table.

11. In all cases, assign distinct numbers to the items, and use a random-number table.

12. Answers vary. Use single digits with even corresponding to true and odd corresponding to false.

13. Answers vary. Use single digits with correct answer placed in corresponding position.

14. (a) This technique is stratified sampling. The population was divided into strata (4 categories of length of hospital stay), then a simple random sample was drawn from each stratum.

(b) This technique is simple random sampling. Every sample of size n from the population has an equal chance of being selected and every member of the population has an equal chance of being included in the sample.

(c) This technique is cluster sampling. There are 5 geographic regions and a random sample of hospitals is selected from each region. Then, for each selected hospital, all patients on the discharge list are surveyed to create the patient satisfaction profiles. Within each hospital, the degree of satisfaction varies patient to patient. The sampling units (the hospitals) are clusters of individuals who will be studied.

(d) This technique is systematic sampling. Every k^{th} element is included in the sample.

(e) This technique is convenience sampling. This technique uses results or data that are conveniently and readily obtained.

15. (a) This technique is simple random sampling. Every sample of size n from the population has an equal chance of being selected and every member of the population has an equal chance of being included in the sample.

(b) This technique is cluster sampling. The state, Hawaii, is divided into regions using, say, the first 3 digits of the Zip code. Within each region a random sample of 10 Zip code areas is selected using, say, all 5 digits of the Zip code. Then, within each selected Zip codes, all businesses are surveyed. The sampling units, defined by 5 digit Zip codes, are clusters of businesses, and within each selected Zip code, the benefits package the businesses offer their employees differs business to business.

(c) This technique is convenience sampling. This technique uses results or data that are conveniently and readily obtained.

(d) This technique is systematic sampling. Every k^{th} element is included in the sample.

(e) This technique is stratified sampling. The population was divided into strata (10 business types), then a simple random sample was drawn from each stratum.

Section 1.3

1. (a) This is an observational study because observations and measurements of individuals are conducted in a way that doesn't change the response or the variable being measured.

(b) This is an experiment because a treatment is deliberately imposed on the individuals in order to observe a possible change in the response or variable being measured.

(c) This is an experiment because a treatment is deliberately imposed on the individuals in order to observe a possible change in the response or variable being measured.

(d) This is an observational study because observations and measurements of individuals are conducted in a way that doesn't change the response or the variable being measured.

2. (a) A census was used because data for all the games were used.

(b) An experiment was used. A treatment is deliberately imposed on the individuals in order to observe change in the response or variable being measured.

(c) A simulation was used because computer imaging of runners was used.

(d) Sampling was used because measurements from a representative part of the population were used.

3. (a) Sampling was used because measurements from a representative part of the population were used.

(b) A simulation was used because computer programs that mimic actual flight were used.

(c) A census was used because data for all scores are available.

(d) An experiment was used. A treatment is deliberately imposed on the individuals in order to observe change in the response or variable being measured.

4. **(a)** No, "over the last few years" could mean the last three years to some and the last five years to others, etc.; answers vary.

(b) Yes. The response to doubling fines would be affected by whether the responder had ever run a stop sign.

(c) Answers vary.

5. **(a)** Use random selection to pick 10 calves to inoculate. Then test all calves to see if there is a difference in resistance to infection between the two groups. There is no placebo being used.

(b) Use random selection to pick 9 schools to visit. Then survey all the schools to see if there is a difference in views between the two groups. There is no placebo being used.

(c) Use random selection to pick 40 volunteers for skin patch with drug. Then survey all volunteers to see if a difference exists between the two groups. A placebo for the remaining 35 volunteers in the second group is used.

6. **(a)** Use random selection to pick 25 cars for high-temperature bond tires. Then examine tires of all the cars to see if a difference exists between the two groups. This is a double-blind experiment because neither the individuals in the study nor the observers know which subjects are receiving the new tires.

(b) Use random selection to pick 10 bags. Then send all bags through the security check. This is not a double-blind experiment because the agent carrying the bag knows whether or not the bag contains a weapon.

(c) Use random selection to pick 35 patients for new eye drops. Then measure eye pressure for all patients to see if a difference exists between the two groups. This is a double-blind experiment because neither the patients nor the doctors know which subjects are receiving the new drops.

Chapter 1 Review

1. Answers vary.

2. The implied population is the opinions of all the listeners. The variable is the opinion of a caller. There is probably bias in the selection of the sample because those with the strongest opinions are most likely to call in.

3. Essay

4. Name, social security number, color of hair and eyes, address, phone number, place of birth, and college major are all nominal because the data consist of names or qualities only. Letter grade on test is ordinal because the data may be arranged in order. Year of birth is interval because the data may be arranged in order and differences are meaningful. Height, age, and distance from home to college are ratio because the data may be arranged in order, differences and ratios are meaningful, and 0 is the starting point for all measurements.

5. In the random number table use groups of 2 digits. Select the first six distinct groups of 2 digits that fall in the range from 01 to 42. Choices vary according to the starting place in the random number table.

6. **(a)** Cluster sampling was used because a random sample of 10 telephone prefixes was selected and all households in the selected prefixes were included in the sample.

(b) Convenience sampling was used because it uses results or data that are conveniently and readily obtained.

(c) Systematic sampling was used because every k^{th} element is included in the sample.

(d) Random sampling was used because every sample of size 30 from the population has an equal chance of being selected and every member of the population has an equal chance of being included.

(e) Stratified sampling was used because the population was divided into strata (three age categories), then a simple random sample was drawn from each stratum.

7. (a) This is an observational study because observations and measurements of individuals are conducted in a way that doesn't change the response or the variable being measured.

(b) This is an experiment because a treatment is deliberately imposed on the individuals in order to observe a possible change in the response or variable being measured.

8. (a) Use random selection to pick half to solicit by mail. Then compute the percentage of donors in each group. Compare the results. No placebo was used.

(b) Use random selection to pick 43 volunteers to be given whitening gel. Evaluate tooth whiteness for all participants. Compare the results. A placebo was used with the remaining 42 in the second group. The experiment could be double-blind if the observers did not know which subjects were receiving the tooth whitening chemicals.

9. This is a good problem for class discussion. Some items such as age and grade point average might be sensitive information. You could ask the class to design a data form that can be filled out anonymously. Other issues to discuss involve the accuracy and honesty of the responses.

10. Students may easily spend several hours at this Web site.

Chapter 2 Organizing Data

Section 2.1

1. Highest Level of Education and Average Annual Household Income (in thousands of dollars)

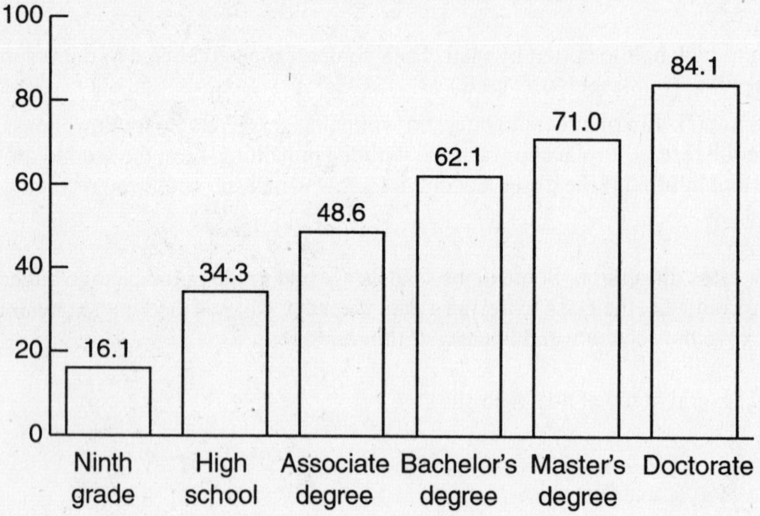

2. Annual Number of Deaths from Injuries per 100,000 Children (Ages 1 to 14)

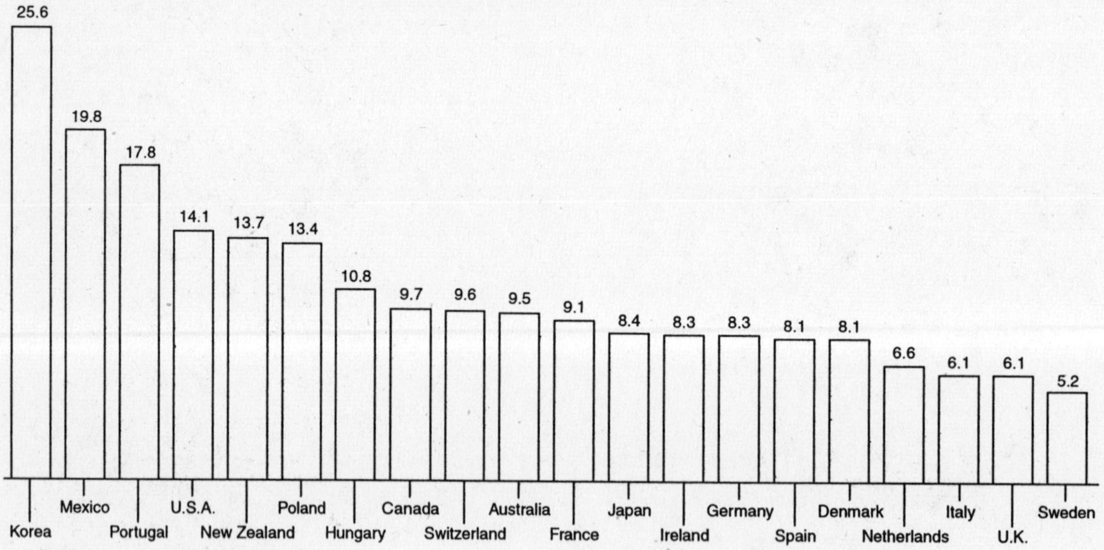

3. Number of People Who Died in a Calendar Year from Listed Causes—Pareto Chart

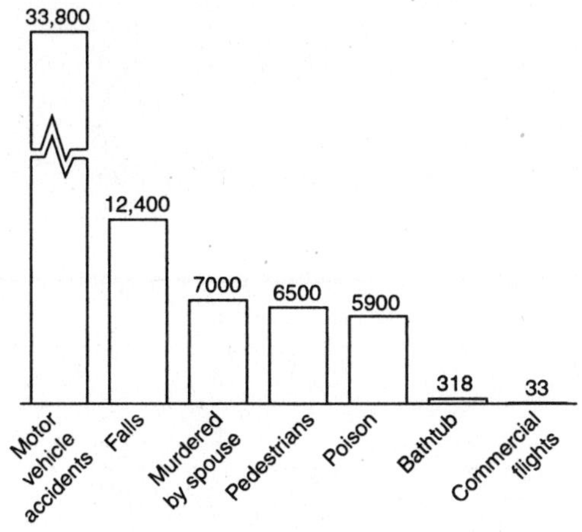

4. (a) Since 88% of those surveyed cited internal problems, 100% – 88% = 12% cited external factors as the leading cause of business failure. Among the internal causes, 88% – 13% – 13% – 18% – 29% = 15% must have listed various other internal factors for the leading cause of business failure.

Causes for Business Failure—Pareto Chart

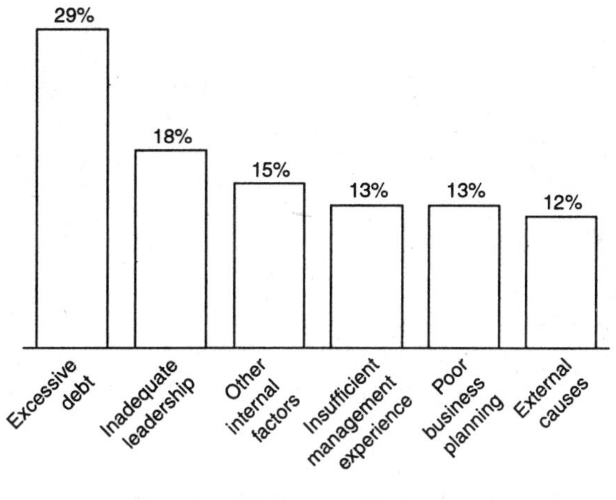

(b) As shown in part (a), 15% of those interviewed cited other internal factors as the leading cause of business failure. Excessive debt was the most commonly cited (internal and overall) cause for business failure.

Cause of Business Failure	Percentage	Frequency
Insufficient management experiences	13%	13% × 1300 = 169
Poor business planning	13%	13% × 1300 = 169
Inadequate leadership	18%	18% × 1300 = 234
Excessive debt	29%	29% × 1300 = 377
Other internal factors	15%	15% × 1300 = 195
External factors	12%	12% × 1300 = 156
Total	100%	1300

5.

Hiding place	Percentage	Number of Degrees
In the closet	68%	$68\% \times 360° \approx 245°$
Under the bed	23%	$23\% \times 360° \approx 83°$
In the bathtub	6%	$6\% \times 360° \approx 22°$
In the freezer	3%	$3\% \times 360° \approx 11°$
Total	100%	361°*

*Total does not add to 360° due to rounding.

Where We Hide the Mess

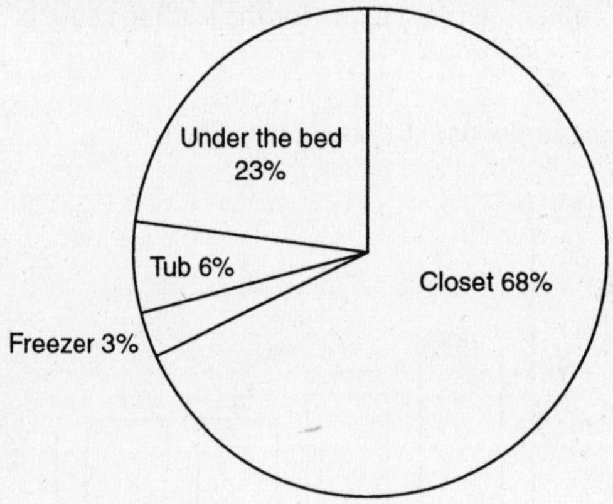

6.

Meal	Percentage	Number of Degrees
Lunch	48.9%	$48.9\% \times 360° \approx 176°$
Breakfast	7.7%	$7.7\% \times 360° \approx 28°$
Dinner	31.6%	$31.6\% \times 360° \approx 114°$
Snack	10.0%	$10.0\% \times 360° = 36°$
Don't know	1.8%	$1.8\% \times 360° \approx 6°$
Total	100.0%	360°

Meals We Are Most Likely to Eat in a Fast-Food Restaurant

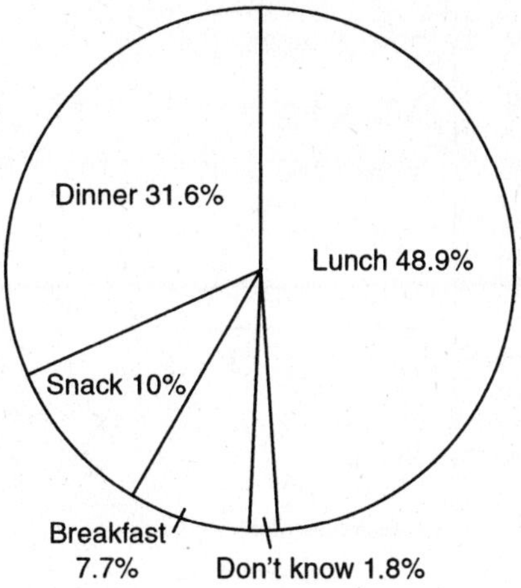

7.

Professional Activity	Percentage	Number of Degrees
Teaching	51%	$51\% \times 360° \approx 184°$
Research	16%	$16\% \times 360° \approx 58°$
Professional growth	5%	$5\% \times 360° = 18°$
Community service	11%	$11\% \times 360° \approx 40°$
Service to the college	11%	$11\% \times 360° \approx 40°$
Consulting outside the college	6%	$6\% \times 360° \approx 22°$
Total	100%	362°*

* Total does not add to 360° due to rounding.

How College Professors Spend Time

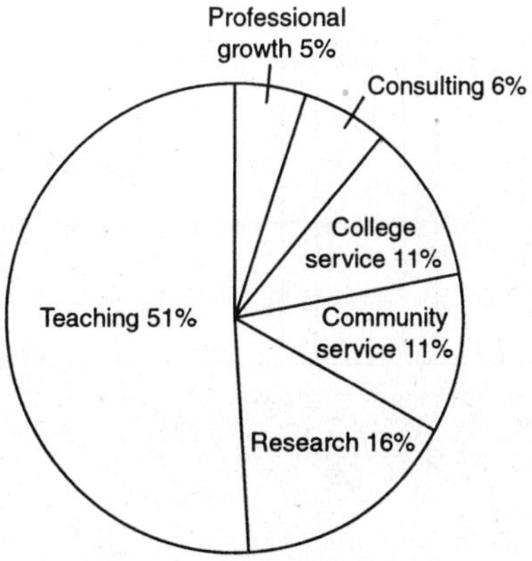

8.

Age	Percentage	Number of Degrees
Under 35 years	8%	$8\% \times 360° \approx 29°$
35–44 years	29%	$29\% \times 360° \approx 104°$
45–54 years	37%	$37\% \times 360° \approx 133°$
55–59 years	13%	$13\% \times 360° \approx 47°$
60–64 years	9%	$9\% \times 360° \approx 32°$
65 years and over	4%	$4\% \times 360° \approx 14°$
Total	100%	359°*

*Total does not add to 360° due to rounding.

Age Distribution of Professors

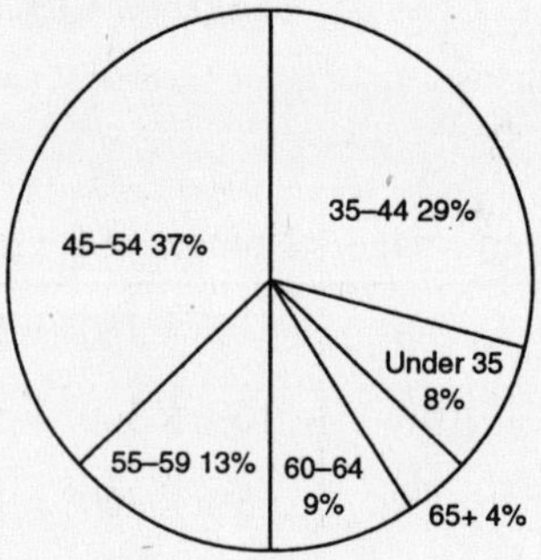

9. Percentage of Households with Telephone Gadgets

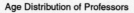

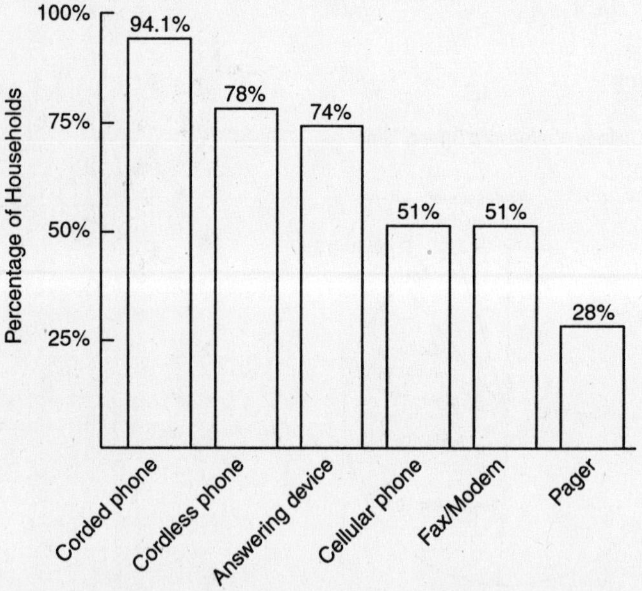

No. Since household can report having more than one telephone gadget, the percentages will not add to 100%.

10. The following Pareto Chart shows the percentage of drivers for each stated complaint.

Driving Problems—Pareto Chart

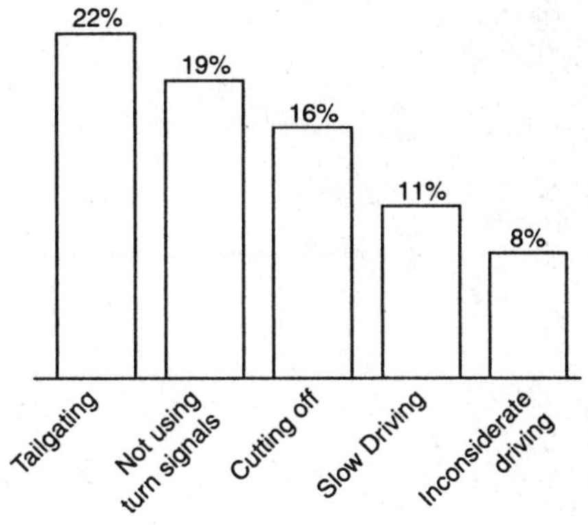

By subtraction, 100% – 22% – 19% – 16% – 11% – 8% = 24% of the respondents citied other bad habits.

Bad Habit	Percentage	Frequency
Tailgating	22%	22% × 500 = 110
Not using turn signals	19%	19% × 500 = 95
Cutting off other drivers	16%	16% × 500 = 80
Driving too slowly	11%	11% × 500 = 55
Being inconsiderate	8%	8% × 500 = 40
Other	24%	24% × 500 = 120
Total	100%	500

As reported, the percentages add to 76%, not the 100% needed for a circle graph. However, if there was only one response per person, knowing that the company surveyed 500 drivers tells us that 120 drivers, or 24%, had other bad driving complaints. Using this fact, a circle graph could be used.

11. Elevation of Pyramid Lake Surface—Time Plot

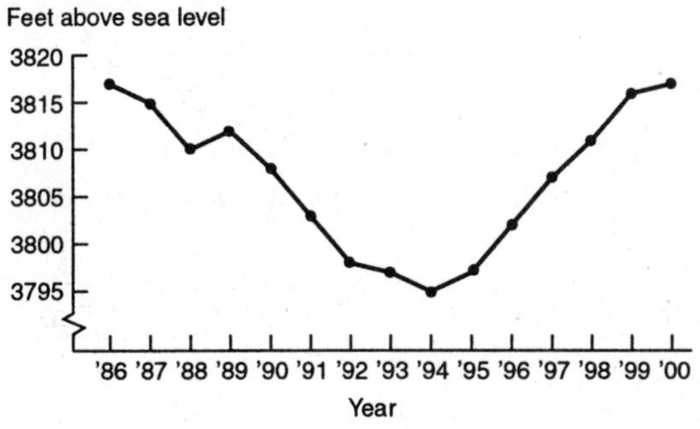

12.

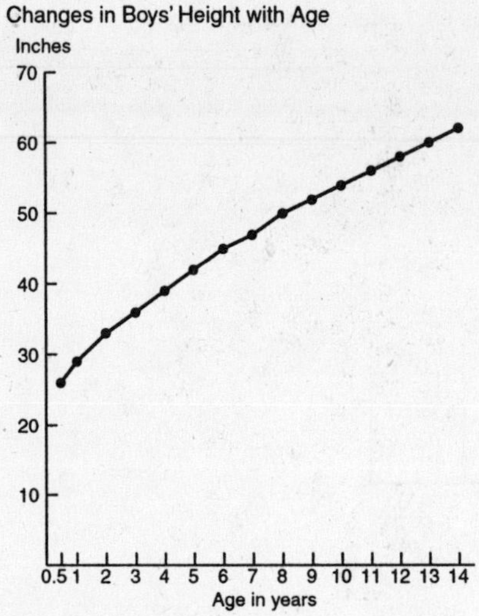

Section 2.2

1. **(a)** largest data value = 360

smallest data value = 236

number of classes specified = 5

class width $= \frac{360-236}{5} = 24.8$, increased to next whole number, 25

(b) The lower class limit of the first class in the smallest value, 236.

The lower class limit of the next class is the previous class's lower class limit plus the class width; for the second class, this is 236 + 25 = 261.

The upper class limit is one value less than lower class limit of the next class; for the first class, the upper class limit is 261 – 1 = 260.

The class boundaries are the halfway points between (i.e., the average of) the (adjacent) upper class limit of one class and the lower class limit of the next class. The lower class boundary of the first class is the lower class limit minus one-half unit. The upper class boundary for the last class is the upper class limit plus one-half unit. For the first class, the class boundaries are $236 - \frac{1}{2} = 235.5$ and $\frac{260+261}{2} = 260.5$. For the last class, the class boundaries are $\frac{335+336}{2} = 335.5$ and $360 + \frac{1}{2} = 360.5$.

The class mark or midpoint is the average of the class limits for that class. For the first class, the midpoint is $\frac{236+260}{2} = 248$.

The class frequency is the number of data values that belong to that class; call this value f.

The relative frequency of a class is the class frequency, f, divided by the total number of data values, i.e., the overall sample size, n.

For the first class, $f = 4$, $n = 57$, and the relative frequency is $f/n = \frac{4}{57} \approx 0.07$.

Class Limits	Boundaries	Midpoint	Frequency	Relative Frequency
236–260	235.5–260.5	248	4	0.07
261–285	260.5–285.5	273	9	0.16
286–310	285.5–310.5	298	25	0.44
311–335	310.5–335.5	323	16	0.28
336–360	335.5–360.5	348	3	0.05

(c) The histogram plots the class frequencies on the y-axis and the class boundaries on the x-axis. Since adjacent classes share boundary values, the bars touch each other. [Alternatively, the bars may be centered over the class marks (midpoints).

(d) The relative frequency histogram is exactly the same shape as the frequency histogram, but the vertical scale is relative frequency, f/n, instead of actual frequency, f.

The following figure shows the histogram and relative-frequency histogram for (c), and (d). Note that two vertical scales are shown.

Hours to Complete the Iditarod
Histogram and Relative-Frequency Histogram

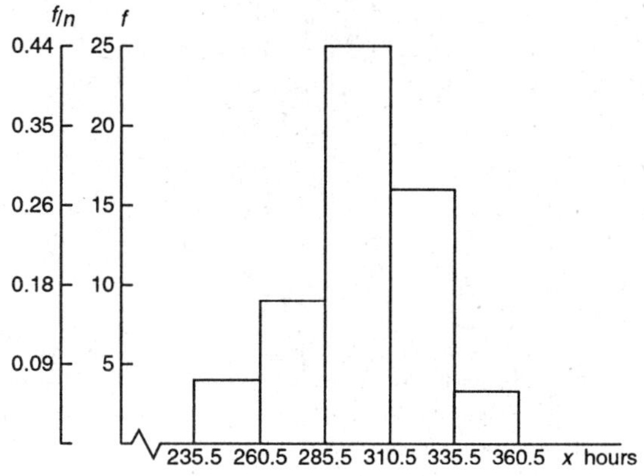

2. (a) largest data value = 65

smallest data value = 20

number of classes specified = 5

class width $= \dfrac{65-20}{5} = 9$, increased to next whole number, 10

(b) The lower class limit of the first class in the smallest value, 20.

The lower class limit of the next class is the previous class's lower class limit plus the class width; for the second class, this is 20 + 10 = 30.

The upper class limit is one value less than lower class limit of the next class; for the first class, the upper class limit is 30 – 1 = 29.

The class boundaries are the halfway points between (i.e., the average of) the (adjacent) upper class limit of one class and the lower class limit of the next class. The lower class boundary of the first class is the lower class limit minus one-half unit. The upper class boundary for the last class is the upper class limit plus one-half unit. For the first class, the class boundaries are $20-\frac{1}{2}=19.5$ and $\frac{29+30}{2}=29.5$. For the last class, the class boundaries are $\frac{59+60}{2}=59.5$ and $69+\frac{1}{2}=69.5$.

The class mark or midpoint is the average of the class limits for that class. For the first class, the midpoint is $\frac{20+29}{2}=24.5$.

The class frequency is the number of data values that belong to that class; call this value f.

The relative frequency of a class is the class frequency, f, divided by the total number of data values, i.e., the overall sample size, n.

For the first class, $f = 3$, $n = 35$, and the relative frequency is $f/n = 3/35 \approx 0.0857$.

Percent Difficult Ski Terrain

Class Limits	Class Boundaries	Midpoint	Frequency	Relative Frequency
20–29	19.5–29.5	24.5	3	0.0857
30–39	29.5–39.5	34.5	6	0.1714
40–49	39.5–49.5	44.5	13	0.3714
50–59	49.5–59.5	54.5	9	0.2571
60–69	59.5–69.5	64.5	4	0.1143

(c) The histogram plots the class frequencies on the y-axis and the class boundaries on the x-axis. Since adjacent classes share boundary values, the bars touch each other. [Alternatively, the bars may be centered over the class marks (midpoints).]

(d) The relative frequency histogram is exactly the same shape as the frequency histogram, but the vertical scale is relative frequency, *f/n*, instead of actual frequency, *f*.

The following figure shows the histogram and relative-frequency histogram for (c) and (d). Note that two vertical scales are shown.

Percent Difficult Ski Terrain
Histogram and Relative-Frequency Histogram

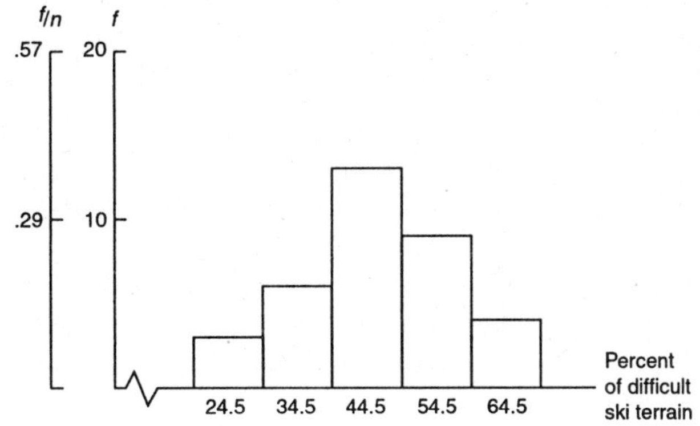

3. (a) largest data value = 53

smallest data value = 5

number of classes specified = 7

class width $= \dfrac{53-5}{7} \approx 6.86$, increased to next whole number, 7

(b) The lower class limit of the first class in the smallest value, 5.

The lower class limit of the next class is the previous class's lower class limit plus the class width; for the second class, this is 5 + 7 = 12.

The upper class limit is one value less than lower class limit of the next class; for the first class, the upper class limit is 12 – 1 = 11.

The class boundaries are the halfway points between (i.e., the average of) the (adjacent) upper class limit of one class and the lower class limit of the next class. The lower class boundary of the first class is the lower class limit minus one-half unit. The upper class boundary for the last class is the upper class limit plus one-half unit. For the first class, the class boundaries are $5-\dfrac{1}{2}=4.5$ and $\dfrac{11+12}{2}=11.5$. For the last class, the class boundaries are $\dfrac{46+47}{2}=46.5$ and $53+\dfrac{1}{2}=53.5$.

The class mark or midpoint is the average of the class limits for that class. For the first class, the midpoint is $\dfrac{5+11}{2}=8$.

The class frequency is the number of data values that belong to that class; call this value *f*.

The relative frequency of a class is the class frequency, *f*, divided by the total number of data values, i.e., the overall sample size, *n*.

For the first class, $f = 4$, $n = 50$, and the relative frequency is $f/n = 4/50 \approx 0.08$.

Class Limits	Class Boundaries	Midpoint	Frequency	Relative Frequency
5–11	4.5–11.5	8	4	0.08
12–18	11.5–18.5	15	7	0.14
19–25	18.5–25.5	22	12	0.24
26–32	25.5–32.5	29	12	0.24
33–39	32.5–39.5	36	12	0.24
40–46	39.5–46.5	43	2	0.04
47–53	46.5–53.5	50	1	0.02

(c) The histogram plots the class frequencies on the *y*-axis and the class boundaries on the *x*-axis. Since adjacent classes share boundary values, the bars touch each other. [Alternatively, the bars may be centered over the class marks (midpoints).]

(d) The relative frequency histogram is exactly the same shape as the frequency histogram, but the vertical scale is relative frequency, *f/n*, instead of actual frequency, *f*.

The following figure shows the histogram and relative-frequency histogram for (c) and (d). Note that two vertical scales are shown.

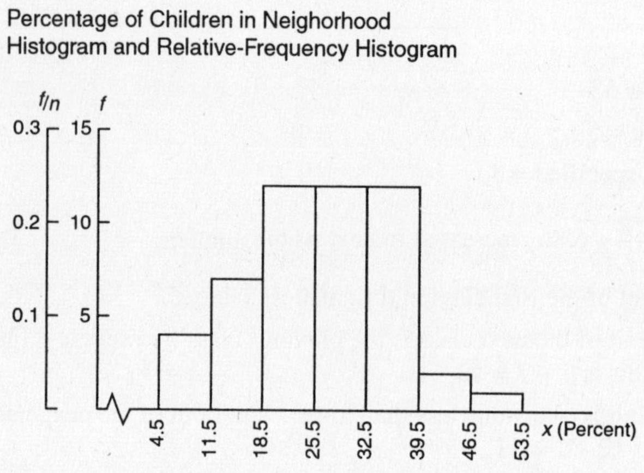

4. (a) largest data value = 75

smallest data value = 5

number of classes specified = 5

class width $= \dfrac{75-5}{5} = 14$, increased to next whole number, 15

(b) The lower class limit of the first class in the smallest value, 5.

The lower class limit of the next class is the previous class's lower class limit plus the class width; for the second class, this is 5 + 15 = 20.

The upper class limit is one value less than lower class limit of the next class; for the first class, the upper class limit is 20 – 1 = 19.

The class boundaries are the halfway points between (i.e., the average of) the (adjacent) upper class limit of one class and the lower class limit of the next class. The lower class boundary of the first class is the lower class limit minus one-half unit. The upper class boundary for the last class is the upper class limit plus one-half unit. For the first class, the class boundaries are $5-\frac{1}{2}=4.5$ and $\frac{19+20}{2}=19.5$. For the last class, the class boundaries are $\frac{64+65}{2}=64.5$ and $79+\frac{1}{2}=79.5$.

The class mark or midpoint is the average of the class limits for that class. For the first class, the midpoint is $\frac{5+19}{2}=12$.

The class frequency is the number of data values that belong to that class; call this value f.

The relative frequency of a class is the class frequency, f, divided by the total number of data values, i.e., the overall sample size, n.

For the first class, $f = 21$, $n = 63$, and the relative frequency is $f/n = 21/63 \approx 0.3333$.

Fast-Food Franchise Fees (in thousands)

Class Limits	Class Boundaries	Midpoint	Frequency	Relative Frequency
5–19	4.5–19.5	12	21	0.3333
20–34	19.5–34.5	27	35	0.5556
35–49	34.5–49.5	42	5	0.0794
50–64	49.5–64.5	57	1	0.0159
65–79	64.5–79.5	72	1	0.0159

(c) The histogram plots the class frequencies on the y-axis and the class boundaries on the x-axis. Since adjacent classes share boundary values, the bars touch each other. [Alternatively, the bars may be centered over the class marks (midpoints).]

(d) The relative frequency histogram is exactly the same shape as the frequency histogram, but the vertical scale is relative frequency, f/n, instead of actual frequency, f.

The following figure shows the histogram and relative-frequency histogram for (c) and (d). Note that two vertical scales are shown.

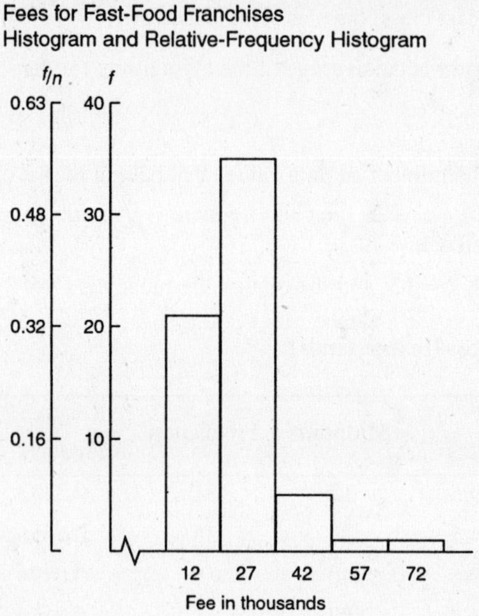

5. (a) largest data value = 102

smallest data value = 18

number of classes specified = 5

class width $= \dfrac{102-18}{5} = 16.8$, increased to next whole number, 17

(b) The lower class limit of the first class in the smallest value, 18.

The lower class limit of the next class is the previous class's lower class limit plus the class width; for the second class, this is 18 + 17 = 35.

The upper class limit is one value less than lower class limit of the next class; for the first class, the upper class limit is 35 – 1 = 34.

The class boundaries are the halfway points between (i.e., the average of) the (adjacent) upper class limit of one class and the lower class limit of the next class. The lower class boundary of the first class is the lower class limit minus one-half unit. The upper class boundary for the last class is the upper class limit plus one-half unit. For the first class, the class boundaries are $18-\dfrac{1}{2}=17.5$ and $\dfrac{34+35}{2}=34.5$. For the last class, the class boundaries are $\dfrac{85+86}{2}=85.5$ and $102+\dfrac{1}{2}=102.5$.

The class mark or midpoint is the average of the class limits for that class. For the first class, the midpoint is $\dfrac{18+34}{2}=26$.

The class frequency is the number of data values that belong to that class; call this value f.

The relative frequency of a class is the class frequency, f, divided by the total number of data values, i.e., the overall sample size, n.

For the first class, $f = 1$, $n = 35$, and the relative frequency is $f/n = 1/35 \approx 0.03$.

Number of Room Calls per Night

Class Limits	Class Boundaries	Midpoint	Frequency	Relative Frequency
18–34	17.5–34.5	26	1	0.03
35–51	34.5–51.5	43	2	0.06
52–68	51.5–68.5	60	5	0.14
69–85	68.5–85.5	77	15	0.43
86–102	85.5–102.5	94	12	0.34

(c) The histogram plots the class frequencies on the y-axis and the class boundaries on the x-axis. Since adjacent classes share boundary values, the bars touch each other. [Alternatively, the bars may be centered over the class marks (midpoints).]

(d) The relative frequency histogram is exactly the same shape as the frequency histogram, but the vertical scale is relative frequency, f/n, instead of actual frequency, f.

The following figure shows the histogram and relative-frequency histogram for (c) and (d). Note that two vertical scales are shown.

Number of Room Calls per Night
Histogram and Relative-Frequency Histogram

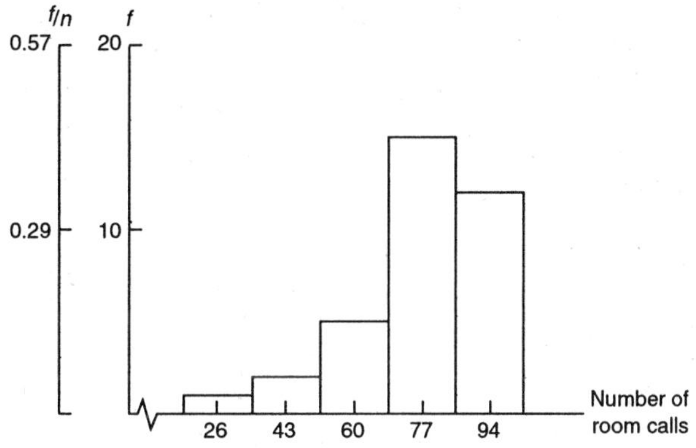

6. (a) largest data value = 43

smallest data value = 0

number of classes specified = 8

$$\text{class width } = \frac{43-0}{8} = 5.375, \text{ increased to next whole number, 6}$$

(b) The lower class limit of the first class in the smallest value, 0.

The lower class limit of the next class is the previous class's lower class limit plus the class width; for the second class, this is 0 + 6 = 6.

The upper class limit is one value less than lower class limit of the next class; for the first class, the upper class limit is 6 – 1 = 5.

The class boundaries are the halfway points between (i.e., the average of) the (adjacent) upper class limit of one class and the lower class limit of the next class. The lower class boundary of the first class is the lower class limit minus one-half unit. The upper class boundary for the last class is the upper class limit plus one-half unit. For the first class, the class boundaries are $0-\frac{1}{2}=-0.5$ and $\frac{5+6}{2}=5.5$.

For the last class, the class boundaries are $\frac{41+42}{2}=41.5$ and $47+\frac{1}{2}=47.5$.

The class mark or midpoint is the average of the class limits for that class. For the first class, the midpoint is $\frac{0+5}{2}=2.5$.

The class frequency is the number of data values that belong to that class; call this value f.

The relative frequency of a class is the class frequency, f, divided by the total number of data values, i.e., the overall sample size, n.

For the first class, $f = 13$, $n = 55$, and the relative frequency is $f/n = 13/55 \approx 0.24$.

Words of Three Syllables or More

Class Limits	Class Boundaries	Midpoint	Frequency	Relative Frequency
0–5	0.5–5.5	2.5	13	0.24
6–11	5.5–11.5	8.5	15	0.27
12–17	11.5–17.5	14.5	11	0.20
18–23	17.5–23.5	20.5	3	0.05
24–29	23.5–29.5	26.5	6	0.11
30–35	29.5–35.5	32.5	4	0.07
36–41	35.5–41.5	38.5	2	0.04
42–47	41.5–47.5	44.5	1	0.02

(c) The histogram plots the class frequencies on the y-axis and the class boundaries on the x-axis. Since adjacent classes share boundary values, the bars touch each other. [Alternatively, the bars may be centered over the class marks (midpoints).]

(d) The relative frequency histogram is exactly the same shape as the frequency histogram, but the vertical scale is relative frequency, *f/n*, instead of actual frequency, *f*.

The following figure shows the histogram and relative-frequency histogram for (c) and (d). Note that two vertical scales are shown.

Words of Three Syllables or More
Histogram and Relative-Frequency Histogram

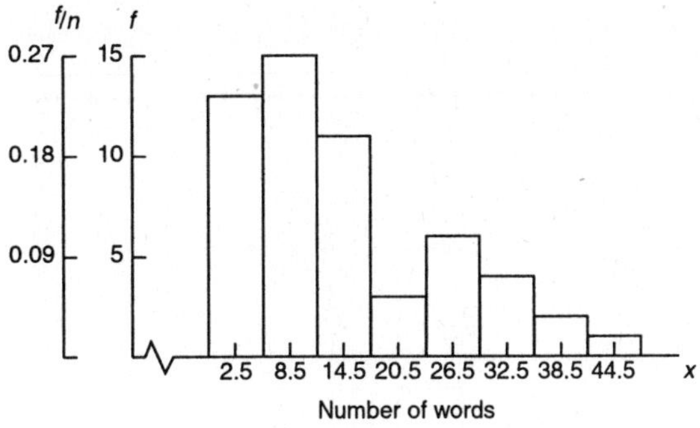

7. (a)

	Largest value	Smallest value	Class width
Food Companies	11	−3	$\frac{11-(-3)}{5} = 2.8$; use 3
Electronic Companies	16	−6	$\frac{16-(-6)}{5} = 4.4$; use 5

Profit as Percent of Sales—Food Companies

Class	Frequency	Midpoint
−3 to −1	2	−2
0–2	16	1
3–5	10	4
6–8	9	7
9–11	2	10

Profit as Percent of Sales—Electronic Companies

Class	Frequency	Midpoint
−6 to −2	3	−4
−1 to 3	13	1
4–8	20	6
9–13	7	11
14–18	1	16

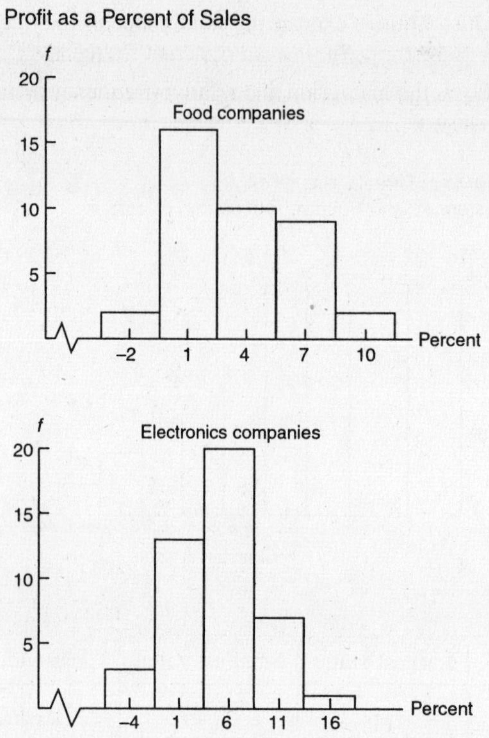

(b) Because the classes and class widths are different for the two company types, it is difficult to compare profits as a percentage of sales. We can notice that for the electronic companies the 16 profits as a percentage of sales extends as high as 18, while for the food companies the highest profit as a percentage of sales is 11. On the other hand, some of the electronic companies also have greater losses than the food companies. Had we made the class limits the same for both company types and overlaid the histograms, it would be easier to compare the data.

8. (a)

	Largest value	Smallest value	Class width
Miami Dolphins	295	175	$\frac{295-175}{6} = 20$; use 21
San Diego Charges	310	119	$\frac{310-119}{6} \approx 31.8$; use 32

Weights of Football Players:
Miami Dolphins

Class	Midpoint	Frequency
175–195	185	13
196–216	206	7
217–237	227	19
238–258	248	8
259–279	269	11
280–300	290	12

San Diego Chargers

Class	Midpoint	Frequency
119–150	134.5	1
151–182	166.5	4
183–214	198.5	27
215–246	230.5	15
247–278	262.5	14
279–310	294.5	11

Weights of Football Players—Miami Dolphins

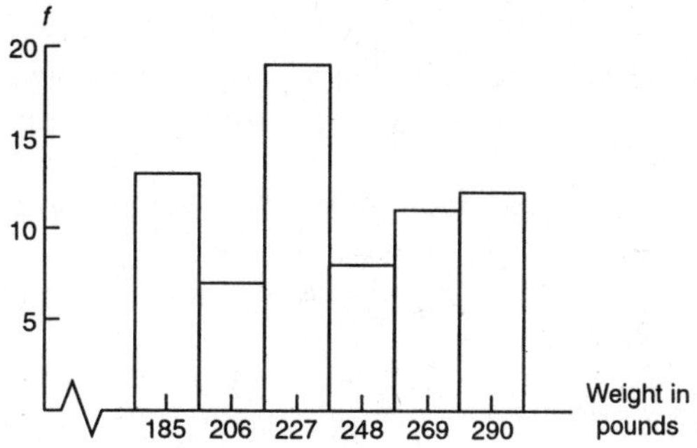

Weights of Football Players—San Diego Chargers

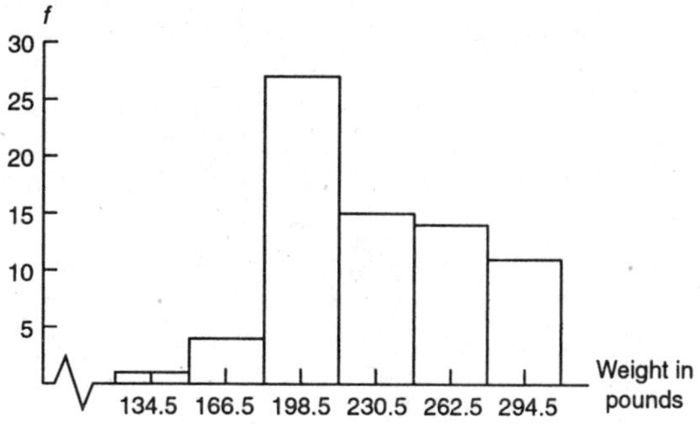

(b) Because the class widths are different, it is difficult to compare the histograms. However, San Diego has 4 players who are smaller than the smallest Miami player, and 4 players who are larger than the largest player.

It would be easier to compare the teams' weights if the histograms had common classes and were overlaid.

9. (a) Uniform is rectangular, symmetric looks like mirror images on each side of the middle, bimodal has two modes (peaks), and skewed distributions have long tails on one side, and are skewed in the direction of the tail ("skew, few"). (Note that uniform distributions are also symmetric, but "uniform" is more descriptive.)

(a) skewed left; (b) uniform, (c) symmetric, (d) bimodal, (e) skewed right.

(b) Answers vary. Students would probably like (a) since there are many high scores and few low scores. Students would probably dislike (e) since there are few high scores but lots of lows scores. (b) is designed to give approximately the same number of As, Bs, etc. (d) has more Bs and Ds, say. (c) is the way many tests are designed: As and Fs for the exceptionally high and low scores with most students receiving Cs.

10. (a) Uniform is rectangular, symmetric looks like mirror images on each side of the middle, bimodal has two modes (peaks), and skewed distributions have long tails on one side, and are skewed in the direction of the tail ("skew, few"). (Note that uniform distributions are also symmetric, but "uniform" is more descriptive.)

(a) uniform, (b) skewed right, (c) bimodal, (d) bimodal, (e) symmetric. [Note that (c) has a major and a minor mode. "Tails" in a distribution's shape "tail off," i.e., get thinner, and do not have "bumps" in them as (c) does.]

(b) Answers vary. Ads should target the largest number of potential buyers, so ads should be aimed at the income levels with the greatest concentration (frequency) of households.

(c) Answers vary. Since warranty/registration cards are returned voluntarily, the income data are most likely not representative of the buying public in general, and probably are not even representative of those buying the specific product. Also, people tend to inflate their income levels on most forms, except those sent to the IRS.

11. (a) $2.71 \times 100 = 271$, $1.62 \times 100 = 162, \ldots, 0.70 \times 100 = 70$.

(b) largest value = 282, smallest value = 46

$$\text{class width} = \frac{282-46}{6} \approx 39.3\text{; use } 40$$

Class Limits	Class Boundaries	Midpoint	Frequency
46–85	45.5–85.5	65.5	4
86–125	85.5–125.5	105.5	5
126–165	125.5–165.5	145.5	10
166–205	165.5–205.5	185.5	5
206–245	205.5–245.5	225.5	5
246–285	245.5–285.5	265.5	3

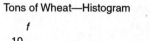

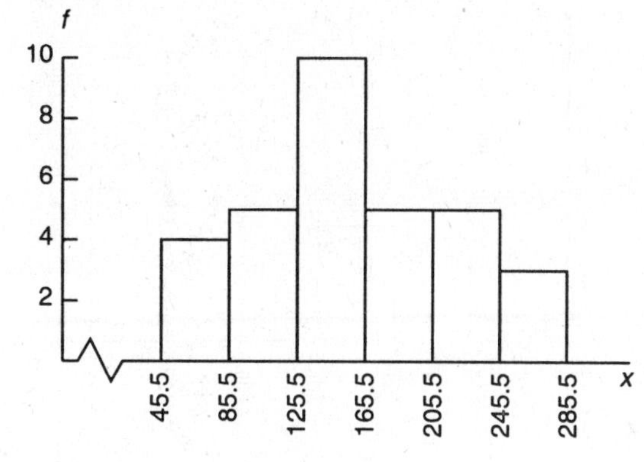

(c) class width is $\frac{40}{100} = 0.40$

Class Limits	Class Boundaries	Midpoint	Frequency
0.46–0.85	0.455–0.855	0. 655	4
0.86–1.25	0.855–1.255	1.055	5
1.26–1.65	1.255–1.655	1.455	10
1.66–2.05	1.655–2.055	1.855	5
2.06–2.45	2.055–2.455	2.255	5
2.46–2.85	2.455–2.855	2.655	3

12. (a) $0.194 \times 1000 = 194, 0.258 \times 1000 = 258, \ldots, 0.200 \times 1000 = 200.$

(b) largest value = 317, smallest value = 107

class width $= \frac{317-107}{5} = 42$, use 43

Class Limits	Class Boundaries	Midpoint	Frequency
107–149	106.5–149.5	128	3
150–192	149.5–192.5	171	4
193–235	192.5–235.5	214	3
236–278	235.5–278.5	257	10
279–321	278.5–321.5	300	6

Baseball Batting Averages—Histogram

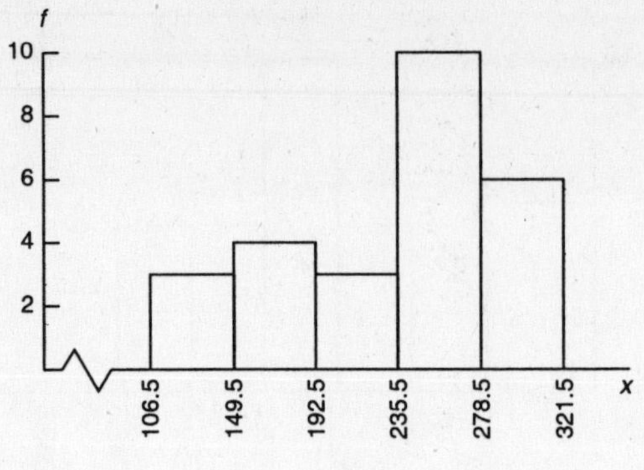

(c) class width = $\dfrac{43}{1000} = 0.043$

Class Limits	Class Boundaries	Midpoint	Frequency
0.107–0.149	0.1065–0.1495	0.128	3
0.150–0.192	0.1495–0.1925	0.171	4
0.193–0.235	0.1925–0.2355	0.214	3
0.236–0.278	0.2355–0.2785	0.257	10
0.279–0.321	0.2785–0.3215	0.300	6

13. (a) There is one dot below 600, so 1 state has 600 or fewer licensed drivers per 1000 residents.

(b) 5 values are close to 800; $\dfrac{5}{51} \approx 0.0980 \approx 9.8\%$

(c) 9 values below 650
37 values between 650 and 750
5 values above 750
From either the counts or the dotplot, the interval from 650 to 750 licensed drivers per 1000 residents has the most "states."

14. The dotplot shows some of the characteristics of the histogram such as more dot density from, say 280 to 340, corresponding roughly to the histogram bars of heights 25 and 16.
However, they are somewhat difficult to compare since the dotplot can be thought of as a histogram with one value, the class mark, i.e., the data value, per class.
Because the definitions of the classes and, therefore, the class widths, differ, it is difficult to compare the two figures.

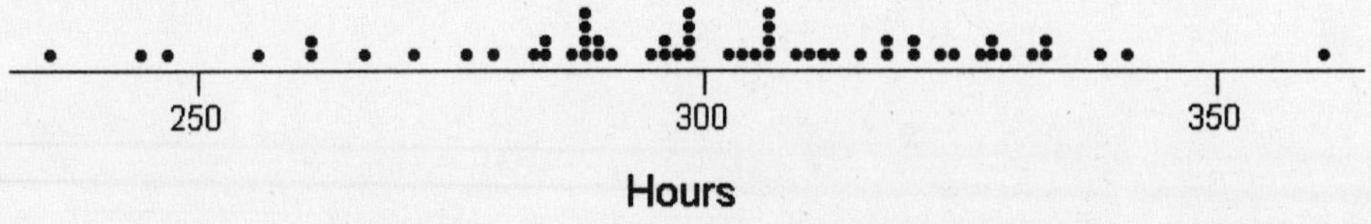

15. The dotplot shows some of the characteristics of the histogram, such as the concentration of most of the data from, say, 20 to 40; this corresponds roughly to the 3 histogram bars of height 12. There are more data (dots) below 20 than above 40, which corresponds to the histogram bars of heights 4 and 7, and the bars of heights 2 and 1, respectively.

However, they are somewhat difficult to compare since the dotplot can be thought of as a histogram with one value, the class mark, i.e., the data value, per class.

Because the definitions of the classes and, therefore, the class widths, differ, it is difficult to compare the two figures.

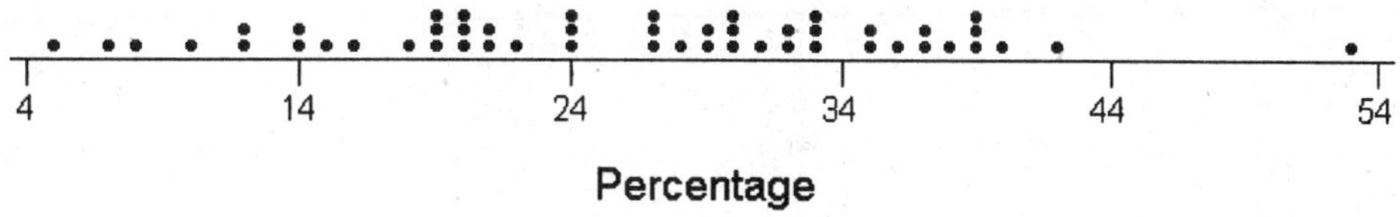

Section 2.3

1. (a) The smallest value is 47 and the largest is 97, so we need stems 4, 5, 6, 7, 8, and 9. Use the tens digit as the stem and the ones digit as the leaf.

Longevity of Cowboys

4	7 = 47 years
4	7
5	2 7 8 8
6	1 6 6 8 8
7	0 2 2 3 3 5 6 7
8	4 4 4 5 6 6 7 9
9	0 1 1 2 3 7

(b) Yes, certainly these cowboys lived long lives, as evidenced by the high frequency of leaves for stems 7, 8, and 9 (i.e., 70-, 80-, and 90-year olds).

2. The largest value is 91 (percent of wetlands lost) and the smallest value is 9 (percent), which is coded as 09. We need stems 0 to 9. Use the tens digit as the stem and the ones digit as the leaf. The percentages are concentrated from 20 to 50 percent. The distribution is asymmetrical but not skewed because of the "bump" in the 80s. If we smoothed the shape, we might consider this bimodal. There is a gap showing none of the lower 48 states has lost from 10 to 19% of its wetlands.

Percent of Wetlands Lost

4	0 = 40%
0	9
1	
2	0 3 4 7 7 8
3	0 1 3 5 5 5 6 7 8 8 9
4	2 2 6 6 6 8 9 9
5	0 0 0 2 2 4 6 6 9 9
6	0 7
7	2 3 4
8	1 5 7 7 9
9	0 1

3. The longest average length of stay is 11.1 days in North Dakota and the shortest is 5.2 days in Utah. We need stems from 5 to 11. Use the digit(s) to the left of the decimal point as the stem, and the digit to the right as the leaf.

Average Length of Hospital Stay

5	2 = 5.2 days
5	2 3 5 5 6 7
6	0 2 4 6 6 7 7 8 8 8 8 9 9
7	0 0 0 0 0 0 1 1 1 2 2 2 3 3 3 3 4 4 5 5 6 6 8
8	4 5 7
9	4 6 9
10	0 3
11	1

The distribution is skewed right.

4. Number of Hospitals per State

0	8 = 8 hospitals		
0	8	15	
1	1 2 5 6 9	16	2
2	1 7 7	17	5
3	5 7 8	18	
4	1 2 7	19	3
5	1 2 3 9	20	9
6	1 6 8	21	
7	1	22	7
8	8	23	1 6
9	0 2 6 8		
10	1 2 7	42	1
11	3 3 7 9	43	
12	2 3 9	44	0
13	3 3 6		
14	8		

Texas and California have the highest number of hospitals, 421 and 440, respectively. Both states have large populations and large areas. The four largest states by area are Alaska, Texas, California, and Montana; however, both Alaska and Montana have small populations, but the population tends to cluster at their largest cities, thus reducing the number of hospitals needed.

5. (a) The longest time during 1961–1980 is 23 minutes (i.e., 2:23) and the shortest time is 9 minutes (2:09). We need stems 0, 1, and 2, which we'll write as 0*, 0•, 1*, 1•, 2*, and 2•. (We can eliminate 0* since no time was 2:04 or less and 2• because no winning time was 2:25 or more. We'll use the tens digit as the stem and the ones digit as the leaf, placing leaves 0, 1, 2, 3, and 4 on the "* stem" and leaves 5, 6, 7, 8, and 9 on the "• stem."

Minutes Beyond 2 Hours (1961-1980)

0	9 = 9 minutes past 2 hours
0•	9 9
1*	0 0 2 3 3
1•	5 5 6 6 7 8 8 9
2*	0 2 3 3

(b) The longest time during the period 1981–2000 was 14 (2:14), and the shortest was 7 (2:07), so we'll need stems 0• and 1* only.

Minutes Beyond 2 Hours (1981-2000)

0	7 = 7 minutes past 2 hours
0•	7 7 7 8 8 8 8 9 9 9 9 9 9 9 9
1*	0 0 1 1 4

(c) In more recent times, the winning times have been closer to 2 hours, with all 20 times between 7 and 14 minutes over two hours. In the earlier period, more than half the times (12 or 20) were more than 2 hours and 14 minutes.

6. (a) The largest (worst) score in the first round was 75; the smallest (best) score was 65. We need stems 6• and both 7* and 7•; leaves 0 to 4 go on the "* stem" and leaves 5–9 belong on the "• stem."

First Round Scores

6	5 = score of 65
6•	5 6 7 7
7*	0 1 1 1 1 1 1 1 1 1 1 2 2 2 3 3 3 3 4 4 4
7•	5 5 5 5 5 5 5

(b) the largest score in the fourth round was 74 and the smallest was 68. Here we need stems 6• and 7*, we don't need 7• because no scores were over 74.

Fourth Round Scores

6	8 = score of 68
6•	8 9 9 9 9 9
7*	0 0 0 0 1 1 1 1 1 1 1 1 2 2 2 2 2 2 2 3 3 3 3 3 4 4 4

(c) Scores are lower in the fourth round. In the first round both the low and high scores were more extreme than in the fourth round.

7. The largest value in the data is 29.8 mg. Of tar per cigarette smoked, and the smallest value is 1.0. We will need stems from 1 to 29, and we will use the numbers to the right of the decimal point as the leaves.

Milligrams of Tar per Cigarette

1	0 = 1.0 mg tar
1	0
2	
3	
4	1 5
5	
6	
7	3 8
8	0 6 8
9	0
10	
11	4
12	0 4 8
13	7
14	1 5 9
15	0 1 2 8
16	0 6
17	0
29	8

8. The largest value in the data set is 23.5 mg Carbon monoxide per cigarette smoked, and the smallest is 1.5. We need stems from 1 to 23, and we'll use the numbers to the right of the decimal point as leaves.

Milligrams of Carbon Monoxide

1	5 = 1.5 mg CO
1	5
2	
3	
4	9
5	4
6	
7	
8	5
9	0 5
10	0 2 2 6
11	
12	3 6
13	0 6 9
14	4 9
15	0 4 9
16	3 6
17	5
18	5
23	5

9. The largest value in the data set is 2.03 mg nicotine per cigarette smoked. The smallest value is 0.13. We will need stems 0*, $0^{\bullet}$, 1*, $1^{\bullet}$, and 2*. Leaves 0 to 4 belong on the * stems and leaves 5 to 9 belong on the $^{\bullet}$ stems. We will use the number to the left of the decimal point as the stem and the first number to the right of the decimal point as the leaf. The number 2 places to the right of the decimal point (the hundredths digit) will be truncated (chopped off; not rounded off).

Milligrams of Nicotine per Cigarette

0	1 = 0.1 milligram
0*	1 4 4
$0^{\bullet}$	5 6 6 6 7 7 7 8 8 9 9 9
1*	0 0 0 0 0 0 0 1 2
$1^{\bullet}$	
2*	0

10. **(a)** For Site I, read the values in Figure 2-27 from the center (stem) to the left to find the least depth is 25 cm and the greatest depth is 110 cm. For Site II, read the values from the center (stem) to the right to find the least depth is 20 cm and the greatest depth is 125 cm.

(b) The Site I depth distribution is, smoothed out, fairly symmetrical around approximately 70 cm. Site II, however, is fairly uniform in shape except that it has a huge gap with no artifacts from about 70 to 100 cm.

(c) It would appear that Site II was probably unoccupied during the time period associated with 70 cm to 100 cm.

11. **(a)** Average salaries in California range from $49,000 to $126,000. Salaries in New York range from $45,000 to $120,000.

(b) New York has a greater number of average salaries in the $60,000 than California, but California has more average salaries than New York in the $70,000 range.

(c) The California data appear to be similar in shape to the New York data, but California's distribution has been shifted up approximately $10,000. It is also heavier in the upper tail and shows no gap in average salaries, unlike New York which has no salaries in the $110,000 range. California has higher average salaries.

Chapter 2 Review

1. Figure 2-1 (a) (in the text) is essentially a bar graph with a "horizontal" axis showing years and a "vertical" axis showing miles per gallon. However, in depicting the data as a highway and showing it in perspective, the ability to correctly compare bar heights visually has been lost. For example, determining what would appear to be the bar heights by measuring from the white line on the road to the edge of the road along a line drawn from the year to its mpg value, we get the bar height for 1983 to be approximately 7/8 inch and the bar height for 1985 to be approximately 1 3/8 inches (i.e., 11/8 inches). Taking the ratio of the given bar heights, we see that the bar for 1985 should be $\frac{27.5}{26} \approx 1.06$ times the length of the 1983 bar.

However, the measurements show a ratio of $\frac{\frac{11}{8}}{\frac{7}{8}} = \frac{11}{7} \approx 1.60$, i.e., the 1985 bar is (visually) 1.6 times the length of the 1983 bar. Also, the years are evenly spaced numerically, but the figure shows the more recent years to be more widely spaced due to the use of perspective.

Figure 2-1(b) is a time plot, showing the years on the x-axis and miles per gallon on the y-axis. Everything is to scale and not distorted visually by the use of perspective. It is easy to see the mpg standards for each year, and you can also see how fuel economy standards for new cars have changed over the eight years shown (i.e., a steep increase in the early years and a leveling off in the later years).

2. (a) By reading the y-coordinate of the dot associated with the year, we estimate the 1980 prison population at approximately 140 prisoners per 100,000, and the 1997 population at approximately 440 prisoners per 100,000 people.

(b) The number of inmates per 100,000 increased.

(c) The population 266,574,000 is 2,665.74 × 100,000, and 444 per 100,000 is $\frac{444}{100,000}$.

So $\frac{444}{100,000} \times (2{,}665.74 \times 100{,}000) \approx 1{,}183{,}589$ prisoners.

The projected 2020 population is 323,724,000, or 3,237.24 × 100,000.

So $\frac{444}{100,000} \times (3{,}237.24 \times 100{,}000) \approx 1{,}437{,}335$ prisoners.

3.

Most Difficult Task	Percentage	Degrees
IRS jargon	43%	$0.43 \times 360° \approx 155°$
Deductions	28%	$0.28 \times 360° \approx 101°$
Right form	10%	$0.10 \times 360° = 36°$
Calculations	8%	$0.08 \times 360° \approx 29°$
Don't know	10%	$0.10 \times 360° = 36°$

Note: Degrees do not total 360° due to rounding.

Problems with Tax Returns

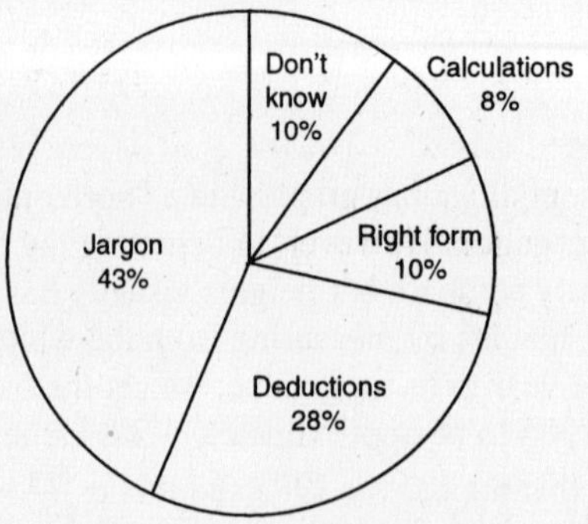

4. (a) Since the ages are two digit numbers, use the tens digit as the stem and the ones digit as the leaf.

Age of DUI Arrests

1	6 = 16 years
1	6 8
2	0 1 1 2 2 2 3 4 4 5 6 6 6 7 7 7 9
3	0 0 1 1 2 3 4 4 5 5 6 7 8 9
4	0 0 1 3 5 6 7 7 9 9
5	1 3 5 6 8
6	3 4

(b) The largest age is 64 and the smallest is 16, so the class with for 7 classes is $\frac{64-16}{7} \approx 6.86$; use 7. The lower class limit for the first class is 16; the lower class limit for the second class is 16 + 7 = 23. The total number of data points is 50, so calculate the relative frequency by dividing the class frequency by 50.

Age Distribution of DUI Arrests

Class Limits	Class Boundaries	Midpoint	Frequency	Relative Frequency	Cumulative Frequency
16–22	15.5–22.5	19	8	0.16	8
23–29	22.5–29.5	26	11	0.22	19
30–36	29.5–36.5	33	11	0.22	30
37–43	36.5–43.5	40	7	0.14	37
44–50	43.5–50.5	47	6	0.12	43
51–57	50.5–57.5	54	4	0.08	47
58–64	57.5–64.5	61	3	0.06	50

The class boundaries are the average of the upper class limit of the next class. The midpoint is the average of the class limits for that class.

(c) The class boundaries are shown in (b).

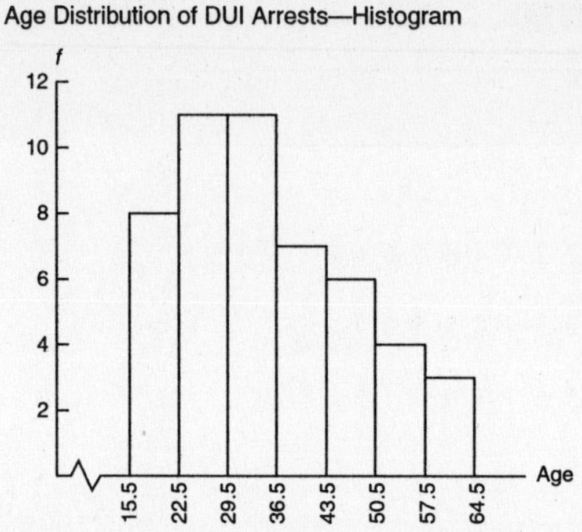

5. (a) The largest value is 96 mg of glucose per 100 ml of blood, and the smallest value is 59. For 7 classes we need a class width of $\frac{96-59}{7} \approx 5.3$; use 6. The lower class limit of the first class is 59, and the lower class limit of the second class is $59 + 6 = 65$.

The class boundaries are the average of the upper class limit of one class and the lower class limit of the next higher class. The midpoint is the average of the class limits for that class. There are 53 data values total so the relative frequency is the class frequency divided by 53.

Class Limits	Class Boundaries	Midpoint	Frequency	Relative Frequency
59–64	58.5–64.5	61.5	1	0.02
65–70	64.5–70.5	67.5	7	0.13
71–76	70.5–76.5	73.5	6	0.11
77–82	76.5–82.5	79.5	13	0.25
83–88	82.5–88.5	85.5	18	0.34
89–94	88.5–94.5	91.5	7	0.13
95–100	94.5–100.5	97.5	1	0.02

(b) The histogram shows the bars centered over the midpoints of each class.

(c) The frequency histogram and the relative frequency histogram are the same except in the latter, the vertical scale is relative frequency, not frequency.

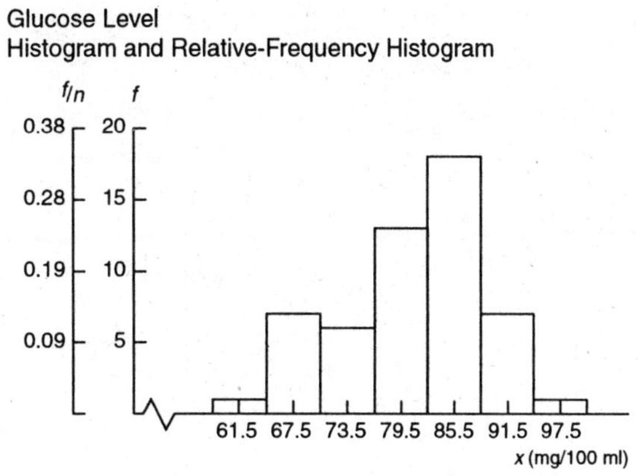

6. (a) A pareto chart is similar to a bar chart, except the bars are in decreasing order by frequency.

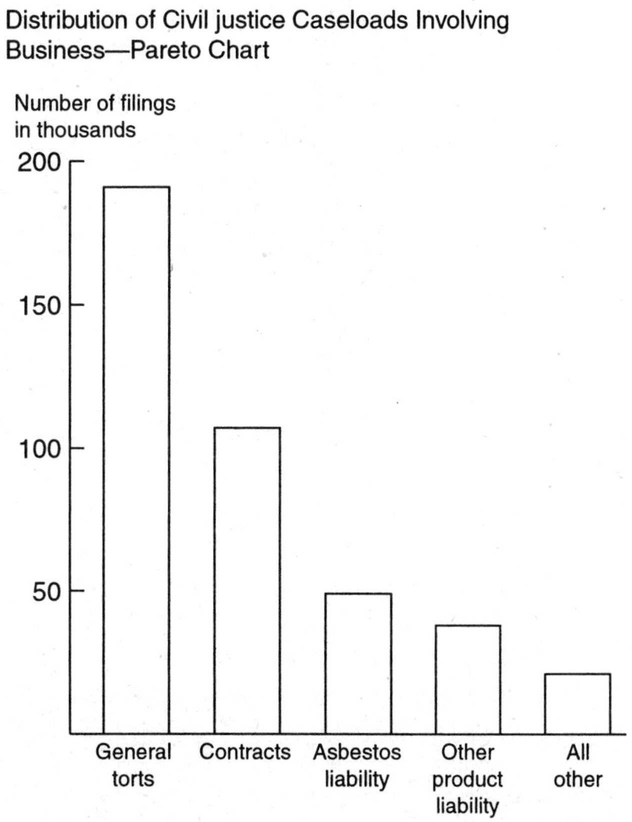

The general torts (personal injury) lawsuits occur with the greatest frequency.

(b) The total number of filings shown is 406 (thousand).

Case Type	Percentage	Degrees
Contracts	107/406 ≈ 26%	0.26 × 360° ≈ 94°
General torts	191/406 ≈ 47%	0.47 × 360° ≈ 169°
Asbestos liability	49/406 ≈ 12%	0.12 × 360° ≈ 43°
Other product liability	38/406 ≈ 9%	0.09 × 360° ≈ 32°
All other	21/406 ≈ 5%	0.05 × 360° = 18°

Note: Percentages do not add to 100% due to rounding. Similarly, the degrees do not add to 360° due to rounding.

Distribution of Civil justice Caseloads Involving Business—Pie Chart

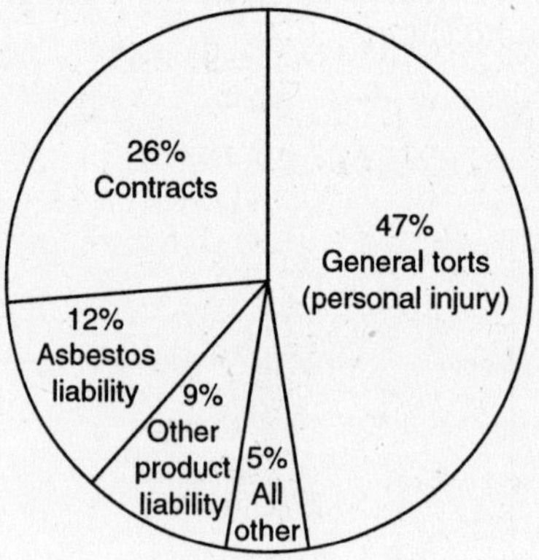

7. (a) To determine the decade which contained the most samples, count <u>both</u> rows (if shown) of leaves; recall leaves 0–4 belong on the first line and 5–9 belong on the second line when two lines per stem are used. The greatest number of leaves is found on stem 124, i.e., the 1240s (the 40s decade in the 1200s), with 40 samples.

(b) The number of samples with tree ring dates 1200 A.D. to 1239 A.D. is 28 + 3 + 19 + 25 = 75.

(c) The dates of the longest interval with no sample values are 1204 through 1211 A.D. This might mean that for these eight years, the pueblo was unoccupied (thus no new or repaired structures) or that the population remained stable (no new structures needed) or that, say, weather conditions were favorable these years so existing structures didn't need repair. If relatively few new structures were built or repaired during this period, their tree rings might have been missed during sample selection.

8. (a) It has a long tail on the left, so it is skewed left.

(b) The class width is the difference between any two adjacent midpoints. Here, for example, the class width is 4 – 3.5 = 0.5 grade points. The average of any two adjacent midpoints is the boundary value between the two midpoints classes*. So, for midpoints 1 and 1.5, the boundary value is $1+\frac{1.5}{26}=1.25$.

The difference between any two adjacent boundary values is also the class width, so the other class boundary values within the histogram are 1.25 + 0.5 = 1.75, 1.75 + 0.5 = 2.25, 2.25 + 0.5 = 2.75, 2.75 + 0.5 = 3.25, and 3.25 + 0.5 = 3.75; 3.75 is the lower class boundary for the class, so its upper class boundary is 3.75 + 0.5 = 4.25. Similarly, the upper class boundary of the first class was 1.25, so its lower class boundary is 1.25 – 0.5 = 0.75. The class boundaries are, therefore, 0.75, 1.25, 1.75, 2.25, 2.75, 3.25, 3.75 and 4.25 (from left to right).

*Recall that the average of a and b is $\frac{a+b}{2}$ which is also the value halfway between a and b.

(c) The relative frequencies are f/n, so if we multiply this decimal value by 100, we have the relative frequency expressed as a percent. The relative frequencies, expressed as percents, are 1%, 1%, 2%, 8%, 17%, 27%, and 44%, from left to right. The GPA of 3.25 is a boundary value, so to find the percentage of college graduates who had high school GPAs less than 3.25 is the sum of the relative frequency percentages for bars at or below 3.25: 1% + 1% + 2% + 8% + 17% = 29%. A high school GPA of 3.75 is the next boundary value above 3.25, so if we take the percentage of students with GPAs less 3.25 (29%), and add the percentage of students with GPAs between 3.25 and 3.75 (27%), we find 29% + 27% = 56% of college graduates had high school GPAs of less than 3.75. (Recall that, technically, boundary values are not values the data can take on. They are values between the upper class limit of one class and the lower class limit of the next class, and the class limits specify the largest and smallest data values, respectively, that can be put in those classes. Traditionally, the boundary values are specified to one more decimal place than the data, and that is the case here: the data are reported to one decimal place, but the boundaries are reported to two decimal places.)

Class Midpoints	Class Boundaries	Relative Frequency	Relative Frequency
1	0.75–1.25	0.01	1%
1.5	1.25–1.75	0.01	1%
2	1.75–2.25	0.02	2%
2.5	2.25–2.75	0.08	8%
3	2.75–3.25	0.17	17%
3.5	3.25–3.75	0.27	27%
4	3.75–4.25	0.44	44%

Chapter 3 Averages and Variation

Section 3.1

1. Mean $= \bar{x} = \frac{\Sigma x}{n} = \frac{156+161+152+\cdots+157}{12}$

$= \frac{1876}{12}$

$= 156.33$

The mean is 156.33.

Organize the data from smallest to largest.

144	148	152	153	156	157
157	157	161	161	162	168

To find the median, add the two middle values and divide by 2 since there is an even number of values.

$$\text{Median} = \frac{157+157}{2} = 157$$

The median is 157.

The mode is 157 because it is the value that occurs most frequently.

A gardener in Colorado should look at seed and plant descriptions to determine if the plant can thrive and mature in the designated number of frost-free days. The mean, median, and mode are all close. About half the locations have 157 or fewer frost-free days.

2. Mean $= \bar{x} = \frac{\Sigma x}{n} = \frac{11+29+54+\cdots+46}{12}$

$= \frac{542}{12}$

$= 45.17$

The mean is 45.17.

Organize the data from smallest to largest.

11	29	41	46	46	46
47	49	54	54	59	60

To find the median, add the two middle values and divide by 2 since there is an even number of values.

$$\text{Median} = \frac{46+47}{2} = 46.5$$

The median is 46.5.

The mode is 46 because it is the value that occurs most frequently.

3. Mean $= \bar{x} = \frac{\Sigma x}{n} = \frac{146+152+168+\cdots+144}{14}$

$= \frac{2342}{14}$

$= 167.3$

The mean is 167.3°F.

Organize the data from smallest to largest.

144	146	152	152	165	168	168
174	178	178	178	179	180	180

To find the median, add the two middle values and divide by 2 since there is an even number of values.

$$\text{Median} = \frac{168+174}{2} = 171$$

The median is 171° F.

The mode is 178° F because it is the value that occurs most frequently.

4. (a) Mean $= \bar{x} = \frac{\Sigma x}{n} = \frac{2723}{20} = 136.15$

The mean is \$136.15.

The median is \$66.50.

The mode is \$60.

(b) 5% of 20 is 1. Eliminate one data value from the bottom and one from the top of the ordered data. In this case eliminate \$40 and \$500.

$$\text{Mean} = \bar{x} = \frac{\Sigma x}{n} = \frac{2183}{18} \approx 121.28$$

The 5% trimmed mean is \$121.28.

Yes, the trimmed mean more accurately reflects the general level of the daily rental cost, but is still higher than the median.

(c) Median. The low and high prices would be helpful also.

5. First organize the data from smallest to largest. Then compute the mean, median, and mode.

(a) Upper Canyon

1	1	1	2	3	3	3	3	4	6	9

Mean $= \bar{x} = \frac{\Sigma x}{n} = \frac{36}{11} \approx 3.27$

Median = 3 (middle value)

Mode = 3 (occurs most frequently)

(b) Lower Canyon

0	0	1	1	1	1	2	2	3	6	7	8	13	14

$$\text{Mean} = \bar{x} = \frac{\Sigma x}{n} = \frac{59}{14} \approx 4.21$$

$$\text{Median} = \frac{2+2}{2} = 2$$

Mode = 1 (occurs most frequently)

(c) The mean for the Lower Canyon is greater than that of the Upper Canyon. However, the median and mode for the Lower Canyon are less than those of the Upper Canyon.

(d) 5% of 14 is 0.7 which rounds to 1. So, eliminate one data value from the bottom of the list and one from the top. Then compute the mean of the remaining 12 values.

$$5\%\text{ trimmed mean} = \frac{\Sigma x}{n} = \frac{45}{12} = 3.75$$

Now this value is closer to the Upper Canyon mean.

6. (a) First arrange the data from smallest to largest. Then compute the mean, median, and mode.

$$\text{Mean} = \bar{x} = \frac{\Sigma x}{n} = \frac{1050}{40} \approx 26.3$$

The mean is 26.3 yr.

$$\text{Median} = \frac{25+26}{2} = 25.5$$

The median is 25.5 yr.

Mode = 25

The mode is 25 yr.

(b) The median may represent the age most accurately. The answers are very close.

7. (a) $\text{Mean} = \bar{x} = \frac{\Sigma x}{n} = \frac{93+80+15+\cdots+13}{12}$

$$= \frac{346}{12}$$

$$\approx 28.83$$

The mean is 28.83 thousand dollars.

(b) $\text{Median} = \frac{18+19}{2} = 18.5$

The median is 18.5 thousand dollars.

The median best describes the salary of the majority of employees, since the mean is influenced by the high salaries of the president and vice president.

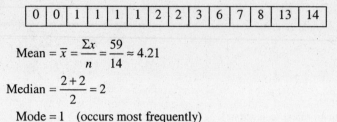

(c) Mean $= \bar{x} = \frac{\Sigma x}{n} = \frac{15+25+14+\cdots+13}{10}$

$= \frac{173}{10}$

$= 17.3$

The mean is 17.3 thousand dollars.

Median $= \frac{16+18}{2} = 17$

The median is 17 thousand dollars.

(d) Without the salaries for the two executives, the mean and the median are closer, and both reflect the salary of most of the other workers more accurately. The mean changed quite a bit, while the median did not, a difference that indicates that the mean is more sensitive to the absence or presence of extreme values.

8. (a) Since this data is at the ratio level of measurement, the mean, median, and mode (if it exists) can be used to summarize the data.

(b) Since this data is at the nominal level of measurement, only the mode (if it exists) can be used to summarize the data.

(c) Since this data is at the ratio level of measurement, the mean, median, and mode (if it exists) can be used to summarize the data.

9. (a) Since this data is at the nominal level of measurement, only the mode (if it exists) can be used to summarize the data.

(b) Since this data is at the ratio level of measurement, the mean, median, and mode (if it exists) can be used to summarize the data.

(c) The mode can be used (if it exists). If a 24-hour clock is used, then the data is at the ratio level of measurement, so the mean and median may be used as well.

10. Discussion question.

11. (a) If the largest data value is *replaced* by a larger value, the mean will increase because the sum of the data values will increase, but the number of them will remain the same. The median will not change. The same value will still be in the eighth position when the data are ordered.

(b) If the largest value is replaced by a value that is smaller (but still higher than the median), the mean will decrease because the sum of the data values will decrease. The median will not change. The same value will be in the eighth position in increasing order.

(c) If the largest value is replaced by a value that is smaller than the median, the mean will decrease because the sum of the data values will decrease. The median also will decrease because the former value in the eighth position will move to the ninth position in increasing order. The median will be the new value in the eighth position.

12. Answers will vary according to data collected.

Section 3.2

1. (a) Range = largest value − smallest value

$$= 58 - 4 = 54$$

The range is 54 deer/km^2.

$$\bar{x} = \frac{\Sigma x}{n} = \frac{251}{12} \approx 20.9$$

The sample mean is 20.9 deer/km^2.

$$s^2 = \frac{\Sigma(x-\bar{x})^2}{n-1} = \frac{2474.9}{11} \approx 225.0$$

The sample variance is 225.0.

$$s = \sqrt{s^2} = \sqrt{225.0} = 15.0$$

The sample standard deviation is 15.0 deer/km^2.

(b) $CV = \frac{s}{\bar{x}} \cdot 100 = \frac{15.0}{20.9} \cdot 100 \approx 71.8\%$

s is 71.8% of $\bar{x}$.

Since the standard deviation is about 71.8% of the mean, there is considerable variation in the distribution of deer from one part of the park to another.

2. (a) Range = largest value − smallest value

$$= 78.6 - 17.8 = 60.8$$

The range is 60.8%.

$$\bar{x} = \frac{\Sigma x}{n} = \frac{540.8}{10} \approx 54.1$$

The mean is 54.1%.

(b) $s^2 = \frac{\Sigma(x-\bar{x})^2}{n-1} = \frac{3400}{9} \approx 377.78$

The sample variance is 377.78.

$$s = \sqrt{s^2} = \sqrt{377.78} \approx 19.44$$

The standard deviation is 19.44%.

(c) $CV = \frac{s}{\bar{x}} \cdot 100 = \frac{19.44}{54.1} \cdot 100 \approx 35.9\%$

s is 35.9% of $\bar{x}$.

3. (a) Range $= 90.3 - 12.7 = 77.6$

The range is 77.6%.

$$\bar{x} = \frac{\Sigma x}{n} = \frac{556.7}{10} \approx 55.7$$

The mean is 55.7%.

(b) $$s^2 = \frac{\Sigma(x-\bar{x})^2}{n-1} = \frac{4833}{9} \approx 537$$

The sample variance is 537.

$$s = \sqrt{s^2} = \sqrt{537} \approx 23.17$$

The standard deviation is 23.17%.

(c) $$CV = \frac{s}{\bar{x}} \cdot 100 = \frac{23.17}{55.7} \cdot 100 \approx 41.6\%$$

s is 41.6% of $\bar{x}$.

This *CV* is larger than the *CV* for geese. So, nesting success rates for ducks have greater relative variability.

4. (a) Range $= 14.1 - 6.8 = 7.3$

$$\bar{x} = \frac{\Sigma x}{n} = \frac{63.7}{7} = 9.1$$

$$s^2 = \frac{\Sigma(x-\bar{x})^2}{n-1} = \frac{53.28}{6} = 8.88$$

$$s = \sqrt{s^2} = \sqrt{8.88} \approx 2.98$$

$$CV = \frac{s}{\bar{x}} \cdot 100 = \frac{2.98}{9.1} \cdot 100 = 32.7\%$$

(b) Range $= 31.0 - 19.1 = 11.9$

$$\bar{x} = \frac{\Sigma x}{n} = \frac{182.9}{7} = 26.1$$

$$s^2 = \frac{\Sigma(x-\bar{x})^2}{n-1} = \frac{118.71}{6} \approx 19.79$$

$$s = \sqrt{s^2} = \sqrt{19.79} \approx 4.45$$

$$CV = \frac{s}{\bar{x}} \cdot 100 = \frac{4.45}{26.1} \cdot 100 = 17.0\%$$

(c) More relatively consistent productivity at a higher average level.

5. (a) $$\text{Pax } CV = \frac{s}{\bar{x}} \cdot 100 = \frac{11.56}{11.69} \cdot 100 \approx 98.9\%$$

$$\text{Vanguard } CV = \frac{s}{\bar{x}} \cdot 100 = \frac{12.50}{5.61} \cdot 100 \approx 222.8\%$$

Pax World Balanced seems less risky.

(b) Pax: $\bar{x}-2s=11.69-2(11.56)=-11.43$

$\bar{x}+2s=11.69+2(11.56)=34.81$

At least 75% of the data fall in the interval −11.43% to 34.81%.

Vanguard: $\bar{x}-2s=5.61-2(12.50)=-19.39$

$\bar{x}+2s=5.61+2(12.50)=30.61$

At least 75% of the data fall in the interval −19.39% to 30.61%.

The performance range for Pax seems better than for Vanguard (based on these historical data).

6. (a) Results round to answers given.

(b) $\bar{x}-2s=730-2(172)=386$

$\bar{x}+2s=730+2(172)=1074$

We expect at least 75% of the years to have between 386 and 1074 tornados.

(c) $\bar{x}-3s=730-3(172)=214$

$\bar{x}+3s=730+3(172)=1246$

We expect at least 88.9% of the years to have between 214 and 1246 tornados.

7. (a) Range $=956-219=737$

$$\bar{x}=\frac{\Sigma x}{n}=\frac{3968}{7}\approx 566.9$$

(b) $$s^2=\frac{\Sigma(x-\bar{x})^2}{n-1}=\frac{427{,}213}{6}\approx 71{,}202$$

$$s=\sqrt{s^2}=\sqrt{71{,}202}\approx 266.8$$

(c) $$CV=\frac{s}{\bar{x}}\cdot 100=\frac{266.8}{566.9}\cdot 100=47.1\%$$

s is 47.1% of $\bar{x}$.

(d) $\bar{x}-2s=566.9-2(266.8)\approx 33$

$\bar{x}+2s=566.9+2(266.8)\approx 1100$

We expect at least 75% of the artifact counts for all such excavation sites to fall in the interval 33 to 1100.

8. $$CV=\frac{s}{\bar{x}}\cdot 100$$

$$\frac{\bar{x}\cdot CV}{100}=s$$

$$s=\frac{\bar{x}\cdot CV}{100}$$

$$s=\frac{2.2(1.5)}{100}$$

$$s=0.033$$

9. (a) Students verify results.

(b) Wal-Mart $CV = \frac{s}{\bar{x}} \cdot 100 = \frac{1.06}{52.03} \cdot 100 \approx 2\%$

Disney $CV = \frac{s}{\bar{x}} \cdot 100 = \frac{0.98}{32.23} \cdot 100 \approx 3\%$

Yes, since the CV's are approximately equal, they appear to be equally attractive.

(c) Wal-Mart:

$\bar{x} - 3s = 52.03 - 3(1.06) = 48.85$

$\bar{x} + 3s = 52.03 + 3(1.06) = 55.21$

Disney:

$\bar{x} - 3s = 32.23 - 3(0.98) = 29.29$

$\bar{x} + 3s = 32.23 + 3(0.98) = 35.17$

The support is \$48.85 and resistance is \$55.21 for Wal-Mart.
The support is \$29.29 and the resistance is \$35.17 for Disney.

10. Answers vary.

11.

Class	f	x	xf	$x-\bar{x}$	$(x-\bar{x})^2$	$(x-\bar{x})^2 f$
21–30	260	25.5	6630	−10.3	106.09	27,583.4
31–40	348	35.5	12,354	−0.3	0.09	31.3
41 and over	287	45.5	13,058.5	9.7	94.09	27,003.8
	$n = \Sigma f = 895$		$\Sigma xf = 32{,}042.5$			$\Sigma(x-\bar{x})^2 f = 54{,}619$

$$\bar{x} = \frac{\Sigma xf}{n} = \frac{32{,}042.5}{895} \approx 35.80$$

$$s^2 = \frac{\Sigma(x-\bar{x})^2 \cdot f}{n-1} = \frac{54{,}619}{894} \approx 61.1$$

$$s = \sqrt{61.1} \approx 7.82$$

12.

Class	f	x	xf	$x-\bar{x}$	$(x-\bar{x})^2$	$(x-\bar{x})^2 f$
1–10	34	5.5	187	−10.6	112.36	3820.24
11–20	18	15.5	279	−0.6	0.36	6.48
21–30	17	25.5	433.5	9.4	88.36	1502.12
31 and over	11	35.5	390.5	19.4	376.36	4139.96
	$n = \Sigma f = 80$		$\Sigma xf = 1290$			$\Sigma(x-\bar{x})^2 f = 9468.8$

$$\bar{x} = \frac{\Sigma xf}{n} = \frac{1290}{80} \approx 16.1$$

$$s^2 = \frac{\Sigma(x-\bar{x})^2 f}{n-1} = \frac{9468.8}{79} \approx 119.9$$

$$s = \sqrt{119.9} \approx 10.95$$

13.

Class	f	x	xf	$x-\bar{x}$	$(x-\bar{x})^2$	$(x-\bar{x})^2 f$
8.6–12.5	15	10.55	158.25	−5.05	25.502	382.537
12.6–16.5	20	14.55	291.00	−1.05	1.102	22.050
16.6–20.5	5	18.55	92.75	2.95	8.703	43.513
20.6–24.5	7	22.55	157.85	6.95	48.303	338.118
24.6–28.5	3	26.55	79.65	10.95	119.903	359.708
	$n=\Sigma f=50$		$\Sigma xf=779.5$			$\Sigma(x-\bar{x})^2 f=1145.9$

$$\bar{x}=\frac{\Sigma xf}{n}=\frac{779.5}{50}\approx 15.6$$

$$s^2=\frac{\Sigma(x-\bar{x})^2 f}{n-1}=\frac{1145.9}{49}\approx 23.4$$

$$s=\sqrt{23.4}\approx 4.8$$

14.

Class	f	x	xf	$x-\bar{x}$	$(x-\bar{x})^2$	$(x-\bar{x})^2 f$
18–24	78	21.0	1638.0	−18.12	328.33	25610.1
25–34	75	29.5	2212.5	−9.62	92.54	6940.8
35–44	48	39.5	1896.0	0.38	0.14	6.9
45–54	33	49.5	1633.5	10.38	107.74	3555.6
55–64	33	59.5	1963.5	20.38	415.34	13706.4
65–80	33	72.5	2392.5	33.38	1114.22	36769.4
	$n=\Sigma f=300$		$\Sigma xf=11{,}736$			$\Sigma(x-\bar{x})^2 f=86{,}589$

$$\bar{x}=\frac{\Sigma xf}{n}=\frac{11{,}736}{300}=39.12$$

$$s=\sqrt{\frac{\Sigma(x-\bar{x})^2 f}{n-1}}=\sqrt{\frac{86{,}589}{299}}\approx 17.02$$

$$CV=\frac{s}{\bar{x}}\cdot 100=\frac{17.02}{39.12}\cdot 100\approx 43.5\%$$

15.

x	f	xf	$x^2 f$
3.5	2	7	24.5
4.5	2	9	40.5
5.5	4	22	121.0
6.5	22	143	929.5
7.5	64	480	3600.0
8.5	90	765	6502.5
9.5	14	133	1263.5
10.5	2	21	220.5
	$\Sigma f=200$	$\Sigma xf=1580$	$\Sigma x^2 f=12{,}702$

$$\bar{x} = \frac{\Sigma xf}{n} = \frac{1580}{200} = 7.9$$

$$SS_x = \Sigma x^2 f - \frac{(\Sigma xf)^2}{n} = 12{,}702 - \frac{(1580)^2}{200} = 220$$

$$s = \sqrt{\frac{SS_x}{n-1}} = \sqrt{\frac{220}{199}} \approx 1.05$$

$$CV = \frac{s}{\bar{x}} \cdot 100 = \frac{1.05}{7.9} \cdot 100 \approx 13.29\%$$

Section 3.3

1. 82% or more of the scores were at or below her score. 100% −82% = 18% or less of the scores were above her score. Note: This answer is correct, but it relies on a more precise definition than that given in the text on page 124. An adequate answer, matching the definition in the text would be: 82% of the scores were at or below her score, and (100 – 82)% = 18% of the scores were at or above her score.

2. The upper quartile is the 75th percentile. Therefore, the minimal percentile rank must be the 75th.

3. No, the score 82 might have a percentile rank less than 70.

4. Timothy performed better because a percentile rank of 72 is greater than a percentile rank of 70.

5. Order the data from smallest to largest.

Lowest value = 2
Highest value = 42

There are 20 data values.

$$\text{Median} = \frac{23+23}{2} = 23$$

There are 10 values less than the Q_2 position and 10 values greater than the Q_2 position.

$$Q_1 = \frac{8+11}{2} = 9.5$$

$$Q_3 = \frac{28+29}{2} = 28.5$$

$$IQR = Q_3 - Q_1 = 28.5 - 9.5 = 19$$

Nurses' Length of Employment (months)

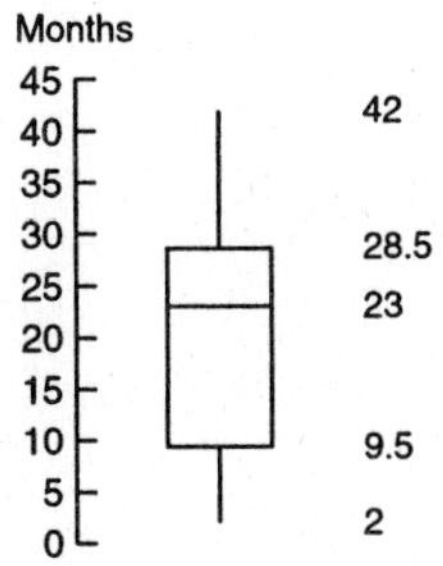

6. (a) Order the data from smallest to largest.

Lowest value = 3
Highest value = 72

There are 20 data values.

$$\text{Median} = \frac{22+24}{2} = 23$$

There are 10 values less than the median and 10 values greater than the median.

$$Q_1 = \frac{15+17}{2} = 16$$

$$Q_3 = \frac{29+31}{2} = 30$$

$$IQR = Q_3 - Q_1 = 30 - 16 = 14$$

Clerical Staff Length of Employment (months)

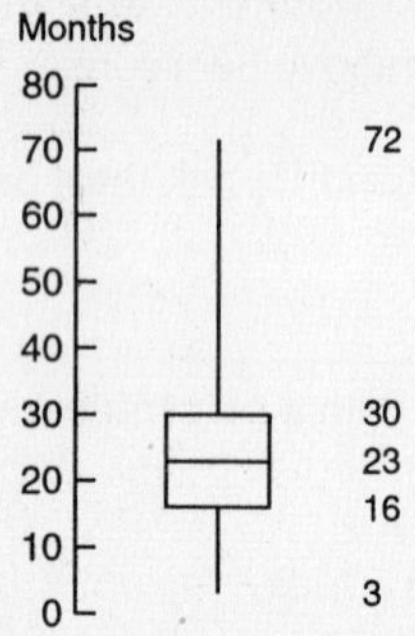

(b) The medians are the same (23) and the *IQR*'s are similar. However, the distances from Q_1 to the minimum value and from Q_3 to the maximum value are greater here than in Problem 7.

7. (a) Order the data from smallest to largest.

Lowest value = 17
Highest value = 38

There are 50 data values.

$$\text{Median} = \frac{24+24}{2} = 24$$

There are 25 values above and 25 values below the Q_2 position.

$Q_1 = 22$
$Q_3 = 27$
$IQR = 27 - 22 = 5$

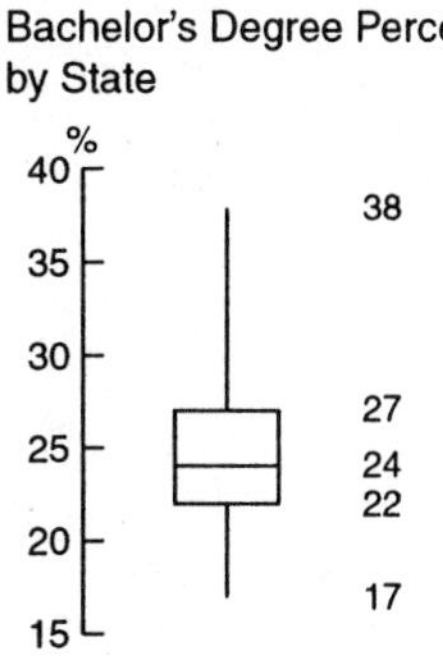

(b) 26% is in the 3rd quartile, since it is between the median and Q_3.

8. (a) Order the data from smallest to largest.

Lowest value = 5
Highest value = 15

There are 50 data values.

$$\text{Median} = \frac{10+10}{2} = 10$$

There are 25 values above and 25 values below the Q_2 position.

$$Q_1 = 9$$
$$Q_3 = 12$$
$$IQR = 12 - 9 = 3$$

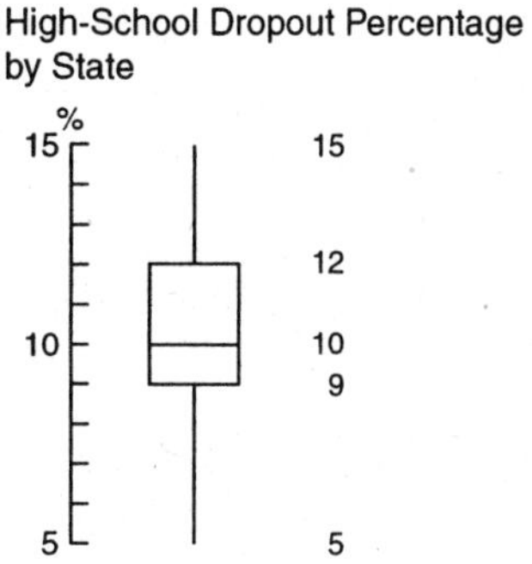

(b) 7% is in the 1st quartile, since it is below Q_1.

9. (a) California has the lowest premium since its left whisker is farthest to the left. Pennsylvania has the highest premium since its right whisker is farthest to the right.

(b) Pennsylvania has the highest median premium since its line in the middle of the box is farthest to the right.

(c) California has the smallest range of premiums since the distance between the ends of the whiskers is the smallest. Texas has the smallest interquartile range since the distance between the ends of the boxes is the smallest.

(d) Based on the answers to (a)-(c) above, we can determine that part (a) of Figure 3-13 is for Texas, part (b) of Figure 3-13 is for Pennsylvania, and part (c) of Figure 3-13 is for California.

10. (a) Order the data from smallest to largest.

Lowest value = 4
Highest value = 80

There are 24 data values.

$$\text{Median} = \frac{65+66}{2} = 65.5$$

There are 12 values above and 12 values below the median.

$$Q_1 = \frac{61+62}{2} = 61.5$$

$$Q_3 = \frac{71+72}{2} = 71.5$$

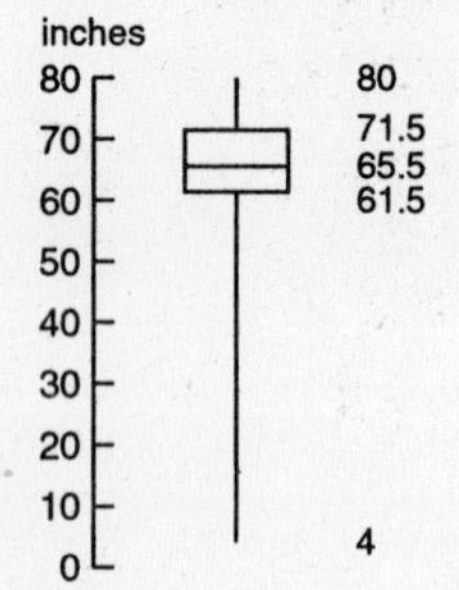

(b) $IQR = Q_3 - Q_1 = 71.5 - 61.5 = 10$

(c) $1.5(10) = 15$
Lower limit: $Q_1 - 1.5(IQR) = 61.5 - 15 = 46.5$
Upper limit: $Q_3 + 1.5(IQR) = 71.5 + 15 = 86.5$

(d) Yes, the value 4 is below the lower limit and so is an outlier; it is probably an error. Our guess is that one of the students is 4 feet tall and listed height in feet instead of inches. There are no values above the upper limit.

11. (a) Assistant had the smallest median percentage salary increase since the bar in the middle of the box is the lowest. Associate had the single highest salary increase since it has the highest asterisk.

(b) Instructor had the largest spread between the first and third quartiles since the distance between the ends of the box is greatest.

(c) Assistant had the smallest spread for the lower 50% of the percentage salary increases since the distance between the bar in the box and the maximum value is the smallest.

(d) Professor had the most symmetric percentage salary increases because there are no outliers and the bar representing the median is close to the center of the box.
Yes, if the outliers for the associate professors were omitted, that distribution would appear to be symmetric.

(e) Associate professor:

$$IQR = 5.075 - 2.350 = 2.725$$
$$Q_3 + 1.5(IQR) = 5.075 + 1.5(2.725) \approx 9.16$$

Yes, since 17.7 is greater than 9.16, there is at least one outlier.

Instructor:

$$IQR = 5.800 - 2.850 = 2.950$$
$$Q_3 + 1.5(IQR) = 5.800 + 1.5(2.950) \approx 10.23$$

Yes, since 13.4 is greater than 10.23, there is at least one outlier.

Chapter 3 Review

1. (a) $\bar{x} = \frac{\Sigma x}{n} = \frac{876}{8} = 109.5$

$$s = \sqrt{\frac{\Sigma(x-\bar{x})^2}{n-1}} = \sqrt{\frac{7044}{7}} = \sqrt{1006.3} \approx 31.7$$

$$CV = \frac{s}{\bar{x}} \cdot 100 = \frac{31.7}{109.5} \cdot 100 \approx 28.9\%$$

range = maximum value − minimum value
= 142 − 73 = 69

(b) $\bar{x} = \frac{\Sigma x}{n} = \frac{881}{8} = 110.125$

$$s = \sqrt{\frac{\Sigma(x-\bar{x})^2}{n-1}} = \sqrt{\frac{358.87}{7}} \approx 7.2$$

$$CV = \frac{s}{\bar{x}} \cdot 100 = \frac{7.2}{110.125} \cdot 100 \approx 6.5\%$$

range = maximum value − minimum value
= 120 − 100 = 20

(c) The means are about the same. The first distribution has greater spread. The standard deviation, *CV*, and range for the first set of measurements are greater than those for the second set of measurements.

2. (a) Mean $= \bar{x} = \frac{\Sigma x}{n} = \frac{1.9 + 2.8 + \cdots + 7.2}{8}$

$$= \frac{36.2}{8}$$

$$= 4.525$$

Order the data from smallest to largest.

1.9 1.9 2.8 3.9 4.2 5.7 7.2 8.6

Median $= \frac{3.9 + 4.2}{2} = 4.05$

The mode is 1.9 because it is the value that occurs most frequently.

(b) $s = \sqrt{\dfrac{\Sigma(x-\bar{x})^2}{n-1}} = \sqrt{\dfrac{42.395}{7}} \approx 2.46$

$CV = \dfrac{s}{\bar{x}} \cdot 100 = \dfrac{2.46}{4.525} \cdot 100 \approx 54.4\%$

Range $= 8.6 - 1.9 = 6.7$

3. (a) Order the data from smallest to largest.

Lowest value = 31
Highest value = 68

There are 60 data values.

$$\text{Median} = \frac{45+45}{2} = 45$$

There are 30 values above and 30 values below the Q_2 position.

$$Q_1 = \frac{40+40}{2} = 40$$

$$Q_3 = \frac{52+53}{2} = 52.5$$

$$IQR = 52.5 - 40 = 12.5$$

Percentage of Democratic Vote by Counties in Georgia

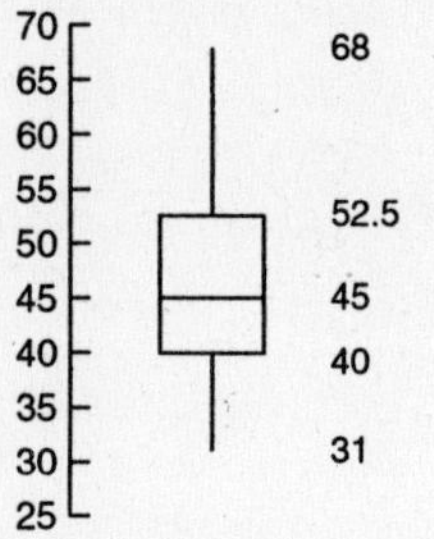

(b) Class width = 8

Class	x Midpoint	f	xf	x^2f
31–38	34.5	11	379.5	13,092.8
39–46	42.5	24	1020	43,350.0
47–54	50.5	15	757.5	38,253.8
55–62	58.5	7	409.5	23,955.8
63–70	66.5	3	199.5	13,266.8
		$n = \Sigma f = 60$	$\Sigma xf = 2766$	$\Sigma x^2 f = 131,919$

$$\bar{x} = \frac{\Sigma xf}{n} = \frac{2766}{60} = 46.1$$

$$SS_x = \Sigma x^2 f - \frac{(\Sigma xf)^2}{n} = 131{,}919 - \frac{(2766)^2}{60} = 4406.4$$

$$s = \sqrt{\frac{SS_x}{n-1}} = \sqrt{\frac{4406.4}{59}} \approx 8.64$$

$\bar{x} - 2s = 46.1 - 2(8.64) = 28.82$
$\bar{x} + 2s = 46.1 + 2(8.64) = 63.38$
We expect at least 75% of the data to fall in the interval 28.82 to 63.38.

(c) $\bar{x} = 46.15,\ s \approx 8.63$

4. (a) Order the data from smallest to largest.

Lowest value = 6
Highest value = 16

There are 50 data values.

$$\text{Median} = \frac{11+11}{2} = 11$$

There are 25 values above and 25 values below the Q_2 position.

$Q_1 = 10$

$Q_3 = 13$

$IQR = Q_3 - Q_1 = 13 - 10 = 3$

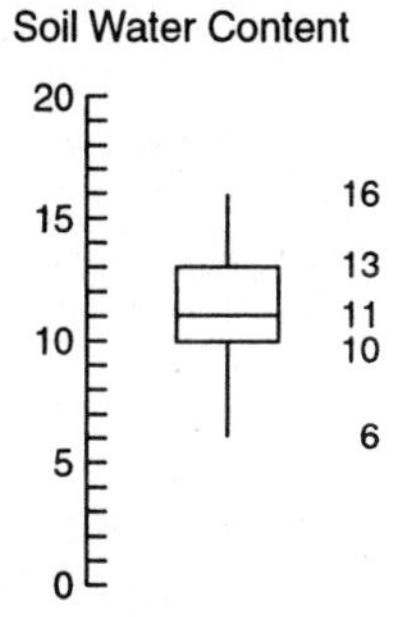

(b)

Class	x Midpoint	f	xf	$x^2 f$
6–8	7	4	28	196
9–11	10	24	240	2400
12–14	13	15	195	2535
15–17	16	7	112	1792
		$n = \Sigma f = 50$	$\Sigma xf = 575$	$\Sigma x^2 f = 6923$

$$\bar{x} = \frac{\Sigma xf}{n} = \frac{575}{50} = 11.5$$

$$SS_x = \Sigma x^2 f - \frac{(\Sigma xf)^2}{n} = 6923 - \frac{(575)^2}{50} = 310.5$$

$$s = \sqrt{\frac{SS_x}{n-1}} = \sqrt{\frac{310.5}{49}} \approx 2.52$$

$$\bar{x} - 2s = 11.5 - 2(2.52) = 6.46$$
$$\bar{x} + 2s = 11.5 + 2(2.52) = 16.54$$

We expect at least 75% of the data to fall in the interval 6.46 to 16.54.

(c) $\bar{x} \approx 11.48$; $s \approx 2.44$

5. (a) Mean $= \bar{x} = \frac{\Sigma x}{n} = \frac{10.1 + 6.2 + \cdots + 5.7}{6} = \frac{47}{6} \approx 7.83$

$$s = \sqrt{\frac{\Sigma(x-\bar{x})^2}{n-1}} = \sqrt{\frac{26.913}{5}} \approx 2.32$$

$$CV = \frac{s}{\bar{x}} \cdot 100 = \frac{2.32}{7.83} \cdot 100 \approx 29.6\%$$

Range = largest value − smallest value
$= 10.1 - 5.3 = 4.8$

(b) Mean $= \bar{x} = \frac{\Sigma x}{n} = \frac{10.2 + 9.7 + \cdots + 10.1}{6} = \frac{59.7}{6} = 9.95$

$$s = \sqrt{\frac{\Sigma(x-\bar{x})^2}{n-1}} = \sqrt{\frac{0.415}{5}} \approx 0.29$$

$$CV = \frac{s}{\bar{x}} \cdot 100 = \frac{0.29}{9.95} \cdot 100 \approx 2.9\%$$

Range = largest value − smallest value
$= 10.3 - 9.6 = 0.7$

(c) Second line has more consistent performance as reflected by the smaller standard deviation, *CV*, and range.

6. Order the data from smallest to largest.

Lowest value = 45
Highest value = 109

There are 70 data values.

$$\text{Median} = \frac{80+80}{2} = 80$$

There are 35 values above and 35 values below the Q_2 position.

$$Q_1 = 71$$
$$Q_3 = 84$$
$$IQR = 84 - 71 = 13$$

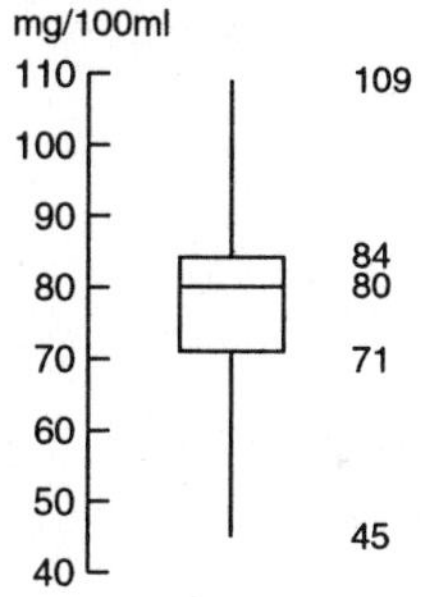

7. $\text{Mean weight} = \frac{2500}{16} = 156.25$

The mean weight is 156.25 lb.

8. **(a)** It is possible for the range and the standard deviation to be the same. For instance, for data values that are all the same, such as 1, 1, 1, 1, 1, the range and standard deviation are both 0.

(b) It is possible for the mean, median, and mode to be all the same. For instance, the data set 1, 2, 3, 3, 3, 4, 5 has mean, median, and mode all equal to 3. The averages can all be different, as in the data set 1, 2, 3, 3. In this case, the mean is 2.25, the median is 2.5, and the mode is 3.

Chapter 4 Regression and Correlation

Section 4.1

1. The points seem close to a straight line, so there is moderate or low linear correlation.

2. No straight line is realistically a good fit, so there is no linear correlation.

3. The points seem very close to a straight line, so there is high linear correlation.

4. The points seem close to a straight line, so there is moderate or low linear correlation.

5. The points seem very close to a straight line, so there is high linear correlation.

6. No straight line is realistically a good fit, so there is no linear correlation.

7. (a)

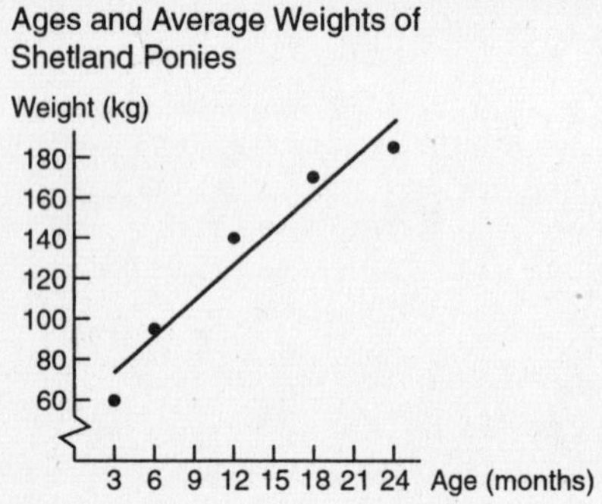

(b) Draw the line you think fits best. (Method to find equation is in Section 4.2.)

(c) Since the points are very close to a straight line, the correlation is high.

8. (a)

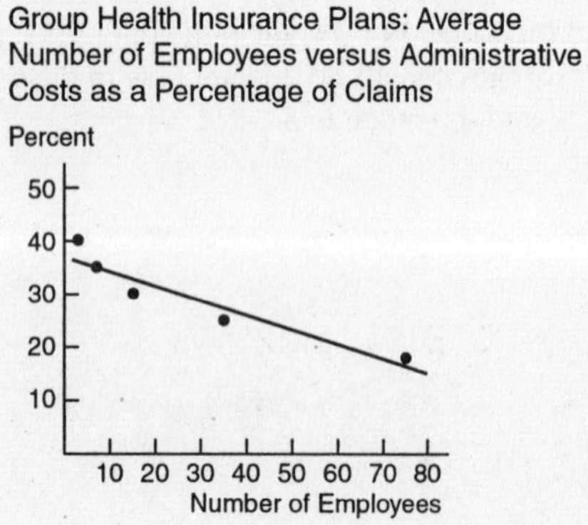

(b) Draw the line you think fits best. (Method to find equation is in Section 4.2.)

(c) Since the points are fairly close to a straight line, the correlation is moderate.

9. (a) Change in Wages and in Consumer Prices in Various Countries (%)

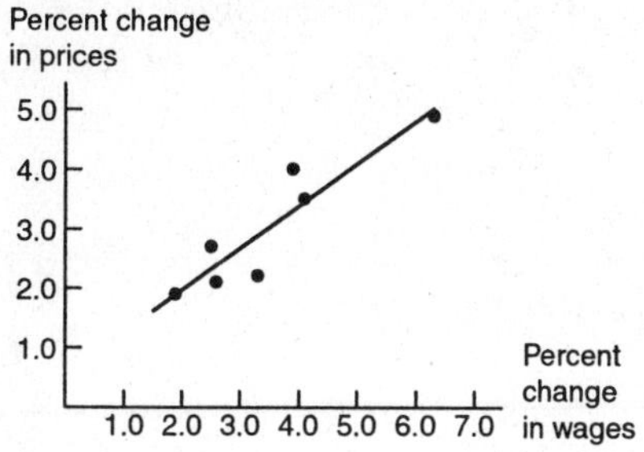

(b) Draw the line you think fits best. (Method to find equation is in Section 4.2.)

(c) Since the points are fairly close to a straight line, the correlation is moderate.

10. (a) Magnitude (Richter Scale) and Depth (km) of Earthquakes

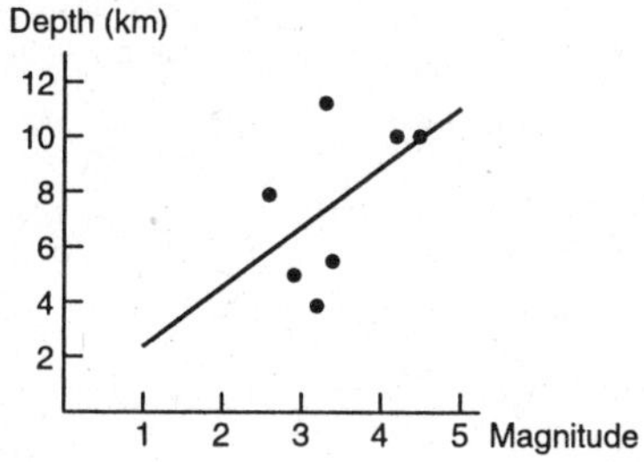

(b) Draw the line you think fits best. (Method to find equation is in Section 4.2.)

(c) Since the points are not close to a straight line, the correlation is low.

Note: One possible reason why there appears to be little, if any, linear relationship is that the Richter scale is logarithmic. An increase of 1 on the Richter scale represents a 60-fold increase in energy.

11. (a) Body Diameter and Weight of Prehistoric Pottery

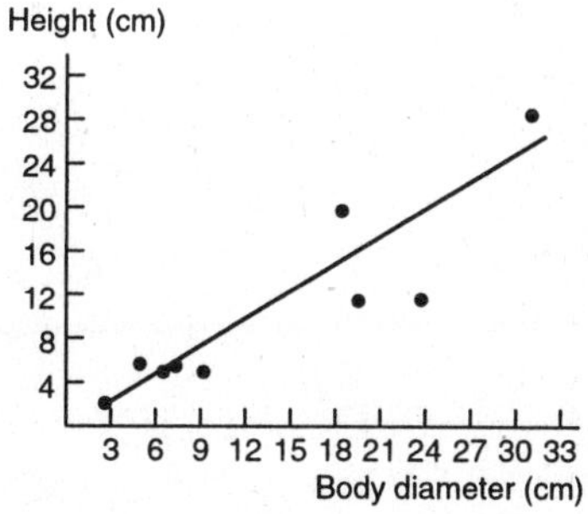

(b) Draw the line you think fits best. (Method to find equation is in Section 4.2.)

(c) Since the points are fairly close to a straight line, the correlation is moderate.

12. (a) Body Weight and Metabolic Rate of Children

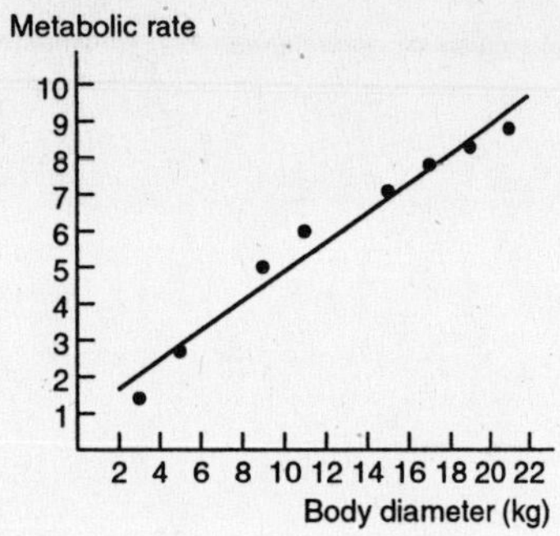

(b) Draw the line you think fits best. (Method to find equation is in Section 4.2.)

(c) Since the points are very close to a straight line, the correlation is high.

13. (a) Unit Length on *y* Same as That on *x*

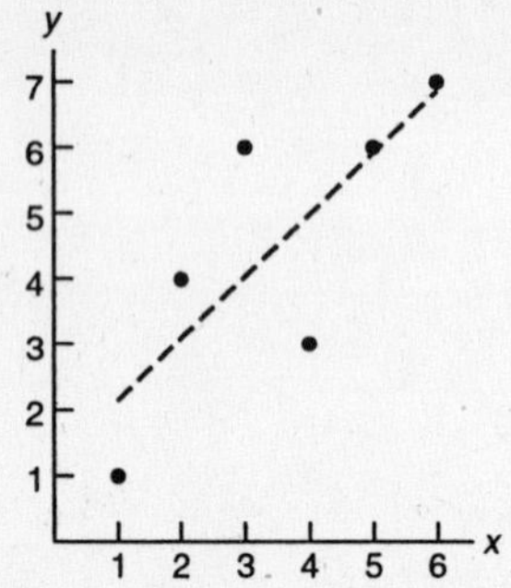

(b) Unit Length on *y* Twice That on *x*

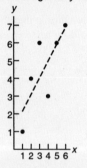

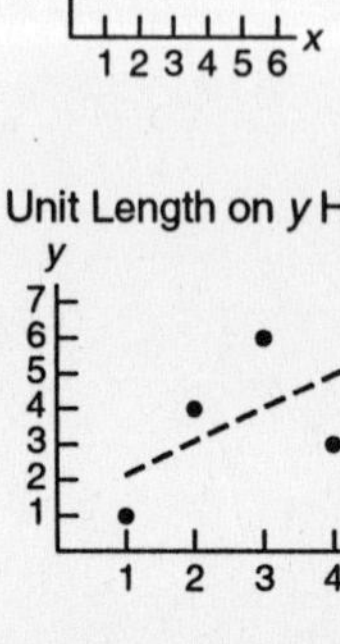

(c) Unit Length on *y* Half That on *x*

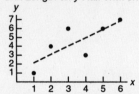

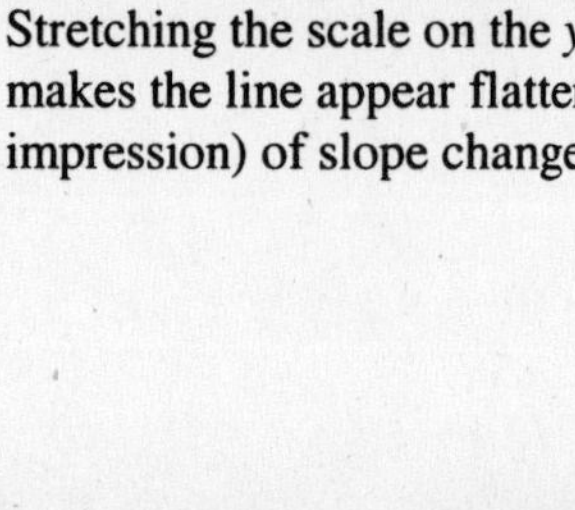

(d) Draw the lines you think best fit the data points.
Stretching the scale on the *y*-axis makes the line appear steeper. Shrinking the scale on the *y*-axis makes the line appear flatter. The slope of the line does not change. Only the appearance (visual impression) of slope changes as the scale of the *y*-axis changes.

Section 4.2

Note: In this section and the next two, answers may vary slightly, depending on how many significant digits are used throughout the calculations.

1. (a) Absenteeism and Number of Assembly Line Defects

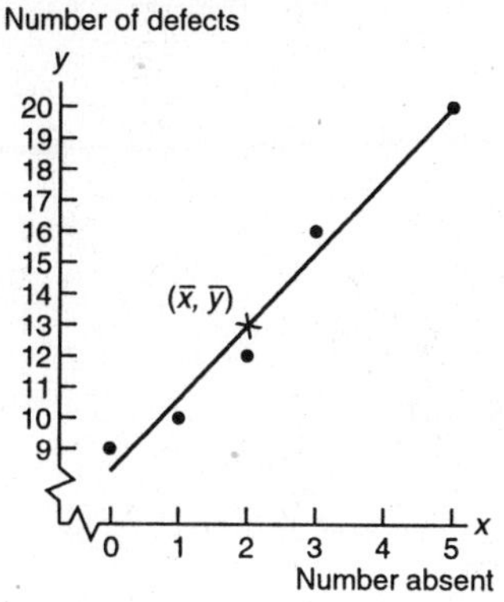

(b) $\bar{x} = \dfrac{\Sigma x}{n} = \dfrac{11}{5} = 2.2$

$\bar{y} = \dfrac{\Sigma y}{n} = \dfrac{67}{5} = 13.4$

$b = \dfrac{SS_{xy}}{SS_x} = \dfrac{34.6}{14.8} = 2.3378$

$a = \bar{y} - b\bar{x} = 13.4 - 2.3378(2.2) = 8.26$

$y = a + bx$ or $y = 8.26 + 2.338x$

(c) See figure of part (a).

(d) Use $x = 4$.

$y_p = 8.26 + 2.338(4) = 17.6$ defects

2. (a) Age and Weight of Healthy Calves

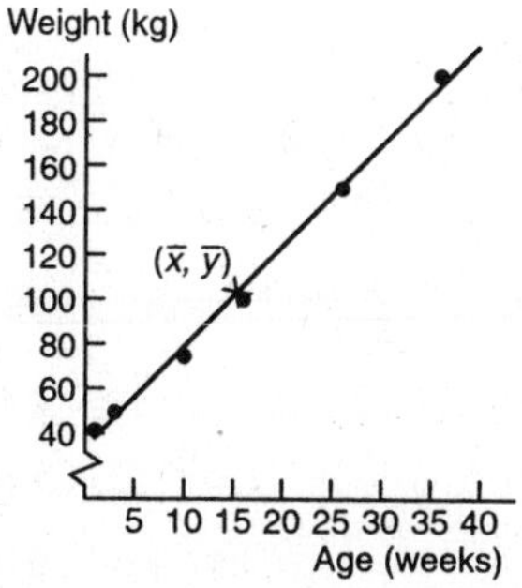

(b) $\bar{x} = \frac{\sum x}{n} = \frac{92}{6} = 15.33$

$\bar{y} = \frac{\sum y}{n} = \frac{617}{6} = 102.83$

$b = \frac{SS_{xy}}{SS_x} = \frac{4181.3}{927.3} = 4.509$

$a = \bar{y} - b\bar{x} = 102.83 - 4.509(15.33) = 33.70$

$y = a + bx$ or $y = 33.70 + 4.51x$

(c) See figure of part (a).

(d) Use $x = 12$.

$y_p = 33.70 + 4.51(12) = 87.8$ kg

3. (a)

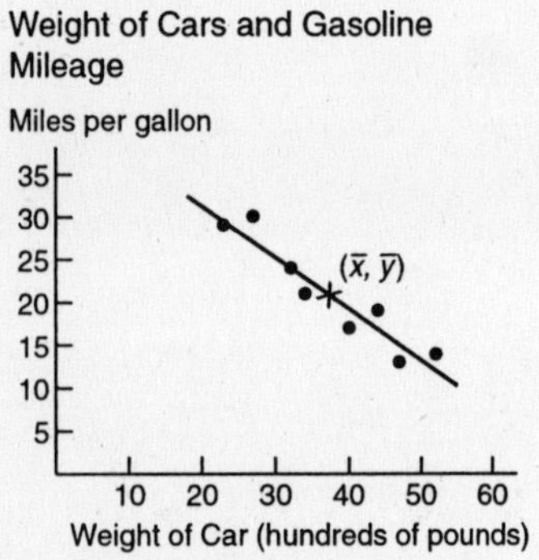

(b) $\bar{x} = \frac{\sum x}{n} = \frac{299}{8} = 37.375$

$\bar{y} = \frac{\sum y}{n} = \frac{167}{8} = 20.875$

$b = \frac{SS_{xy}}{SS_x} = \frac{-427.625}{711.875} = -0.6007$

$a = \bar{y} - b\bar{x} = 20.875 - (-0.6007)(37.375) = 43.3263$

$y = a + bx$ or $y = 43.3263 - 0.6007x$

(c) See figure of part (a).

(d) Use $x = 38$.

$y_p = 43.3263 - 0.6007(38) = 20.5$ mpg

4. (a)

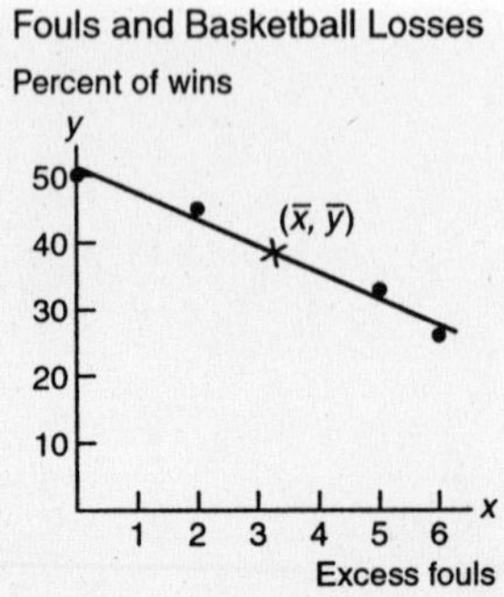

(b) $\bar{x} = \frac{\sum x}{n} = \frac{13}{4} = 3.25$

$\bar{y} = \frac{\sum y}{n} = \frac{154}{4} = 38.5$

$b = \frac{SS_{xy}}{SS_x} = \frac{-89.5}{22.75} = -3.934$

$a = \bar{y} - b\bar{x} = 38.5 - (-3.934)(3.25) = 51.29$

$y = a + bx$ or $y = 51.29 - 3.934x$

(c) See figure of part (a).

(d) Use $x = 4$.

$y_p = 51.29 - 3.934(4) = 35.55\%$

5. (a) Education and Income in Small Cities

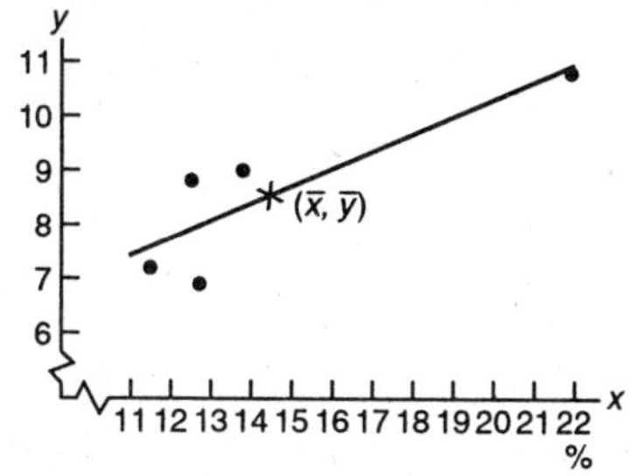

(b) $\bar{x} = \frac{\sum x}{n} = \frac{72.4}{5} = 14.48$

$\bar{y} = \frac{\sum y}{n} = \frac{42.7}{5} = 8.54$

$b = \frac{SS_{xy}}{SS_x} = \frac{22.854}{71.448} = 0.320$

$a = \bar{y} - b\bar{x} = 8.54 - 0.320(14.48) = 3.91$

$y = a + bx$ or $y = 3.91 + 0.320x$

(c) See figure of part (a).

Note that the regression line would be much steeper if (21.9, 10.8) were eliminated from the data set [which would also affect $(\bar{x}, \bar{y})$]. Not all outliers (this point is an outlier in both x (probably) and y) have this effect; however, when the parameter estimates a and b depend heavily on a particular observation, as is the case here, the point is called "influential," and conclusions drawn are shaky at best when influential observations remain in the data. For further information, refer to a more advanced textbook such as Applied Regression Analysis by Draper and Smith.

(d) Use $x = 20$.

$y_p = 3.91 + 0.320(20) = 10.31$ i.e., 10.31 thousand dollars

6. (a) Percentage of 16 to 19-Year-Olds Not in School and per Capita Income (thousands of dollars)

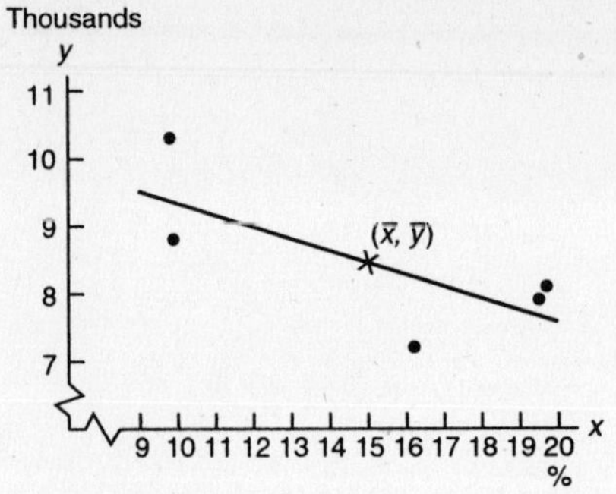

(b) $\bar{x} = \frac{\sum x}{n} = \frac{75.1}{5} = 15.02$

$\bar{y} = \frac{\sum y}{n} = \frac{42.3}{5} = 8.46$

$b = \frac{SS_{xy}}{SS_x} = \frac{-17.026}{96.828} = -0.1758$

$a = \bar{y} - b\bar{x} = 8.46 - (-0.1758)15.02 = 11.10$

$y = a + bx$ or $y = 11.10 - 0.176x$

(c) See figure of part (a).

(d) Use $x = 17$.

$y_p = 11.10 - 0.176(17) = 8.11$ thousand dollars

7. (a) Cultural Affiliation and Elevation of Achaeological Sites

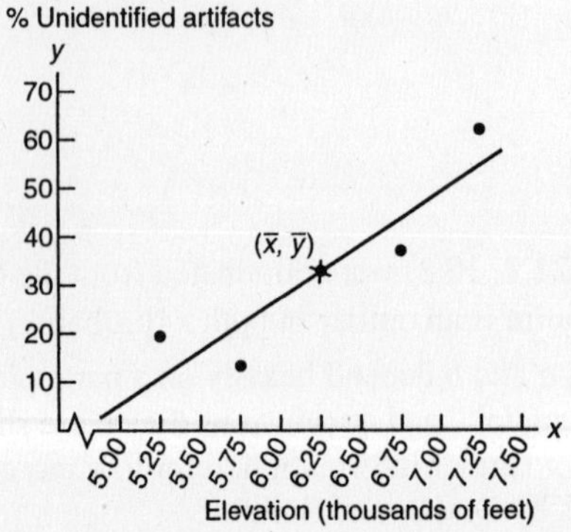

(b) $\bar{x} = \frac{\sum x}{n} = \frac{31.25}{5} = 6.25$

$\bar{y} = \frac{\sum y}{n} = \frac{164}{5} = 32.8$

$b = \frac{SS_{xy}}{SS_x} = \frac{55}{2.5} = 22.0$

$a = \bar{y} - b\bar{x} = 32.8 - 22.0(6.25) = -104.7$

$y = a + bx$ or $y = -104.7 + 22.0x$

(c) See figure of part (a).

(d) Use $x = 6.5$.

$y_p = -104.7 + 22.0(6.5) = 38.3$ percent

8. (a)

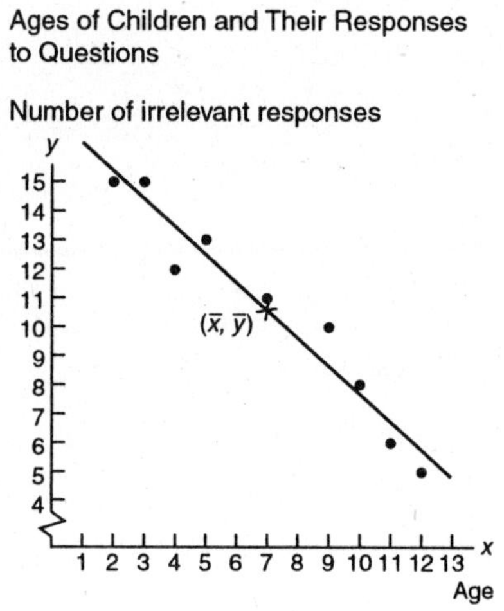

(b) $\bar{x} = \frac{\sum x}{n} = \frac{63}{9} = 7.0$

$\bar{y} = \frac{\sum y}{n} = \frac{95}{9} = 10.56$

$b = \frac{SS_{xy}}{SS_x} = \frac{-104}{108} = -0.96296$

$a = \bar{y} - b\bar{x} = 10.56 - (-0.96296)(7.0) = 17.30$

$y = a + bx$ or $y = 17.30 - 0.963x$

(c) See figure of part (a).

(d) Use $x = 9.5$.

$y_p = 17.30 - 0.963(9.5) = 8.15$ irrelevant responses

9. (a)

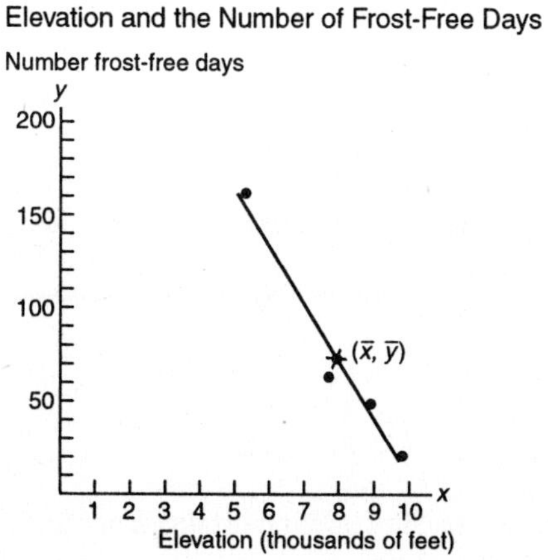

(b) $\bar{x} = \frac{\Sigma x}{n} = \frac{39.6}{5} = 7.92$

$\bar{y} = \frac{\Sigma y}{n} = \frac{368}{5} = 73.6$

$b = \frac{SS_{xy}}{SS_x} = \frac{-352.26}{11.408} = -30.8783$

$a = \bar{y} - b\bar{x} = 73.6 - (-30.8783)(7.92) = 318.16$

$y = a + bx$ or $y = 318.16 - 30.878x$

(c) See figure of part (a).

Note: Compare this figure to that in Problem 5 above, the point (5.3, 162) is an outlier (possibly in x, definitely in y) but it is more or less along the regression line that would be drawn if it were eliminated from the data set. Thus, this is not an "influential" observation.

(d) Use $x = 6$.

$y_p = 318.16 - 30.878(6) = 132.89$ days

10. (a) Results checks.

(b) Results checks.

(c) Yes.

(d) $y = 0.143 + 1.071x$

$$y - 0.143 = 1.071x$$

$$\frac{y - 0.143}{1.071} = x$$

$$\frac{1}{1.071}y - \frac{0.143}{1.071} = x$$

or

$x = 0.9337y - 0.1335$

The equation $x = 0.9337y - 0.1335$ does not match part (b), with the symbols x and y exchanged.

(e) In general, switching x and y values produces a different least-squares equation. It is important that when you perform a linear regression, you know which variable is the explanatory variable and which is the response variable.

Section 4.3

1. (a) No, high positive correlation does not mean causation.

(b) An increase in the population is a third factor that might cause traffic accidents and the number of safety stickers to increase together.

2. (a) No, high positive correlation does not mean causation.

(b) There is an increase in buying power due to increase in salaries.

3. (a) No, strong negative correlation does not mean causation.

(b) Better medical treatment is a third factor that might be decreasing infant mortalities and at the same time increasing life span.

4. (a) No, strong positive correlation does not mean causation.

(b) An increase in population could account for increases both in consumption of soda pop and in number of traffic accidents.

5. (a)

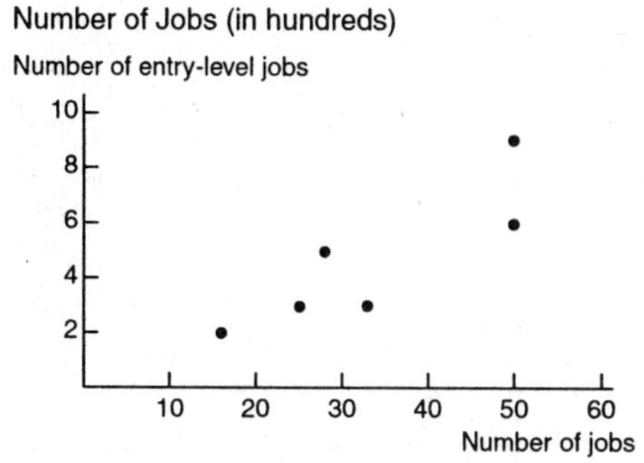

(b) r should be close to 1 because the points seem to be clustered fairly close to a straight line going up from left to right.

(c) $r = \frac{SS_{xy}}{\sqrt{SS_x SS_y}} = \frac{153.\overline{3}}{\sqrt{953.\overline{3}(33.\overline{3})}} = 0.860$

$r^2 = (0.860)^2 = 0.740$

This means that 74.0% of the variation in y = number of entry-level jobs can be explained by the corresponding variation in x = total number of jobs using the least squares line. 100% – 74.0% = 26.0% of the variation is unexplained.

6. (a)

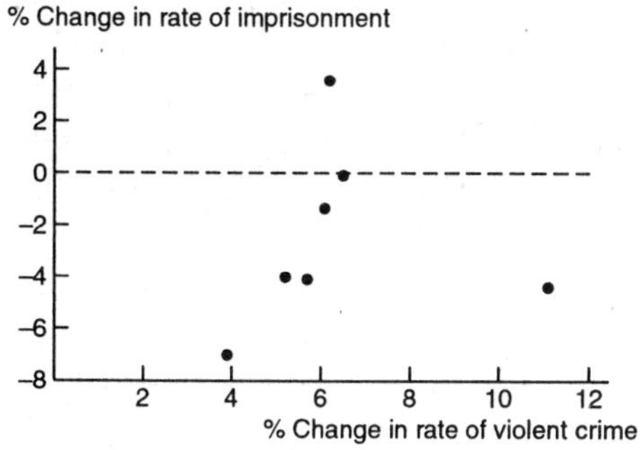

(b) r should be close to 0 because the points are not all clustered around a straight line, due to (11.1, –4.4) (which is an influential observation).

(c) $r = \frac{SS_{xy}}{\sqrt{SS_x SS_y}} = \frac{3.9314}{\sqrt{30.4086(72.8486)}} = 0.084$

$r^2 = (0.084)^2 = 0.007$

This means that 0.7% of the variation in y = percent change in the rate of imprisonment can be explained by the corresponding variation in x = percent change in the rate of violent crime using the least squares line. 100% – 0.7% = 99.3% of the variation is unexplained.

7. (a)

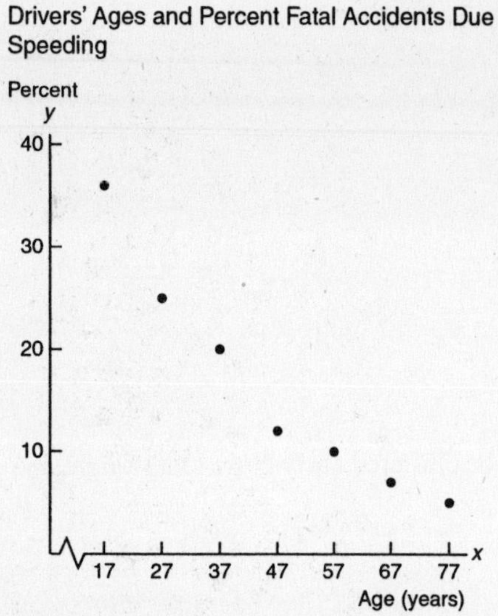

(b) r should be closer to –1 because the points are clustered very close to a straight line going down from left to right. (Note also that the data values fall nicely on a curve.)

(c) $$r = \frac{SS_{xy}}{\sqrt{SS_x SS_y}} = \frac{-1390}{\sqrt{2800(749.714)}} = -0.959$$

$r^2 = (-0.959)^2 = 0.920$

This means that 92% of the variation in y = percentage of all fatal accidents due to speeding can be explained by the corresponding variation in x = age in years of a licensed automobile driver using the least squares line. 100% – 92% = 8% of the variation is unexplained.

8. (a)

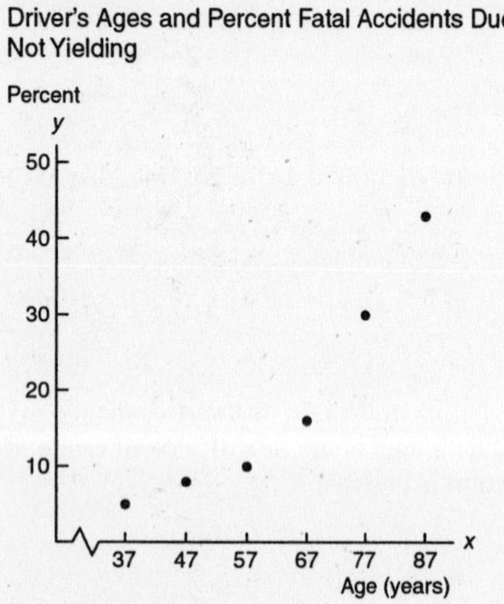

(b) r should be closer to 1 because the points are clustered very close to a straight line going up from left to right. (Note also that the data follow a curve.)

(c) $$r = \frac{SS_{xy}}{\sqrt{SS_x SS_y}} = \frac{1310}{\sqrt{1750(1103.\overline{3})}} = 0.943$$

$r^2 = (0.943)^2 = 0.889$

This means that 88.9% of the variation in y = percentage of fatal accidents due to failure to yield the right of way can be explained by the corresponding variation in x = age of a licensed driver in years using the least squares line. 100% – 88.9% = 11.1% of the variation is unexplained.

9. (a) Body Height and Bone Size

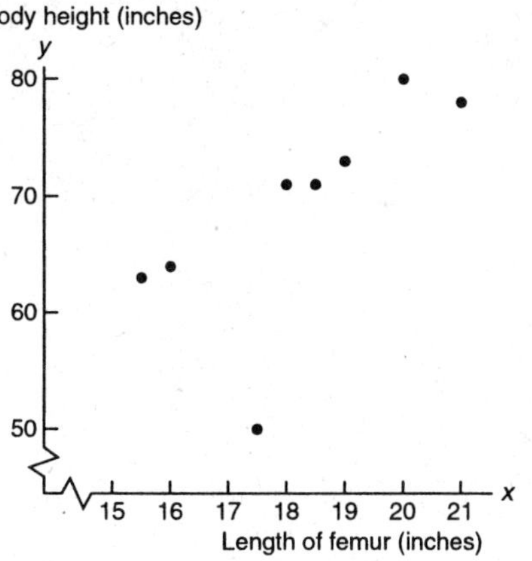

(b) r should be closer to 1 because the points are clustered close to a straight line going up from left to right.

(c) $$r = \frac{SS_{xy}}{\sqrt{SS_x SS_y}} = \frac{88.875}{\sqrt{24.4688(647.5)}} = 0.7061$$

$r^2 = (0.7061)^2 = 0.499$

This means that 49.9% of the variation in y = body height can be explained by the corresponding variation in x = length of femur using the least squares line. 100% – 49.9% = 50.1% of the variation is unexplained.

10. (a) Lowest Barometric Pressure and Maximum Wind Speed for Tropical Cyclones

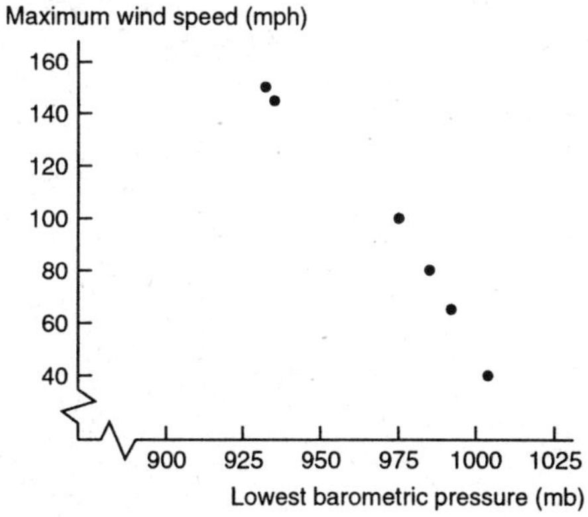

(b) r should be closer to –1 because the points are clustered very close to a straight line going down from left to right.

(c) $$r = \frac{SS_{xy}}{\sqrt{SS_x SS_y}} = \frac{-6575}{\sqrt{4557.5(9683.\overline{3})}} = -0.9897$$

$r^2 = (-0.9897)^2 = 0.9795$ or 0.98

This means that 98% of the variation in y = maximum wind speed of the cyclone can be explained by the corresponding variation in x = lowest pressure as a cyclone approaches using the least-squares line. 100% – 98% = 2% of the variation is unexplained.

11. (a) We get the same result.

$SS_{xy} = SS_{yx}$

(b) We get the same result.

(c) We get the same result.

(d) First set: $r = \dfrac{SS_{xy}}{\sqrt{SS_x SS_y}} = \dfrac{5}{\sqrt{4.\overline{6}(14)}} = 0.618590$

Second set: $r = \dfrac{SS_{xy}}{\sqrt{SS_x SS_y}} = \dfrac{5}{\sqrt{14(4.\overline{6})}} = 0.618590$

$r = 0.618590$ in both cases.

The least-squares equations are not necessarily the same.

Chapter 4 Review Problems

1. (a) Age and Mortality Rate for Bighorn Sheep

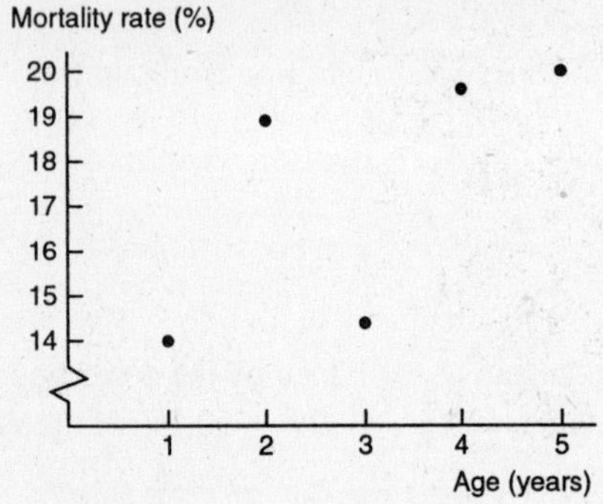

(b) $\overline{x} = \dfrac{\sum x}{n} = \dfrac{15}{5} = 3$

$\overline{y} = \dfrac{\sum y}{n} = \dfrac{86.9}{5} = 17.38$

$b = \dfrac{SS_{xy}}{SS_x} = \dfrac{12.7}{10} = 1.27$

$a = \overline{y} - b\overline{x} = 17.38 - 1.27(3) = 13.57$

$y = a + bx$ or $y = 13.57 + 1.27x$

(c) $r = \dfrac{SS_{xy}}{\sqrt{SS_x SS_y}} = \dfrac{12.7}{\sqrt{10(34.408)}} = 0.685$

$r^2 = (0.685)^2 = 0.469$

The correlation coefficient r measures the strength of the linear relationship between a bighorn sheep's age and the mortality rate. The coefficient of determination, r^2, means that 46.9% of the variation in y = mortality rate in this age groups can be explained by the corresponding variation in x = age of a bighorn sheep using the least-squares line.

2. (a) Annual Salary (thousands) and Number of Job Changes

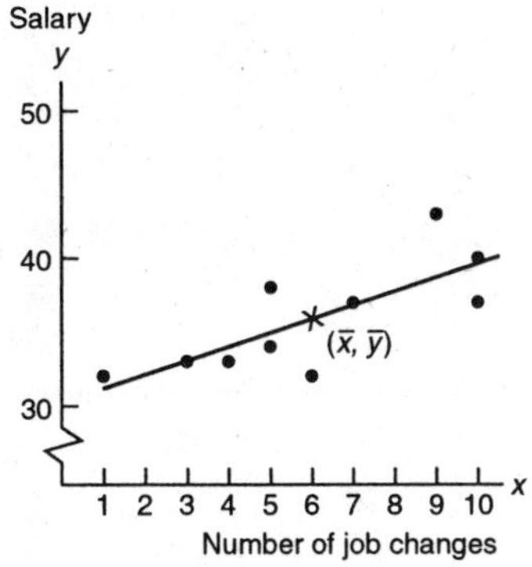

(b) $\bar{x} = \frac{\sum x}{n} = \frac{60}{10} = 6.0$

$\bar{y} = \frac{\sum y}{n} = \frac{359}{10} = 35.9$

$b = \frac{SS_{xy}}{SS_x} = \frac{77}{82} = 0.939024$

$a = \bar{y} - b\bar{x} = 35.9 - 0.939024(6.0) = 30.266$

$y = a + bx$ or $y = 30.266 + 0.939x$

(c) See the figure in part (a).

(d) Let $x = 2$.

$y_p = 30.266 + 0.939(2) = 32.14$

The predicted salary is $32,140.

(e) The correlation coefficient will be positive because the points are clustered around a straight line going up from left to right.

(f) $r = \frac{SS_{xy}}{\sqrt{SS_x SS_y}} = \frac{77}{\sqrt{82(124.9)}} = 0.761$

$r^2 = (0.761)^2 = 0.579$

This means that 57.9% of the variation in y = salary can be explained by the corresponding variation in x = number of job changes using the least-squares line.

3. (a) Weight of One-Year-Old versus Weight of Adult

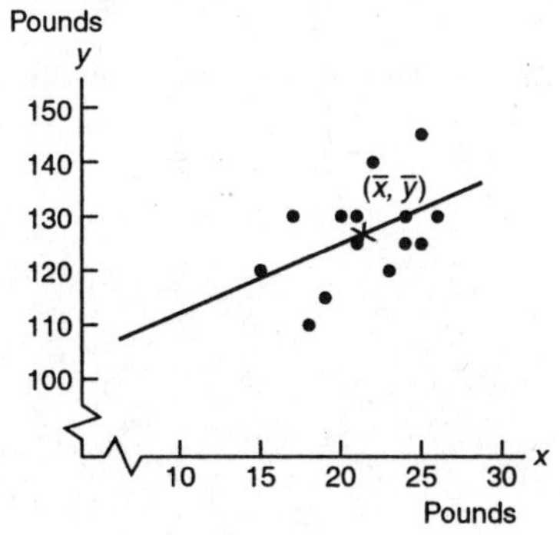

(b) $\bar{x} = \frac{\sum x}{n} = \frac{300}{14} = 21.43$

$\bar{y} = \frac{\sum y}{n} = \frac{1775}{14} = 126.79$

$b = \frac{SS_{xy}}{SS_x} = \frac{184.2857}{143.4286} = 1.285$

$a = \bar{y} - b\bar{x} = 126.79 - (1.285)(21.43) = 99.25$

$y = a + bx$ or $y = 99.25 + 1.285x$

(c) See the figure in part (a).

(d) Let $x = 20$.

$y_p = 99.25 + 1.285(20) = 124.95$

The predicted weight is 124.95 pounds.

(e) The correlation coefficient will be positive because the points are clustered around a straight line going up from left to right.

(f) $r = \frac{SS_{xy}}{\sqrt{SS_x SS_y}} = \frac{184.2857}{\sqrt{143.4286(1080.36)}} = 0.468$

$r^2 = (0.468)^2 = 0.219$

The correlation coefficient r measures the strength of the linear relationship between a woman's weight at age 1 and at age 30. The coefficient of determination r^2 means that 21.9% of the variation in y = weight of a mature adult (30 years old) can be explained by the corresponding variation in x = weight of a 1-year-old baby using the least-squares line.

4. (a) Number of Insurance Sales and Number of Visits

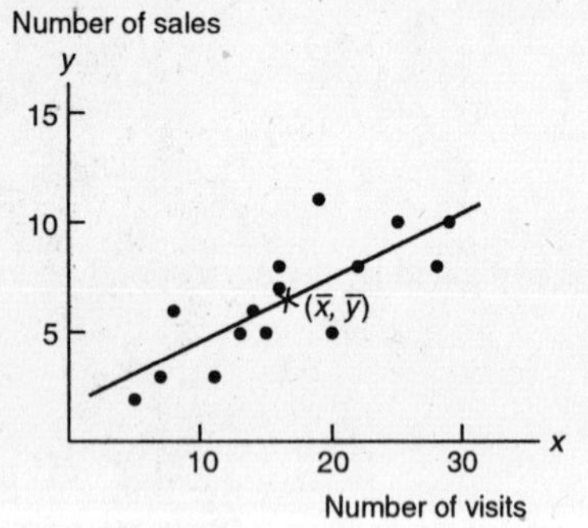

(b) $\bar{x} = \frac{\sum x}{n} = \frac{248}{15} = 16.5\bar{3} \approx 16.53$

$\bar{y} = \frac{\sum y}{n} = \frac{97}{15} = 6.4\bar{6} \approx 6.47$

$b = \frac{SS_{xy}}{SS_x} = \frac{221.2\bar{6}}{755.7\bar{3}} = 0.292784$

$a = \bar{y} - b\bar{x} = 6.4\bar{6} - 0.292784(16.5\bar{3}) = 1.626$

$y = a + bx$ or $y = 1.626 + 0.293x$

(c) See the figure in part (a).

(d) Let $x = 18$.

$y_p = 1.626 + 0.293(18) = 6.9$

The predicted number of sales is 6.9 or 7.

(e) $r = \dfrac{SS_{xy}}{\sqrt{SS_x SS_y}} = \dfrac{221.2\overline{6}}{\sqrt{755.7\overline{3}(103.7\overline{3})}} = 0.790$

$r^2 = (0.790)^2 = 0.624$

This means that 62.4% of the variation in y = number of people who bought insurance that week can be explained by the corresponding variation in x = number of visits made each week using the least-squares line.

5. (a)

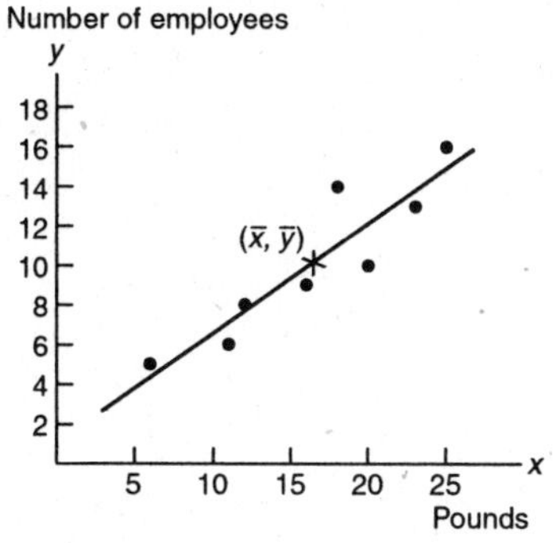

(b) $\bar{x} = \dfrac{\Sigma x}{n} = \dfrac{131}{8} = 16.375 \approx 16.38$

$\bar{y} = \dfrac{\Sigma y}{n} = \dfrac{81}{8} = 10.125 \approx 10.13$

$b = \dfrac{SS_{xy}}{SS_x} = \dfrac{160.625}{289.875} = 0.554118 \approx 0.554$

$a = \bar{y} - b\bar{x} = 10.125 - 0.554118(16.375) = 1.051$

$y = a + bx$ or $y = 1.051 + 0.544x$

(c) See the figure in part (a).

(d) Use $x = 15$.

$y_p = 1.051 + 0.544(15) = 9.36$

About 9 or 10 employees should be assigned mail duty.

(e) $r = \dfrac{SS_{xy}}{\sqrt{SS_x SS_y}} = \dfrac{160.625}{\sqrt{289.875(106.875)}} = 0.913$

$r^2 = (0.913)^2 = 0.834$

The correlation coefficient r measures the strength of the linear association between weight of incoming mail and number of employees assigned to answer it. The coefficient of determination, r^2, means that 83.4% of the variation in y = number of employees can be explained by the corresponding variation in x = weight of incoming mail using the least-squares line.

6. (a) Percent Population Change

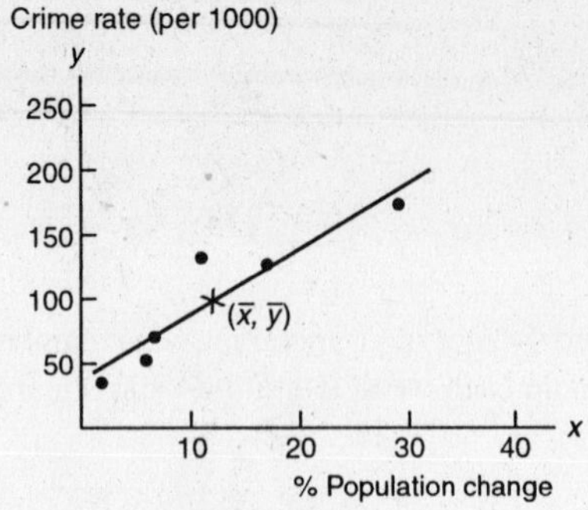

(b) $\bar{x} = \frac{\sum x}{n} = \frac{72}{6} = 12.0$

$\bar{y} = \frac{\sum y}{n} = \frac{589}{6} = 98.1\bar{6} \approx 98.17$

$b = \frac{SS_{xy}}{SS_x} = \frac{2431}{476} = 5.1071 \approx 5.11$

$a = \bar{y} - b\bar{x} = 98.1\bar{6} - 5.1071(12.0) = 36.881 \approx 36.9$

$y = a + bx$ or $y = 36.9 + 5.11x$

See the figure in part (a).

(c) Let $x = 12$

$y = 36.9 + 5.11(12) = 98.2$

The predicted crime rate is 98.2 crimes per thousand.

(d) $r = \frac{SS_{xy}}{\sqrt{SS_x SS_y}} = \frac{2431}{\sqrt{476(14456.8\bar{3})}} = 0.927$

$r^2 = (0.927)^2 = 0.859$

$H_0: \rho = 0$

$H_1: \rho \neq 0$

$t = \frac{r\sqrt{n-2}}{\sqrt{1-r^2}} = \frac{0.927\sqrt{6-2}}{\sqrt{1-(0.927)^2}} = 4.94$

$d.f. = n - 2 = 6 - 2 = 4$

At 1% level of significance, $t_0 = \pm 4.604$.

Since 4.94 > 4.604, we reject H_0 and conclude that the sample evidence supports a significant correlation coefficient.

(e) High correlation does not guarantee a "cause-and-effect" situation. Before causation is established, more work needs to be done taking other variables into account.

High correlation is simply an indication of a mathematical relationship between variables.

Chapter 5 Elementary Probability Theory

Section 5.1

1. Answers vary. Probability is a number between 0 and 1, inclusive, that expresses the likelihood that a specific event will occur. Three ways to find or assign a probability to an event are (1) through intuition (subjective probability), (2) by considering the long-term relative frequency of recurrence of an event in repeated independent trials (empirical probability), and (3) by computing the ratio of the number of favorable outcomes to the total number of possible outcomes, assuming all outcomes are equally likely (classical probability).

2. Answers vary. Probability in business: market research; in medicine: drug tests to determine if a new drug is more effective than the standard treatment; in social science: determining which characteristics to use in creating a profile to detect terrorists; in natural sciences: predicting the likely path and location of landfall for a hurricane.
Statistics is the science of collecting, analyzing, and interpreting quantitative data in such a way that the reliability of the conclusions based on the data can be evaluated objectively. Probability is used in determining the reliability of the results.

3. These are not probabilities: (b) because it is greater than 1, (d) because it is less than zero (negative), (h) 150% = 1.50, because it is greater than 1.

4. Remember $0 \leq$ probability of an event ≤ 1

(a) $-0.41 < 0$

(b) $1.21 > 1$

(c) $120\% = 1.2 > 1$

(d) yes, $0 \leq 0.56 \leq 1$

5. Answers vary. The result is a sample, although not necessarily a good one, showing the relative frequency of people able to wiggle their ears.

6. Answers vary. The results are one example (not necessarily a good one) of the relative frequency of occurrence of raising one eyebrow.

7. (a) $P(\text{no similar preferences}) = P(0) = \frac{15}{375}$, $P(1) = \frac{71}{375}$, $P(2) = \frac{124}{375}$, $P(3) = \frac{131}{375}$, $P(4) = \frac{34}{375}$

(b) $\frac{15+71+124+131+34}{375} = \frac{375}{375} = 1$, yes

Personality types were classified into 4 main preferences; all possible numbers of shared preferences were considered. The sample space is 0, 1, 2, 3, and 4 shared preferences.

8. (a) $P(\text{couple not engaged}) = \frac{200}{1000} = 0.20$, $P(\text{dated less than 1 year}) = \frac{240}{1000} = 0.24$, $P(\text{dated 1 to 2 years}) = \frac{210}{1000} = 0.21$, $P(\text{dated more than 2 years}) = \frac{350}{1000} = 0.35$, based on the number of favorable outcomes divided by the total number of outcomes (1000 couples' engagement status)

(b) $\frac{200+240+210+350}{1000} = \frac{1000}{1000} = 1$, yes

They should add to 1 because all possible outcomes were considered. The sample space is never engaged, engaged less than 1 year, engaged 1 to 2 years, engaged more than 2 years.

9. **(a)** Note: "includes the left limit but not the right limit" means 6 A.M. $\le$ time $t <$ noon, noon $\le t <$ 6 P.M., 6 P.M. $\le t <$ midnight, midnight $\le t <$ 6 A.M. P(best idea 6 A.M.–12 noon) $= \frac{290}{966} \approx 0.30$; P(best idea 12 noon–6 P.M.) $= \frac{135}{966} \approx 0.14$; P(best idea 6 P.M.–12 midnight) $\frac{319}{966} \approx 0.33$; P(best idea from 12 midnight to 6 A.M.) $= \frac{222}{966} \approx 0.23$.

(b) The probabilities add up to 1. They should add up to 1 provided that the intervals do not overlap and each inventor chose only one interval. The sample space is the set of four time intervals.

10. **(a)** $P(\text{germinate}) = \dfrac{\text{number germinated}}{\text{number planted}} = \dfrac{2430}{3000} = 0.81$

(b) $P(\text{not germinate}) = \dfrac{3000 - 2430}{3000} = \dfrac{570}{3000} = 0.19$

(c) The sample space is 2 outcomes, germinate and not germinate.
$P(\text{germinate}) + P(\text{not germinate}) = 0.81 + 0.19 = 1$
The probabilities of all the outcomes in the sample space should and do sum to 1.

(d) no; $P(\text{germinate}) = 0.81$, $P(\text{not germinate}) = 0.19$
If they were equally likely, each would have probability $\frac{1}{2} = 0.5$.

11. Make a table showing the information known about the 127 people who walked by the store: [Example 6 in Section 4.2 uses this technique.]

	Buy	Did not buy	Row Total
Came into the store	25	58 – 25 = 33	58
Did not come in	0	69	127 – 58 = 69
Column Total	25	102	127

If 58 came in, 69 didn't; 25 of the 58 bought something, so 33 came in but didn't buy anything. Those who did not come in, couldn't buy anything. The row entries must sum to the row totals; the column entries must sum to the column totals; and the row totals, as well as the column totals, must sum to the overall total, i.e., the 127 people who walked by the store. Also, the four inner cells must sum to the overall total: $25 + 33 + 0 + 69 = 127$.

This kind of problem relies on formula (2), $P(\text{event } A) = \dfrac{\text{number outcomes favorable to } A}{\text{total number of outcomes}}$. The "trick" is to decide what belongs in the denominator *first*. If the denominator is a row total, stay in that row. If the denominator is a column total, stay in that column. If the denominator is the overall total, the numerator can be a row total, a column total, or the number in any one of the four "cells" inside the table.

(a) total outcomes: people walking by, overall total, 127
favorable outcomes: enter the store, row total, 58 (that's all we know about them)
$P(A) = \frac{58}{127} \approx 0.46$

(b) total outcomes: people who walk into the store, row total 58
favorable outcomes: staying in the row, those who buy: 25
$P(A) = \frac{25}{58} \approx 0.43$

(c) total outcomes: people walking by, overall total 127
favorable outcomes: people coming in *and* buying, the cell at the *intersection* of the "coming in" row and the "buying" column (the upper left corner), 25 (Recall from set theory that "and" means both things happen, that the two sets *intersect*: >)

$P(A) = \frac{25}{127} \approx 0.20$

(d) total outcomes: people coming into the store, row total, 58
favorable outcomes: staying in the row, those who do not buy, 33

$P(A) = \frac{33}{58} \approx 0.57$

$\left(\text{alternate method: this is the complement to (b): } P(A) = 1 - \frac{25}{58} = \frac{33}{58} \approx 0.57\right)$

Section 5.2

1. (a) Green and blue are mutually exclusive because each M&M candy is only 1 color.
$P(\text{green } or \text{ blue}) = P(\text{green}) + P(\text{blue}) = 10\% + 10\% = 20\%$

(b) Yellow and red are mutually exclusive, again, because each candy is only one color, and if the candy is yellow, it can't be red, too.
$P(\text{yellow } or \text{ red}) = P(\text{yellow}) + P(\text{red}) = 20\% + 20\% = 40\%$

(c) It is faster here to use the complementary event rule than to add up the probabilities of all the colors except purple.
$P(not \text{ purple}) = 1 - P(\text{purple}) = 1 - 0.20 = 0.80$, or 80%

2. (a) Green and blue are mutually exclusive because each M&M candy is only 1 color.
$P(\text{green } or \text{ blue}) = P(\text{green}) + P(\text{blue}) = 20\% + 10\% = 30\%$

(b) Yes, mutually exclusive colors
$P(\text{yellow } or \text{ red}) = P(\text{yellow}) + P(\text{red}) = 20\% + 20\% = 40\%$

(c) $P(not \text{ purple}) = 1 - P(\text{purple}) = 1 - 0.20 = 0.80 = 80\%$
Since the percentage of green M&Ms is 10% for plain and 20% for almond, I expect the results for part (a) to be different. Parts (b) and (c) should be the same because the percentages for these colors are the same.

3. (a) Green and blue are mutually exclusive because each M&M candy is only 1 color.
$P(\text{green } or \text{ blue}) = P(\text{green}) + P(\text{blue}) = 16.6\% + 16.6\% = 33.2\%$

(b) Mutually exclusive: $P(\text{yellow } or \text{ red}) = P(\text{yellow}) + P(\text{red}) = 16.6\% + 16.6\% = 33.2\%$

(c) $P(not \text{ purple}) = 1 - P(\text{purple}) = 1 - 0 = 1 = 100\%$ (no purple)
Since the color distributions differ for plain and Dulce de Leche~Carmel M&Ms, I expect the results for all parts to be different. If the answers were the same, it would only be by coincidence.

4. The total number of arches tabled is 288. Arch heights are mutually exclusive because if the height is 12 feet, it can't be 42 feet as well.

(a) $P(3 \text{ to } 9) = \frac{111}{288}$

(b) $P(30 \text{ } or \text{ taller}) = P(30 \text{ to } 49) + P(50 \text{ to } 74) + P(75 \text{ and higher}) = \frac{30}{288} + \frac{33}{288} + \frac{18}{288} = \frac{81}{288}$

(c) $P(3 \text{ to } 49) = P(3-9) + P(10-29) + P(30-49) = \frac{111}{288} + \frac{96}{288} + \frac{30}{288} = \frac{237}{288}$

(d) $P(10 \text{ to } 74) = P(10-29) + P(30-49) + P(50-74) = \frac{96}{288} + \frac{30}{288} + \frac{33}{288} = \frac{159}{288}$

(e) $P(75 \text{ } or \text{ taller}) = \dfrac{18}{288}$

Hint for Problems 5–8: Refer to Figure 4–1 if necessary. (Without loss of generality, let the red die be the first die and the green die be the second die in Figure 4–1.) Think of the outcomes as an (x, y) ordered pair. Then, without loss of generality, (1, 6) means 1 on the red die and 6 on the green die. (We are "ordering" the dice for convenience only — which is first and which is second have no bearing on this problem.) The only important fact is that they are distinguishable outcomes, so that (1 on red, 2 on green) is different from (2 on red, 1 on green).

5. (a) Yes, the outcome of the red die does not influence the outcome of the green die.

(b) $P(5 \text{ on green } and \text{ } 3 \text{ on red}) = P(5 \text{ on green}) \cdot P(3 \text{ on red}) = \left(\dfrac{1}{6}\right)\left(\dfrac{1}{6}\right) = \dfrac{1}{36} \approx 0.028$ because they are independent.

(c) $P(3 \text{ on green } and \text{ } 5 \text{ on red}) = P(3 \text{ on green}) \cdot P(5 \text{ on red}) = \left(\dfrac{1}{6}\right)\left(\dfrac{1}{6}\right) = \dfrac{1}{36} \approx 0.028$

(d) $P[(5 \text{ on green } and \text{ } 3 \text{ on red}) \text{ } or \text{ } (3 \text{ on green } and \text{ } 5 \text{ on red})]$
$= P(5 \text{ on green } and \text{ } 3 \text{ on red}) + P(3 \text{ on green } and \text{ } 5 \text{ on red})]$
$= \dfrac{1}{36} + \dfrac{1}{36} = \dfrac{2}{36} = \dfrac{1}{18} \approx 0.056$ [because they are mutually exclusive outcomes]

6. (a) Yes, the outcome of the red die does not influence the outcome of the green die.

(b) $P(1 \text{ on green } and \text{ } 2 \text{ on red}) = P(1 \text{ on green}) \cdot P(2 \text{ on red}) = \left(\dfrac{1}{6}\right)\left(\dfrac{1}{6}\right) = \dfrac{1}{36}$

(c) $P(2 \text{ on green } and \text{ } 1 \text{ on red}) = P(2 \text{ on green}) \cdot P(1 \text{ on red}) = \left(\dfrac{1}{6}\right)\left(\dfrac{1}{6}\right) = \dfrac{1}{36}$

(d) $P[(1 \text{ on green } and \text{ } 2 \text{ on red}) \text{ } or \text{ } (2 \text{ on green } and \text{ } 1 \text{ on red})]$
$= P(1 \text{ on green } and \text{ } 2 \text{ on red}) + P(2 \text{ on green } and \text{ } 1 \text{ on red})]$
$= \dfrac{1}{36} + \dfrac{1}{36} = \dfrac{2}{36} = \dfrac{1}{18}$ [because they are mutually exclusive outcomes]

7. (a) $1 + 5 = 6, 2 + 4 = 6, 3 + 3 = 6, 4 + 2 = 6, 5 + 1 = 6$
$P(\text{sum} = 6) = P[(1, 5) \text{ } or \text{ } (2, 4) \text{ } or \text{ } (3 \text{ on red}, 3 \text{ on green}) \text{ } or \text{ } (4, 2) \text{ } or \text{ } (5, 1)]$
$= P(1, 5) + P(2, 4) + P(3, 3) + P(4, 2) + P(5, 1)$
since the (red, green) outcomes are mutually exclusive
$= \left(\dfrac{1}{6}\right)\left(\dfrac{1}{6}\right) + \left(\dfrac{1}{6}\right)\left(\dfrac{1}{6}\right) + \left(\dfrac{1}{6}\right)\left(\dfrac{1}{6}\right) + \left(\dfrac{1}{6}\right)\left(\dfrac{1}{6}\right) + \left(\dfrac{1}{6}\right)\left(\dfrac{1}{6}\right)$
because the red die outcome is independent of the green die outcome
$= \dfrac{1}{36} + \dfrac{1}{36} + \dfrac{1}{36} + \dfrac{1}{36} + \dfrac{1}{36} = \dfrac{5}{36}$

(b) $1 + 3 = 4, 2 + 2 = 4, 3 + 1 = 4$
$P(\text{sum is } 4) = P[(1, 3) \text{ } or \text{ } (2, 2) \text{ } or \text{ } (3, 1)]$
$= P(1, 3) + P(2, 2) + P(3, 1)$
because the (red, green) outcomes are mutually exclusive
$= \left(\dfrac{1}{6}\right)\left(\dfrac{1}{6}\right) + \left(\dfrac{1}{6}\right)\left(\dfrac{1}{6}\right) + \left(\dfrac{1}{6}\right)\left(\dfrac{1}{6}\right)$
because the red die outcome is independent of the green die outcome
$= \dfrac{1}{36} + \dfrac{1}{36} + \dfrac{1}{36} = \dfrac{3}{36} = \dfrac{1}{12}$

(c) Since a sum of six can't simultaneously be a sum of 4, these are mutually exclusive events;
$P(\text{sum of } 6 \text{ } or \text{ } 4) = P(\text{sum of } 6) + P(\text{sum of } 4) = \dfrac{5}{36} + \dfrac{3}{36} = \dfrac{8}{36} = \dfrac{2}{9}$

8. (a) $1+6=7, 2+5=7, 3+4=7, 4+3=7, 5+2=7, 6+1=7$

$P(\text{sum is } 7) = P[(1, 6) \text{ or } (2, 5) \text{ or } (3, 4) \text{ or } (4, 3) \text{ or } (5, 2) \text{ or } (6, 1)]$

$= P(1, 6)+P(2, 5)+P(3, 4)+P(4, 3)+P(5, 2)+P(6, 1)$

because the (red, green) outcomes are mutually exclusive

$= \left(\frac{1}{6}\right)\left(\frac{1}{6}\right)+\left(\frac{1}{6}\right)\left(\frac{1}{6}\right)+\left(\frac{1}{6}\right)\left(\frac{1}{6}\right)+\left(\frac{1}{6}\right)\left(\frac{1}{6}\right)+\left(\frac{1}{6}\right)\left(\frac{1}{6}\right)+\left(\frac{1}{6}\right)\left(\frac{1}{6}\right)$

because the red die outcome is independent of the green die outcome

$= \frac{1}{36}+\frac{1}{36}+\frac{1}{36}+\frac{1}{36}+\frac{1}{36}+\frac{1}{36}=\frac{6}{36}=\frac{1}{6}$

(b) $5+6=11, 6+5=11$

$P(\text{sum is } 11) = P[(5, 6) \text{ or } (6, 5)]$

$= P(5, 6)+P(6, 5)$

because the (red, green) outcomes are mutually exclusive

$= \left(\frac{1}{6}\right)\left(\frac{1}{6}\right)+\left(\frac{1}{6}\right)\left(\frac{1}{6}\right)$

because the red die outcome is independent of the green die outcome

$= \frac{1}{36}+\frac{1}{36}=\frac{2}{36}=\frac{1}{18}$

(c) Since a sum of can't be both 7 and 11, they are mutually exclusive

$P(\text{sum is 7 } or \text{ 11}) = P(\text{sum is 7}) + P(\text{sum is 11}) = \frac{6}{36}+\frac{2}{36}=\frac{8}{36}=\frac{2}{9}$

9. (a) No, the key idea here is "without replacement," which means the draws are dependent, because the outcome of the second card drawn depends on what the first card drawn was. Let the card draws be represented by an (x, y) ordered pair. For example, (K, 6) means the first card drawn was a king and the second card drawn was a 6. Here the order of the cards *is* important.

(b) $P(\text{ace on 1st } and \text{ king on second}) = P(\text{ace, king}) = \left(\frac{4}{52}\right)\left(\frac{4}{51}\right)=\frac{16}{2652}=\frac{4}{663}$

There are 4 aces and 4 kings in the deck. Once the first card is drawn and not replaced, there are only 51 cards left to draw from, but all the kings are still there.

(c) $P(\text{king, ace}) = \left(\frac{4}{52}\right)\left(\frac{4}{51}\right)=\frac{16}{2652}=\frac{4}{663}$

There are 4 kings and 4 aces in the deck. Once the first card is drawn and not replaced, there are only 51 cards left to draw from, but all the aces are still there.

(d) $P(\text{ace } and \text{ king in either order})$

$= P[(\text{ace, king}) \text{ or } (\text{king, ace})]$

$= P(\text{ace, king}) + P(\text{king, ace})$ because these two outcomes are mutually exclusive

$= \frac{16}{2652}+\frac{16}{2652}=\frac{32}{2652}=\frac{8}{663}$

10. (a) No, the key idea here is "without replacement," which means the draws are dependent, because the outcome of the second card drawn depends on what the first card drawn was. Let the card draws be represented by an (x, y) ordered pair. For example, (K, 6) means the first card drawn was a king and the second card drawn was a 6. Here the order of the cards *is* important.

(b) $P(3, 10) = P[(3 \text{ on 1st}) \text{ and } (10 \text{ on 2nd, given 3 on 1st})]$

$= P(3 \text{ on 1st}) \cdot P(10 \text{ on 2nd, given 3 on 1st})$

$= \left(\frac{4}{52}\right)\left(\frac{4}{51}\right)=\frac{16}{2652}=\frac{4}{663}\approx 0.006$

(c) $P(10, 3) = P[(10 \text{ on 1st}) \textit{ and } (3 \text{ on 2nd, } \textit{given} \text{ 10 on 1st})]$

$= P(10 \text{ on 1st}) \cdot P(3 \text{ on 2nd, } \textit{given} \text{ 10 on 1st})$

$= \left(\frac{4}{52}\right)\left(\frac{4}{51}\right) = \frac{16}{2652} = \frac{4}{663} \approx 0.006$

(d) $P[(3, 10) \textit{ or } (10, 3)] = P(3, 10) + P(10, 3)$ since these 2 outcomes are mutually exclusive

$= \frac{4}{663} + \frac{4}{663} = \frac{8}{663} \approx 0.012$

11. (a) Yes; the key idea here is "with replacement." When the first card drawn is replaced, the sample space is the same when the second card is drawn as it was when the first card was drawn and the second card is in no way influenced by the outcome of the first draw; in fact, it is possible to draw the same card twice. Let the card draws be represented by an (x, y) ordered pair; for example (K, 6) means a king was drawn first, replaced, and then the second card, a "6," was drawn independently of the first.

(b) $P(\text{A, K}) = P(\text{A}) \cdot P(\text{K})$ because they are independent

$= \left(\frac{4}{52}\right)\left(\frac{4}{52}\right) = \frac{16}{2704} = \frac{1}{169}$

(c) $P(\text{K, A}) = P(\text{K}) \cdot P(\text{A})$ because they are independent

$= \left(\frac{4}{52}\right)\left(\frac{4}{52}\right) = \frac{16}{2704} = \frac{1}{169}$

(d) $P[(\text{A, K}) \textit{ or } (\text{K, A})] = P(\text{A, K}) + P(\text{K, A})$ since the 2 outcomes are mutually exclusive when we consider the order

$= \frac{1}{169} + \frac{1}{169} = \frac{2}{169}$

12. (a) Yes; the key idea here is "with replacement." When the first card drawn is replaced, the sample space is the same when the second card is drawn as it was when the first card was drawn and the second card is in no way influenced by the outcome of the first draw; in fact, it is possible to draw the same card twice. Let the card draws be represented by an (x, y) ordered pair; for example (K, 6) means a king was drawn first, replaced, and then the second card, a "6," was drawn independently of the first.

(b) $P(3, 10) = P(3) \cdot P(10)$ because draws are independent

$= \left(\frac{4}{52}\right)\left(\frac{4}{52}\right) = \frac{16}{2704} = \frac{1}{169} \approx 0.0059$

(c) $P(10, 3) = P(10) \cdot P(3)$ because of independence

$= \left(\frac{4}{52}\right)\left(\frac{4}{52}\right) = \frac{16}{2704} = \frac{1}{169} \approx 0.0059$

(d) $P[(3, 10) \textit{ or } (10, 3)] = P(3, 10) + P(10, 3)$ because these outcomes are mutually exclusive

$= \frac{1}{169} + \frac{1}{169} = \frac{2}{169} \approx 0.0118$

13. (a) $P(6 \textit{ or } \text{older}) = P[(6 \text{ to } 9) \textit{ or } (10 \text{ to } 12) \textit{ or } (13 \text{ and over})]$

$= P(6-9) + P(10-12) + P(13+)$ because they are mutually exclusive age groups – no child is both 7 and 11 years old.

$= 27\% + 14\% + 22\% = 63\% = 0.63$

(b) $P(12 \text{ or younger}) = 1 - P(13 \text{ and over}) = 1 - 0.22 = 0.78$

(c) $P(\text{between 6 and 12}) = P[(6 \text{ to } 9) \textit{ or } (10 \text{ to } 12)]$

$= P(6 \text{ to } 9) + P(10 \text{ to } 12)$ because the age groups are mutually exclusive

$= 27\% + 14\% = 41\% = 0.41$

(d) $P(\text{between 3 and 9}) = P[(3 \text{ to } 5) \textit{ or } (6 \text{ to } 9)]$

$= P(3 \text{ to } 5) + P(6 \text{ to } 9)$ because age categories are mutually exclusive

$= 22\% + 27\% = 49\% = 0.49$

Answers vary; however, category 10–12 years covers only 3 years while 13 and over covers many more years and many more people, including adults who buy toys for themselves.

14. What we know: $P(\text{seniors get flu}) = 0.14$,
$P(\text{younger people get flu}) = 0.24$
$P(\text{senior}) = 0.125$
Let S denote seniors, so *not* S denotes younger people. Let F denote flu and *not* F denote did not get the flu. So $P(\text{F, } \textit{given } \text{S}) = 0.14$, $P(\text{F, } \textit{given not } \text{S}) = 0.24$ and $P(\text{S}) = 0.125$ so $P(\textit{not } \text{S}) = 1 - 0.125 = 0.875$. Note the phrases 14% *of* seniors, i.e., they were already seniors, so this is a given condition; and 24% *of* people under 65, i.e., these people were already under 65, so under 65 (younger) is a given condition.

(a) $P(\text{person is senior } \textit{and} \text{ will get flu}) = P(\text{S } \textit{and} \text{ F})$

$= P(\text{S}) \cdot P(\text{F, } \textit{given } \text{S}) = (0.125)(0.14) = 0.0175$

conditional probability rule

(b) $P(\text{person is } \textit{not} \text{ senior } \textit{and} \text{ will get flu}) = P[(\textit{not } \text{S}) \textit{ and } \text{F}]$

$= P(\textit{not } \text{S}) \cdot P(\text{F, } \textit{given not } \text{S}) = 0.875(0.24) = 0.21$

(c) Here, $P(\text{S}) = 0.95$ so $P(\textit{not } \text{S}) = 1 - 0.95 = 0.05$
(a) $P(\text{S } \textit{and} \text{ F}) = P(\text{S}) \cdot P(\text{F, } \textit{given } \text{S}) = (0.95)(0.14) = 0.133$
(b) $P(\textit{not } \text{S } \textit{and} \text{ F}) = P(\textit{not } \text{S}) \cdot P(\text{F, } \textit{given not } \text{S}) = (0.05)(0.24) = 0.012$

(d) Here, $P(\text{S}) = P(\textit{not } \text{S}) = 0.50$
(a) $P(\text{S } \textit{and} \text{ F}) = P(\text{S}) \cdot P(\text{F, } \textit{given } \text{S}) = 0.50(0.14) = 0.07$
(b) $P(\textit{not } \text{S } \textit{and} \text{ F}) = P(\textit{not } \text{S}) \cdot P(\text{F, } \textit{given not } \text{S}) = 0.50(0.24) = 0.12$

15. What we know: $P(\text{polygraph says "lying" when person is lying}) = 72\%$
$P(\text{polygraph says "lying" when person is not lying}) = 7\%$
Let L denote that the polygraph results show lying and *not* L denote that the polygraph results show the person is not lying. Let T denote that the person is telling the truth and let *not* T denote that the person is not telling the truth, so $P(\text{L, } \textit{given not } \text{T}) = 72\%$
$P(\text{L, } \textit{given } \text{T}) = 7\%$.
We are told whether the person is telling the truth or not; what we know is what the polygraph results are, given the case where the person tells the truth, and given the situation where the person is not telling the truth.

(a) $P(\text{T}) = 0.90$ so $P(\textit{not } \text{T}) = 0.10$
$P(\text{polygraph says lying and person tells truth})$
$= P(\text{L } \textit{and} \text{ T}) = P(\text{T}) \cdot P(\text{L, } \textit{given } \text{T})$
$= (0.90)(0.07) = 0.063 = 6.3\%$

(b) $P(\textit{not } \text{T}) = 0.10$ so $P(\text{T}) = 0.90$
$P(\text{polygraph says lying and person is not telling the truth})$
$= P(\text{L } \textit{and not } \text{T}) = P(\textit{not } \text{T}) \cdot P(\text{L, } \textit{given not } \text{T})$
$= (0.10)(0.72) = 0.072 = 7.2\%$

(c) $P(\text{T}) = P(\textit{not } \text{T}) = 0.50$
(a) $P(\text{L } \textit{and} \text{ T}) = P(\text{T}) \cdot P(\text{L, } \textit{given } \text{T})$
$= (0.50)(0.07) = 0.035 = 3.5\%$
(b) $P(\text{L } \textit{and not } \text{T}) = P(\textit{not } \text{T}) \cdot P(\text{L, } \textit{given not } \text{T})$
$= (0.50)(0.72) = 0.36 = 36\%$

(d) $P(\text{T}) = 0.15$ so $P(not\ \text{T}) = 1 - P(\text{T}) = 1 - 0.15 = 0.85$

(a) $\text{P}(\text{L}\ and\ \text{T}) = P(\text{T}) \cdot P(\text{L},\ given\ \text{T})$

$= (0.15)(0.07) = 0.0105 = 1.05\%$

(b) $\text{P}(\text{L}\ and\ not\ \text{T}) = P(not\ \text{T}) \cdot P(\text{L},\ given\ not\ \text{T})$

$= (0.85)(0.72) = 0.612 = 61.2\%$

16. What we know: P(polygraph says "lying" when person is lying) = 72%
P(polygraph says "lying" when person is not lying) = 7%
Let L denote that the polygraph results show lying and *not* L denote that the polygraph results show the person is not lying. Let T denote that the person is telling the truth and let *not* T denote that the person is not telling the truth, so $P(\text{L},\ given\ not\ \text{T}) = 72\%$

$P(\text{L},\ given\ \text{T}) = 7\%$

We are told whether the person is telling the truth or not; what we know is what the polygraph results are, given the case where the person tells the truth, and given the situation where the person is not telling the truth.

(a) P(polygraph reports "lying") = $P(\text{L}) = 30\%$
We want to find P(person is lying) = $P(not\ \text{T})$
There are two possibilities when the polygraph says the person is lying: either the polygraph is right, or the polygraph is wrong. If the polygraph is right, the polygraph results show "lying" and the person is not telling the truth, i.e., $P(\text{L}\ and\ not\ \text{T})$. If the polygraph is wrong, then the polygraph results show "lying" but, in fact, the person is telling the truth, i.e., $P(\text{L}\ and\ \text{T})$. (This is the basic "trick" to this problem, and the idea comes directly from set theory.)

So $P(\text{L}) = P(\text{L}\ and\ not\ \text{T}) + P(\text{L}\ and\ \text{T})$

$= [P(not\ \text{T}) \cdot P(\text{L},\ given\ not\ \text{T})] + [P(\text{T}) \cdot P(\text{L},\ given\ \text{T})]$

using conditional probability rules

$= [P(not\ \text{T}) \cdot P(\text{L},\ given\ not\ \text{T})] + \{[1 - P(not\ \text{T})] \cdot P(\text{L},\ given\ \text{T})\}$

using the complementary event rule to rewrite $P(\text{T})$ as $1 - P(not\ \text{T})$

$0.30 = [P(not\ \text{T})] \cdot (0.72)] + [(1 - P(not\ \text{T})] \cdot (0.07)$

substituting in the known values as given in # 15, and as given above

$= (0.72) \cdot P(not\ \text{T}) + [0.07 - (0.07) \cdot P(not\ \text{T})]$

$0.30 - 0.07 = (0.72) \cdot P(not\ \text{T}) - (0.07) \cdot P(not\ \text{T})$

$0.23 = P(not\ \text{T})(0.72 - 0.07) = P(not\ \text{T})(0.65)$

$\frac{0.23}{0.65} = P(not\ \text{T})$, or $P(not\ \text{T}) \approx 0.354 = 35.4\%$

(b) Here, $P(\text{L}) = 70\% = 0.70$
This is the same as (a) except for the new $P(\text{L})$. Starting from the step in (a) just before we substituted in the numerical values we knew:

$P(\text{L}) = [P(not\ \text{T}) \cdot P(\text{L},\ given\ not\ \text{T})] + \{[1 - P(not\ \text{T})] \cdot P(\text{L},\ given\ \text{T})\}$

$0.70 = P(not\ \text{T}) \cdot (0.72) + [1 - P(not\ \text{T})] \cdot (0.07)$

$0.70 = (0.72) \cdot P(not\ \text{T}) + [0.07 - 0.07 \cdot P(not\ \text{T})]$

$0.70 - 0.07 = (0.72 - 0.07) \cdot P(not\ \text{T})$

$0.63 = 0.65P(not\ \text{T})$

so $P(not\ \text{T}) = \frac{0.63}{0.65} \approx 0.969 = 96.9\%$.

17. We have $P(A) = \frac{580}{1160}$, $P(Pa) = \frac{580}{1160} = P(not\ A)$, $P(S) = \frac{686}{1160}$, $P(N) = \frac{474}{1160} = P(not\ S)$

(a) $P(S) = \frac{686}{1160}$

$P(S,\ given\ A) = \frac{270}{580}$ (*given A* means stay in the *A*, aggressive row)

$P(S,\ given\ Pa) = \frac{416}{580}$ (staying in row *Pa*)

(b) $P(S) = \dfrac{686}{1160} = \dfrac{343}{580}$

$P(S, given\ Pa) = \dfrac{416}{580}$

They are not independent since the probabilities are not the same.

(c) $P(A\ and\ S) = P(A) \cdot P(S,\ given\ A)$

$= \left(\dfrac{580}{1160}\right)\left(\dfrac{270}{580}\right) = \dfrac{270}{1160}$

$P(Pa\ and\ S) = P(Pa) \cdot P(S,\ given\ Pa)$

$= \left(\dfrac{580}{1160}\right)\left(\dfrac{416}{580}\right) = \dfrac{416}{1160}$

(d) $P(N) = \dfrac{474}{1160}$

$P(N,\ given\ A) = \dfrac{310}{580}$ (stay in the A row)

(e) $P(N) = \dfrac{474}{1160} = \dfrac{237}{580}$

$P(N,\ given\ A) = \dfrac{310}{580}$

Since the probabilities are not the same, N and A are not independent.

(f) $P(A\ or\ S) = P(A) + P(S) - P(A\ and\ S)$

$= \dfrac{580}{1160} + \dfrac{686}{1160} - \dfrac{270}{1160} = \dfrac{996}{1160}$

18. **(a)** $P(+,$ *given* condition present$) = \dfrac{110}{130}$ (stay in "condition present" row)

(b) $P(-,$ *given* condition present$) = \dfrac{20}{130}$ (stay in "condition present" row)

[(a) and (b) are complementary events]

(c) $P(-,$ *given* condition absent$) = \dfrac{50}{70}$ (stay in the row or column of the "*given*")

(d) $P(+,$ *given* condition absent$) = \dfrac{20}{70}$

(e) $P($condition present *and* $+) = P($condition present$) \cdot P(+,$ *given* condition present$)$

$= \left(\dfrac{130}{200}\right)\left(\dfrac{110}{130}\right) = \dfrac{110}{200}$

(f) $P($condition present *and* $-) = P($condition present$) \cdot P(-,$ *given* condition present$)$

$= \left(\dfrac{130}{200}\right)\left(\dfrac{20}{130}\right) = \dfrac{20}{200}$

19. Let C denote the condition is present, and *not* C denote the condition is absent.

(a) $P(+,$ *given* C$) = \dfrac{72}{154}$ (stay in C column)

(b) $P(-,$ *given* C$) = \dfrac{82}{154}$ (stay in C column)

(c) $P(-,$ *given not* C$) = \dfrac{79}{116}$ (stay in *not* C column)

(d) $P(+,$ *given not* C$) = \dfrac{37}{116}$ (stay in *not* C column)

(e) $P(\text{C } and +) = P(\text{C}) \cdot P(+, \textit{given } \text{C}) = \left(\frac{154}{270}\right)\left(\frac{72}{154}\right) = \frac{72}{270}$

(f) $P(\text{C } and -) = P(\text{C}) \cdot P(-, \textit{given } \text{C}) = \left(\frac{154}{270}\right)\left(\frac{82}{154}\right) = \frac{82}{270}$

20. First determine the denominator. If it is a row or column total, the numerator will be in the body (inside) of the table in that same row or column. If the denominator is the grand total the numerator can be one or more row totals, one or more column totals, or a body-of-the-table cell entry. A cell entry is usually indicated when the problem mentions both a row category and a column category, in which case the desired cell is the one where the row and column intersect.

(a) customer at random, denominator is grand total, 2008; loyal 10–14 years, numerator is column total, 291; $\frac{291}{2008}$

(b) given: customer is from the East, so denominator is row total, 452; loyal 10–14 years: the cell entry in that row, 77; $\frac{77}{452}$

(c) no qualifiers on the customers, so denominator is grand total, 2008; at least 10 years: need entry for 10–14 years and 15+ years, so numerator is the sum of these 2 column totals, 291 + 535 = 826; $\frac{826}{2008}$

(d) given: from the West means the denominator is the West row total, 373; loyal at least 10 years means we sum the numbers in the West row for 10–14 and 15+ years, 45 + 86 = 131; $\frac{131}{373}$

(e) given: loyal less than 1 year means the denominator is column total, 157; from the West means the numerator is the cell entry for West in that column, 41; $\frac{41}{157}$

(f) given: loyal < 1 year, so denominator is 157; from South, so numerator is at the intersection of the < 1 year column and the South row, 53; $\frac{53}{157}$

(g) given: East, so denominator is 452; loyal 1+ years: either add up all the entries except < 1 year in the East row, or use the complementary event rule (less work!);

$$P(\text{loyal } 1+ \text{ years, } \textit{given} \text{ East}) = 1 - P(\text{loyal} < 1 \text{ year, } \textit{given} \text{ East})$$
$$= 1 - \frac{32}{452} = \frac{420}{452}$$

(h) given: West, so denominator is West row total, 373; loyal 1+ years is either the sum of all the West row entries except < 1 year, or apply the complementary event rule in the probability calculation;

$$P(\text{loyal } 1+ \text{ years, } \textit{given} \text{ West}) = 1 - P(\text{loyal} < 1 \text{ year, } \textit{given} \text{ West})$$
$$= 1 - \frac{41}{373} = \frac{332}{373}$$

(i) $P(\text{East}) = \frac{452}{2008} \approx 0.2251$

$P(\text{loyal 15+ years}) = \frac{535}{2008} \approx 0.2664$

$P(\text{East, } \textit{given} \text{ 15+ years}) = \frac{118}{535} \approx 0.2206$

$P(\text{loyal 15+ years, } \textit{given} \text{ East}) = \frac{118}{452} \approx 0.2611$

If they are independent, $P(\text{East}) = P(\text{East, } \textit{given} \text{ 15+ years})$ but $0.2251 \neq 0.2206$, and if they are independent, $P(\text{loyal 15+ years}) = P(\text{loyal 15+ years, } \textit{given} \text{ East})$ but $0.2664 \neq 0.2611$, so they aren't independent. (If you use decimal approximations, and the 2 probabilities are quite close, it's time to reduce fractions or use the least common denominator to get an accurate comparison.)
Note: independence is symmetric, i.e., if *A* is independent of *B*, then *B* is independent of *A*; this means you don't have to do *both* independence checks; one is sufficient.

Section 5.3

1. (a) Outcomes for Tossing a Coin Three Times

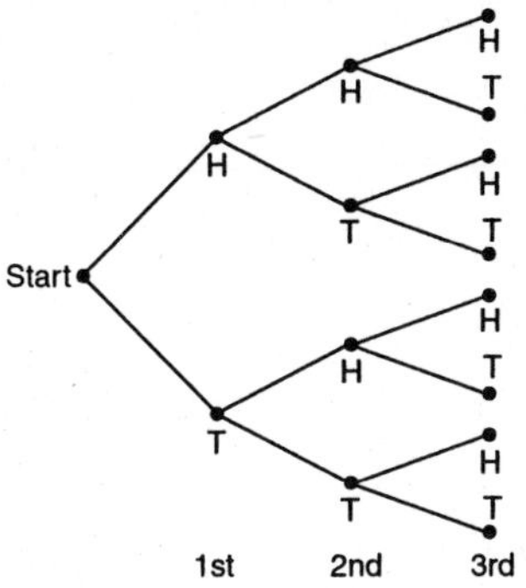

(b) HHT, HTH, THH: 3

(c) 8 possible outcomes, 3 with exactly 2 Hs: $\frac{3}{8}$

2. (a) Outcomes of Tossing a Coin and Throwing a Die

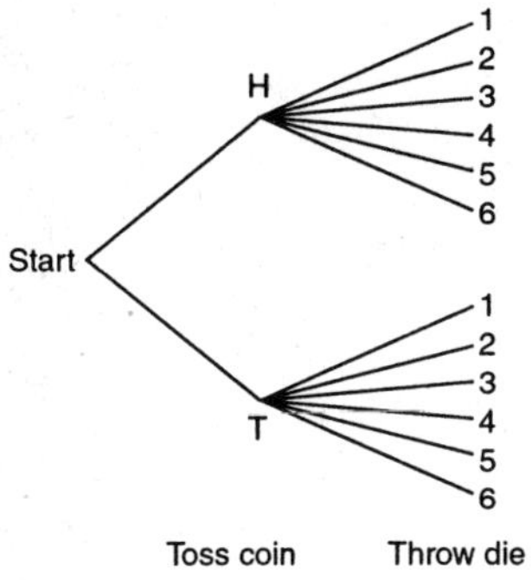

(b) outcomes with H and > 4
H5, H6:2

(c) 12 outcomes, two with H and > 4: $\frac{2}{12} = \frac{1}{6}$

3. (a) Outcomes for Drawing Two Balls (without replacement)

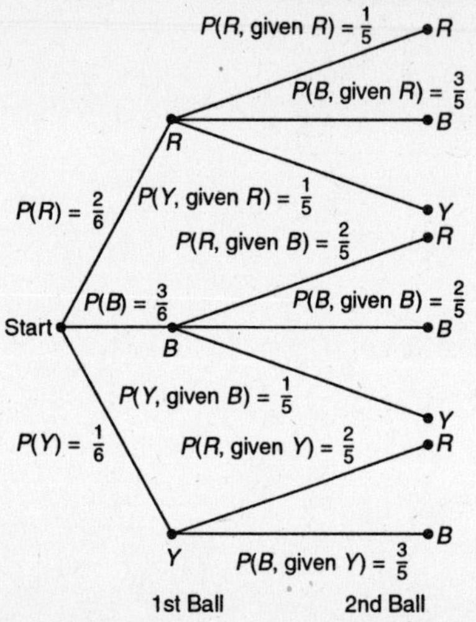

Because we drew without replacement the number of available balls drops to 5 and one of the colors drops by 1. Note that if the yellow ball is drawn first, there are only two possibilities for the second draw: red and blue; the yellow balls are exhausted.

(b) $P(R, R) = \left(\frac{2}{6}\right)\left(\frac{1}{5}\right) = \frac{2}{30} = \frac{1}{15}$

$P(R, B) = \left(\frac{2}{6}\right)\left(\frac{3}{5}\right) = \frac{6}{30} = \frac{1}{5}$

$P(R, Y) = \left(\frac{2}{6}\right)\left(\frac{1}{5}\right) = \frac{2}{30} = \frac{1}{15}$

$P(B, R) = \left(\frac{3}{6}\right)\left(\frac{2}{5}\right) = \frac{6}{30} = \frac{1}{5}$

$P(B, B) = \left(\frac{3}{6}\right)\left(\frac{2}{5}\right) = \frac{6}{30} = \frac{1}{5}$

$P(B, Y) = \left(\frac{3}{6}\right)\left(\frac{1}{5}\right) = \frac{3}{30} = \frac{1}{10}$

$P(Y, R) = \left(\frac{1}{6}\right)\left(\frac{2}{5}\right) = \frac{2}{30} = \frac{1}{15}$

$P(Y, B) = \left(\frac{1}{6}\right)\left(\frac{3}{5}\right) = \frac{3}{30} = \frac{1}{10}$

where $P(x, y)$ is the probability the first ball is color x, and the second ball is color y. Multiply the branch probability values along each branch from start to finish. Observe the sum of the probabilities is 1.

4. (a) Outcomes of Three Multiple-Choice Questions

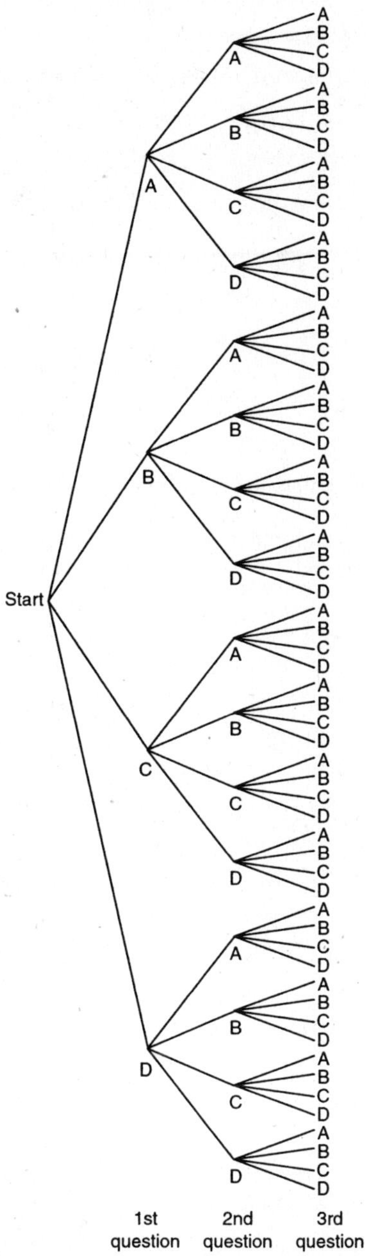

This is a gaudier version of problems 1 and 5 where there are 3 questions, but now there are 4 responses (A, B, C, D) for each question at each step.

(b) If the outcomes are equally likely, then $P(\text{all 3 correct}) = \left(\frac{1}{4}\right)\left(\frac{1}{4}\right)\left(\frac{1}{4}\right) = \frac{1}{64}$.

5. 4 wire choices for the first leaves 3 wire choices for the second, 2 for the third, and only 1 wire choice for the fourth wire connection: $4 \cdot 3 \cdot 2 \cdot 1 = 4! = 24$.

6. 4 choices for his first stop, 3 for the second, 2 for the third, and only 1 city for his (last) fourth stop: $4 \cdot 3 \cdot 2 \cdot 1 = 4! = 24$. This problem is identical to problem 7 except wires were changed to cities.

7. (a) Choose 1 card from each deck. The number of pairs (one card from the first deck and one card from the second) is $52 \cdot 52 = 52^2 = 2704$.

(b) There are 4 kings in the first deck and four in the second, so $4 \cdot 4 = 16$.

(c) There are 16 ways to draw a king from each deck, and 2704 ways to draw a card from each deck, so $\frac{16}{2704} = \frac{1}{169} \approx 0.006$.

8. (a) The die rolls are independent, so multiply the 6 ways the first die can land by the 6 ways the second die can land: $6 \cdot 6 = 36$.

(b) Even numbers are 2, 4, and 6, three possibilities per die, so $3 \cdot 3 = 9$.

(c) $P(\text{even, even}) = \frac{9}{36} = \frac{1}{4} = 0.25$

using $P(\text{event}) = \dfrac{\text{number of favorable outcomes}}{\text{total number of outcomes}}$

Problems 9–12 deal with permutations, $P_{n,r} = \dfrac{n!}{(n-r)!}$. This counts the number of ways r objects can be selected from n when the order of the result is important. For example, if we choose two people from a group, the first of which is to be the group's chair, and the second, the assistant chair, then (John, Mary) is distinct from (Mary, John).

9. $P_{5,2}: n = 5,\ r = 2$

$$P_{5,2} = \frac{5!}{(5-2)!} = \frac{5\cdot 4\cdot 3\cdot 2\cdot 1}{3!} = 20$$

10. $P_{8,3}: n = 8,\ r = 3$

$$P_{8,3} = \frac{8!}{(8-3)!} = \frac{8\cdot 7\cdot 6\cdot 5\cdot 4\cdot 3\cdot 2\cdot 1}{5!} = \frac{8\cdot 7\cdot 6\cdot 5!}{5!} = 336$$

11. $P_{7,7}: n = r = 7$

$$P_{7,7} = \frac{7!}{(7-7)!} = \frac{7!}{0!} = 7! = 5040 \text{ (recall } 0! = 1)$$

In general, $P_{n,n} = \dfrac{n!}{(n-n)!} = \dfrac{n!}{0!} = \dfrac{n!}{1} = n!$.

12. $P_{9,9}: n = r = 9$

$$P_{9,9} = \frac{9!}{(9-9)!} = \frac{9!}{0!} = \frac{9!}{1} = 362{,}880$$

In general, $P_{n,n} = \dfrac{n!}{(n-n)!} = \dfrac{n!}{0!} = \dfrac{n!}{1} = n!$.

Problems 13–16 deal with combination, $C_{n,r} = \dfrac{n!}{r!(n-r)!}$. This counts the number of ways r items can be selected from among n items when the order of the result doesn't matter. For example, when choosing two people from an office to pick up coffee and doughnuts, (John, Mary) is the same as (Mary, John) — both get to carry the goodies back to the office.

13. $C_{5,2}$: $n = 5$, $r = 2$

$$C_{5,2} = \frac{5!}{2!(5-2)!} = \frac{5!}{2!3!} = \frac{5\cdot4\cdot3\cdot2\cdot1}{2\cdot1\cdot3\cdot2\cdot1} = \frac{20}{2} = 10$$

14. $C_{8,3}$: $n = 8$, $r = 3$

$$C_{8,3} = \frac{8!}{3!(8-3)!} = \frac{8!}{3!5!} = \frac{8\cdot7\cdot6\cdot5!}{3\cdot2\cdot1\cdot5!} = 56$$

15. $C_{7,7}$: $n = r = 7$

$$C_{7,7} = \frac{7!}{7!(7-7)!} = \frac{7!}{7!0!} = \frac{7!}{7!(1)} = 1 \text{ (recall } 0! = 1)$$

In general, $C_{n,n} = \dfrac{n!}{n!(n-n)!} = \dfrac{n!}{n!0!} = \dfrac{n!}{n!(1)} = 1$. There is only 1 way to choose all n objects without regard to order.

16. $C_{8,8}$: $n = r = 8$

$$P_{9,9} = \frac{8!}{8!(8-8)!} = \frac{8!}{8!0!} = \frac{8!}{8!(1)} = 1 \text{ (recall } 0! = 1)$$

In general, $C_{n,n} = \dfrac{n!}{n!(n-n)!} = \dfrac{n!}{n!0!} = \dfrac{n!}{n!(1)} = 1$. There is only 1 way to choose all n objects without regard to order.

17. Since the order matters (first is day supervisor, second is night supervisor, and third is coordinator), this is a permutation of 15 nurse candidates to fill 3 positions.

$$P_{15,3} = \frac{15!}{(15-3)!} = \frac{15!}{12!} = \frac{15\cdot14\cdot13\cdot12!}{12!} = 2730$$

18. The order of the software packages selected doesn't matter, since all three are going home with the customer. (Assume the software packages are of equal interest to the customer.)

$$C_{10,3} = \frac{10!}{3!(10-3)!} = \frac{10!}{3!7!} = \frac{10\cdot9\cdot8\cdot7!}{3!7!} = \frac{720}{6} = 120$$

19. The order of trainee selection doesn't matter, since they are all going to be trained the same.

$$C_{15,5} = \frac{15!}{5!(15-5)!} = \frac{15!}{5!10!} = \frac{15\cdot14\cdot13\cdot12\cdot11\cdot10!}{5!10!} = \frac{15\cdot14\cdot13\cdot12\cdot11}{5\cdot4\cdot3\cdot2\cdot1} = 3003$$

20. It doesn't matter in which order the professor grades the problems, the 5 selected problems all get graded.

(a) $C_{12,5} = \dfrac{12!}{5!(12-5)!} = \dfrac{12!}{5!7!} = \dfrac{12\cdot11\cdot10\cdot9\cdot8\cdot7!}{5!7!} = \dfrac{12\cdot11\cdot10\cdot9\cdot8}{5\cdot4\cdot3\cdot2\cdot1} = 792$

(b) Jerry must have the very same 5 problems as the professor selected to grade, so

$P(\text{Jerry chose the right problems}) = \dfrac{1}{792} \approx 0.001$. (Jerry is pushing his luck.)

(c) Silvia did seven problems, which have $C_{7,5}$ subsets of 5 problems which would be among 792 subsets of 5 the professor selected from.

$$C_{7,5} = \frac{7!}{5!(7-5)!} = \frac{7!}{5!2!} = \frac{7 \cdot 6 \cdot 5!}{5!2!} = \frac{7 \cdot 6}{2 \cdot 1} = \frac{42}{2} = 21$$

$$P(\text{Silvia lucked out}) = \frac{21}{792} \approx 0.027$$

(Silvia is pushing her luck, too, but she increased her chances by a factor of 21, compared to Jerry, just by doing two more problems. Now, if these two had just done all the problems, or even split them half and half, ...)

21. (a) Six applicants are selected from among 12 without regard to order.

$$C_{12,6} = \frac{12!}{6!6!} = \frac{479{,}001{,}600}{(720)^2} = 924$$

(b) There are 7 women and 5 men. This problem really asks, in how many ways can 6 women be selected from among 7, and zero men be selected from 5?

$$(C_{7,6})(C_{5,0}) = \left(\frac{7!}{6!(7-6)!}\right)\left(\frac{5!}{0!(5-0)!}\right) = \frac{7!}{6!1!} \cdot \frac{5!}{0!5!} = 7$$

Since the zero men are "selected" be default, all positions being filled. This problem reduces to, in how many ways can 6 applicants be selected from 7 women?

(c) $P(\text{event A}) = \dfrac{\text{number of favorable outcomes}}{\text{total number of outcomes}}$

$$P(\text{all hired are women}) = \frac{7}{924} = \frac{1}{132} \approx 0.008$$

22. It doesn't matter in which order you or the state select the 6 numbers each. It only matters that you and the state pick the *same* six numbers. While you spend your zillion dollars, you can always reorder your numbers if you want to.

(a) $C_{42,6} = \dfrac{42!}{6!(42-6)!} = \dfrac{42!}{6!36!} = \dfrac{42 \cdot 41 \cdot 40 \cdot 39 \cdot 38 \cdot 37 \cdot 36!}{6!36!} = \dfrac{3{,}776{,}965{,}920}{720} = 5{,}245{,}786$

(Most calculators will handle numbers through 69! But if you hate to see numbers like 1.771×10^{98}, cancel out the common factorial factors, such as 36! here.)

(b) This problem asks, what is the chance you choose the very same 6 numbers the state chose.

$$P(\text{winning ticket}) = \frac{1}{5{,}245{,}786} \approx 0.000000191$$

(c) What is the chance one of your 10 tickets is the winning ticket? (We'll assume each ticket is different from the other 9 you have, but, it really doesn't matter much ...)

$$P(\text{win}) = \frac{10}{5{,}245{,}786} = \frac{5}{2{,}622{,}893} \approx 0.0000019$$

Chapter 5 Review

1. $P(\text{asked}) = 24\% = 0.24$
$P(\text{received, } \textit{given} \text{ asked}) = 45\% = 0.45$
$P(\text{asked } \textit{and} \text{ received}) = P(\text{asked}) \cdot P(\text{received, } \textit{given} \text{ asked}) = (0.24)(0.45) = 0.108 = 10.8\%$

2. $P(\text{asked}) = 20\% = 0.20$
$P(\text{received, } \textit{given} \text{ asked}) = 59\% = 0.59$
$P(\text{asked } \textit{and} \text{ received}) = P(\text{asked}) \cdot P(\text{received, } \textit{given} \text{ asked}) = (0.20)(0.59) = 0.118 = 11.8\%$

3. (a) If the first card is replaced before the second is chosen (sampling with replacement), they are independent. If the sampling is without replacements they are dependent.

(b) $P(\text{heart}) = \frac{13}{52} = \frac{1}{4}$
with replacement, independent
$P(\text{H on both}) = \left(\frac{1}{4}\right)\left(\frac{1}{4}\right) = \frac{1}{16} = 0.0625 \approx 0.063$

(c) without replacement, dependent
$P(\text{H on first } \textit{and} \text{ H on second}) = \frac{13}{52} \cdot \frac{12}{51} = \frac{156}{2652} \approx 0.059$

4. (a) There are 11 other outcomes besides 3H. The sample space is the 12 outcomes shown.

1H	1T
2H	2T
(3H)	3T
4H	4T
5H	5T
6H	6T

(b) Yes; the die and the coin are independent. Each outcome has probability $\left(\frac{1}{6}\right)\left(\frac{1}{2}\right) = \frac{1}{12}$.

(c) $P(\text{H } \textit{and} \text{ number} < 3) = P[\text{H } \textit{and} \text{ (1 } \textit{or} \text{ 2)}] = P(\text{1H } \textit{or} \text{ 2H}) = P(\text{1H}) + P(\text{2H}) = \frac{1}{12} + \frac{1}{12} = \frac{2}{12} = \frac{1}{6} \approx 0.167$

5. (a) Throw a large number of similar thumbtacks, or one thumbtack a large number of times, and record the frequency of occurrence of the various outcomes. Assume the thumbtack falls either flat side down (i.e., point up), or tilted (with the point down, resting on the edge of the flat side). (We will

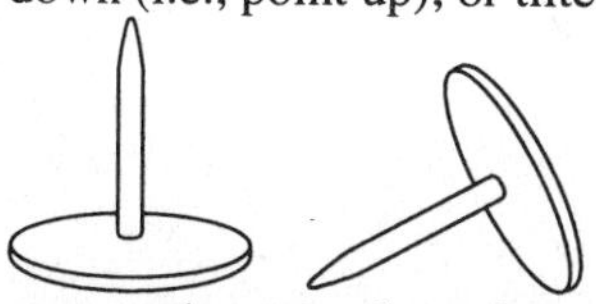

assume these are the only two ways the tack can land.) To estimate the probability the tack lands on its flat side with the point up, find the relative frequency of this occurrence, dividing the number of times this occurred by the total number of thumbtack tosses.

(b) The sample space is the two outcomes flat side down (point up) and tilted (point down).

(c) $P(\text{flat side down, point up}) = \frac{340}{500} = 0.68$
$P(\text{tilted, point down}) = 1 - 0.68 = 0.32$

6. **(a)** $P(N) = \frac{470}{1000} = 0.470$

$P(M) = \frac{390}{1000} = 0.390$

$P(S) = \frac{140}{1000} = 0.140$

(b) $P(N, \textit{given } W) = \frac{420}{500} = 0.840$

$P(S, \textit{given } W) = \frac{20}{500} = 0.040$

(c) $P(N, \textit{given } A) = \frac{50}{500} = 0.100$

$P(S, \textit{given } A) = \frac{120}{500} = 0.240$

(d) $P(N \textit{ and } W) = P(W) \cdot P(N, \textit{ given } W)$

$= \left(\frac{500}{1000}\right)(0.840) = 0.420$

$P(M \textit{ and } W) = P(W) \cdot P(M, \textit{ given } W)$

$= \left(\frac{500}{1000}\right)\left(\frac{60}{500}\right) = 0.060$

(e) $P(N \textit{ or } M) = P(N) + P(M)$ if mutually exclusive

$= \left(\frac{470}{1000}\right) + \left(\frac{390}{1000}\right) = \frac{860}{1000} = 0.860$

They are mutually exclusive because the reactions are defined into 3 distinct, mutually exclusive categories, a reaction can't be both mild and non-existent.

(f) If N and W were independent, $P(N \text{ and } W) = P(N) \cdot P(W) = (0.470)(0.500) = 0.235$. However, from (d), we have $P(N \text{ and } W) = 0.420$. They are not independent.

7. (a) possible values for x, the sum of the two dice faces, is 2, 3, 4, 5, 6, 7, 8, 9, 10, 11, and 12

(b) 2:1 and 1
3:1 and 2, or 2 and 1
4:1 and 3, 2 and 2, 3 and 1
5:1 and 4, 2 and 3, 3 and 2, 4 and 1
6:1 and 5, 2 and 4, 3 and 3, 4 and 2, 5 and 1
7:1 and 6, 2 and 5, 3 and 4, 4 and 3, 5 and 2, 6 and 1
8:2 and 6, 3 and 5, 4 and 4, 5 and 3, 6 and 2
9:3 and 6, 4 and 5, 5 and 4, 6 and 3
10:4 and 6, 5 and 5, 6 and 4
11:5 and 6, 6 and 5
12:6 and 6

x	$P(x)$
2	$\frac{1}{36} \approx 0.028$
3	$\frac{2}{36} \approx 0.056$
4	$\frac{3}{36} \approx 0.083$
5	$\frac{4}{36} \approx 0.111$
6	$\frac{5}{36} \approx 0.139$
7	$\frac{6}{36} \approx 0.167$
8	$\frac{5}{36} \approx 0.139$
9	$\frac{4}{36} \approx 0.111$
10	$\frac{3}{36} \approx 0.083$
11	$\frac{2}{36} \approx 0.056$
12	$\frac{1}{36} \approx 0.028$

Where there are (6)(6) = 36 possible, equally likely outcomes (the sums, however, are not equally likely).

8. $P(\text{pass } 101) = 0.77$
$P(\text{pass } 102, \textit{given} \text{ pass } 101) = 0.90$
$P(\text{pass } 101 \textit{ and } \text{pass } 102) = P(\text{pass } 101) \cdot P(\text{pass } 102, \textit{given} \text{ pass } 101) = 0.77(0.90) = 0.693$

9. $C_{8,2} = \frac{8!}{2!6!} = \frac{8 \cdot 7 \cdot 6!}{(2 \cdot 1)6!} = \frac{56}{2} = 28$

10. (a) $P_{7,2} = \frac{7!}{(7-2)!} = \frac{7!}{5!} = 7(6) = 42$

(b) $C_{7,2} = \frac{7!}{2!5!} = \frac{7 \cdot 6}{2} = 21$

(c) $P_{3,3} = \frac{3!}{(3-3)!} = \frac{3!}{0!} = 6$

(d) $C_{4,4} = \frac{4!}{4!(4-4)!} = \frac{4!}{4!0!} = 1$

11. $3 \cdot 2 \cdot 1 = 6$

12. Ways to Satisfy Literature, Social Science, and Philosophy Requirements

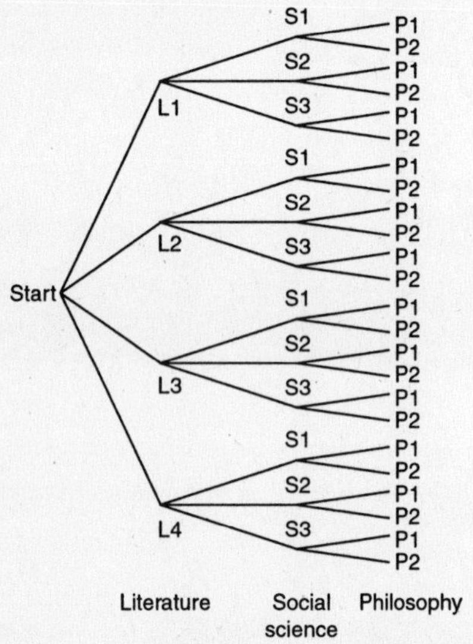

Let Li, $i = 1, \cdots, 4$ denote the 4 literature courses.
Let Si, $i = 1, 2, 3$ denote the 3 social science courses.
Let Pi, $i = 1, 2$ denote the 2 philosophy courses.
There are $4 \cdot 3 \cdot 2 = 24$ possible course combinations.

13. 5 multiple choice questions, each with 4 possible answers (A, B, C, or D), so 4 answers for first question; and for each of those, 4 answers for the second question; and for each of those, 4 answers for the third question; and for each of those, 4 answers for the fifth question. There are $4 \cdot 4 \cdot 4 \cdot 4 \cdot 4 = 4^5 = 1024$ possible sequences, such as A, D, B, B, C or C, B, A, D, D, etc.

$P(\text{getting the correct sequence}) = \frac{1}{1024} \approx 0.00098$

14. Two possible outcomes per coin toss; 6 tosses to get a sequence such as THTHHT
$2 \cdot 2 \cdot 2 \cdot 2 \cdot 2 \cdot 2 = 2^6 = 64$ possible sequences.

Chapter 6 The Binomial Probability Distribution and Related Topics

Section 6.1

1. (a) The number of traffic fatalities can be only a whole number. This is a discrete random variable.

(b) Distance can assume any value, so this is a continuous random variable.

(c) Time can take on any value, so this is a continuous random variable.

(d) The number of ships can be only a whole number. This is a discrete random variable.

(e) Weight can assume any value, so this is a continuous random variable.

2. (a) Speed can assume any value, so this is a continuous random variable.

(b) Age can take on any value, so this is a continuous random variable.

(c) Number of books can be only a whole number. This is a discrete random variable.

(d) Weight can assume any value, so this is a continuous random variable.

(e) Number of lightning strikes can be only a whole number. This is a discrete random variable.

3. (a) $\Sigma P(x) = 0.25 + 0.60 + 0.15 = 1.00$

Yes, this is a valid probability distribution because a probability is assigned to each distinct value of the random variable and the sum of these probabilities is 1.

(b) $\Sigma P(x) = 0.25 + 0.60 + 0.20 = 1.05$

No, this is not a probability distribution because the probabilities total to more than 1.

4. (a) $\Sigma P(x) = 0.07 + 0.44 + 0.24 + 0.4 + 0.11 = 1.00$

Yes, this is a valid probability distribution because the events are distinct and the probabilities total to 1.

(b) Age of Promotion Sensitive Shoppers

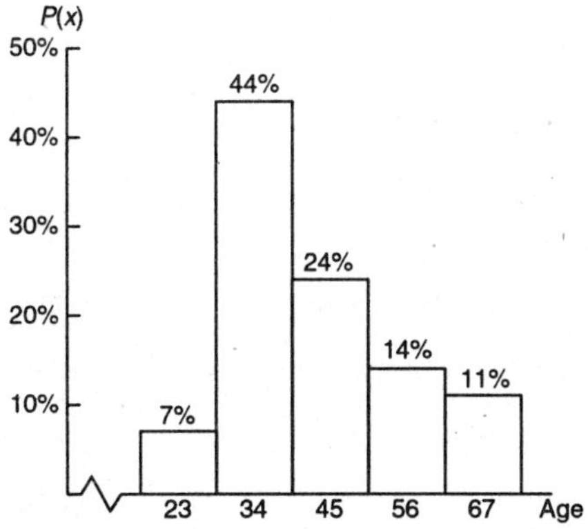

(c) $\mu = \Sigma xP(x)$

$= 23(0.07) + 34(0.44) + 45(0.24) + 56(0.14) + 67(0.11)$

$= 42.58$

(d) $\sigma = \sqrt{\Sigma(x-\mu)^2 P(x)}$

$= \sqrt{(-19.58)^2(0.07)+(-8.58)^2(0.44)+(2.42)^2(0.24)+(13.42)^2(0.14)+(24.42)^2(0.11)}$

$= \sqrt{151.44}$

≈ 12.31

5. (a) $\Sigma P(x) = 0.21+0.14+0.22+0.15+0.20+0.08 = 1.00$

Yes, this is a valid probability distribution because the events are distinct and the probabilities total to 1.

(b) Income Distribution ($1000)

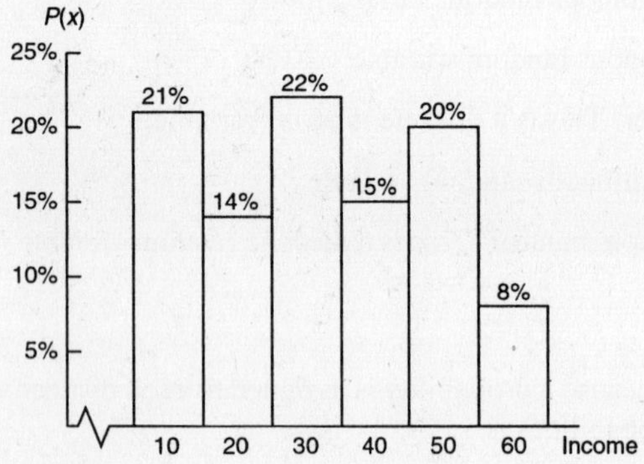

(c) $\mu = \Sigma xP(x)$

$= 10(0.21)+20(0.14)+30(0.22)+40(0.15)+50(0.20)+60(0.08)$

$= 32.3$

(d) $\sigma = \sqrt{\Sigma(x-\mu)^2 P(x)}$

$= \sqrt{(-22.3)^2(0.21)+(-12.3)^2(0.14)+(-2.3)^2(0.22)+(7.7)^2(0.15)+(17.7)^2(0.20)+(27.7)^2(0.08)}$

$= \sqrt{259.71}$

≈ 16.12

6. (a) $\Sigma P(x) = 0.057+0.097+0.195+0.292+0.250+0.091+0.018$

$= 1.000$

Yes, this is a valid probability distribution because the outcomes are distinct and the probabilities total to 1.

(b) Age of Nurses

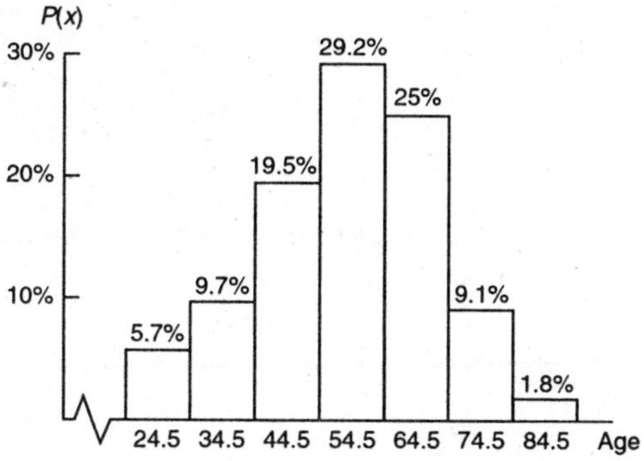

(c) $P(60 \text{ years of age or older}) = P(64.5) + P(74.5) + P(84.5)$

$$= 0.250 + 0.091 + 0.018$$

$$= 0.359$$

The probability is 35.9%.

(d) $\mu = \Sigma xP(x)$

$$= 24.5(0.057) + 34.5(0.097) + 44.5(0.195) + 54.5(0.292) + 64.5(0.250) + 74.5(0.091) + 84.5(0.018)$$

$$= 53.76$$

(e) $\sigma = \sqrt{\Sigma(x-\mu)^2 P(x)}$

$$= \sqrt{(-29.26)^2(0.057) + (-19.26)^2(0.097) + (-9.26)^2(0.195) + (0.74)^2(0.292) + (10.74)^2(0.250) + (20.74)^2(0.091) + (30.74)^2(0.018)}$$

$$= \sqrt{186.65}$$

$$\approx 13.66$$

7. (a) Number of Fish Caught in a 6-Hour Period at Pyramid Lake, Nevada

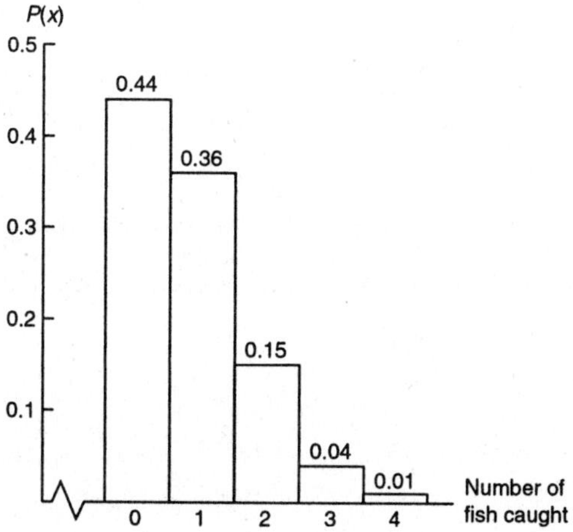

(b) $P(1 \text{ or more}) = 1 - P(0)$
$= 1 - 0.44$
$= 0.56$

(c) $P(2 \text{ or more}) = P(2) + P(3) + P(4 \text{ or more})$
$= 0.15 + 0.04 + 0.01$
$= 0.20$

(d) $\mu = \Sigma xP(x)$
$= 0(0.44) + 1(0.36) + 2(0.15) + 3(0.04) + 4(0.01)$
$= 0.82$

(e) $\sigma = \sqrt{\Sigma (x-\mu)^2 P(x)}$

$= \sqrt{(-0.82)^2 (0.44) + (0.18)^2 (0.36) + (1.18)^2 (0.15) + (2.18)^2 (0.04) + (3.18)^2 (0.01)}$

$= \sqrt{0.8076}$

≈ 0.899

8. $\Sigma P(x) \neq 1.000$ due to rounding

(a) $P(1 \text{ or more}) = 1 - P(0)$
$= 1 - 0.237$
$= 0.763$

This is the complement of the probability that none of the parolees will be repeat offenders.

(b) $P(2 \text{ or more}) = P(2) + P(3) + P(4) + P(5)$
$= 0.264 + 0.088 + 0.015 + 0.001$
$= 0.368$

(c) $P(4 \text{ or more}) = P(4) + P(5)$
$= 0.015 + 0.001$
$= 0.016$

(d) $\mu = \Sigma xP(x)$
$= 0(0.237) + 1(0.396) + 2(0.264) + 3(0.088) + 4(0.015) + 5(0.001)$
$= 1.253$

(e) $\sigma = \sqrt{\Sigma (x-\mu)^2 P(x)}$

$$= \sqrt{\begin{aligned}&(-1.253)^2 (0.237) + (-0.253)^2 (0.396) + (0.747)^2 (0.264) + (1.747)^2 (0.088) \\ &+ (2.747)^2 (0.015) + (3.747)^2 (0.001)\end{aligned}}$$

$= \sqrt{0.941}$

≈ 0.97

9. (a) $P(\text{win}) = \dfrac{15}{719} \approx 0.021$

$P(\text{not win}) = \dfrac{719-15}{719} = \dfrac{704}{719} \approx 0.979$

(b) Expected earnings $= (\text{value of dinner})(\text{probability of winning})$

$$= \$35\left(\frac{15}{719}\right)$$

$$\approx \$0.73$$

Lisa's expected earnings are \$0.73.

contribution $= \$15 - \$0.73 = \$14.27$

Lisa effectively contributed \$14.27 to the hiking club.

10. (a) $P(\text{win}) = \dfrac{6}{2852} \approx 0.0021$

$P(\text{not win}) = \dfrac{2852-6}{2852} = \dfrac{2846}{2852} \approx 0.9979$

(b) Expected earnings $= (\text{value of cruise})(\text{probability of winning})$

$$\approx \$2000(0.0021)$$

$$\approx \$4.20$$

Kevin spent 6(\$5) = \$30 for the tickets. His expected earnings are less than the amount he paid.

contribution $= \$30 - \$4.20 = \$25.80$

Kevin effectively contributed \$25.80 to the homeless center.

11. (a) $P(60 \text{ years}) = 0.01191$

Expected loss $= \$50{,}000(0.01191) = \$595.50$

The expected loss for Big Rock Insurance is \$595.50.

(b)

Probability	Expected Loss
$P(61) = 0.01292$	$\$50,000(0.01292) = \$646$
$P(62) = 0.01396$	$\$50,000(0.01396) = \$698$
$P(63) = 0.01503$	$\$50,000(0.01503) = \$751.50$
$P(64) = 0.01613$	$\$50,000(0.01613) = \$806.50$

Expected loss $= \$595.50 + \$646 + \$698 + \$751.50 + \$806.50$
$= \$3497.50$

The total expected loss is \$3497.50.

(c) \$3497.50 + \$700 = \$4197.50

They should charge \$4197.50.

(d) $\$5000 - \$3497.50 = \$1502.50$

They can expect to make \$1502.50.

Comment: losses are usually denoted by negative numbers such as –\$50,000.

12. (a) $P(60 \text{ years}) = 0.00756$

Expected loss $= \$50{,}000(0.00756) = \$378$

The expected loss for Big Rock Insurance is \$378.

(b)

Probability	Expected Loss
$P(61) = 0.00825$	$\$50{,}000(0.00825) = \$412.50$
$P(62) = 0.00896$	$\$50{,}000(0.00896) = \$448$
$P(63) = 0.00965$	$\$50{,}000(0.00965) = \$482.50$
$P(64) = 0.01035$	$\$50{,}000(0.01035) = \$517.50$

$$\begin{aligned}\text{expected loss} &= \$378 + \$412.50 + \$448 + \$482.50 + \$517.50 \\ &= \$2238.50\end{aligned}$$

The total expected loss is \$2238.50.

(c) $\$2238.50 + \$700 = \$2938.50$

They should charge \$2938.50.

(d) $\$5000 - \$2238.50 = \$2761.50$

They can expect to make \$2761.50.

Section 6.2

1. A trial is one flip of a fair quarter. Success = head. Failure = tail.

$n = 3$, $p = 0.5$, $q = 1 - 0.5 = 0.5$

(a)
$$\begin{aligned}P(3) &= C_{3,3}(0.5)^3(0.5)^{3-3} \\ &= 1(0.5)^3(0.5)^0 \\ &= 0.125\end{aligned}$$

To find this value in Table 3 of Appendix II, use the group in which $n = 3$, the column headed by $p = 0.5$, and the row headed by $r = 3$.

(b)
$$\begin{aligned}P(2) &= C_{3,2}(0.5)^2(0.5)^{3-2} \\ &= 3(0.5)^2(0.5)^1 \\ &= 0.375\end{aligned}$$

To find this value in Table 3 of Appendix II, use the group in which $n = 3$, the column headed by $p = 0.5$, and the row headed by $r = 2$.

(c)
$$\begin{aligned}P(r \geq 2) &= P(2) + P(3) \\ &= 0.125 + 0.375 \\ &= 0.5\end{aligned}$$

(d) The probability of getting exactly three tails is the same as getting exactly zero heads.

$$\begin{aligned} P(0) &= C_{3,0}(0.5)^0(0.5)^{3-0} \\ &= 1(0.5)^0(0.5)^3 \\ &= 0.125 \end{aligned}$$

To find this value in Table 3 of Appendix II, use the group in which $n = 3$, the column headed by $p = 0.5$, and the row headed by $r = 0$.

The results from Table 3 of Appendix II are the same.

In the problems that follow, there are often other ways the solve the problems than those shown. As long as you get the same answer, your method is probably correct.

2. A trial is answering a question on the quiz. Success = correct answer. Failure = incorrect answer.

$n = 10,\ p = \frac{1}{5} = 0.2,\ q = 1 - 0.2 = 0.8$

(a)
$$\begin{aligned} P(10) &= C_{10,10}(0.2)^{10}(0.8)^{10-10} \\ &= 1(0.2)^{10}(0.8)^0 \\ &= 0.000 \text{ (to three digits)} \end{aligned}$$

(b) 10 incorrect is the same as 0 correct.

$$\begin{aligned} P(0) &= C_{10,0}(0.2)^0(0.8)^{10-0} \\ &= 1(0.2)^0(0.8)^{10} \\ &= 0.107 \end{aligned}$$

(c) First method:

$$\begin{aligned} P(r \ge 1) &= P(1)+P(2)+P(3)+P(4)+P(5)+P(6)+P(7)+P(8)+P(9)+P(10) \\ &= 0.268+0.302+0.201+0.088+0.026+0.006+0.001+0.000+0.000+0.000 \\ &= 0.892 \end{aligned}$$

Second method:

$$\begin{aligned} P(r \ge 1) &= 1 - P(0) \\ &= 1 - 0.107 \\ &= 0.893 \end{aligned}$$

The two results should be equal, but because of rounding error, they differ slightly.

(d)
$$\begin{aligned} P(r \ge 5) &= P(5)+P(6)+P(7)+P(8)+P(9)+P(10) \\ &= 0.026+0.006+0.001+0.000+0.000+0.000 \\ &= 0.033 \end{aligned}$$

3. (a) A trial is a man's response to the question, "Would you marry the same woman again?"
Success = a positive response. Failure = a negative response.
$n=10,\ p=0.80,\ q=1-0.80=0.20$
Using values in Table 3 of Appendix II:

$$P(r\geq 7)=P(7)+P(8)+P(9)+P(10)$$
$$=0.201+0.302+0.268+0.107$$
$$=0.878$$

$$P(r \text{ is less than half of } 10)=P(r<5)$$
$$=P(0)+P(1)+P(2)+P(3)+P(4)$$
$$=0.000+0.000+0.000+0.001+0.006$$
$$=0.007$$

(b) A trial is a woman's response to the question, "Would you marry the same man again?"
Success = a positive response. Failure = a negative response.
$n=10,\ p=0.5,\ q=1-0.5=0.5$
Using values in Table 3 of Appendix II:

$$P(r\geq 7)=P(7)+P(8)+P(9)+P(10)$$
$$=0.117+0.044+0.010+0.001$$
$$=0.172$$

$$P(r<5)=P(0)+P(1)+P(2)+P(3)+P(4)$$
$$=0.001+0.010+0.044+0.117+0.205$$
$$=0.377$$

4. A trial is a one-time fling. Success = has done a one-time fling. Failure = has not done a one-time fling.
$n=7,\ p=0.10,\ q=1-0.10=0.90$

(a) $P(0)=C_{7,0}(0.10)^0(0.90)^{7-0}$
$$=1(0.10)^0(0.90)^7$$
$$=0.478$$

(b) $P(r\geq 1)=1-P(0)$
$$=1-0.478$$
$$=0.522$$

(c) $P(r\leq 2)=P(0)+P(1)+P(2)$
$$=0.478+0.372+0.124$$
$$=0.974$$

5. A trial consists of a woman's response regarding her mother-in-law. Success = dislike. Failure = like.
$n=6,\ p=0.90,\ q=1-0.90=0.10$

(a) $P(6)=C_{6,6}(0.90)^6(0.10)^{6-6}$
$$=1(0.90)^6(0.10)^0$$
$$=0.531$$

(b) $P(0) = C_{6,0}(0.90)^0(0.10)^{6-0}$

$= 1(0.90)^0(0.10)^6$

≈ 0.000 (to 3 digits)

(c) $P(r \geq 4) = P(4) + P(5) + P(6)$

$= 0.098 + 0.354 + 0.531$

$= 0.983$

(d) $P(r \leq 3) = 1 - P(r \geq 4)$

$\approx 1 - 0.983$

$= 0.017$

From the table:

$P(r \leq 3) = P(0) + P(1) + P(2) + P(3)$

$= 0.000 + 0.000 + 0.001 + 0.015$

$= 0.016$

6. A trial is how a businessman wears a tie. Success = too tight. Failure = not too tight.
$n = 20,\ p = 0.10,\ q = 1 - 0.10 = 0.90$

(a) $P(r \geq 1) = 1 - P(r = 0)$

$= 1 - 0.122$

$= 0.878$

(b) $P(r > 2) = 1 - P(r \leq 2)$

$= 1 - [P(r = 0) + P(r = 1) + P(r = 2)]$

$= 1 - P(r = 0) - P(r = 1) - P(r = 2)$

$= [1 - P(r = 0)] - P(r = 1) - P(r = 2)$

$= P(r \geq 1) - P(r = 1) - P(r = 2)$

$= 0.878 - 0.270 - 0.285$ using (a)

$= 0.323$

(c) $P(r = 0) = 0.122$

(d) At least 18 are not too tight is the same as at most 2 are too tight. (To see this, note that at least 18 failures is the same as 18 or 19 or 20 failures, which is 2, 1, or 0 successes; i.e., at most 2 successes.)

$P(r \leq 2) = 1 - P(r > 2)$

$= 1 - 0.323$ using (b)

$= 0.677$

7. A trial consists of taking a polygraph examination. Success = pass. Failure = fail.
$n = 9,\ p = 0.85,\ q = 1 - 0.85 = 0.15$

(a) $P(9) = 0.232$

(b) $P(r \geq 5) = P(5) + P(6) + P(7) + P(8) + P(9)$

$= 0.028 + 0.107 + 0.260 + 0.368 + 0.232$

$= 0.995$

(c) $P(r \le 4) = 1 - P(r \ge 5)$
$= 1 - 0.995$
$= 0.005$

From the table:
$P(r \le 4) = P(0) + P(1) + P(2) + P(3) + P(4)$
$= 0.000 + 0.000 + 0.000 + 0.001 + 0.005$
$= 0.006$

The two results should be equal, but because of rounding error, they differ slightly.

(d) All students fail is the same as no students pass.
$P(0) = 0.000$ (to 3 digits)

8. A trial consists of checking the gross receipts of the store for one business day. Success = gross over \$850. Failure = gross is at or below \$850. $p = 0.6$, $q = 1 - p = 0.4$

(a) $n = 5$
$P(r \ge 3) = P(3) + P(4) + P(5)$
$= 0.346 + 0.259 + 0.078$
$= 0.683$

(b) $n = 10$
$P(r \ge 6) = P(6) + P(7) + P(8) + P(9) + P(10)$
$= 0.251 + 0.215 + 0.121 + 0.040 + 0.006$
$= 0.633$

(c) $n = 10$
$P(r < 5) = P(0) + P(1) + P(2) + P(3) + P(4)$
$= 0.000 + 0.002 + 0.011 + 0.042 + 0.111$
$= 0.166$

(d) $n = 20$
$P(r < 6) = P(0) + P(1) + P(2) + P(3) + P(4) + P(5)$
$= 0.000 + 0.000 + 0.000 + 0.000 + 0.000 + 0.001$
$= 0.001$

Yes. If p were really 0.60, then the event of a 20-day period with gross income exceeding \$850 fewer than 6 days would be very rare. If it happened again, we would suspect that $p = 0.60$ is too high.

(e) $n = 20$
$P(r > 17) = P(18) + P(19) + P(20)$
$= 0.003 + 0.000 + 0.000$
$= 0.003$

Yes. If p were really 0.60, then the event of a 20-day period with gross income exceeding \$850 more than 17 days would be very rare. If it happened again, we would suspect that $p = 0.60$ is too low.

9. A trial is catching and releasing a pike. Success = pike dies. Failure = pike lives.
$n = 16$, $p = 0.05$, $q = 1 - 0.05 = 0.95$

(a) $P(0) = 0.440$

(b) $P(r < 3) = P(0) + P(1) + P(2)$

$= 0.440 + 0.371 + 0.146$

$= 0.957$

(c) All of the fish lived is the same as none of the fish died.

$P(0) = 0.440$

(d) More than 14 fish lived is the same as less than 2 fish died.

$P(r < 2) = P(0) + P(1)$

$= 0.440 + 0.371$

$= 0.811$

10. A trial is tasting coffee. Success = choose Tasty Bean. Failure = do not choose Tasty Bean.

$n = 4,\ p = \frac{1}{5} = 0.2,\ q = 1 - 0.2 = 0.8$

(a) $P(4) = 0.002$

(b) $P(0) = 0.410$

(c) $P(r \geq 3) = P(3) + P(4)$

$= 0.026 + 0.002$

$= 0.028$

11. (a) A trial consists of using the Meyers-Briggs instrument to determine if a person in marketing is an extrovert. Success = extrovert. Failure = not extrovert.

$n = 15,\ p = 0.75,\ q = 1 - 0.75 = 0.25$

$P(r \geq 10) = P(10) + P(11) + P(12) + P(13) + P(14) + P(15)$

$= 0.165 + 0.225 + 0.225 + 0.156 + 0.067 + 0.013$

$= 0.851$

$P(r \geq 5) = P(5) + P(6) + P(7) + P(8) + P(9) + P(r \geq 10)$

$= 0.001 + 0.003 + 0.013 + 0.039 + 0.092 + 0.851$

$= 0.999$

$P(15) = 0.013$

(b) A trial consists of using the Meyers-Briggs instrument to determine if a computer programmer is an introvert. Success = introvert. Failure = not introvert.

$n = 5,\ p = 0.60,\ q = 1 - 0.60 = 0.40$

$P(0) = 0.010$

$P(r \geq 3) = P(3) + P(4) + P(5)$

$= 0.346 + 0.259 + 0.078$

$= 0.683$

$P(5) = 0.078$

12. A trial consists of a man's response regarding welcoming a woman taking the initiative in asking for a date. Success = yes. Failure = no.

$n = 20,\ p = 0.70,\ q = 1 - 0.70 = 0.30$

(a) $P(r \geq 18) = P(18) + P(19) + P(20)$
$= 0.028 + 0.007 + 0.001$
$= 0.036$

(b) $P(r < 3) = P(0) + P(1) + P(2)$
$= 0.000 + 0.000 + 0.000$
$= 0.000$ (to 3 digits)

(c) $P(0) = 0.000$ (to 3 digits)

(d) At least 5 say no is the same as at most 15 say yes.

$P(r \leq 15) = 1 - P(r \geq 16)$
$= 1 - [P(16) + P(17) + P(18) + P(19) + P(20)]$
$= 1 - (0.130 + 0.072 + 0.028 + 0.007 + 0.001)$
$= 1 - 0.238$
$= 0.762$

13. A trial is checking the development of hypertension in patients with diabetes. Success = yes. Failure = no.

$n = 10,\ p = 0.40,\ q = 1 - 0.40 = 0.60$

(a) $P(0) = 0.006$

(b) $P(r < 5) = P(0) + P(1) + P(2) + P(3) + P(4)$
$= 0.006 + 0.040 + 0.121 + 0.215 + 0.251$
$= 0.633$

A trial is checking the development of an eye disease in patients with diabetes. Success = yes. Failure = no.

$n = 10,\ p = 0.30,\ q = 1 - 0.30 = 0.70$

(c) $P(r \leq 2) = P(0) + P(1) + P(2)$
$= 0.028 + 0.121 + 0.233$
$= 0.382$

(d) At least 6 will never develop a related eye disease is the same as at most 4 will never develop a related eye disease.

$P(r \leq 4) = P(0) + P(1) + P(2) + P(3) + P(4)$
$= 0.028 + 0.121 + 0.233 + 0.267 + 0.200$
$= 0.849$

14. A trial consists of the response of adults regarding their concern that employers are monitoring phone calls. Success = yes. Failure = no.

$p = 0.37,\ q = 1 - 0.37 = 0.63$

(a) $n = 5$

$$P(0) = C_{5,0}(0.37)^0(0.63)^{5-0}$$
$$= 1(0.37)^0(0.63)^5$$
$$\approx 0.099$$

(b) $n = 5$

$$P(5) = C_{5,5}(0.37)^5(0.63)^{5-5}$$
$$= 1(0.37)^5(0.63)^0$$
$$\approx 0.007$$

(c) $n = 5$

$$P(3) = C_{5,3}(0.37)^3(0.63)^{5-3}$$
$$= 10(0.37)^3(0.63)^2$$
$$\approx 0.201$$

15. A trial consists of the response of adults regarding their concern that Social Security numbers are used for general identification. Success = concerned that SS numbers are being used for identification. Failure = not concerned that SS numbers are being used for identification.

$n = 8,\ p = 0.53,\ q = 1 - 0.53 = 0.47$

(a) $P(r \le 5) = P(0) + P(1) + P(2) + P(3) + P(4) + P(5)$

$$= 0.002381 + 0.021481 + 0.084781 + 0.191208 + 0.269521 + 0.243143$$
$$= 0.812515$$

$P(r \le 5) = 0.81251$ from the cumulative probability is the same, truncated to 5 digits.

(b) $P(r > 5) = P(6) + P(7) + P(8)$

$$= 0.137091 + 0.044169 + 0.006726$$
$$= 0.187486$$

$$P(r > 5) = 1 - P(r \le 5)$$
$$= 1 - 0.81251$$
$$= 0.18749$$

Yes, this is the same result rounded to 5 digits.

16. $n = 3,\ p = 0.0228,\ q = 1 - p = 0.9772$

(a) $P(2) = C_{3,2}p^2q^{3-2} = 3(0.0228)^2(0.9772)^1 = 0.00152$

(b) $P(3) = C_{3,3}p^3q^{3-3} = 1(0.0228)^3(0.9772)^0 = 0.00001$

(c) $P(2 \text{ or } 3) = P(2) + P(3) = 0.00153$

17. (a) $p = 0.30,\ P(3) = 0.132$
$p = 0.70,\ P(2) = 0.132$
They are the same.

(b) $p = 0.30,\ P(r \geq 3) = 0.132 + 0.028 + 0.002 = 0.162$
$p = 0.70,\ P(r \leq 2) = 0.002 + 0.028 + 0.132 = 0.162$
They are the same.

(c) $p = 0.30,\ P(4) = 0.028$
$p = 0.70,\ P(1) = 0.028$
$r = 1$

(d) The column headed by $p = 0.80$ is symmetrical with the one headed by $p = 0.20$.

Section 6.3

1. (a) Binomial Distribution
The distribution is symmetrical.

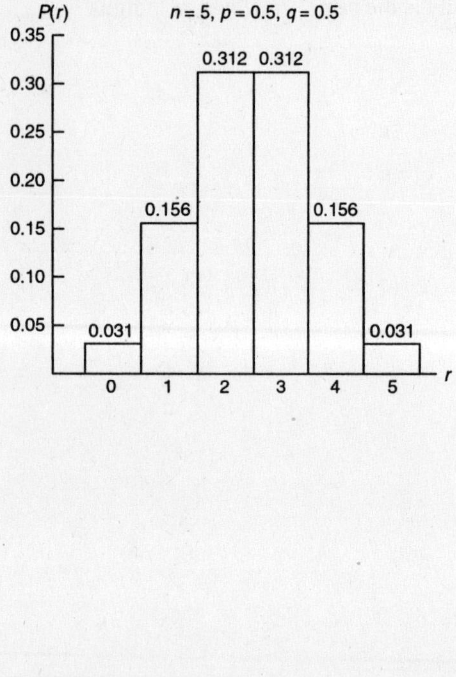

(b) Binomial Distribution
The distribution is skewed right.

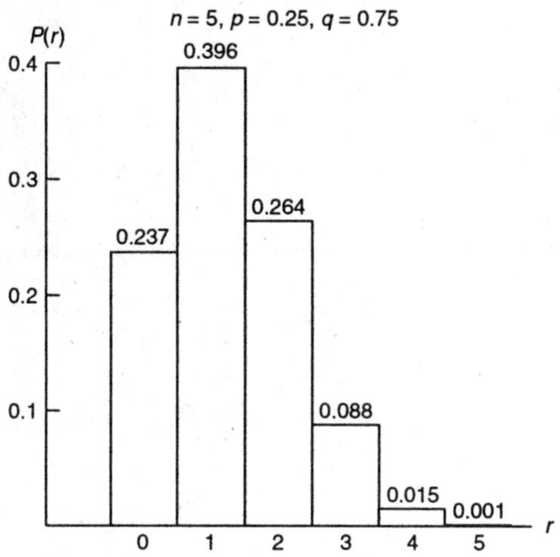

(c) Binomial Distribution
The distribution is skewed left.

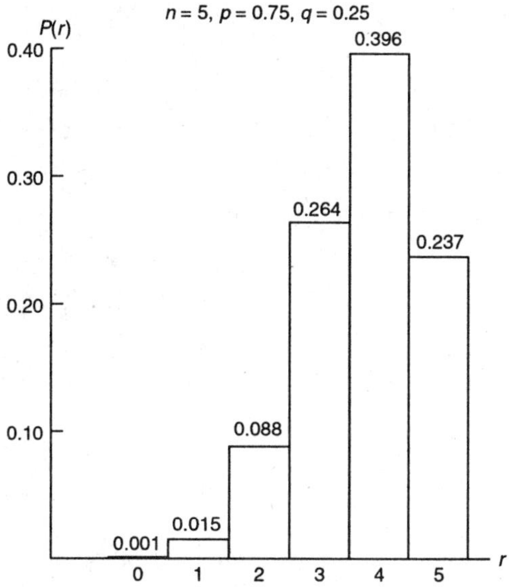

(d) The distributions are mirror images of one another.

(e) The distribution would be skewed left for $p = 0.73$ because the more likely number of successes are to the right of the middle.

2. (a) $p = 0.30$ goes with graph II since it is slightly skewed right.

(b) $p = 0.50$ goes with graph I since it is symmetrical.

(c) $p = 0.65$ goes with graph III since it is slightly skewed left.

(d) $p = 0.90$ goes with graph IV since it is drastically skewed left.

(e) The graph is more symmetrical when p is close to 0.5. The graph is skewed left when p is close to 1 and skewed right when p is close to 0.

3. The probabilities can be taken directly from Table 3 in Appendix II.

(a) $n = 10,\ p = 0.80$

Households with Children Under 2 That Buy Film

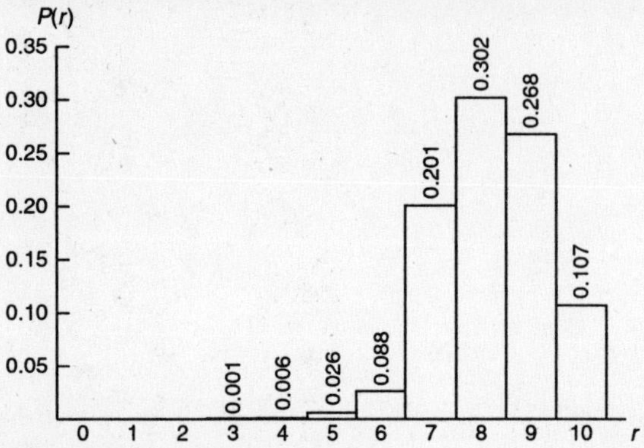

$\mu = np = 10(0.8) = 8$

$\sigma = \sqrt{npq} = \sqrt{10(0.8)(0.2)} \approx 1.26$

(b) $n = 10,\ p = 0.5$

Households with No Children Under 21 that Buy Film

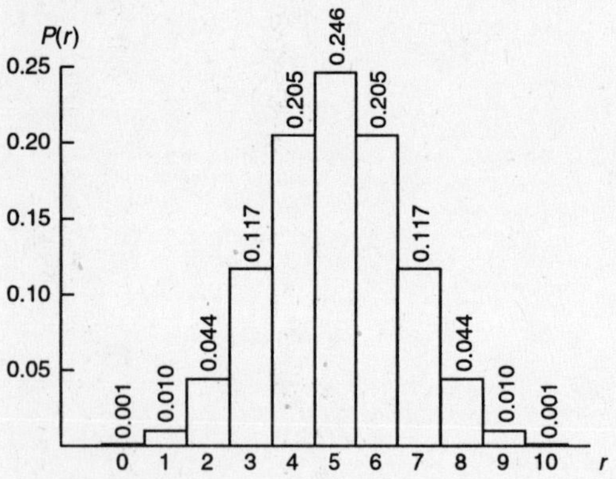

$\mu = np = 10(0.5) = 5$

$\sigma = \sqrt{npq} = \sqrt{10(0.5)(0.5)} \approx 1.58$

(c) Yes; since the graph in part (a) is skewed left, it supports the claim that more households buy film that have children under 2 years than households that have no children under 21 years.

4. (a) $n = 8,\ p = 0.01$

The probabilities can be taken directly from Table 3 in Appendix II.

Binomial Distribution for Number of Defective Syringes

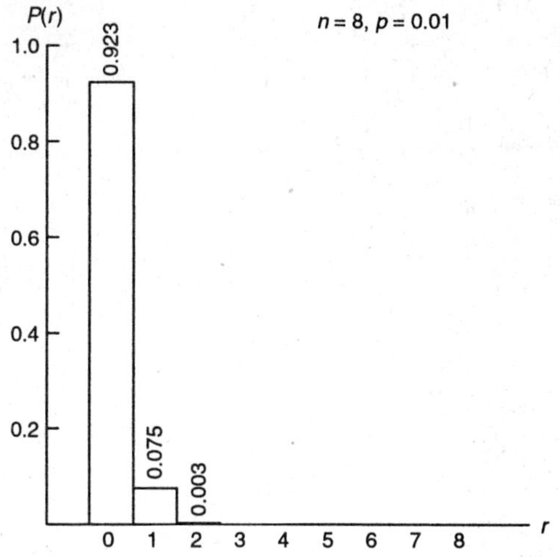

(b) $\mu = np = 8(0.01) = 0.08$

The expected number of defective syringes the inspector will find is 0.08.

(c) The batch will be accepted if less than 2 defectives are found.

$$\begin{aligned} P(r<2) &= P(0)+P(1) \\ &= 0.923+0.075 \\ &= 0.998 \end{aligned}$$

(d) $\sigma = \sqrt{npq} = \sqrt{8(0.01)(0.99)} \approx 0.281$

5. (a) $n = 8,\ p = 0.25$

The probabilities can be taken directly from Table 3 in Appendix II.

Binomial Distribution for Number of Gullible Customers

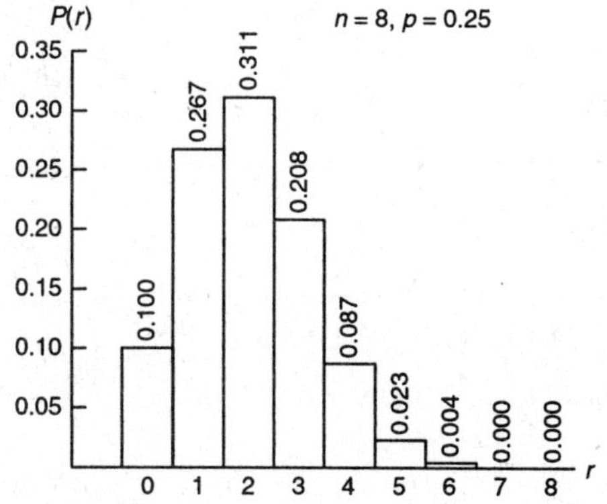

(b) $\mu = np = 8(0.25) = 2$

$$\sigma = \sqrt{npq} = \sqrt{8(0.25)(0.75)} \approx 1.225$$

The expected number of people in this sample who believe the product is improved is 1.4.

(c) Find n such that

$$P(r \geq 1) = 0.99.$$

Try $n = 16$.

$$\begin{aligned} P(r \geq 1) &= 1 - P(0) \\ &= 1 - 0.01 \\ &= 0.99 \end{aligned}$$

Sixteen people are needed in the marketing study to be 99% sure that at least one person believes the product to be improved.

6. (a) $n = 5,\ p = 0.85$

The probabilities can be taken directly from Table 3 in Appendix II.

Binomial Distribution for Number of Automobile Damage Claims by People Under Age 25

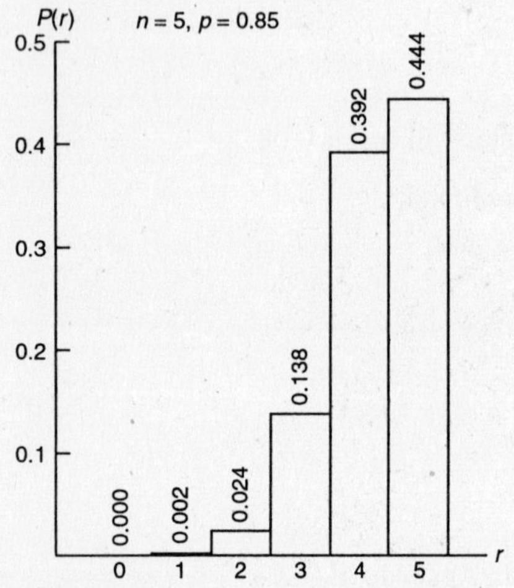

(b) $\mu = np = 5(0.85) = 4.25$

$$\sigma = \sqrt{npq} = \sqrt{5(0.85)(0.15)} \approx 0.798$$

For samples of size 5, the expected number of claims made by people under 25 years of age is about 4.

7. (a) Japan: $n = 7, p = 0.95$

$$P(7) = 0.698$$

United States: $n = 7, p = 0.60$

$$P(7) = 0.028$$

(b) Japan: $n = 7, p = 0.95$

$$\mu = np = 7(0.95) = 6.65$$

$$\sigma = \sqrt{npq} = \sqrt{7(0.95)(0.05)} \approx 0.58$$

United States: $n = 7, p = 0.60$

$$\mu = np = 7(0.60) = 4.2$$

$$\sigma = \sqrt{npq} = \sqrt{7(0.60)(0.40)} \approx 1.30$$

The expected number of verdicts in Japan is 6.65 and in the United States is 4.2.

(c) United States: $p = 0.60$

Find n such that

$$P(r \geq 2) = 0.99.$$

Try $n = 8$.

$$\begin{aligned} P(r \geq 2) &= 1 - [P(0) + P(1)] \\ &= 1 - (0.001 + 0.008) \\ &= 0.991 \end{aligned}$$

Japan: $p = 0.95$

Find n such that

$$P(r \geq 2) = 0.99.$$

Try $n = 3$.

$$\begin{aligned} P(r \geq 2) &= P(2) + P(3) \\ &= 0.135 + 0.857 \\ &= 0.992 \end{aligned}$$

Cover 8 trials in the U.S. and 3 trials in Japan.

8. $n = 6$, $p = 0.45$

(a) $P(6) = 0.008$

(b) $P(0) = 0.028$

(c)
$$\begin{aligned} P(r \geq 2) &= P(2) + P(3) + P(4) + P(5) + P(6) \\ &= 0.278 + 0.303 + 0.186 + 0.061 + 0.008 \\ &= 0.836 \end{aligned}$$

(d) $\mu = np = 6(0.45) = 2.7$

The expected number is 2.7.

$$\sigma = \sqrt{npq} = \sqrt{6(0.45)(0.55)} \approx 1.219$$

(e) Find n such that

$$P(r \geq 3) = 0.90.$$

Try $n = 10$.

$$\begin{aligned} P(r \geq 3) &= 1 - [P(0) + P(1) + P(2)] \\ &= 1 - (0.003 + 0.021 + 0.076) \\ &= 1 - (0.100) \\ &= 0.900 \end{aligned}$$

You need to interview 10 professors to be at least 90% sure of filling the quota.

9. (a) Since $p = 0.47$, the probability of success is close to 50%.

The histogram is more symmetrical.

$\mu = np = (0.47) = 5.17$

Yes, the expected value μ is close to the center of the graph.

(b) Since $p = 0.36$, the probability of success is less than 50%.

Yes, the histogram is slightly skewed to the right.

$\mu = np = 11(0.36) = 3.96$

The expected value is on the left side of the graph.

(c) $p = 1 - 0.36 = 0.64$

The distribution is skewed to the left.

$\mu = np = 11(0.64) = 7.04$

The expected value lies on the right side of the graph.

(d) The entries are the same.

(e) For part (b),

$r = \sqrt{npg} = \sqrt{11(0.36)(0.64)} = 1.59$

For part (c),

$r = \sqrt{npq} = \sqrt{11(0.64)(0.36)} = 1.59$

Yes, the standard deviations are the same. We would expect this because the same values are being used to compute r although the probability of success for parts (b) and (c) are opposites.

Chapter 6 Review

1. (a) $\mu = \Sigma xP(x)$

$= 18.5(0.127) + 30.5(0.371) + 42.5(0.285) + 54.5(0.215) + 66.5(0.002)$

$= 37.628$

≈ 37.63

The expected lease term is about 38 months.

$\sigma = \sqrt{\Sigma (x-\mu)^2 P(x)}$

$= \sqrt{(-19.13)^2 (0.127) + (-7.13)^2 (0.371) + (4.87)^2 (0.285) + (16.87)^2 (0.215) + (28.87)^2 (0.002)}$

$\approx \sqrt{134.95}$

≈ 11.6 (using $\mu = 37.63$ in the calculations)

(b) Leases in Months

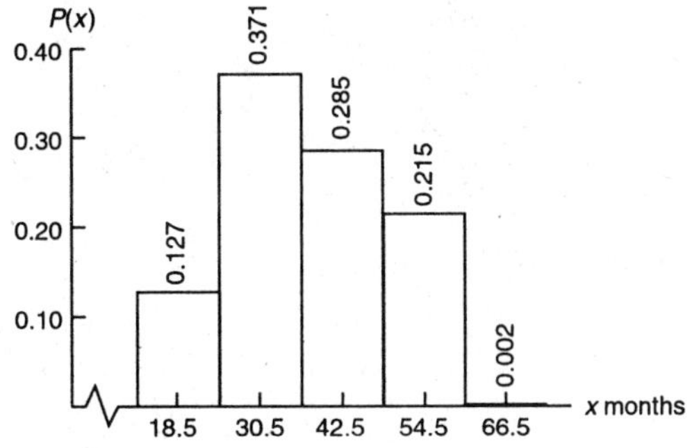

2. (a)

Number killed by Wolves	Relative Frequency	$P(x)$
112	112/296	0.378
53	53/296	0.179
73	73/296	0.247
56	56/296	0.189
2	2/296	0.007

(b) $\mu = \Sigma xP(x)$

$= 0.5(0.378) + 3(0.179) + 8(0.247) + 13(0.189) + 18(0.007)$

≈ 5.28 yr

$\sigma = \sqrt{\Sigma (x-\mu)^2 P(x)}$

$= \sqrt{(-4.78)^2 (0.378) + (-2.28)^2 (0.179) + (2.72)^2 (0.247) + (7.72)^2 (0.189) + (12.72)^2 (0.007)}$

$= \sqrt{23.8}$

≈ 4.88 yr

3. This is a binomial experiment with 10 trials. A trial consists of a claim.

Success = submitted by a male under 25 years of age.
Failure = not submitted by a male under 25 years of age.

(a) The probabilities can be taken directly from Table 3 in Appendix II, $n = 10$, $p = 0.55$.

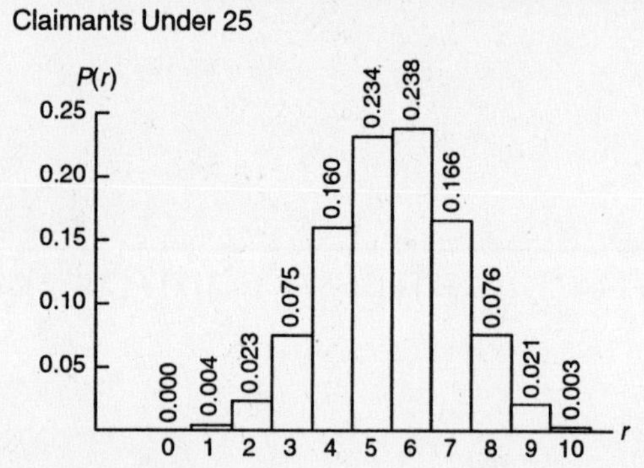

(b) $P(r \geq 6) = P(6) + P(7) + P(8) + P(9) + P(10)$

$= 0.238 + 0.166 + 0.076 + 0.021 + 0.003$

$= 0.504$

(c) $\mu = np = 10(0.55) = 5.5$

The expected number of claims made by males under age 25 is 5.5.

$\sigma = \sqrt{npq} = \sqrt{10(0.55)(0.45)} \approx 1.57$

4. **(a)** $n = 20, p = 0.05$

$P(r \leq 2) = P(0) + P(1) + P(2)$

$= 0.358 + 0.377 + 0.189$

$= 0.924$

(b) $n = 20, p = 0.15$

Probability accepted:

$P(r \leq 2) = P(0) + P(1) + P(2)$

$= 0.039 + 0.137 + 0.229$

$= 0.405$

Probability not accepted:

$1 - 0.405 = 0.595$

5. $n = 16, p = 0.50$

(a) $P(r \geq 12) = P(12) + P(13) + P(14) + P(15) + P(16)$

$= 0.028 + 0.009 + 0.002 + 0.000 + 0.000$

$= 0.039$

(b) $P(r \le 7) = P(0) + P(1) + P(2) + P(3) + P(4) + P(5) + P(6) + P(7)$

$= 0.000 + 0.000 + 0.002 + 0.009 + 0.028 + 0.067 + 0.122 + 0.175$

$= 0.403$

(c) $\mu = np = 16(0.50) = 8$

The expected number of inmates serving time for drug dealing is 8.

6. $n = 200, p = 0.80$

$$\mu = np = 200(0.80) = 160$$

The expected number that will arrive on time is 160 flights.

$$\sigma = \sqrt{npq} = \sqrt{200(0.80)(0.20)} \approx 5.66$$

The standard deviation is 5.66 flights.

7. $n = 10, p = 0.75$

(a) The probabilities can be obtained directed from Table 3 in Appendix II.

Number of Good Grapefruit

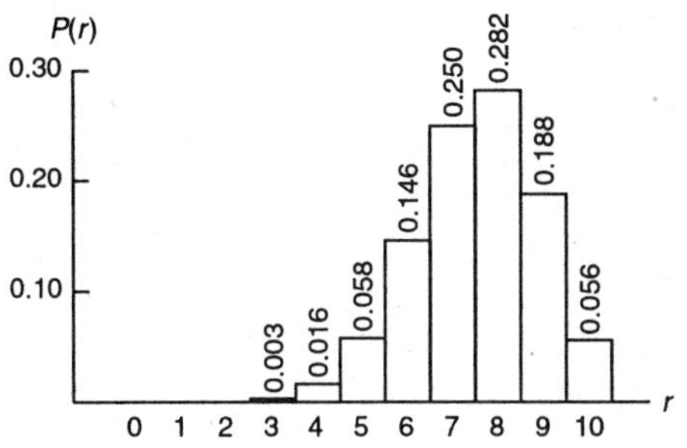

(b) No more than one bad is the same as at least nine good.

$P(r \ge 9) = P(9) + P(10)$

$= 0.188 + 0.056$

$= 0.244$

$P(r \ge 1) = P(1) + P(2) + P(3) + P(4) + P(5) + P(6) + P(7) + P(8) + P(9) + P(10)$

$= 0.000 + 0.000 + 0.003 + 0.016 + 0.058 + 0.146 + 0.250 + 0.282 + 0.188 + 0.056$

$= 0.999$

(c) $\mu = np = 10(0.75) = 7.5$

The expected number of good grapefruit in a sack is 7.5.

(d) $\sigma = \sqrt{npq} = \sqrt{10(0.75)(0.25)} \approx 1.37$

8. Let success = show up, then $p = 0.95$, $n = 82$.

$$\mu = np = 82(0.95) = 77.9$$

If 82 party reservations have been made, 77.9 or about 78 can be expected to show up.

$$\sigma = \sqrt{npq} = \sqrt{82(0.95)(0.05)} \approx 1.97$$

9. $p = 0.85, n = 12$

$$P(r \le 2) = P(0) + P(1) + P(2)$$
$$= 0.000 + 0.000 + 0.000$$
$$= 0.000 \quad \text{(to 3 digits)}$$

The data seem to indicate that the percent favoring the increase in fees is less than 85%.

Chapter 7 Normal Distributions

Section 7.1

1. **(a)** not normal; left skewed instead of symmetric

   (b) not normal; curve touches and goes below
   x-axis instead of always being above the x-axis and being asymptotic to the x-axis in the tails

   (c) not normal; not bell-shaped, not unimodal

   (d) not normal; not a smooth curve

2. $\mu = 16$, $\sigma = 2$, $\mu + \sigma = 16 + 2 = 18$
   (The mean is located directly below the peak; one standard deviation from the mean is the x-value under the point of inflection [the transition point between the curve cupping upward and cupping downward].)

3. The mean is the x-value directly below the peak; in Figure 6-16, $\mu = 10$; in Figure 6-17, $\mu = 4$. Assuming the two figures are drawn on the same scale, Figure 6-16, being shorter and more spread out, has the larger standard deviation.

4. **(a)**

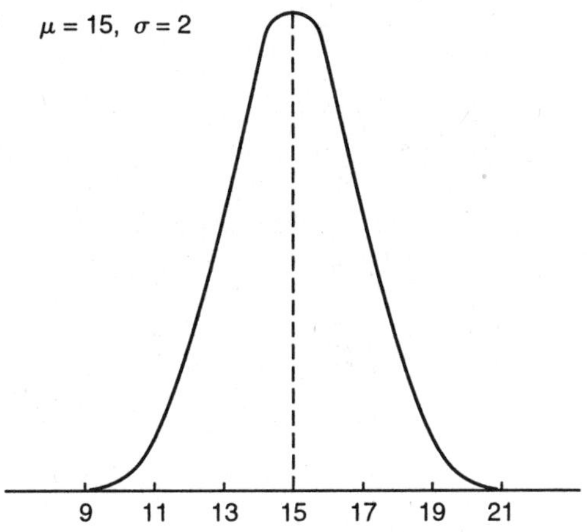

   (b)

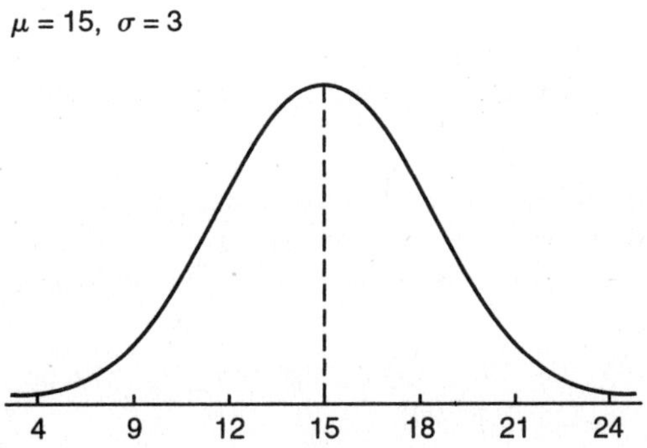

(c)

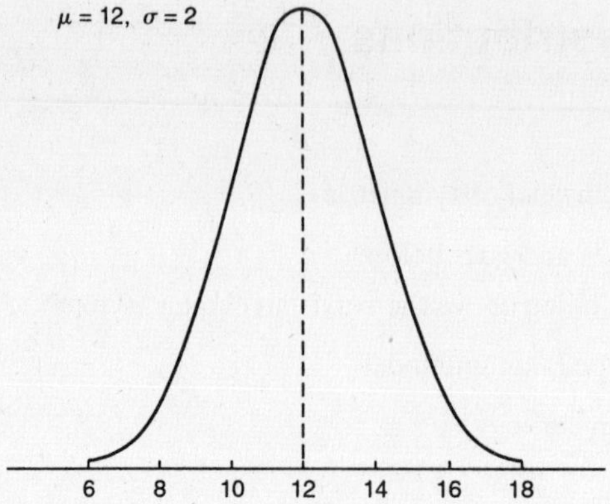

(d)

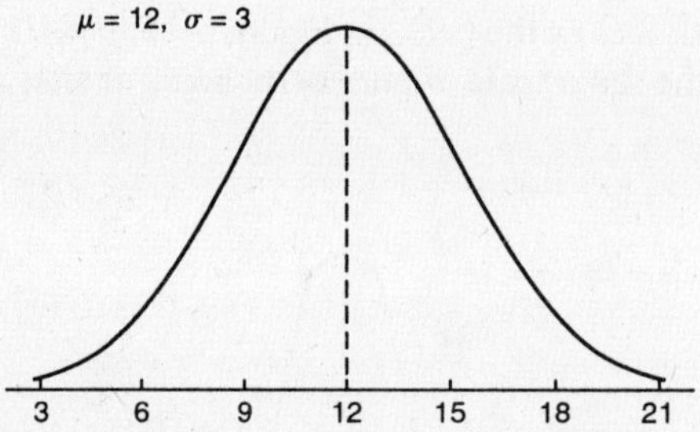

(e) No; the mean μ and the standard deviation σ are independent of one another. If $\mu_1 > \mu_2$, then $\sigma_1 > \sigma_2$, $\sigma_1 = \sigma_2$, and $\sigma_1 < \sigma_2$ are all possible.

5. (a) 50%; the normal curve is symmetric about μ

(b) 68%

(c) 99.7%

6. (a) 50%; the normal curve is symmetric about μ

(b) 95%

(c) $\frac{1}{2}(100\% - 99.7\%) = \frac{1}{2}(0.3\%) = 0.15\%$,

99.7% lies between $\mu - 3\sigma$ and $\mu + 3\sigma$, so 0.3% lies in the tails, and half of that is in the upper tail.

7. (a) $\mu = 65$, so 50% are taller than 65 in.

(b) $\mu = 65$, so 50% are shorter than 65 in.

(c) $\mu - \sigma = 65 - 2.5 = 62.5$ in. and $\mu + \sigma = 65 + 2.5 = 67.5$ in. so 68% of college women are between 62.5 in. and 67.5 in. tall.

(d) $\mu - 2\sigma = 65 - 2(2.5) = 65 - 5 = 60$ in. and $\mu - 2\sigma = 65 + 2(2.5) = 65 + 5 = 70$ in. so 95% of college women are between 60 in. and 70 in. tall.

8. (a) $\mu - 2\sigma = 21 - 2(1) = 19$ days and $\mu + 2\sigma = 21 + 2(1) = 23$ days so 95% of 1000 eggs, or 950 eggs will hatch between 19 and 23 days of incubation.

(b) $\mu - \sigma = 21 - 1 = 20$ days and $\mu + \sigma = 21 + 1 = 22$ days so 68% of 1000 eggs, or 680 eggs, will hatch between 20 and 22 days of incubation.

(c) $\mu = 21$, so 50%, or 500, of the eggs will hatch in at most 21 days.

(d) $\mu - 3\sigma = 21 - 3(1) = 21 - 3 = 18$ days and $\mu + 3\sigma = 21 + 3(1) = 21 + 3 = 24$ days so 99.7%, or 997, eggs will hatch between 18 and 24 days of incubation.

9. (a) $\mu - \sigma = 1243 - 36 = 1207$ and $\mu + \sigma = 1243 + 36 = 1279$ so about 68% of the tree rings will date between 1207 and 1279 AD.

(b) $\mu - 2\sigma = 1243 - 2(36) = 1171$ and $\mu + 2\sigma = 1243 + 2(36) = 1315$ so about 95% of the tree rings will date between 1171 and 1315 AD.

(c) $\mu - 3\sigma = 1243 - 3(36) = 1135$ and $\mu + 3\sigma = 1243 + 3(36) = 1351$ so 99.7% (almost all) of the tree rings will date between 1135 and 1351 AD.

10. (a) $\mu + \sigma = 7.6 + 0.4 = 8.0$

Since 68% of the cups filled will fall into the $\mu \pm \sigma$ range, 100% – 68% = 32% will fall outside that range and $\frac{32\%}{2} = 16\% = 0.16$ will be over $\mu + \sigma = 8$ oz. Approximately 16% of the time, the cups will overflow.

(b) 100% – 16% = 84% = 0.84 so, by the complementary event rule, 84% of the time the cups will not overflow.

(c) Since 16% of 850 = 136, we can expect approximately 136 of the cups filled by this machine will overflow.

11. (a) $\mu - \sigma = 3.15 - 1.45 = 1.70$ and $\mu + \sigma = 3.15 + 1.45 = 4.60$ so 68% of the experimental group will have millamperes pain thresholds between 1.70 and 4.60 mA.

(b) $\mu - 2\sigma = 3.15 - 2(1.45) = 0.25$ and $\mu + 2\sigma = 3.15 + 2(1.45) = 6.05$ so 95% of the experimental group will have pain thresholds between 0.25 and 6.05 mA.

12. (a)

Visitors Treated Each Day by YPMS (first 10 day period)

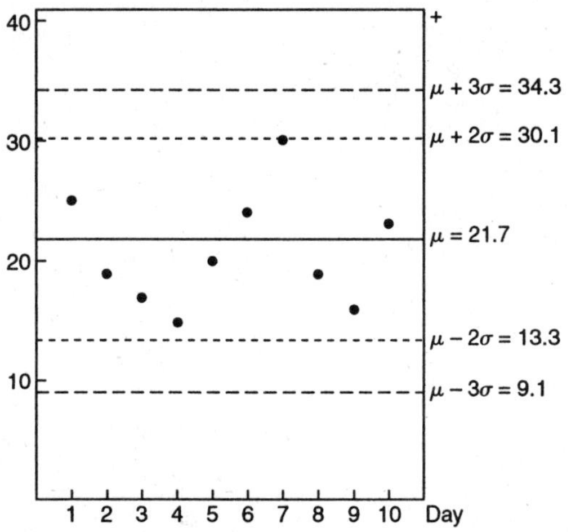

The data indicate the process is in control; none of the out-of-control warning signals are present.

(b)

Visitors Treated Each Day by YPMS (second 10 day period)

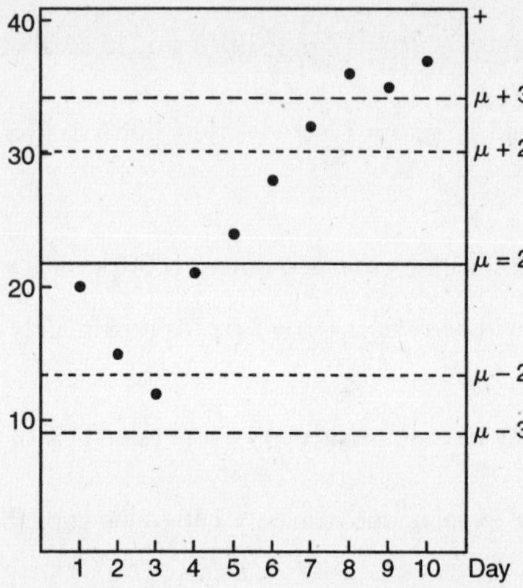

Three points fall beyond $\mu + 3\sigma = 34.3$. Four consecutive points lie beyond $\mu + 2\sigma = 30.1$. Out-of-control warning signals I and III are present; the data indicate the process is out-of-control. Under the conditions or time period (say, July 4) represented by the second 10-day period, YPMS probably needs (temporary) extra help to provide timely emergency health care for park visitors.

13. (a)

Tri-County Bank Monthly Loan Request—First Year (thousands of dollars)

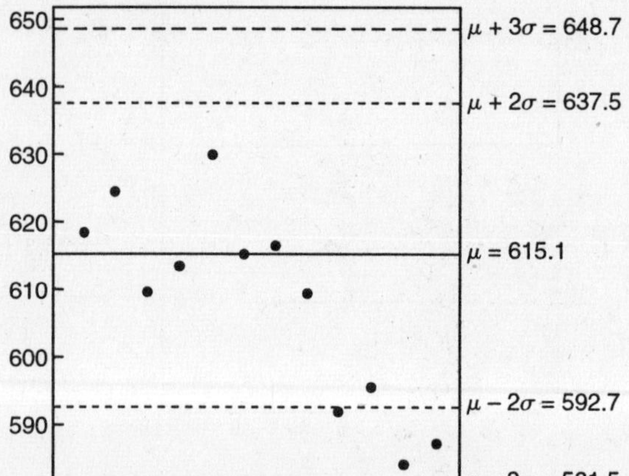

The economy would appear to be cooling off as evidenced by an overall downward trend. Out-of-control warning signal III is present: 2 of the last 3 consecutive points are below $\mu - 2\sigma = 592.7$.

(b)

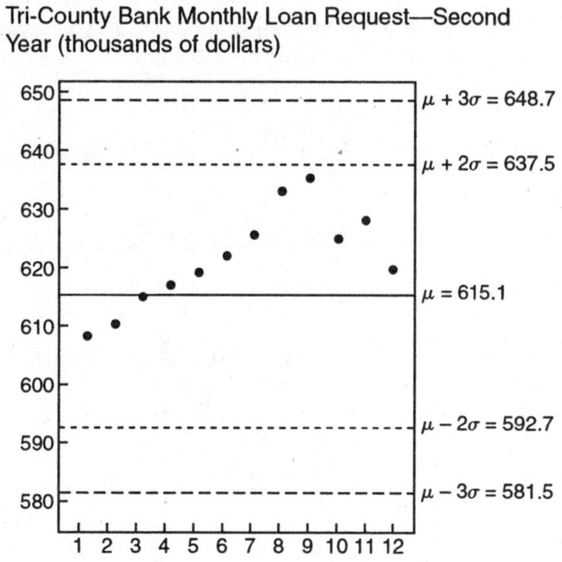

Here, it looks like the economy was heating up during months 1–9 and perhaps cooling off during months 10–12. Out-of-control warning signal II is present: there is a run of 9 consecutive points above $\mu = 615.1$.

14. (a) The room rental rate has been unusually high. There are eleven values in a row above the center line. Since there are more than nine points in a row, the out-of-control signals is Type II. The out-of-control signal may be the result of a convention, special tours, or a vacation period.

(b) The room rental rate is mixed with some high values and some low values. There is one out-of-control. Signal of Type I on the high side. There are two out-of-control signals of Type III on the low side. Low occupancy rates could be caused by weather or by high cancellation rates. High occupancy could be caused by tours, conventions, or holidays.

15.

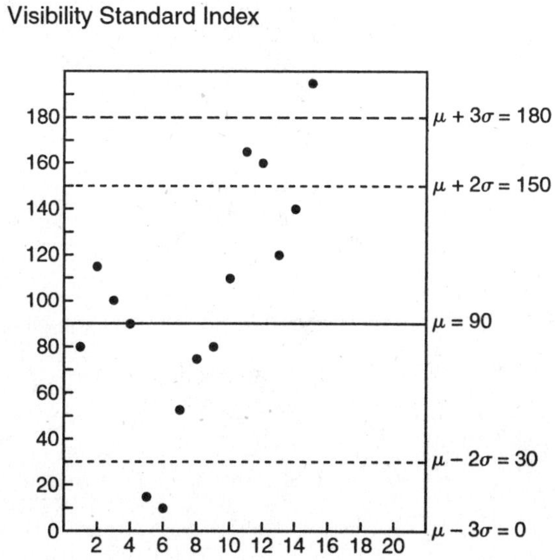

Out-of-control warning signals I and III are present. Day 15's VSI exceeds $\mu + 3\sigma$. Two of 3 consecutive points (days 10, 11, 12 or days 11, 12, 13) are about $\mu + 2\sigma = 150$, and 2 of 3 consecutive points (days 4, 5, 6 or days 5, 6, 7) are below $\mu - 2\sigma = 30$. Days 10–15 all show above average air pollution levels; days 11, 12, and 15 triggered out-of-control signals, indicating pollution abatement procedures should be in place.

Section 7.2

1. **(a)** z-scores > 0 indicate the student scored above the mean: Robert, Jane, and Linda

(b) z-scores = 0 indicates the student scored at the mean: Joel

(c) z-scores < 0 indicate the student scored below the mean: John and Susan

(d) $z = \dfrac{x-\mu}{\sigma}$ so $x = \mu + z\sigma$

In this case, if the student's score is x, $x = 150 + z(20)$.

Robert:	$x = 150 + 1.10(20) = 172$
Joel:	$x = 150 + 0(20) = 150$
Jan:	$x = 150 + 1.70(20) = 184$
John:	$x = 150 - 0.80(20) = 134$
Susan:	$x = 150 - 2.00(20) = 110$
Linda:	$x = 150 + 1.60(20) = 182$

2. Use $z = \dfrac{x-\mu}{\sigma}$. In this case, $z = \dfrac{x-0.51}{0.25}$ with x expressed as a decimal or $z = \dfrac{x-51\%}{25\%}$ with x expressed in percent.

(a) $z = \dfrac{0.45-0.51}{0.25} = -0.24$

(b) $z = \dfrac{0.72-0.51}{0.25} = 0.84$

(c) $z = \dfrac{0.75-0.51}{0.25} = 0.96$

(d) $z = \dfrac{65\%-51\%}{25\%} = 0.56$

(e) $z = \dfrac{33\%-51\%}{25\%} = -0.72$

(f) $z = \dfrac{55\%-51\%}{25\%} = 0.16$

3. Use $z = \frac{x-\mu}{\sigma}$. In this case, $z = \frac{x-73}{5}$.

(a) $53°F < x < 93°F$

$\frac{53-73}{5} < \frac{x-73}{5} < \frac{93-73}{5}$ Subtract $\mu = 73°F$ from each piece; divide result by $\sigma = 5°F$.

$-\frac{20}{5} < z < \frac{20}{5}$

$-4.00 < z < 4.00$

(b) $x < 65°F$

$x - 73 < 65 - 73$ Subtract $\mu = 73°F$.

$\frac{x-73}{5} < \frac{65-73}{5}$ Divide both sides by $\sigma = 5°F$.

$z < -1.6$

(c) $78°F < x$

$\frac{78-73}{5} < \frac{x-73}{5}$ Subtract $\mu = 73°F$ from each side; divide by $\sigma = 5°F$.

$\frac{5}{5} < z$

$1.00 < z$ (or $z > 1.00$)

Since $z = \frac{x-73}{5}$, $x = 73 + 5z$.

(d) $1.75 < z$

$5(1.75) < 5z$ Multiply both sides by $\sigma = 5°F$.

$73 + 5(1.75) < 73 + 5z$ Add $\mu = 73°F$ to both sides.

$81.75°F < x$ (or $x > 81.75$)

(e) $z < -1.90$

$5z < 5(-1.90)$ Multiply both sides by $\sigma = 5°F$.

$73 + 5z < 73 + 5(-1.90)$ Add $\mu = 73°F$ to both sides.

$x < 63.5°F$

(f) $-1.80 < z < 1.65$

$5(-1.80) < 5z < 5(1.65)$ Multiply each part by $\sigma = 5°F$.

$73 + 5(-1.80) < 73 + 5z < 73 + 5(1.65)$ Add $\mu = 73°F$ to each part of the inequality.

$64°F < x < 81.25°F$

4. $z = \frac{x-\mu}{\sigma}$; here, $z = \frac{x-27.2}{4.3}$

(a) $x < 30$ kg

$x - 27.2 < 30 - 27.2$ Subtract $\mu = 27.2$ kg from each side.

$\frac{x-27.2}{4.3} < \frac{30-27.2}{4.3}$ Divide both sides by $\sigma = 4.3$ kg.

$z < 0.65$ (rounded to 2 decimal places)

(b) $19 \text{ kg} < x$

$19 - 27.2 < x - 27.2$ Subtract $\mu = 27.2$ kg from each side.

$\frac{19-27.2}{4.3} < \frac{x-27.2}{4.3}$ Divide both sides by $\sigma = 4.3$ kg.

$-1.91 < z$ (rounded)

(c) $32 \text{ kg} < x < 35 \text{ kg}$

$32 - 27.2 < x - 27.2 < 35 - 27.2$ Subtract $\mu = 27.2$ kg from each part.

$\frac{32-27.2}{4.3} < \frac{x-27.2}{4.3} < \frac{35-27.2}{4.3}$ Divide each part by $\sigma = 4.3$ kg.

$1.12 < z < 1.81$ (rounded)

Since $z = \frac{x-27.2}{4.3}$, $x = 27.2 + 4.3z$ kg.

(d) $-2.17 < z$

$(4.3)(-2.17) < 4.3z$ Multiply both sides by $\sigma = 4.3$ kg.

$27.2 + 4.3(-2.17) < 27.2 + 4.3z$ Add $\mu = 27.2$ kg to each side.

$17.9 \text{ kg} < x$, or $x > 17.9$ kg (rounded)

(e) $z < 1.28$

$4.3z < 4.3(1.28)$ Multiply both sides by $\sigma = 4.3$ kg.

$27.2 + 4.3z < 27.2 + 4.3(1.28)$ Add $\mu = 27.2$ kg to both sides.

$x < 32.7$ kg (rounded)

(f) $-1.99 < z < 1.44$

$4.3(-1.99) < 4.3z < 4.3(1.44)$ Multiply each part by $\sigma = 4.3$ kg.

$27.2 + 4.3(-1.99) < 27.2 + 4.3z < 27.2 + 4.3(1.44)$ Add $\mu = 27.2$ kg to each part.

$18.6 \text{ kg} < x < 33.4 \text{ kg}$ (rounded)

(g) 14 kg is an unusually low weight for a fawn

$z = \frac{x-27.2}{4.3} = \frac{14-27.2}{4.3} = -3.07$ (rounded)

(note $\mu = 27.2$ kg)

(h) An unusually large fawn would have a large positive z, such as 3.

5. $z = \frac{x-\mu}{\sigma}$, here $z = \frac{x-4400}{620}$

(a) $3300 < x$

$3300 - 4400 < x - 4400$ Subtract $\mu = 4400$ deer.

$\frac{3300-4400}{620} < \frac{x-4400}{620}$ Divide by $\sigma = 620$ deer.

$-1.77 < z$

(b) $x < 5400$

$x - 4400 < 5400 - 4400$ Subtract $\mu = 4400$ deer.

$\frac{x-4400}{620} < \frac{5400-4400}{620}$ Divide by $\sigma = 620$ deer.

$z < 1.61$

(c) $3500 < x < 5300$

$3500 - 4400 < x - 4400 < 5300 - 4400$ Subtract $\mu = 4400$.

$\frac{3500\text{-}4400}{620} < \frac{x-4400}{620} < \frac{5300-4400}{620}$ Divide by $\sigma = 620$.

$-1.45 < z < 1.45$

Since $z = \frac{x-4400}{620}$, $x = 4400 + 620z$ deer.

(d)
$$-1.12 < z < 2.43$$
$$620(-1.12) < 620z < 620(2.43) \quad \text{Multiply by } \sigma = 620.$$
$$4400 + 620(-1.12) < 4400 + 620z < 4400 + 620(2.43) \quad \text{Add } \mu = 4400 \text{ to each part.}$$
$$3706 \text{ deer } < x < 5907 \text{ deer (rounded)}$$

(e)
$$z < 1.96$$
$$620z < 620(1.96) \quad \text{Multiply by } \sigma.$$
$$4400 + 620z < 4400 + 620(1.96) \quad \text{Add } \mu.$$
$$x < 5615 \text{ deer}$$

(f)
$$2.58 < z$$
$$620(2.58) < 620z \quad \text{Multiply by } \sigma.$$
$$4400 + 620(2.58) < 4400 + 620z \quad \text{Add } \mu.$$
$$6000 \text{ deer } < x$$

(g) If $x = 2800$ deer, $z = \dfrac{2800 - 4400}{620} = -2.58$.

This is a small z-value, so 2800 deer is quite low for the fall deer population.

If $x = 6300$ deer, $z = \dfrac{6300 - 4400}{620} = 3.06$.

This is a very large z-value, so 6300 deer would be an unusually large fall population size.

6. $z = \dfrac{x - \mu}{\sigma}$ so in this case, $z = \dfrac{x - 7500}{1750}$.

(a)
$$9000 < x$$
$$\frac{9000 - 7500}{1750} < \frac{x - 7500}{1750} \quad \text{Subtract } \mu\text{; divide by } \sigma.$$
$$0.86 < z$$

(b)
$$x < 6000$$
$$\frac{x - 7500}{1750} < \frac{6000 - 7500}{1750} \quad \text{Subtract } \mu\text{; divide by } \sigma.$$
$$z < -0.86$$

(c)
$$3500 < x < 4500$$
$$\frac{3500 - 7500}{1750} < \frac{x - 7500}{1750} < \frac{4500 - 7500}{1750} \quad \text{Subtract } \mu\text{; divide by } \sigma.$$
$$-2.29 < z < -1.71$$

Since $z = \dfrac{x - 7500}{1750}$, $x = 7500 + 1750z$.

(d)
$$z < 1.15$$
$$7500 + 1750z < 7500 + 1750(1.15) \quad \text{Multiply by } \sigma\text{; add } \mu.$$
$$x < 9513$$

(e)
$$2.19 < z$$
$$7500 + 1750(2.19) < 7500 + 1750(z) \quad \text{Multiply by } \sigma\text{; add } \mu.$$
$$11{,}333 < x$$

(f)
$$0.25 < z < 1.25$$
$$7500 + 1750(0.25) < 7500 + 1750z < 7500 + 1750(1.25) \quad \text{Multiply by } \sigma\text{; add } \mu.$$
$$7938 < x < 9688$$

(g) Since $\mu = 7500$, $x = 2500$ is quite low.

$$z = \frac{x-\mu}{\sigma} = \frac{2500-7500}{1750} = -2.86 \text{ (a very small } z\text{)}$$

7. $z = \frac{x-\mu}{\sigma}$ so in this case, $z = \frac{x-4.8}{0.3}$.

(a)
$$4.5 < x$$
$$\frac{4.5-4.8}{0.3} < \frac{x-4.8}{0.3} \quad \text{Subtract } \mu\text{; divide by } \sigma.$$
$$-1.00 < z$$

(b)
$$x < 4.2$$
$$\frac{x-4.8}{0.3} < \frac{4.2-4.8}{0.3} \quad \text{Subtract } \mu\text{; divide by } \sigma.$$
$$z < -2.00$$

(c)
$$4.0 < x < 5.5$$
$$\frac{4.0-4.8}{0.3} < \frac{x-4.8}{0.3} < \frac{5.5-4.8}{0.3} \quad \text{Subtract } \mu\text{; divide by } \sigma.$$
$$-2.67 < z < 2.33$$

Since $z = \frac{x-4.8}{0.3}$, $x = 4.8 + 0.3z$.

(d)
$$z < -1.44$$
$$0.3z < 0.3(-1.44) \quad \text{Multiply by } \sigma.$$
$$4.8 + 0.3z < 4.8 + 0.3(-1.44) \quad \text{Add } \mu.$$
$$x < 4.4$$

(e)
$$1.28 < z$$
$$0.3(1.28) < 0.3z \quad \text{Multiply by } \sigma.$$
$$4.8 + 0.3(1.28) < 4.8 + 0.3z \quad \text{Add } \mu.$$
$$5.2 < x$$

(f)
$$-2.25 < z < -1.00$$
$$0.3(-2.25) < 0.3z < 0.3(-1.00) \quad \text{Multiply by } \sigma.$$
$$4.8 + 0.3(-2.25) < 4.8 + 0.3z < 4.8 + 0.3(-1.00) \quad \text{Add } \mu.$$
$$4.1 < x < 4.5$$

(g) If the RBC was 5.9 or higher, that would be an unusually high red blood cell count.

$$x \geq 5.9$$
$$\frac{x-4.8}{0.3} \geq \frac{5.9-4.8}{0.3}$$
$$z \geq 3.67 \text{ (a very large } z\text{-value)}$$

8. (a) $z = \frac{x - \mu}{\sigma}$

Site 1: $z_1 = \frac{x_1 - \mu_1}{\sigma_1}$

$z_1 = \frac{x_1 - 1272}{35}$

so for $x_1 = 1250$

$z_1 = \frac{1250 - 1272}{35} = -0.63$

Site 2: $z_2 = \frac{x_2 - \mu_2}{\sigma_2}$

$z_2 = \frac{x_2 - 1122}{40}$

so for $x_2 = 1234$

$z_2 = \frac{1234 - 1122}{40} = 2.80$

(b) x_2, the object dated 1234 AD, is more unusual at its site, since $z_2 = 2.8$ vs. $z_1 = -0.63$.

For problems 9–48, refer to the following sketch patterns for guidance in calculations

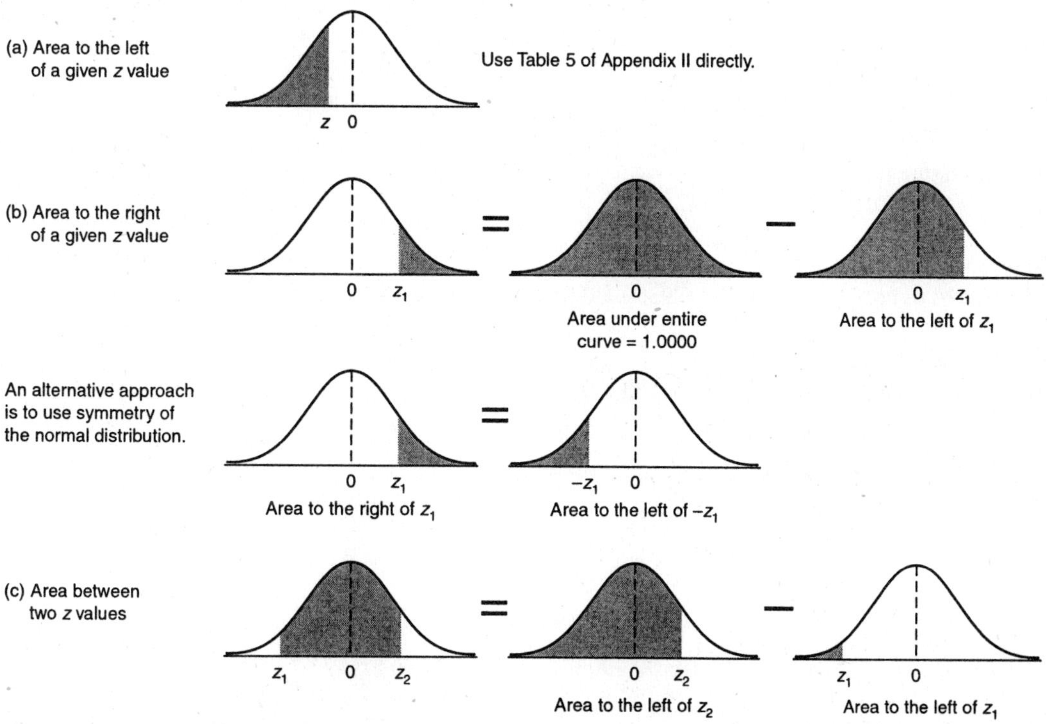

Using the left-tail style standard normal distribution table (see figures above)

(a) For areas to the *left* of a specified z value, use the table entry directly.

(b) For areas to the *right* of a specified z value, look up the table entry for z and subtract the table value from 1. (This is the complementary event rule as applied to area as probability.)

OR: Use the fact that the normal curve is symmetric about the mean, 0. The area in the right tail above a z-value is the same as the area in the left tail below the value of $-z$. So, to find the area to the right of z, look up the table value for $-z$.

(c) For areas *between* two z-values, z_1 and z_2, where $z_1 < z_2$, subtract the tabled value for z_1, from the tabled value for z_2.

These sketches and rules for finding the area for probability from the standard normal table apply for *any* z: $-\infty < z < +\infty$.

Student sketches should resemble those indicated with negative z-values to left of 0 and positive z-values to the right of zero.

9. Refer to figure (b).
The area to the right of $z = 0$ is 1 – area to left of $z = 0$, or 1 – 0.5000 = 0.5000.

10. Refer to figure (a).
The area to the left of $z = 0$ is 0.5000 (direct read).

11. Refer to figure (a).
The area to the left of $z = -1.32$ is 0.0934.

12. Refer to figure (a).
The area to the left of $z = -0.47$ is 0.3192.

13. Refer to figure (a).
The area to left of $z = 0.45$ is 0.6736.

14. Refer to figure (a).
The area to left of $z = 0.72$ is 0.7642.

15. Refer to figure (b).
The area to right of $z = 1.52$ is 1 – 0.9357 = 0.0643.

16. Refer to figure (b).
The area to right of $z = 0.15$ is 1 – 0.5596 = 0.4404.

17. Refer to figure (b).
The area to right of $z = -1.22$ is 1 – 0.1112 = 0.8888.

18. Refer to figure (b).
The area to right of $z = -2.17$ is 1 – 0.0150 = 0.9850.

19. Refer to figure (c).
The area between $z = 0$ and $z = 3.18$ is 0.9993 – 0.5000 = 0.4993.

20. Refer to figure (c).
The area between $z = 0$ and $z = 2.92$ is 0.9982 – 0.5000 = 0.4982.

21. Refer to figure (c).
The area between $z = 0$ and $z = -2.01$ is 0.5000 – 0.0222 = 0.4778.

22. Refer to figure (c).
The area between $z = 0$ and $z = -1.93$ is 0.5000 – 0.0268 = 0.4732.

23. Refer to figure (c).
The area between $z = -2.18$ and $z = 1.34$ is 0.9099 – 0.0146 = 0.8953.

24. Refer to figure (c).
The area between $z = -1.40$ and $z = 2.03$ is $0.9788 - 0.0808 = 0.8980$.

25. Refer to figure (c).
The area between $z = 0.32$ and $z = 1.92$ is $0.9726 - 0.6255 = 0.3471$.

26. Refer to figure (c).
The area between $z = 1.42$ and $z = 2.17$ is $0.9850 - 0.9222 = 0.0628$.

27. Refer to figure (c).
The area between $z = -2.42$ and $z = -1.77$ is $0.0384 - 0.0078 = 0.0306$.

28. Refer to figure (c).
The area between $z = -1.98$ and $z = -0.03$ is $0.4880 - 0.0239 = 0.4641$.

29. Refer to figure (a).
$P(z \le 0) = -0.5000$

30. Refer to figure (b).
$P(z \ge 0) = 1 - P(z < 0) = 1 - 0.5000 = 0.5000$

31. Refer to figure (a).
$P(z \le -0.13) = 0.4483$ (direct read)

32. Refer to figure (a).
$P(z \le -2.15) = 0.0158$

33. Refer to figure (a).
$P(z \le 1.20) = 0.8849$

34. Refer to figure (a).
$P(z \le 3.20) = 0.9993$

35. Refer to figure (b).
$P(z \ge 1.35) = 1 - P(z < 1.35) = 1 - 0.9115 = 0.0885$

36. Refer to figure (b).
$P(z \ge 2.17) = 1 - P(z < 2.17) = 1 - 0.9850 = 0.0150$

37. Refer to figure (b).
$P(x \ge -1.20) = 1 - P(z < -1.20) = 1 - 0.1151 = 0.8849$

38. Refer to figure (b).
$P(z \ge -1.50) = 1 - P(z < -1.50) = 1 - 0.0668 = 0.9332$

39. Refer to figure (c).
$P(-1.20 \le z \le 2.64) = P(z \le 2.64) - P(z < -1.20) = 0.9959 - 0.1151 = 0.8808$

40. Refer to figure (c).
$P(-2.20 \le z \le 1.04) = P(z \le 1.04) - P(z < -2.20) = 0.8508 - 0.0139 = 0.8369$

41. Refer to figure (c).
$P(-2.18 \le z \le -0.42) = P(z \le -0.42) - P(z < -2.18) = 0.3372 - 0.0146 = 0.3226$

42. Refer to figure (c).
$P(-1.78 \le z \le -1.23) = P(z \le -1.23) - P(z < -1.78) = 0.1093 - 0.0375 = 0.0718$

43. Refer to figure (c).
$P(0 \le z \le 1.62) = P(z \le 1.62) - P(z < 0) = 0.9474 - 0.5000 = 0.4474$

44. Refer to figure (c).
$P(0 \le z \le 0.54) = P(z \le 0.54) - P(z < 0) = 0.7054 - 0.5000 = 0.2054$

45. Refer to figure (c).
$P(-0.82 \le z \le 0) = P(z \le 0) - P(z < -0.82) = 0.5000 - 0.2061 = 0.2939$

46. Refer to figure (c).
$P(-2.37 \le z \le 0) = P(z \le 0) - P(z < -2.37) = 0.5000 - 0.0089 = 0.4911$

47. Refer to figure (c).
$P(-0.45 \le z \le 2.73) = P(z \le 2.73) - P(z < -0.45) = 0.9968 - 0.3264 = 0.6704$

48. Refer to figure (c).
$P(-0.73 \le z \le 3.12) = P(z \le 3.12) - P(z < -0.73) = 0.9991 - 0.2327 = 0.7664$

Section 7.3

1. We are given $\mu = 4$ and $\sigma = 2$. Since $z = \dfrac{x-\mu}{\sigma}$, we have $z = \dfrac{x-4}{2}$.

$P(3 \le x \le 6)$
$= P(3-4 \le x-4 \le 6-4)$ Subtract $\mu = 4$ from each part of the inequality.
$= P\left(\dfrac{3-4}{2} \le \dfrac{x-4}{2} \le \dfrac{6-4}{2}\right)$ Divide each part by $\sigma = 2$.
$= P\left(-\dfrac{1}{2} \le z \le \dfrac{2}{2}\right)$
$= P(-0.5 \le z \le 1)$
$= P(z \le 1) - P(z < -0.5)$ Refer to sketch (c) in the solutions for Section 7.2.
$= 0.8413 - 0.3085$
$= 0.5328$

2. We are given $\mu = 15$ and $\sigma = 4$. Since $z = \frac{x-\mu}{\sigma}$, we have $z = \frac{x-15}{4}$.

$$
\begin{aligned}
&P(10 \le x \le 26)\\
&= P(10-15 \le x-15 \le 26-15) && \text{Subtract } \mu = 15.\\
&= P\left(\frac{10-15}{4} \le \frac{x-15}{4} \le \frac{26-15}{4}\right) && \text{Divide each part of the inequality by } \sigma = 4.\\
&= P(-1.25 \le z \le 2.75)\\
&= P(z \le 2.75) - P(z < -1.25) && \text{Refer to sketch (c) in solutions for Section 7.2.}\\
&= 0.9970 - 0.1056\\
&= 0.8914
\end{aligned}
$$

3. We are given $\mu = 40$ and $\sigma = 15$. Since $z = \frac{x-\mu}{\sigma}$, we have $z = \frac{x-40}{15}$.

$$
\begin{aligned}
&P(50 \le x \le 70)\\
&= P(50-40 \le x-40 \le 70-40) && \text{Subtract } \mu = 40.\\
&= P\left(\frac{50-40}{15} \le \frac{x-40}{15} \le \frac{70-40}{15}\right) && \text{Divide by } \sigma = 15.\\
&= P(0.67 \le z \le 2)\\
&= P(z \le 2) - P(z < 0.67)\\
&= 0.9772 - 0.7486 = 0.2286
\end{aligned}
$$

4. We are given $\mu = 5$ and $\sigma = 1.2$. Since $z = \frac{x-\mu}{\sigma}$, we have $z = \frac{x-5}{1.2}$.

$$
\begin{aligned}
&P(7 \le x \le 9)\\
&= P(7-5 \le x-5 \le 9-5) && \text{Subtract } \mu = 5.\\
&= P\left(\frac{7-5}{1.2} \le \frac{x-5}{1.2} \le \frac{9-5}{1.2}\right) && \text{Divide by } \sigma = 1.2.\\
&= P(1.67 \le z \le 3.33)\\
&= P(z \le 3.33) - P(z < 1.67)\\
&= 0.9996 - 0.9525 = 0.0471
\end{aligned}
$$

5. We are given $\mu = 15$ and $\sigma = 3.2$. Since $z = \frac{x-\mu}{\sigma}$, we have $z = \frac{x-15}{3.2}$.

$$
\begin{aligned}
&P(8 \le x \le 12)\\
&= P(8-15 \le x-15 \le 12-15) && \text{Subtract } \mu = 15.\\
&= P\left(\frac{8-15}{3.2} \le \frac{x-15}{3.2} \le \frac{12-15}{3.2}\right) && \text{Divide by } \sigma = 3.2.\\
&= P(-2.19 \le z \le -0.94)\\
&= P(z \le -0.94) - P(z < -2.19)\\
&= 0.1736 - 0.0143 = 0.1593
\end{aligned}
$$

6. We are given $\mu = 50$ and $\sigma = 15$. Since $z = \dfrac{x-\mu}{\sigma}$, we have $z = \dfrac{x-50}{15}$.

$$\begin{aligned}
&P(40 \le x \le 47)\\
&= P(40-50 \le x-50 \le 47-50) && \text{Subtract } \mu = 50.\\
&= P\left(\frac{40-50}{15} \le \frac{x-50}{15} \le \frac{47-50}{15}\right) && \text{Divide by } \sigma = 15.\\
&= P(-0.67 \le z \le -0.20)\\
&= P(z \le -0.20) - P(z < -0.67)\\
&= 0.4207 - 0.2514 = 0.1693
\end{aligned}$$

7. We are given $\mu = 20$ and $\sigma = 3.4$. Since $z = \dfrac{x-\mu}{\sigma}$, we have $z = \dfrac{x-20}{3.4}$.

$$\begin{aligned}
&P(x \ge 30)\\
&= P(x-20 \ge 30-20) && \text{Subtract } \mu = 20.\\
&= P\left(\frac{x-20}{3.4} \ge \frac{30-20}{3.4}\right) && \text{Divide by } \sigma = 3.4.\\
&= P(z \ge 2.94)\\
&= 1 - P(z < 2.94) && \text{Refer to sketch (b) in Section 7.2.}\\
&= 1 - 0.9984 = 0.0016
\end{aligned}$$

8. We are given $\mu = 100$ and $\sigma = 15$. Since $z = \dfrac{x-\mu}{\sigma}$, we have $z = \dfrac{x-100}{15}$.

$$\begin{aligned}
&P(x \ge 120)\\
&= P(x-100 \ge 120-100) && \text{Subtract } \mu.\\
&= P\left(\frac{x-100}{15} \ge \frac{120-100}{15}\right) && \text{Divide by } \sigma.\\
&= P(z \ge 1.33) && \text{Refer to sketch (b) in Section 7.2.}\\
&= 1 - P(z < 1.33)\\
&= 1 - 0.9082 = 0.0918
\end{aligned}$$

9. We are given $\mu = 100$ and $\sigma = 15$. Since $z = \dfrac{x-\mu}{\sigma}$, we have $z = \dfrac{x-100}{15}$.

$$\begin{aligned}
&P(x \ge 90)\\
&= P\left(\frac{x-100}{15} \ge \frac{90-100}{15}\right) && \text{Subtract } \mu\text{; divide by } \sigma.\\
&= P(z \ge -0.67)\\
&= 1 - P(z < -0.67) = 1 - 0.2514 = 0.7486
\end{aligned}$$

10. We are given $\mu = 3$ and $\sigma = 0.25$. Since $z = \dfrac{x-\mu}{\sigma}$, we have $z = \dfrac{x-3}{0.25}$.

$$\begin{aligned}
&P(x \ge 2)\\
&= P\left(\frac{x-3}{0.25} \ge \frac{2-3}{0.25}\right) && \text{Subtract } \mu\text{; divide by } \sigma.\\
&= P(z \ge -4)\\
&= 1 - P(z < -4) \approx 1 - 0 = 1
\end{aligned}$$

For problems 11–20, refer to the following sketch patterns for guidance in calculation.

(a) **Left-tail case:**
The given area A is to the left of z.

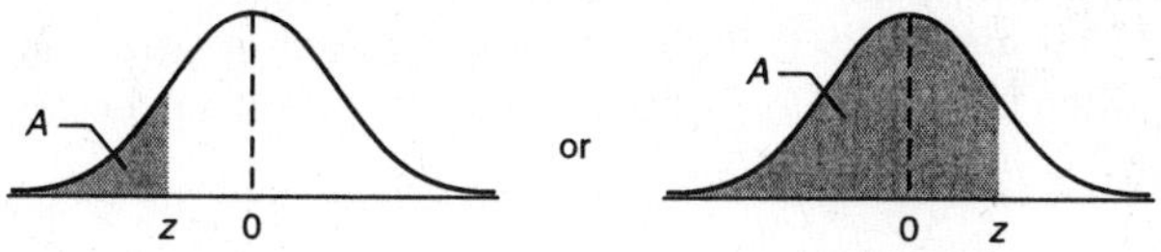

For the left-tail case, look up the number A in the body of the table and use the corresponding z value.

(b) **Right-tail case:**
The given area A is to the right of z.

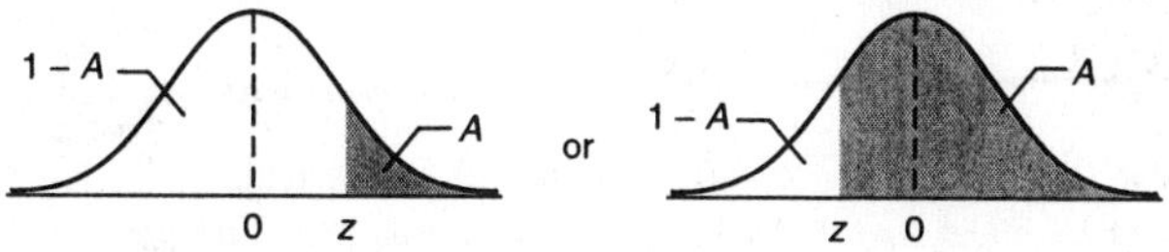

For the right-tail case, look up the number $1 - A$ in the body of the table and use the corresponding z value.

(c) **Center case:**
The given area A is symmetric and centered above $z = 0$. Half of A lies to the left and half lies to the right of $z = 0$.

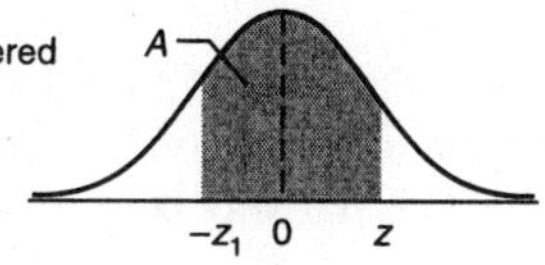

For the center case, look up the number $\frac{1-A}{2}$ in the body of the table and use the corresponding $\pm z$ value.

Student sketches should resemble the figures above, with negative z-values to the left of zero and positive z-values to the right of zero, and A written as a decimal.

11. Refer to figure (a).
Find z so that the area A to the left of z is 6% = 0.06. Since $A = 0.06$ is less than 0.5000, look for a negative z value. A to left of –1.55 is 0.0606 and A to left of –1.56 is 0.0594. Since 0.06 is in the middle of 0.0606 and 0.0594, for our z-value we will use the average of –1.55 and –1.56:

$$\frac{-1.55+(-1.56)}{2} = -1.555.$$

12. Refer to figure (a).
Find z so that the area A to the left of z is 5.2% = 0.052. Since $A = 0.052 < 0.5000$, look for a negative z value. A to the left of –1.63 is 0.0516, which is closer to 0.052 than is A to the left of –1.62 (0.0526), so $z = -1.63$.

13. Refer to figure (a).
Find z so that the area A to the left of z is 55% = 0.55. Since $A = 0.55 > 0.5000$, look for a positive z-value. The area to the left of 0.13 is 0.5517, so $z = 0.13$.

14. Refer to figure (a).
Find z so that the area A to the left of z is 97.5% = 0.975. Since $A = 0.975 > 0.5000$, look for a positive z. A to left of $z = 1.96$ is 0.9750.

15. Refer to figure (b).
Find z so that the area A to the right of z is 8% = 0.08. Since A to the right of z is 0.08, $1 - A = 1 - 0.08 = 0.92$ is to the left of z-value. The area to the left of 1.41 is 0.9207.

16. Refer to figure (b).
Find z so that the area A to the right of z is 5% = 0.05. Since A to the right of z is 0.05, $1 - A = 1 - 0.05 = 0.95$ is to the left of z. Since $1 - A = 0.95 > 0.5000$, look for a positive z-value. The area to the left of 1.64 is 0.9495, and the area to the left of 1.65 is 0.9505. Since 0.95 is halfway between 0.9495 and 0.9505, we average the two z values.

$$\frac{1.64+1.65}{2} = 1.645$$

17. Refer to figure (b).
Find z so that the area A to the right of z is 82% = 0.82. Since A to the right of z, $1 - A = 1 - 0.82 = 0.18$ is to the left of z. Since $1 - A = 0.18 < 0.5000$, look for a negative z value. The area to the left of $z = -0.92$ is 0.1788.

18. Refer to figure (b).
Find z so that the area A to the right of z is 95% = 0.95. Since A to the right of z is 0.95, $1 - A = 1 - 0.95 = 0.05$ is to the left of z. Because $1 - A = 0.05 < 0.5000$, look for a negative z value. The area to the left of −1.64 is 0.0505. The area to the left of −1.65 is 0.0495. Since 0.05 is halfway between these two area values we average the two z-values.

$$\frac{-1.64+(-1.65)}{2} = -1.645$$

19. Refer to figure (c).
Find z such that the area A between $-z$ and z is 98% = 0.98. Since A is between $-z$ and z, $1 - A = 1 - 0.98 = 0.02$ lies in the tails, and since we need $\pm z$, half of $1 - A$ lies in each tail. The area to the left of $-z$ is $\frac{1-A}{2} = \frac{0.02}{2} = 0.01$. The area to the left of −2.33 is 0.0099. Thus $-z = -2.33$ and $z = 2.33$.

20. Refer to figure (c).
Find z such that the area A between $-z$ and z is 95% = 0.95. If A between $-z$ and $z = 0.95$, then $1 - A = 1 - 0.95 = 0.05$ is the area in the tails, and that is split evenly between the two tails. Thus, the area to the left of $-z$ is $\frac{1-A}{2} = \frac{0.05}{2} = 0.025$. The area to the left of −1.96 is 0.0250, so $-z$ is −1.96 and $z = 1.96$.

21. x is approximately normal with $\mu = 85$ and $\sigma = 25$. Since $z = \frac{x-\mu}{\sigma}$, we have $z = \frac{x-85}{25}$.

(a) $P(x > 60)$

$$= P\left(\frac{x-85}{25} > \frac{60-85}{25}\right) = P(z > -1)$$
$$= 1 - P(z \le -1) = 1 - 0.1587 = 0.8413$$

(b) $P(x < 110) = P\left(\frac{x-85}{25} < \frac{110-85}{25}\right) = P(z < 1) = 0.8413.$

(c) $P(60 < x < 110)$

$$= P(-1 < z < 1) \quad \text{using (a) and (b)}$$
$$= P(z < 1) - P(z \le -1) = 0.8413 - 0.1587 = 0.6826$$

(i.e., approximately 68% of the blood glucose measurements lie within $\mu \pm \sigma$)

(d) $P(x > 140)$

$$= P\left(\frac{x-85}{25} > \frac{140-85}{25}\right) = P(z > 2.2)$$
$$= 1 - P(z \le 2.2) = 1 - 0.9861 = 0.0139$$

22. Maintenance cost, x, is approximately normal with $\mu = 615$ and $\sigma = 42$ dollars.

(a) $P(x > 646) = P\left(z > \frac{646-615}{42}\right) = P(z > 0.74) = 1 - P(z \le 0.74) = 1 - 0.7704 = 0.2296$

(b) Find x_0 such that $P(x > x_0) = 0.10$.
But, if the actual cost exceeds the budgeted amount 10% of the time, the actual cost must be within the budgeted amount 90% of the time. The problem can be rephrased as how much should be budgeted so that the probability the actual cost is less than or equal to the budgeted amount is 0.90, or find x_0 such that $P(x \le x_0) = 0.90$. First, find z_0 such that $P(z \le z_0) = 0.90$. $P(z \le 1.28) = 0.8997$, so let $z_0 = 1.28$. Since $z_0 = \frac{x_0 - \mu}{\sigma}$, $x_0 = \mu + z_0\sigma = 615 + 1.28(42) = \$668.76 \approx \$669$.

23. SAT scores, x, are normal with $\mu_x = 500$ and $\sigma_x = 100$. Since $z = \frac{x - \mu_x}{\sigma_x}$, we have $x = \frac{x-500}{100}$.

(a) $P(x > 675) = P\left(\frac{x-500}{100} > \frac{675-500}{100}\right) = P(z > 1.75) = 1 - P(z \le 1.75) = 1 - 0.9599 = 0.0401$

(b) $P(x < 450) = P\left(\frac{x-500}{100} < \frac{450-500}{100}\right) = P(z < -0.5) = 0.3085$

(c) $P(450 \le x \le 675)$

$= P(-0.5 \le z \le 1.75)$ using (a), (b)

$= P(z \le 1.75) - P(z < -0.5) = 0.9599 - 0.3085$

$= 0.6514$ using work in (a), and (b)

ACT scores, y, are normal with $\mu_y = 18$ and $\sigma_y = 6$. Since $z = \frac{y - \mu_y}{\sigma_y}$, we have $z = \frac{y-18}{6}$.

(d) $P(y > 28) = P\left(\frac{y-18}{6} > \frac{28-18}{6}\right) = P(z > 1.67) = 1 - P(z \le 1.67) = 1 - 0.9525 = 0.0475$

(e) $P(y > 12) = P\left(\frac{y-18}{6} > \frac{12-18}{6}\right) = P(z > -1) = 1 - P(z \le -1) = 1 - 0.1587 = 0.8413$

(f) $P(12 \le y \le 28)$

$= P(-1 \le z \le 1.67)$ using (a), (b)

$= P(z \le 1.67) - P(z < -1) = 0.9525 - 0.1587 = 0.7938$

24. SAT scores, x, are normal with $\mu_x = 500$ and $\sigma_x = 100$; ACT scores, y, are normal with $\mu_y = 18$ and $\sigma_x = 6$. Since $z_0 = \frac{x_0 - \mu_x}{\sigma_x}$, a little algebra shows $x_0 = \mu_x + z_0\sigma_x$ and, similarly, $y_0 = \mu_y + z_0\sigma_y$.

(a) Find the SAT score, x_0, such that $P(x \ge x_0) = 10\% = 0.10$.

$$P(x \geq x_0) = P\left(\frac{x-500}{100} \geq \frac{x_0-500}{100}\right) = P(z \geq z_0) = 0.10$$

(that is, find the value z_0 such that 10% of the standard normal curve lies to the right of z_0). Since 10% is to the right of z_0, $1 - 0.10 = 0.90 = 90\%$ is to the left of z_0. Because $0.90 > 0.5000$, z_0 will be a positive number.

$$P(z \leq 1.28) = 0.8997, \text{ so } z_0 = 1.28$$

$x_0 = \mu_x + z_0\sigma_x$ so here, $x_0 = 500 + 1.28(100) = 628$ students scoring 628 points or more on the SAT math exam are in the top 10%.

Similarly, find y_0 such that $P(y \geq y_0) = 0.10$. Since 10% of the standard normal curve is to the right of z_0, $100\% - 10\% = 90\% = 0.90$ is to the left of z_0. $P(z \leq 1.28) = 0.8997$, so $z_0 = 1.28$. Then $y_0 = \mu_y + z_0\sigma_y = 18 + 1.28(6) = 25.68 \approx 26$. Students scoring 26 or more points on the ACT math test are in the top 10%.

(b) Find the SAT score, x_0, and the ACT score, y_0, such that $P(x \geq x_0) = P(y \geq y_0) = 20\% = 0.20$. First, find z_0 such that $P(z \geq z_0) = 0.20$, or $P(z < z_0) = 1 - 0.20 = 0.80$. $P(z < 0.84) = 0.7995$, so $z_0 = 0.84$. Then $x_0 = \mu_x + z_0\sigma_x = 500 + 0.84(100) = 584$ and $y_0 = \mu_y + z_0\sigma_y = 18 + 0.84(6) = 23.04 \approx 23$. Students scoring at least 584 on the SAT math test, or at least 23 on the ACT math test, are in the top 20%.

(c) Find x_0, y_0, and z_0 such that $P(x \geq x_0) = P(y \geq y_0) = P(z \geq z_0) = 60\% = 0.60$. First, z_0: $P(z < z_0) = 1 - 0.60 = 0.40$. $P(z < -0.25) = 0.4013$, so $z_0 = -0.25$. Then $x_0 = \mu_x + z_0\sigma_x = 500 + (-0.25)(100) = 475$ and $y_0 = \mu_y + z_0\sigma_y = 18 + (-0.25)(6) = 16.5$. So students scoring at least 475 on the SAT test or at least 16.5 on the ACT test are in the top 60%.

25. Lifetime, x, is normally distributed with $\mu = 45$ and $\sigma = 8$ months.

(a) $P(x \leq 36) = P\left(\frac{x-45}{8} \leq \frac{36-45}{8}\right) = P(z \leq -1.125) \approx P(z \leq -1.13) = 0.1292$

The company will have to replace approximately 13% of its batteries.

(b) Find x_0 such that $P(x \leq x_0) = 10\% = 0.10$. First, find z_0 such that $P(z \leq z_0) = 0.10$. $P(z \leq -1.28) = 0.1003$, so $z_0 = -1.28$. Then $x_0 = \mu + z_0\sigma = 45 + (-1.28)(8) = 34.76 \approx 35$. The company should guarantee the batteries for 35 months.

26. Lifetime, x, is normally distributed with $\mu = 28$ and $\sigma = 5$ months.

(a) 2 years = 24 months

$P(x \leq 24) = P\left(z \leq \frac{24-28}{5}\right) = P(z \leq -0.8) = 0.2119$

The company should expect to replace about 21.2% of its watches.

(b) Find x_0 such that $P(x \leq x_0) = 12\% = 0.12$. First, find z_0 such that $P(z \leq z_0) = 0.12$. $P(z \leq -1.17) = 0.1210$ and $P(z \leq -1.18) = 0.1190$. Since 0.12 is halfway between 0.1210 and 0.1190, we will average the z-values: $z_0 = \frac{-1.17 + (-1.18)}{2} = -1.175$.

So $x_0 = \mu + z_0\sigma = 28 + (-1.175)(5) = 22.125$.

The company should guarantee its watches for 22 months.

27. Age at replacement, x, is approximately normal with $\mu = 8$ and range = 6 years.

(a) The empirical rule says that about 95% of the data are between $\mu - 2\sigma$ and $\mu + 2\sigma$, or about 95% of the data are in a $(\mu + 2\sigma) - (\mu - 2\sigma) = 4\sigma$ range (centered around μ). Thus, the range $\approx 4\sigma$, or $\sigma \approx$ range/4. Here, we can approximate σ by 6/4 = 1.5 years.

(b) $P(x > 5) = P\left(z > \frac{5-8}{1.5}\right) = P(z > -2)$ using the estimate of σ from (a)
$= 1 - P(z \le -2) = 1 - 0.0228 = 0.9772$

(c) $P(x < 10) = P\left(z < \frac{10-8}{1.5}\right) = P(z < 1.33) = 0.9082$

(d) Find x_0 so that $P(x \le x_0) = 10\% = 0.10$. First, find z_0 such that $P(z \le z_0) = 0.10$.
$P(z \le -1.28) = 0.1003$, so $z_0 = -1.28$. Then $x_0 = \mu + z_0\sigma = 8 + (-1.28)(1.5) = 6.08$.
The company should guarantee their TVs for about 6.1 years.

28. Resting heart rate, x, is approximately normal with $\mu = 46$ and (95%) range from 22 to 70 bpm.

(a) From Problem 31(a), range $\approx 4\sigma$, or $\sigma \approx$ range/4. Here range = 70 – 22 = 48, so $\sigma \approx 48/4 = 12$ bpm.

(b) $P(x < 25) = P\left(z < \frac{25-46}{12}\right) = P(z < -1.75) = 0.0401$

(c) $P(x > 60) = P\left(z > \frac{60-46}{12}\right) = P(z > 1.17) = 1 - P(z \le 1.17) = 1 - 0.8790 = 0.1210$

(d) $P(25 \le x \le 60) = P(-1.75 \le z \le 1.17)$ using (b), (c)
$= P(z \le 1.17) - P(z < -1.75)$
$= 0.8790 - 0.0401$
$= 0.8389$

(e) Find x_0 such that $P(x > x_0) = 10\% = 0.10$. First, find z_0 such that $P(z > z_0) = 0.10$.
$P(z \le z_0) = 1 - 0.10 = 0.90$
$P(z \le 1.28) = 0.8997 \approx 0.90$, so let $z_0 = 1.28$.
When $x_0 = \mu + z_0\sigma = 46 + 1.28(12) = 61.36$, so horses with resting rates of 61 bpm or more may need treatment.

29. Life expectancy x is normal with $\mu = 90$ and $\sigma = 3.7$ months.

(a) The insurance company wants 99% of the microchips to last <u>longer</u> than x_0. Saying this another way: the insurance company wants to pay the \$50 million at most 1% of the time. So, find x_0 such that $P(x \le x_0) = 1\% = 0.01$. First, find z_0 such that $P(z \le z_0) = 0.01$. $P(z \le -2.33) = 0.0099 \approx 0.01$, so let $z_0 = -2.33$. Since $z_0 = \frac{x_0 - \mu}{\sigma}$, $x_0 = \mu + z_0\sigma = 90 + (-2.33)(3.7) = 81.379 \approx 81$ months.

(b) $P(x \le 84) = P\left(z \le \frac{84-90}{3.7}\right) = P(z \le -1.62) = 0.0526 \approx 5\%$.

(c) The "expected loss" is 5.26% [from (b)] of the \$50 million, or 0.0526(50,000,000) = \$2,630,000.

(d) Profit is the difference between the amount of money taken in (here, \$3 million), and the amount paid out (here, \$2.63 million, from (c)). So the company expects to profit 3,000,000 – 2,630,000 = \$370,000.

30. (Questions 1–6 in the text will be labeled (a)–(f) below.) Daily attendance, x, is normally distributed with $\mu = 8000$ and $\sigma = 500$ people.

(a) $P(x < 7200) = P\left(z < \frac{7200-8000}{500}\right) = P(z < -1.6) = 0.0548$

(b) $P(x > 8900) = P\left(z > \frac{8900-8000}{500}\right) = P(z > 1.8) = 1 - P(z \le 1.8) = 1 - 0.9641 = 0.0359$

(c) $P(7200 \le x \le 8900) = P(-1.6 \le z \le 1.8) = P(z \le 1.8) - P(z < -1.6) = 0.9641 - 0.0548 = 0.9093$

Arrival times are normal with $\mu = 3$ hours, 48 minutes and $\sigma = 52$ minutes after the doors open. Convert μ to minutes: $(3 \times 60) + 48 = 228$ minutes.

(d) Find x_0 such that $P(x \le x_0) = 90\% = 0.90$. First, find z_0 such that $P(z \le z_0) = 0.90$.
$P(z \le 1.28) = 0.8997 \approx 0.90$, so let $z_0 = 1.28$. Since $z_0 = \frac{x_0 - \mu}{\sigma}$, $x_0 = \mu + z_0\sigma = 228 + (1.28)(52)$
$= 294.56$ minutes, or $294.56/60 = 4.9093 \approx 4.9$ hours after the doors open.

(e) Find x_0 such that $P(x \le x_0) = 15\% = 0.15$. First, find z_0 such that $P(z \le z_0) = 0.15$.
$P(z \le -1.04) = 0.1492 \approx 0.15$, so let $z_0 = -1.04$. Then $x_0 = \mu + z_0\sigma = 228 + (-1.04)(52)$
$= 173.92$ minutes, or $173/60 = 2.899 \approx 2.9$ hours after the doors open.

(f) Answers vary. Most people have Saturday off, so many may come early in the day. Most people work Friday, so most people would probably come after 5 P.M. There is no reason to think weekday and weekend arrival times would have the same distribution.

Section 7.4

Answers may vary slightly due to rounding.

1. Previously, $p = 88\% = 0.88$; now, $p = 9\% = 0.09$; $n = 200$; $r = 50$
Let a success be defined as a child with a high blood-lead level.

(a) $P(r \ge 50) = P(50 \le r) = P(49.5 \le x)$
$np = 200(0.88) = 176$; $nq = n(1-p) = 200(0.12) = 24$
Since both np and nq are greater than 5, we will use the normal approximation to the binomial with $\mu = np = 176$ and $\sigma = \sqrt{npq} = \sqrt{200(0.88)(0.12)} = \sqrt{21.12} = 4.60$.
So, $P(r \ge 50) = P(49.5 \le x) = P\left(\frac{49.5-176}{4.6} \le z\right) = P(-27.5 \le z)$.
Almost every z value will be greater than or equal to –27.5, so this probability is approximately 1. It is almost certain that 50 or more children a decade ago had high blood-lead levels.

(b) $P(r \geq 50) = P(50 \leq r) = P(49.5 \leq x)$

In this case, $np = 200(0.09) = 18$ and $nq = 200(0.91) = 182$, so both are greater than 5. Use the normal approximation with $\mu = np = 18$ and $\sigma = \sqrt{npq} = \sqrt{200(0.09)(0.91)} = \sqrt{16.38} = 4.05$.

So $P(49.5 \leq x) = P\left(\frac{49.5-18}{4.05} \leq z\right) = P(7.78 \leq z)$.

Almost no z values will be larger than 7.78, so this probability is approximately 0. Today, it is almost impossible that a sample of 200 children would include at least 50 with high blood-lead levels.

2. We are given $p = 0.40$ and $n = 128$. Let a success be defined as an insurance claim inflated (padded) to cover the deductible.

(a) $\frac{1}{2}(128) = 64$

$P(r \geq 64) = P(64 \leq r) = P(63.5 \leq x)$ 64 is a left endpoint $np = 128(0.4) = 51.2$ and $nq = 128(0.6) = 76.8$ are both greater than 5, so we will use the normal approximation to the binomial with $\mu = np = 51.2$ and $\sigma = \sqrt{npq} = \sqrt{128(0.4)(0.6)} = \sqrt{30.72} = 5.54$.

$$P(63.5 \leq x) = P\left(\frac{63.5-51.2}{5.54} \leq z\right) = P(2.22 \leq z) = P(z \geq 2.22) = 1 - P(z < 2.22) = 1 - 0.9868 = 0.0132$$

(b) $P(r < 45) = P(r \leq 44) = P(x \leq 44.5)$ 44 is a right endpoint.

$$= P\left(z \leq \frac{44.5-51.2}{5.54}\right) = P(z \leq -1.21) = 0.1131$$

(c) $P(40 \leq r \leq 64) = P(39.5 \leq x \leq 64.5)$

$$\begin{aligned} &= P\left(\frac{39.5-51.2}{5.54} \leq z \leq \frac{64.5-51.2}{5.54}\right) \\ &= P(-2.11 \leq z \leq 2.40) \\ &= P(z \leq 2.40) - P(z < -2.11) \\ &= 0.9918 - 0.0174 \\ &= 0.9744 \end{aligned}$$

(d) More than 80 *not* padded = 81 or more *not* padded, i.e., 128 – 81 = 47 or fewer *are padded*.

Method 1:

$$P(r \leq 47) = P(x \leq 47.5) = P\left(z \leq \frac{47.5-51.2}{5.54}\right) = P(z \leq -0.67) = 0.2514$$

Method 2:

Success is now *redefined* to mean an insurance claim that has not been padded, and p is not $1 - 0.40 = 0.60$.

$P(r \geq 81) = P(81 \leq r) = P(80.5 \leq x)$. 81 is a left endpoint. The normal approximation is still valid, since what was np in (a) is now nq and vice versa. The standard deviation is still the same, but now $\mu = np = 128(0.60) = 76.8$. So,

$$P(80.5 \leq x) = P\left(\frac{80.5-76.8}{5.54} \leq z\right) = P(0.67 \leq z) = P(z \geq 0.67) = 1 - P(z < 0.67) = 1 - 0.7486 = 0.2514.$$

3. We are given $n = 125$ and $p = 17\% = 0.17$. Let a success be defined as the police receiving enough information to locate and arrest a fugitive within 1 week.

(a) $P(r \geq 15) = P(15 \leq r) = P(14.5 \leq x)$. 15 is a left endpoint. $np = 125(0.17) = 21.25$ and $nq = 125(1-0.17) = 125(0.83) = 103.75$, which are both greater than 5, so we can use the normal approximation with $\mu = np = 21.25$ and $\sigma = \sqrt{npq} = \sqrt{125(0.17)(0.83)} = \sqrt{17.6375} = 4.20$. So

$$P(14.5 \leq x) = P\left(\frac{14.5-21.25}{4.2} \leq z\right) = P(-1.61 \leq z) = P(z \geq -1.61) = 1 - P(z < -1.61) = 1 - 0.0537 = 0.9463.$$

(b) $P(r \geq 28) = P(28 \leq r)$ 28 is a left endpoint.

$$\begin{aligned} &= P(27.5 \leq x) \\ &= P\left(\frac{27.5-21.25}{4.2} \leq z\right) \\ &= P(1.49 \leq z) \\ &= P(z \geq 1.49) \\ &= 1 - P(z < 1.49) \\ &= 1 - 0.9319 \\ &= 0.0681 \end{aligned}$$

(c) Remember, r "between" a and b is $a \leq r \leq b$.

$P(15 \leq r \leq 28) = P(14.5 \leq x \leq 28.5)$ 15 is a left endpoint and 28 is a right endpoint.

$$\begin{aligned} &= P\left(\frac{14.5-21.25}{4.2} \leq z \leq \frac{28.5-21.25}{4.2}\right) \\ &= P(-1.61 \leq z \leq 1.73) \\ &= P(z \leq 1.73) - P(z < -1.61) \\ &= 0.9582 - 0.0537 \\ &= 0.9045. \end{aligned}$$

(d) $n = 125, p = 0.17, q = 1 - p = 1 - 0.17 = 0.83$.

np and nq are both greater than 5, so the normal approximation is appropriate.

4. $n = 316, p = 11\% = 0.11$; a success occurs when the book sold is a romance novel.

(a) $P(r < 40) = P(r \leq 39) = P(x \leq 39.5)$ 39 is a right endpoint $np = 316(0.11) = 34.76$, $nq = n(1-p) = 316(1-0.11) = 316(0.89) = 281.24$, both of which are greater than 5, so we can apply the normal approximation with $\mu = np = 34.76$ and $\sigma = \sqrt{npq} = \sqrt{316(0.11)(0.89)} = \sqrt{30.9364} = 5.56$.

$$P(x \leq 39.5) = P\left(z \leq \frac{39.5-34.76}{5.56}\right) = P(z \leq 0.85) = 0.8023$$

(b) $P(r \geq 25) = P(25 \leq r)$ 25 is a left endpoint.

$$\begin{aligned} &= P(24.5 \leq x) \\ &= P\left(\frac{24.5-34.76}{5.56} \leq z\right) \\ &= P(-1.85 \leq z) \\ &= P(z \geq -1.85) \\ &= 1 - P(z < -1.85) \\ &= 1 - 0.0322 \\ &= 0.9678 \end{aligned}$$

(c) $P(25 \le r \le 40) = P(24.5 \le x \le 40.5)$

$$= P\left(\frac{24.5 - 34.76}{5.56} \le z \le \frac{40.5 - 34.76}{5.56}\right)$$
$$= P(-1.85 \le z \le 1.03)$$
$$= P(z \le 1.03) - P(z < -1.85)$$
$$= 0.8485 - 0.0322$$
$$= 0.8163$$

(d) $n = 316,\ p = 0.11,\ q = 1 - p = 0.89$

np and nq are both greater than 5, so the normal approximation to the binomial is appropriate. (See (a) above.)

5. We are given $n = 753$ and $p = 3.5\% = 0.035$; $q = 1 - p = 1 - 0.035 = 0.965$.
Let a success be a person living past age 90.

(a) $P(r \ge 15) = P(15 \le r) = P(14.5 \le x)$ 15 is a left endpoint.

Here, $np = 753(0.035) = 26.355$, and $nq = 753(0.965) = 726.645$, both of which are greater than 5; the normal approximation is appropriate, using $\mu = np = 26.355$ and

$\sigma = \sqrt{npq} = \sqrt{753(0.035)(0.965)} = \sqrt{25.4326} = 5.0431.$

$$P(14.5 \le x) = P\left(\frac{14.5 - 26.355}{5.0431} \le z\right)$$
$$= P(-2.35 \le z)$$
$$= P(z \ge -2.35)$$
$$= 1 - P(z < -2.35)$$
$$= 1 - 0.0094$$
$$= 0.9906$$

(b) $P(r \ge 30) = P(30 \le r)$

$$= P(29.5 \le x)$$
$$= P\left(\frac{29.5 - 26.355}{5.0431} \le z\right)$$
$$= P(0.62 \le z)$$
$$= P(z \ge 0.62)$$
$$= 1 - P(z < 0.62)$$
$$= 1 - 0.7324$$
$$= 0.2676$$

(c) $P(25 \le r \le 35) = P(24.5 \le x \le 35.5)$

$$= P\left(\frac{24.5 - 26.355}{5.0431} \le z \le \frac{35.5 - 26.355}{5.0431}\right)$$
$$= P(-0.37 \le z \le 1.81)$$
$$= P(z \le 1.81) - P(z < -0.37)$$
$$= 0.9649 - 0.3557$$
$$= 0.6092$$

(d) $P(r > 40) = P(r \geq 41)$
$= P(41 \leq r)$
$= P(40.5 \leq x)$
$= P\left(\dfrac{40.5 - 26.355}{5.0431} \leq z\right)$
$= P(2.80 \leq z)$
$= P(z \geq 2.80)$
$= 1 - P(z < 2.80)$
$= 1 - 0.9974$
$= 0.0026$

6. $n = 24$, $p = 44\% = 0.44$, $q = 1 - p = 1 - 0.44 = 0.56$
A success occurs when a billfish striking the line is caught.

(a) $P(r \leq 12) = P(x \leq 12.5)$
$np = 24(0.44) = 10.56$ and $nq = 24(0.56) = 13.44$, both of which are greater than 5, so the normal approximation is appropriate. Here, $\mu = np = 10.56$ and

$\sigma = \sqrt{npq} = \sqrt{24(0.44)(0.56)} = \sqrt{5.9136} = 2.4318.$

$P(x \leq 12.5) = P\left(z \leq \dfrac{12.5 - 10.56}{2.4318}\right) = P(z \leq 0.80) = 0.7881$

(b) $P(r \geq 5) = P(5 \leq r)$
$= P(4.5 \leq x)$
$= P\left(\dfrac{4.5 - 10.56}{2.4318} \leq z\right)$
$= P(-2.49 \leq z)$
$= P(z \geq -2.49)$
$= 1 - P(z < -2.49)$
$= 1 - 0.0064$
$= 0.9936$

(c) $P(5 \leq r \leq 12) = P(4.5 \leq x \leq 12.5)$
$= P(-2.49 \leq z \leq 0.80)$
$= P(z \leq 0.80) - P(z < -2.49)$
$= 0.7881 - 0.0064$
$= 0.7817$

(d) $n = 24$, $p = 0.44$, $q = 0.56$
Both np and $nq > 5$, so the normal approximation to the binomial is appropriate.

7. $n = 66, p = 80\% = 0.80, q = 1 - p = 1 - 0.80 = 0.20$
A success is when a new product fails within 2 years.

(a) $P(r \geq 47) = P(47 \leq r) = P(46.5 \leq x)$
$np = 66(0.80) = 52.8$, and $nq = 66(0.20) = 13.3$. Both exceed 5, so the normal approximation with $\mu = np = 52.8$ and $\sigma = \sqrt{npq} = \sqrt{66(0.8)(0.2)} = \sqrt{10.56} = 3.2496$ is appropriate.

$$\begin{aligned} P(46.5 \leq x) &= P\left(\frac{46.5-52.8}{3.2496} \leq z\right) \\ &= P(-1.94 \leq z) \\ &= P(z \geq -1.94) \\ &= 1 - P(z < -1.94) \\ &= 1 - 0.0262 \\ &= 0.9738 \end{aligned}$$

(b) $P(r \leq 58) = P(x \leq 58.5) = P\left(z \leq \frac{58.5-52.8}{3.2496}\right) = P(z \leq 1.75) = 0.9599$

For (c) and (d), note we are interested now in products succeeding, so a success is redefined to be a new product staying on the market for 2 years. Here, $n = 66$, p is now 0.20 with q is now 0.80 (p and q above have been switched. Now $np = 13.2$ and $nq = 52.8$, $\mu = 13.2$, and σ stays equal to 3.2496.

(c)
$$\begin{aligned} P(r \geq 15) &= P(15 \leq r) \\ &= P(14.5 \leq x) \\ &= P\left(\frac{14.5-13.2}{3.2496} \leq z\right) \\ &= P(0.40 \leq z) \\ &= P(z \geq 0.40) \\ &= 1 - P(z < 0.40) \\ &= 1 - 0.6554 \\ &= 0.3446 \end{aligned}$$

(d) $P(r < 10) = P(r \leq 9) = P(x \leq 9.5) = P\left(z \leq \frac{9.5-13.2}{3.2496}\right) = P(z \leq -1.14) = 0.1271$

8. $n = 63, p = 64\% = 0.64, q = 1 - p = 1 - 0.64 = 0.36$
A success is when the murder victim knows the murderer.

(a) $P(r \geq 35) = P(35 \leq r) = P(34.5 \leq x)$
$np = 63(0.64) = 40.32$ and $nq = 63(0.36) = 22.68$
Since both np and nq are greater than 5, the normal approximation is appropriate. Use $\mu = np = 40.32$ and $\sigma = \sqrt{npq} = \sqrt{63(0.64)(0.36)} = \sqrt{14.5152} = 3.8099$. So

$$P(34.5 \leq x) = P\left(\frac{34.5-40.32}{3.8099} \leq z\right) = P(-1.53 \leq z) = 1 - P(z < -1.53) = 1 - 0.0630 = 0.9370$$

(b) $P(r \leq 48) = P(x \leq 48.5) = P\left(z \leq \frac{48.5-40.32}{3.8099}\right) = P(z \leq 2.15) = 0.9842$

(c) If fewer than 30 victims, i.e., 29 or fewer did not know their murderer, then 63 – 29 = 34 or more victims did know their murderer.

$$\begin{aligned}P(r \ge 34) &= P(34 \le r)\\ &= P(33.5 \le x)\\ &= P\left(\frac{33.5-40.32}{3.8099} \le z\right)\\ &= P(-1.79 \le z)\\ &= 1-P(z<-1.79)\\ &= 1-0.0367\\ &= 0.9633\end{aligned}$$

(d) If more than 20, i.e., 21 or more, victims did not know their murderer, then 63 – 21 = 42 or fewer victims did know their murderer.

$$P(r \le 42) = P(x \le 42.5) = P\left(z \le \frac{42.5-40.32}{3.8099}\right) = P(z \le 0.57) = 0.7157$$

9. $n = 430$, $p = 70\% = 0.70$, $q = 1 - p = 1 - 0.70 = 0.30$
A success is finding the address or lost acquaintances.

(a) $P(r > 280) = P(r \ge 281) = P(281 \le r) = P(280.5 \le x)$
$np = 430(0.7) = 301$ and $nq = 430(0.3) = 129$
Since both np and nq are greater than 5, the normal approximation with $\mu = np = 301$ and $\sigma = \sqrt{npq} = \sqrt{430(0.7)(0.3)} = \sqrt{90.3} = 9.5026$ is appropriate.

$$P(280.5 \le x) = P\left(\frac{280.5-301}{9.5026} \le z\right) = P(-2.16 \le z) = 1-P(z<-2.16) = 1-0.0154 = 0.9846$$

(b)
$$\begin{aligned}P(r \ge 320) &= P(320 \le r)\\ &= P(319.5 \le x)\\ &= P\left(\frac{319.5-301}{9.5026} \le z\right)\\ &= P(1.95 \le z)\\ &= 1-P(z>1.95)\\ &= 1-0.9744\\ &= 0.0256\end{aligned}$$

(c)
$$\begin{aligned}P(280 \le r \le 320) &= P(279.5 \le x \le 320.5)\\ &= P\left(\frac{279.5-301}{9.5026} \le z \le \frac{320.5-301}{9.5026}\right)\\ &= P(-2.26 \le z \le 2.05)\\ &= P(z \le 2.05) - P(z < -2.26)\\ &= 0.9798 - 0.0119\\ &= 0.9679\end{aligned}$$

(d) $n = 430$, $p = 0.7$, $q = 0.3$
Both np and nq are greater than 5 so the normal approximation is appropriate, See (a).

10. $n = 8641, p = 61\% = 0.61, q = 1 - p = 0.39$
A success is when a pottery shard is Santa Fe black on white.

(a) $P(r < 5200) = P(r \le 5199) = P(x \le 5199.5)$
$np = 8641(0.61) = 5271.01$ and $nq = 3369.99$
Since both np and nq are greater than 5, we can use the normal approximation with
$\mu = np = 5271.01$ and $\sigma = \sqrt{npq} = \sqrt{8641(0.61)(0.39)} = \sqrt{2055.6939} = 45.3398$

$$P(x \le 5199.5) = P\left(z \le \frac{5199.5 - 5271.01}{45.3398}\right) = P(z \le -1.58) = 0.0571$$

(b)
$$\begin{aligned} P(r > 5400) &= P(r \ge 5401) \\ &= P(5401 \le r) \\ &= P(5400.5 \le x) \\ &= P\left(\frac{5400.5 - 5271.01}{45.3398} \le z\right) \\ &= P(2.86 \le z) \\ &= 1 - P(z < 2.86) \\ &= 1 - 0.9979 \\ &= 0.0021 \end{aligned}$$

(c)
$$\begin{aligned} P(5200 \le r \le 5400) &= P(5199.5 \le x \le 5400.5) \\ &= P\left(-1.58 \le z \le \frac{5400.5 - 5271.01}{45.3398}\right) \\ &= P(-1.58 \le z \le 2.86) \\ &= P(z \le 2.86) - P(z < -1.58) \\ &= 0.9979 - 0.0571 \\ &= 0.9408 \end{aligned}$$

(d) $n = 8641, p = 0.61, q = 0.39, np = 5271.01, nq = 3369.99.$

11. $n = 850, p = 57\% = 0.57, q = 0.43$
Success = pass Ohio bar exam

(a) $P(r \ge 540) = P(540 \le r) = P(539.5 \le x)$

$np = 484.5, nq = 365.5, \mu = np = 484.5, \sigma = \sqrt{npq} = \sqrt{208.335} = 14.4338$

Since both np and nq are greater than 5, use normal approximation with μ and σ as above.

$$P(539.5 \le x) = P\left(\frac{539.5 - 484.5}{14.4338} \le z\right) = P(3.81 \le z) \approx 0$$

(b) $P(r \le 500) = P(x \le 500.5) = P\left(z \le \frac{500.5 - 484.5}{14.4338}\right) = P(z \le 1.11) = 0.8665$

(c)
$$\begin{aligned} P(485 \le r \le 525) &= P(484.5 \le x \le 525.5) \\ &= P\left(0 \le z \le \frac{525.5 - 484.5}{14.4338}\right) \\ &= P(0 \le z \le 2.84) \\ &= P(z \le 2.84) - P(z < 0) \\ &= 0.9977 - 0.5 \\ &= 0.4977 \end{aligned}$$

12. $n = 5000, p = 3.2\% = 0.032, q = 0.968$
Success = coupon redeemed

(a) $P(100 < r) = P(101 \le r) = P(100.5 \le x)$

$np = 160, nq = 4840, \sigma = \sqrt{npq} = \sqrt{154.88} = 12.4451$

Since both np and nq are greater than 5, use normal approximation with $\mu = np$ and σ as shown.

$$P(100.5 \le x) = P\left(\frac{100.5-160}{12.4451} \le z\right) = P(-4.78 \le z) \approx 1$$

(b)
$$\begin{aligned} P(r < 200) &= P(r \le 199) \\ &= P(x \le 199.5) \\ &= P\left(z \le \frac{199.5-160}{12.4451}\right) \\ &= P(z \le 3.17) \\ &= 0.9992 \end{aligned}$$

(c)
$$\begin{aligned} P(100 \le r \le 200) &= P(99.5 \le x \le 200.5) \\ &= P\left(\frac{99.5-160}{12.4451} \le z \le \frac{200.5-160}{12.4451}\right) \\ &= P(-4.86 \le z \le 3.25) \\ &\approx P(z \le 3.25) \\ &= 0.9994 \end{aligned}$$

Chapter 7 Review

1. (a) $P(0 \le z \le 1.75) = P(z \le 1.75) - P(z < 0) = 0.9599 - 0.5 = 0.4599$

(b) $P(-1.29 \le z \le 0) = P(z \le 0) - P(z < -1.29) = 0.5 - 0.0985 = 0.4015$

(c) $P(1.03 \le z \le 1.21) = P(z \le 1.21) - P(z < 1.03) = 0.8869 - 0.8485 = 0.0384$

(d) $P(z \ge 2.31) = 1 - P(z < 2.31) = 1 - 0.9896 = 0.0104$

(e) $P(z \le -1.96) = 0.0250$

(f) $P(z \le 1) = 0.8413$

2. (a) $P(0 \le z \le 0.75) = P(z \le 0.75) - P(z < 0) = 0.7734 - 0.5 = 0.2734$

(b) $P(-1.50 \le z \le 0) = P(z \le 0) - P(z < -1.50) = 0.5 - 0.0668 = 0.4332$

(c) $P(-2.67 \le z \le -1.74) = P(z \le -1.74) - P(z < -2.67) = 0.0409 - 0.0038 = 0.0371$

(d) $P(z \ge 1.56) = 1 - P(z < 1.56) = 1 - 0.9406 = 0.0594$

(e) $P(z \le -0.97) = 0.1660$

(f) $P(z \le 2.01) = 0.9778$

3. x is normal with $\mu = 47$ and $\sigma = 6.2$

(a) $P(x \le 60) = P\left(z \le \frac{60-47}{6.2}\right) = P(z \le 2.10) = 0.9821$

(b) $P(x \geq 50) = P\left(z \geq \frac{50-47}{6.2}\right) = P(z \geq 0.48) = 1 - P(z < 0.48) = 1 - 0.6844 = 0.3156$

(c) $P(50 \leq x \leq 60) = P(0.48 \leq z \leq 2.10) = P(z \leq 2.10) - P(z < 0.48) = 0.9821 - 0.6844 = 0.2977$

4. x is normal with $\mu = 110$, $\sigma = 12$

(a) $P(x \leq 120) = P\left(z \leq \frac{120-110}{12}\right) = P(z \leq 0.83) = 0.7967$

(b) $P(x \geq 80) = P\left(z \geq \frac{80-110}{12}\right) = P(z \geq -2.5) = 1 - P(z < -2.5) = 1 - 0.0062 = 0.9938$

(c) $P(108 \leq x \leq 117) = P\left(\frac{108-110}{12} \leq z \leq \frac{117-110}{12}\right) = P(-0.17 \leq z \leq 0.58)$
$= P(z \leq 0.58) - P(z < -0.17) = 0.7190 - 0.4325 = 0.2865$

5. Find z_0 such that $P(z \geq z_0) = 5\% = 0.05$. Same as find z_0 such that $P(z < z_0) = 0.95$
$P(z < 1.645) = 0.95$, so $z_0 = 1.645$

6. Find z_0 such that $P(z \leq z_0) = 1\% = 0.01$.
$P(z \leq -2.33) = 0.0099$, so $z_0 = -2.33$

7. Find z_0 such that $P(-z_0 \leq z \leq +z_0) = 0.95$
Same as 5% of area outside $[-z_0, +z_0]$; split in half:
$P(z \leq -z_0) = 0.05/2 = 0.025$
$P(z \leq -1.96) = 0.0250$ so $-z_0 = -1.96$ and $+z_0 = 1.96$

8. Find z_0 so $P(-z_0 \leq z \leq +z_0) = 0.99$
Same as 1% outside $[-z_0, +z_0]$; divide in half.
$P(z \leq -z_0) = 0.01/2 = 0.005$.
$P(z \leq -2.575) = 0.0050$, so $\pm z_0 = \pm 2.575$, or ± 2.58

9. $\mu = 79$, $\sigma = 9$

(a) $z = \frac{x-\mu}{\sigma} = \frac{87-79}{9} = 0.89$

(b) $z = \frac{79-79}{9} = 0$

(c) $P(x > 85) = P\left(z > \frac{85-79}{9}\right) = P(z > 0.67) = 1 - P(z \leq 0.67) = 1 - 0.7486 = 0.2514$

10. $\mu = 270$, $\sigma = 35$

(a) $z = \frac{x-\mu}{\sigma}$ so $x = \mu + z\sigma$
here, $x = 270 + 1.9(35) = 336.5$

(b) $x = \mu + z\sigma$; here, $x = 270 + (-0.25)(35) = 261.25$

(c) $P(200 \leq x \leq 340) = P\left(\frac{200-270}{35} \leq z \leq \frac{340-270}{35}\right) = P(-2 \leq z \leq 2)$
$= P(z \leq 2) - P(z < -2) = 0.9772 - 0.0228 = 0.9544$

11. Binomial with $n = 400$, $p = 0.70$, and $q = 0.30$.
Success = can recycled

(a) $P(r \geq 300) = P(300 \leq r) = P(299.5 \leq x)$

$np = 280 > 5,\ nq = 120 > 5,\ \sqrt{npq} = \sqrt{84} = 9.1652$

Use normal approximation with $\mu = np$ and $\sigma = \sqrt{npq}$.

$$P(299.5 \leq x) = P\left(\frac{299.5 - 280}{9.1652} \leq z\right) = P(2.13 \leq z) = 1 - P(z < 2.13) = 1 - 0.9834 = 0.0166$$

(b)
$$\begin{aligned} P(260 \leq r \leq 300) &= P(259.5 \leq x \leq 300.5) \\ &= P\left(\frac{259.5 - 280}{9.1652} \leq z \leq \frac{300.5 - 280}{9.1652}\right) \\ &= P(-2.24 \leq z \leq 2.24) \\ &= P(z \leq 2.24) - P(z < -2.24) \\ &= 0.9875 - 0.0125 \\ &= 0.9750 \end{aligned}$$

12. Lifetime x is normally distributed with $\mu = 5000$ and $\sigma = 450$ hours.

(a) $P(x \leq 5000) = P(z \leq 0) = 0.5000$

(b) Find x_0 such that $P(x \leq x_0) = 0.05$. First, find z_0 so that $P(z \leq z_0) = 0.05$.
$P(z \leq -1.645) = 0.05$, so $z_0 = -1.645$
$x_0 = \mu + z_0\sigma = 5000 + (-1.645)(450) = 4259.75$
Guarantee the CD player for 4260 hours.

13. Delivery time x is normal with $\mu = 14$ and $\sigma = 2$ hours.

(a) $P(x \leq 18) = P\left(z \leq \frac{18 - 14}{2}\right) = P(z \leq 2) = 0.9772$

(b) Find x_0 such that $P(x \leq x_0) = 0.95$.
Find z_0 so that $P(z \leq z_0) = 0.95$.
$P(z \leq 1.645) = 0.95$, so $z_0 = 1.645$.
$x_0 = \mu + z_0\sigma = 14 + 1.645(2) = 17.29 \approx 17.3$ hours

14. (a)

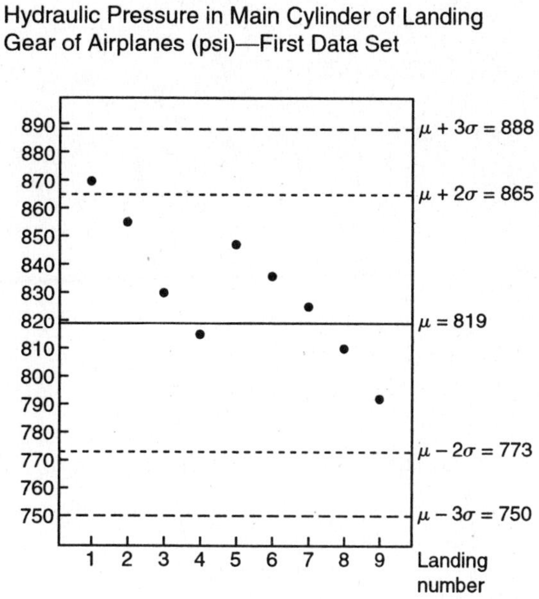

The pressure is "in control;" none of the 3 warning signals is present.

(b)

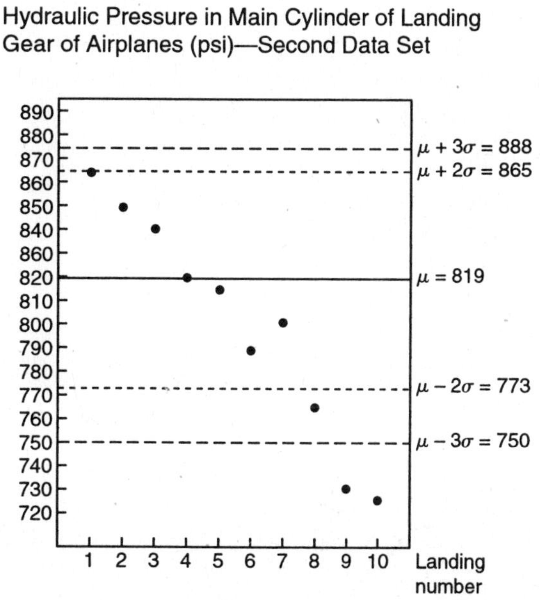

The last 2 points are below $\mu - 3\sigma$. The last 3 (consecutive) points are all below $\mu - 2\sigma$. Since warning signals I and III are present, the pressure is "out-of-control."

15. Scanner price errors in the store's favor are mound-shaped with $\mu = \$2.66$ and $\sigma = \$0.85$.

(a) 68% of the errors should be in the range $\mu \pm 1\sigma$, approximately, or $2.66 \pm 1(0.85)$ which is \$1.81 to \$3.51.

(b) Approximately 95% of the errors should be in the range $\mu \pm 2\sigma$, or $2.66 \pm 2(0.85)$, which is \$0.96 to \$4.36.

(c) Almost all (99.7%) of the errors should lie in the range $\mu \pm 3\sigma$, or $2.66 \pm 3(0.85)$, which is \$0.11 to \$5.21.

16. Time spent on a customer's complaint, x, is normally distributed with $\mu = 9.3$ and $\sigma = 2.5$ minutes.

(a) $P(x<10) = P\left(z<\dfrac{10-9.3}{2.5}\right) = P(z<0.28) = 0.6103$

(b) $P(x>5) = P\left(z>\dfrac{5-9.3}{2.5}\right) = P(z>-1.72) = 1-P(z\le -1.72) = 1-0.0427 = 0.9573$

(c) $P(8\le x\le 15) = P\left(\dfrac{8-9.3}{2.5}\le z\le \dfrac{15-9.3}{2.5}\right)$
$= P(-0.52\le z\le 2.28)$
$= P(z\le 2.28) - P(z<-0.52)$
$= 0.9887 - 0.3015$
$= 0.6872$

17. Response time, x, is normally distributed with $\mu = 42$ and $\sigma = 8$ minutes.

(a) $P(30\le x\le 45) = P\left(\dfrac{30-42}{8}\le z\le \dfrac{45-42}{8}\right)$
$= P(-1.5\le z\le 0.375)$
$= P(z\le 0.38) - P(z<-1.5)$
$= 0.6480 - 0.0668$
$= 0.5812$

(b) $P(x<30) = P(z<-1.5) = 0.0668$

(c) $P(x>60) = P\left(z>\dfrac{60-42}{8}\right) = P(z>2.25) = 1-P(z\le 2.25) = 1-0.9878 = 0.0122$

18. Success = unlisted phone number
$n = 150$, $p = 68\% = 0.68$, $q = 1 - p = 0.32$
$np = 150(0.68) = 102$, $nq = 48$, $npq = 150(0.68)(0.32) = 32.64$

(a) $P(r\ge 100) = P(100\le r) = P(99.5\le x)$

Since np and nq are both greater than 5, we can use the normal approximation with $\mu = np = 102$ and $\sigma = \sqrt{npq} = \sqrt{32.64} = 5.7131$.

$P(99.5\le x) = P\left(\dfrac{99.5-102}{5.7131}\le z\right) = P(-0.44\le z) = 1-P(z<-0.44) = 1-0.3300 = 0.6700$

(b) $P(r<100) = P(r\le 99) = P(x\le 99.5) = P\left(z\le \dfrac{99.5-102}{5.7131}\right) = P(z\le -0.44) = 0.3300$

(c) Success is redefined to be a listed phone number, so $n = 150$, $p = 0.32$, $q = 0.68$, $np = 48$, $nq = 102$, and $\sqrt{npq} = \sqrt{32.64} = 5.7131 = \sigma$; μ is now 48; normal approximation is still appropriate.

$P(50\le r\le 65) = P(49.5\le x\le 65.5)$
$= P\left(\dfrac{49.5-48}{5.7131}\le z\le \dfrac{65.5-48}{5.7131}\right)$
$= P(0.26\le z\le 3.06)$
$= P(z\le 3.06) - P(z<0.26)$
$= 0.9989 - 0.6026$
$= 0.3963$

19. Success = having blood type AB
$n = 250$, $p = 3\% = 0.03$, $q = 1 - p = 0.97$, $np = 7.5$, $nq = 242.5$, $npq = 7.275$

(a) $P(5 \le r) = P(4.5 \le x)$

$np > 7.5$ and $\sigma = \sqrt{npq} = \sqrt{7.275} = 2.6972$

$$P(4.5 \le x) = P\left(\frac{4.5-7.5}{2.6972} \le z\right) = P(-1.11 \le z) = 1 - P(z < -1.11) = 1 - 0.1335 = 0.8665$$

(b)
$$\begin{aligned} P(5 \le r \le 10) &= P(4.5 \le x \le 10.5) \\ &= P\left(-1.11 \le z \le \frac{10.5-7.5}{2.6972}\right) \\ &= P(-1.11 \le z \le 1.11) \\ &= 1 - 2P(z < -1.11) \\ &= 1 - 2(0.1335) \\ &= 0.7330 \end{aligned}$$

Chapter 8 Introduction to Sampling Distributions

Section 8.1

1. Answers vary. Students should identify the individuals (subjects) and variable involved. Answers may include: A population is a set of measurements or counts either existing or conceptual. For example, the population of all ages of all people in Colorado; the population of weights of all students in your school; the population count of all antelope in Wyoming.

2. See Section 1.2. Answer may include:

   A simple random sample of n measurements from a population is a subset of the population selected in a manner such that

   (a) every sample of size n from the population has an equal chance of being selected and

   (b) every member of the population has an equal chance of being included in the sample.

3. A population parameter is a numerical descriptive measure of a population, such as μ, the population mean; σ, the population standard deviation; σ^2, the population variance; p, the population proportion of success in a binomial distribution.

4. A sample statistic is a numerical descriptive measure of a sample such as $\bar{x}$, the sample mean; s, the sample standard deviation; s^2, the sample variance; $\hat{p}$, the sample proportion; r, the sample correlation coefficient for those who have already studied linear regression from Chapter 10.

5. A statistical inference is a conclusion about the value of a population parameter based on information about the corresponding sample statistic and probability. We will do both estimation and testing.

6. A sampling distribution is a probability distribution for a sample statistic.

7. They help us visualize the sampling distribution by using tables and graphs that approximately represent the sampling distribution.

8. Relative frequencies can be thought of as a measure or estimate of the likelihood of a certain statistic falling within the class bounds.

9. We studied the sampling distribution of mean trout lengths based on samples of size 5. Other such sampling distributions abound. Notice that the sample size remains the same for each sample in a sampling distribution.

Section 8.2

Note: Answers may vary slightly depending on the number of digits carried in the standard deviation.

1. (a) $\mu_{\bar{x}} = \mu = 15$

$$\sigma_{\bar{x}} = \frac{\sigma}{\sqrt{n}} = \frac{14}{\sqrt{49}} = 2.0$$

Because $n = 49 \geq 30$, by the central limit theorem, we can assume that the distribution of $\bar{x}$ is approximately normal.

$$z = \frac{\bar{x} - \mu}{\sigma_{\bar{x}}} = \frac{\bar{x} - 15}{2.0}$$

$\bar{x} = 15$ converts to $z = \frac{15 - 15}{2.0} = 0$

$\bar{x} = 17$ converts to $z = \frac{17 - 15}{2.0} = 1$

$$\begin{aligned} P(15 \leq \bar{x} \leq 17) &= P(0 \leq z \leq 1) \\ &= P(z \leq 1) - P(z \leq 0) \\ &= 0.8413 - 0.5000 \\ &= 0.3413 \end{aligned}$$

(b) $\mu_{\bar{x}} = \mu = 15$

$$\sigma_{\bar{x}} = \frac{\sigma}{\sqrt{n}} = \frac{14}{\sqrt{64}} = 1.75$$

Because $n = 64 \geq 30$, by the central limit theorem, we can assume that the distribution of $\bar{x}$ is approximately normal.

$$z = \frac{\bar{x} - \mu}{\sigma_{\bar{x}}} = \frac{\bar{x} - 15}{1.75}$$

$\bar{x} = 15$ converts to $z = \frac{15 - 15}{1.75} = 0$

$\bar{x} = 17$ converts to $z = \frac{17 - 15}{1.75} = 1.14$

$$\begin{aligned} P(15 \leq \bar{x} \leq 17) &= P(0 \leq z \leq 1.14) \\ &= P(z \leq 1.14) - P(z \leq 0) \\ &= 0.8729 - 0.5000 \\ &= 0.3729 \end{aligned}$$

(c) The standard deviation of part (b) is smaller because of the larger sample size. Therefore, the distribution about $\mu_{\bar{x}}$ is narrower in part (b).

2. (a) $\mu_{\bar{x}} = \mu = 100$

$$\sigma_{\bar{x}} = \frac{\sigma}{\sqrt{n}} = \frac{48}{\sqrt{81}} = 5.33$$

Because $n = 81 \geq 30$, by the central limit theorem, we can assume that the distribution of $\bar{x}$ is approximately normal.

$$z = \frac{\bar{x} - \mu}{\sigma_{\bar{x}}} = \frac{\bar{x} - 100}{5.33}$$

$\bar{x} = 92$ converts to $z = \frac{92 - 100}{5.33} = -1.50$

$\bar{x} = 100$ converts to $z = \frac{100 - 100}{5.33} = 0$

$$\begin{aligned} P(92 \le \bar{x} \le 100) &= P(-1.50 \le z \le 0) \\ &= P(z \le 0) - P(z \le -1.50) \\ &= 0.5000 - 0.0668 \\ &= 0.4332 \end{aligned}$$

(b) $\mu_{\bar{x}} = \mu = 100$

$$\sigma_{\bar{x}} = \frac{\sigma}{\sqrt{n}} = \frac{48}{\sqrt{121}} = 4.36$$

Because $n = 121 \ge 30$, by the central limit theorem, we can assume that the distribution of $\bar{x}$ is approximately normal.

$$z = \frac{\bar{x} - \mu}{\sigma_{\bar{x}}} = \frac{\bar{x} - 100}{4.36}$$

$\bar{x} = 92$ converts to $z = \frac{92 - 100}{4.36} = -1.83$

$\bar{x} = 100$ converts to $z = \frac{100 - 100}{4.36} = 0$

$$\begin{aligned} P(92 \le \bar{x} \le 100) &= P(-1.83 \le z \le 0) \\ &= P(z \le 0) - P(z \le -1.83) \\ &= 0.5000 - 0.0336 \\ &= 0.4664 \end{aligned}$$

(c) The probability of part (b) is greater than that of part (a). The standard deviation of part (b) is smaller because of the larger sample size. Therefore, the distribution about $\mu_{\bar{x}}$ is narrower in part (b).

3. (a) No, we cannot say anything about the distribution of sample means because the sample size is only 9 and so it is too small to apply the central limit theorem.

(b) Yes, now we can say that the $\bar{x}$ distribution will also be normal with

$$\mu_{\bar{x}} = \mu = 25 \text{ and } \sigma_{\bar{x}} = \frac{\sigma}{\sqrt{n}} = \frac{3.5}{\sqrt{9}} = 1.17.$$

$$z = \frac{\bar{x} - \mu}{\sigma_{\bar{x}}} = \frac{\bar{x} - 25}{1.17}$$

$$\begin{aligned} P(23 \le \bar{x} \le 26) &= P\left(\frac{23 - 25}{1.17} \le z \le \frac{26 - 25}{1.17}\right) \\ &= P(-1.71 \le z \le 0.86) \\ &= P(z \le 0.86) - P(z \le -1.71) \\ &= 0.8051 - 0.0436 \\ &= 0.7615 \end{aligned}$$

4. **(a)** No, we cannot say anything about the distribution of sample means because the sample size is only 16 and so it is too small to apply the central limit theorem.

(b) Yes, now we can say that the $\overline{x}$ distribution will also be normal with

$$\mu_{\overline{x}} = \mu = 72 \text{ and } \sigma_{\overline{x}} = \frac{\sigma}{\sqrt{n}} = \frac{8}{\sqrt{16}} = 2.$$

$$z = \frac{\overline{x} - \mu}{\sigma_{\overline{x}}} = \frac{\overline{x} - 72}{2}$$

$$\begin{aligned} P(68 \le \overline{x} \le 73) &= P\left(\frac{68-72}{2} \le z \le \frac{73-72}{2}\right) \\ &= (-2 \le z \le 0.5) \\ &= P(z \le 0.5) - P(z \le -2) \\ &= 0.6915 - 0.0228 \\ &= 0.6687 \end{aligned}$$

5. **(a)** $\mu = 75,\ \sigma = 0.8$

$$\begin{aligned} P(x < 74.5) &= P\left(z < \frac{74.5-75}{0.8}\right) \\ &= P(z < -0.63) \\ &= 0.2643 \end{aligned}$$

(b) $\mu_{\overline{x}} = 75,\ \sigma_{\overline{x}} = \frac{\sigma}{\sqrt{n}} = \frac{0.8}{\sqrt{20}} = 0.179$

$$\begin{aligned} P(\overline{x} < 74.5) &= P\left(z < \frac{74.5-75}{0.179}\right) \\ &= P(z < -2.79) \\ &= 0.0026 \end{aligned}$$

(c) No. If the weight of only one car were less than 74.5 tons, we cannot conclude that the loader is out of adjustment. If the mean weight for a sample of 20 cars were less than 74.5 tons, we would suspect that the loader is malfunctioning. As we see in part (b), the probability of this happening is very low if the loader is correctly adjusted.

6. **(a)** $\mu = 68,\ \sigma = 3$

$$\begin{aligned} P(67 \le x \le 69) &= P\left(\frac{67-68}{3} \le z \le \frac{69-68}{3}\right) \\ &= P(-0.33 \le z \le 0.33) \\ &= P(z \le 0.33) - P(z \le -0.33) \\ &= 0.6293 - 0.3707 \\ &= 0.2586 \end{aligned}$$

(b) $\mu_{\bar{x}} = 68,\ \sigma_{\bar{x}} = \dfrac{\sigma}{\sqrt{n}} = \dfrac{3}{\sqrt{9}} = 1$

$$\begin{aligned} P(67 \le \bar{x} \le 69) &= P\left(\frac{67-68}{1} \le z \le \frac{69-68}{1}\right) \\ &= P(-1 \le z \le 1) \\ &= P(z \le 1) - P(z \le -1) \\ &= 0.8413 - 0.1587 \\ &= 0.6826 \end{aligned}$$

(c) The probability in part (b) is much higher because the standard deviation is smaller for the $\bar{x}$ distribution.

7. (a) $\mu = 85,\ \sigma = 25$

$$\begin{aligned} P(x < 40) &= P\left(z < \frac{40-85}{25}\right) \\ &= P(z < -1.8) \\ &= 0.0359 \end{aligned}$$

(b) The probability distribution of $\bar{x}$ is approximately normal with $\mu_{\bar{x}} = 85;\ \sigma_{\bar{x}} = \dfrac{\sigma}{\sqrt{n}} = \dfrac{25}{\sqrt{2}} = 17.68.$

$$\begin{aligned} P(\bar{x} < 40) &= P\left(z < \frac{40-85}{17.68}\right) \\ &= P(z < -2.55) \\ &= 0.0054 \end{aligned}$$

(c) $\mu_{\bar{x}} = 85,\ \sigma_{\bar{x}} = \dfrac{\sigma}{\sqrt{n}} = \dfrac{25}{\sqrt{3}} = 14.43$

$$\begin{aligned} P(\bar{x} < 40) &= P\left(z < \frac{40-85}{14.43}\right) \\ &= P(z < -3.12) \\ &= 0.0009 \end{aligned}$$

(d) $\mu_{\bar{x}} = 85,\ \sigma_{\bar{x}} = \dfrac{\sigma}{\sqrt{n}} = \dfrac{25}{\sqrt{5}} = 11.2$

$$\begin{aligned} P(\bar{x} < 40) &= P\left(z < \frac{40-85}{11.2}\right) \\ &= P(z < -4.02) \\ &\approx 0 \end{aligned}$$

(e) Yes; If the average value based on five tests were less than 40, the patient is almost certain to have excess insulin.

8. $\mu = 7500,\ \sigma = 1750$

(a) $$P(x < 3500) = P\left(z < \frac{3500 - 7500}{1750}\right)$$
$$= P(z < -2.29)$$
$$= 0.0110$$

(b) The probability distribution of $\bar{x}$ is approximately normal with

$$\mu_{\bar{x}} = 7500;\ \sigma_{\bar{x}} = \frac{\sigma}{\sqrt{n}} = \frac{1750}{\sqrt{2}} = 1237.44.$$

$$P(\bar{x} < 3500) = P\left(z < \frac{3500 - 7500}{1237.44}\right)$$
$$= P(z < -3.23)$$
$$= 0.0006$$

(c) $$\mu_{\bar{x}} = 7500,\ \sigma_{\bar{x}} = \frac{\sigma}{\sqrt{n}} = \frac{1750}{\sqrt{3}} = 1010.36$$

$$P(\bar{x} < 3500) = P\left(z < \frac{3500 - 7500}{1010.36}\right)$$
$$= P(z < -3.96)$$
$$\approx 0$$

(d) The probabilities decreased as n increased. It would be an extremely rare event for a person to have two or three tests below 3500 purely by chance; the person probably has leukopenia.

9. (a) $\mu = 63.0,\ \sigma = 7.1$

$$P(x < 54) = P\left(z < \frac{54 - 63.0}{7.1}\right)$$
$$= P(z < -1.27)$$
$$= 0.1020$$

(b) The expected number undernourished is 2200(0.1020) = 224.4, or about 224.

(c) $$\mu_{\bar{x}} = 63.0,\ \sigma_{\bar{x}} = \frac{\sigma}{\sqrt{n}} = \frac{7.1}{\sqrt{50}} = 1.004$$

$$P(\bar{x} < 60) = P\left(z < \frac{60 - 63.0}{1.004}\right)$$
$$= P(z < -2.99)$$
$$= 0.0014$$

(d) $\mu_{\bar{x}} = 63.0,\ \sigma_{\bar{x}} = 1.004$

$$P(\bar{x} < 64.2) = P\left(z < \frac{64.2 - 63.0}{1.004}\right)$$
$$= P(z < 1.20)$$
$$= 0.8849$$

Since the sample average is above the mean, it is quite unlikely that the doe population is undernourished.

10. **(a)** Blue for sample size 49; red for sample size 100

(b) Blue

(c) Yes, because by the Central Limit Theorem the standard deviation is larger for smaller sample sizes. A larger standard deviation results in the data being more spread out around the mean.

11. **(a)** The random variable x is itself an average based on the number of stocks or bonds in the fund. Since x itself represents a sample mean return based on a large (random) sample of stocks or bonds, x has a distribution that is approximately normal (Central Limit Theorem).

(b) $\mu_{\bar{x}} = 1.6\%,\ \sigma_{\bar{x}} = \dfrac{\sigma}{\sqrt{n}} = \dfrac{0.9\%}{\sqrt{6}} = 0.367\%$

$$\begin{aligned} P(1\% \le \bar{x} \le 2\%) &= P\left(\frac{1\% - 1.6\%}{0.367\%} \le z \le \frac{2\% - 1.6\%}{0.367\%}\right) \\ &= P(-1.63 \le z \le 1.09) \\ &= P(z \le 1.09) - P(z \le -1.63) \\ &= 0.8621 - 0.0516 \\ &= 0.8105 \end{aligned}$$

Note: It does not matter whether you solve the problem using percents or their decimal equivalents as long as you are consistent.

(c) Note: 2 years = 24 months; x is <u>monthly</u> percentage return.

$\mu_{\bar{x}} = 1.6\%,\ \sigma_{\bar{x}} = \dfrac{\sigma}{\sqrt{n}} = \dfrac{0.9\%}{\sqrt{24}} = 0.1837\%$

$$\begin{aligned} P(1\% \le \bar{x} \le 2\%) &= P\left(\frac{1\% - 1.6\%}{0.1837\%} \le z \le \frac{2\% - 1.6\%}{0.1837\%}\right) \\ &= P(-3.27 \le z \le 2.18) \\ &= P(z \le 2.18) - P(z \le -3.27) \\ &= 0.9854 - 0.0005 \\ &= 0.9849 \end{aligned}$$

(d) Yes. The probability increases as the standard deviation decreases. The standard deviation decreases as the sample size increases.

(e) $\mu_{\bar{x}} = 1.6\%,\ \sigma_{\bar{x}} = 0.1837\%$

$$\begin{aligned} P(\bar{x} < 1\%) &= P\left(z < \frac{1\% - 1.6\%}{0.1837\%}\right) \\ &= P(z < -3.27) \\ &= 0.0005 \end{aligned}$$

This is very unlikely if $\mu = 1.6\%$. One would suspect that μ has slipped below 1.6%.

12. **(a)** By the Central Limit Theorem, the sampling distribution of $\bar{x}$ is approximately normal with mean $\mu_{\bar{x}} = \mu = \$20$ and standard error $\sigma_{\bar{x}} = \dfrac{\sigma}{\sqrt{n}} = \dfrac{\$7}{\sqrt{100}} = \$0.70$. It is not necessary to make any assumption about the x distribution because n is large.

(b) $\mu_{\bar{x}} = \$20,\ \sigma_{\bar{x}} = \$0.70$

$$P(\$18 \le \bar{x} \le \$22) = P\left(\frac{\$18 - \$20}{\$0.70} \le z \le \frac{\$22 - \$20}{\$0.70}\right)$$
$$= P(-2.86 \le z \le 2.86)$$
$$= P(z \le 2.86) - P(z \le -2.86)$$
$$= 0.9979 - 0.0021$$
$$= 0.9958$$

(c) $\mu_x = \$20,\ \sigma = \$7$

$$P(\$18 \le x \le \$22) = P\left(\frac{\$18 - \$20}{\$7} \le z \le \frac{\$22 - \$20}{\$7}\right)$$
$$= P(-0.29 \le z \le 0.29)$$
$$= 0.6141 - 0.3859$$
$$= 0.2282$$

(d) We expect the probability in part (b) to be much higher than the probability in part (c) because the standard deviation is smaller for the $\bar{x}$ distribution than it is for the x distribution. By the Central Limit Theorem, the sampling distribution of $\bar{x}$ will be approximately normal as n increases, and its standard deviation, $\sigma/\sqrt{n}$, will decrease as n increases. The standard deviation of $\bar{x}$, a.k.a. the standard error of $\bar{x}$, measures the spread of the $\bar{x}$ values; the smaller $\sigma/\sqrt{n}$ is, the less variability there is in the $\bar{x}$ values. The less variability there is in the values of $\bar{x}$, the more reliable $\bar{x}$ is as an estimate or predictor of μ. For large n, approximately 95% of the possible values of $\bar{x}$ are within $2\sigma/\sqrt{n}$ of μ. The amount x a typical customer spends on impulse buys also estimates μ (recall $\mu_{\bar{x}} = \mu_x = \mu$), but approximately 95% of individual impulse buys x are within 2σ of μ (using either the Empirical Rule for somewhat mound-shaped data, or assuming x has a distribution that is approximately normal). For a fixed interval, such as \$18 to \$22, centered at the mean, \$20 in this case, the proportion of the possible $\bar{x}$ values within the interval will be greater than the proportion of the possible x values within the same interval.

13. (a) The sample size should be 30 or more.

(b) No. If the distribution of x is normal, the distribution of $\bar{x}$ is also normal, regardless of the sample size.

Section 8.3

1. Step 1:

Sample 1	Sample 2	Sample 3	Sample 4	Sample 5
5	2	3	4	5
3	5	4	6	5
2	5	5	5	2
6	1	2	6	1
4	3	1	3	6
1	5	5	5	6
2	5	1	1	5
3	2	1	2	6
3	6	2	3	6
2	4	1	2	2

Step 2: $\bar{x}_1 = 3.1$ $\quad \bar{x}_2 = 3.8$ $\quad \bar{x}_3 = 2.5$ $\quad \bar{x}_4 = 3.7$ $\quad \bar{x}_5 = 4.4$

$s_1 = 1.524$ $\quad s_2 = 1.687$ $\quad s_3 = 1.650$ $\quad s_4 = 1.767$ $\quad s_5 = 1.955$

Step 3: $\bar{x}_{\bar{x}} = \dfrac{3.1+3.8+2.5+3.7+4.4}{5} = 3.5$

$$S_{\bar{x}} = \sqrt{\frac{(3.1^2+\ldots+4.4^2)-\frac{1}{5}(3.1+\ldots+4.4)^2}{4}} = 0.725$$

Step 4: Intervals:

$\bar{x}-3s$ to $\bar{x}-2s = 3.5-3(0.725)$ to $3.5-2(0.725)$
$= 1.325$ to 2.05

$\bar{x}-2s$ to $\bar{x}-s = 3.5-2(0.725)$ to $3.5-0.725$
$= 2.05$ to 2.775

$\bar{x}-s$ to $\bar{x} = 3.5-0.725$ to 3.5
$= 2.775$ to 3.5

$\bar{x}$ to $\bar{x}+s = 3.5$ to $3.5+0.725$
$= 3.5$ to 4.225

$\bar{x}+s$ to $\bar{x}+2s = 3.5+0.725$ to $3.5+2(0.725)$
$= 4.225$ to 4.95

$\bar{x}+2s$ to $\bar{x}+3s = 3.5+2(0.725)$ to $3.5+3(0.705)$
$= 4.95$ to 5.675

Interval	Frequency	Percent	Hypothetical Normal Dist.
1.325 – 2.05	0	0.0	2 – 3%
2.05 – 2.775	1	0.2	13 or 14%
2.775 – 3.5	1	0.2	About 34%
3.5 – 4.225	2	0.4	About 34%
4.225 – 4.95	1	0.2	13 or 14%
4.95 – 5.675	0	0.0	2 or 3%

Step 5:

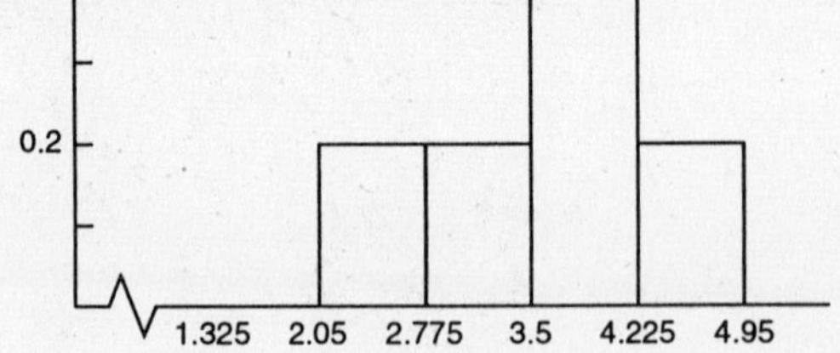

No, the shape does not seem to be approximately mound-shaped and symmetrical. It does not seem to give the general outline of a normal curve.

2. "The larger the sample size, the more likely it is that the sample mean $\bar{x}$ is close to the population mean μ of the distribution."

The larger the sample we are able to collect, the more information we gather about the population. Therefore, as the size of our sample increases, the value of the sample mean $\bar{x}$ should get closer to the value of the population mean μ.

Chapter 8 Review

1. **(a)** The $\bar{x}$ distribution approaches a normal distribution.

   (b) The mean $\mu_{\bar{x}}$ of the $\bar{x}$ distribution equals the mean μ of the x distribution, regardless of the sample size.

   (c) The standard deviation $\sigma_{\bar{x}}$ of the sampling distribution equals $\frac{\sigma}{\sqrt{n}}$, where σ is the standard deviation of the x distribution and n is the sample size.

   (d) They will both be approximately normal with the same mean, but the standard deviations will be $\frac{\sigma}{\sqrt{50}}$ and $\frac{\sigma}{\sqrt{100}}$ respectively.

2. All the $\bar{x}$ distributions will be normal with mean $\mu_{\bar{x}} = \mu = 15$. The standard deviations will be:

$$n = 4:\ \sigma_{\bar{x}} = \frac{\sigma}{\sqrt{n}} = \frac{3}{\sqrt{4}} = \frac{3}{2}$$

$$n = 16:\ \sigma_{\bar{x}} = \frac{\sigma}{\sqrt{n}} = \frac{3}{\sqrt{16}} = \frac{3}{4}$$

$$n = 100:\ \sigma_{\bar{x}} = \frac{\sigma}{\sqrt{n}} = \frac{3}{\sqrt{100}} = \frac{3}{10}$$

3. **(a)** $\mu = 35,\ \sigma = 7$

$$P(x \ge 40) = P\left(z \ge \frac{40-35}{7}\right) = P(z \ge 0.71) = 0.2389$$

   (b) $\mu_{\bar{x}} = \mu = 35,\ \sigma_{\bar{x}} = \frac{\sigma}{\sqrt{n}} = \frac{7}{\sqrt{9}} = \frac{7}{3}$

$$P(\bar{x} \ge 40) = P\left(z \ge \frac{40-35}{\frac{7}{3}}\right) = P(z \ge 2.14) = 0.0162$$

4. (a) $\mu = 38,\ \sigma = 5$

$$\begin{aligned} P(x \le 35) &= P\left(z \le \frac{35-38}{5}\right) \\ &= P(z \le -0.6) \\ &= 0.2743 \end{aligned}$$

(b) $\mu_{\bar{x}} = \mu = 38,\ \sigma_{\bar{x}} = \dfrac{\sigma}{\sqrt{n}} = \dfrac{5}{\sqrt{10}} = 1.58$

$$\begin{aligned} P(\bar{x} \le 35) &= P\left(z \le \frac{35-38}{1.58}\right) \\ &= P(z \le -1.90) \\ &= 0.0287 \end{aligned}$$

(c) The probability in part (b) is much smaller because the standard deviation is smaller for the $\bar{x}$ distribution.

5. $\mu_{\bar{x}} = \mu = 100,\ \sigma_{\bar{x}} = \dfrac{\sigma}{\sqrt{n}} = \dfrac{15}{\sqrt{100}} = 1.5$

$$\begin{aligned} P(100-2 \le \bar{x} \le 100+2) &= P(98 \le \bar{x} \le 102) \\ &= P\left(\frac{98-100}{1.5} \le z \le \frac{102-100}{1.5}\right) \\ &= P(-1.33 \le z \le 1.33) \\ &= P(z \le 1.33) - P(z \le -1.33) \\ &= 0.9082 - 0.0918 \\ &= 0.8164 \end{aligned}$$

6. $\mu_{\bar{x}} = \mu = 15,\ \sigma_{\bar{x}} = \dfrac{\sigma}{\sqrt{n}} = \dfrac{2}{\sqrt{36}} = 0.333$

$$\begin{aligned} P(15-0.5 \le \bar{x} \le 15+0.5) &= P(14.5 \le \bar{x} \le 15.5) \\ &= P\left(\frac{14.5-15}{0.333} \le z \le \frac{15.5-15}{0.333}\right) \\ &= P(-1.5 \le z \le 1.5) \\ &= P(z \le 1.5) - P(z \le -1.5) \\ &= 0.9332 - 0.0668 \\ &= 0.8664 \end{aligned}$$

7. $\mu_{\bar{x}} = \mu = 750,\ \sigma_{\bar{x}} = \dfrac{\sigma}{\sqrt{n}} = \dfrac{20}{\sqrt{64}} = 2.5$

(a)
$$\begin{aligned} P(\bar{x} \ge 750) &= P\left(z \ge \frac{750-750}{2.5}\right) \\ &= P(z \ge 0) \\ &= 0.5000 \end{aligned}$$

(b) $P(745 \le \overline{x} \le 755) = P\left(\frac{745-750}{2.5} \le z \le \frac{755-750}{2.5}\right)$

$= P(-2 \le z \le 2)$

$= P(z \le 2) - P(z \le -2)$

$= 0.9772 - 0.0228$

$= 0.9544$

8. (a) Miami: $\mu = 76$, $\sigma = 1.9$

$$P(x < 77) = P\left(z < \frac{77-76}{1.9}\right) = P(z < 0.53) = 0.7019$$

Fairbanks: $\mu = 0$, $\sigma = 5.3$

$$P(x < 3) = P\left(z < \frac{3-0}{5.3}\right) = P(z < 0.57) = 0.7157$$

(b) Since x has a normal distribution, the sampling distribution of $\overline{x}$ is also normal regardless of the sample size.

Miami: $\mu_{\overline{x}} = \mu = 76$, $\sigma_{\overline{x}} = \frac{\sigma}{\sqrt{n}} = \frac{1.9}{\sqrt{7}} = 0.718$

$$P(\overline{x} < 77) = P\left(z < \frac{77-76}{0.718}\right) = P(z < 1.39) = 0.9177$$

Fairbanks: $\mu_{\overline{x}} = \mu = 0$, $\sigma_{\overline{x}} = \frac{\sigma}{\sqrt{n}} = \frac{5.3}{\sqrt{7}} = 2.003$

$$P(\overline{x} < 3) = P\left(z < \frac{3-0}{2.003}\right) = P(z < 1.50) = 0.9332$$

(c) We cannot say anything about the probability distribution of $\overline{x}$, because the sample size is not 30 or greater. Consider using all 31 days.

Miami: $\mu_{\overline{x}} = \mu = 76$, $\sigma_{\overline{x}} = \frac{\sigma}{\sqrt{n}} = \frac{1.9}{\sqrt{31}} = 0.341$

$$P(\overline{x} < 77) = P\left(z < \frac{77-76}{0.341}\right)$$
$$= P(z < 2.93)$$
$$= 0.9983$$

Fairbanks: $\mu_{\overline{x}} = \mu = 0$, $\sigma_{\overline{x}} = \frac{\sigma}{\sqrt{n}} = \frac{5.3}{\sqrt{31}} = 0.952$

$$P(\overline{x} < 3) = P\left(z < \frac{3-0}{0.952}\right)$$
$$= P(z < 3.15)$$
$$= 0.9992$$

Chapter 9 Estimation

Section 9.1

Answers may vary slightly due to rounding.

1. $n = 196,\ \overline{x} = 11.9,\ s = 4.30,\ c = 95\%,\ z_c = 1.96$

Since $n = 196 \geq 30$, we can use s to estimate σ.

$$E \approx \frac{z_c s}{\sqrt{n}} = \frac{1.96(4.30)}{\sqrt{196}} = 0.6$$
$$(\overline{x} - E) < \mu < (\overline{x} + E)$$
$$(11.9 - 0.6) < \mu < (11.9 + 0.6)$$
$$11.3 \text{ mg/liter} < \mu < 12.5 \text{ mg/liter}$$

2. $n = 99,\ \overline{x} = 10.5,\ s = 3.2,\ c = 95\%,\ z_c = 1.96$

Since $n = 99 \geq 30$, we can use s to estimate σ.

$$E \approx \frac{z_c s}{\sqrt{n}} = \frac{1.96(3.2)}{\sqrt{99}} = 0.63$$
$$(\overline{x} - E) < \mu < (\overline{x} + E)$$
$$(10.5 - 0.63) < \mu < (10.5 + 0.63)$$
$$9.87 \text{ mg/liter} < \mu < 11.13 \text{ mg/liter}$$

3. $n = 36,\ \overline{x} = 16{,}000,\ s = 2400,\ c = 90\%,\ z_c = 1.645$

Since $n = 36 \geq 30$, we can use s to estimate σ.

$$E \approx \frac{z_c s}{\sqrt{n}} = \frac{1.645(2400)}{\sqrt{36}} = 658$$
$$(\overline{x} - E) < \mu < (\overline{x} + E)$$
$$(16{,}000 - 658) < \mu < (16{,}000 + 658)$$
$$15{,}342 \text{ cars} < \mu < 16{,}658 \text{ cars}$$

4. $n = 56,\ \overline{x} = 97^\circ\text{C},\ s = 17^\circ\text{C}$

Since $n = 56 \geq 30$, we can use s to estimate σ.

(a) $c = 95\%,\ z_c = 1.96$

$$E \approx \frac{z_c s}{\sqrt{n}} = \frac{1.96(17)}{\sqrt{56}} = 4.5$$
$$(\overline{x} - E) < \mu < (\overline{x} + E)$$
$$(97 - 4.5) < \mu < (97 + 4.5)$$
$$92.5^\circ\text{C} < \mu < 101.5^\circ\text{C}$$

(b) If the temperature rises, the hot air in the balloon rises, so the balloon would go up. The upper limit of the confidence interval (101.5°C) is an estimate of the (maximum) temperature at which the balloon is at equilibrium (neither going up nor down).

5. (a) $n = 102,\ \bar{x} = 1.2,\ s = 0.4,\ c = 99\%,\ z_c = 2.58$

Since $n = 102 \geq 30$, we can use s to estimate σ.

$$E \approx \frac{z_c s}{\sqrt{n}} = \frac{2.58(0.4)}{\sqrt{102}} = 0.10$$
$(\bar{x} - E) < \mu < (\bar{x} + E)$
$(1.2 - 0.10) < \mu < (1.2 + 0.10)$
1.10 seconds $< \mu < 1.30$ seconds

(b) $\bar{x} = 609,\ s = 248,\ c = 95\%,\ z_c = 1.96,\ n$ still 102

$$E \approx \frac{z_c s}{\sqrt{n}} = \frac{1.96(248)}{\sqrt{102}} = 48.13$$
$(\bar{x} - E) < \mu < (\bar{x} + E)$
$(609 - 48.13) < \mu < (609 + 48.13)$
560.87 Hz $< \mu <$ 657.13 Hz

6. $n = 50 \geq 30$, so we can use s to estimate σ.

(a) $\bar{x} = 5.55,\ s = 0.57,\ c = 85\%,\ z_c = 1.44$

$$E \approx \frac{z_c s}{\sqrt{n}} = \frac{1.44(0.57)}{\sqrt{50}} = 0.12$$
$(\bar{x} - E) < \mu < (\bar{x} + E)$
$(5.55 - 0.12) < \mu < (5.55 + 0.12)$
5.43 cm $< \mu <$ 5.67 cm

(b) $\bar{x} = 2.03,\ s = 0.27,\ c = 90\%,\ z_c = 1.645$

$$E \approx \frac{z_c s}{\sqrt{n}} = \frac{1.645(0.27)}{\sqrt{50}} = 0.06$$
$(\bar{x} - E) < \mu < (\bar{x} + E)$
$(2.03 - 0.06) < \mu < (2.03 + 0.06)$
1.97 cm $< \mu <$ 2.09 cm

7. (a) $n = 42,\ \bar{x} = \frac{\Sigma x_i}{n} = \frac{1511.8}{42} = 35.9952 \approx 36.0$, as stated

$$s^2 = \frac{\Sigma x_i^2 - n\bar{x}^2}{n-1} = \frac{58{,}714.96 - 42(36.0)^2}{41} = 104.4624$$
$s^2 = \sqrt{104.4624} = 10.2207 \approx 10.2$, as stated

Since $n = 42 \geq 30$, we can use s to approximate σ.

(b) $c = 75\%,\ z_c = 1.15$

$$E \approx \frac{z_c s}{\sqrt{n}} = \frac{1.15(10.2)}{\sqrt{42}} = 1.81$$
$(\bar{x} - E) < \mu < (\bar{x} + E)$
$(36.0 - 1.81) < \mu < (36.0 + 1.81)$
$34.19 < \mu < 37.81$ thousand dollars per employee profit

(c) Since \$30 thousand per employee profit is less than the lower limit of the confidence interval (34.19), your bank profits are low, compared to other similar financial institutions.

(d) Since \$40 thousand per employee profit exceeds the upper limit of the confidence interval (37.81), your bank profit is higher than other similar financial institutions.

(e) $c = 90\%,\ z_c = 1.645$

$$E \approx \frac{z_c s}{\sqrt{n}} = \frac{1.645(10.2)}{\sqrt{42}} = 2.59$$

$(\bar{x} - E) < \mu < (\bar{x} + E)$

$(36.0 - 2.59) < \mu < (36.0 + 2.59)$

$33.41 < \mu < 38.59$ thousand dollars per employee profit

\$30 thousand is less than the lower limit of the confidence interval (33.44), so your bank's profit is less than that of other financial institutions.

\$40 thousand is more than the upper limit of the confidence interval (38.59), so your bank is doing better (profit-wise) than other financial institutions.

8. (a) $n = 35,\ \bar{x} = 5.1029 \approx 5.1$, as stated

$s = 3.7698 \approx 3.8$, as stated

Since $n = 35 \geq 30$, we can use s to estimate σ.

(b) $c = 80\%,\ z_c = 1.28$

$$E \approx \frac{z_c s}{\sqrt{n}} = \frac{1.28(3.8)}{\sqrt{35}} = 0.82$$

$(\bar{x} - E) < \mu < (\bar{x} + E)$

$(5.1 - 0.82) < \mu < (5.1 + 0.82)$

$4.28 < \mu < 5.92$ thousand dollars per employee profit

(c) Yes. \$3 thousand per employee profit is less than the lower limit of the confidence interval (4.28).

(d) Yes. \$6.5 thousand dollars profit per employee is larger than the upper limit of the confidence interval (5.92).

(e) $c = 95\%,\ z_c = 1.96$

$$E \approx \frac{z_c s}{\sqrt{n}} = \frac{1.96(3.8)}{\sqrt{35}} = 1.26$$

$(\bar{x} - E) < \mu < (\bar{x} + E)$

$(5.1 - 1.26) < \mu < (5.1 + 1.26)$

$3.84 < \mu < 6.36$ thousand dollars per employee profit

Yes. \$3 thousand per employee profit is less than the lower limit of the confidence interval (3.84).

Yes. \$6.5 thousand dollars profit per employee is larger than the upper limit of the confidence interval (6.36).

9. (a) $\bar{x} = \frac{\Sigma x_i}{n} = \frac{5128}{35} = 146.5143 \approx 146.5,$

$$s^2 = \frac{\Sigma x_i^2 - \frac{(\Sigma x_i)^2}{n}}{n-1} = \frac{756{,}820 - \frac{(5128)^2}{35}}{34} = 161.6101$$

$s = \sqrt{s^2} = \sqrt{161.6101} = 12.7126 \approx 12.7$ as stated

Since $n = 35 \geq 30$, we can use s to approximate σ.

(b) $c = 80\%$ so $z_c = 1.28$

$$\left(\bar{x} - \frac{z_c s}{\sqrt{n}}\right) < \mu < \left(\bar{x} + \frac{z_c s}{\sqrt{n}}\right),\ \left(146.5 - \frac{1.28(12.7)}{\sqrt{35}}\right) < \mu < \left(146.5 + \frac{1.28(12.7)}{\sqrt{35}}\right)$$

$(146.5 - 2.7) < \mu < (146.5 + 2.7)$

143.8 calories $< \mu < 149.2$ calories

(c) $c = 90\%$ so $z_c = 1.645$

$$E \approx \frac{z_c s}{\sqrt{n}} = \frac{1.645(12.7)}{\sqrt{35}} \approx 3.5$$

$(\bar{x} - E) < \mu < (\bar{x} + E)$

$(146.5 - 3.5) < \mu < (146.5 + 3.5)$

143.0 calories $< \mu <$ 150.0 calories

(d) $c = 99\%$ so $t_c = 2.58$

$$E \approx \frac{z_c s}{\sqrt{n}} = \frac{2.58(12.7)}{\sqrt{35}} \approx 5.5; \ (\bar{x} - E) < \mu < (\bar{x} + E)$$

$(146.5 - 5.5) < \mu < (146.5 + 5.5)$

141.0 calories $< \mu <$ 152.0 calories

(e)

c	z_c	Length of confidence interval
80%	1.28	149.2 – 143.8 = 5.4
90%	1.645	150.0 – 143.0 = 7.0
99%	2.58	152.0 – 141.0 = 11.0

As the confidence level, c, increases, so does z_c; therefore, all else being the same, the length of the confidence interval increases, too. We can be more confident the interval captures μ if the interval is longer.

10. (a) $n = 30, \bar{x} = 15.71$ inches, $s = 4.63$ inches, since $n = 30 \geq 30$, we can use s to approximate σ

$c = 95\%$ so $z_c = 1.96$

$$E \approx \frac{z_c s}{\sqrt{n}} = \frac{1.96(4.63)}{\sqrt{30}} = 1.66$$

$(\bar{x} - E) < \mu < (\bar{x} + E)$

$(15.71 - 1.66) < \mu < (15.71 + 1.66)$

14.05 inches $< \mu <$ 17.37 inches

(b) $n = 90$, $\bar{x} = 15.58$ inches, $s = 4.61$ inches

Since $n = 90 \geq 30$, we can use s to approximate σ.

$c = 95\%$ so $z_c = 1.96$

$$E \approx \frac{z_c s}{\sqrt{n}} = \frac{1.96(4.61)}{\sqrt{90}} = 0.95$$

$(\bar{x} - E) < \mu < (\bar{x} + E)$

$(15.58 - 0.95) < \mu < (15.58 + 0.95)$

14.63 inches $< \mu <$ 16.53 inches

(c) $n = 300$, $\bar{x} = 15.59$ inches, $s = 4.62$ inches

Since $n = 300 \geq 30$, we can use s to approximate σ.

$c = 95\%$ so $z_c = 1.96$

$$E \approx \frac{z_c s}{\sqrt{n}} = \frac{1.96(4.62)}{\sqrt{300}} = 0.52$$

$(\bar{x} - E) < \mu < (\bar{x} + E)$

$(15.59 - 0.52) < \mu < (15.59 + 0.52)$

15.07 inches $< \mu <$ 16.11 inches

(d)

Sample size, n	Length of confidence interval
30	17.37 – 14.05 = 3.32
90	16.53 – 14.63 = 1.90
300	16.11 – 15.07 = 1.04

As n increases, so does $\sqrt{n}$, which appears in the denominator of E. All else being the same, the length of the confidence interval decreases as n increases.

11. (a) $n = 38,\ \bar{x} = 2.5,\ s = 0.7,\ c = 90\%,\ z_c = 1.645$

Since $n = 38 \geq 30$, we can use s to estimate σ.

$$E \approx \frac{z_c s}{\sqrt{n}} = \frac{1.645(0.7)}{\sqrt{38}} = 0.2$$
$$(\bar{x} - E) < \mu < (\bar{x} + E)$$
$$(2.5 - 0.2) < \mu < (2.5 + 0.2)$$

2.3 minutes $< \mu <$ 2.7 minutes; length = 2.7 – 2.3 = 0.4 minute

(b) $\bar{x} = 15.2,\ s = 4.8,\ n = 38,\ c = 90\%,\ z_c = 1.645$

$$E \approx \frac{z_c s}{\sqrt{n}} = \frac{1.645(4.8)}{\sqrt{38}} = 1.3$$
$$(\bar{x} - E) < \mu < (\bar{x} + E)$$
$$(15.2 - 1.3) < \mu < (15.2 + 1.3)$$

13.9 min $< \mu <$ 16.5 min; length = 16.5 – 13.9 = 2.6 min

(c) $\bar{x} = 25.7,\ s = 8.3,\ n = 38,\ c = 90\%,\ z_c = 1.645$

$$E \approx \frac{z_c s}{\sqrt{n}} = \frac{1.645(8.3)}{\sqrt{38}} = 2.2$$
$$(\bar{x} - E) < \mu < (\bar{x} + E)$$
$$(25.7 - 2.2) < \mu < (25.7 + 2.2)$$

23.5 min $< \mu <$ 27.9 min, length = 27.9 – 23.5 = 4.4 min

(d)

length	s
0.4	0.7
2.6	4.8
4.4	8.3

As s increases, so does the length of the interval. This is because s is in the numerator of E and the length of any confidence interval is $(\bar{x} + E) - (\bar{x} - E) = 2E$.

12. (a) $n = 40,\ \bar{x} = 27.775 \approx 27.8$ as stated, $s = 3.7994 \approx 3.8$, as stated.

(b) $c = 80\%,\ z_c = 1.28$

$$E \approx \frac{z_c s}{\sqrt{n}} = \frac{1.28(3.8)}{\sqrt{40}} = 0.77$$
$$(\bar{x} - E) < \mu < (\bar{x} + E)$$
$$(27.8 - 0.77) < \mu < (27.8 + 0.77)$$

27.03 years $< \mu <$ 28.57 years

(c) Yes, he would be somewhat old for such a position because 33 is higher than the upper limit of the confidence interval (28.57).

(d) $c = 99\%,\ z_c = 2.58$

$$E \approx \frac{z_c s}{\sqrt{n}} = \frac{2.58(3.8)}{\sqrt{40}} = 1.55$$

$(\bar{x} - E) < \mu < (\bar{x} + E)$

$(27.8 - 1.55) < \mu < (27.8 + 1.55)$

26.25 years $< \mu <$ 29.35 years

33 years is still quite "old," it is above the upper limit of 29.35 years.

Section 9.2

Answers may vary slightly due to rounding.

1. $n = 18$ so $d.f. = n - 1 = 18 - 1 = 17$, $c = 0.95$

$t_c = t_{0.95} = 2.110$

2. $n = 4$ so $d.f. = n - 1 = 4 - 1 = 3$, $c = 0.99$

$t_c = t_{0.99} = 5.841$

3. $n = 22$ so $d.f. = n - 1 = 22 - 1 = 21$, $c = 0.90$

$t_c = t_{0.90} = 1.721$

4. $n = 12$ so $d.f. = n - 1 = 12 - 1 = 11$, $c = 0.95$

$t_c = t_{0.95} = 2.201$

5. $n = 9$ so $d.f. = n - 1 = 9 - 1 = 8$

(a) $\bar{x} = \frac{\Sigma x}{n} = \frac{11,450}{9} \approx 1272$, as stated

$$s^2 = \frac{\Sigma x_i^2 - \frac{(\Sigma x_i)^2}{n}}{n-1} = \frac{14,577,854 - \frac{(11,450)^2}{9}}{8} = 1363.6944$$

$s = \sqrt{1363.6944} = 36.9282 \approx 37$, as stated

(b) $c = 90\%,\ t_c = t_{0.90}$ with 8 $d.f. = 1.860$

$$E = \frac{t_c s}{\sqrt{n}} = \frac{1.86(37)}{\sqrt{9}} = 22.94 \approx 23$$

$(\bar{x} - E) < \mu < (\bar{x} + E)$

$(1272 - 23) < \mu < (1272 + 23)$

1249 A.D. $< \mu <$ 1295 A.D.

6. $n = 6$ so $d.f. = n - 1 = 5$

(a) $\bar{x} = 91.0$, as stated

$s = 30.7181 \approx 30.7$, as stated

(b) $c = 75\%$, so $t_{0.75}$ with 5 $d.f. = 1.301$

$$E = \frac{t_c s}{\sqrt{n}} = \frac{1.301(30.7)}{\sqrt{6}} \approx 16.3$$

$(\bar{x} - E) < \mu < (\bar{x} + E)$

$(91.0 - 16.3) < \mu < (91.0 + 16.3)$

74.7 pounds $< \mu <$ 107.3 pounds

7. $n = 8$, so $d.f. = n - 1 = 7$

(a) $\bar{x} = 12.3475 \approx 12.35$, as stated

$s = 2.2487 \approx 2.25$, as stated

(b) $c = 90\%$, so $t_{0.90}$ with 7 $d.f. = 1.895$

$$E = \frac{t_c s}{\sqrt{n}} = \frac{1.895(2.25)}{\sqrt{8}} = 1.5075 \approx 1.51$$

$(\bar{x} - E) < \mu < (\bar{x} + E)$

$(12.35 - 1.51) < \mu < (12.35 + 1.51)$

$\$10.84 < \mu < \$13.86$

8. $n = 9$, $d.f. = n - 1 = 8$

$\bar{x} = 106.8889 \approx 106.9$, as stated

$s = 29.4425 \approx 29.4$, as stated

$c = 90\%$, so $t_{0.90}$ with 8 $d.f. = 1.860$

$$E = \frac{t_c s}{\sqrt{n}} = \frac{1.860(29.4)}{\sqrt{9}} = 18.228 \approx 18.2$$

$(\bar{x} - E) < \mu < (\bar{x} + E)$

$(106.9 - 18.2) < \mu < (106.9 + 18.2)$

$\$88.7$ thousand $< \mu < \$125.1$ thousand

9. (a) $n = 19$, $d.f. = n - 1 = 18$, $c = 90\%$, $t_{0.90}$ with 18 $d.f. = 1.734$

$\bar{x} = 9.8421 \approx 9.8$, as stated

$s = 3.3001 \approx 3.3$, as stated

$$E = \frac{t_c s}{\sqrt{n}} = \frac{1.734(3.3)}{\sqrt{19}} \approx 1.3$$

$(\bar{x} - E) < \mu < (\bar{x} + E)$

$(9.8 - 1.3) < \mu < (9.8 + 1.3)$

8.5 inches $< \mu <$ 11.1 inches

(b) $n = 7$, $d.f. = n - 1 = 6$, $c = 80\%$, $t_{0.80}$ with 6 $d.f. = 1.440$

$\bar{x} = 17.0714 \approx 17.1$, as stated

$s = 2.4568 \approx 2.5$, as stated

$$E = \frac{t_c s}{\sqrt{n}} = \frac{1.440(2.5)}{\sqrt{7}} \approx 1.4$$

$(\bar{x} - E) < \mu < (\bar{x} + E)$

$(17.1 - 1.4) < \mu < (17.1 + 1.4)$

15.7 inches $< \mu <$ 18.5 inches

10. (a) $n = 12$, $d.f. = n - 1 = 6$, $c = 90\%$, $t_{0.90}$ with 11 $d.f. = 1.796$

$\bar{x} = 4.70$, as stated

$s = 1.9781 \approx 1.98$, as stated

$$E = \frac{t_c s}{\sqrt{n}} = \frac{1.796(1.98)}{\sqrt{12}} \approx 1.03$$

$(\bar{x} - E) < \mu < (\bar{x} + E)$

$(4.70 - 1.03) < \mu < (4.70 + 1.03)$

$3.67\% < \mu < 5.73\%$

(b) $n = 14$, $d.f. = n - 1 = 13$, $c = 90\%$, $t_{0.90}$ with 13 $d.f. = 1.771$

$\bar{x} = 3.1786 \approx 3.18$, as stated

$s = 1.3429 \approx 1.34$, as stated

$$E = \frac{t_c s}{\sqrt{n}} = \frac{1.771(1.34)}{\sqrt{12}} \approx 0.63$$

$(\bar{x} - E) < \mu < (\bar{x} + E)$

$(3.18 - 0.63) < \mu < (3.18 + 0.63)$

$2.55\% < \mu < 3.81\%$

11. (a) $n = 8$, $d.f. = n - 1 = 7$, $c = 85\%$, $t_{0.85}$ with 7 $d.f. = 1.617$

$\bar{x} = 33.125 \approx 33.1$, as stated

$s = 6.3852 \approx 6.4$, as stated

$$E = \frac{t_c s}{\sqrt{n}} = \frac{1.617(6.4)}{\sqrt{8}} = 3.65885 \approx 3.7$$

$(\bar{x} - E) < \mu < (\bar{x} + E)$

$(33.1 - 3.7) < \mu < (33.1 + 3.7)$

\$29.4 thousand $< \mu <$ \$36.8 thousand

(b) $n = 9$, $d.f. = n - 1 = 8$, $c = 99\%$, $t_{0.99}$ with 8 $d.f. = 3.355$

$\bar{x} = 20.8222 \approx 20.8$, as stated

$s = 2.5228 \approx 2.5$, as stated

$$E = \frac{t_c s}{\sqrt{n}} = \frac{3.355(2.5)}{\sqrt{9}} \approx 2.8$$

$(\bar{x} - E) < \mu < (\bar{x} + E)$

$(20.8 - 2.8) < \mu < (20.8 + 2.8)$

\$18.0 thousand $< \mu <$ \$23.6 thousand

12. (a) $n = 15$, $d.f. = n - 1 = 14$, $c = 99\%$, $t_{0.99}$ with 14 $d.f. = 2.977$

$\bar{x} = 7.2867 \approx 7.3$, as stated

$s = 0.7954 \approx 0.8$, as stated

$$E = \frac{t_c s}{\sqrt{n}} = \frac{2.977(0.8)}{\sqrt{15}} \approx 0.6$$

$(\bar{x} - E) < \mu < (\bar{x} + E)$

$(7.3 - 0.6) < \mu < (7.3 + 0.6)$

6.7 days $< \mu <$ 7.9 days

(b) $n = 10$, $d.f. = n - 1 = 9$, $c = 90\%$, $t_{0.90}$ with 9 $d.f. = 1.833$

$\bar{x} = 62.26 \approx 62.3$, as stated

$s = 8.0185 \approx 8.0$, as stated

$$E = \frac{t_c s}{\sqrt{n}} = \frac{1.833(8.0)}{\sqrt{10}} \approx 4.6$$

$(\bar{x} - E) < \mu < (\bar{x} + E)$

$(62.3 - 4.6) < \mu < (62.3 + 4.6)$

$57.7\% < \mu < 66.9\%$

13. Notice that the four figures are drawn on different scales.

(a) The box plots differ in range (distance between whisker ends), in interquartile range (distance between box ends), in medians (line through boxes), in symmetry (indicated by the placement of the median within the box, and by the placement of the median relative to the whisker endpoints), in whisker lengths, and in the presence/absence of outliers. These differences are to be expected since each box plot represents a different sample of size 20. (Although the data sets were all selected as samples of size $n = 20$ from a normal distribution with $\mu = 68$ and $\sigma = 3$, it is very interesting that Sample 2, figure (b), shows 2 outliers.)

(b)

Sample	Confidence interval width	Includes $\mu = 68$?
1	69.407 – 66.692 = 2.715	Yes
2	69.426 – 66.490 = 2.936	Yes
3	69.211 – 66.741 = 2.470	Yes
4	68.050 – 65.766 = 2.284	Yes (barely)

The intervals differ in length; all 4 cover/capture/enclose $\mu = 68$. If many additional samples of size 20 were generated from this distribution, we would expect about 95% of the confidence intervals created from these samples to cover/capture/enclose the number 68; in approximately 5% of the intervals, 68 would be outside the confidence interval, i.e., 68 would be less than the lower limit or greater than the upper limit. Drawing all these samples (at least conceptually) and checking whether or not the confidence intervals include the number 68, keeping track of the percentage that do, is an illustration of the definition of (95%) confidence intervals.

Section 9.3

Answers may vary slightly due to rounding.

1. $r = 39,\ n = 62,\ \hat{p} = \dfrac{r}{n} = \dfrac{39}{62},\ \hat{q} = 1 - \hat{p} = \dfrac{23}{62}$

(a) $\hat{p} = \dfrac{39}{62} = 0.6290$

(b) $c = 95\%, z_c = z_{0.95} = 1.96$
$E \approx z_c\sqrt{\hat{p}\hat{q}/n} = 1.96\sqrt{(0.6290)(1-0.6290)/62} = 0.1202$
$(\hat{p} - E) < p < (\hat{p} + E), (0.6290 - 0.1202) < p < (0.6290 + 0.1202)$
$0.5088 < p < 0.7492$ or approximately 0.51 to 0.75.
We are 95% confident that the true proportion of actors who are extroverts is between 0.51 and 0.75, approximately. In repeated sampling from the same population, approximately 95% of the samples would generate confidence intervals that would cover/capture/enclose the true value of p.

(c) $np \approx n\hat{p} = r = 39$
$nq \approx n\hat{q} = n - r = 62 - 39 = 23$
It is quite likely that np and $nq > 5$, since their estimates are much larger than 5. If np and $nq > 5$ then $\hat{p}$ is approximately normal with $\mu = p$ and $\sigma = \sqrt{pq/n}$. This forms the basis for the large sample confidence interval derivation.

2. $n = 519,\ r = 285$

(a) $\hat{p} = \dfrac{r}{n} = \dfrac{285}{519} = 0.5491$

(b) $c = 99\%, z_c = 2.58, \hat{q} = 1 - \hat{p} = 0.4509$
$E \approx z_c\sqrt{\hat{p}\hat{q}/n} = 2.58\sqrt{(0.5491)(0.4509)/519} = 0.0564$
$(\hat{p} - E) < p < (\hat{p} + E), (0.5491 - 0.0564) < p < (0.5491 + 0.0564)$
$0.4927 < p < 0.6055$ or approximately 0.49 to 0.61.
In repeated sampling, approximately 99% of the intervals generated from the samples would include p, the proportion of judges who are introverts.

(c) $np \approx n\hat{p} = n\left(\frac{r}{n}\right) = r = 285 > 5$

$nq, n\hat{q} = n(1 - \hat{p}) = n\left(1 - \frac{r}{n}\right) = n - r = 234 > 5$

Since the estimates of np and nq are substantially greater than 5, it is quite likely np, $nq > 5$. This allows us to use the normal approximation to the distribution of $\hat{p}$, with $\mu = p$ and $\sigma = \sqrt{pq/n}$.

3. $n = 5222$, $r = 1619$

(a) $\hat{p} = \frac{r}{n} = \frac{1619}{5222} = 0.3100$
so $\hat{q} = 1 - \hat{p} = 0.6900$

(b) $c = 99\%$, so $z_c = 2.58$
$E \approx z_c\sqrt{\hat{p}\hat{q}/n} = 2.58\sqrt{(0.3100)(0.6900)/5222} = 0.0165$
$(\hat{p} - E) < p < (\hat{p} + E), (0.3100 - 0.0165) < p < (0.3100 + 0.0165)$
$0.2935 < p < 0.3265$ or approximately 0.29 to 0.33.
In repeated sampling, approximately 99% of the confidence intervals generated from the samples would include p, the proportion of judges who are hogans.

(c) $np \approx n\hat{p} = r = 1619, nq \approx n\hat{q} = n(1 - \hat{p}) = n - r = 3603$
Since the estimates of np and nq are much greater than 5, it is reasonable to assume np, $nq > 5$. Then we can use the normal distribution with $\mu = p$ and $\sigma = \sqrt{pq/n}$ to approximate the distribution of $\hat{p}$.

4. $n = 592$, $r = 360$

(a) $\hat{p} = \frac{r}{n} = \frac{360}{592} = 0.6081$
so $\hat{q} = 1 - \hat{p} = 0.3919$

(b) $c = 95\%$, $z_c = 1.96$
$E \approx z_c\sqrt{\hat{p}\hat{q}/n} = 1.96\sqrt{(0.6081)(0.3919)/592} = 0.0393$
$(\hat{p} - E) < p < (\hat{p} + E)$
$(0.6081 - 0.0393) < p < (0.6081 + 0.0393)$
$0.5688 < p < 0.6474$, or approximately 0.57 to 0.65.
In repeated sampling from this population, approximately 95% of the samples would generate confidence intervals capturing p, the population proportion of Santa Fe black on white potsherds at the excavation site.

(c) $np \approx n\hat{p} = r = 360 > 5$
$nq \approx n\hat{q} = n(1 - \hat{p}) = n - r = 592 - 360 = 232 > 5$
Since the estimates of np and nq are both greater than 5, it is reasonable to assume that np, nq are also greater than 5. We can then approximate the sampling distribution of $\hat{p}$ with a normal distribution with $\mu = p$ and $\sigma = \sqrt{pq/n}$.

5. $n = 5792, r = 3139$

(a) $\hat{p} = \frac{r}{n} = \frac{3139}{5792} = 0.5420$

so $\hat{q} = 1 - \hat{p} = 0.4580$

(b) $c = 99\%$, so $z_c = 2.58$

$E \approx z_c\sqrt{\hat{p}\hat{q}/n} = 2.58\sqrt{(0.5420)(0.4580)/5792} = 0.0169$

$(\hat{p} - E) < p < (\hat{p} + E)$

$(0.5420 - 0.0169) < p < (0.5420 + 0.0169)$

$0.5251 < p < 0.5589$, or approximately 0.53 to 0.56.

If we drew many samples of size 5792 physicians from those in Colorado, and generated a confidence interval from each sample, we would expect approximately 99% of the intervals to include the true proportion of Colorado physicians providing at least some charity care.

(c) $np \approx n\hat{p} = r = 3139 > 5; nq \approx n\hat{q} = n - r = 2653 > 5.$

Since the estimates of *np* and *nq* are much larger than 5, it is reasonable to assume *np* and *nq* are both greater than 5. Under the circumstances, it is appropriate to approximate the distribution of $\hat{p}$ with a normal distribution with $\mu = p$ and $\sigma = \sqrt{pq/n}$.

6. $n = 10{,}351, r = 7867$

(a) $\hat{p} = \frac{r}{n} = \frac{7867}{10{,}351} = 0.7600$

so $\hat{q} = 1 - \hat{p} = 0.2400$

(b) $c = 99\%$, so $z_c = 2.58$

$E \approx z_c\sqrt{\hat{p}\hat{q}/n} = 2.58\sqrt{(0.7600)(0.2400)/10{,}351} = 0.0108$

$(\hat{p} - E) < p < (\hat{p} + E)$

$(0.7600 - 0.0108) < p < (0.7600 + 0.0108)$

$0.7492 < p < 0.7708$, or approximately 0.75 to 0.77.

In repeated sampling from the population of convicts who escaped from U.S. prisons, approximately 99% of the confidence intervals created from those samples would include *p*, the proportion of recaptured escaped convicts.

(c) $np \approx n\hat{p} = r = 7867 > 5; nq \approx n\hat{q} = n(1 - \hat{p}) = n - r = 2484 > 5.$

Since the estimates of *np* and *nq* are each considerably greater than 5, it is reasonable to assume that *np* and *nq* > 5. When *np* and *nq* > 5, the distribution of $\hat{p}$ can be quite accurately approximated by a normal distribution with $\mu = p$ and $\sigma = \sqrt{pq/n}$.

7. $n = 855, r = 26$

(a) $\hat{p} = \frac{r}{n} = \frac{26}{855} = 0.0304$

so $\hat{q} = 1 - \hat{p} = 0.9696$

(b) $c = 99\%$, so $z_c = 2.58$
$E \approx z_c\sqrt{\hat{p}\hat{q}/n} = 2.58\sqrt{(0.0304)(0.9696)/855} = 0.0151$
$(\hat{p} - E) < p < (\hat{p} + E)$
$(0.0304 - 0.0151) < p < (0.0304 + 0.0151)$
$0.0153 < p < 0.0455$, or approximately 0.02 to 0.05.
If many additional samples of size n = 855 were drawn from this fish population, and a confidence interval was created from each such sample, approximately 99% of those confidence intervals would contain p, the catch-and-release mortality rate (barbless hooks removed).

(c) $np \approx n\hat{p} = r = 26 > 5; nq \approx n\hat{q} = n - r = 829 > 5$.
Based on the estimates of np and nq, it is safe to assume both np and $nq > 5$. When np and $nq > 5$, the distribution of $\hat{p}$ can be accurately approximated by a normal distribution with $\mu = p$ and $\sigma = \sqrt{pq/n}$.

8. (a) For a 95% confidence interval, use $z = 1.96$. The point estimate for p is $\hat{p}$ is 85%.
The standard error is

$$\sqrt{\frac{\hat{p}(1-\hat{p})}{n}} = \sqrt{\frac{0.85(0.15)}{100}} = \sqrt{0.001275} = 0.03571 \approx 0.036$$

Thus, the maximal value of $E = \pm\,1.96(0.036) = \pm\,0.071$. Thus, $\hat{p} - E = 0.85 - 0.071 = 0.779$ and $\hat{p} + E = 0.85 + 0.071 = 0.921$, So, we are 95% sure that the population proportion is between 0.779 and 0.921 or 77.9% and 92.1%.
Since $n\hat{p} = 100(0.85) = 85$ and $n\hat{q} = 100(0.15) = 15$ are both greater than 5, the use of the normal distribution in approximating the binomial is justified because the sampling distribution is approximately normal.

(b) The point estimate for p is $\hat{p} = 85\%$. The standard error is

$$\sqrt{\frac{\hat{p}(1-\hat{p})}{n}} = \sqrt{\frac{0.85(0.15)}{1000}} = \sqrt{0.0001275} = 0.0113$$

The maximal value of $E = \pm\,1.96(0.0113) = \pm\,0.02221$.
So, the endpoints of the interval are $\hat{p} - E = 0.85 - 0.0221 = 0.8279$ and

$$\begin{aligned}\hat{p} + E &= 0.85 + 0.221 \\ &= 0.8721\end{aligned}$$

Thus, if our sample has 1000 pharmacists, we are 95% sure that the population proportion is between 0.8279 and 0.8721.

(c) For part (a), the confidence interval has a width of 0.921 – 0.779 = 0.142. However, in part (b), the width is 0.8721 – 0.8279 = 0.0442. When we increase the sample size, the standard error decreases which in turn decreases E, the maximal error of estimate.

9. **(a)** For a 95% confidence interval, use $z = 1.96$. The point estimate for p is $\hat{p}$ is 26.6%. The standard error is

$$\sqrt{\frac{\hat{p}(1-\hat{p})}{n}} = \sqrt{\frac{0.266(0.734)}{1000}} = \sqrt{0.000195} = 0.01397 \approx 0.014.$$

Thus, maximal value of $E = \pm 1.96(0.014) = \pm 0.027$. Thus, $\hat{p} - E = 0.266 - 0.027 = 0.239$ and $\hat{p} + E = 0.266 + 0.027 = 0.293$. So, we are 95% sure that the population proportion is between 0.239 and 0.293 or 23.9% and 29.3%. Since $n\hat{p} = 1000(0.266) = 266$ and $n\hat{q} = 1000(0.734) = 734$ are both greater than 5, the use of the normal distribution in approximately the binomial is justified because the sampling distribution is approximately normal.

(b) The point estimate for p is $\hat{p} = 85\%$. The standard error is

$$\sqrt{\frac{\hat{p}(1-\hat{p})}{n}} = \sqrt{\frac{0.266(0.734)}{10{,}000}} = \sqrt{0.0000195} = 0.0044.$$

The maximal value of $E = \pm 1.96(0.0044) = \pm 0.0087$. So the endpoints of the interval are

$\hat{p} - E = 0.266 - 0.0087 = 0.2573$ and $\hat{p} + E = 0.266 + 0.0087 = 0.2747$.

Thus, if our sample has 10,000 orthopedic cases, we are 95% sure that the population proportion is between 0.2573 and 0.2747.

(c) For part (a), the confidence interval has a width of 0.293 – 0.239 = 0.054. However, in part (b), the width is 0.2747 – 0.2573 = 0.0174. When we increase the sample size, the standard error decreases, which in turn decreases E, the maximal error of estimate.

10. From Problem 8, we see the calculation for the margin of error.
The standard error is

$$\sqrt{\frac{\hat{p}(1-\hat{p})}{n}} = \sqrt{\frac{0.85(0.15)}{100}} = \sqrt{0.001275}$$
$$= 0.03571 \approx 0.036$$

Thus, the margin of error is $E = \pm 1.96(0.036)$
$= \pm 0.071$.

11. $n = 730$, $r = 628$, $n - r = 102$; both np and $nq > 5$.

(a) $\hat{p} = \frac{r}{n} = \frac{628}{730} = 0.8603$, so $\hat{q} = 0.1397$

(b) $c = 95\%$, so $z_c = 1.96$
$E \approx z_c\sqrt{\hat{p}\hat{q}/n} = 1.96\sqrt{(0.8603)(0.1397)/730} = 0.0251$
$(\hat{p} - E) < p < (\hat{p} + E)$
$(0.8603 - 0.0251) < p < (0.8603 + 0.0251)$
$0.8352 < p < 0.8854$, or about 0.84 to 0.89.

In repeated sampling, approximately 95% of the intervals created from the samples would include p, the proportion of loyal women shoppers.

(c) Margin of error $= E \approx 2.5\%$
A recent study by the Food Marketing Institute showed that about 86% of women shoppers remained loyal to their favorite supermarket last year. The study's margin of error is 2.5 percentage points.

12. $n = 1001$, $r = 273$, $n - r = 728$; both np and $nq > 5$.

(a) $\hat{p} = \frac{r}{n} = \frac{273}{1001} = 0.2727$, so $\hat{q} = 0.7273$

(b) $c = 95\%$, so $z_c = 1.96$
$E \approx z_c\sqrt{\hat{p}\hat{q}/n} = 1.96\sqrt{(0.2727)(0.7273)/1001} = 0.0276$
$(\hat{p} - E) < p < (\hat{p} + E)$
$(0.2727 - 0.0276) < p < (0.2727 + 0.0276)$
$0.2451 < p < 0.3003$, or about 0.25 to 0.30.

If many additional samples of size $n = 1001$ were drawn from this population, about 95% of the confidence intervals created from these samples would include p, the proportion of shoppers who stock up on bargains.

(c) Margin of error $= E \approx 2.8\%$
The Food Marketing Institute reported that, based on a recent study, 27.3% of shoppers stock up on an item when it is a real bargain. The study had a margin of error of 2.8 percentage points.

13. $n = 1000$, $r = 250$, $n - r = 750$; both np and $nq > 5$.

(a) $\hat{p} = \frac{r}{n} = \frac{250}{1000} = 0.2500$, so $\hat{q} = 0.7500$

(b) $c = 95\%$, so $z_c = 1.96$
$E \approx z_c\sqrt{\hat{p}\hat{q}/n} = 1.96\sqrt{(0.2500)(0.7500)/1000} = 0.0268$
$(\hat{p} - E) < p < (\hat{p} + E)$
$(0.2500 - 0.0268) < p < (0.2500 + 0.0268)$
$0.2232 < p < 0.2768$, or about 0.22 to 0.28.

(c) Margin of error $= E \approx 2.7\%$
In a survey reported in *USA Today*, 25% of large corporations interviewed admitted that, given a choice between equally qualified applicants, they would offer the job to the nonsmoker. The survey's margin of error was 2.7 percentage points.

14. **(a)** For a 90% confidence interval, use $z = 1.645$. The point estimate for p is $\hat{p} = \frac{11}{35} = 0.31$.

The standard error is

$$\sqrt{\frac{\hat{p}(1-\hat{p})}{n}} = \sqrt{\frac{0.31(0.69)}{35}} = \sqrt{0.006111} = 0.07818$$
$$\approx 0.078$$

Thus, the maximal value of $E = \pm 1.645(0.078) = \pm 0.128$. Thus, $\hat{p} - E = 0.31 - 0.128 = 0.182$ and $\hat{p} + E = 0.31 + 0.128 = 0.438$. So, we are 90% sure that the population proportion is between 0.182 and 0.438 or 18.2% and 43.8%.

(b) Since $n\hat{p} = 35(0.31) = 10.85$ and $n\hat{q} = 35(0.69) = 24.15$ are both greater than 5, the use of the normal distribution in approximating the binomial is justified because the sampling distribution is approximately normal.

(c) Yes, 0.28 is in the confidence interval computed in part (a). Since the national proportion lies within the confidence interval, it does not seem that the Cherry Creek neighborhood is much different from the population of all U.S. households.

Section 9.4

1. The goal is to estimate μ, the population mean number of new lodgepole pine saplings in a 50 square meter plot in Yellowstone National Park. Use $n = (z_c\sigma / E)^2$, where $c = 95\%$, $z_c = 1.96$, $\sigma = 44$, and $E = 10$.

$$n \approx \left[\frac{1.96(44)}{10}\right]^2 = 74.37, \text{ “round up” to 75 plots.}$$

2. The goal is to estimate the mean root depth, μ, in glacial outwash soil.
Use $n = (z_c\sigma / E)^2$ with $c = 90\%$, $z_c = 1.645, E = 0.5$, and $\sigma = 8.94$.
$n \approx [1.645(8.94)/0.5]^2 = 865.10$; “roundup” to 866 plants.

3. The goal is to find μ, the mean player weight. Preliminary, $n = 56$.
Use $n = (z_c\sigma / E)^2$. $s = 26.58, c = 90\%, z_c = 1.645, E = 4$.
Then $n \approx [1.645(26.58)/4]^2 = 119.5$, or 120 players. However, since 56 players have already been drawn to estimate σ, we need only $120 - 56 = 64$ additional players.

4. The goal is to estimate μ, the mean height of NBA players; use $n = (z_c\sigma / E)^2$. Preliminary, $n = 41$.

$s = 3.32$, $c = 95\%$, $z_c = 1.96, E = 0.75$.

Then $n \approx [1.96(3.32)/0.75]^2 = 75.3$, or 76 players. However, the preliminary sample had 41 players in it, so we need to sample only 76 – 41 = 35 additional players.

5. The goal is to estimate μ, the mean reconstructed clay vessel diameter, use $n = (z_c\sigma / E)^2$.

Preliminary $n = 83$, $s = 5.5$, $c = 95\%$, $z_c = 1.96$, $E = 1.0$

$n \approx [1.96(5.5)/1.0]^2 = 116.2 \approx 117$ clay pots.

Since 83 pots were already measured, we need 117 – 83 = 34 additional reconstructed clay pots.

6. The goal is to estimate the mean weight, μ, of bighorn sheep; use $n = (z_c\sigma / E)^2$.

Preliminary $n = 37, s = 15.8, c = 90\%, z_c = 1.645, E = 2.5$.

$n \approx [1.645(15.8)/2.5]^2 = 108.1 \approx 109$ sheep.

Since 37 bighorn sheep have already been weighed, we need 109 – 37 = 72 additional bighorn sheep for the sample.

7. The goal is to estimate μ, the average time phone customers are on hold; use $n = (z_c\sigma / E)^2$.

Preliminary sample of size $n = 167, s = 3.8$ minutes, $c = 99\%, z_c = 2.58, E = 30$ seconds = 0.5 minute (all time figures must be in the same units).

$n \approx [2.58(3.8)/0.5]^2 = 384.5 \approx 385$ phone calls.

Since the airline already measured the time on hold for 167 calls, it needs to measure the time on hold for 385 – 167 = 218 more phone calls.

8. **(a)** Since we have no value of $\hat{p}$, we will use $n = \frac{1}{4}\left(\frac{Zc}{E}\right)^2$. The confidence level is 90% so $z = 1.645$.

We are given that $E = 0.05$.

Thus, $n = \frac{1}{4}\left(\frac{1.534}{0.05}\right)^2 = 270.60$.

Rounding up to the nearest whole number, $n = 271$ people.

(b) We use the preliminary estimate $\hat{p} = 0.52$.

So, $\hat{q} = 1 - \hat{p} = 0.48$. Thus, $n = (0.52)(0.48)\left(\frac{1.645}{0.05}\right)^2 = 270.17$.

Rounding up the nearest whole number, $n = 271$ people.

Note that since the product (0.52)(0.48) = 0.2496, it is not surprising that the sample size requirement is essentially the same with or without a preliminary sample.

9. We are estimating the proportion of p drivers who admit that they exceed the speed limit.

(a) Since we have no value of $\hat{p}$, we will use $n = \frac{1}{4}\left(\frac{Zc}{E}\right)^2$. The confidence level is 95% so $z = 1.96$. We are given that $E = 0.05$. Thus

$n = \frac{1}{4}\left(\frac{0.96}{0.05}\right)^2 = 384.16$.

Rounding up to the nearest whole number, $n = 386$ drivers.

(b) We use the preliminary estimate $\hat{p} = 0.14$. So, $\hat{q} = 0.86$. Thus,

$$n = (0.14)(0.86)\left(\frac{1.96}{0.05}\right)^2 = 185.01.$$

Rounding up to the nearest whole number, $n = 186$ drivers.

10. The goal is to estimate p, the proportion of callers who reach a business person by phone on the first call, so use $n = pq(z_c / E)^2$.

(a) $c = 80\%, z_c = 1.28, E = 0.03$

Since there is no preliminary estimate of p, we will use $\hat{p} = \hat{q} = 1/2$, the "worst case" estimate; no other choice of $\hat{p}$ would give us a larger sample size.

When $n = pq(z_c / E)^2$ becomes $n \approx 0.5 \cdot 0.5(z_c / E)^2$ or $n \approx 0.25(z_c / E)^2 = 0.25(1.28/0.03)^2 = 455.1$, or 456. We need to have a sample size of 456 business phone calls.

(b) $\hat{p} = 17\% = 0.17$, so $\hat{q} = 1 - \hat{p} = 1 - 0.17 = 0.83$

$n = pq(z_c / E)^2 \approx (0.17)(0.83)(1.28/0.03)^2 = 256.9$, or 257 business phone calls.

11. The goal is to estimate the proportion of women students; use $n = pq(z_c / E)^2$.

(a) $c = 99\%, z_c = 2.58, E = 0.05$

Since there is no preliminary estimate of p, we'll use the "worst case" estimate, $\hat{p} = \frac{1}{2}, \hat{q} = 1 - \hat{p} = \frac{1}{2}$.

$n \approx 0.5 \cdot 0.5(z_c / E)^2 = 0.25(z_c / E)^2 = 0.25(2.58/0.05)^2 = 665.6$, or 666 students.

(b) Preliminary estimate of p: $\hat{p} = 54\% = 0.54, \hat{q} = 1 - \hat{p} = 0.46$.

$n \approx (0.54)(0.46)(2.58/0.05)^2 = 661.4 \approx 662$ students.

(There is very little difference between (a) and (b) because (a) uses $\hat{p} = \frac{1}{2} = 0.50$ and (b) uses $\hat{p} = 0.54$, and the $\hat{p}$s are approximately the same.)

12. The goal is to estimate p, the proportion of small businesses declaring bankruptcy; use $n = pq(z_c / E)^2$.

(a) $c = 95\%, z_c = 1.96, E = 0.10$

Since there is no preliminary estimate of p available, use the worst case estimate, $\hat{p} = \frac{1}{2}; \hat{q} = 1 - \hat{p} = \frac{1}{2}$.

$n \approx 0.5 \cdot 0.5(z_c / E)^2 = 0.25(z_c / E)^2 = 0.25(1.96/0.10)^2 = 96.04 \approx 97$ small businesses.

(b) Preliminary sample of $n = 38$ found $r = 6$ small businesses had declared bankruptcy, so $\hat{p} = r/n = 6/38 = 0.1579, \hat{q} = 1 - \hat{p} = 0.8421$.

$n \approx (0.1579)(0.8421)(1.96/0.10)^2 = 51.08 \approx 52$ small businesses, total.

Since 38 have already been surveyed, the National Council of Small Businesses needs to survey an additional 52 – 38 = 14 small businesses.

13. The goal is to estimate the proportion, p, of votes for the Democratic presidential candidate. Use $n = pq(z_c / E)^2$.

(a) $E = 0.001, c = 99\%, z_c = 2.58$

Since there is no preliminary estimate of p, use $\hat{p} = \hat{q} = \frac{1}{2}$.

$$n \approx 0.5 \cdot 0.5(z_c / E)^2 = 0.25(z_c / E)^2 = 0.25(2.58/0.001)^2 = 1{,}664{,}100 \text{ votes.}$$

(b) No. If the preliminary estimate of p was $\hat{p} = 0.5$, the same as the no information, worst case estimate of p, the sample size would be exactly the same as in (a). The general formula $n = pq(z_c / E)^2$ with preliminary estimates of p (and q) = 0.5 is the same as the no-information-about-p formula, $n = \frac{1}{4}(z_c / E)^2$.

[The above solutions have repeatedly used this fact, demonstrating that using $\hat{p} = \hat{q} = 0.5$ (the worst case estimates of p and q, giving the largest possible sample size that meets the stated criteria) in the special no-information-about-p formula, $n = \frac{1}{4}(z_c / E)^2$, directly.]

14. Note: all sample size calculations, after "rounding up," give the <u>minimum</u> sample size required to meet the c, E, etc., criteria.
Recall from text that the margin of error is the maximal error E of a <u>95%</u> confidence interval for p.
The problem, therefore, implies $E = 3\% = 0.03$, $c = 95\%$, $z_c = 1.96$, and since no preliminary estimate or p is given, use $\hat{p} = \frac{1}{2}; \hat{q} = 1 - \hat{p} = \frac{1}{2}$.
Because the goal is to estimate p, the proportion of registered voters who favor using lottery proceeds for park improvements, use

$$n = pq(z_c / E)^2 \approx 0.5 \cdot 0.5(z_c / E)^2 = 0.25(z_c / E)^2 = 0.25(1.96/0.03)^2 = 1067.1 \approx 1068 \text{ registered voters.}$$

Chapter 9 Review

1. point estimate: a single number used to estimate a population parameter

critical value: the x-axis values (arguments) of a probability density function (such as the standard normal or Student's t) which cut off an area of $c, 0 \le c \le 1$, under the curve between them. Examples: the area under the standard normal curve between $-z_c$ and $+z_c$ is c; the area under the curve of a Student's t distribution between $-t_c$ and $+t_c$ as c. The area is symmetric about the curve's mean, μ.

maximal error of estimate, E: the largest distance ("error") between the point estimate and the parameter it estimates that can be tolerated under certain circumstances; E is the half-width of a confidence interval.

confidence level, c: A measure of the reliability of an (interval) estimate: c denotes the proportion of all possible confidence interval estimates of a parameter (or difference between 2 parameters) that will cover/capture/enclose the true value being estimated. It is a statement about the probability the <u>procedure</u> being used has of capturing the value of interest; it <u>cannot</u> be considered a measure of the reliability of a <u>specific</u> interval, because any specific interval is either right or wrong-either it captures the parameter value, or it does not, period.

confidence interval: a procedure designed to give a range of values as an (interval) estimate of an unknown parameter value; compare point estimate. What separates confidence interval estimates from any other interval estimate (such as 4, give or take 2.8) is that the <u>reliability</u> of the procedure can be determined: if $c = 0.90 = 90\%$, for example, a 90% confidence interval (estimate) for μ says that it all possible samples of size n were drawn, and a 90% confidence interval for μ was created for each such sample using the prescribed method (such as $\bar{x} \pm z_c\sigma / \sqrt{n}$), then if the true value of μ became known, 90% of the confidence intervals so created would cover/capture/enclose the value of μ.

large/small samples: a large sample in our context is one that is of sufficient size to warrant using a normal approximation to the exact method, i.e., large enough that the central limit theorem can reasonably be applied and that the approximation results in sufficiently accurate estimates of the exact method results. We have said that $n \geq 30$ is large enough, in all but the most extreme cases, to say that the Student's *t*-distribution can be approximated by the normal and that s^2 can be used to estimate σ^2. Similarly, if np_i and nq_i are both greater than 5, the normal distribution can be used to approximate the exact binomial calculations of the probability of *r* successes.

Small samples are those of a size where using a normal approximation instead of the exact method would give unreliable results; the difference between the exact and appropriate answers is too large to be tolerated. For Student's *t* distribution, if $n < 30$, the normal approximation results are considered too crude to be useful. For the central limit theorem to be applied when estimating *p* or $p^1 - p^2$, or, specifically, for the normal approximation to the binomial to be applied we have said np_i and nq_i must both be >5.* Here the criteria are the <u>products</u> np_i and nq_i, not just the size of *n*. A sample of size $n = 300$ seems large enough for just about any purpose, but if $p = 0.01$, $np_i = 300(0.01) = 3 \leq 5$, and $n = 300$ is not large enough for a normal approximation to be accurate enough. In general, if *p* (or *q*) is near 0 or near 1, *n* must be quite large before the normal approximation can be used.

*Some textbooks use other criteria or rules of thumb, such as np_i and $nq_i > 10$.

2. For a 0.90 confidence interval, use $zc = 1.645$. The sample size $n = 370$ is large enough that we may approximate *r* as $s = \$150$.
Therefore,
$$E \approx Zc\left(\frac{s}{\sqrt{n}}\right) = 1.645\left(\frac{150}{\sqrt{370}}\right) = 12.83$$
The 90% confidence interval for μ is
$$\bar{x} - E < \mu < \bar{x} + E$$
$$1{,}750 - 12.83 < \mu < 1{,}750 + 12.83$$
$$1{,}737.17 < \mu < 1{,}762.83$$
We conclude with 90% confidence that the population mean μ of claim payments is between \$1,737.17 and \$1,762.83.
For a 0.95 confidence interval, use $zc = 2.58$.
Therefore, $E \approx Zc\left(\frac{s}{\sqrt{n}}\right) = 2.58\left(\frac{150}{\sqrt{370}}\right) = 20.12$
The 95% confidence interval for μ is
$$\bar{x} - E < \mu < \bar{x} + E$$
$$1{,}750 - 20.12 < \mu < 1{,}750 + 20.12$$
$$1{,}729.88 < \mu < 1{,}770.12$$
We conclude with 95% confidence that the population mean μ of claim payments is between \$1,729.88 and \$1,770.12.

3. $n = 73 \geq 30$; use a large sample procedure to estimate μ.
$\bar{x} = 178.70, s = 7.81, c = 95\%, z_c = 1.96$
$$E \approx \frac{z_c s}{\sqrt{n}} = \frac{1.96(7.81)}{\sqrt{73}} = 1.7916 \approx 1.79$$
$$(\bar{x} - E) < \mu < (\bar{x} + E)$$
$$(178.70 - 1.79) < \mu < (178.70 + 1.79)$$
$$176.91 < \mu < 180.49$$

4. $c = 99\%, z_c = 2.58, E = 2, s = 7.81$ from Problem 3 above

$$n = \left(\frac{z_c s}{E}\right)^2 = \left[\frac{2.58(7.81)}{2}\right]^2 = 101.5036 \approx 102$$

5. **(a)** $\overline{x} = 74.2, s = 18.2530 \approx 18.3$, as indicated

(b) $c = 95\%, n = 15$ so use small sample procedure to estimate μ

$t_{0.95}$ with $n - 1 = 14$ $d.f. = 2.145$

$$E \approx \frac{t_c s}{\sqrt{n}} = \frac{2.145(18.3)}{\sqrt{15}} = 10.1352 \approx 10.1$$

$(\overline{x} - E) < \mu < (\overline{x} + E)$

$(74.2 - 10.1) < \mu < (74.2 + 10.1)$

64.1 centimeters $< \mu <$ 84.3 centimeters

6. **(a)** $n = 10, \overline{x} = 15.78 \approx 15.8, s = 3.4608 \approx 3.5$, as indicated

(b) estimate μ with small sample procedure because $n = 10 < 30$

$c = 80\%, n - 1 = 9\ d.f., t_{0.80}$ with $9\ d.f. = 1.383$

$$E \approx \frac{t_c s}{\sqrt{n}} = \frac{1.383(3.5)}{\sqrt{10}} = 1.5307 \approx 1.53$$

$(\overline{x} - E) < \mu < (\overline{x} + E)$

$(15.8 - 1.53) < \mu < (15.8 + 1.53)$

14.27 centimeters $< \mu <$ 17.33 centimeters

7. $n = 2958, r = 1538, \hat{p} = \frac{r}{n} = \frac{1538}{2958} = 0.5199 \approx 0.52$

$\hat{q} = 1 - \hat{p} = 0.4801, n - r = 1420, c = 90\%, z_c = 1.645$

$np \approx n\hat{p} = r = 1538 > 5, nq \approx n\hat{q} = n - r = 1420 > 5$, so use large sample method to estimate p

$$E \approx z_c\sqrt{\frac{\hat{p}\hat{q}}{n}} = 1.645\sqrt{\frac{(0.5199)(0.4801)}{2958}} = 0.0151 \approx 0.02$$

$(\hat{p} - E) < p < (\hat{p} + E)$

$(0.52 - 0.02) < p < (0.52 + 0.02)$

$0.50 < p < 0.54$

8. $95\%, z_c = 1.96$, preliminary estimate $\hat{p} = 0.52$

$E = 0.01, \hat{q} = 1 - \hat{p} = 0.48$

$$n = \left(\frac{z_c}{E}\right)^2 \hat{p}\hat{q} = \left(\frac{1.96}{0.01}\right)^2 (0.52)(0.48) = 9{,}588.6336$$

sample size of 9589

9. $n = 167, r = 68$

(a) $\hat{p} = \frac{r}{n} = \frac{68}{167} = 0.4072, \hat{q} = 0.5928, n - r = 99$

(b) $n\hat{p} = r = 68 > 5$ and $n\hat{q} = n - r = 99 > 5$, so use large sample method to estimate p

$c = 95\%, z_c = 1.96$

$$E \approx z_c\sqrt{\frac{\hat{p}\hat{q}}{n}} = 1.96\sqrt{\frac{(0.4072)(0.5928)}{167}} = 0.0745$$

$(\hat{p} - E) < p < (\hat{p} + E)$

$(0.4072 - 0.0745) < p < (0.4072 + 0.0745)$

$0.3327 < p < 0.4817$, or about 0.333 to 0.482

10. $c = 95\%, z_c = 1.96, E = 0.06, \hat{p} = 0.4072, \hat{q} = 0.5928$ from Problem 9

$$n \approx \left(\frac{z_c}{E}\right)^2 \hat{p}\hat{q} = \left(\frac{1.96}{0.06}\right)^2 (0.4072)(0.5928) = 257.5880 \approx 258 \text{ potsherds}$$

Since Problem 9 says 167 potsherds have already been collected, we need 258 – 167 = 91 additional potsherds to be collected.

11. **(a)** Mean and standard deviation round to results shown.

(b) Since n is small, use $t = 2.093$ for a 95% confidence interval. Therefore,

$$E \approx t\left(\frac{s}{\sqrt{n}}\right) = 2.093\left(\frac{2.72}{\sqrt{20}}\right) = 1.27$$

The 95% confidence interval for μ is

$$\bar{x} - E < \mu < \bar{x} + E$$

$$9.55 - 1.27 < \mu < 9.55 + 1.27$$

$$8.3 < \mu < 10.8$$

(c) We conclude with 95% confidence that the population mean μ of times between commercials is between 8.3 and 10.8 minutes.

12. First we find that $\bar{x} = 0.526$ and $s = 0.0445$

Since n is small, use t = 2,776. Therefore,

$$E \approx t\left(\frac{s}{\sqrt{n}}\right) = 2.776\left(\frac{0.0445}{\sqrt{5}}\right) = 0.055$$

The 95% confidence interval for μ is

$$\bar{x} - E < \mu < \bar{x} + E$$

$$0.526 - 0.055 < \mu < 0.526 + 0.055$$

$$0.471 < \mu < 0.581$$

We conclude with 95% confidence that the population mean μ of coyote density is between 0.471 and 0.581 coyote per square kilometer.

Chapter 10 Hypothesis Testing

Section 10.1

1. See text for definitions. Essay may include:

(a) A working hypothesis about the population parameter in question is called the null hypothesis. The value specified in the null hypothesis is often a historical value, a claim, or a production specification.

(b) Any hypothesis that differs from the null hypothesis is called an alternate hypothesis.

(c) If we reject the null hypothesis when it is in fact true, we have an error that is called a type I error. On the other hand, if we accept (i.e., fail to reject) the null hypothesis when it is in fact false, we have made an error that is called a type II error.

(d) The probability with which we are willing to risk a type I error is called the level of significance of a test. The probability of making a type II error is denoted by β.

2. The alternate hypothesis is used to determine which type of critical region is used. An alternate hypothesis is constructed in such a way that it is the one to be accepted when the null hypothesis must be rejected.

3. No, if we fail to reject the null hypothesis, we have not proven it to be true beyond all doubt. The evidence is not sufficient to merit rejecting H_0.

4. No, if we reject the null hypothesis, we have not proven it to be false beyond all doubt. The test was conducted with a level of significance, α, which is the probability with which we are willing to risk a type I error (rejecting H_0 when it is in fact true).

5. (a) The claim is $\mu = 60$ kg, so you would use $H_0: \mu = 60$ kg.

(b) We want to know if the average weight is less than 60 kg, so you would use $H_1: \mu < 60$ kg.

(c) We want to know if the average weight is greater than 60 kg, so you would use $H_1: \mu > 60$ kg.

(d) We want to know if the average weight is different from (more or less than) 60 kg, so you would use $H_1: \mu \neq 60$ kg.

(e) Since part (b) is a left-tailed test, the critical region is on the left. Since part (c) is a right-tailed test, the critical region is on the right. Since part (d) is a two-tailed test, the critical region is on both sides of the mean.

6. (a) The claim is $\mu = 8.3$ min, so you would use $H_0: \mu = 8.3$ min. If you believe the average is less than 8.3 min, then you would use $H_1: \mu < 8.3$ min. Since this is a left-tailed test, the critical region is on the left side of the mean.

(b) The claim is $\mu = 8.3$ min, so you would use $H_0: \mu = 8.3$ min. If you believe the average is different from 8.3 min, then you would use $H_1: \mu \neq 8.3$ min. Since this is a two-tailed test, the critical region is on both sides of the mean.

(c) The claim is $\mu = 4.5$ min, so you would use $H_0: \mu = 4.5$ min. If you believe the average is more than 4.5 min, then you would use $H_1: \mu > 4.5$ min. Since this is a right-tailed test, the critical region is on the right side of the mean.

(d) The claim is $\mu = 4.5$ min, so you would use $H_0: \mu = 4.5$ min. If you believe the average is different from 4.5 min, then you would use $H_1: \mu \neq 4.5$ min. Since this is a two-tailed test, the critical region is on both sides of the mean.

7. (a) The claim is $\mu = 16.4$ ft, so $H_0: \mu = 16.4$ ft.

(b) You want to know if the average is getting larger, so $H_1: \mu > 16.4$ ft.

(c) You want to know if the average is getting smaller, so $H_1: \mu < 16.4$ ft.

(d) You want to know if the average is different from 16.4 ft, so $H_1: \mu \neq 16.4$ ft.

(e) Since part (b) is a right-tailed test, the critical region is on the right. Since part (c) is a left-tailed test, the critical region is on the left. Since part (d) is a two-tailed test, the critical region is on both sides of the mean.

8. (a) The claim is $\mu = 8.7$ sec, so $H_0: \mu = 8.7$ sec.

(b) You want to know if the average is longer (larger), so $H_1: \mu > 8.7$ sec.

(c) You want to know if the average is reduced (smaller), so $H_1: \mu < 8.7$ sec.

(d) Since part (b) is a right-tailed test, the critical region is on the right. Since part (c) is a left-tailed test, the critical region is on the left.

9. (a) The claim is $\mu = 288$ lb, so $H_0: \mu = 288$ lb.

(b) If you want to know if the average is higher (larger), $H_1: \mu > 288$ lb. Since this is a right-tailed test, the critical region is on the right.

If you want to know if the average is lower (smaller), $H_1: \mu < 288$ lb. Since this is a left-tailed test, the critical region is on the left.

If you want to know if the average is different, $H_1: \mu \neq 288$ lb. Since this is a two-tailed test, the critical region is in both tails.

Section 10.2

1. $H_0: \mu = 16.4$ ft
$H_1: \mu < 16.4$ ft

Since < is in H_1, a left-tailed test is used.
Since the sample size $n = 36$ is large, the sampling distribution of $\bar{x}$ is approximately normal by the central limit theorem, and we can estimate σ by s.

For $\alpha = 0.01$, the critical value is $z_0 = -2.33$.

$$z = \frac{\bar{x} - \mu}{\sigma/\sqrt{n}} = \frac{15.1 - 16.4}{3.2/\sqrt{36}} = -2.44$$

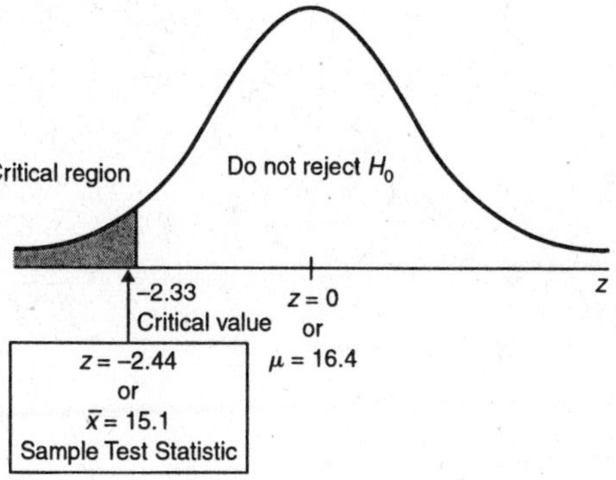

The sample test statistic falls in the critical region (−2.44 < −2.33). Therefore, we reject H_0. We conclude that the storm is lessening. The data are statistically significant.

2. H_0: $\mu = 38$ hr

H_1: $\mu < 38$ hr

Since < is in H_1, a left-tailed test is used.

Since the sample size $n = 47$ is large, the sampling distribution of $\bar{x}$ is approximately normal by the central limit theorem, and we can estimate σ by s.

For $\alpha = 0.01$, the critical value is $z_0 = -2.33$.

$$z = \frac{\bar{x} - \mu}{\sigma/\sqrt{n}} = \frac{37.5 - 38}{1.2/\sqrt{47}} = -2.86$$

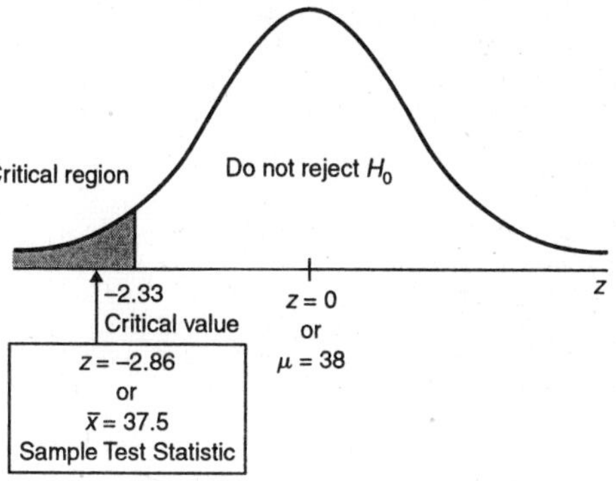

The sample test statistic falls in the critical region (−2.86 < −2.33). Therefore, we reject H_0. We conclude that the average assembly time is less. The data are statistically significant.

3. H_0: $\mu = 31.8$ calls/day

H_1: $\mu \neq 31.8$ calls/day

Since ≠ is in H_1, a two-tailed test is used.

Since the sample size $n = 63$ is large, the sampling distribution of $\bar{x}$ is approximately normal by the central limit theorem, and we can estimate σ by s.

For $\alpha = 0.01$, the critical value is $z_0 = \pm 2.58$.

$$z = \frac{\bar{x} - \mu}{\sigma/\sqrt{n}} = \frac{28.5 - 31.8}{10.7/\sqrt{63}} = -2.45$$

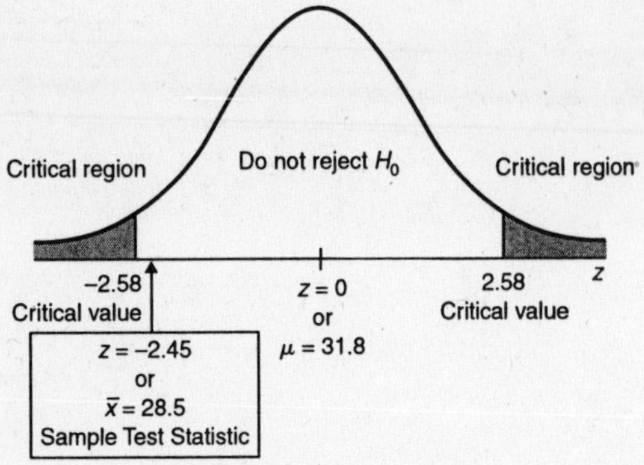

The sample test statistic does not fall in the critical region (−2.58 < −2.45 < 2.58). Therefore, we do not reject H_0. There is not enough evidence to conclude that the mean number of messages has changed. The data are not statistically significant.

4. H_0: $\mu = 3218$

H_1: $\mu > 3218$

Since > is in H_1, a right-tailed test is used.

Since the sample size $n = 42$ is large, the sampling distribution of $\bar{x}$ is approximately normal by the central limit theorem, and we can estimate σ by s.

For $\alpha = 0.01$, the critical value is $z_0 = 2.33$.

$$z = \frac{\bar{x} - \mu}{\sigma/\sqrt{n}} = \frac{3392 - 3218}{287/\sqrt{42}} = 3.93$$

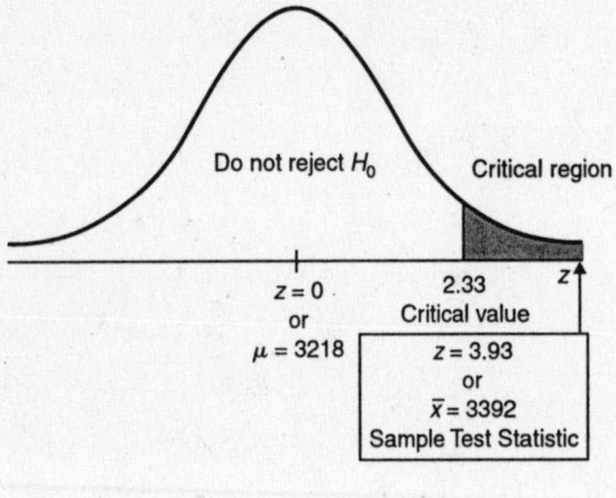

The sample test statistic falls in the critical region (3.93 > 2.33). Therefore, we reject H_0. We conclude the average number of people entering the store each day has increased. The data are statistically significant.

5. H_0: $\mu = \$4.75$

H_1: $\mu > \$4.75$

Since > is in H_1, a right-tailed test is used.

Since the sample size $n = 52$ is large, the sampling distribution of $\bar{x}$ is approximately normal by the central limit theorem, and we can estimate σ by s.

For $\alpha = 0.01$, the critical value is $z_0 = 2.33$.

$$z = \frac{\bar{x} - \mu}{\sigma/\sqrt{n}} = \frac{5.25 - 4.75}{1.15/\sqrt{52}} = 3.14$$

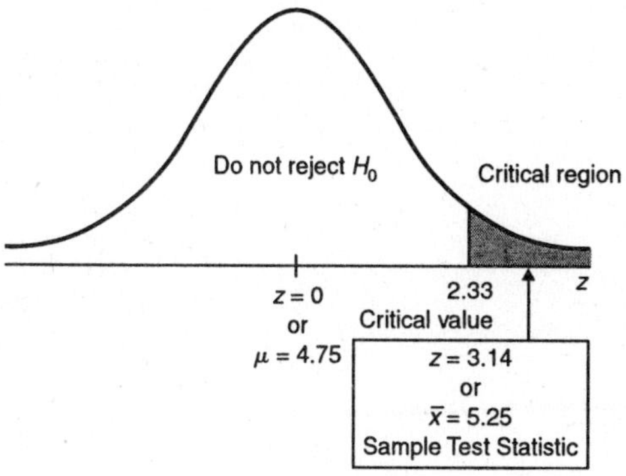

The sample test statistic falls in the critical region (3.14 > 2.33). Therefore, we reject H_0. We conclude that her average tip is more than \$4.75. The data are statistically significant.

6. H_0: $\mu = 17.2\%$

   H_1: $\mu < 17.2\%$

   Since < is in H_1, a left-tailed test is used.

   Since the sample size $n = 50$ is large, the sampling distribution of $\bar{x}$ is approximately normal by the central limit theorem, and we can estimate σ by s.

   For $\alpha = 0.05$, the critical value is $z_0 = -1.645$.

   $$z = \frac{\bar{x} - \mu}{\sigma/\sqrt{n}} = \frac{15.8 - 17.2}{5.3/\sqrt{50}} = -1.87$$

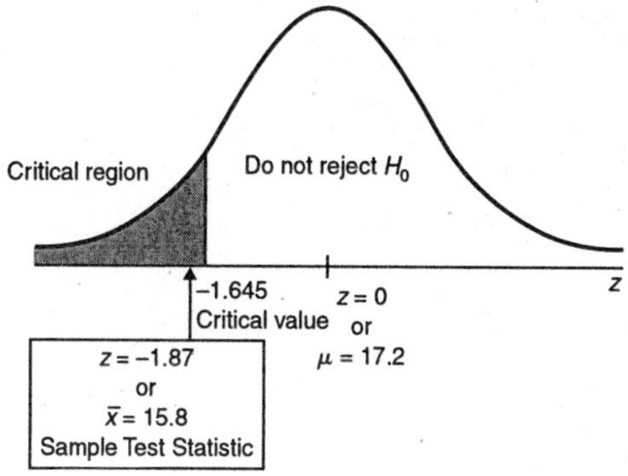

   The sample test statistic falls in the critical region (−1.87 < −1.645). Therefore, we reject H_0. We conclude that this hay has lower protein content. The data are statistically significant.

7. H_0: $\mu = 10.2$ sec

   H_1: $\mu < 10.2$ sec

   Since < is in H_1, a left-tailed test is used.

   Since the sample size $n = 41$ is large, the sampling distribution of $\bar{x}$ is approximately normal by the central limit theorem, and we can estimate σ by s.

   For $\alpha = 0.05$, the critical value is $z_0 = -1.645$.

   $$z = \frac{\bar{x} - \mu}{\sigma/\sqrt{n}} = \frac{9.7 - 10.2}{2.1/\sqrt{41}} = -1.52$$

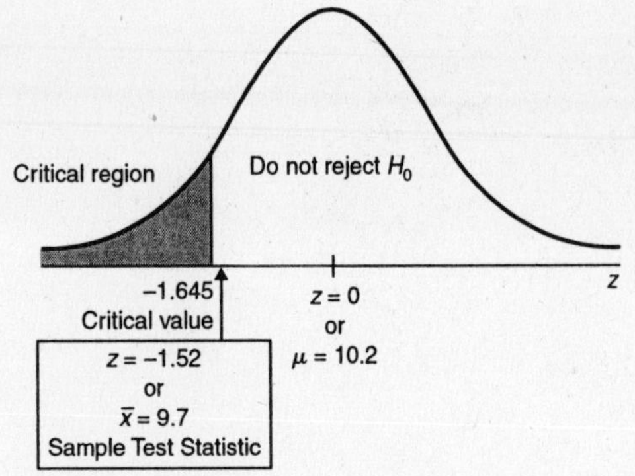

The sample test statistic does not fall in the critical region (−1.645 < −1.52). Therefore, we do not reject H_0. There is not enough evidence to conclude that the mean acceleration time is less. The data are not statistically significant.

8. H_0: $\mu = 159$ ft

H_1: $\mu < 159$ ft

Since < is in H_1, a left-tailed test is used.

Since the sample size $n = 45$ is large, the sampling distribution of $\bar{x}$ is approximately normal by the central limit theorem, and we can estimate σ by s.

For $\alpha = 0.01$, the critical value is $z_0 = -2.33$.

$$z = \frac{\bar{x} - \mu}{\sigma/\sqrt{n}} = \frac{148 - 159}{23.5/\sqrt{45}} = -3.14$$

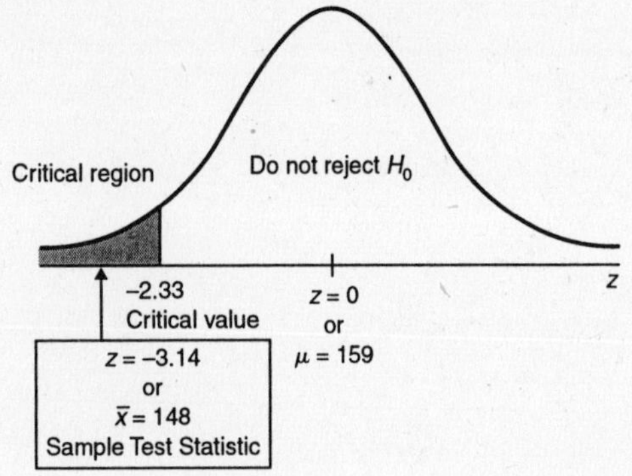

The sample test statistic falls in the critical region (−3.14 < −2.33). Therefore, we reject H_0. We conclude the mean braking distance is reduced for the new tire tread. The data are statistically significant.

9. H_0: $\mu = 19.0$ ml/dl

H_1: $\mu > 19.0$ ml/dl

Since > is in H_1, a right-tailed test is used.

Since the sample size $n = 48$ is large, the sampling distribution of $\bar{x}$ is approximately normal by the central limit theorem, and we can estimate σ by s.

For $\alpha = 0.01$, the critical value is $z_0 = 2.33$.

$$z = \frac{\bar{x} - \mu}{\sigma/\sqrt{n}} = \frac{20.7 - 19.0}{9.9/\sqrt{48}} = 1.19$$

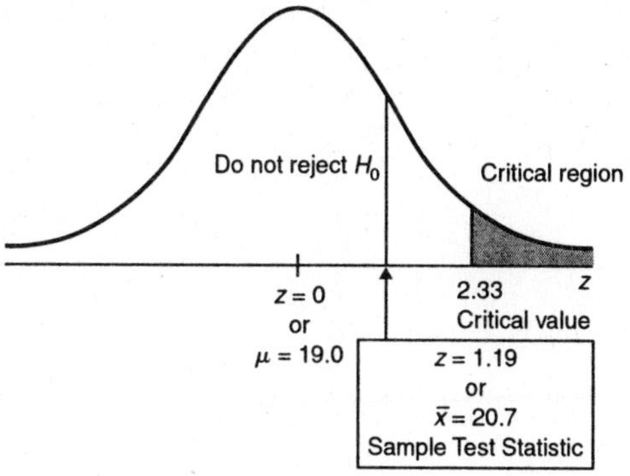

The sample test statistic does not fall in the critical region (1.19 < 2.33). Therefore, we do not reject H_0. There is not sufficient evidence to conclude that the average oxygen capacity has increased. The data are not statistically significant.

10. H_0: $\mu = \$1,789,556$

H_1: $\mu \neq \$1,789,556$

Since ≠ is in H_1, a two-tailed test is used.

Since the sample size $n = 35$ is large, the sampling distribution of $\overline{x}$ is approximately normal by the central limit theorem, and we can estimate σ by s.

For α = 0.01, the critical values are $z_0 = \pm 2.58$.

$$z = \frac{\overline{x} - \mu}{\sigma/\sqrt{n}} = \frac{1,621,726 - 1,789,556}{591,218/\sqrt{35}} = -1.68$$

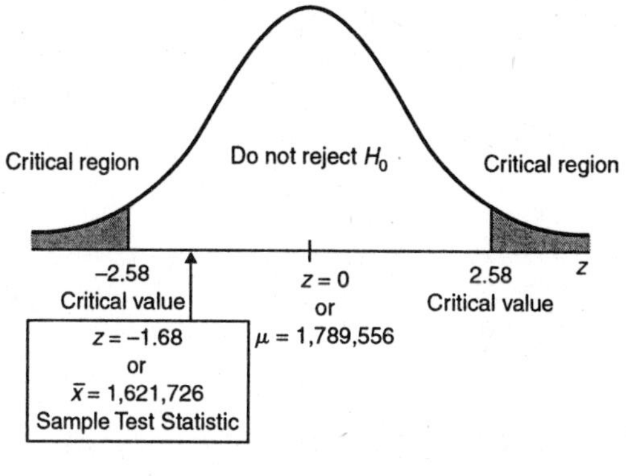

The sample test statistic does not fall in the critical region (−2.58 < −1.68 < 2.58). Therefore, we do not reject H_0. There is not sufficient evidence to conclude that the average salary of major league baseball players in Florida is different from the national average. The data are not statistically significant.

11. H_0: $\mu = 7.4$ pH

H_1: $\mu \neq 7.4$ pH

Since ≠ is in H_1, a two-tailed test is used.

Since the sample size $n = 33$ is large, the sampling distribution of $\overline{x}$ is approximately normal by the central limit theorem, and we can estimate σ by s.

For α = 0.05, the critical values are $z_0 = \pm 1.96$.

$$z = \frac{\overline{x} - \mu}{\sigma/\sqrt{n}} = \frac{8.1 - 7.4}{1.9/\sqrt{33}} = 2.12$$

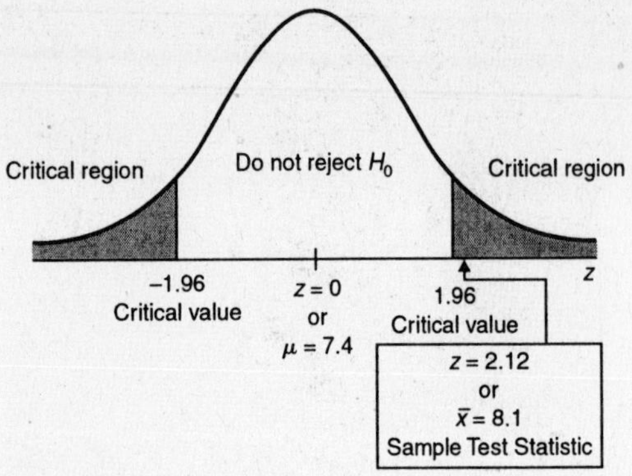

The sample test statistic falls in the critical region $(2.12 > 1.96)$. Therefore, we reject H_0. We conclude that the drug has changed the mean pH of the blood. The data are statistically significant.

12. H_0: $\mu = 0.25$ gal
H_1: $\mu > 0.25$ gal

Since > is in H_1, a right-tailed test is used.

Since the sample size $n = 100$ is large, the sampling distribution of $\bar{x}$ is approximately normal by the central limit theorem, and we can estimate σ by s.

For $\alpha = 0.05$, the critical value is $z_0 = 1.645$.

$$z = \frac{\bar{x} - \mu}{\sigma/\sqrt{n}} = \frac{0.28 - 0.25}{0.10/\sqrt{100}} = 3.00$$

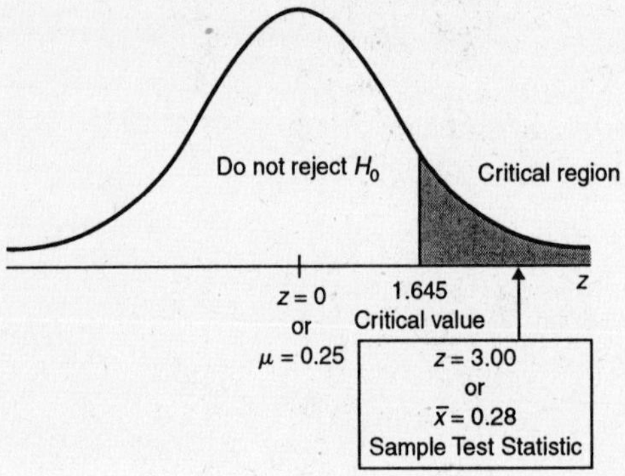

The sample test statistic falls in the critical region $(3.00 > 1.645)$. Therefore, we reject H_0. We conclude that the supplier's claim is too low. The data are statistically significant.

13. The mean and the standard deviation round to the values given.

H_0: $\mu = \$13.9$ thousand

H_1: $\mu > \$13.9$ thousand

Since > is in H_1, a right-tailed test is used.

Since the sample size $n = 50$ is large, the sampling distribution of $\overline{x}$ is approximately normal by the central limit theorem, and we can estimate σ by s.

For $\alpha = 0.05$, the critical value is $z_0 = 1.645$.

$$z = \frac{\overline{x} - \mu}{\sigma/\sqrt{n}} = \frac{16.58 - 13.9}{8.11/\sqrt{50}} = 2.34$$

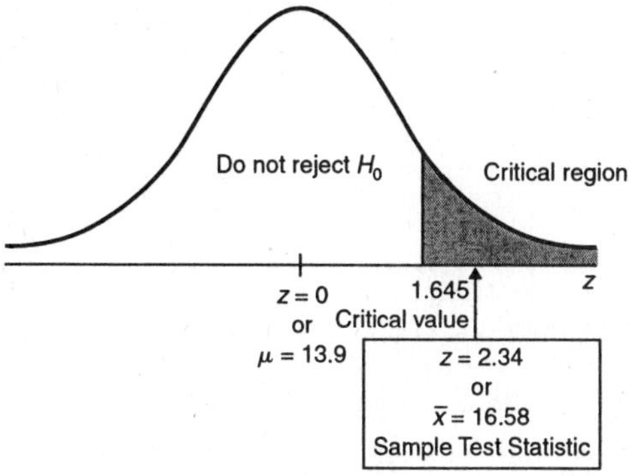

The sample test statistic falls in the critical region (2.34 > 1.645). Therefore, we reject H_0. We conclude that the mean franchise costs for pizza businesses are higher. The data are statistically significant.

14. **(a)** One-tailed test; if the null hypothesis is rejected, then one has the additional information from the test that the true population mean is larger/smaller than stated in H_0. Also, the critical value for a one-tailed test is nearer to zero than the critical values of a two-tailed test, meaning that *if* one is able to correctly surmise the direction of H_1, there is a greater chance of rejecting H_0. For example, for a right-tailed test at $\alpha = 0.05$, $z = 1.7$ would result in rejecting H_0, but in a two-tailed test at $\alpha = 0.05$ for the same $z = 1.7$, one would fail to reject H_0.

(b) Two-tailed test; for a given α-level the absolute value of the critical value for a one-tailed test is less than that of a two-tailed test, making the one-tailed test more likely to reject H_0.

(c) Yes. The rejection regions are different for one- and two-tailed tests.

15. Essay or class discussion.

16. **(a)** H_0: $\mu = 20$

H_1: $\mu \neq 20$

For $\alpha = 0.01$, $c = 1 - 0.01 = 0.99$, $s = 4$, and the critical value $z_c = 2.58$.

$$E \approx z_c \frac{s}{\sqrt{n}} = 2.58 \frac{4}{\sqrt{36}} = 1.72$$

$$\overline{x} - E < \mu < \overline{x} + E$$

$$22 - 1.72 < \mu < 22 + 1.72$$

$$20.28 < \mu < 23.72$$

The hypothesized mean $\mu = 20$ is not in the interval. Therefore, we reject H_0.

(b) For $\alpha = 0.01$, the two-tailed test's critical values are $z_0 = \pm 2.58$. Because $n = 36$ is large, the sampling distribution of $\bar{x}$ is approximately normal by the central limit theorem, and we can estimate σ by s.

$$z = \frac{\bar{x} - \mu}{\sigma/\sqrt{n}} = \frac{22 - 20}{4/\sqrt{36}} = 3.00$$

Since the sample test statistic falls inside the critical region (3.00 > 2.58), we reject H_0. The results are the same.

17. (a) H_0: $\mu = 21$

H_1: $\mu \neq 21$

For $\alpha = 0.01$, $c = 1 - 0.01 = 0.99$, $s = 4$, and the critical value $z_c = 2.58$.

$$E \approx z_c \frac{s}{\sqrt{n}} = 2.58 \frac{4}{\sqrt{36}} = 1.72$$

$$\bar{x} - E < \mu < \bar{x} + E$$

$$22 - 1.72 < \mu < 22 + 1.72$$

$$20.28 < \mu < 23.72$$

The hypothesized mean $\mu = 21$ falls into the confidence interval. Therefore, we do not reject H_0.

(b) For $\alpha = 0.01$, the two-tailed test's critical values are $z_0 = \pm 2.58$. Because $n = 36$ is large, the sampling distribution of $\bar{x}$ is approximately normal by the central limit theorem, and we can estimate σ by s.

$$z = \frac{\bar{x} - \mu}{\sigma/\sqrt{n}} = \frac{22 - 21}{4/\sqrt{36}} = 1.50$$

Since the sample test statistic falls outside the critical region (−2.58 < 1.50 < 2.58), we do not reject H_0. The results are the same.

Section 10.3

1. H_0: $\mu = 5$

H_1: $\mu > 5$

$$z = \frac{\bar{x} - \mu}{\sigma/\sqrt{n}} = \frac{6.1 - 5}{2.5/\sqrt{40}} = 2.78$$

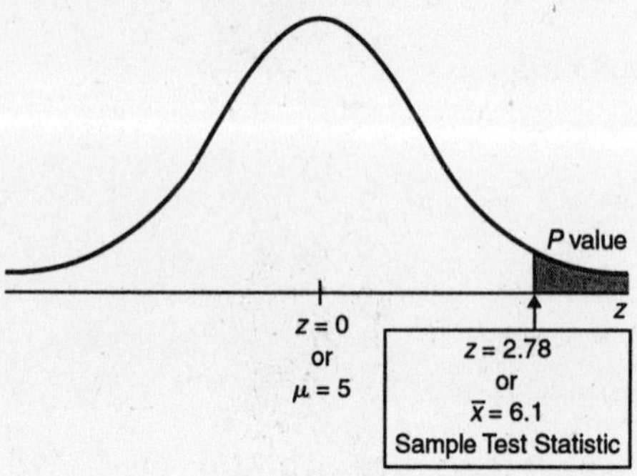

$$\begin{aligned} P \text{ value} &= P(\bar{x} \ge 6.1) \\ &= P(z \ge 2.78) \\ &= 1 - 0.9973 \\ &= 0.0027 \end{aligned}$$

Since 0.0027 < 0.01, the data are significant at the 1% level.

2. H_0: $\mu = 53.1$

H_1: $\mu < 53.1$

$$z = \frac{\bar{x} - \mu}{\sigma/\sqrt{n}} = \frac{52.7 - 53.1}{4.5/\sqrt{41}} = -0.57$$

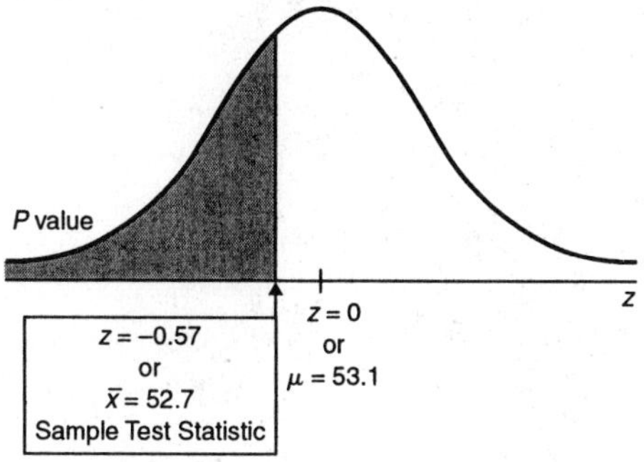

$$\begin{aligned} P \text{ value} &= P(\bar{x} \le 52.7) \\ &= P(z \le -0.57) \\ &= 0.2843 \end{aligned}$$

Since 0.2843 > 0.01, the data are not significant at the 1% level.

3. H_0: $\mu = 21.7$

H_1: $\mu \ne 21.7$

$$z = \frac{\bar{x} - \mu}{\sigma/\sqrt{n}} = \frac{20.5 - 21.7}{6.8/\sqrt{45}} = -1.18$$

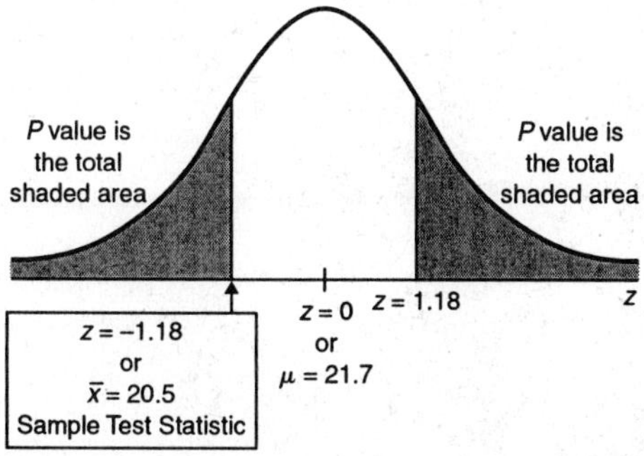

$$\begin{aligned}P\text{ value} &= 2P(\overline{x} \le 20.5)\\ &= 2P(z \le -1.18)\\ &= 2(0.1190)\\ &= 0.2380\end{aligned}$$

Since 0.2380 > 0.05, the data are not significant at the 5% level.

4. H_0: $\mu = 18.7$
H_1: $\mu \ne 18.7$

$$z = \frac{\overline{x} - \mu}{\sigma/\sqrt{n}} = \frac{19.1 - 18.7}{5.2/\sqrt{32}} = 0.44$$

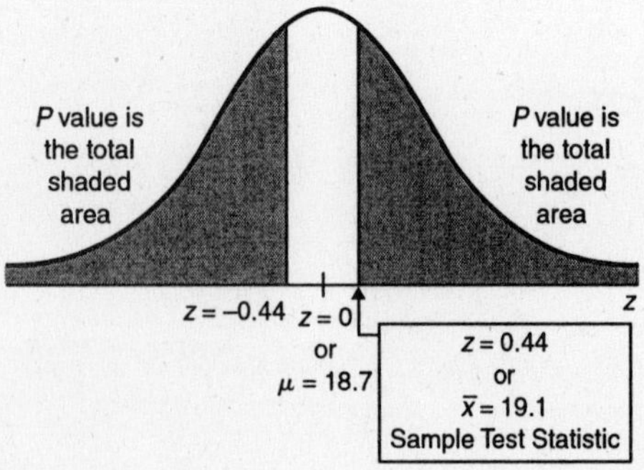

$$\begin{aligned}P\text{ value} &= 2P(\overline{x} \ge 19.1)\\ &= 2P(z \ge 0.44)\\ &= 2(1 - 0.6700)\\ &= 2(0.3300)\\ &= 0.6600\end{aligned}$$

Since 0.6600 > 0.05, the data are not significant at the 5% level.

5. H_0: $\mu = 1.75$ yr
H_1: $\mu > 1.75$ yr

$$z = \frac{\overline{x} - \mu}{\sigma/\sqrt{n}} = \frac{2.05 - 1.75}{0.82/\sqrt{68}} = 3.02$$

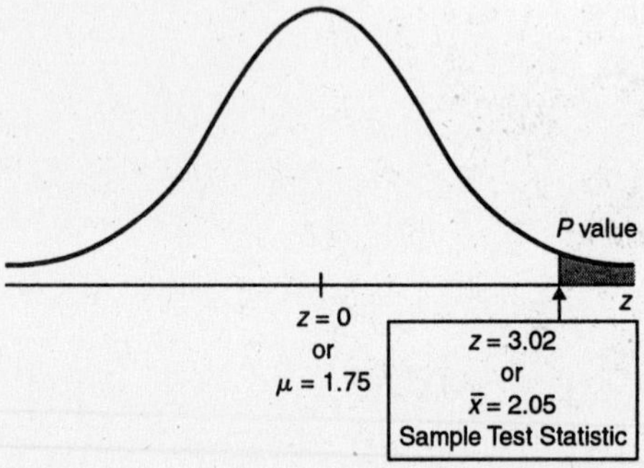

$$\begin{aligned} P \text{ value} &= P(\bar{x} \geq 2.05) \\ &= P(z \geq 3.02) \\ &= 1 - 0.9987 \\ &= 0.0013 \end{aligned}$$

Since $0.0013 < 0.01$, the data are significant at the 1% level. Coyotes in this region appear to live longer.

6. H_0: $\mu = 12 \text{ m}^2$

H_1: $\mu \neq 12\ m^2$

$$z = \frac{\bar{x} - \mu}{\sigma/\sqrt{n}} = \frac{12.3 - 12}{3.4/\sqrt{56}} = 0.66$$

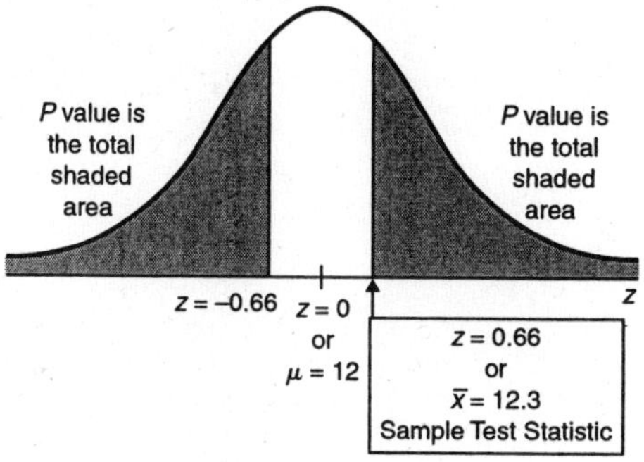

$$\begin{aligned} P \text{ value} &= 2P(\bar{x} \geq 12.3) \\ &= 2P(z \geq 0.66) \\ &= 2(1 - 0.7454) \\ &= 2(0.2546) \\ &= 0.5092 \end{aligned}$$

Since $0.5092 > 0.01$, the data are not significant at the 1% level. The evidence does not support the idea that Mesa Verdi Rivas have an average floor space size different from 12 square meters.

7. H_0: $\mu = 19$ in.

H_1: $\mu < 19$ in.

$$z = \frac{\bar{x} - \mu}{\sigma/\sqrt{n}} = \frac{18.7 - 19}{3.2/\sqrt{73}} = -0.80$$

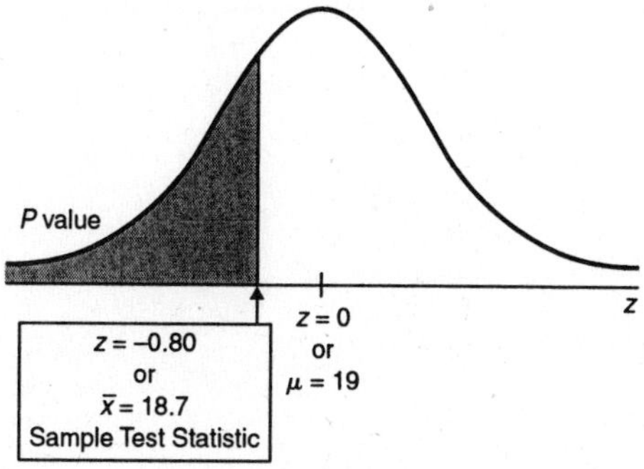

$$P \text{ value} = P(\bar{x} \le 18.7)$$
$$= P(z \le -0.80)$$
$$= 0.2119$$

Since 0.2119 > 0.05, the data are not significant at the 5% level. The data support the new hypothesis that average trout length is 19 inches.

8. H_0: $\mu = \$61{,}400$

H_1: $\mu < \$61{,}400$

$$z = \frac{\bar{x} - \mu}{\sigma/\sqrt{n}} = \frac{55{,}200 - 61{,}400}{18{,}800/\sqrt{34}} = -1.92$$

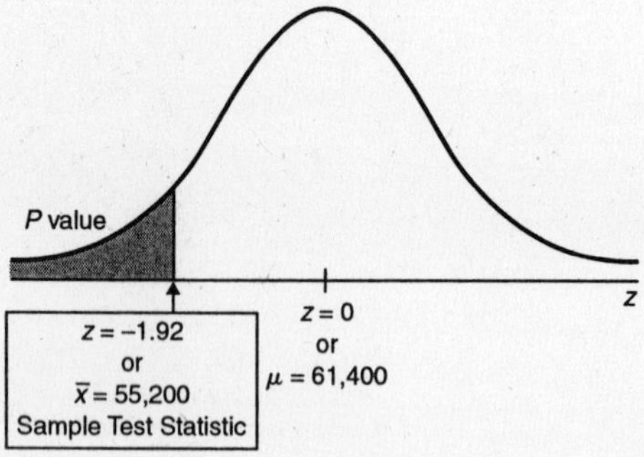

$$P \text{ value} = P(\bar{x} \le 55{,}200)$$
$$= P(z \le -1.92)$$
$$= 0.0274$$

Since 0.0274 < 0.05, the data are significant at the 5% level. The data tend to support a smaller start-up cost than $61,400.

9. H_0: $\mu = \$15.35$

H_1: $\mu < \$15.35$

$$z = \frac{\bar{x} - \mu}{\sigma/\sqrt{n}} = \frac{11.85 - 15.35}{6.21/\sqrt{34}} = -3.29$$

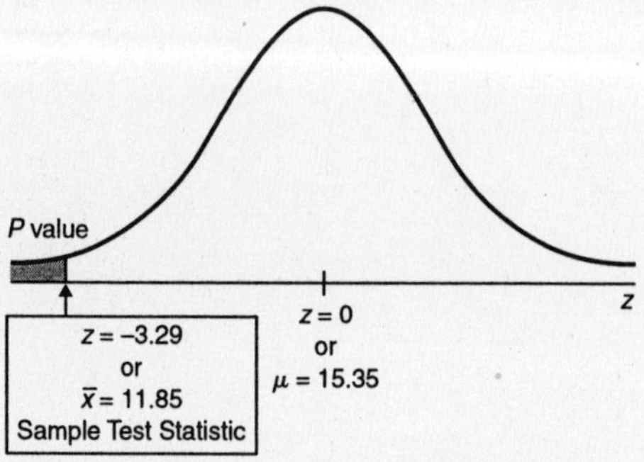

$$P \text{ value} = P(\bar{x} \le 11.85)$$
$$= P(z \le -3.29)$$
$$= 0.0005$$

Since 0.0005 < 0.05, the data are significant at the 5% level. The data support the claim that college students have lower daily ownership costs.

10. (a) H_0: $\mu = 19.5$ mpg

H_1: $\mu < 19.5$ mpg

The sample mean is $\bar{x} = 18.750$.
The P value is 0.047.
Reject H_0 for all $\alpha \ge 0.047$.

(b) H_0: $\mu = 19.5$ mpg

H_1: $\mu \ne 19.5$ mpg

The sample mean is $\bar{x} = 18.750$.
The P value is 0.094.
Reject H_0 for all $\alpha \ge 0.094$.

(c) The P value for a two-tailed test is twice that of the one-tailed test $\left[2(0.047) = 0.094\right]$.

Section 10.4

1. In this case we use the column headed by $\alpha' = 0.05$ and the row headed by $d.f. = n - 1 = 9 - 1 = 8$. This gives us $t = 1.860$. For a left-tailed test, we use symmetry of the distribution to get $t_0 = -1.860$.

2. In this case we use the column headed by $\alpha' = 0.01$ and the row headed by $d.f. = n - 1 = 13 - 1 = 12$. The critical value is $t_0 = 2.681$.

3. In this case we use the column headed by $\alpha'' = 0.01$ and the row headed by $d.f. = n - 1 = 24 - 1 = 23$. By the symmetry of the curve, the critical values are $t_0 = \pm 2.807$.

4. In this case we use the column headed by $\alpha' = 0.05$ and the row headed by $d.f. = n - 1 = 18 - 1 = 17$. This gives $t = 1.740$. For a left-tailed test, we use symmetry of the distribution to get $t_0 = -1.740$.

5. In this case we use the column headed by $\alpha'' = 0.05$ and the row headed by $d.f. = n - 1 = 12 - 1 = 11$. By the symmetry of the curve, the critical values are $t_0 = \pm 2.201$.

6. In this case we use the column headed by $\alpha' = 0.01$ and the row headed by $d.f. = n - 1 = 29 - 1 = 28$. The critical value is $t_0 = 2.467$.

7. (a) Answers are used in part (b).

(b) H_0: $\mu = 4.8$

H_1: $\mu < 4.8$

Since < is in H_1, a left-tailed test is used. Since the sample size is small and the data distribution is approximately normal, critical values are found using the Student's *t* distribution (use Table 5 in Appendix I). For a one-tailed test, look in the column headed by $\alpha' = 0.05$ and the row headed by $d.f. = 6 - 1 = 5$. The critical value is $t_0 = -2.015$.

$$t = \frac{\overline{x} - \mu}{s/\sqrt{n}} = \frac{4.07 - 4.8}{0.44/\sqrt{6}} = -4.06$$

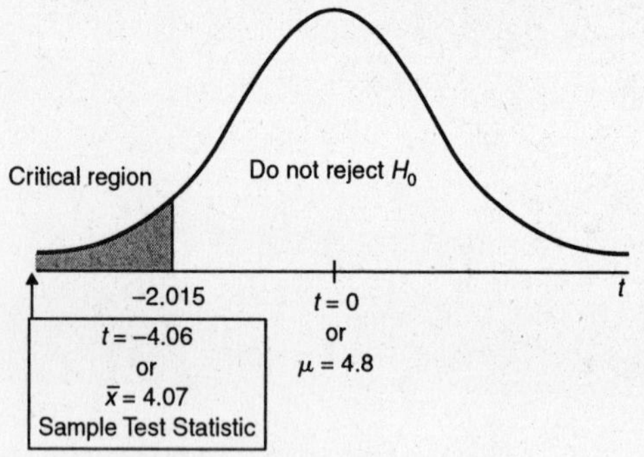

To find the *P*-value interval, use the α' value since our test is one-tailed and look in the row headed by $d.f. = 5$. We find that the sample *t* value $t = -4.06$ falls to the left of −4.032. Therefore, *P* value < 0.005.

The sample test statistic falls in the critical region (−4.06 < −2.015) and the *P* value is less than the level of significance $\alpha = 0.05$. Therefore, we reject H_0. We conclude at the 5% significance level that this patient's average red blood cell count is less than 4.8.

8. (a) Answers are used in part (b).

(b) H_0: $\mu = 14$

H_1: $\mu > 14$

Since > is in H_1, a right-tailed test is used. Since the sample size is small and the data distribution is approximately normal, critical values are found using the Student's *t* distribution (use Table 5 in Appendix I). For a one-tailed test, look in the column headed by $\alpha' = 0.01$ and the row headed by $d.f. = 12 - 1 = 11$. The critical value is $t_0 = 2.718$.

$$t = \frac{\overline{x} - \mu}{s/\sqrt{n}} = \frac{18.33 - 14}{2.71/\sqrt{12}} = 5.53$$

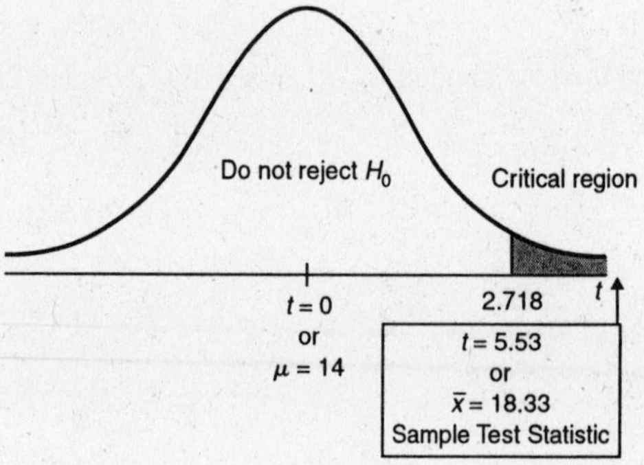

To find the P-value interval, use the α' values since our test is one-tailed and look in the row headed by $d.f. = 11$. We find that the sample t value, $t = 5.53$, falls to the right of 3.106. Therefore, P value < 0.005.

The sample test statistic falls in the critical region ($5.53 > 2.718$) and the P value is less than the level of significance $\alpha = 0.01$. Therefore, we reject H_0. We conclude that the population average HC for this patient is higher than 14.

9. (a) Answers are used in part (b).

(b) $H_0: \mu = 16.5$ days

$H_1: \mu \neq 16.5$ days

Since $\neq$ is in H_1, a two-tailed test is used. Since the sample size is small and the data distribution is approximately normal, critical values are found using the Student's t distribution (use Table 5 in Appendix I). For a two-tailed test, look in the column headed by $\alpha'' = 0.05$ and the row headed by $d.f. = 18 - 1 = 17$. The critical values are $t_0 = \pm 2.110$.

$$t = \frac{\overline{x} - \mu}{s/\sqrt{n}} = \frac{17.83 - 16.5}{2.20/\sqrt{18}} = 2.565$$

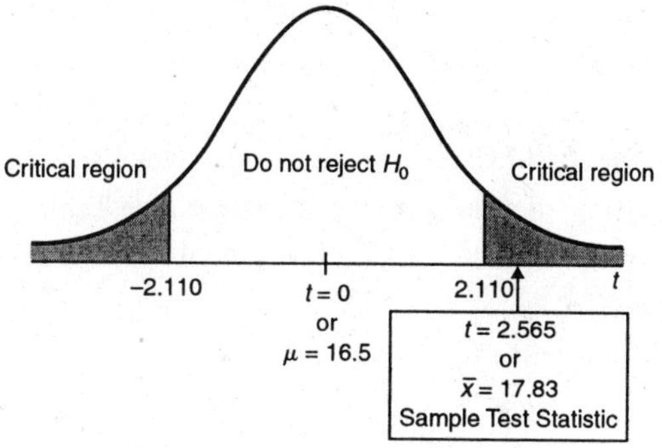

To find the P-value interval, use the α'' values since our test is two-tailed and look in the row headed by $d.f. = 17$. We find that the sample t value $t = 2.565$ falls between 2.110 and 2.567. Therefore, $0.02 < P$ value < 0.05.

The sample test statistic falls in the critical region ($2.565 > 2.110$) and the P value is less than the level of significance $\alpha = 0.05$. Therefore, we reject H_0. We conclude at the 5% significance level that the mean incubation time above 8000 feet is different from 16.5 days.

10. (a) Answers are used in part (b).

(b) $H_0: \mu = 8.8$

$H_1: \mu \neq 8.8$

Since $\neq$ is in H_1, a two-tailed test is used. Since the sample size is small and the data distribution is approximately normal, critical values are found using the Student's t distribution (use Table 5 in Appendix I). For a two-tailed test, look in the column headed by $\alpha'' = 0.05$ and the row headed by $d.f. = 14 - 1 = 13$. The critical values are $t_0 = \pm 2.160$.

$$t = \frac{\overline{x} - \mu}{s/\sqrt{n}} = \frac{7.36 - 8.8}{4.03/\sqrt{14}} = -1.34$$

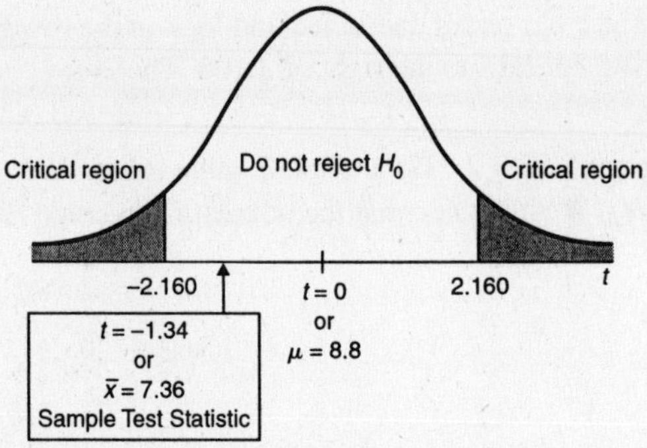

To find the *P*-value interval, use the α'' values since our test is two-tailed and look in the row headed by *d.f.* = 13. We find that the sample *t* value $t = -1.34$ falls between −1.204 and −1.350. Therefore, $0.200 < P$ value < 0.250.

The sample test statistic does not fall in the critical region ($-2.160 < -1.34 < 2.160$) and the *P* value is greater than the level of significance $\alpha = 0.05$. Therefore, we fail to reject H_0. We cannot conclude that the average catch is different from 8.8 fish per day.

11. (a) Answers are used in part (b).

(b) $H_0: \mu = 1300$

$H_1: \mu \neq 1300$

Since ≠ is in H_1, a two-tailed test is used. Since the sample size is small and the data distribution is approximately normal, critical values are found using the Student's *t* distribution (use Table 5 in Appendix I). For a two-tailed test, look in the column headed by $\alpha'' = 0.01$ and the row headed by *d.f.* $= 10 - 1 = 9$. The critical values are $t_0 = \pm 3.250$.

$$t = \frac{\bar{x} - \mu}{s/\sqrt{n}} = \frac{1268 - 1300}{37.29/\sqrt{10}} = -2.71$$

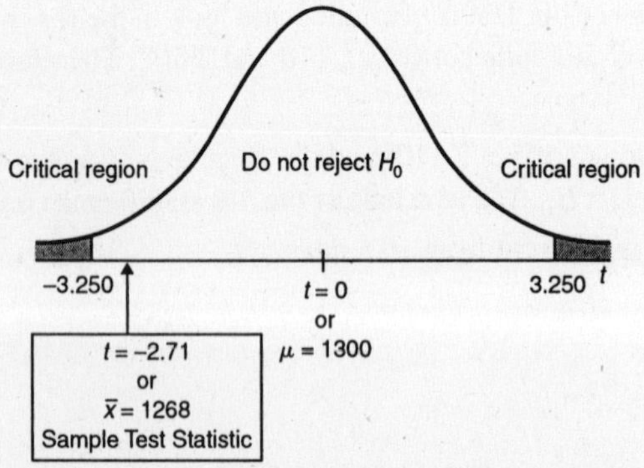

To find the *P*-value interval, use the α'' values since our test is two-tailed and look in the row headed by *d.f.* = 9. We find that the sample *t* value, $t = -2.71$, falls between −2.262 and −2.821. Therefore, $0.020 < P$ value < 0.050.

The sample test statistic does not fall in the critical region ($-3.250 < -2.71 < 3.250$) and the *P* value is greater than the level of significance $\alpha = 0.01$. Therefore, do not reject H_0. There is not enough evidence to conclude that the population mean of tree ring dates is different from 1300.

12. (a) Answers are used in part (b).

(b) $H_0: \mu = 67$

$H_1: \mu \neq 67$

Since $\neq$ is in H_1, a two-tailed test is used. Since the sample size is small and the data distribution is approximately normal, critical values are found using the Student's t distribution (use Table 5 in Appendix I). For a two-tailed test, look in the column headed by $\alpha'' = 0.01$ and the row headed by $d.f. = 16 - 1 = 15$. The critical values are $t_0 = \pm 2.947$.

$$t = \frac{\bar{x} - \mu}{s/\sqrt{n}} = \frac{61.8 - 67}{10.6/\sqrt{16}} = -1.962$$

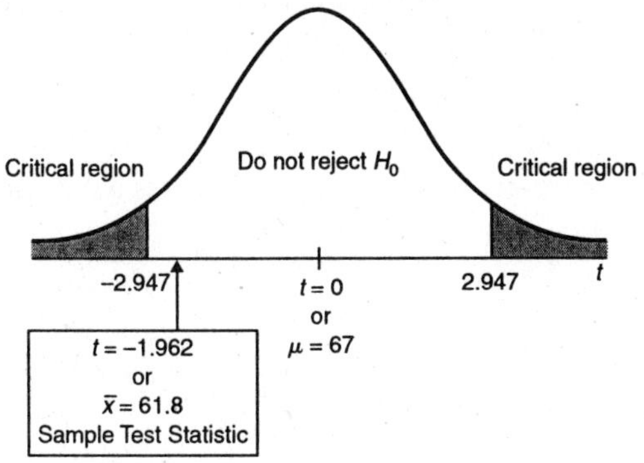

To find the P-value interval, use the α'' values since our test is two-tailed and look in the row headed by $d.f. = 15$. We find that the sample t value, $t = -1.962$, falls between -1.753 and -2.131. Therefore, $0.050 < P \text{ value} < 0.100$.

The sample test statistic does not fall in the critical region ($-2.947 < -1.962 < 2.947$) and the P value is greater than the level of significance $\alpha = 0.01$. Therefore, do not reject H_0. There is not enough evidence to conclude that the average thickness of slab avalanches in Vail is different from those in Canada.

13. (a) Answers are used in part (b).

(b) $H_0: \mu = 77$ yr

$H_1: \mu < 77$ yr

Since < is in H_1, a left-tailed test is used. Since the sample size is small and the data distribution is approximately normal, critical values are found using the Student's t distribution (use Table 5 in Appendix I). For a one-tailed test, look in the column headed by $\alpha' = 0.05$ and the row headed by $d.f. = 20 - 1 = 19$. The critical value is $t_0 = -1.729$.

$$t = \frac{\bar{x} - \mu}{s/\sqrt{n}} = \frac{74.45 - 77}{18.09/\sqrt{20}} = -0.6304$$

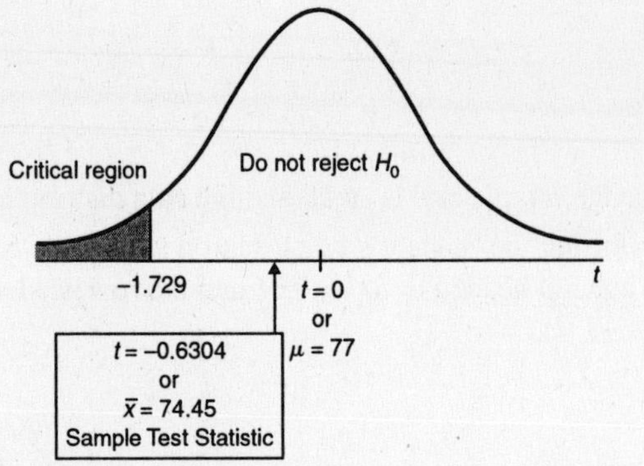

To find the P-value interval, use the α' values since our test is one-tailed and look in the row headed by $d.f. = 19$. We find that the sample t value, $t = -0.6304$, falls to the right of −1.187. Therefore, P value > 0.125.

The sample test statistic does not fall in the critical region (−1.729 < −0.6304) and the P value is greater than the level of significance $\alpha = 0.05$. Therefore, do not reject H_0. There is not enough evidence to conclude that the population mean life span is less than 77 years.

14. (a) Answers are used in part (b).

(b) $H_0: \mu = 40$

$H_1: \mu \neq 40$

Since ≠ is in H_1, a two-tailed test is used. Since the sample size is small and the data distribution is approximately normal, critical values are found using the Student's t distribution (use Table 5 in Appendix I). For a two-tailed test, look in the column headed by $\alpha'' = 0.05$ and the row headed by $d.f. = 6 - 1 = 5$. The critical values are $\pm t_0 = \pm 2.571$.

$$t = \frac{\bar{x} - \mu}{s/\sqrt{n}} = \frac{36.5 - 40}{4.2/\sqrt{6}} = -2.04$$

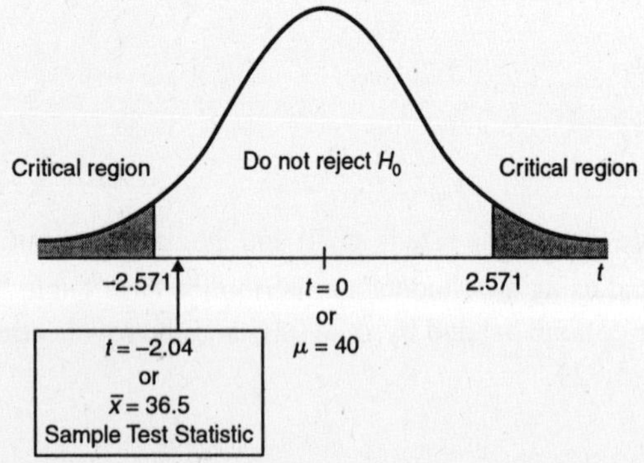

To find the P-value interval, use the α'' values since our test is two-tailed and look in the row headed by $d.f. = 5$. We find that the sample t value, $t = -2.04$, falls between −2.015 and −2.571. Therefore, 0.050 < P value < 0.100.

The sample test statistic does not fall in the critical region (−2.571 < −2.04 < 2.571) and the P value is greater than the level of significance $\alpha = 0.05$. Therefore, do not reject H_0. There is not enough evidence to conclude that the population average heart rate of the lion is different from 40 beats per minute.

15. (a) Answers are used in part (b).

(b) H_0: $\mu = 7.3$

H_1: $\mu > 7.3$

Since > is in H_1, a right-tailed test is used. Since the sample size is small and the data distribution is approximately normal, critical values are found using the Student's t distribution (use Table 5 in Appendix I). For a one-tailed test, look in the column headed by $\alpha' = 0.05$ and the row headed by $d.f. = 20 - 1 = 19$. The critical value is $t_0 = 1.729$.

$$t = \frac{\bar{x} - \mu}{s/\sqrt{n}} = \frac{8.1 - 7.3}{1.4/\sqrt{20}} = 2.556$$

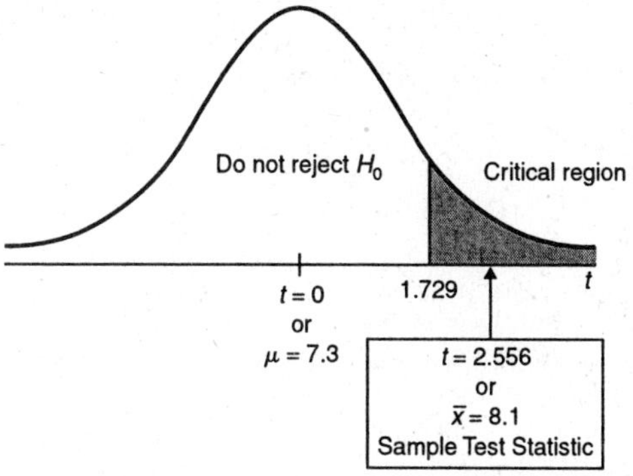

To find the P-value interval, use the α' values since our test is one-tailed and look in the row headed by $d.f. = 19$. We find that the sample t value, $t = 2.556$, falls between 2.539 and 2.861. Therefore, $0.005 < P \text{ value} < 0.010$.

The sample test statistic falls in the critical region (2.555 > 1.729) and the P value is less than the level of significance $\alpha = 0.05$. Therefore, reject H_0. We conclude that the evidence supports the claim that the average time women with children spend shopping in houseware stores in Cherry Creek Mall is higher than the national average.

Section 10.5

1. H_0: $p = 0.70$

H_1: $p \neq 0.70$

Since ≠ is in H_1, a two-tailed test is used. The $\hat{p}$ distribution is approximately normal when n is sufficiently large, which it is here, since $np = 32(0.7) = 22.4$ and $nq = 32(0.3) = 9.6$ are both > 5. For $\alpha = 0.01$, the critical values are $z_0 = \pm 2.58$.

$$\hat{p} = \frac{r}{n} = \frac{24}{32} = 0.75$$

$$z = \frac{\hat{p} - p}{\sqrt{\frac{pq}{n}}} = \frac{0.75 - 0.70}{\sqrt{\frac{0.70(0.30)}{32}}} = 0.62$$

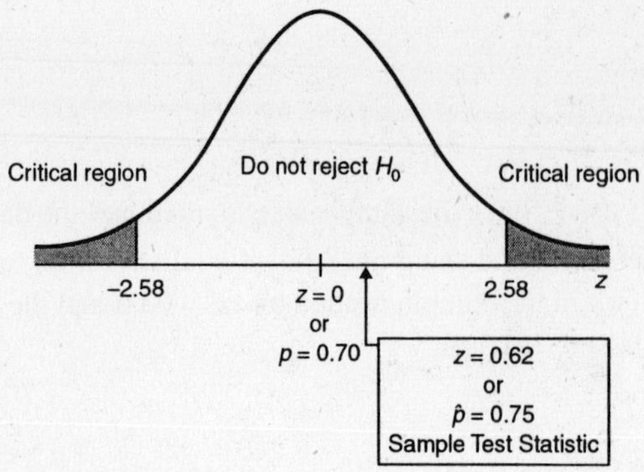

Next, find the P value associated with $z = 0.62$ and a two-tailed test.

$$\begin{aligned} P \text{ value} &= 2P(z \geq 0.62) \\ &= 2(1-0.7324) \\ &= 2(0.2676) \\ &= 0.5352 \end{aligned}$$

Since the sample z value falls outside the critical region and the P value is greater than the level of significance, $\alpha = 0.01$, we do not reject H_0. There is not enough evidence to conclude that the population proportion of such arrests is different from 0.70.

2. $H_0: p = 0.67$
$H_1: p < 0.67$

Since < is in H_1, a left-tailed test is used. The $\hat{p}$ distribution is approximately normal when n is sufficiently large, which it is here, since $np = 38(0.67) = 25.46$ and $nq = 38(0.33) = 12.54$ are both > 5. For $\alpha = 0.05$, the critical value is $z_0 = -1.645$.

$$\hat{p} = \frac{r}{n} = \frac{21}{38} = 0.5526$$

$$z = \frac{\hat{p} - p}{\sqrt{\frac{pq}{n}}} = \frac{0.5526 - 0.67}{\sqrt{\frac{0.67(0.33)}{38}}} = -1.54$$

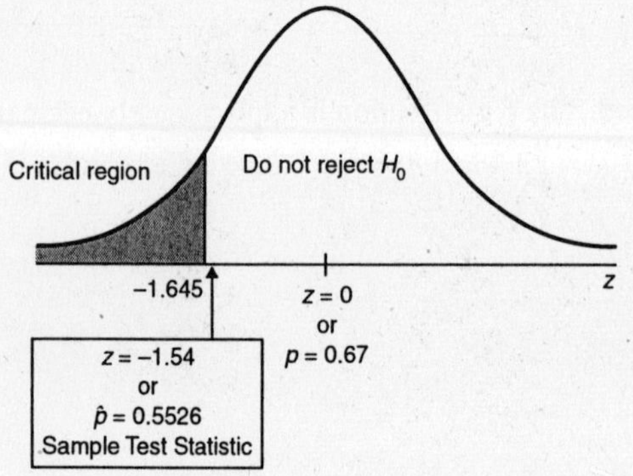

Next, find the P value associated with $z = -1.54$ and a one-tailed test.

$$P \text{ value} = P(z \le -1.54)$$
$$= 0.0618$$

Since the sample z value falls outside the critical region and the P value is greater than the level of significance, $\alpha = 0.05$, we do not reject H_0. There is not enough evidence to conclude that the population proportion of women athletes who graduate at CU Boulder is now less than 67%.

3. $H_0\colon p = 0.77$

$H_1\colon p < 0.77$

Since < is in H_1, a left-tailed test is used. The $\hat{p}$ distribution is approximately normal when n is sufficiently large, which it is here, because $np = 27(0.77) = 20.79$ and $nq = 27(0.23) = 6.21$ are both > 5. For $\alpha = 0.01$, the critical value is $z_0 = -2.3$.

$$\hat{p} = \frac{r}{n} = \frac{15}{27} = 0.5556$$

$$z = \frac{\hat{p} - p}{\sqrt{\frac{pq}{n}}} = \frac{0.5556 - 0.77}{\sqrt{\frac{0.77(0.23)}{27}}} = -2.65$$

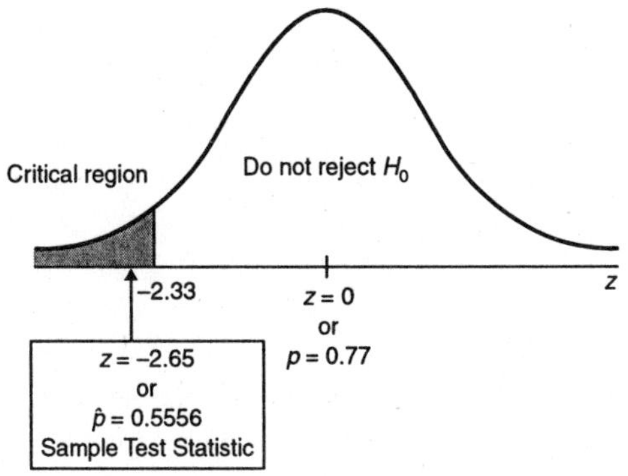

Next, find the P value associated with $z = -2.65$ and a one-tailed test.

$$P \text{ value} = P(z \le -2.65)$$
$$= 0.0040$$

Since the sample z value falls inside the critical region and the P value is less than the level of significance, $\alpha = 0.01$, we reject H_0. We conclude the population proportion of driver fatalities related to alcohol is less than 77%.

4. $H_0: p = 0.73$

$H_1: p > 0.73$

Since > is in H_1, a right-tailed test is used. The $\hat{p}$ distribution is approximately normal when n is sufficiently large, which it is here, because $np = 41(0.73) = 29.93$ and $nq = 41(0.27) = 11.07$ are both > 5. For $\alpha = 0.05$, the critical value is $z_0 = 1.645$.

$$\hat{p} = \frac{r}{n} = \frac{33}{41} = 0.8049$$

$$z = \frac{\hat{p} - p}{\sqrt{\frac{pq}{n}}} = \frac{0.8049 - 0.73}{\sqrt{\frac{0.73(0.27)}{41}}} = 108$$

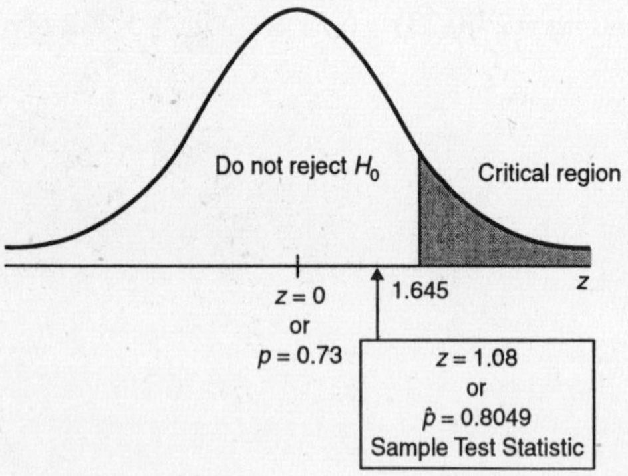

Next, find the P value associated with $z = 1.08$ and a one-tailed test.

$$\begin{aligned} P \text{ value} &= P(z \geq 1.08) \\ &= 1 - 0.8599 \\ &= 0.1401 \end{aligned}$$

Since the sample z value falls outside the critical region and the P value is greater than the level of significance, $\alpha = 0.05$, we do not reject H_0. There is not enough evidence to conclude that the population proportion of such accidents is higher than 73% in the Fargo district.

5. $H_0: p = 0.50$

$H_1: p < 0.50$

Since < is in H_1, a left-tailed test is used. The $\hat{p}$ distribution is approximately normal when n is sufficiently large, and it is here because $np = 34(0.50) = 17$ and $nq = 17$ are both > 5. For $\alpha = 0.01$, the critical value is $z_0 = -2.33$.

$$\hat{p} = \frac{r}{n} = \frac{10}{34} = 0.2941$$

$$z = \frac{\hat{p} - p}{\sqrt{\frac{pq}{n}}} = \frac{0.2941 - 0.50}{\sqrt{\frac{0.5(0.5)}{34}}} = -2.40$$

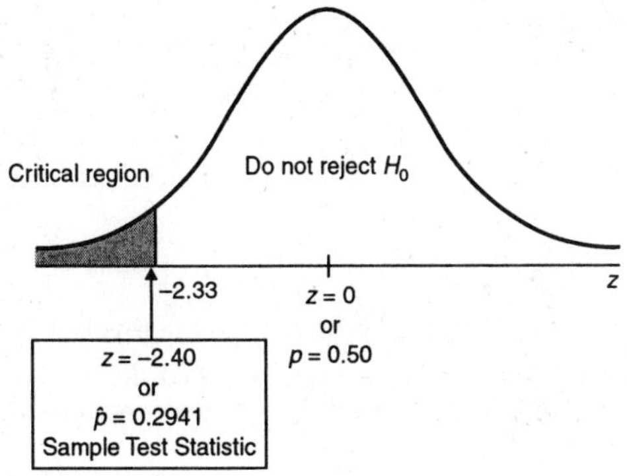

Next, find the P value associated with $z = -2.40$ and a one-tailed test.

$$P \text{ value} = P(z \le -2.40) = 0.0082$$

Since the sample z value falls inside the critical region and the P value is less than the level of significance, $\alpha = 0.01$, we reject H_0. We conclude that the population proportion of female wolves is less than 50%.

6. $H_0: p = 0.75$
$H_1: p \ne 0.75$

Since $\ne$ is in H_1, a two-tailed test is used. The $\hat{p}$ distribution is approximately normal when n is sufficiently large, which it is here, because $np = 83(0.75) = 62.25$ and $nq = 83(0.25) = 20.75$ are both > 5. For $\alpha = 0.05$, the critical values are $z_0 = \pm 1.96$.

$$\hat{p} = \frac{r}{n} = \frac{64}{83} = 0.7711$$

$$z = \frac{\hat{p} - p}{\sqrt{\frac{pq}{n}}} = \frac{0.7711 - 0.75}{\sqrt{\frac{0.75(0.25)}{83}}} = 0.44$$

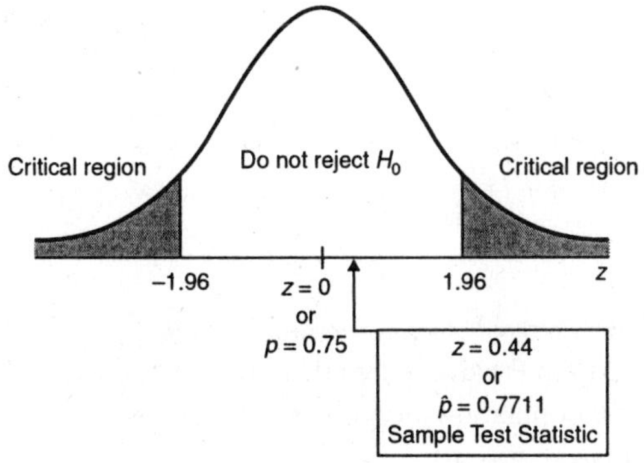

Next, find the P value associated with $z = 0.44$ and a two-tailed test.

$$\begin{aligned} P \text{ value} &= 2P(z \geq 0.44) \\ &= 2(1 - 0.6700) \\ &= 2(0.3300) \\ &= 0.6600 \end{aligned}$$

Since the sample z value falls outside the critical region and the P value is greater than the level of significance, $\alpha = 0.05$, we do not reject H_0. There is insufficient evidence to conclude that the population proportion is different from 75%.

7. $H_0: p = 0.261$

$H_1: p \neq 0.261$

Since $\neq$ is in H_1, a two-tailed test is used. The $\hat{p}$ distribution is approximately normal when n is sufficiently large, which it is here, because $np = 317(0.261) = 82.737$ and $nq = 317(0.739) = 234.263$ are both > 5. For $\alpha = 0.01$, the critical values are $z_0 = \pm 2.58$.

$$\hat{p} = \frac{r}{n} = \frac{61}{317} = 0.1924$$

$$z = \frac{\hat{p} - p}{\sqrt{\frac{pq}{n}}} = \frac{0.1924 - 0.261}{\sqrt{\frac{0.261(0.739)}{317}}} = -2.78$$

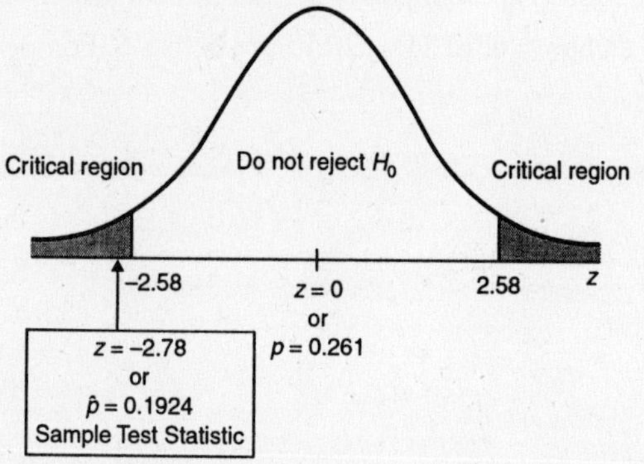

Next, find the P value associated with $z = -2.78$ and a two-tailed test.

$$\begin{aligned} P \text{ value} &= 2P(z \leq -2.78) \\ &= 2(0.0027) \\ &= 0.0054 \end{aligned}$$

Since the sample z value falls inside the critical region and the P value is less than the level of significance, $\alpha = 0.01$, we reject H_0. We conclude that the population proportion of this type of five-syllable sequence is significantly different from that of Plato's *Republic*.

8. $H_0: p = 0.214$

$H_1: p > 0.214$

Since > is in H_1, a right-tailed test is used. The $\hat{p}$ distribution is approximately normal when n is sufficiently large, which it is here, because $np = 493(0.214) = 105.502$ and $nq = 493(0.786) = 387.498$ are both > 5. For $\alpha = 0.01$, the critical value is $z_0 = 2.33$.

$$\hat{p} = \frac{r}{n} = \frac{136}{493} = 0.2759$$

$$z = \frac{\hat{p} - p}{\sqrt{\frac{pq}{n}}} = \frac{0.2759 - 0.214}{\sqrt{\frac{0.214(0.786)}{493}}} = 3.35$$

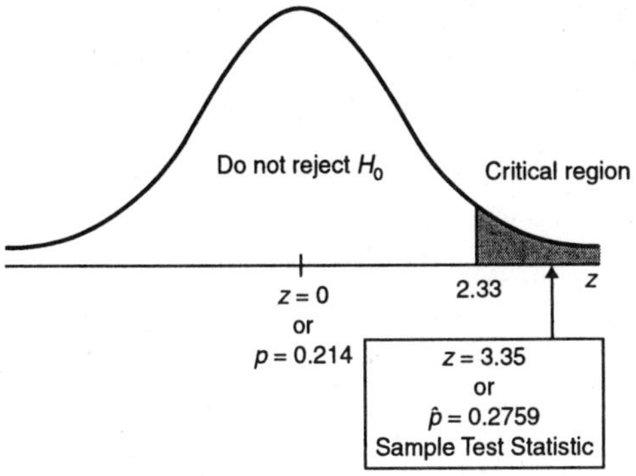

Next, find the P value associated with $z = 3.35$ and a one-tailed test.

$$P \text{ value} = P(z \geq 3.35)$$
$$= 1 - 0.9996$$
$$= 0.0004$$

Since the sample z value falls inside the critical region and the P value is less than the level of significance, $\alpha = 0.01$, we reject H_0. We conclude that the population proportion is higher than 21.4%

9. $H_0: p = 0.47$

$H_1: p > 0.47$

Since > is in H_1, a right-tailed test is used. The $\hat{p}$ distribution is approximately normal when n is sufficiently large, which it is here, because $np = 1006(0.47) = 472.82$ and $nq = 1006(0.53) = 533.18$ are both > 5. For $\alpha = 0.01$, the critical value is $z_0 = 2.33$.

$$\hat{p} = \frac{r}{n} = \frac{490}{1006} = 0.4871$$

$$z = \frac{\hat{p} - p}{\sqrt{\frac{pq}{n}}} = \frac{0.4871 - 0.47}{\sqrt{\frac{0.47(0.53)}{1006}}} = 1.09$$

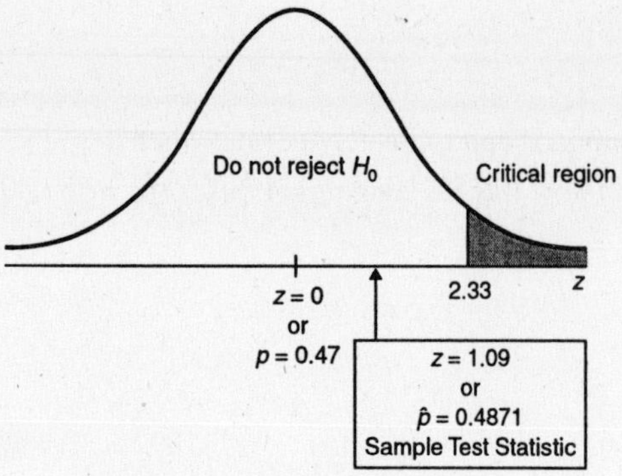

Next, find the P value associated with $z = 1.09$ and a one-tailed test.

$$\begin{aligned} P \text{ value} &= P(z \geq 1.09) \\ &= 1 - 0.8621 \\ &= 0.1379 \end{aligned}$$

Since the sample z value falls outside the critical region and the P value is greater than the level of significance, $\alpha = 0.01$, we do not reject H_0. There is insufficient evidence to conclude that the population proportion is more than 47%.

10. $H_0: p = 0.80$

$H_1: p < 0.80$

Since < is in H_1, a left-tailed test is used. The $\hat{p}$ distribution is approximately normal when n is sufficiently large, which it is here, because $np = 115(0.8) = 92$ and $nq = 115(0.2) = 23$ are both > 5. For $\alpha = 0.05$, the critical value is $z_0 = -1.645$.

$$\hat{p} = \frac{r}{n} = \frac{88}{115} = 0.7652$$

$$z = \frac{\hat{p} - p}{\sqrt{\frac{pq}{n}}} = \frac{0.7652 - 0.80}{\sqrt{\frac{0.80(0.20)}{115}}} = -0.93$$

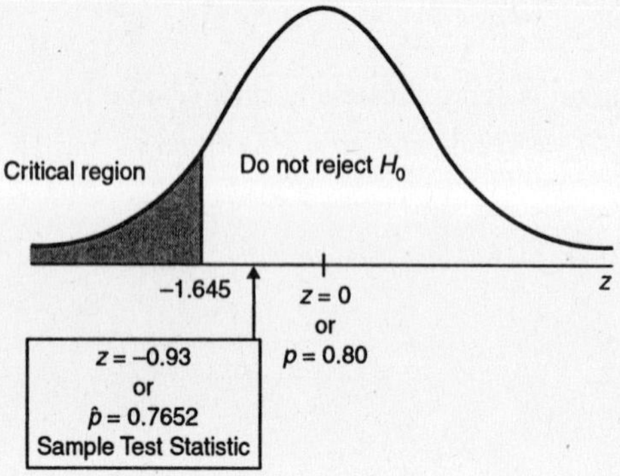

Next, find the P value associated with $z = -0.93$ and a one-tailed test.

$$P \text{ value} = P(z \le -0.93) = 0.1762$$

Since the sample z value falls outside the critical region and the P value is greater than the level of significance, $\alpha = 0.05$, we do not reject H_0. There is insufficient evidence to conclude less than 80% of the prices in the store end in the digits 9 or 5.

11. $H_0: p = 0.092$

$H_1: p > 0.092$

Since > is in H_1, a right-tailed test is used. The $\hat{p}$ distribution is approximately normal when n is sufficiently large, which it is here, because $np = 196(0.092) = 18.032$ and $nq = 196(0.908) = 177.968$ are both > 5. For $\alpha = 0.05$, the critical value is $z_0 = 1.645$.

$$\hat{p} = \frac{r}{n} = \frac{29}{196} = 0.1480$$

$$z = \frac{\hat{p} - p}{\sqrt{\frac{pq}{n}}} = \frac{0.1480 - 0.092}{\sqrt{\frac{0.092(0.908)}{196}}} = 2.71$$

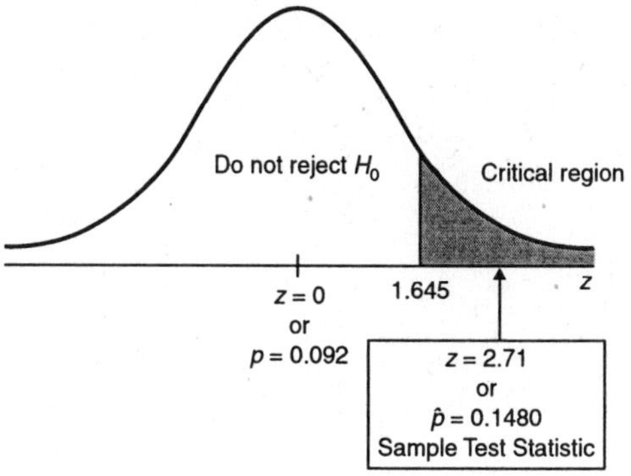

Next, find the P value associated with $z = 2.71$ and a one-tailed test.

$$P \text{ value} = P(z \ge 2.71) = 1 - 0.9966 = 0.0034$$

Since the sample z value falls inside the critical region and the P value is less than the level of significance, $\alpha = 0.05$, we reject H_0. We conclude that the population proportion of students with hypertension during final exams week is higher than 9.2%.

12. $H_0: p = 0.12$
$H_1: p < 0.12$

Since < is in H_1, a left-tailed test is used. The $\hat{p}$ distribution is approximately normal when n is sufficiently large, which it is here, because $np = 209(0.12) = 25.08$ and $nq = 209(0.88) = 183.92$ are both > 5. For $\alpha = 0.01$, the critical value is $z_0 = -2.33$.

$$\hat{p} = \frac{r}{n} = \frac{16}{209} = 0.0766$$

$$z = \frac{\hat{p} - p}{\sqrt{\frac{pq}{n}}} = \frac{0.0766 - 0.12}{\sqrt{\frac{0.12(0.88)}{209}}} = -1.93$$

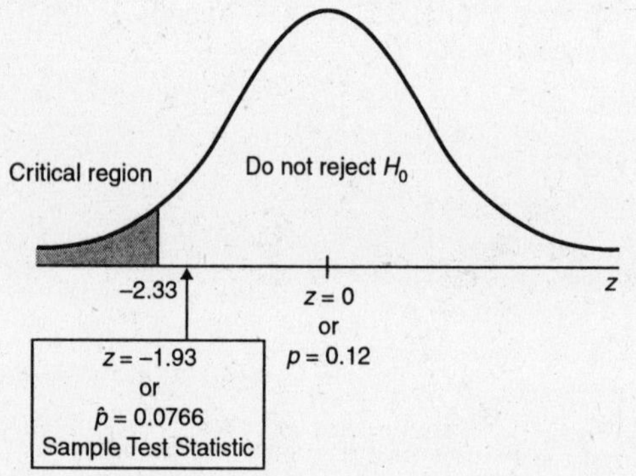

Next, find the P value associated with $z = -1.93$ and a one-tailed test.

$$P \text{ value} = P(z \le -1.93) = 0.0268$$

Since the sample z value falls outside the critical region and the P value is greater than the level of significance, $\alpha = 0.01$, we do not reject H_0. There is insufficient evidence to conclude that there has been a reduction in the population proportion of patients having headaches.

13. $H_0: p = 0.82$
$H_1: p \ne 0.82$

Since ≠ is in H_1, a two-tailed test is used. The $\hat{p}$ distribution is approximately normal when n is sufficiently large, which it is here, because $np = 73(0.82) = 59.86$ and $nq = 73(0.18) = 13.14$ are both > 5. For $\alpha = 0.01$, the critical values are $z_0 = \pm 2.58$.

$$\hat{p} = \frac{r}{n} = \frac{56}{73} = 0.7671$$

$$z = \frac{\hat{p} - p}{\sqrt{\frac{pq}{n}}} = \frac{0.7671 - 0.82}{\sqrt{\frac{0.82(0.18)}{73}}} = -1.18$$

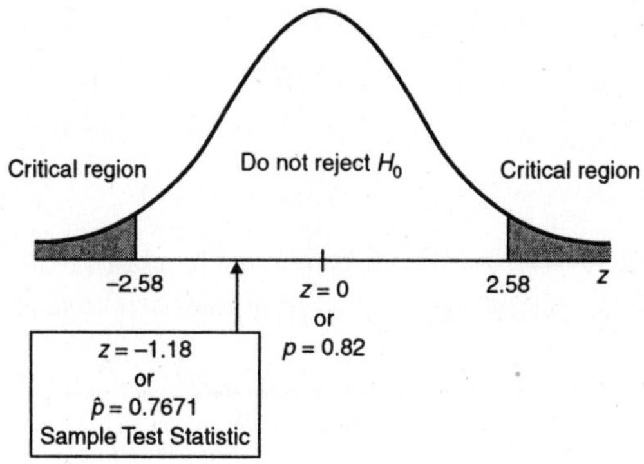

Next, find the P value associated with $z = -1.18$ and a two-tailed test.

$$\begin{aligned} P \text{ value} &= 2P(z \le -1.18) \\ &= 2(0.1190) \\ &= 0.2380 \end{aligned}$$

Since the sample z value falls outside the critical region and the P value is greater than the level of significance, $\alpha = 0.01$, we do not reject H_0. There is insufficient evidence to conclude that the population proportion is different from 82%.

14. H_0: $p = 0.28$

H_1: $p > 0.28$

Since > is in H_1, a right-tailed test is used. The $\hat{p}$ distribution is approximately normal when n is sufficiently large, which it is here, because $np = 48(0.28) = 13.44$ and $nq = 48(0.72) = 34.56$ are both > 5. For $\alpha = 0.05$, the critical value is $z_0 = 1.645$.

$$\hat{p} = \frac{r}{n} = \frac{19}{48} = 0.3958$$

$$z = \frac{\hat{p} - p}{\sqrt{\frac{pq}{n}}} = \frac{0.3958 - 0.28}{\sqrt{\frac{0.28(0.72)}{48}}} = 1.79$$

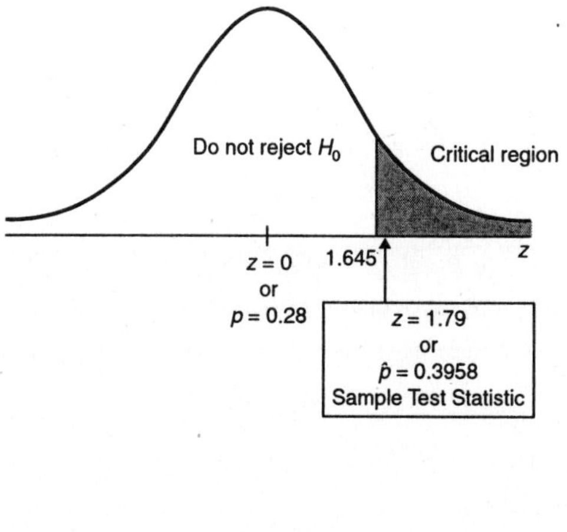

Next, find the P value associated with $z = 1.79$ and a one-tailed test.

$$P \text{ value} = P(z \geq 1.79)$$
$$= 1 - 0.9633$$
$$= 0.0367$$

Since the sample z value falls inside the critical region and the P value is less than the level of significance, $\alpha = 0.05$, we reject H_0. We conclude that the population proportion of interstate truckers who believe NAFTA benefits America is higher than 28%.

15. $H_0: p = 0.76$

$H_1: p \neq 0.76$

Since $\neq$ is in H_1, a two-tailed test is used. The $\hat{p}$ distribution is approximately normal when n is sufficiently large, which it is here, because $np = 59(0.76) = 44.84$ and $nq = 59(0.24) = 14.16$ are both > 5. For $\alpha = 0.01$, the critical values are $z_0 = \pm 2.58$.

$$\hat{p} = \frac{r}{n} = \frac{47}{59} = 0.7966$$

$$z = \frac{\hat{p} - p}{\sqrt{\frac{pq}{n}}} = \frac{0.7966 - 0.76}{\sqrt{\frac{0.76(0.24)}{59}}} = 0.66$$

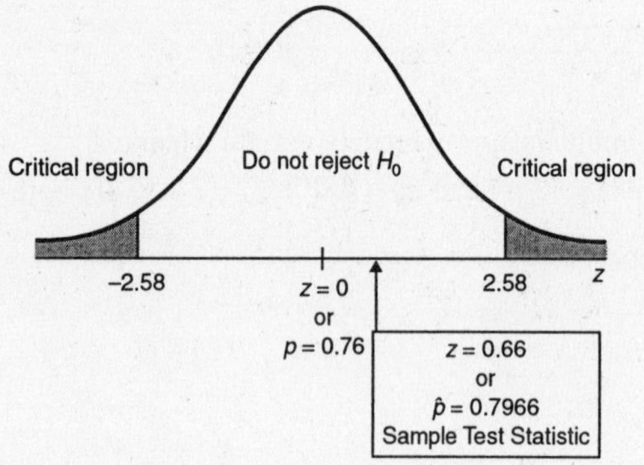

Next, find the P value associated with $z = 0.66$ and a two-tailed test.

$$P \text{ value} = 2P(z \geq 0.66)$$
$$= 2(1 - 0.7454)$$
$$= 2(0.2546)$$
$$= 0.5092$$

Since the sample z value falls outside the critical region and the P value is greater than the level of significance, $\alpha = 0.01$, we do not reject H_0. There is insufficient evidence to conclude that the population proportion of professors in Colorado who would choose the career again is different from 76%.

Chapter 10 Review

1. We are testing a single mean with a large sample.
$H_0: \mu = 11.3$
$H_1: \mu \neq 11.3$
Since $\neq$ is in H_1, a two-tailed test is used.
For $\alpha = 0.05$, the critical values are $z_0 = \pm 1.96$.

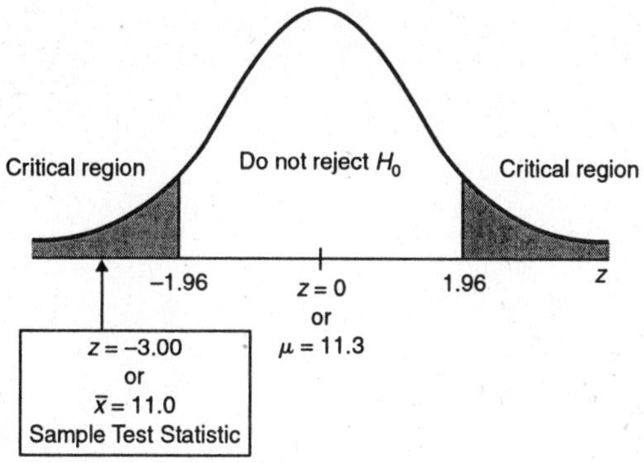

Next, find the P value associated with $z = -3.00$ and a two-tailed test.

$$\begin{aligned} P\text{-value} &= 2P(z \leq -3.00) \\ &= 2(0.0013) \\ &= 0.0026 \end{aligned}$$

Since the sample z falls inside the critical region and the P value is less than $\alpha = 0.05$, we reject H_0. We conclude that the average number of miles driven per vehicle in Chicago is different from the national average.

2. We are testing a single proportion with a large sample.
$H_0: p = 0.25$
$H_1: p \neq 0.25$
Since $\neq$ is in H, a two-tailed test is used. For $\alpha = 0.01$, the critical value is $z_0 = 2.58$.

$$\hat{p} = \frac{r}{n} = \frac{18}{50} = 0.36$$

$$z = \frac{\hat{p} - p}{\sqrt{\frac{pq}{n}}} = \frac{0.36 - 0.25}{\sqrt{\frac{(0.25)(0.75)}{50}}} = 1.80$$

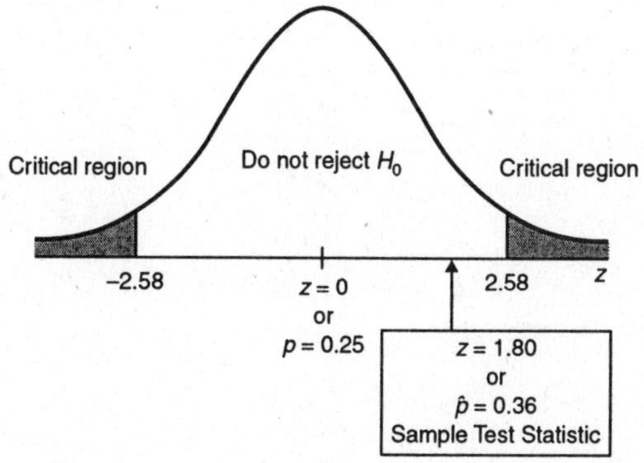

Next, find the P value associated with $z = 1.80$ and a two-tailed test.

$$\begin{aligned} P \text{ value} &= 2P(z > 1.80) \\ &= 2(0.0359) \\ &= 0.0718 \end{aligned}$$

Since the sample z does not fall in the critical region and the P value is greater than $\alpha = 0.01$, we do not reject H_0. There is insufficient evidence to conclude that the proportion is different from 25%.

3. We are testing a single mean with a small sample.

$H_0: \mu = 10.8$
$H_1: \mu < 10.8$

Since < is in H_1, a left-tailed test is used. For $\alpha^1 = 0.01$ and $d.f. = 20 - 1 = 19$, the critical value is $t_0 = 2.539$.

$$t = \frac{\bar{x} - \mu}{\frac{s}{\sqrt{n}}} = \frac{9.2 - 10.8}{\frac{3}{\sqrt{20}}} = -2.39$$

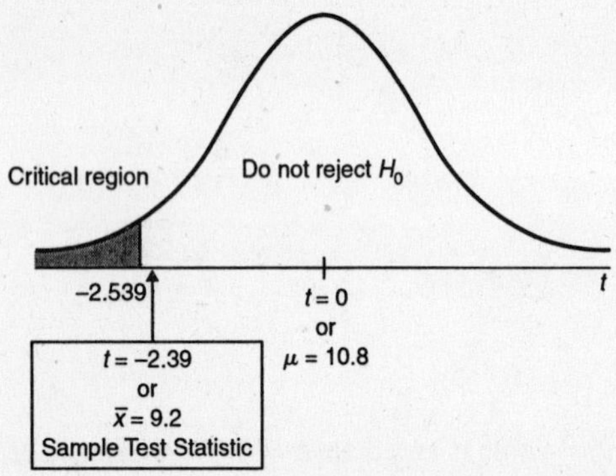

Next, find the P value interval using α' values and the row headed by $d.f. = 19$. We find that the sample t value, $t = 2.39$, falls to the right of -2.539. Therefore, $0.01 < P < 0.025$.

Since the sample t falls outside the critical region and the P value is greater than $\alpha = 0.01$, we do not reject H_0. There is insufficient evidence to conclude that the average wait is less than 10.8 days.

4. We are testing a single proportion with a large sample.

$H_0: p = 0.35$
$H_1: p > 0.35$

Since > is in H_1, a right-tailed test is used. For $\alpha = 0.05$, the critical value is $z_0 = 1.645$.

$$\hat{p} = \frac{r}{n} = \frac{39}{81} = 0.4815$$

$$z = \frac{\hat{p} - p}{\sqrt{\frac{pq}{n}}} = \frac{0.4815 - 0.35}{\sqrt{\frac{0.35(0.65)}{81}}} = 2.48$$

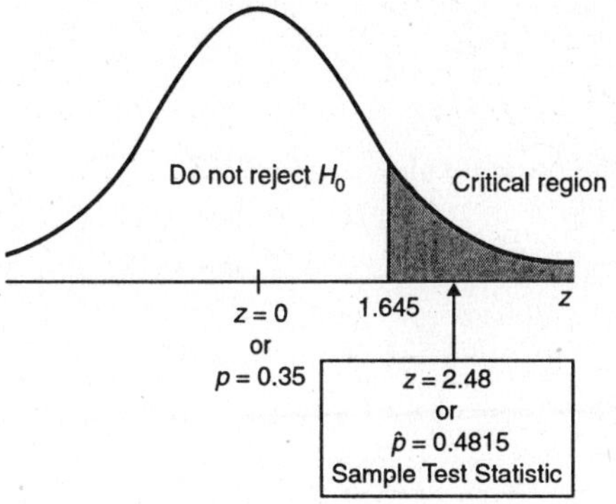

Next, find the P value associated with $z = 2.48$ and a one-tailed test.

$$P \text{ value} = P(z \geq 2.48) = 0.0066$$

Since the sample z falls inside the critical region and the P value is less than $\alpha = 0.05$, we reject H_0. We conclude that more than 35% of the students have jobs.

5. We are testing a single mean with a large sample.
$H_0 : \mu = 88.9$
$H_1 : \mu > 88.9$
Since > is in H_1, a right-tailed test is used. For $\alpha = 0.05$, the critical value is $z_0 = 1.645$.

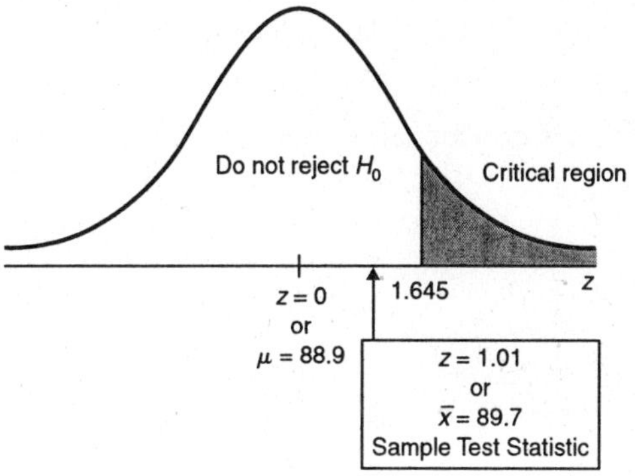

Next, find the P value associated with $z = 1.01$ and a right-tailed test.

$$P \text{ value} = P(z \geq 1.01) = 0.1562$$

Since the sample z falls outside the critical region and the P value is greater than $\alpha = 0.05$, we do not reject H_0. There is insufficient evidence to conclude that the average annual sales is greater than 88.9 thousand dollars.

6. We are testing a single mean with a large sample.
$H_0: \mu = 131.4$
$H_1: \mu \neq 131.4$
Since $\neq$ is in H_1, a two-tailed test is used. For $\alpha = 0.01$, the critical value is $z_0 = 2.58$

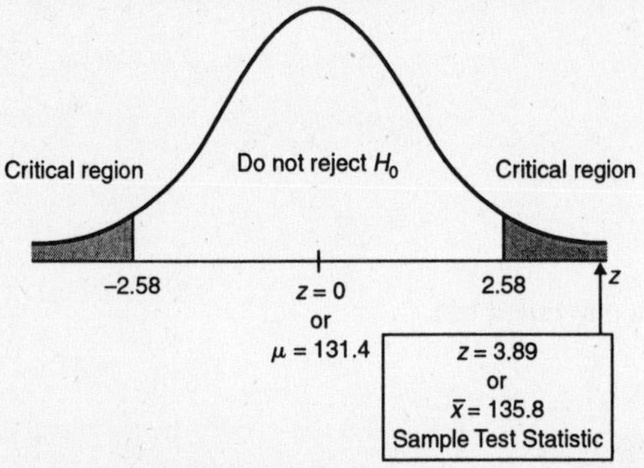

Next, find the P value associated with $z = 3.89$ and a two-tailed test.
P value $= 2P(z > 3.89)$
$= 2(0.0001) = 0.0002$
Since the sample z falls inside the critical region and the P value is less than $\alpha = 0.01$, we reject H_0.
We conclude that the average annual sales per Great Foods employee is different from 131.4 thousand.

7. We are testing a single proportion with a large sample.
$H_0: p = 0.846$
$H_1: p < 0.846$
Since < is in H_1, a left-tailed test is used. For $\alpha = 0.01$, the critical value $z_0 = -2.33$.

$$\hat{p} = \frac{r}{n} = \frac{36}{50} = 0.72$$

$$z = \frac{\hat{p} - p}{\sqrt{\frac{pq}{n}}} = \frac{0.72 - 0.846}{\sqrt{\frac{0.846(0.154)}{50}}} = -2.47$$

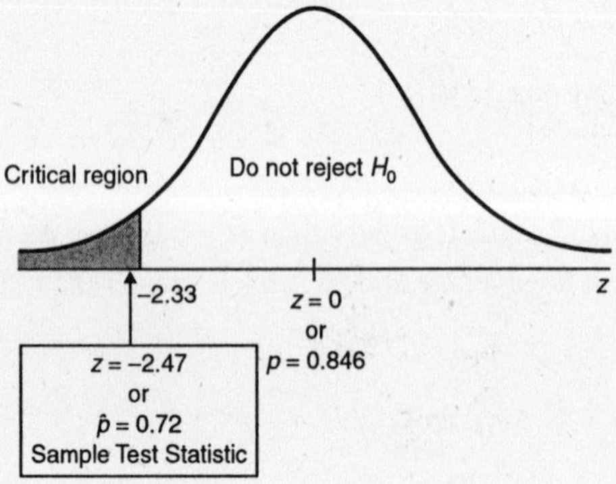

Next, find the P value associated with $z = -2.47$
P value $= P(z < -2.47)$
$= 0.0068$
Since the sample z falls inside the critical region and the P value is less than $\alpha = 0.01$, we reject H_0.
We conclude that the proportion of people in the 25-to-34 age group who believe everyone has a perfect match is less than 84.6%.

8. We are testing a single proportion with a large sample.

$H_0: p = 0.60$
$H_1: p < 0.60$

Since < is in H_1, a left-tailed test is used. For $\alpha = 0.01$, the critical value is $z_0 = -2.33$.

$$\hat{p} = \frac{r}{n} = \frac{40}{90} = 0.4444$$

$$z = \frac{\hat{p} - p}{\sqrt{\frac{pq}{n}}} = \frac{0.4444 - 0.60}{\sqrt{\frac{0.60(0.40)}{90}}} = -3.01$$

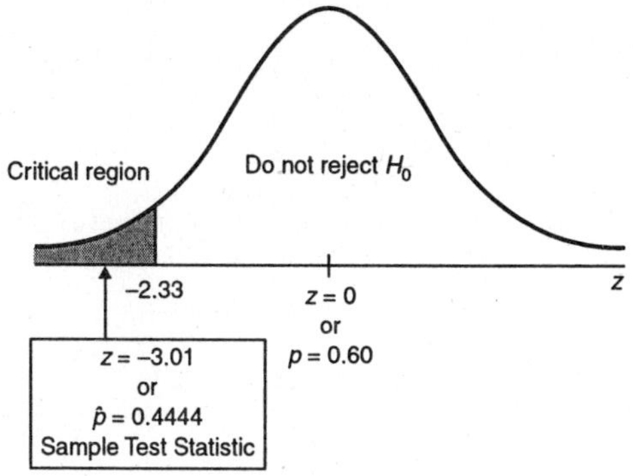

Next, find the P value associated with $z = 3.01$ and a one-tailed test.

$$P \text{ value} = P(z \le -3.01) = 0.0013$$

Since the sample z falls inside the critical region and the P value is less than $\alpha = 0.01$, we reject H_0. We conclude that the population mortality rate has dropped.

9. We are testing a single proportion with a large sample.

$H_0: p = 0.36$
$H_1: p < 0.36$

Since < is in H_1, a left-tailed test is used. For $\alpha = 0.05$, the critical value is $z_0 = -1.645$.

$$\hat{p} = \frac{r}{n} = \frac{33}{120} = 0.275$$

$$z = \frac{\hat{p} - p}{\sqrt{\frac{pq}{n}}} = \frac{0.275 - 0.36}{\sqrt{\frac{0.36(0.64)}{120}}} = -1.94$$

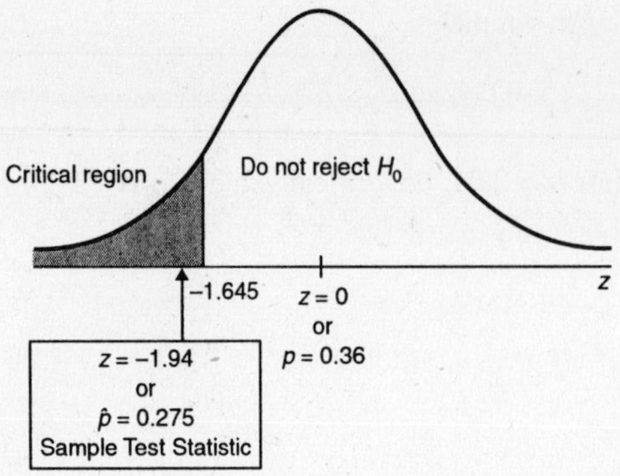

Next, find the P value associated with $z = -1.94$ and a one-tailed test.

$$P \text{ value} = P(z \le -1.94)$$
$$= 0.0262$$

Since the sample z falls inside the critical region and the P value is less than $\alpha = 0.05$, we reject H_0. We conclude that the percentage is less than 36%.

10. We are testing a single mean with a small sample.

H_0: $\mu = 7.0$

H_1: $\mu \neq 7.0$

Since ≠ is in H_1, a two-tailed test is used. For $\alpha'' = 0.05$ and $d.f. = 8 - 1 = 7$, the critical values are $t_0 = \pm 2.365$.

$$t = \frac{\bar{x} - \mu}{s/\sqrt{n}} = \frac{7.3 - 7.0}{0.5/\sqrt{8}} = 1.697$$

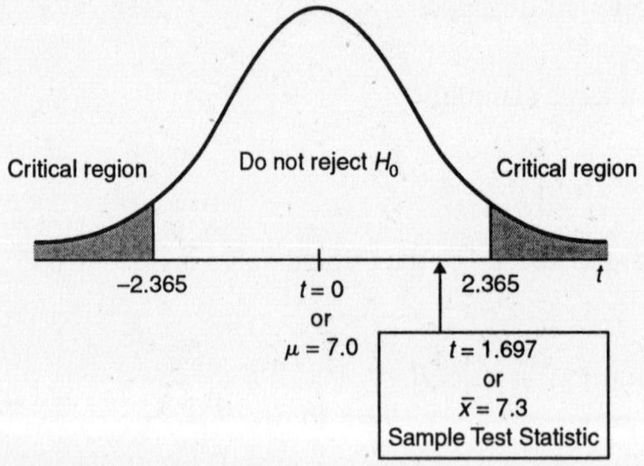

Next, find the P value interval using the t values and the row headed by $d.f. = 7$. We find that the sample t value, $t = 1.697$, falls between 1.617 and 1.895. Therefore, $0.10 < P < 0.15$.

Since the sample t falls outside the critical region and the P value is greater than $\alpha'' = 0.05$, we do not reject H_0. There is not enough evidence to conclude that the machine has slipped out of adjustment.

11. (a) The mean and standard deviation round to the results given.

(b) We are testing a single mean with a small sample.

H_0: $\mu = 48$

H_1: $\mu < 48$

Since < is in H_1, a left-tailed test is used. For $\alpha' = 0.05$ and $d.f. = 10 - 1 = 9$, the critical value is $t_0 = -1.833$.

$$t = \frac{\bar{x} - \mu}{s/\sqrt{n}} = \frac{46.2 - 48}{10.85/\sqrt{10}} = -0.525$$

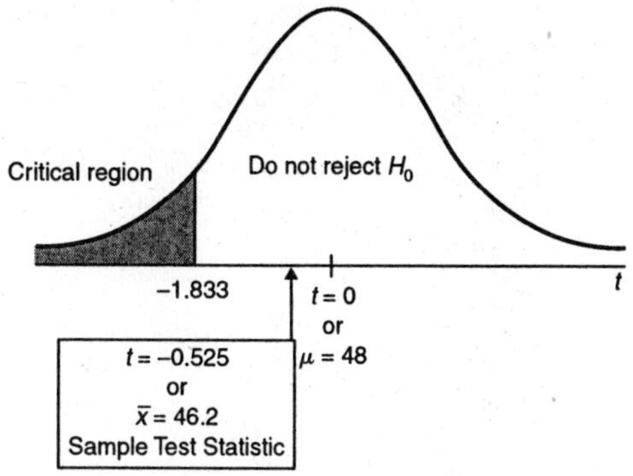

Next, find the P value interval using the α' values and the row headed by $d.f. = 9$. We find that the sample t value, $t = -0.525$, falls to the right of -1.230. Therefore, P value > 0.125.

Since the sample t falls outside the critical region and the P value is greater than $\alpha' = 0.05$, we do not reject H_0. There is insufficient evidence to conclude that the average is less than 48 months.

Chapter 11 Inferences about Differences

Section 11.1

Note: In the following problems, we will make the assumption that the data are (approximately) normally distributed.

1. H_0: $\mu_d = 0$
   H_1: $\mu_d \neq 0$

   Since $\neq$ is in H_1, a two-tailed test is used. Since the sample size is small, critical values are found using the Student's t distribution (use Table 6 in Appendix II). For a two-tailed test, look in the column headed by $\alpha'' = 0.05$ and the row headed by $d.f. = 8 - 1 = 7$. The critical values are $t_0 = \pm 2.365$.

$$\bar{d} = 2.25, s_d = 7.78$$
$$t = \frac{\bar{d} - \mu_d}{s_d/\sqrt{n}} = \frac{2.25 - 0}{7.78/\sqrt{8}} = 0.818$$

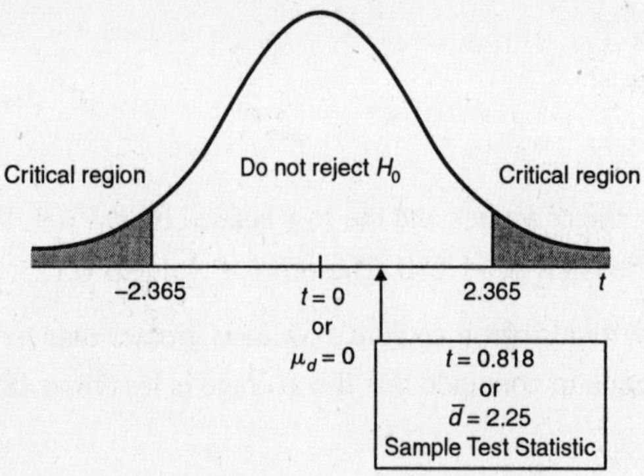

   To find the P value interval, use the α'' values since the test is two-tailed and look in the row headed by $d.f. = 7$. We find that the sample t value, $t = 0.818$, falls to the left of 1.254. Therefore, P value > 0.250.

   Since the sample test statistic falls outside the critical region and the P value is greater than the level of significance, $\alpha = 0.05$, we do not reject H_0. There is not enough evidence to conclude that there is a significant difference between the population mean percentage increase in corporate revenue and the population mean percentage increase in CEO salary.

2. H_0: $\mu_d = 0$
   H_1: $\mu_d \neq 0$

   Since $\neq$ is in H_1, a two-tailed test is used. Since the sample size is small, critical values are found using the Student's t distribution (use Table 6 in Appendix II). For a two-tailed test, look in the column headed by $\alpha'' = 0.01$ and the row headed by $d.f. = 7 - 1 = 6$. The critical values are $t_0 = \pm 3.707$.

$$\bar{d} = 0.37, s_d = 0.47$$
$$t = \frac{\bar{d} - \mu_d}{s_d/\sqrt{n}} = \frac{0.37 - 0}{0.47/\sqrt{7}} = 2.08$$

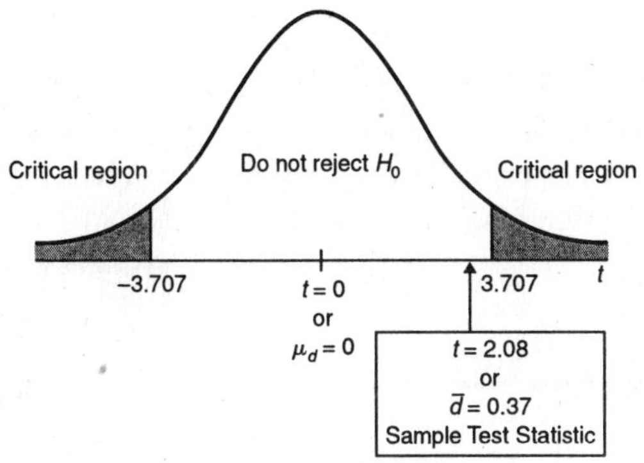

To find the P value interval, use the α'' values since the test is two-tailed and look in the row headed by $d.f. = 6$. We find that the sample t value, $t = 2.08$, falls between 1.943 and 2.447. Therefore, $0.050 < P \text{ value} < 0.100$.

Since the sample test statistic falls outside the critical region and the P value is greater than the level of significance, $\alpha = 0.01$, we do not reject H_0. There is not enough evidence to conclude that there is a difference in the population mean hours per fish using a boat compared to the population mean hours per fish fishing from the shore.

3. $H_0: \mu_d = 0$

$H_1: \mu_d > 0$

Since > is in H_1, a right-tailed test is used. Since the sample size is small, critical values are found using the Student's t distribution (use Table 6 in Appendix II). For a one-tailed test, look in the column headed by $\alpha' = 0.01$ and the row headed by $d.f. = 5 - 1 = 4$. The critical value is $t_0 = 3.747$.

$$\bar{d} = 12.6, s_d = 22.66$$

$$t = \frac{\bar{d} - \mu_d}{s_d/\sqrt{n}} = \frac{12.6 - 0}{22.66/\sqrt{5}} = 1.243$$

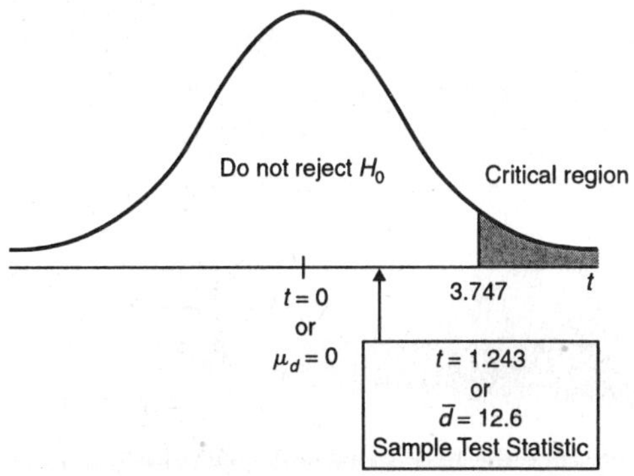

To find the P value interval, use the α' values since the test is one-tailed and look in the row headed by $d.f. = 4$. We find that the sample t value, $t = 1.243$, falls to the left of 1.344. Therefore, $P \text{ value} > 0.125$.

Since the sample test statistic falls outside the critical region and the P value is greater than the level of significance, $\alpha = 0.01$, we do not reject H_0. There is insufficient evidence to conclude that, on average, peak wind gusts are higher in January than they are in April.

4. $H_0: \mu_d = 0$

$H_1: \mu_d > 0$

Since > is in H_1, a right-tailed test is used. Since the sample size is small, critical values are found using the Student's t distribution (use Table 6 in Appendix II). For a one-tailed test, look in the column headed by $\alpha' = 0.01$ and the row headed by $d.f. = 10 - 1 = 9$. The critical value is $t_0 = 2.821$.

$$\bar{d} = 0.08, \; s_d = 1.701$$

$$t = \frac{\bar{d} - \mu_d}{s_d/\sqrt{n}} = \frac{0.08 - 0}{1.701/\sqrt{10}} = 0.1487$$

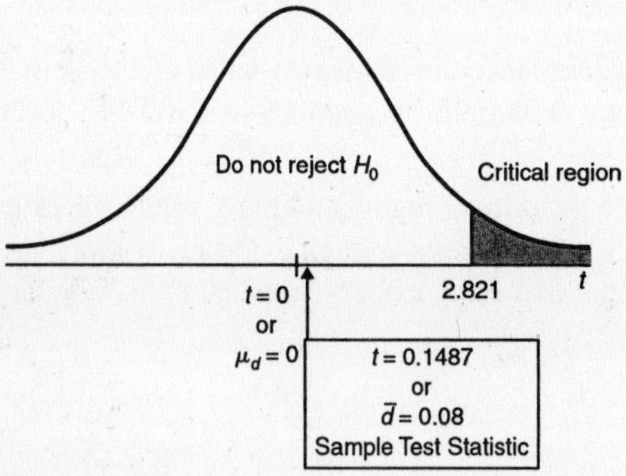

To find the P value interval, use the α' values since the test is one-tailed and look in the row headed by $d.f. = 9$. We find that the sample t value, $t = 0.1487$, falls to the left of 1.230. Therefore, P value > 0.125.

Since the sample test statistic falls outside the critical region and the P value is greater than the level of significance, $\alpha = 0.01$, we do not reject H_0. There is insufficient evidence to conclude that the January population mean has dropped.

5. $H_0: \mu_d = 0$

$H_1: \mu_d > 0$

Since > is in H_1, right-tailed test is used. Since the sample size is small, critical values are found using the Student's t distribution (use Table 6 in Appendix II). For a one-tailed test, look in the column headed by $\alpha' = 0.05$ and the row headed by $d.f. = 8 - 1 = 7$. The critical value is $t_0 = 1.895$.

$$\bar{d} = 6.125, \; s_d = 9.83$$

$$t = \frac{\bar{d} - \mu_d}{s_d/\sqrt{n}} = \frac{6.125 - 0}{9.83/\sqrt{8}} = 1.76$$

To find the P value interval, use the α' values since the test is one-tailed and look in the row headed by $d.f. = 7$. We find that the sample t value, $t = 1.76$, falls between 1.617 and 1.895. Therefore, $0.050 < P \text{ value} < 0.075$.

Since the sample test statistic falls outside the critical region and the P value is greater than the level of significance, $\alpha = 0.05$, we do not reject H_0. There is not enough evidence to conclude that average percentage of males in a wolf pack is higher in winter.

6. $H_0: \mu_d = 0$

$H_1: \mu_d \neq 0$

Since ≠ is in H_1, a two-tailed test is used. Since the sample size is small, critical values are found using the Student's t distribution (use Table 6 in Appendix II). For a two-tailed test, look in the column headed by $\alpha'' = 0.01$ and the row headed by $d.f. = 12 - 1 = 11$. The critical values are $t_0 = \pm 3.106$.

$$\bar{d} = -0.84,\ s_d = 3.57$$

$$t = \frac{\bar{d} - \mu_d}{s_d/\sqrt{n}} = \frac{-0.84 - 0}{3.57/\sqrt{12}} = -0.815$$

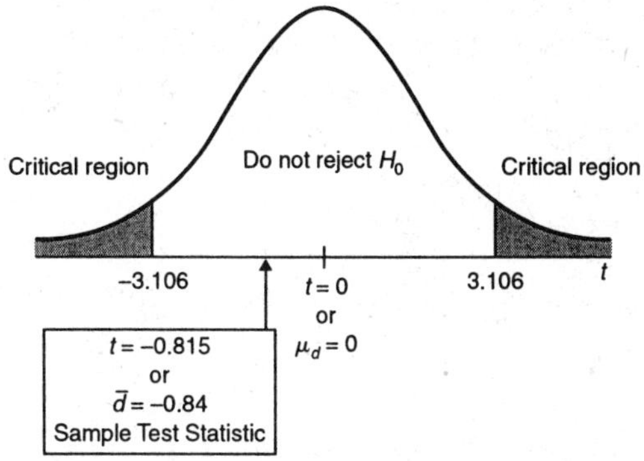

To find the P value interval, use the α'' values since the test is two-tailed and look in the row headed by $d.f. = 11$. We find that the sample t value, $t = -0.815$, falls to the right of −1.214. Therefore, P value > 0.250.

Since the sample test statistic falls outside the critical region and the P value is greater than the level of significance, $\alpha = 0.01$, we do not reject H_0. There is insufficient evidence to conclude that the average temperature in Miami is different from that in Honolulu.

7. $H_0: \mu_d = 0$

$H_1: \mu_d > 0$

Since > is in H_1, a right-tailed test is used. Since the sample size is small, critical values are found using the Student's t distribution (use Table 6 in Appendix II). For a one-tailed test, look in the column headed by $\alpha' = 0.05$ and the row headed by $d.f. = 8 - 1 = 7$. The critical value is $t_0 = 1.895$.

$$\bar{d} = 6,\ s_d = 21.5$$

$$t = \frac{\bar{d} - \mu_d}{s_d/\sqrt{n}} = \frac{6 - 0}{21.5/\sqrt{8}} = 0.789$$

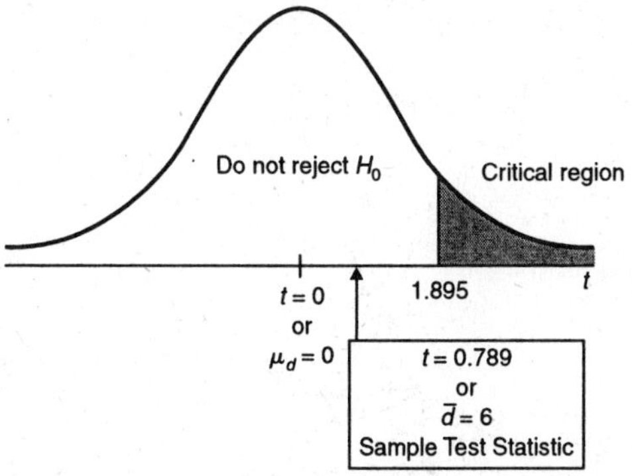

To find the P value interval, use the α' values since the test is one-tailed and look in the row headed by $d.f. = 7$. We find that the sample t value, $t = 0.789$, falls to the left of 1.254. Therefore, P value > 0.125.

Since the sample test statistic falls outside the critical region and the P value is greater than the level of significance, $\alpha = 0.05$, we do not reject H_0. We do not have enough evidence to conclude that the average number of inhabited houses is greater than the average number of inhabited hogans on the Navajo Reservation.

8. $H_0: \mu_d = 0$

$H_1: \mu_d > 0$

Since > is in H_1, a right-tailed test is used. Since the sample size is small, critical values are found using the Student's t distribution (use Table 6 in Appendix II). For a one-tailed test, look in the column headed by $\alpha' = 0.05$ and the row headed by $d.f. = 7 - 1 = 6$. The critical value is $t_0 = 1.943$.

$$\bar{d} = 0.0, \quad s_d = 8.76$$

$$t = \frac{\bar{d} - \mu_d}{s_d/\sqrt{n}} = \frac{0.0 - 0}{8.76/\sqrt{7}} = 0.000$$

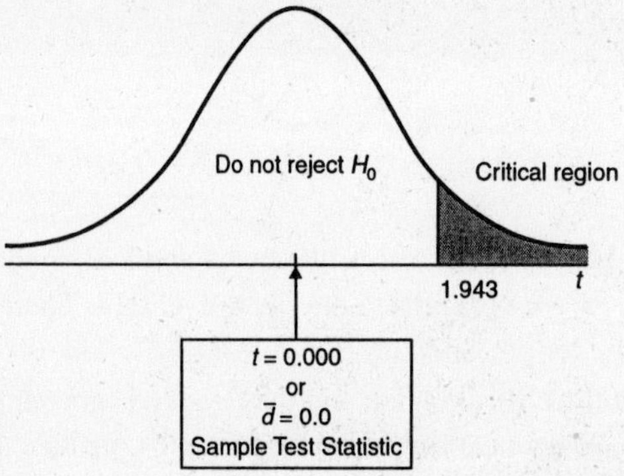

To find the P value interval, use the α' values since the test is one-tailed and look in the row headed by $d.f. = 6$. We find that the sample t value, $t = 0$, falls to the left of 1.273. Therefore, P value > 0.125.

Since the sample test statistic falls outside the critical region and the P value is greater than the level of significance, $\alpha = 0.05$, we do not reject H_0. We do not have enough evidence to conclude that there tend to be more flaked stone tools than nonflaked stone tools at this excavation site. Note: In fact, there is no reason to do a hypothesis test or "find" the P value for this data. Since d = the hypothesized mean, there is absolutely no evidence for H_1.

9. $H_0: \mu_d = 0$

$H_1: \mu_d \neq 0$

Since ≠ is in H_1, a two-tailed test is used. Since the sample size is small, critical values are found using the Student's t distribution (use Table 6 in Appendix II). For a two-tailed test, look in the column headed by $\alpha'' = 0.05$ and the row headed by $d.f. = 5 - 1 = 4$. The critical values are $t_0 = \pm 2.776$.

$$\bar{d} = 1.0, \quad s_d = 5.24$$

$$t = \frac{\bar{d} - \mu_d}{s_d/\sqrt{n}} = \frac{1.0 - 0}{5.24/\sqrt{5}} = 0.427$$

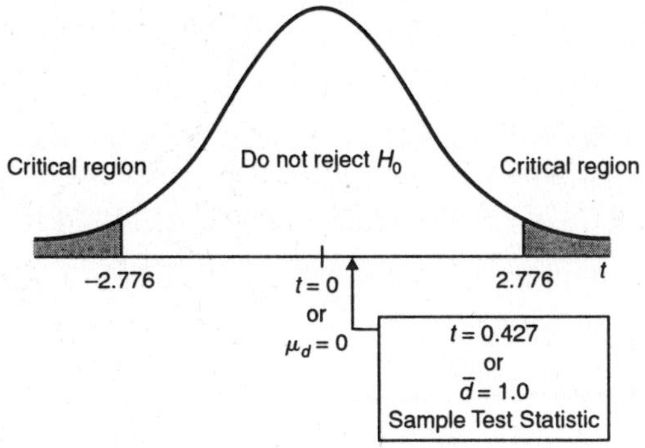

To find the P value interval, use the α'' values since the test is two-tailed and look in the row headed by $d.f. = 4$. We find that the sample t value, $t = 0.427$, falls to the left of 1.344. Therefore, P value > 0.250.

Since the sample test statistic falls outside the critical region and the P value is greater than the level of significance, $\alpha = 0.05$, we do not reject H_0. There is not enough evidence to conclude that there is a difference in the average number of service ware sherds in subarea 1 compared to subarea 2.

10. $H_0: \mu_d = 0$

$H_1: \mu_d > 0$

Since > is in H_1, a right-tailed test is used. Since the sample size is small, critical values are found using the Student's t distribution (use Table 6 in Appendix II). For a one-tailed test, look in the column headed by $\alpha' = 0.05$ and the row headed by $d.f. = 8 - 1 = 7$. The critical value is $t_0 = 1.895$.

$$\bar{d} = 1.25, \ s_d = 1.91$$

$$t = \frac{\bar{d} - \mu_d}{s_d/\sqrt{n}} = \frac{1.25 - 0}{1.91/\sqrt{8}} = 1.851$$

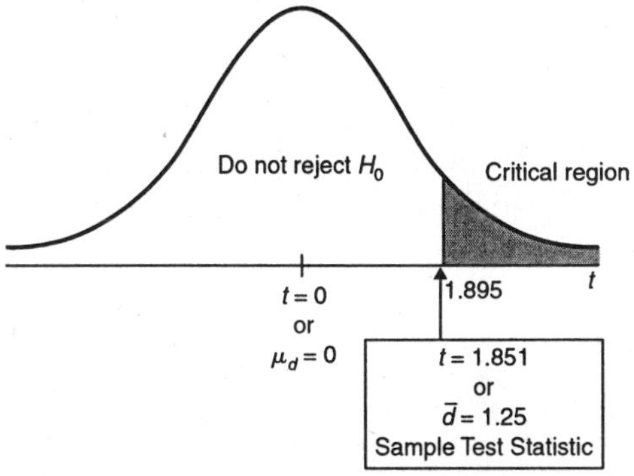

To find the P value interval, use the α' values since the test is one-tailed and look in the row headed by $d.f. = 7$. We find that the sample t value, $t = 1.851$, falls between 1.617 and 1.895. Therefore, $0.050 < P$ value < 0.075.

Since the sample test statistic falls outside the critical region and the P value is greater than the level of significance, $\alpha = 0.05$, we do not reject H_0. We do not have sufficient evidence to conclude that the mothers are more successful in picking out their own babies when a hunger cry is involved.

11. H_0: $\mu_d = 0$

H_1: $\mu_d < 0$

Since < is in H_1, a left-tailed test is used. Since the sample size is small, critical values are found using the Student's t distribution (use Table 6 in Appendix II). For a one-tailed test, look in the column headed by $\alpha' = 0.05$ and the row headed by $d.f. = 6 - 1 = 5$. The critical value is $t_0 = -2.015$.

$$\bar{d} = -10.5, \; s_d = 5.17$$

$$t = \frac{\bar{d} - \mu_d}{s_d/\sqrt{n}} = \frac{-10.5 - 0}{5.17/\sqrt{6}} = -4.97$$

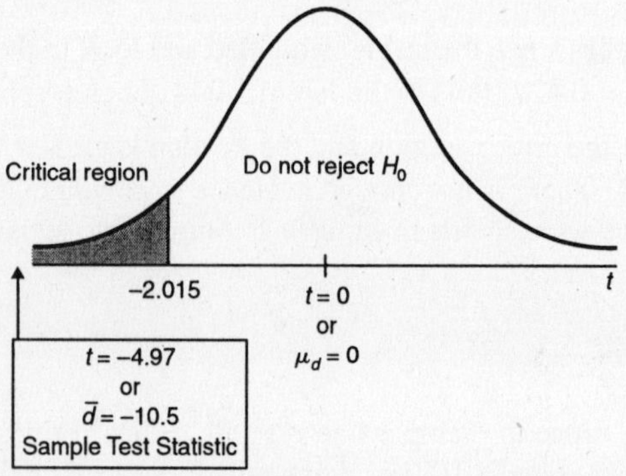

To find the P value interval, use the α' values since the test is one-tailed and look in the row headed by $d.f. = 5$. We find that the sample t value, $t = -4.97$, falls to the left of -4.032. Therefore, P value > 0.005.

Since the sample test statistic falls inside the critical region and the P value is less than the level of significance, $\alpha = 0.05$, we reject H_0. We conclude that the population mean heart rate after the test is higher than that before the test.

12. H_0: $\mu_d = 0$

H_1: $\mu_d \neq 0$

Since ≠ is in H_1, a two-tailed test is used. Since the sample size is small, critical values are found using the Student's t distribution (use Table 6 in Appendix II). For a two-tailed test, look in the column headed by $\alpha'' = 0.05$ and the row headed by $d.f. = 6 - 1 = 5$. The critical values are $t_0 = \pm 2.571$.

$$\bar{d} = -3.33, \; s_d = 7.34$$

$$t = \frac{\bar{d} - \mu_d}{s_d/\sqrt{n}} = \frac{-3.33 - 0}{7.34/\sqrt{6}} = -1.111$$

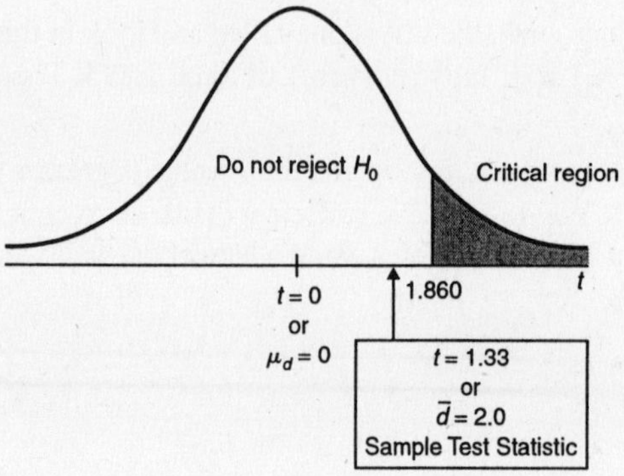

To find the P value interval, use the α'' values since the test is two-tailed and look in the row headed by $d.f. = 5$. We find that the sample t value, $t = -1.111$, falls to the right of -1.301. Therefore, P value > 0.250.

Since the sample test statistic falls outside the critical region and the P value is greater than the level of significance, $\alpha = 0.05$, we do not reject H_0. There is insufficient evidence to conclude that the population mean systolic blood pressure is different before and 6 minutes after the treadmill test.

13. $H_0: \mu_d = 0$

$H_1: \mu_d > 0$

Since > is in H_1, a right-tailed test is used. Since the sample size is small, critical values are found using the Student's t distribution (use Table 6 in Appendix II). For a one-tailed test, look in the column headed by $\alpha' = 0.05$ and the row headed by $d.f. = 9 - 1 = 8$. The critical value is $t_0 = 1.860$.

$$\bar{d} = 2.0, \; s_d = 4.5$$

$$t = \frac{\bar{d} - \mu_d}{s_d/\sqrt{n}} = \frac{2.0 - 0}{4.5/\sqrt{9}} = 1.33$$

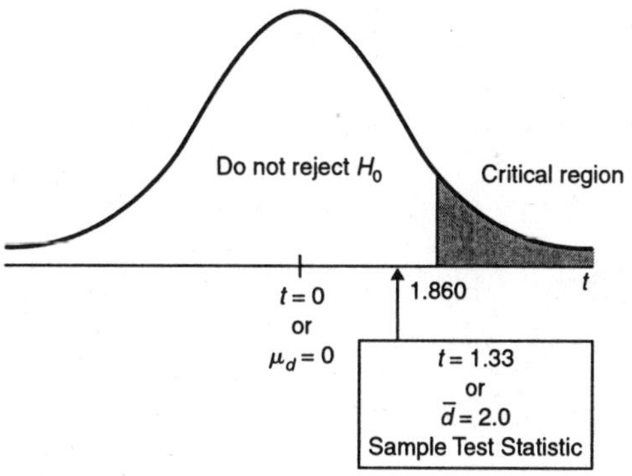

To find the P value interval, use the α' values since the test is one-tailed and look in the row headed by $d.f. = 8$. We find that the sample t value, $t = 1.33$, falls between 1.240 and 1.397. Therefore, $0.100 < P$ value < 0.125.

Since the sample test statistic falls outside the critical region and the P value is greater than the level of significance, $\alpha = 0.05$, we do not reject H_0. There is insufficient evidence to conclude that the population mean score on the last round is significantly higher than the population mean score on the first round.

14. **(a)** The null hypothesis for this test is $H_0 : \mu_d = 0$.

(b) Consider the test statistic formula (t stat)

$$t = \frac{\bar{d} - \mu_d}{\frac{s_d}{\sqrt{n}}}$$

If t stat is positive, then the value $\bar{d}$ must be positive since we know from part (a) that μ_d is zero and the denominator is always positive.

(c) Since the P value is 0.4756, we do not reject H_0 at $\alpha = 0.05$.

(d) For $\alpha = 0.05$ and $d.f. = 9 - 1 = 8$, the critical value for a one-tailed test is $t_0 = 1.86$. Since the sample test statistic for a one-tailed test 0.063 is less than 1.86, we do not reject H_0. This result is consistent with the result in part (c).

(e) Based on the results stated in parts (c) and (d), there is not sufficient evidence to conclude that, in small and medium-sized colleges in the western United States, the population mean number of male assistant professors is greater than that of female assistant professors.

(f) For a two-tailed test, $H_1: \mu_d \neq 0$. For part (c), we do not reject H_0 since the P value 0.951 is greater than $\alpha = 0.05$. For part (d), the critical value for a two-tailed test and $\alpha = 0.05$ is $t_0 = \pm 2.306$. Since the sample test statistic is $t = 0.063$ is less than the critical value, we do not reject H_0. This result is consistent with part (c) for this part of the problem. For part (e), there is not sufficient evidence to conclude that, in small and medium-sized colleges in the western United states, the population mean number of male assistant professors is greater than that of female assistant professors.

Selected Solutions for Section 11.2

1. (a) The hypotheses are $H_0: \mu_1 - \mu_2 = 0$ and $H_1: \mu_1 - \mu_2 > 0$. The test is right-tailed. The level of significance is $\alpha = 0.01$.

(b) We will use a normal distribution because independent samples are used and both samples are large. The appropriate critical value is $z_0 = 2.33$

(c)

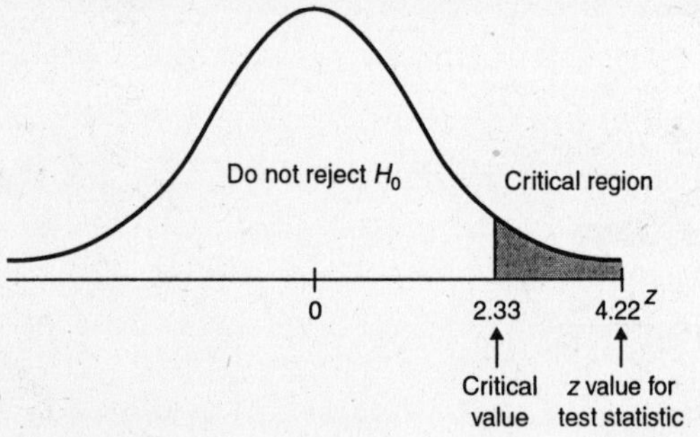

(d) The appropriate z value is

$$z = \frac{(2.6-1.9)-0}{\sqrt{\frac{(.5)^2}{33}+\frac{(.8)^2}{32}}} = \frac{0.7}{0.1661} = 4.22.$$

(e) The P value for $z = 4.22$ is less than the P value for $z = 4.0$ which is $0.5 - 0.4999 = 0.0001$. Assuming there is no difference in the population means, the probability of getting a greater difference between the means of the two samples than 0.7 is less than 0.0001. Since the P value is less than α, we reject the null hypothesis.

(f) We conclude that there is sufficient evidence to choose the hypothesis that ten year old children tend to have significantly more REM sleep than 50 year old adults.

2. (a) The hypotheses are $H_0: \mu_1 - \mu_2 = 0$ and $H_1: \mu_1 - \mu_2 \neq 0$. The test is two-tailed. The level of significance is $\alpha = 0.01$.

(b) We will use a normal distribution because independent samples are used and both samples are large. The appropriate critical value is $z_0 = \pm 2.58$.

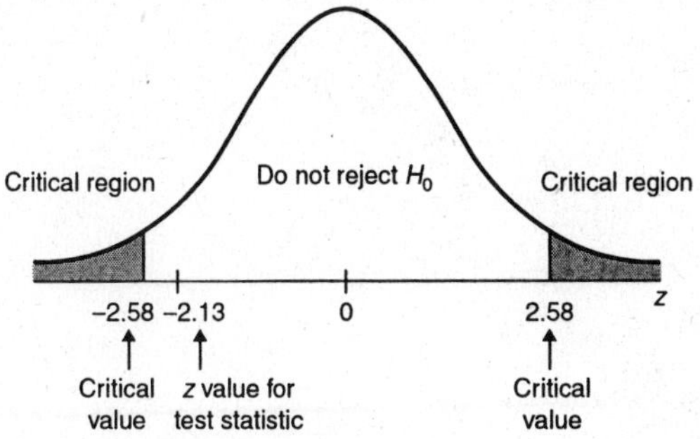

(c)

(d) The appropriate z value is

$$z = \frac{(83{,}000 - 91{,}000) - 0}{\sqrt{\frac{(17{,}000)^2}{62} + \frac{(22{,}000)^2}{51}}} = \frac{-8000}{3761.8461} = -2.13$$

(e) The area under the left tail to the left of $z = -2.13$ is $0.5 - 0.4834 = 0.0166$. Since this is a two-tailed test, the P value is $P = 2(0.0166) = 0.0332$. Assuming there is no difference in the population means, the probability of getting a greater difference between the means of the two samples is 0.0332. This is greater than α. Thus, we fail to reject the null hypothesis.

(f) We conclude that there is not sufficient evidence to choose the hypothesis that there is a difference in the mean startup costs of small clothing stores compared with small bakeries.

3. **(a)** The hypotheses are $H_0 : \mu_1 - \mu_2 = 0$ and $H_1 : \mu_1 - \mu_2 > 0$. The test is right-tailed. The level of significance is $\alpha = 0.01$.

(b) We will use a normal distribution because independent samples are used and both samples are large. The appropriate critical value is $z_0 = 2.33$.

(c)

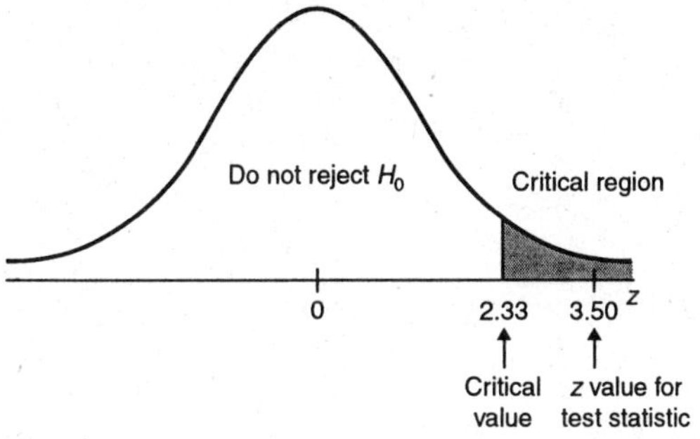

(d) The appropriate z value is

$$z = \frac{(43 - 30) - 0}{\sqrt{\frac{(22)^2}{45} + \frac{(12)^2}{47}}} = \frac{3}{3.7174} = 3.50$$

(e) The P value for $z = 3.50$ is $0.5 - 0.4998 = 0.0002$. Since the P value is less than α, we reject the null hypothesis. Assuming there is no difference in the population means, the probability of getting a greater difference between the means of the two samples is 0.0002.

(f) We conclude that there is sufficient evidence to choose the hypothesis that the mean pollution index for Englewood is less than that for Denver in the winter.

4. **(a)** The hypotheses are $H_0 : \mu_1 - \mu_2 = 0$ and $H_1 : \mu_1 - \mu_2 \neq 0$. The test is two-tailed. The level of significance is $\alpha = 0.01$.

(b) We will use a normal distribution because independent samples are used and both samples are large. The appropriate critical value is $z_0 = \pm 2.58$.

(c)

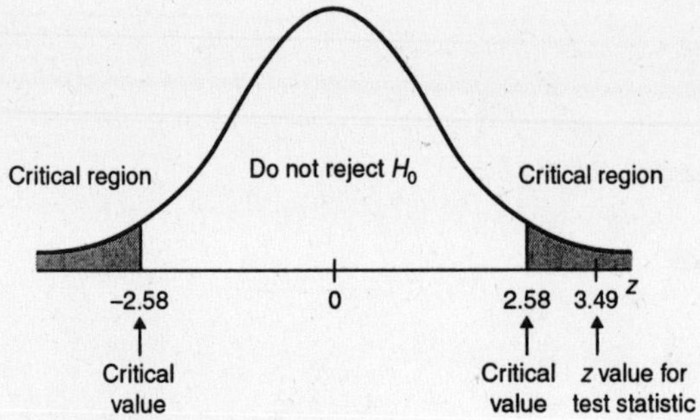

(d) The appropriate z value is

$$z = \frac{(4.7-4.2)-0}{\sqrt{\frac{(1.1)^2}{201}+\frac{(1.4)^2}{135}}} = \frac{0.5}{0.1433} = 3.49$$

(e) The area under the right tail to the right of 3.49 is 0.5 – 0.4998 = 0.0002. Since this is a two-tailed test, the P value is $P = 2(0.0002) = 0.0004$. Assuming there is no difference in the population means, the probability of getting a greater difference between the means of the two samples is 0.0004. This is less than α. Thus, we reject the null hypothesis.

(f) We conclude that there is sufficient evidence to choose the hypothesis that there is a difference regarding preference for camping or preference for fishing as an outdoor activity.

5. **(a)** The hypotheses are $H_0: \mu_1 - \mu_2 = 0$ and $H_1: \mu_1 - \mu_2 \neq 0$. The test is two-tailed. The level of significance is $\alpha = 0.05$

(b) We will use a normal distribution because independent samples are used and both samples are large. The appropriate critical values are $z = \pm 1.96$.

(c)

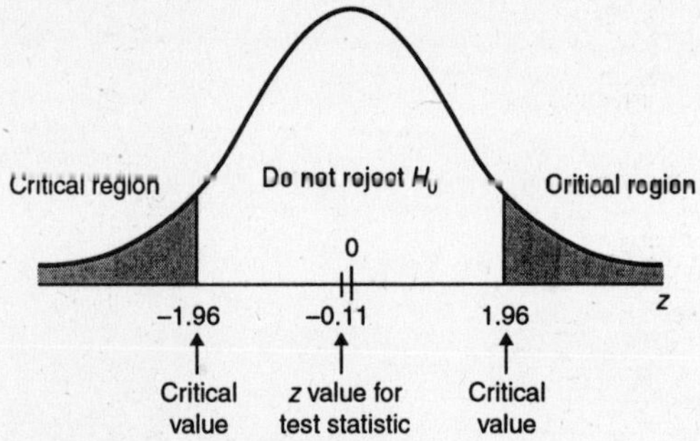

(d) The appropriate z value is

$$z = \frac{(344.5-345.9)-0}{\sqrt{\frac{(49.1)^2}{30}+\frac{(50.9)^2}{30}}} = \frac{-1.4}{12.9120} = -0.11.$$

(e) The area under the left tail to the left of $z = -0.11$ is $0.5 - 0.0438 = 0.4562$. Since this is a two-tailed test, the P value is $P = 2(0.4562) = 0.9124$. Assuming there is no difference in the population means, the probability of getting a greater difference between the means of the two samples is 0.9124. This is greater than α. Thus, we fail to reject the null hypothesis.

(f) We conclude that there is no evidence of a difference in the vocabulary scores of the two groups before the instruction began.

6. **(a)** The hypotheses are $H_0: \mu_1 - \mu_2 = 0$ and $H_1: \mu_1 - \mu_2 > 0$. The test is right-tailed. The level of significance is $\alpha = 0.01$.

(b) We will use a normal distribution because independent samples are used and both samples are large. The appropriate critical value is $z_0 = 2.33$.

(c)

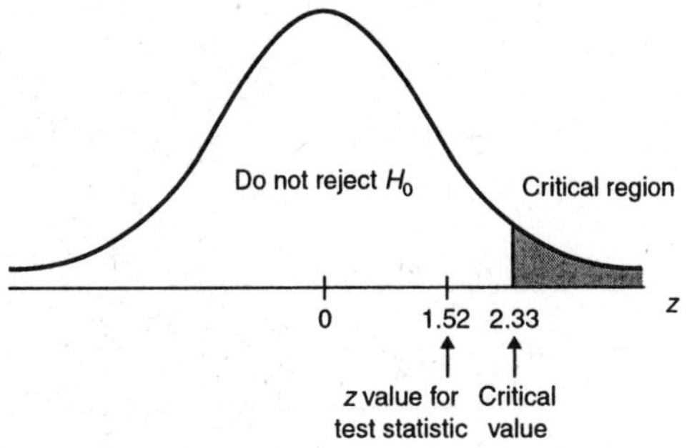

(d) The appropriate z value is

$$z = \frac{(368.4 - 349.2) - 0}{\sqrt{\frac{(39.5)^2}{30} + \frac{(56.6)^2}{30}}} = \frac{19.2}{12.6013} = 1.52$$

(e) The P value for $z = 1.52$ is $0.5 - 0.4357 = 0.0643$. Assuming there is no difference in the population means, the probability of getting a greater difference between the means of the two samples is 0.0643. Since the P value is greater than α, we fail to reject the null hypothesis.

(f) We conclude that there is not sufficient evidence to choose the hypothesis that the experimental group performed better than the control group.

7. **(a)** From the Excel output, the z value of the sample test statistic is $z = -3.72$.

(b) The alternate hypothesis for a two-tailed test is $H_1 : \mu_1 - \mu_2 \neq 0$. The P value for a two-tailed test is 0.0002. Since the P value $= 0.0002$ is less than $\alpha = 0.05$, we reject the null hypothesis and conclude that the heights of football and basketball players differ.

8. **(a)** Based on the printout, the level of confidence used for the confidence interval is 95%.

(b) At the 95% confidence level, we see that the difference between average petal length is positive, with that for *Iris virginica* longer than that for *Iris setosa.*

9. **(a)** This problem calls for the formation of a confidence interval for the difference in population means using large samples. We are given $n_1 = 32, x_1 = 69.44$ and $s_1 = 11.69$ (the data for mothers). Furthermore, $n_2 = 32$, $\bar{x}_2 = 59.00$, and $s_2 = 11.60$ (the data for fathers). The distribution of sample differences is normal with a mean difference of 0 and a standard error of

$$\sqrt{\frac{(11.69)^2}{32} + \frac{(11.60)^2}{32}} = \sqrt{8.4755} = 2.9113.$$

The point estimate for the difference is $\bar{x}_1 - \bar{x}_2 = 69.44 - 59.00 = 10.44$. The maximal value for the error of estimate is

$$E = \pm 2.58\sqrt{\frac{(11.69)^2}{32} + \frac{(11.60)^2}{32}} = \pm 2.58\sqrt{8.4755} = \pm 7.51.$$

The endpoints are $10.44 - 7.51 = 2.93$ and $10.44 + 7.51 = 17.95$. Thus, we are 99% sure that the difference in the mean empathy scores of mothers and fathers is between 2.93 and 17.95.

(b) All values for this confidence interval are positive. This indicates that we can be 99% certain that mothers have a higher empathy score than fathers.

10. **(a)** This problem calls for the formulation of a confidence interval for the difference in population means using large samples. We are given that $n_1 = 9340$, $\bar{x}_1 = 63.3$ and $s_1 = 9.17$ (the data for 1948 to 1952). Furthermore, $n_2 = 25{,}111$, $\bar{x}_2 = 72.1$, and $s_2 = 12.67$ (the data for 1983 to 1987). The distribution of sample differences is normal with a mean difference of 0 and a standard error of

$$\sqrt{\frac{(9.17)^2}{9340} + \frac{(12.67)^2}{25{,}111}} = \sqrt{0.0154} = 0.1241.$$

The point estimate for the difference is $\bar{x}_1 - \bar{x}_2 = 63.3 - 72.1 = -8.8$. The maximal value for the error of estimate is

$$E = \pm 2.58\sqrt{\frac{(9.17)^2}{9340} + \frac{(12.67)^2}{25{,}111}} = \pm 2.58\sqrt{0.0154}$$
$$= \pm 0.3202$$

The endpoints are $-8.8 - 0.3202 = -9.12$ and $-8.8 + 0.3202 = -8.48$. Thus, we are 99% sure that the difference in the mean interval length between intervals is between -9.12 and -8.48.

(b) All values for this confidence interval are negative. Yes, it does appear that a change in the interval length between eruptions has occurred since the interval does not contain the hypothesized mean difference value of 0.

11. **(a)** This problem calls for the formation of a confidence interval for the difference in population means using large samples. We are given that $n_1 = 51$, $\bar{x}_1 = 74.04$, and $s_1 = 17.19$ (the data for the Cache la Poudre Region). Furthermore, $n_2 = 36$, $\bar{x}_2 = 94.53$, and $s_2 = 19.66$ (the data for the Mesa Verde Region). The distribution of sample differences is normal with a mean difference of 0 and standard error of

$$\sqrt{\frac{(17.19)^2}{51} + \frac{(19.66)^2}{36}} = \sqrt{16.5306} = 4.0658.$$

The point estimate for the difference is $\bar{x}_1 - \bar{x}_2 = 74.04 - 94.53 = -20.49$. The maximal value for the error of estimate is $E = \pm 1.96\sqrt{\frac{(17.19)^2}{51} + \frac{(19.66)^2}{36}} = \pm 1.96\sqrt{16.5306} = \pm 7.9689$

The endpoints are $-20.49 - 7.9689 = -28.46$ and $-20.49 + 7.9689 = -12.52$. Thus, we are 95% sure that the difference in the mean weight of all bucks in the Cache la Poudre Region and the Mesa Verde Region is between -28.46 and -12.52.

(b) All values in this confidence interval are negative. This indicates that we can be 95% certain that the mean weight of bucks for the Cache la Poudre region is less than those in the Mesa Verde Region.

12. **(a)** This problem calls for the formation of a confidence interval for the difference in population means using large samples. We are given that $n_1 = 47$, $\bar{x}_1 = 483.43$, and $s_1 = 126.62$ (the data for males). Furthermore, $n_2 = 51$, $\bar{x}_2 = 414.43$, and $s_2 = 105.99$ (the data for females). The distribution of sample differences is normal with a mean difference of 0 and a standard error of

$$\sqrt{\frac{(126.62)^2}{47} + \frac{(105.99)^2}{51}} = \sqrt{561.3918} = 23.6937.$$

The point estimate for the difference is $\bar{x}_1 - \bar{x}_2 = 483.43 - 414.43 = 69$. The maximal value for the error of estimate is

$$E = \pm 1.645\sqrt{\frac{(126.62)^2}{47} + \frac{(105.99)^2}{51}} = \pm 1.645\sqrt{561.3918}$$
$$= \pm 38.9761.$$

The endpoints are $69 - 38.9761 = 30.02$ and $69 + 38.9761 = 107.98$. Thus, we are 90% sure that the difference in the mean annual premium among males and females is between \$30.02 and \$107.98.

(b) All values in this confidence interval are positive. This indicates that we can be 90% certain that males have a higher annual premium than females.

13. For the same data, the 99% confidence interval will contain all the values that a 95% confidence interval will. So, if the 95% interval contained both positive and negative values, the 99% interval will, too. The 90% confidence interval is smaller and as such will not contain all the value that the 95% interval does, so it might not contain both positive and negative values.

Selected Solutions for Section 11.3

1. (a) The means and standard deviations round to the results given.

(b) H_0: $\mu_1 = \mu_2$

H_1: $\mu_1 \neq \mu_2$

Since ≠ is in H_1, a two-tailed test is used. Since the samples are both small, we use the Student's t distribution. For $\alpha'' = 0.05$ and $d.f. = 16 + 15 - 2 = 29$, the critical values are $t_0 = \pm 2.045$.

$$s = \sqrt{\frac{(n_1-1)s_1^2 + (n_2-1)s_2^2}{n_1+n_2-2}} = \sqrt{\frac{15(2.82)^2 + 14(2.43)^2}{16+15-2}} = 2.6389$$

$$\bar{x}_1 - \bar{x}_2 = 4.75 - 3.93 = 0.82$$

$$t = \frac{(\bar{x}_1 - \bar{x}_2) - (\mu_1 - \mu_2)}{s\sqrt{\frac{1}{n_1} + \frac{1}{n_2}}} = \frac{0.82 - 0}{2.6389\sqrt{\frac{1}{16} + \frac{1}{15}}} = 0.865$$

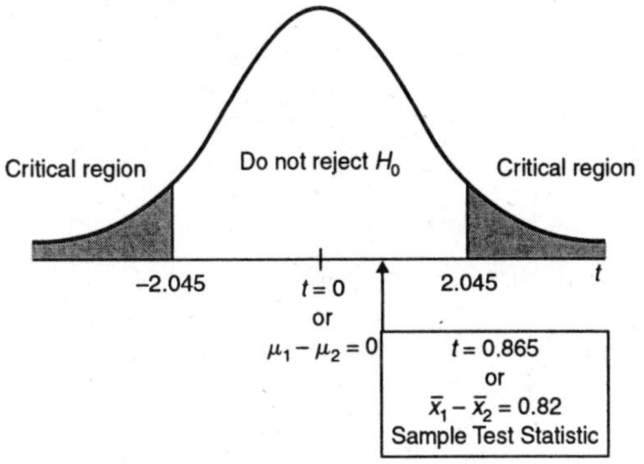

Next, find the P value interval using the α'' values and the row headed by $d.f. = 29$. We find the sample t value, $t = 0.85$, falls to the left of 1.174. Therefore, P value > 0.250.

Since the sample test statistic falls outside the critical region and the P value is greater than the level of significance $\alpha = 0.05$, we do not reject H_0. There is insufficient evidence to conclude a difference exists in the mean number of cases of fox rabies between the two regions.

2. **(a)** The means and standard deviations round to the results given.

(b) H_0: $\mu_1 = \mu_2$

H_1: $\mu_1 > \mu_2$

Since > is in H_1, a right-tailed test is used. Since the samples are both small, we use the Student's t distribution. For $\alpha' = 0.05$ and $d.f. = 14 + 15 - 2 = 27$, the critical value is $t_0 = 1.703$.

$$s = \sqrt{\frac{(n_1-1)s_1^2 + (n_2-1)s_2^2}{n_1+n_2-2}} = \sqrt{\frac{13(2.39)^2 + 14(2.44)^2}{14+15-2}} = 2.416$$

$$\bar{x}_1 - \bar{x}_2 = 12.53 - 10.65 = 1.88$$

$$t = \frac{(\bar{x}_1 - \bar{x}_2) - (\mu_1 - \mu_2)}{s\sqrt{\frac{1}{n_1} + \frac{1}{n_2}}} = \frac{1.88 - 0}{2.416\sqrt{\frac{1}{14} + \frac{1}{15}}} = 2.09$$

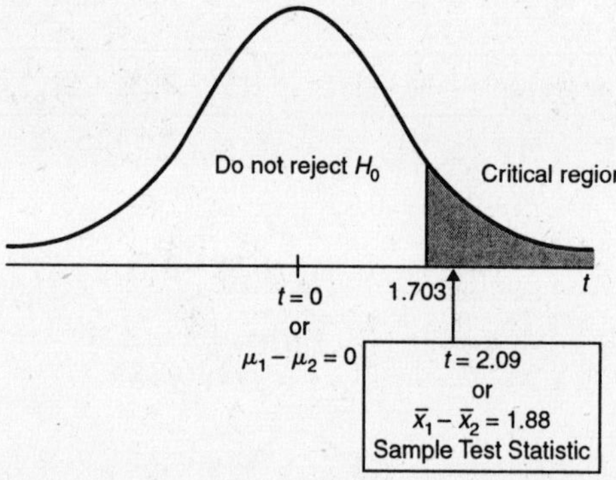

Next, find the P value interval using the α' values and the row headed by $d.f. = 27$. We find the sample t value, $t = 2.09$, falls between 2.052 and 2.473. Therefore, $0.010 < P$ value < 0.025.

Since the sample test statistic falls inside the critical region and the P value is less than the level of significance $\alpha = 0.05$, we reject H_0. We conclude that field A has on average higher soil water content than field B.

3. **(a)** The means and standard deviations round to the results given.

(b) H_0: $\mu_1 = \mu_2$

H_1: $\mu_1 \neq \mu_2$

Since ≠ is in H_1, a two-tailed test is used. Since the samples are both small, we use the Student's t distribution. For $\alpha'' = 0.05$ and $d.f. = 7 + 7 - 2 = 12$, the critical values are $t_0 = \pm 2.179$.

$$s = \sqrt{\frac{(n_1-1)s_1^2 + (n_2-1)s_2^2}{n_1+n_2-2}} = \sqrt{\frac{6(3.18)^2 + 6(4.11)^2}{7+7-2}} = 3.6745$$

$$\bar{x}_1 - \bar{x}_2 = 4.86 - 4.71 = 0.15$$

$$t = \frac{(\bar{x}_1 - \bar{x}_2) - (\mu_1 - \mu_2)}{s\sqrt{\frac{1}{n_1} + \frac{1}{n_2}}} = \frac{0.15 - 0}{3.6745\sqrt{\frac{1}{7} + \frac{1}{7}}} = 0.0764$$

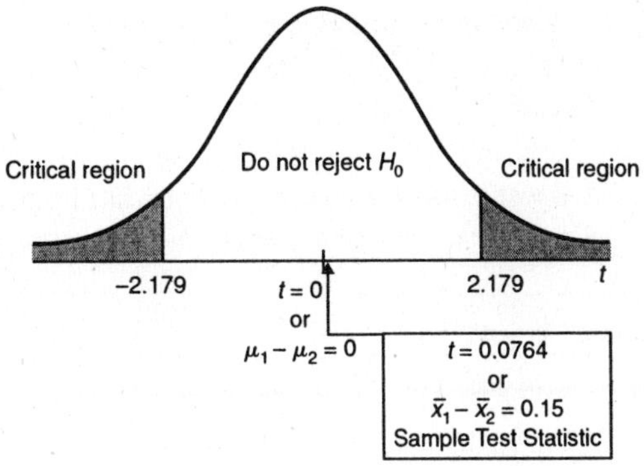

Next, find the P value interval using the α'' values and the row headed by $d.f. = 12$. We find the sample t value, $t = 0.0764$, falls to the left of 1.209. Therefore, P value > 0.250.

Since the sample test statistic falls outside the critical region and the P value is greater than the level of significance $\alpha = 0.05$, we do not reject H_0. There is insufficient evidence to conclude that the population mean time lost for hot tempers is different from that lost due to disputes.

4. (a) The means and standard deviations round to the results given.

(b) $H_0: \mu_1 = \mu_2$

$H_1: \mu_1 < \mu_2$

Since < is in H_1, a left-tailed test is used. Since the samples are both small, we use the Student's t distribution. For $\alpha' = 0.05$ and $d.f. = 7 + 7 - 2 = 12$, the critical value is $t_0 = -1.782$.

$$s = \sqrt{\frac{(n_1-1)s_1^2 + (n_2-1)s_2^2}{n_1+n_2-2}} = \sqrt{\frac{6(2.38)^2 + 6(2.69)^2}{7+7-2}} = 2.5397$$

$$\bar{x}_1 - \bar{x}_2 = 4.00 - 4.29 = -0.29$$

$$t = \frac{(\bar{x}_1 - \bar{x}_2) - (\mu_1 - \mu_2)}{s\sqrt{\frac{1}{n_1} + \frac{1}{n_2}}} = \frac{-0.29 - 0}{2.5397\sqrt{\frac{1}{7} + \frac{1}{7}}} = -0.214$$

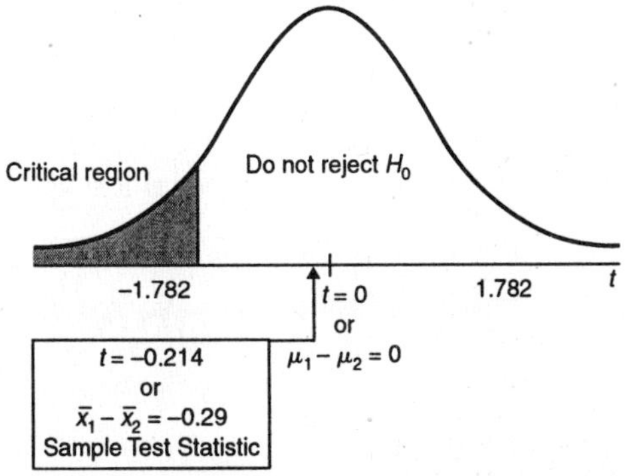

Next, find the P value interval using the α' values and the row headed by $d.f. = 12$. We find the sample t value, $t = -0.214$, falls to the right of -1.209. Therefore, P value > 0.125.

Since the sample test statistic falls outside the critical region and the P value is greater than the level of significance $\alpha = 0.05$, we fail to reject H_0. There is insufficient evidence to conclude that the population mean time lost due to stressors is greater than the population mean time lost due to intimidators.

5. (a)–(b) The means and standard deviations round to the results given.

(c) $H_0: \mu_1 = \mu_2$

$H_1: \mu_1 < \mu_2$

Since < is in H_1, a left-tailed test is used. Since the samples are both small, we use the Student's t distribution. For $\alpha' = 0.05$ and $d.f. = 10 + 12 - 2 = 20$, the critical value is $t_0 = -1.725$.

$$s = \sqrt{\frac{(n_1-1)s_1^2+(n_2-1)s_2^2}{n_1+n_2-2}} = \sqrt{\frac{9(2.7)^2+11(2.5)^2}{10+12-2}} = 2.592$$

$$\bar{x}_1 - \bar{x}_2 = 7.2 - 10.8 = -3.6$$

$$t = \frac{(\bar{x}_1-\bar{x}_2)-(\mu_1-\mu_2)}{s\sqrt{\frac{1}{n_1}+\frac{1}{n_2}}} = \frac{-3.6-0}{2.592\sqrt{\frac{1}{10}+\frac{1}{12}}} = -3.244$$

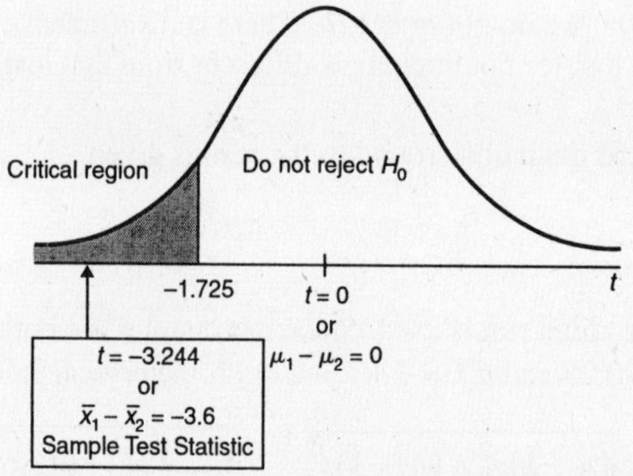

Next, find the P value interval using the α' values and the row headed by $d.f. = 20$. We find the sample t value, $t = -3.244$, falls to the left of -2.845. Therefore, P value < 0.005.

Since the sample test statistic falls inside the critical region and the P value is less than the level of significance $\alpha = 0.05$, we reject H_0. We conclude that the average change in water temperature has increased.

6. $H_0: \mu_1 = \mu_2$

$H_1: \mu_1 < \mu_2$

Since < is in H_1, a left-tailed test is used. Since the samples are both small, we use the Student's t distribution. For $\alpha' = 0.01$ and $d.f. = 11 + 15 - 2 = 24$, the critical value is $t_0 = -2.492$.

$$s = \sqrt{\frac{(n_1-1)s_1^2+(n_2-1)s_2^2}{n_1+n_2-2}} = \sqrt{\frac{10(9)^2+14(7)^2}{11+15-2}} = 7.895$$

$$\bar{x}_1 - \bar{x}_2 = 76 - 82 = -6$$

$$t = \frac{(\bar{x}_1-\bar{x}_2)-(\mu_1-\mu_2)}{s\sqrt{\frac{1}{n_1}+\frac{1}{n_2}}} = \frac{-6-0}{7.895\sqrt{\frac{1}{11}+\frac{1}{15}}} = -1.914$$

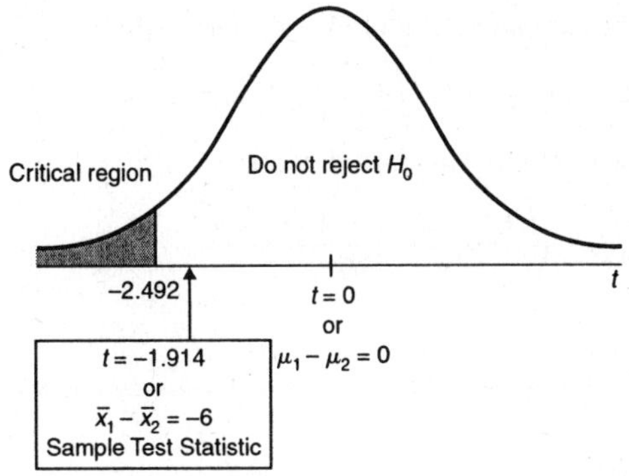

Next, find the P value interval using the α' values and the row headed by $d.f. = 24$. We find the sample t value, $t = -1.914$, falls between -1.711 and -2.064. Therefore, $0.025 < P \text{ value} < 0.050$.

Since the sample test statistic falls outside the critical region and the P value is greater than the level of significance $\alpha = 0.01$, we do not reject H_0. We have insufficient evidence to conclude that airplane pilots are less susceptible to perceptual illusions than the general population.

7. **(a)** Using a calculator, the values of the means and standard deviations are verified.

(b)

Electronic	Food/Drug
$n_1 = 15$	$n_2 = 12$
$\bar{x}_1 = 5.6$	$\bar{x}_2 = 2.5$
$s_1 = 1.0$	$s_2 = 0.9$

The difference between the sample means is $\bar{x}_1 - \bar{x}_2 = 5.6 - 2.5 = 3.1$. The error of estimate formula for a 99% confidence interval is $E = t_c S\sqrt{\frac{1}{n_1} + \frac{1}{n_2}}$. Using 15 + 12 − 2 = 25 $d.f.$, $t_c = 2.787$. Calculate the pooled estimate for the common standard deviation.

$$s = \sqrt{\frac{(n_1 - 1)s_1^2 + (n_2 - 1)s_2^2}{n_1 + n_2 - 2}} = \sqrt{\frac{(15-1)1.0^2 + (12-1)0.9^2}{25}}$$

$$= \sqrt{0.9164} = 0.9573$$

$$E = (2.787)(0.9573)\sqrt{\frac{1}{15} + \frac{1}{12}} = 1.03$$

$$3.1 + E = 3.1 + 1.03 = 4.13$$

$$3.1 - E = 3.1 - 1.03 = 2.07$$

So, the interval estimate for $\mu_1 - \mu_2$ is from 2.07 to 4.13. We are 99% sure the mean profit difference is between 2.07 and 4.13.

(c) All values in this confidence interval are all positive. This indicates that we can be 99% certain that the mean profit for electronic companies is higher than those of food and drug companies.

8. **(a)** Using a calculator, the values of the means and standard deviations are verified.

(b)

Insurance	Healthcare
$n_1 = 12$	$n_2 = 10$
$\bar{x}_1 = 4.4$	$\bar{x}_2 = 4.6$
$s_1 = 1.6$	$s_2 = 1.4$

The difference between the sample means is $\bar{x}_1 - \bar{x}_2 = 4.4 - 4.6 = -0.2$. The error of estimate formula for a 90% confidence interval is $E = t_c s\sqrt{\frac{1}{n_1}+\frac{1}{n_2}}$. Using 12 + 10 − 2 = 20 *d.f.*, $t_c = 1.725$. Calculate the pooled estimate for the common standard deviation.

$$s = \sqrt{\frac{(n_1-1)s_1^2 + (n_2-1)s_2^2}{n_1+n_2-2}} = \sqrt{\frac{(12-1)1.6^2 + (10-1)1.4^2}{20}}$$

$$= \sqrt{2.29} = 1.5133$$

$$E = (1.725)(1.5133)\sqrt{\frac{1}{12}+\frac{1}{10}} = 1.12$$

$$-0.2 + E = -0.2 + 1.12 = 0.9$$

$$-0.2 - E = -0.2 + 1.12 = -1.3$$

So, the interval estimate for $\mu_1 - \mu_2$ is from −1.3 to 0.9. We are 90% sure that the mean annual profit difference is between −1.3 and 0.9.

(c) The values in this confidence interval are of different signs. At the 90% level of confidence we can not say that one industry group has a higher profit.

9.

Competence	Social Acceptance	Attractiveness
$n_1 = 15$	$n_2 = 15$	$n_3 = 15$
$\bar{x}_1 = 19.84$	$\bar{x}_2 = 19.32$	$\bar{x}_3 = 17.88$
$s_1 = 3.07$	$s_2 = 3.62$	$s_3 = 3.74$

(a) The difference between the sample means is $\bar{x}_1 - \bar{x}_2 = 19.84 - 19.32 = 0.52$. The error of estimate formula for an 85% confidence interval is $E = t_c s\sqrt{\frac{1}{n_1}+\frac{1}{n_2}}$. Using 15 + 15 − 2 = 28 *d.f.*, $t_c = 1.480$.

Calculate the pooled estimate for the common standard deviation.

$$s = \sqrt{\frac{(n_1-1)s_1^2 + (n_2-1)s_2^2}{n_1+n_2-2}} = \sqrt{\frac{(15-1)3.07^2 + (15-1)3.62^2}{28}}$$

$$= \sqrt{11.26465} = 3.3563$$

$$E = (1.48)(3.3563)\sqrt{\frac{1}{15}+\frac{1}{15}} \approx 1.81$$

$$0.52 + E = 0.52 + 1.81 = 2.33$$

$$0.52 - E = 0.52 - 1.81 = -1.29$$

So the interval estimate for $\mu_1 - \mu_2$ is from −1.29 to 2.33. The probability that this interval contains the true difference of the means is 0.85. Note that zero is in the confidence interval so there is no evidence of a difference in means of index of self-esteem based on competence and social acceptance.

(b) The difference between the sample means is $\bar{x}_1 - \bar{x}_3 = 19.84 - 17.88 = 1.96$. The error of estimate formula for an 85% confidence interval is $E = t_c s\sqrt{\frac{1}{n_1} + \frac{1}{n_2}}$. Using 15 + 15 – 2 = 28 *d.f.*, $t_c = 1.480$.

Calculate the pooled estimate for the common standard deviation.

$$s = \sqrt{\frac{(n_1 - 1)s_1^2 + (n_3 - 1)s_3^2}{n_1 + n_3 - 2}} = \sqrt{\frac{(15-1)3.07^2 + (15-1)3.74^2}{28}}$$
$$= \sqrt{11.70625} = 3.4214$$
$$E = (1.48)(3.4214)\sqrt{\frac{1}{15} + \frac{1}{15}} \approx 1.85$$
$$1.96 + E = 1.96 + 1.85 = 3.81$$
$$1.96 - E = 1.96 - 1.85 = 0.1$$

So the interval for $u_1 - \mu_3$ is from 0.11 to 3.81. The probability that this interval contains the true difference of the means of 0.85. Since zero is not in the confidence interval we conclude that the mean of index of self-esteem based on competence is larger than the mean of index of self esteem based on physical attractiveness.

(c) The difference between the sample means is $\bar{x}_2 - \bar{x}_3 = 19.32 - 17.88 = 1.44$. The error of estimate formula for an 85% confidence interval is $E = t_c s\sqrt{\frac{1}{n_1} + \frac{1}{n_2}}$. Using 15 + 15 – 2 = 28 *d.f.*, $t_c = 1.480$.

Calculate the pooled estimate for the common standard deviation.

$$s = \sqrt{\frac{(n_2 - 1)s_2^2 + (n_3 - 1)s_3^2}{n_2 + n_3 - 2}} = \sqrt{\frac{(15-1)3.62^2 + (15-1)3.74^2}{28}}$$
$$= \sqrt{13.546} = 3.6805$$
$$E = (1.48)(3.68)\sqrt{\frac{1}{15} + \frac{1}{15}} \approx 1.99$$
$$1.44 + E = 1.44 + 1.99 = 3.43$$
$$1.44 - E = 1.44 - 1.99 = -0.55$$

So the interval estimate for $\mu_2 - \mu_3$ is from –0.55 to 3.43. Since zero is in the confidence interval there is no evidence of a difference in means of index of self-esteem based on social acceptance and attractiveness.

We cannot conclude that there is a difference in mean index of self-esteem based on competence and that based on social acceptance. We cannot conclude that there is a difference in the mean indices based on social acceptance and physical attractiveness.

(d) At the 85% confidence level we can say that the mean index of self-esteem based on competence is greater than the mean index of self-esteem based on physical attractiveness.

10. (a) Using a calculator, the values of the means and standard deviations are verified.

(b)

<u>Chihuahua</u>	<u>Durango</u>
$n_1 = 10$	$n_2 = 18$
$\bar{x}_1 = 75.80$	$\bar{x}_1 = 66.83$
$s_1 = 8.32$	$s_2 = 8.87$

The difference between the sample means is $\bar{x}_1 - \bar{x}_2 = 75.80 - 66.83 = 8.97$. The error of estimate formula for a 85% confidence interval is $E = t_c s\sqrt{\frac{1}{n_1} + \frac{1}{n_2}}$. Using 10 + 18 − 2 = 26 *d.f.*, $t_c = 1.483$.

Calculate the pooled estimate for the common standard deviation.

$$s = \sqrt{\frac{(n_1-1)s_1^2 + (n_2-1)s_2^2}{n_1+n_2-2}} = \sqrt{\frac{(10-1)8.32^2 + (18-1)8.87^2}{26}}$$

$$= \sqrt{75.40419} = 8.6836$$

$$E = (1.483)(8.6836)\sqrt{\frac{1}{10} + \frac{1}{18}} = 5.08$$

$8.97 + E = 8.97 + 5.08 = 14.05$
$8.97 + E = 8.97 - 5.08 = 3.89$

So, the interval estimate for $\mu_1 - \mu_2$ is from 3.89 to 14.05. We are 85% sure that the mean weight difference is between 3.89 and 14.05.

(c) The values in this confidence interval are all positive. This indicates that we can be 85% certain that the mean weight for the Chihuahua region is higher than that of the Durango region.

11. (a) The values in this confidence interval are of different signs. At the 95% confidence level there does not seem to be a difference in the average number of children for low- and high-income families since 0 lies in the confidence interval.

(b) The *P* value for this test is 0.47. Since the *P* value is greater than $\alpha = 0.05$, we fail to reject the null hypothesis. Therefore, we conclude that there is not sufficient evidence to say that family income makes a difference in family size. This result is consistent with the result we found in part (a).

Section 11.4

1. (a) $H_0 : p_1 = p_2$ and $H_1 : p_1 \neq p_2$. The test is two-tailed because we are looking for any differences in the high school dropout rate between Oahu, Hawaii, and Sweetwater, Wyoming. The given level of significance is $\alpha = 0.01$.

(b) We will use the sampling distribution of the difference of two proportions which is a normal distribution. The critical value of z_0 is ± 2.58.

(c)

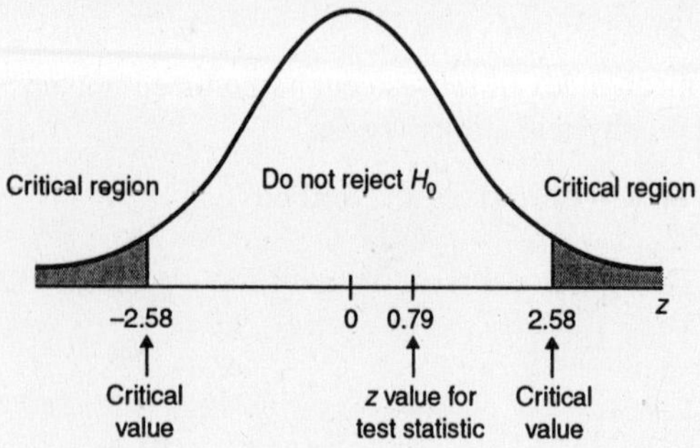

(d) We calculate that $\hat{p}_1 - \hat{p}_2 = 0.0784 - 0.0547 = 0.0237$. Thus, $\hat{p} = \frac{12+7}{153+128} = \frac{19}{281} = 0.0676$.

Remember, $\hat{p}$ represents the dropout rate if there is no difference.

Since $\hat{p} = 0.0676$, $\hat{q} = 1 - \hat{p} = 0.9324$. Then,

$$z = \frac{(\hat{p}_1 - \hat{p}_2) - 0}{\sqrt{\frac{\hat{p}\hat{q}}{n_1} + \frac{\hat{p}\hat{q}}{n_2}}} = \frac{0.0237 - 0}{\sqrt{\frac{(0.0676)(0.9324)}{154} + \frac{(0.0676)(0.9324)}{128}}} = \frac{0.0237}{0.03003} = 0.79$$

Since $0.79 < 2.58$, we fail to reject H_0

(e) The P value for this test is found by looking up 0.79 in the Standard Normal Table and subtracting from 0.5. Since the test is two tailed, we then multiply by 2. So the area under the tail is 0.5 – 0.2852 = 0.2148. Thus, $P = 2(0.2148) = 0.4296$. Since $P > 0.01$, we fail to reject H_0.

(f) This indicates that there is not a significant difference in the dropout rate of Oahu, Hawaii, and Sweetwater, Wyoming.

2. **(a)** $H_0 : p_1 = p_2$ and $H_1 : p_1 < p_2$. The test is left-tailed because we are looking for a negative difference in the proportion of voter turnout between Colorado and California. The given level of significance is $\alpha = 0.05$.

(b) We will use the sampling distribution of the difference of two proportions which is a normal distribution. The critical value of z_0 is -1.645.

(c)

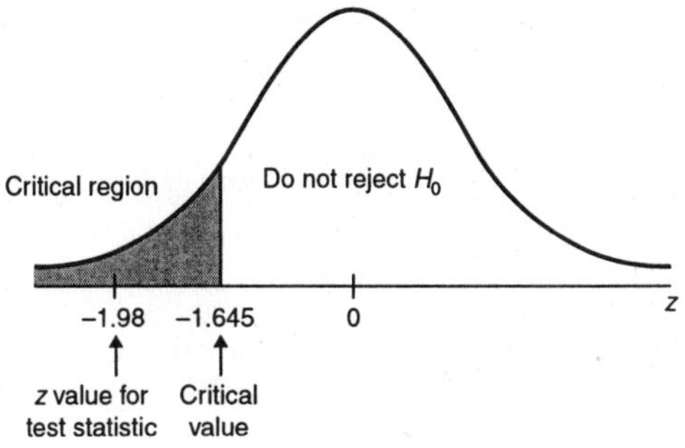

(d) We calculate that $\hat{p}_1 - \hat{p}_2 = 0.4896 - 0.5787 = -0.0891$.

Thus, $\hat{p} = \frac{141+125}{288+216} = \frac{266}{504} = 0.5278$. Remember, $\hat{p}$ represents the voter turnout if there is no difference. Since $\hat{p} = 0.5278, \hat{q} = 1 - \hat{p} = 0.4722$. Then,

$$z = \frac{(\hat{p}_1 - \hat{p}_2) - 0}{\sqrt{\frac{\hat{p}\hat{q}}{n_1} + \frac{\hat{p}q}{n_2}}} = \frac{-0.0891 - 0}{\sqrt{\frac{(0.5278)(0.4722)}{288} + \frac{(.5278)(0.4722)}{216}}} = \frac{-0.0891}{0.0449} = -1.98$$

Since $-1.98 < -1.645$, we reject H_0.

(e) The P value for this test is found by looking up 1.98 (since the distribution is symmetric) in the Standard Normal Table and subtracting from 0.5. So the P value is 0.5 – 0.4761 = 0.0239. Since $P < 0.05$, we reject H_0.

(f) This indicates that there is a significant difference in the voter turnout rate between California and Colorado. The rate is less in California than in Colorado.

3. **(a)** $H_0 : p_1 = p_2$ and $H_1 : p_1 < p_2$. The test is left-tailed because we are looking for a negative difference in the rate of people who believe in extraterrestrials between those who did and did not attend college. The given level of significance is $\alpha = 0.01$. The given level of significance is $\alpha = 0.01$.

(b) We will use the sampling distribution of the difference of two proportions which is a normal distribution. The critical value of z_0 is –2.33.

(c)

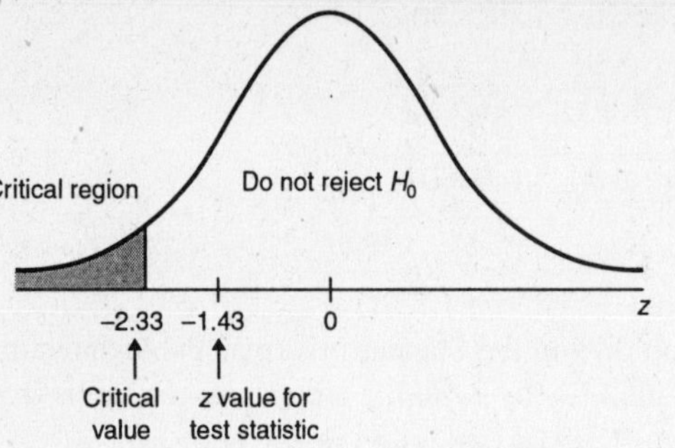

(d) We calculate that $\hat{p}_1 - \hat{p}_2 = 0.37 - 0.47 = -0.10$. Thus, $\hat{p} = \frac{37+47}{100+100} = \frac{84}{200} = 0.42$. Remember, $\hat{p}$ represents the proportion of people who believe in extraterrestrials if there is no difference. Since $\hat{p} = 0.42$, $\hat{q} = 1 - \hat{p} = 0.58$. Then,

$$z = \frac{(\hat{p}_1 - \hat{p}_2) - 0}{\sqrt{\frac{\hat{p}\hat{q}}{n_1} + \frac{\hat{p}\hat{q}}{n_2}}} = \frac{(0.37 - 0.47) - 0}{\sqrt{\frac{(0.42)(0.58)}{100} + \frac{(0.42)(0.58)}{100}}} = \frac{-0.10}{0.0698} = -1.43$$

Since $-1.43 > -2.33$, we fail to reject H_0.

(e) The *P* value for this test is found by looking up 1.43 (since the distribution is symmetric) in the Standard Normal Table and subtracting from 0.5. Thus, *P* = 0.5 0.4236 = 0.0764. Since *P* > 0.01, we fail to reject H_0.

(f) This indicates that there is not a significant difference in the proportion of people who believe in extraterrestrials among those who did and did not attend college.

4. **(a)** $H_0: p_1 = p_2$ and $p_1 > p_2$. The test is a right-tailed test because we are looking for any positive differences in the rate of organ donation between people who would donate a loved one's organs after death and those who would donate their own organs after death. The given level of significance is $\alpha = 0.01$.

(b) We will use the sampling distribution of the difference of two proportions which is a normal distribution. The critical value of z_0 is 2.33.

(c)

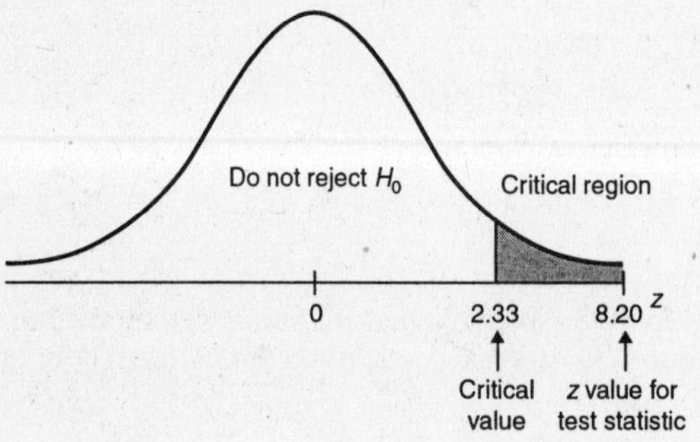

(d) We calculate that $\hat{p}_1 - \hat{p}_2 = 0.78 - 0.20 = 0.58$. Thus, $\hat{p} = \dfrac{78+20}{100+100} = \dfrac{98}{200} = 0.49$. Remember, $\hat{p}$ represents the organ donation rate if there is no difference. Since $\hat{p} = 0.49$, $\hat{q} = 1 - \hat{p} = 0.51$. Then

$$z = \frac{(\hat{p}_1 - \hat{p}_2) - 0}{\sqrt{\frac{\hat{p}\hat{q}}{n_1} + \frac{\hat{p}\hat{q}}{n_2}}} = \frac{(0.78 - 0.20) - 0}{\sqrt{\frac{(0.49)(0.51)}{100} + \frac{(0.49)(0.51)}{100}}} = \frac{0.58}{0.0707} = 8.20$$

Since $8.20 > 2.33$, we reject H_0.

(e) The P value for this test is found by looking up 8.20 in the Standard Normal Table and subtracting from 0.5. Since the table only goes up to 3.69, we know that the P value for 8.20 iess less than that for 3.69. So, the P value for 3.69 is $0.5 - 0.4999 = 0.0001$. Therefore, the P value for 8.20 is less than 0.0001. Since $P < 0.001$, we reject H_0.

(f) This indicates that the proportion of American adults that would donate a loved one's organs after death is greater than the proportion of American adults who would donate their own organs after death.

5. **(a)** $H_0: p_1 = p_2$ and $p_1 \neq p_2$. The test is two-tailed because we are looking for any differences in the proportion of artifacts between the two areas. The given level of significance is $\alpha = 0.05$.

(b) We will use the sampling distribution of the difference of two proportions which is'a normal distribution. The critical value of z_0 is ± 1.96.

(c)

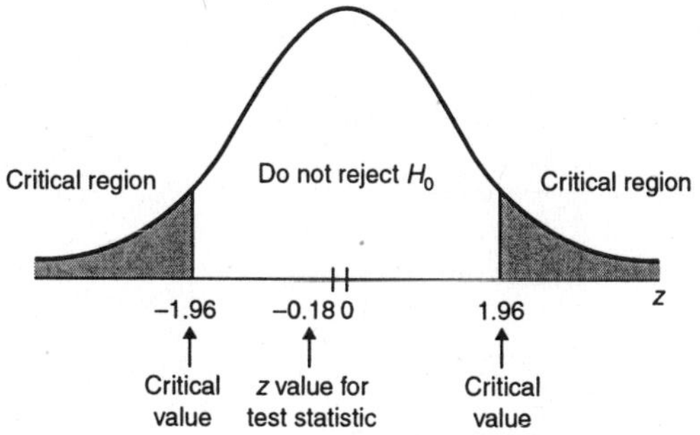

(d) We calculate that $\hat{p}_1 - \hat{p}_2 = 0.0225 - 0.245 = -0.0020$. Thus, $\hat{p} = \dfrac{10+8}{444+326} = \dfrac{18}{770} = 0.0234$.

Remember, $\hat{p}$ represents the proportion of artifacts if there is no difference.

Since $\hat{p} = 0.0234$, $\hat{q} = 1 - \hat{p} = 0.9766$. Then

$$z = \frac{(\hat{p}_1 - \hat{p}_2) - 0}{\sqrt{\frac{\hat{p}\hat{q}}{n_1} + \frac{\hat{p}\hat{q}}{n_2}}} = \frac{-0.002}{\sqrt{\frac{(0.0234)(0.9766)}{444} + \frac{(0.0234)(0.9766)}{326}}} = \frac{-0.002}{0.01103} = -0.18$$

Since $-0.18 > -1.96$, we fail to reject H_0.

(e) The area under the tail is $0.5 - 0.0714 = 0.4286$. Thus, $P = 2(0.4286) = 0.8572$. Since $P > 0.05$, we fail to reject H_0.

(f) This indicates that there is not a significant difference in the proportion of indulgence artifacts found at the two sites.

6. **(a)** $H_0: p_1 = p_2$ and $p_1 > p_2$. The test is right-tailed because we are looking for a positive difference in the proportion of artifacts between the two sites. The given level of significance is $\alpha = 0.05$.

(b) We will use the sampling distribution of the difference of two proportions, which is a normal distribution. The critical value of z_0 is 1.645.

(c)

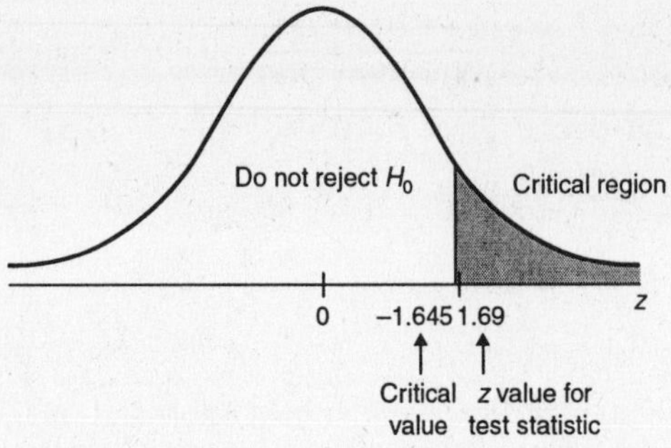

(d) We calculate that $\hat{p}_1 - \hat{p}_2 = 0.0533 - 0.0274 = 0.0259$. Thus, $\hat{p} = \dfrac{18+9}{338+329} = \dfrac{27}{667} = 0.0405$.

Remember, $\hat{p}$ represents the proportion of artifacts if there was no difference.

Since $\hat{p} = 0.0405$, $\hat{q} = 1 - \hat{p} = 0.9595$. Then,

$$z = \frac{(\hat{p}_1 - \hat{p}_2) - 0}{\sqrt{\dfrac{\hat{p}\hat{q}}{n_1} + \dfrac{\hat{p}\hat{q}}{n_2}}} = \frac{(0.0533 - 0.0274) - 0}{\sqrt{\dfrac{(0.0405)(0.9595)}{338} + \dfrac{(0.0405)(0.9595)}{329}}} = \frac{0.0259}{0.0153} = 1.69$$

Since $1.69 > 1.645$, we reject H_0.

(e) The P value for this test is found by looking up 1.69 in the Standard Normal Table and subtracting from 0.5 so the P value is $0.5 - 0.4545 = 0.0455$. Since $P < 0.05$, we reject H_0.

(f) This indicates that the proportion of artifacts found in one site is greater than the proportion of artifacts found in another site.

7. **(a)** First, we calculate that $\hat{p}_1 = \dfrac{289}{375} = 0.7707$ and $\hat{p}_2 = \dfrac{23}{571} = 0.0403$.

Thus, $\hat{q}_1 = 1 - \hat{p}_1 = 0.2293$ and $\hat{q}_2 = 1 - \hat{p}_2 = 0.9597$. Then, calculate that $\hat{p}_1 - \hat{p}_2 = 0.7707 - 0.0403 = 0.7304$. So,

$$\sigma = \sqrt{\frac{(0.7707)(0.2293)}{375} + \frac{(0.0403)(0.9597)}{571}} = 0.0232.$$

The appropriate z value is $z_c = 2.58$. So, $E = z_c\hat{\sigma} = 2.58(0.0232) = 0.0599$.

Finally, the endpoints are $0.7304 + 0.0599 = 0.7903$ and $0.7304 - 0.0599 = 0.6705$.

Thus, $0.6705 \le p_1 - p_2 \le 0.7903$ or $0.67 \le p_1 - p_2 \le 0.79$.

(b) Since the confidence interval is all positive, this indicates that there is a significant difference between the proportion of couples who have two or more personality preferences in common and those who have none. This would indicate that there is a higher proportion of married couples who have two or more personality preferences in common than those who have no personality preferences in common.

8. **(a)** First, we calculate that $\hat{p}_1 = \dfrac{132}{375} = 0.3520$ and $\hat{p}_2 = \dfrac{217}{571} = 0.3800$. Thus,

$\hat{q}_1 = 1 - \hat{p}_1 = 0.6480$ and $\hat{q}_2 = 1 - \hat{p}_2 = 0.6200$. Then, calculate that

$\hat{p}_1 - \hat{p}_2 = 0.3520 - 0.3800 = -0.0280$. So,

$$\hat{\sigma} = \sqrt{\frac{(0.3520)(0.6480)}{375} + \frac{(0.38)(0.62)}{571}} = 0.03195.$$

The appropriate z value is $z_0 = 1.645$. So, $E = z_c\hat{\sigma} = 1.645(0.03195) = 0.0526$.

Finally, the endpoints are $-0.0280 + 0.0526 = 0.0246$ and $-0.0280 - 0.0526 = -0.0806$.

Thus, $-0.0806 \le p_1 - p_2 \le 0.0246$ or $-0.08 \le p_1 - p_2 \le 0.02$.

(b) Since the confidence interval contains both positive and negative values, this indicates that there is no significant difference between the proportion of married couples who have two personality preferences in common and those who have three personality preferences in common (0 is in the confidence interval).

9. **(a)** First, we calculate that $\hat{p}_1 = \dfrac{65}{210} = 0.3095$ and $\hat{p}_2 = \dfrac{18}{152} = 0.1184$. Thus, $\hat{q}_1 = 1 - \hat{p}_1 = 0.6905$ and $\hat{q}_2 = 1 - \hat{p}_2 = 0.8816$. Then, calculate that

$\hat{p}_1 - \hat{p}_2 = 0.3095 - 0.1184 = 0.1911$. So, $\hat{\sigma} = \sqrt{\dfrac{(0.3095)(0.6905)}{210} + \dfrac{(0.1184)(0.8816)}{152}} = 0.0413$.

The appropriate z value is $z_c = 2.58$. So, $E = z_c\hat{\sigma} = 2.58(0.0413) = 0.1066$.

Finally, the endpoints are $0.1911 = 0.1066 = 0.2977$ and $0.1911 - 0.1066 = 0.0845$.

Thus, $0.0845 \le p_1 - p_2 \le 0.2977$ or $p_1 - p_2 \le 0.30$.

(b) Since the confidence interval is all positive, this indicates that there is a significant difference between the two area in proportions of traditional housing (0 is not in the confidence interval). This would indicate that there is a greater proportion of Navajo who follow the traditional culture of their people in the Fort Defiance region.

10. **(a)** First, we calculate that $\hat{p}_1 = \dfrac{69}{112} = 0.6161$ and $\hat{p}_2 = \dfrac{26}{140} = 0.1857$. Thus, $\hat{q}_1 = 1 - \hat{p}_1 = 0.3839$ and $\hat{q}_2 = 1 - \hat{p}_2 = 0.8143$. Then, calculate that

$\hat{p}_1 - \hat{p}_2 = 0.6161 - 0.1857 = 0.4304$. So, $\hat{\sigma} = \sqrt{\dfrac{(0.6161)(0.3839)}{112} + \dfrac{(0.1857)(0.8143)}{140}} = 0.0565$.

The appropriate z value is $z_c = 2.58$. So, $E = z_c\hat{\sigma} = 2.58(0.0565) = 0.1458$.

Finally, the endpoints are $0.4304 + 0.1458 = 0.5762$ and $0.4304 - 0.1458 = 0.2846$.

Thus, $0.2846 \le p_1 - p_2 \le 0.5762$ or $0.28 \le p_1 - p_2 \le 0.58$.

(b) Since the confidence interval is all positive, this indicates that there is a significant difference between the proportion of unidentifiable artifacts at different elevations. (0 is not in the confidence interval.) This would indicate that there is a greater proportion of unidentifiable artifacts at the higher elevation than at the lower elevation.

11. **(a)** The confidence interval contains all positive values. Based on the confidence interval it does appear that group I placement of nesting boxes produces a higher hatch proportion.

(b) The P value for the test is 0.000. Since $p < 0.01$, we reject H_0. This is consistent with the information obtained from the 95% confidence interval.

Chapter 11 Review

1. **(I)** **(a)** Since the commercials are not related in any way, the samples are independent.

(b) Since the samples are both large ($n_1 = 142$ and $n_2 = 47$), the sampling distribution follows a normal distribution.

(c) **(i)** The hypotheses are $H_0 : \mu_1 - \mu_2 = 0$ and $H_1 : \mu_1 - \mu_2 \ne 0$. The test is two-tailed. The level of significance is $\alpha = 0.05$.

(ii) We will use a normal distribution because independent samples are used and both samples are large. The appropriate critical value is $z_0 = \pm 1.96$.

(iii)

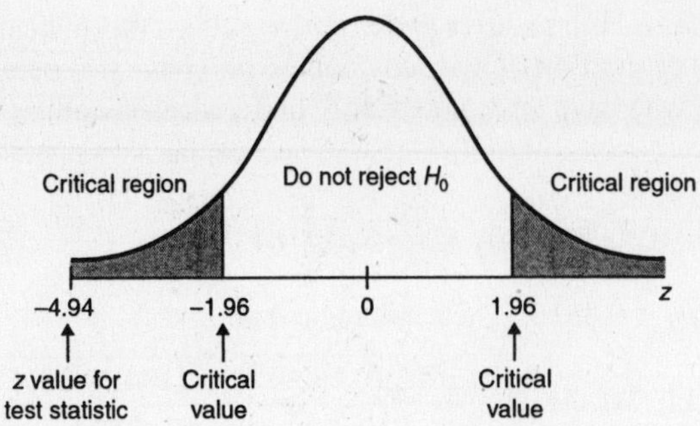

(iv) The appropriate z value is $z = \dfrac{(6.03-7.09)-0}{\sqrt{\dfrac{(1.2)^2}{142}+\dfrac{(1.3)^2}{47}}} = \dfrac{-1.06}{0.2147} = -4.94$

(v) P value $= 2(0.5 - 0.4999) = 0.0002$

(vi) Since $-4.94 < -1.96$ and $P < 0.05$, we reject H_0. We conclude that there is a difference in the ratings for 30-second ads and 60-second ads.

(II) First, we calculate the standard error.

$$\sqrt{\frac{(1.2)^2}{142}+\frac{(1.3)^2}{47}} = 0.2147$$

The point estimate for the difference is $\bar{x}_1 - \bar{x}_2 = 6.03 - 7.09 = -1.06$. The maximal value for the error of estimate is

$$E = \pm 1.96\sqrt{\frac{(1.2)^2}{142}+\frac{(1.3)^2}{47}} = \pm 1.96(0.2147) = \pm 0.4208.$$

The endpoints are $-1.06 + 0.4208 = -0.604$ and $-1.06 - 0.4208 = -1.48$. Thus, we are 95% sure that the difference in the mean ratings between 30-second ads and 60-second ads is between -1.48 and -0.64. All values in this confidence interval are negative. This indicates that we can be 95% certain that the average ratings for the 30-second ads are lower than those for the 60-second ads.

2. **(a)** Since each rat was in the one pellet and five pellet reward program, the samples are dependent.

(b) Since the samples are small and paired, the sampling distribution follows a Student's t distribution.

(c) **(i)** The hypotheses are $H_0 : \mu_d = 0$ and $H_1 : \mu_d > 0$. This is a right-tailed test. The level of significance is $\alpha = 0.05$.

(ii) The appropriate critical value for $6 - 1 = 5$ *d.f.* is $t_0 = 2.015$.

(iii)

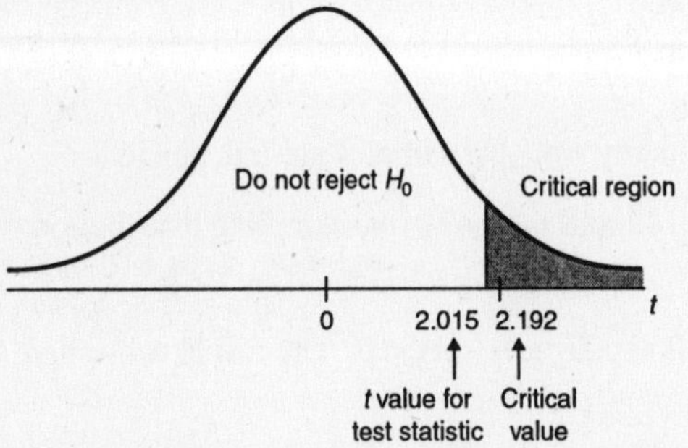

(iv) The appropriate t-value is $t = \dfrac{\bar{d}-0}{\frac{s_d}{\sqrt{n}}} = \dfrac{0.400-0}{\frac{0.447}{\sqrt{6}}} = 2.192$

(v) For 5 *d.f.*, 2.192 lies between 2.015 and 2.571. So, $0.025 < p < 0.05$.

(vi) Since $2.192 > 2.015$ and $p < 0.05$, we reject H_0. We conclude that the rats receiving larger rewards tent to run the maze in less time.

3. **(a)** Since each rat was in the one pellet and five pellet program, the samples are dependent.

(b) Since the samples are small and paired, the sampling distribution follows a Student's t distribution.

(c) **(i)** The hypotheses are $H_0 : \mu_d = 0$ and $H_1 : \mu_d > 0$.

(ii) The appropriate critical for $8 - 1 = 7$ *d.f.* is $t_0 = 1.895$.

(iii)

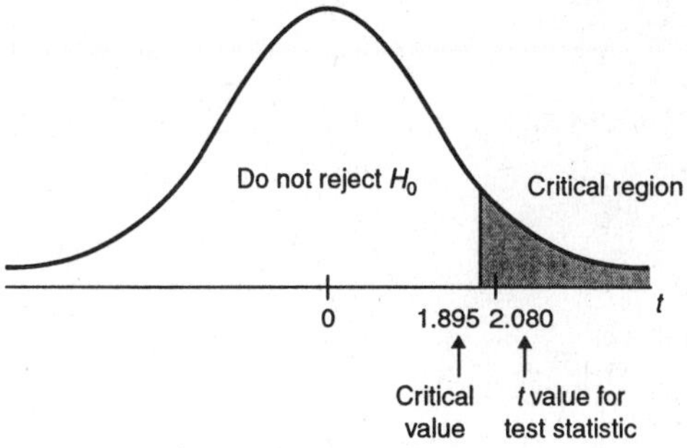

(iv) The appropriate t value is $t = \dfrac{\bar{d} - 0}{\frac{s_d}{\sqrt{n}}} = \dfrac{0.775 - 0}{\frac{1.054}{\sqrt{8}}} = 2.080$

(v) For 7 *d.f.*, 2.080 lies between 1.895 and 2.365. So, $0.025 < p < 0.05$.

(vi) Since $2.080 > 1.895$ and $P < 0.05$, we reject H_0. We conclude that the rats receiving larger rewards tend to perform the ladder climb in less time.

4. **(a)** Since the two bus lines are unrelated, the samples are independent.

(b) Since both samples are large, the sampling distribution follows a normal distribution.

(c) **(i)** The hypotheses are $H_0 : \mu_1 - \mu_2 = 0$ and $H_1 : \mu_1 - \mu_2 \neq 0$. This is a two-tailed test. The level of significance is $\alpha = 0.05$.

(ii) The appropriate critical value is $z_0 = \pm 1.96$.

(iii)

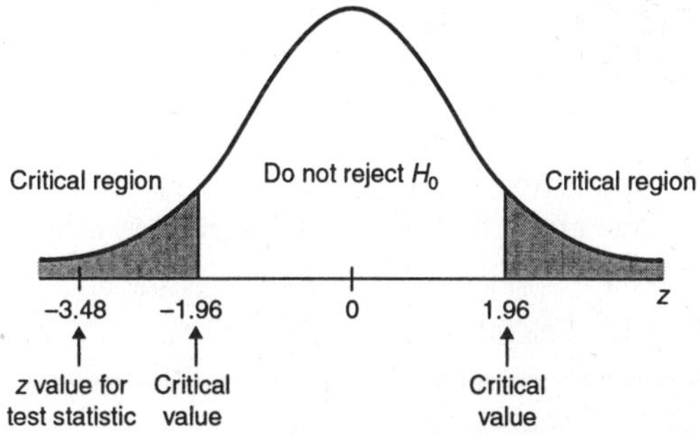

(iv) The appropriate z value is $z = \dfrac{(53 - 62) - 0}{\sqrt{\frac{19^2}{81} + \frac{15^2}{100}}} = \dfrac{-9}{2.5897} = -3.48$.

(v) The P value is $2(0.5 - 0.4997) = 0.0006$.

(vi) Since $-3.48 < -1.96$ and $p < 0.05$, we reject H_0. We conclude that there is a significant difference in the off-schedule times.

5. **(I)** **(a)** Since the people in the survey were not involved in both methods of surveying, the samples are independent.

(b) Since the sample deals with proportions and the samples are large, the sampling distribution follows a normal distribution.

(c) **(i)** The hypotheses are $H_0 : p_1 - p_2 = 0$ and $H_1 : p_1 - p_2 \neq 0$. This is a two-tailed test. The level of significance is $\alpha = 0.01$.

(ii) The appropriate critical value is $z_0 = \pm 2.58$.

(iii)

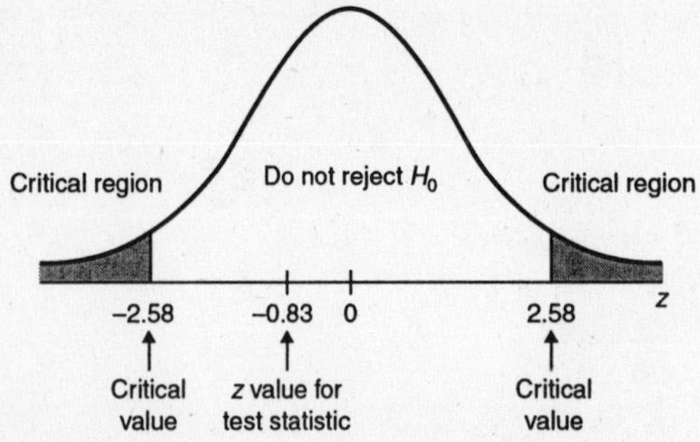

(iv) We first calculate the following statistics.

$$\hat{p}_1 = \frac{79}{93} = 0.8495,\ \hat{q}_1 = 1 - \hat{p}_1 = 0.1505$$

$$\hat{p}_2 = \frac{74}{83} = 0.8916,\ \hat{q}_2 = 1 - \hat{p}_2 = 0.1084$$

$$\hat{p} = \frac{79+74}{93+83} = \frac{153}{176} = 0.8693,\ \hat{q} = 1 - \hat{p} = 0.1307$$

So,

$$z = \frac{(\hat{p}_1 - \hat{p}_2) - 0}{\sqrt{\frac{\hat{p}\hat{q}}{n_1} + \frac{\hat{p}\hat{q}}{n_2}}} = \frac{(0.8495 - 0.8916) - 0}{\sqrt{\frac{(0.8693)(0.1307)}{93} + \frac{(0.8693)(0.1307)}{83}}}$$

$$= \frac{-0.0421}{0.0509} = -0.83.$$

(v) P value $= 0.5 - 0.2967 = 0.2033$

(vi) Since $-0.83 > -1.96$ and $P > 0.01$, we fail to reject H_0. We conclude that there is not sufficient evidence to conclude that the proportion of respondents who answer accurately during face-to-face interviews is different from the proportion who answer accurately during telephone interviews.

(II) First, we find that $\hat{p}_1 - \hat{p}_2 = 0.8495 - 0.8916 = -0.0421$ and

$$\hat{\sigma} = \sqrt{\frac{(0.8495)(0.1505)}{93} + \frac{(0.8916)(0.1084)}{83}} = 0.0504$$

The appropriate z value is $z_0 = 2.58$. So, $E = z_c\hat{\sigma} = 2.58(0.0504) = 0.13003$.

Finally, the endpoints are $-0.0421 + 0.13003 = 0.0879$ and $-0.0421 - 0.13003 = -0.1721$.

Thus, $-0.1721 \leq p_1 - p_2 \leq 0.0879$ or $-0.17 \leq p_1 - p_2 \leq 0.09$.

Since 0 does not lie in this confidence interval we do not detect any differences between the proportion of answering correctly during face-to-face interviews or telephone interviews.

6. **(I)** **(a)** Since the people in the survey were not involved in both methods of surveying, the samples are independent.

(b) Since the sample deals with proportions and the samples are large, the sampling distribution follows a normal distribution.

(c) (i) The hypotheses are $H_0 : p_1 - p_2 = 0$ and $H_1 : p_1 - p_2 \neq 0$. This is a two-tailed test. The level of significance is $\alpha = 0.05$.

(ii) The appropriate critical value is $z_0 = \pm 1.96$.

(iii)

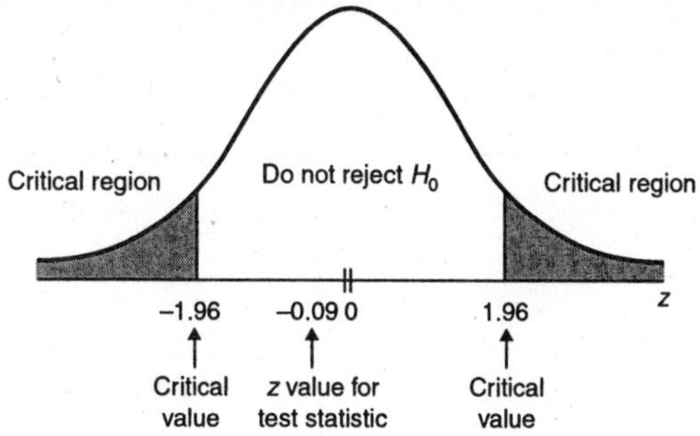

(iv) We first calculate the following statistics.

$$\hat{p}_1 = \frac{16}{30} = 0.5333,\ \hat{q}_1 = 1 - \hat{p}_1 = 0.4667$$

$$\hat{p}_2 = \frac{25}{46} = 0.5435,\ \hat{q}_2 = 1 - \hat{p}_2 = 0.4565$$

$$\hat{p} = \frac{16+25}{30+46} = \frac{41}{76} = 0.5395,\ \hat{q} = 1 - \hat{p} = 0.4605$$

So,

$$z = \frac{(\hat{p}_1 - \hat{p}_2) - 0}{\sqrt{\frac{\hat{p}\hat{q}}{n_1} + \frac{\hat{p}\hat{q}}{n_2}}} = \frac{(0.5333 - 0.5435) - 0}{\sqrt{\frac{(0.5395)(0.4605)}{30} + \frac{(0.5395)(0.4605)}{46}}}$$

$$= \frac{-0.0102}{0.1170} = -0.09$$

(v) P value = 2(0.5 – 0.0359) = 0.9282.

(vi) Since –0.09 > –1.95 and $P > 0.05$, we fail to reject H_0. We conclude that there is not a significant difference between the proportion of respondents, who answer accurately during face-to-face interviews and those who answer accurately during telephone interviews.

(II) First, we find that $\hat{p}_1 - \hat{p}_2 = 0.5333 - 0.5435 = -0.0102$ and

$$\hat{\sigma} = \sqrt{\frac{(0.5333)(0.4667)}{30} + \frac{(0.5435)(0.4565)}{46}} = 0.1170$$

The appropriate z value is $z_0 = 1.96$. So,

$E = z_c \hat{\sigma} = 1.96(0.1170) = 0.2293$

Finally, the endpoints are

–0.0102 + 0.2293 = 0.2191 and –0.0102 – 0.2293 = –0.2395.

Thus, $-0.2395 \leq p_1 - p_2 < 0.2191$ or $-0.24 \leq p_1 - p_2 \leq 0.22$

Since 0 does lie in this confidence interval we do not detect any differences in the proportion of accurate responses from the face-to-face interviews compared with the proportion of accurate responses from the phone interviews.

7. (a)

Burger Queen	McGregor's
$n_1 - 16$	$n_2 = 14$
$\bar{x}_1 = 4.8$	$\bar{x}_2 = 5.2$
$s_1 = 2.0$	$s_2 = 1.8$

The difference between the sample means is $\bar{x}_1 - \bar{x}_2 = 4.8 - 5.2 = -0.40$. The error of estimate formula for a 90% confidence interval is

$E = t_c s\sqrt{\frac{1}{n_1}+\frac{1}{n_2}}$. Using $16+14-2=28\ d.f.$,

$t_c = 1.701$. Calculate the pooled estimate for the common standard deviation.

$$s = \sqrt{\frac{(n_1-1)s_1^2+(n_2-1)s_2^2}{n_1+n_2-2}} = \sqrt{\frac{(16-1)2.0^2+(14-1)1.8^2}{28}} = 1.9097$$

$$E = (1.701)(1.9097)\sqrt{\frac{1}{16}+\frac{1}{14}} = 1.19$$

$-0.40 + E = -0.40 + 1.19 = 0.79$

$-0.40 - E = -0.40 - 1.19 = -1.59$

So, the interval estimate for $\mu_1 - \mu_2$ is from -1.59 to 0.79. We are 90% sure that the difference in mean waiting times is between -1.59 and 0.79.

(b) This confidence interval contains both positive and negative numbers. We conclude that there does not appear to be a difference in average waiting times at the 90% confidence level.

8. We are testing the difference of means with small independent samples.

$H_0: \mu_1 = \mu_2$

$H_1: \mu_1 > \mu_2$

Since > is in H_1, a right-tailed test is used. For $\alpha' = 0.01$ and $d.f. = 12 + 12 - 2 = 22$, the critical value is $t_0 = 2.508$.

$$s = \sqrt{\frac{(n_1-1)s_1^2+(n_2-1)s_2^2}{n_1+n_2-2}} = \sqrt{\frac{11(2.1)^2+11(2.0)^2}{12+12-2}} = 2.0506$$

$\bar{x}_1 - \bar{x}_2 = 9.4 - 6.8 = 2.6$

$$t = \frac{\bar{x}_1 - \bar{x}_2}{s\sqrt{\frac{1}{n_1}+\frac{1}{n_2}}} = \frac{2.6}{2.0506\sqrt{\frac{1}{12}+\frac{1}{12}}} = 3.106$$

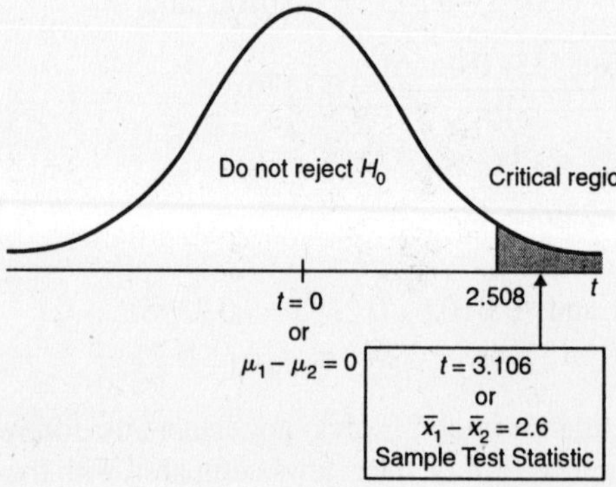

Next, find the P value interval using the α' values and the row headed by $d.f. = 22$. We find the sample t value, $t = 3.106$, falls to the right of 2.819. Therefore, P value < 0.005.

Since the sample t falls inside the critical region and the P value is less than $\alpha = 0.01$, we reject H_0. We conclude that the yellow paint has less visibility after 1 year.

9. We are testing the difference of means with large independent samples.

$$H_0: \mu_1 = \mu_2$$
$$H_1: \mu_1 \neq \mu_2$$

Since ≠ is in H_1, a two-tailed test is used. For $\alpha = 0.05$, the critical values are $z_0 = \pm 1.96$.

$$\bar{x}_1 - \bar{x}_2 = 3.0 - 2.7 = 0.3$$

$$z = \frac{(\bar{x}_1 - \bar{x}_2) - (\mu_1 - \mu_2)}{\sqrt{\frac{\sigma_1^2}{n_1} + \frac{\sigma_2^2}{n_2}}} = \frac{0.3 - 0}{\sqrt{\frac{0.8^2}{55} + \frac{0.9^2}{52}}} = 1.82$$

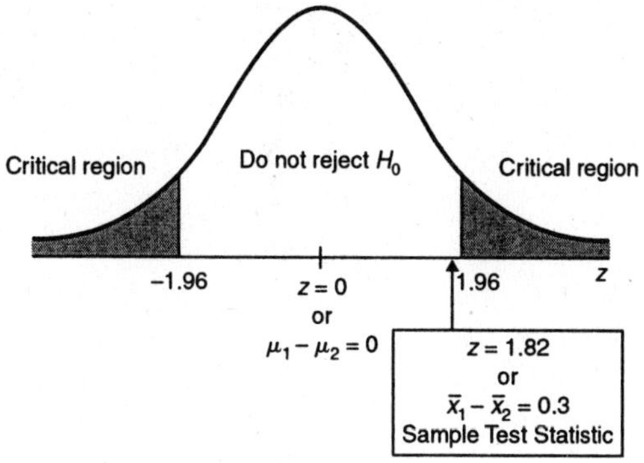

Next, find the P value associated with $z = 1.82$ and a two-tailed test.

$$P \text{ value} = 2P(z \geq 1.82)$$
$$= 2(0.0344)$$
$$= 0.0688$$

Since the sample z falls outside the critical region and the P value is greater than $\alpha = 0.05$, we do not reject H_0. There is insufficient evidence to conclude that a difference exists in the population mean lengths of the two types of projectile points.

Chapter 12 Additional Topics Using Inferences

Section 12.1

1. H_0: Myers-Briggs preference and profession are independent.

H_1: Myers-Briggs preference and profession are not independent.

$$\chi^2 = \sum \frac{(O-E)^2}{E}$$

$$= \frac{(308-241.05)^2}{241.05} + \frac{(226-292.95)^2}{292.95} + \frac{(667-723.61)^2}{723.61} + \frac{(936-879.39)^2}{879.39} + \frac{(112-122.33)^2}{122.33} + \frac{(159-148.67)^2}{148.67}$$

$$= 43.5562$$

Since there are 3 rows and 2 columns, *d.f.* = (3 − 1)(2 − 1) = 2. For $\alpha = 0.01$, the critical value is $\chi^2_{0.01} = 9.21$.

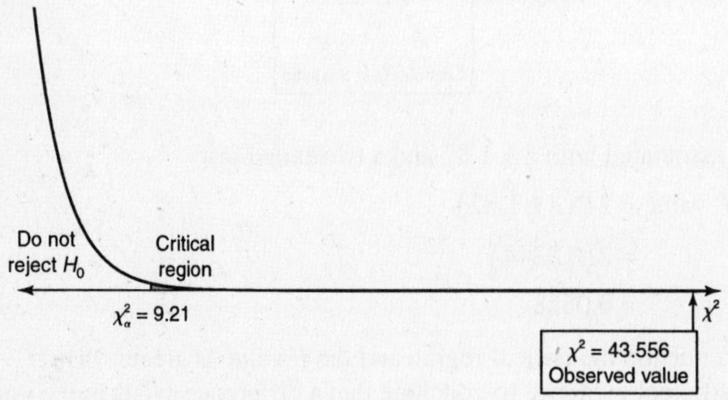

Since the sample statistic falls inside the critical region, we reject H_0. We conclude that Myers-Briggs preference and profession are not independent.

2. H_0: Myers-Briggs preference and profession are independent.

H_1: Myers-Briggs preference and profession are not independent.

$$\chi^2 = \sum \frac{(O-E)^2}{E}$$

$$= \frac{(114-238.39)^2}{238.39} + \frac{(420-295.61)^2}{295.61} + \frac{(785-715.62)^2}{715.62} + \frac{(818-887.37)^2}{887.37} + \frac{(176-120.98)^2}{120.98} + \frac{(95-150.02)^2}{150.02}$$

$$= 174.6$$

Since there are 3 rows and 2 columns, *d.f.* = (3 − 1)(2 − 1) = 2. For $\alpha = 0.01$, the critical value is $\chi^2_{0.01} = 9.21$.

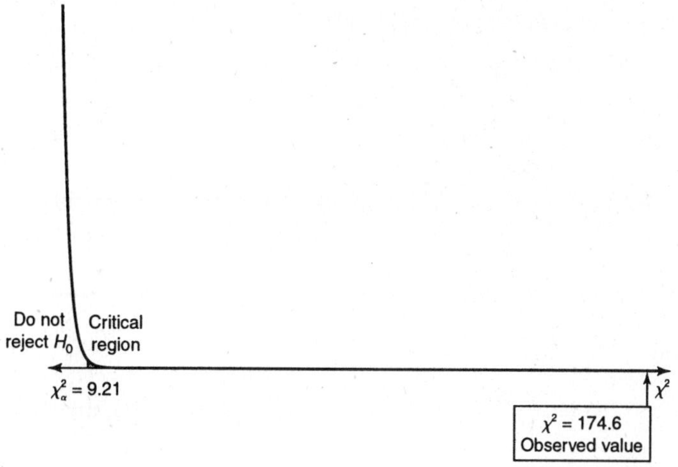

Since the sample statistic falls inside the critical region, we reject H_0. We conclude that Myers-Briggs preference and profession are not independent.

3. H_0: Site type and pottery type are independent.

H_1: Site type and pottery type are not independent.

$$\chi^2 = \sum \frac{(O-E)^2}{E}$$

$$= \frac{(75-74.64)^2}{74.64} + \frac{(61-59.89)^2}{59.89} + \frac{(53-54.47)^2}{54.47} + \frac{(81-84.11)^2}{84.11} + \frac{(70-67.5)^2}{67.5}$$

$$+ \frac{(62-61.39)^2}{61.39} + \frac{(92-89.25)^2}{89.25} + \frac{(68-71.61)^2}{71.61} + \frac{(66-65.14)^2}{65.14}$$

$$= 0.5552$$

Since there are 3 rows and 3 columns, $d.f. = (3-1)(3-1) = 4$. For $\alpha = 0.05$, the critical value is $\chi^2_{0.05} = 9.49$.

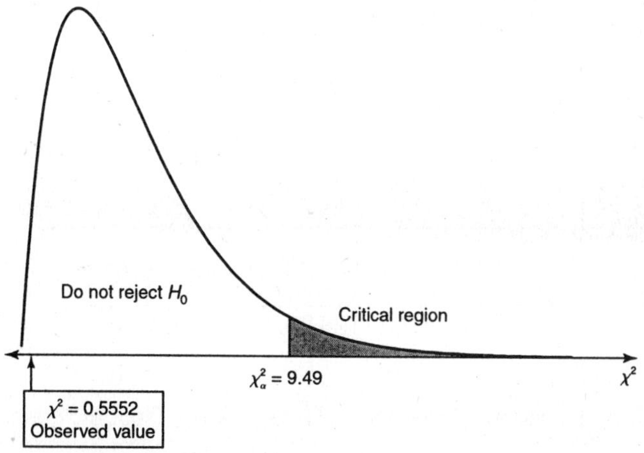

Since the sample statistic falls outside the critical region, do not reject H_0. There is insufficient evidence to conclude that site type and pottery type are not independent.

4. H_0: Ceremonial ranking and pottery type are independent.

H_1: Ceremonial ranking and pottery type are not independent.

$$\chi^2 = \sum \frac{(O-E)^2}{E}$$

$$= \frac{(242-242.61)^2}{242.61} + \frac{(26-25.39)^2}{25.39} + \frac{(658-636.41)^2}{636.41} + \frac{(45-66.59)^2}{66.59} + \frac{(371-391.98)^2}{391.98} + \frac{(62-41.02)^2}{41.02}$$

$$= 19.6079$$

Since there are 3 rows and 2 columns, $d.f. = (3-1)(2-1) = 2$. For $\alpha = 0.05$, the critical value is $\chi^2_{0.05} = 5.99$.

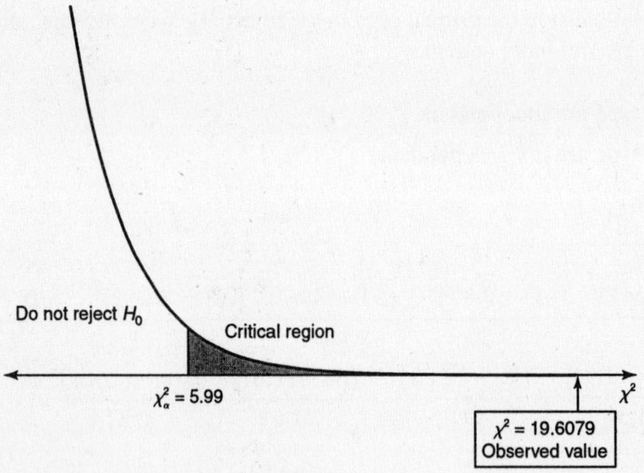

Since the sample statistic falls inside the critical region, we reject H_0. We conclude that ceremonial ranking and pottery type are not independent.

5. H_0: Age distribution and location are independent.

H_1: Age distribution and location are not independent.

$$\chi^2 = \sum \frac{(O-E)^2}{E}$$

$$= \frac{(13-14.08)^2}{14.08} + \frac{(13-12.84)^2}{12.84} + \frac{(15-14.08)^2}{14.08} + \frac{(10-11.33)^2}{11.33} + \frac{(11-10.34)^2}{10.34} + \frac{(12-11.33)^2}{11.33} + \frac{(34-31.59)^2}{31.59} + \frac{(28-28.82)^2}{28.82} + \frac{(30-31.59)^2}{31.59}$$

$$= 0.67$$

Since there are 3 rows and 3 columns, $d.f. = (3-1)(3-1) = 4$. For $\alpha = 0.05$, the critical value is $\chi^2_{0.05} = 9.49$.

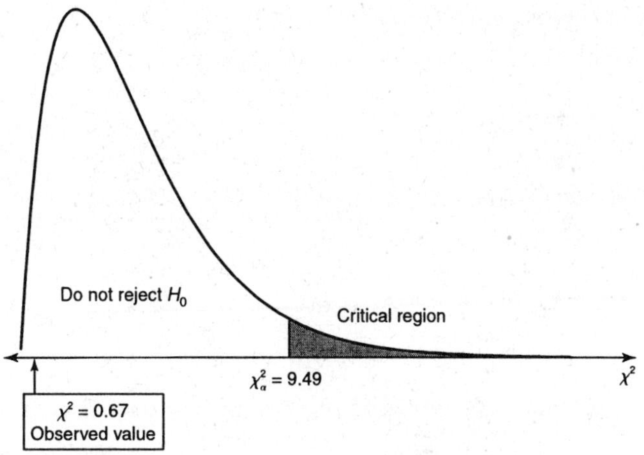

Since the sample statistic falls outside the critical region, we do not reject H_0. There is insufficient evidence to conclude that age distribution and location are not independent.

6. H_0: Type and career choice are independent.

H_1: Type and career choice are not independent.

$$\chi^2 = \sum \frac{(O-E)^2}{E}$$

$$= \frac{(64-53.46)^2}{53.46} + \frac{(15-24.79)^2}{24.79} + \frac{(17-17.76)^2}{17.76} + \frac{(82-85.75)^2}{85.75} + \frac{(42-39.76)^2}{39.76} + \frac{(30-28.49)^2}{28.49}$$
$$+ \frac{(68-64.04)^2}{64.04} + \frac{(35-29.69)^2}{29.69} + \frac{(12-21.27)^2}{21.27} + \frac{(75-85.75)^2}{85.75} + \frac{(42-39.76)^2}{39.76} + \frac{(37-28.49)^2}{28.49}$$
$$= 15.602$$

Since there are 4 rows and 3 columns, $d.f. = (4-1)(3-1) = 6$. For $\alpha = 0.01$, the critical value is $\chi^2_{0.01} = 16.81$.

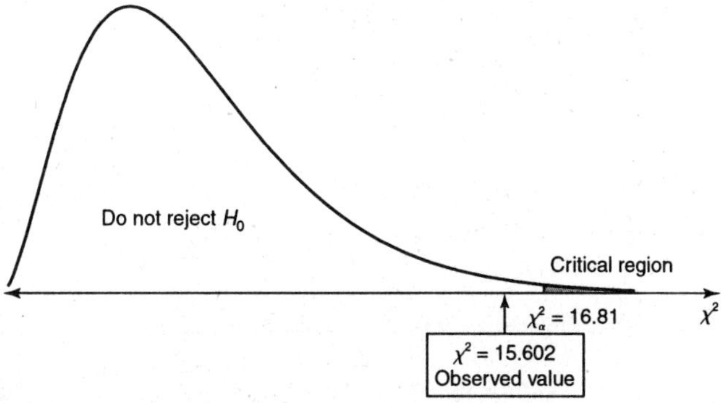

Since the sample statistic falls outside the critical region, we do not reject H_0. There is insufficient evidence to conclude that type and career choice are not independent.

7. H_0: Ages of young adults and movie preferences are independent.

H_1: Ages of young adults and movie preferences are not independent.

$$\chi^2 = \sum \frac{(O-E)^2}{E}$$

$$= \frac{(8-10.60)^2}{10.60} + \frac{(15-12.06)^2}{12.06} + \frac{(11-11.33)^2}{11.33} + \frac{(12-9.35)^2}{9.35} + \frac{(10-10.65)^2}{10.65}$$

$$+ \frac{(8-10.00)^2}{10.00} + \frac{(9-9.04)^2}{9.04} + \frac{(8-10.29)^2}{10.29} + \frac{(12-9.67)^2}{9.67}$$

$$= 3.623$$

Since there are 3 rows and 3 columns, $d.f. = (3-1)(3-1) = 4$. For $\alpha = 0.05$, the critical value is $\chi^2_{0.05} = 9.49$.

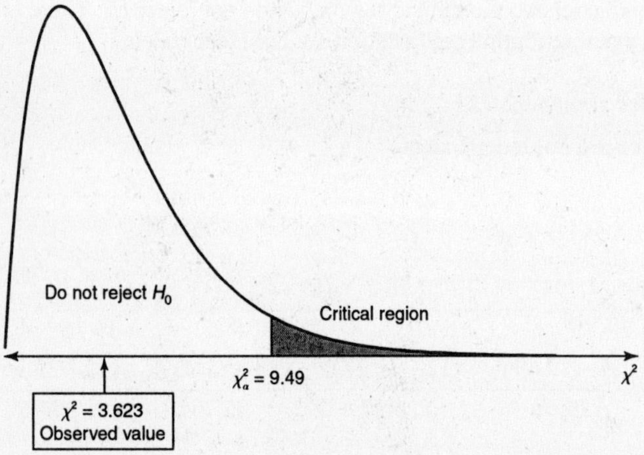

Since the sample statistic falls outside the critical region, we do not reject H_0. There is insufficient evidence to conclude that ages of young adults and movie preferences are not independent.

8. H_0: Contribution and ethnic group are independent.

H_1: Contribution and ethnic group are not independent.

$$\chi^2 = \sum \frac{(O-E)^2}{E}$$

$$= \frac{(310-441.42)^2}{441.42} + \frac{(715-569.96)^2}{569.96} + \frac{(201-244.61)^2}{244.61} + \frac{(105-86.87)^2}{86.87} + \frac{(42-30.13)^2}{30.13}$$

$$+ \frac{(619-501.86)^2}{501.86} + \frac{(511-648.01)^2}{648.01} + \frac{(312-278.10)^2}{278.10} + \frac{(97-98.77)^2}{98.77} + \frac{(22-34.26)^2}{34.26}$$

$$+ \frac{(402-439.17)^2}{439.17} + \frac{(624-567.06)^2}{567.06} + \frac{(217-243.36)^2}{243.36} + \frac{(88-86.43)^2}{86.43} + \frac{(35-29.98)^2}{29.98}$$

$$+ \frac{(544-492.54)^2}{492.54} + \frac{(571-635.97)^2}{635.97} + \frac{(309-272.93)^2}{272.93} + \frac{(79-96.93)^2}{96.93} + \frac{(29-33.62)^2}{33.62}$$

$$= 190.44$$

Since there are 4 rows and 5 columns, $d.f. = (4 - 1)(5 - 1) = 12$. For $\alpha = 0.01$, the critical value is $\chi^2_{0.01} = 26.22$.

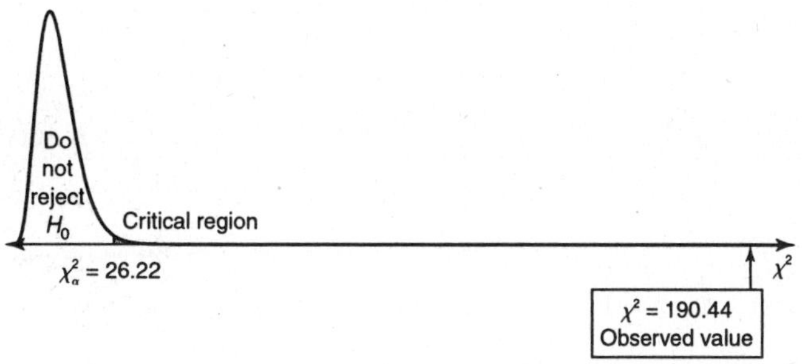

Since the sample statistic falls inside the critical region, we reject H_0. We conclude that contribution and ethnic group are not independent.

9. **(a)** H_0 : Ticket sales and type of billing are independent and H_1 : Ticket sales and type of billing are not independent.

(b) Look in the cell in the first row and fourth column of data. $E = 7.52$.

(c) $x^2 = 1.868$

(d) Since the P value, 0.60, is greater than $\alpha = 0.05$, we do not reject H_0. It appears that type of billing and ticket sales are independent.

Section 12.2

1. H_0: The distributions are the same.
H_1: The distributions are different.

$$\chi^2 = \sum \frac{(O-E)^2}{E}$$
$$= \frac{(47-32.76)^2}{32.76} + \frac{(75-61.88)^2}{61.88} + \frac{(288-305.31)^2}{301.31} + \frac{(45-55.06)^2}{55.06}$$
$$= 11.79$$

$d.f.$ = (number of E entries) $- 1 = 4 - 1 = 3$. The critical value is $\chi^2_{0.05} = 7.81$.

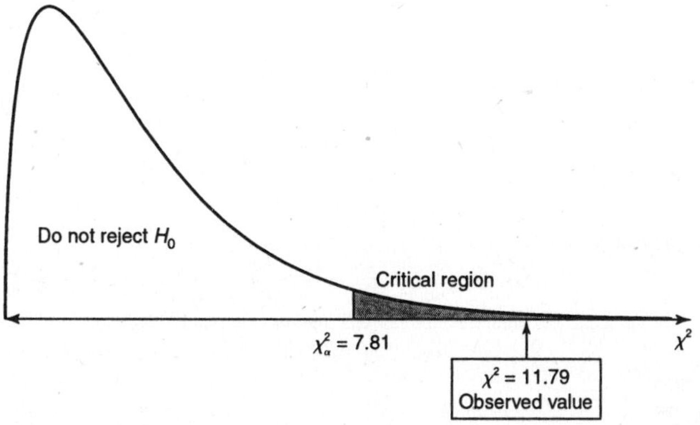

Since the sample statistic falls inside the critical region, we reject H_0. We conclude the distributions are different.

2. H_0: The distributions are the same.

H_1: The distributions are different.

$$\chi^2 = \sum \frac{(O-E)^2}{E}$$

$$= \frac{(102-106.86)^2}{106.86} + \frac{(112-119.19)^2}{119.19} + \frac{(33-36.99)^2}{36.99} + \frac{(96-102.75)^2}{102.75} + \frac{(68-45.21)^2}{45.21}$$

$$= 13.017$$

d.f. = (number of *E* entries) − 1 = 5 − 1 = 4. The critical value is $\chi^2_{0.05} = 9.49$.

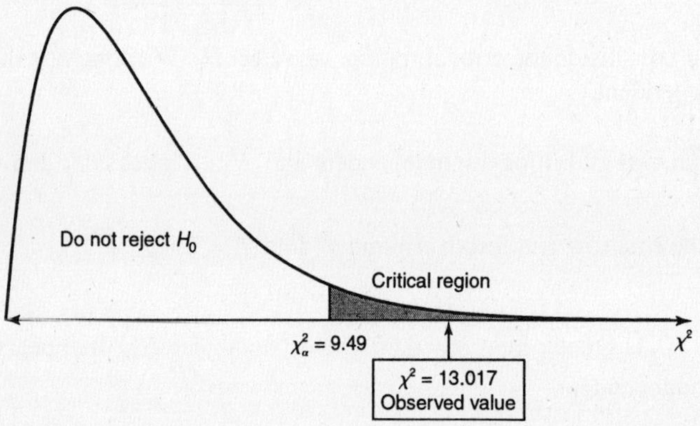

Since the sample statistic falls inside the critical region, we reject H_0. We conclude the distributions are different.

3. H_0: The distributions are the same.

H_1: The distributions are different.

$$\chi^2 = \sum \frac{(O-E)^2}{E}$$

$$= \frac{(906-910.92)^2}{910.92} + \frac{(162-157.52)^2}{157.52} + \frac{(168-169.40)^2}{169.40} + \frac{(197-194.67)^2}{194.67} + \frac{(53-53.50)^2}{53.50}$$

$$= 0.1984$$

d.f. = (number of *E* entries) − 1 = 5 − 1 = 4. The critical value is $\chi^2_{0.01} = 13.28$.

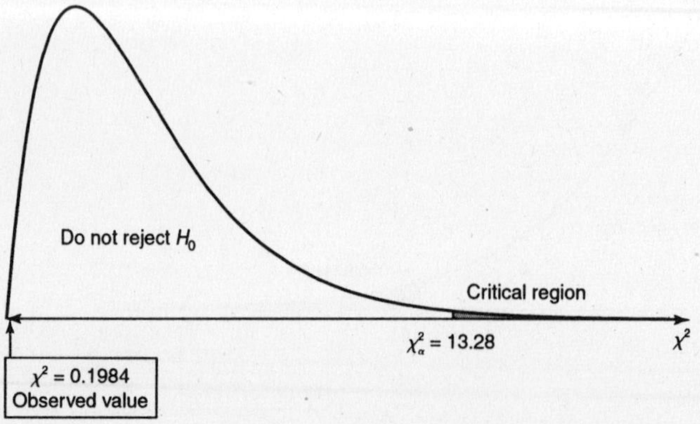

Since the sample statistic falls outside the critical region, we do not reject H_0. There is insufficient evidence to conclude that the distributions are different.

4. H_0: The distributions are the same.

H_1: The distributions are different.

$$\chi^2 = \sum \frac{(O-E)^2}{E}$$

$$= \frac{(102-102.40)^2}{102.40} + \frac{(125-123.84)^2}{123.84} + \frac{(43-38.40)^2}{38.40} + \frac{(27-29.76)^2}{29.76} + \frac{(23-25.60)^2}{25.60}$$

$$= 1.084$$

d.f. = (number of *E* entries) − 1 = 5 − 1 = 4. The critical value is $\chi^2_{0.05} = 9.49$.

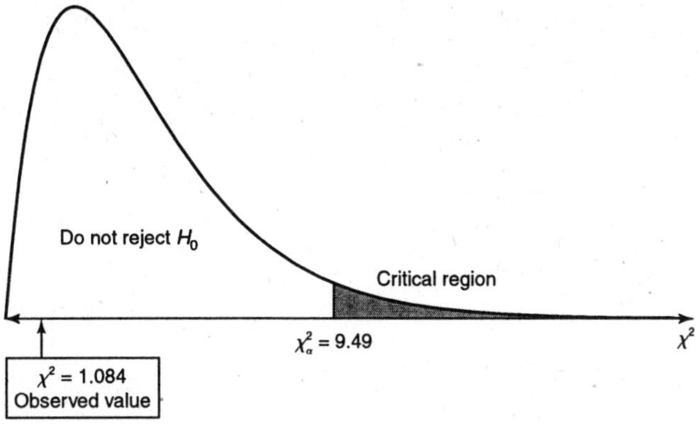

Since the sample statistic falls outside the critical region, we do not reject H_0. There is insufficient evidence to conclude that the distributions are different.

5. (a) Essay.

(b) H_0: The distributions are the same.

H_1: The distributions are different.

$$\chi^2 = \sum \frac{(O-E)^2}{E}$$

$$= \frac{(16-14.57)^2}{14.57} + \frac{(78-83.70)^2}{83.70} + \frac{(212-210.80)^2}{210.80} + \frac{(221-210.80)^2}{210.80}$$

$$+ \frac{(81-83.70)^2}{83.70} + \frac{(12-14.57)^2}{14.57}$$

$$= 1.5716$$

d.f. = (number of *E* entries) − 1 = 6 − 1 = 5. The critical value is $\chi^2_{0.01} = 15.09$.

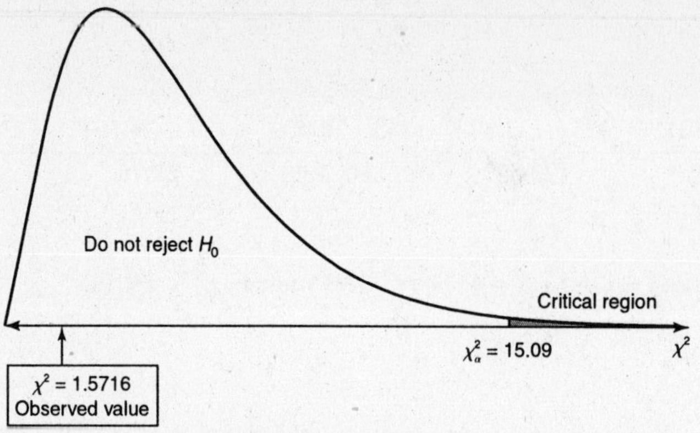

Since the sample statistic falls outside the critical region, we do not reject H_0. There is insufficient evidence to conclude that the distribution is not normal.

6. (a) Essay.

(b) H_0: The distributions are the same.

H_1: The distributions are different.

$$\chi^2 = \sum \frac{(O-E)^2}{E}$$

$$= \frac{(14-14.57)^2}{14.57} + \frac{(86-83.70)^2}{83.70} + \frac{(207-210.80)^2}{210.80} + \frac{(215-210.80)^2}{210.80} + \frac{(83-83.70)^2}{83.70} + \frac{(15-14.57)^2}{14.57}$$

$$= 0.23392$$

d.f. = (number of *E* entries) − 1 = 6 − 1 = 5. The critical value is $\chi^2_{0.01} = 15.09$.

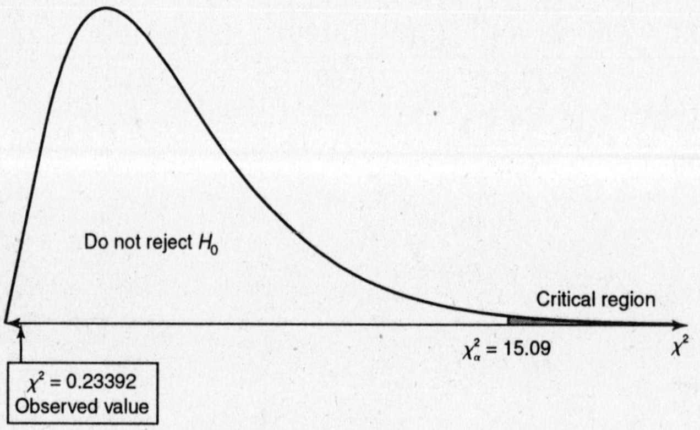

Since the sample statistic falls outside the critical region, we do not reject H_0. There is insufficient evidence to conclude that the distribution is not normal.

7. H_0: The distributions are the same.
H_1: The distributions are different.

$$\chi^2 = \sum \frac{(O-E)^2}{E}$$
$$= \frac{(120-150)^2}{150} + \frac{(85-75)^2}{75} + \frac{(220-200)^2}{200} + \frac{(75-75)^2}{75}$$
$$= 9.333$$

d.f. = (number of *E* entries) − 1 = 4 − 1 = 3. The critical value is $\chi^2_{0.05} = 7.81$.

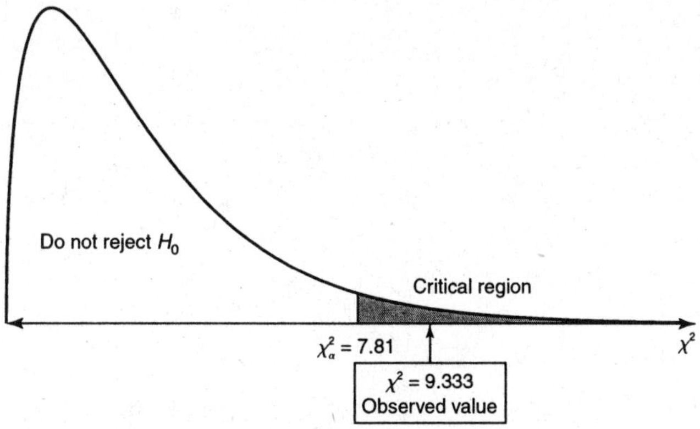

Since the sample statistic falls inside the critical region, we reject H_0. We conclude that the fish distribution has changed.

8. H_0: The distributions are the same.
H_1: The distributions are different.

$$\chi^2 = \sum \frac{(O-E)^2}{E}$$
$$= \frac{(1210-1349.44)^2}{1349.44} + \frac{(956-1054.25)^2}{1054.25} + \frac{(940-843.40)^2}{843.40} + \frac{(814-632.55)^2}{632.55} + \frac{(297-337.36)^2}{337.36}$$
$$= 91.51$$

d.f. = (number of *E* entries) − 1 = 5 − 1 = 4. The critical value is $\chi^2_{0.05} = 9.49$.

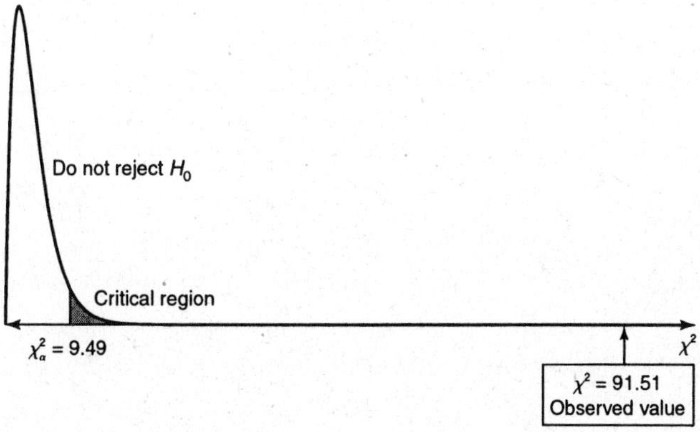

Since the sample statistic falls inside the critical region, we reject H_0. We conclude that the distributions are different.

9. H_0: The distributions are the same.

H_1: The distributions are different.

$$\chi^2 = \sum \frac{(O-E)^2}{E}$$

$$= \frac{(127-121.50)^2}{121.50} + \frac{(40-36.45)^2}{36.45} + \frac{(480-461.70)^2}{461.70}$$

$$+ \frac{(502-498.15)^2}{498.15} + \frac{(56-72.90)^2}{72.90} + \frac{(10-24.30)^2}{24.30}$$

$$= 13.7$$

d.f. = (number of *E* entries) – 1 = 6 – 1 = 5. The critical value is $\chi^2_{0.01} = 15.09$.

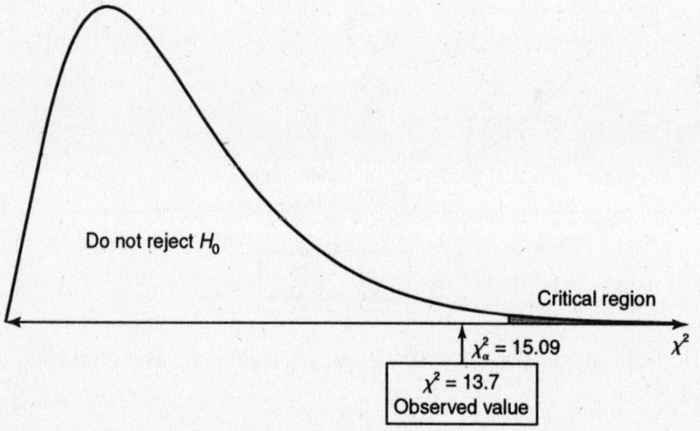

Since the sample statistic falls outside the critical region, we do not reject H_0. There is insufficient evidence to conclude that the distributions are different.

10. H_0: The distributions are the same.

H_1: The distributions are different.

$$\chi^2 = \sum \frac{(O-E)^2}{E}$$

$$= \frac{(88-62.28)^2}{62.28} + \frac{(135-150.51)^2}{150.51} + \frac{(52-57.09)^2}{57.09}$$

$$+ \frac{(40-51.90)^2}{51.90} + \frac{(76-72.66)^2}{72.66} + \frac{(128-124.56)^2}{124.56}$$

$$= 15.65$$

d.f. = (number of *E* entries) − 1 = 6 − 1 = 5. The critical value is $\chi^2_{0.01} = 15.09$.

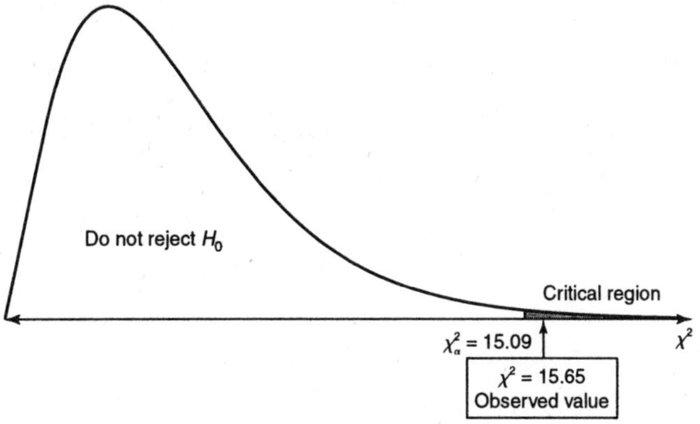

Since the sample statistic falls inside the critical region, we reject H_0. We conclude that the distributions are different.

Section 12.3

1. H_0: $\sigma^2 = 42.3$
H_1: $\sigma^2 > 42.3$

Since > is in H_1, a right-tailed test is used.
For *d.f.* = 23 − 1 = 22, the critical value is $\chi^2_{0.05} = 33.92$.

$$\chi^2 = \frac{(n-1)s^2}{\sigma^2} = \frac{(23-1)46.1}{42.3} = 23.98$$

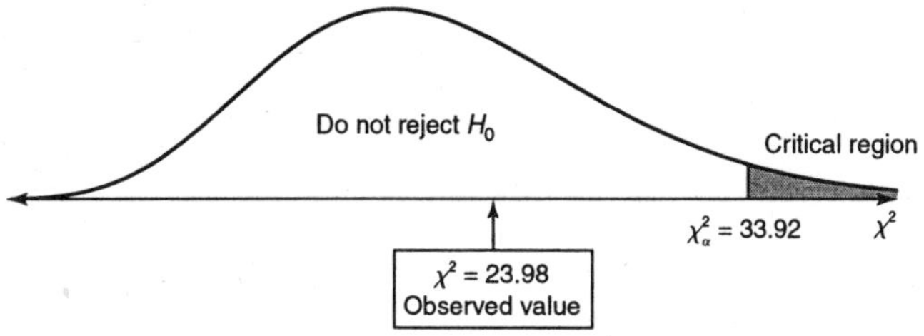

Since the sample statistic falls outside the critical region, we do not reject H_0. We have insufficient evidence to conclude that the variance in the new section is greater than 42.3.

2. H_0: $\sigma^2 = 5.1$
H_1: $\sigma^2 < 5.1$
Since < is in H_1, a left-tailed test is used.
For *d.f.* = 41 − 1 = 40, the critical value is $\chi^2_{0.95} = 26.51$.

$$\chi^2 = \frac{(n-1)s^2}{\sigma^2} = \frac{(41-1)3.3}{5.1} = 25.88$$

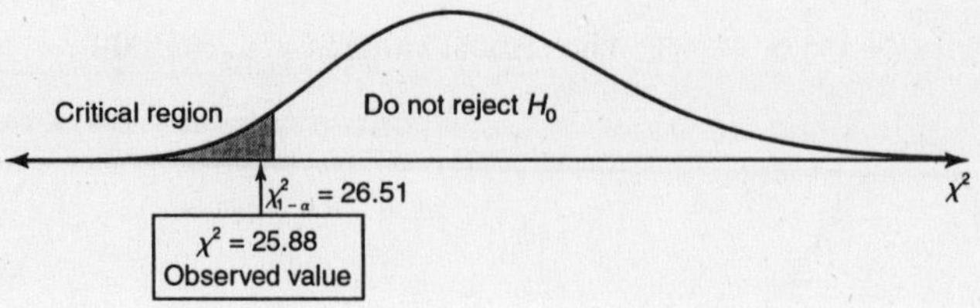

Since the sample statistic falls inside the critical region, we reject H_0. We conclude that the current variance is less than 5.1.

3. H_0: $\sigma^2 = 136.2$
H_1: $\sigma^2 < 136.2$
Since < is in H_1, a left-tailed test is used.
For *d.f.* = 8 − 1 = 7, the critical value is $\chi^2_{0.99} = 1.24$.

$$\chi^2 = \frac{(n-1)s^2}{\sigma^2} = \frac{(8-1)115.1}{136.2} = 5.92$$

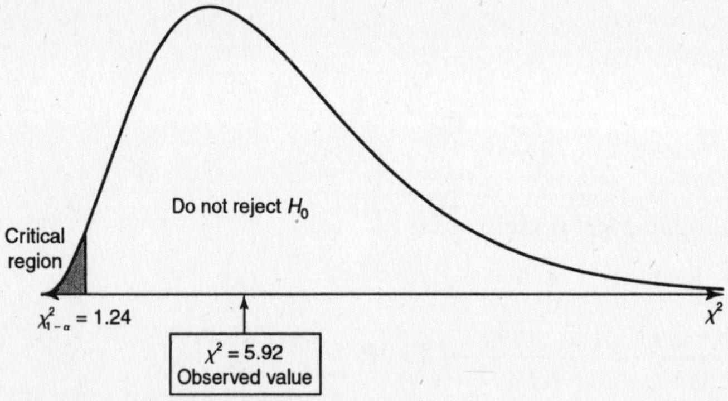

Since the sample statistic falls outside the critical region, we do not reject H_0. We have insufficient evidence to conclude that the recent variance for number of mountain-climber deaths is less than 136.1.

4. H_0: $\sigma^2 = 47.1$
H_1: $\sigma^2 > 47.1$

Since > is in H_1, a right-tailed test is used.
For *d.f.* = 15 − 1 = 14, the critical value is $\chi^2_{0.05} = 23.68$.

$$\chi^2 = \frac{(n-1)s^2}{\sigma^2} = \frac{(15-1)83.2}{47.1} = 24.73$$

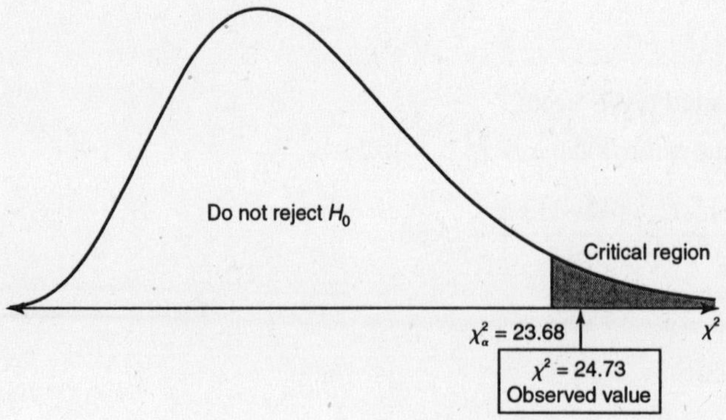

Since the sample statistic falls inside the critical region, we reject H_0. We conclude that the variance for colleges and universities in Kansas is greater than 47.1.

5. H_0: $\sigma^2 = 9$
H_1: $\sigma^2 < 9$
Since < is in H_1, a left-tailed test is used.
For $d.f. = 24 - 1 = 23$, the critical value is $\chi^2_{0.95} = 13.09$.

$$\chi^2 = \frac{(n-1)s^2}{\sigma^2} = \frac{(24-1)(1.9)^2}{3^2} = 9.23$$

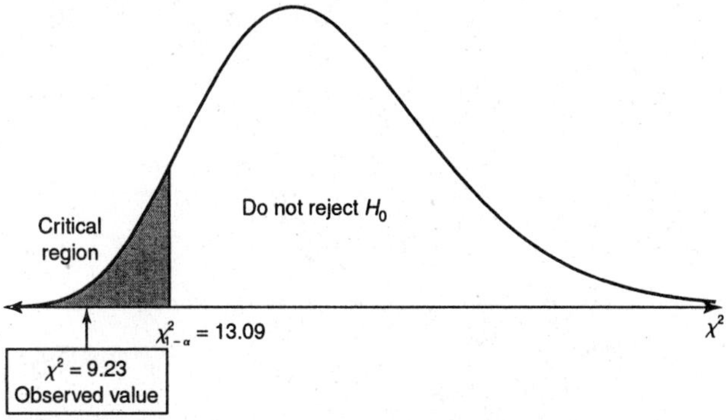

Since the sample statistic falls inside the critical region, we reject H_0. We conclude that the new typhoid shot has a smaller variance of protection times.

6. H_0: $\sigma^2 = 225$
H_1: $\sigma^2 > 225$
Since > is in H_1, a right-tailed test is used.
For $d.f. = 10 - 1 = 9$, the critical value is $\chi^2_{0.01} = 21.67$.

$$\chi^2 = \frac{(n-1)s^2}{\sigma^2} = \frac{(10-1)(24)^2}{(15)^2} = 23.04$$

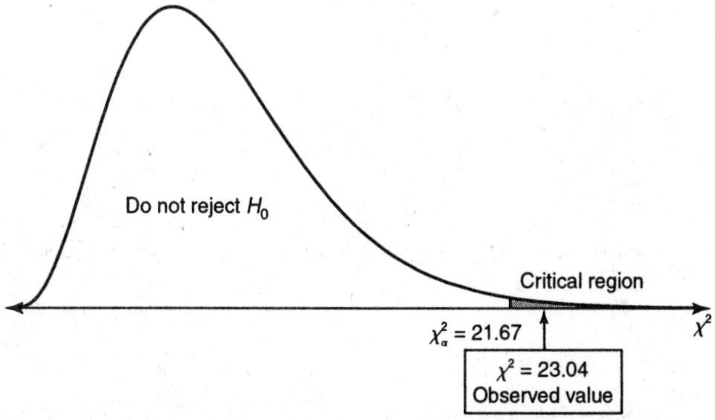

Since the sample statistic falls inside the critical region, we reject H_0. We conclude that the variance is larger than that stated in his journal.

7. H_0: $\sigma^2 = 0.15$
H_1: $\sigma^2 > 0.15$

Since > is in H_1, a right-tailed test is used.

For *d.f.* = 61 − 1 = 60, the critical value is $\chi^2_{0.01} = 88.38$.

$$\chi^2 = \frac{(n-1)s^2}{\sigma^2} = \frac{(61-1)0.27}{0.15} = 108$$

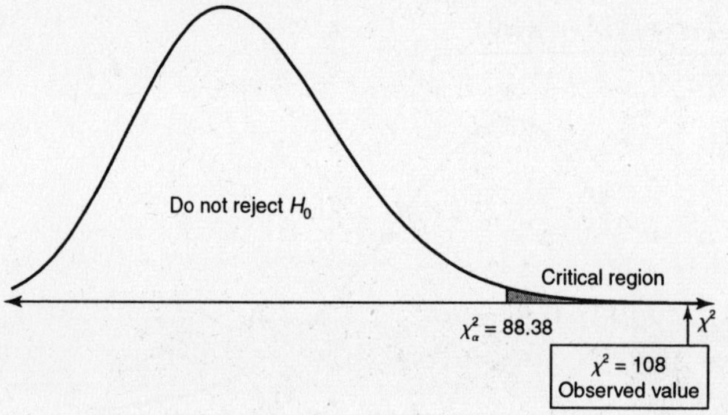

Since the sample statistic falls inside the critical region, we reject H_0. We conclude that all the engine blades must be replaced (i.e., the variance exceeds 0.15).

8. $\left.\begin{matrix} H_0: \sigma^2 = 5625 \\ H_1: \sigma^2 \neq 5625 \end{matrix}\right\}$ these hypotheses are equivalent to $\left\{\begin{matrix} H_0: \sigma = 75 \\ H_1: \sigma \neq 75 \end{matrix}\right.$

Since ≠ is in H_1, a two-tailed test is used.

For *d.f.* = 24 − 1 = 23, the critical values are $\chi^2_{0.005} = 44.18$ and $\chi^2_{0.995} = 9.26$.

$$\chi^2 = \frac{(n-1)s^2}{\sigma^2} = \frac{(24-1)(72)^2}{(75)^2} = 21.20$$

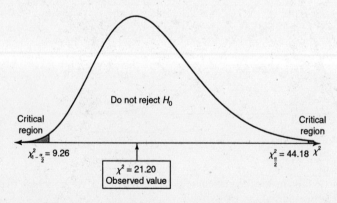

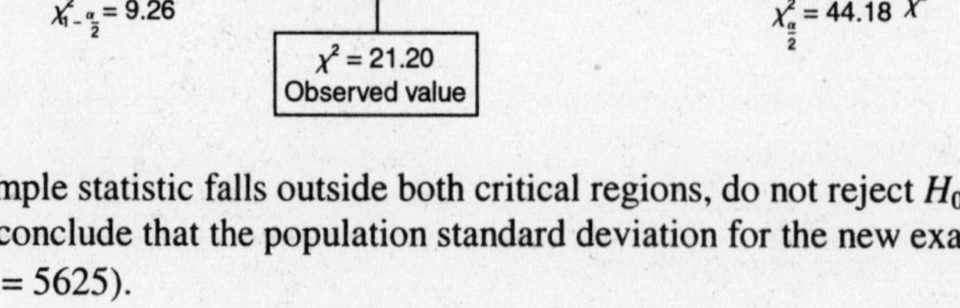

Since the sample statistic falls outside both critical regions, do not reject H_0. There is insufficient evidence to conclude that the population standard deviation for the new examination is different from 75 ($\sigma^2 = 75^2 = 5625$).

Section 12.4

1. (a)

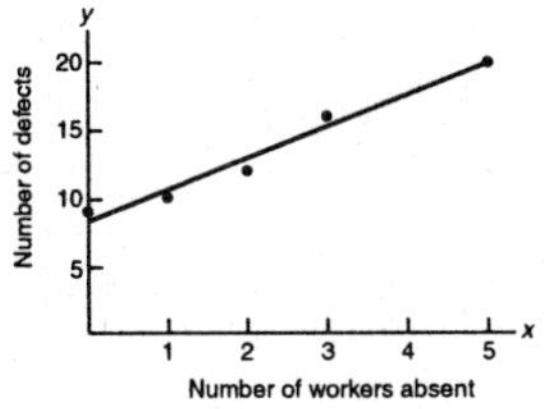

(b)

x	y	xy	x^2	y^2
3	16	48	9	256
5	20	100	25	400
0	9	0	0	81
2	12	24	4	144
1	10	10	1	100
11	67	182	39	981

$$SS_{xy} = \sum xy - \frac{\sum x \sum y}{n} = 182 - \frac{(11)(67)}{5} = 182 - 147.40 = 34.6$$

$$\bar{x} = \frac{11}{5} = 2.2$$

$$\bar{y} = \frac{67}{5} = 13.4$$

$$SS_x = \sum x^2 - \frac{(\sum x)^2}{n} = 39 - \frac{(11)^2}{5} = 14.8$$

$$b = \frac{SS_{xy}}{SS_x} = \frac{34.6}{14.8} = 2.338$$

$$a = \bar{y} - b\bar{x} = 13.4 - (2.338)(2.2) = 8.26$$

$$y = 8.26 + 2.338x$$

(c) $SS_y = \sum y^2 - \frac{(\sum y)^2}{n} = 981 - \frac{(67)^2}{5} = 83.2$

$$S_e = \sqrt{\frac{SSy - b \cdot SS_{xy}}{n-2}} = \sqrt{\frac{83.2(2.338)(34.6)}{5-2}} = \sqrt{\frac{2.305}{3}} = 0.8766$$

(d) Let $x = 4$. Then $y = 8.26 + 2.338(4) = 17.61$.

(e) For a 95% confidence interval with $n - 2 = 5 - 2 = 3$ degrees of freedom, $t_c = 3.182$ (from the t table).

$$E = t_c S_e \sqrt{1 + \frac{1}{n} + \frac{(x - \bar{x})^2}{SS_x}}$$

$$= 3.182(0.8766)\sqrt{1 + \frac{1}{5} + \frac{(4 - 2.2)^2}{14.8}}$$

$$= 2.79\sqrt{1.4189}$$

$$= 3.32$$

$y + E = 17.61 + 3.32 = 20.93$

$y - E = 17.61 - 3.32 = 14.29$

A 95% confident forecast for the number of defects when four workers are absent is between 14.29 and 20.93.

2. **(a)**

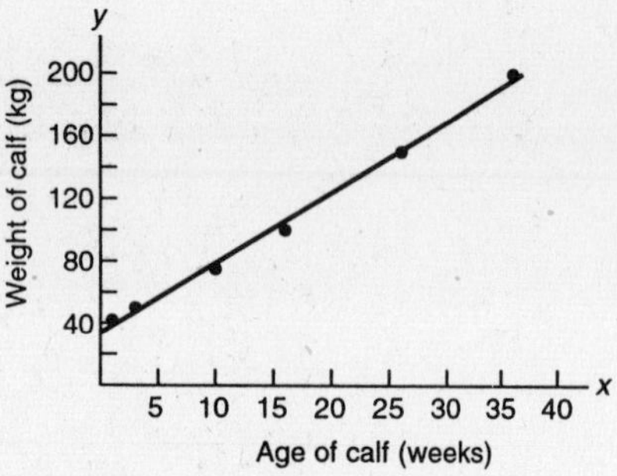

(b)

x	y	xy	x^2	y^2
1	42	42	1	1764
3	50	150	9	2500
10	75	750	100	5625
16	100	1600	256	10,000
26	150	3900	676	22,500
36	200	7200	1296	40,000
92	617	13,642	2338	82,389

$$SS_{xy} = \sum xy - \frac{\sum x \sum y}{n} = 13{,}642 - \frac{(92)(617)}{6} = 13{,}642 - 9460.67$$
$$= 4181.33$$

$$\overline{x} = \frac{92}{6} = 15.33$$

$$\overline{y} = \frac{617}{6} = 102.83$$

$$SS_x = \sum x^2 - \frac{(\sum x)^2}{n} = 2338 - \frac{(92)^2}{6} = 2338 - 1410.67 = 927.33$$

$$b = \frac{SS_{xy}}{SS_x} = \frac{4181.33}{927.33} = 4.509$$

$$a = \overline{y} - b\overline{x} = 102.83 - 4.509(15.33) = 33.707$$

$$y = 33.707 + 4.509x$$

(c) $SS_y = \sum y^2 - \frac{(\sum y)^2}{n} = 82{,}389 - \frac{(617)^2}{6} = 82{,}389 - 63{,}448.17 = 18{,}940$

$$S_e = \sqrt{\frac{Ss_y - b \cdot SS_{xy}}{n-2}} = \sqrt{\frac{18{,}940.83 - (4.509)(4181.33)}{6-2}} = \sqrt{\frac{87.213}{4}} = 4.6694$$

(d) Let $x = 12$. Then $y = 33.707 + 4.509(12) = 87{,}815$.

(e) For a 90% confidence interval with $n - 2 = 6 - 2 = 4$ degrees of freedom, $t_c = 2.132$ (from the t table).

$$E = t_c S_e \sqrt{1 + \frac{1}{n} + \frac{(x - \overline{x})^2}{SS_x}}$$

$$= 2.132(4.6694)\sqrt{1 + \frac{1}{6} + \frac{(12 - 15.33)^2}{927.33}}$$

$$= 9.96\sqrt{1.1786}$$

$$= 10.81$$

$y + E = 87.82 + 10.81 = 98.63$

$y - E = 87.82 - 10.81 = 77.01$

A 90% confident forecast for the weight of calf is between 77.01 kg and 98.63 kg.

3. **(a)**

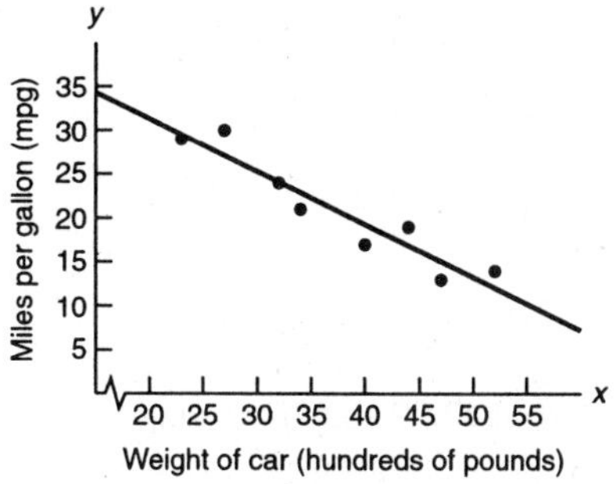

(b)

x	y	xy	x^2	y^2
27	30	810	729	900
44	19	836	1936	361
32	24	768	1024	576
47	13	611	2209	169
23	29	667	529	841
40	17	680	1600	289
34	21	714	1156	441
52	14	728	2704	196
299	167	5814	11,887	3773

$$SS_{xy} = \sum xy - \frac{\sum x \sum y}{n} = 5814 - \frac{(299)(167)}{8} = 5814 - 6241.625$$

$$= -427.625$$

$$\overline{x} = \frac{299}{8} = 37.375$$

$$\overline{y} = \frac{167}{8} = 20.875$$

$$SS_x = \sum x^2 - \frac{\left(\sum x\right)^2 n}{n} = 11{,}887 - \frac{(299)^2}{8} = 11{,}887 - 11{,}175.125$$

$$= 711.875$$

$$b = \frac{SS_{xy}}{SS_x} = \frac{-427.625}{711.875} = -0.601$$

$a = \overline{y} - b\overline{x} = 20.875 - (-0.601)(37.375) = 43.337$

$y = 43{,}337 - 0.601x$

(c) $SS_y = \sum y^2 - \frac{(\sum y)^2}{n} = 3773 - \frac{(167)^2}{8} = 3773 - 3486.125$
$= 286.875$

$$S_e = \sqrt{\frac{SS_y - b \cdot SS_{xy}}{n-2}} = \sqrt{\frac{286.875 - (-0.601)(-427.625)}{8-2}}$$
$$= \sqrt{\frac{29.872}{6}} = 2.2313$$

(d) Let $x = 38$. Then $y = 43.337 - 0.601(38) = 20.499$

(e) For an 80% confidence interval with $n - 2 = 8 - 2 = 6$ degrees of freedom, $t_c = 1.440$.

$$E = t_c S_e \sqrt{1 + \frac{1}{n} + \frac{(x - \bar{x})^2}{SS_x}}$$
$$= 1.440(2.2313)\sqrt{1 + \frac{1}{8} + \frac{(38 - 37.375)^2}{711.875}}$$
$$= 3.213\sqrt{1.1255}$$
$$= 3.41$$

$y + E = 20.499 + 3.41 = 23.91$
$y - E = 20.499 - 3.41 = 17.09$
An 80% confident forecast for the miles per gallon is between 17.09 and 23.91 mpg.

4. (a)

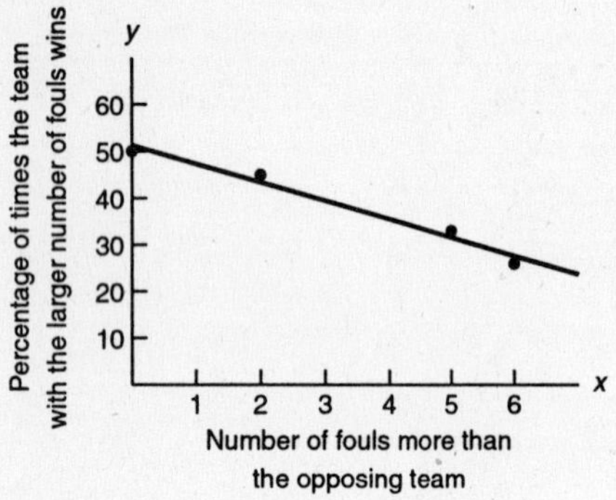

(b)

x	y	xy	x^2	y^2
0	50	0	0	2500
2	45	90	4	2025
5	33	165	25	1089
6	26	156	36	676
13	154	411	65	6290

$$SS_{xy} = \sum xy - \frac{\sum x \sum y}{n} = 411 - \frac{(13)(154)}{4} = 411 - 500.5 = -89.5$$
$$\bar{x} = \frac{13}{4} = 3.25$$
$$\bar{y} = \frac{154}{4} = 38.5$$

$$SS_x = \sum x^2 - \frac{(\sum x)^2}{n} = 65 - \frac{(13)^2}{4} = 65 - 42.25 = 22.75$$

$$b = \frac{SS_{xy}}{SS_x} = \frac{-89.5}{22.75} = -3.934$$

$$a = \overline{y} - b\overline{x} = 38.5 - (-3.934)(3.25) \approx 51.286$$

$$y = 51.286 - 3.934x$$

(c) $$SS_y = \sum y^2 - \frac{(\sum y)^2}{n} = 6290 - \frac{(154)^2}{4} = 361$$

$$S_3 = \sqrt{\frac{SS_y - b \cdot SS_{xy}}{n-2}} = \sqrt{\frac{361-(-3.934)(-89.5)}{4-2}} = \sqrt{\frac{8.907}{2}} = 2.1103$$

(d) Let $x = 4$. Then $y = 51.286 - 3.934(4) = 35.55\%$.

(e) For an 80% confidence interval with $n - 2 = 4 - 2 = 2$ degrees of freedom, $t_c = 1.886$ (from the t table).

$$E = t_c S_e \sqrt{1 + \frac{1}{n} + \frac{(x-\overline{x})^2}{SS_x}}$$

$$= 1.886(2.1103)\sqrt{1 + \frac{1}{4} + \frac{(4-4.35)^2}{22.75}}$$

$$= 3.98\sqrt{1.2747}$$

$$= 4.49$$

$$y + E = 33.55 + 4.49 = 40.04$$

$$y - E = 33.55 - 4.49 = 31.06$$

An 80% confident forecast for the percent of time the team with four more fouls wins is between 31.06% and 40.04%.

5. **(a)**

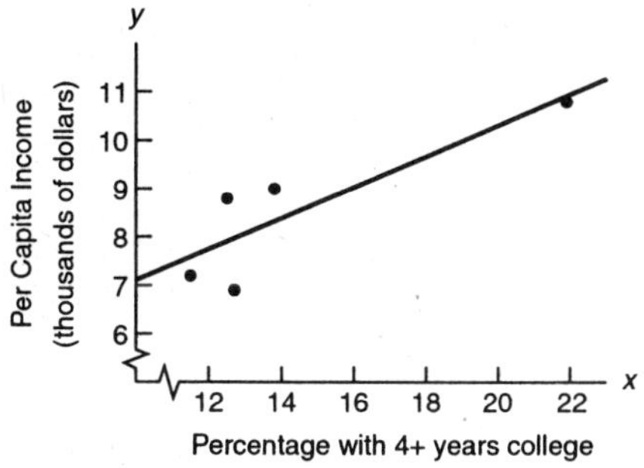

(b)

x	y	xy	x^2	y^2
13.8	9.0	124.20	190.44	81
21.9	10.8	236.52	479.61	116.64
12.5	8.8	110.00	156.25	77.44
12.7	6.9	87.63	161.29	47.61
11.5	7.2	82.8	132.25	51.84
72.4	42.7	641.15	1119.84	374.53

$$SS_{xy} = \sum xy - \frac{\sum x \sum y}{n} = 641.15 - \frac{(72.4)(42.7)}{5} = 22.854$$

$$\overline{x} = \frac{\sum x}{n} = \frac{72.4}{5} = 14.48$$

$$\overline{y} = \frac{\sum y}{n} = \frac{42.7}{5} = 8.54$$

$$SS_x = \sum x^2 - \frac{(\sum x)^2}{n} = 1119.84 - \frac{(72.4)^2}{5} = 1119.84 - 1048.352 = 71.488$$

$$b = \frac{SS_{xy}}{SS_x} = \frac{22.854}{71.488} = 0.3197 \approx 0.320$$

$$a = \overline{y} - b\overline{x} = 8.54 - (0.3197)(14.48) = 3.911 \approx 3.91$$

$$y = 3.91 + 0.320x$$

(c) $SS_y = \sum y^2 - \frac{(\sum y)^2}{n} = 374.53 - \frac{(42.7)^2}{5} = 9.872$

$$S_e = \sqrt{\frac{SS_y - b \cdot SS_{xy}}{n-2}} = \sqrt{\frac{9.872 - (0.320)(22.854)}{5-2}} = \sqrt{\frac{2.55872}{3}} = 0.9235$$

(d) Let $x = 20$. Then $y = 3.91 + 0.320(20) = 10.31 \approx 10.3$ thousand.

(e) For an 80% confidence interval with $n - 2 = 5 - 2 = 3$ degrees of freedom, $t_c = 1.638$ (from the t table).

$$E = t_c S_e \sqrt{1 + \frac{1}{n} + \frac{(x - \overline{x})^2}{SS_x}}$$

$$= 1.638(0.9235)\sqrt{1 + \frac{1}{5} + \frac{(20 - 14.48)^2}{71.488}}$$

$$= 1.5127\sqrt{1.6262}$$

$$= 1.9291 \approx 1.9$$

$$y + E = 10.3 + 1.9 = 12.2$$

$$y - E = 10.3 - 1.9 = 8.4$$

An 80% confident forecast for the income of a town having 20% of its population who had 4 or more years of college education is between 8.4 and 12.2 thousands of dollars.

6. (a)

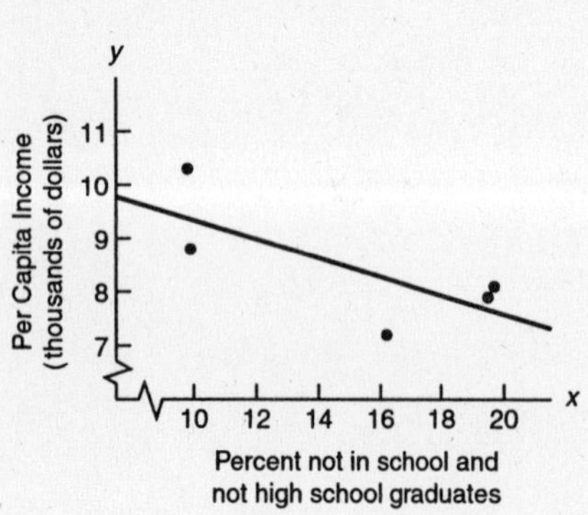

(b)

x	y	xy	x^2	y^2
16.2	7.2	116.64	262.44	51.84
9.9	8.8	87.12	98.01	77.44
19.5	7.9	154.05	380.25	62.41
19.7	8.1	159.57	388.09	65.61
9.8	10.3	100.94	96.04	106.09
75.1	42.3	618.32	1224.83	363.39

$$SS_{xy} = \sum xy - \frac{\sum x \sum y}{n} = 618.32 - \frac{(75.1)(42.3)}{5} = -17.026$$

$$\overline{x} = \frac{\sum x}{n} = \frac{75.1}{5} = 15.02$$

$$\overline{y} = \frac{\sum y}{n} = \frac{42.3}{5} = 8.46$$

$$SS_x = \sum x^2 - \frac{(\sum x)^2}{n} = 1224.83 - \frac{(75.1)^2}{5} = 96.828$$

$$b = \frac{SS_{xy}}{SS_x} = \frac{-17.026}{96.828} = -0.176$$

$$a = \overline{y} - b\overline{x} = 8.46 - (-0.176)(15.02) = 11.1$$

$$y = 11.1 - 0.176x$$

(c) $$SS_y = \sum y^2 - \frac{(\sum y)^2}{n} = 363.39 - \frac{(42.3)^2}{5} = 5.532$$

$$S_e = \sqrt{\frac{SS_y - b \cdot SS_{xy}}{n-2}} = \sqrt{\frac{5.532 - (-0.176)(-17.026)}{5-2}} = \sqrt{\frac{2.5354}{3}}$$

$$= 0.9193$$

(d) Let $x = 17$. Then $y = 11.1 - 0.176(17) = 8.1$ thousand.

(e) For a 75% confidence interval with $n - 2 = 5 - 2 = 3$ degrees of freedom, $t_c = 1.423$ (from the t table).

$$E = t_c S_e \sqrt{1 + \frac{1}{n} + \frac{(x - \overline{x})^2}{SS_x}}$$

$$= 1.423(0.9193)\sqrt{1 + \frac{1}{5} + \frac{(17 - 15.02)^2}{96.828}}$$

$$= 1.3082\sqrt{1.2405}$$

$$= 1.5$$

$y + E = 8.1 + 1.5 = 9.6$

$y - E = 8.1 - 1.5 = 6.6$

A 75% confident forecast for the per capita income is between 6.6 and 9.6 thousands of dollars.

7. **(a)**

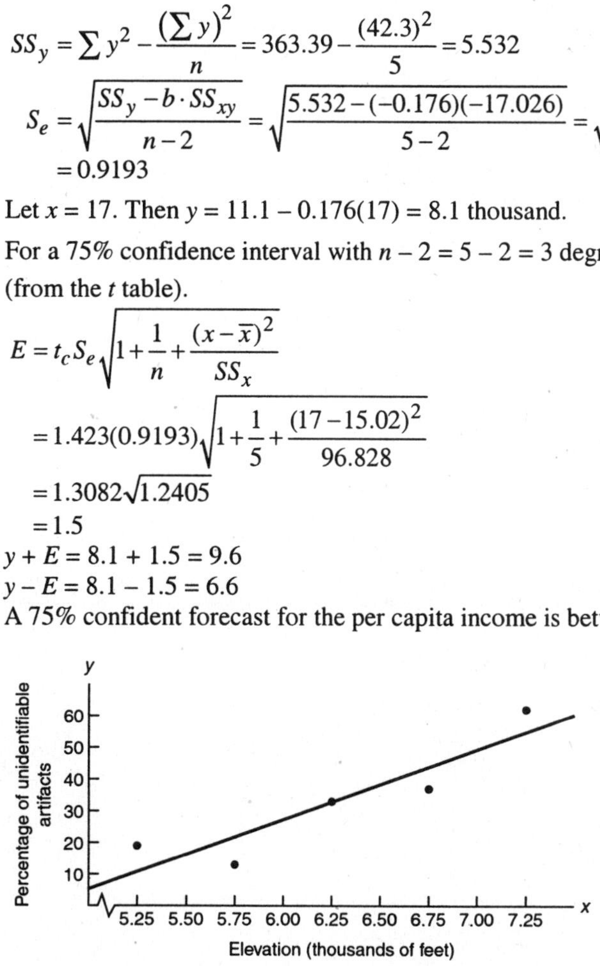

(b)

x	y	xy	x^2	y^2
5.25	19	99.75	27.5625	361
5.75	13	74.75	33.0625	169
6.25	33	206.25	39.0625	1089
6.75	37	249.75	45.5625	1369
7.25	62	449.50	52.5625	3844
31.25	164	1080	197.8125	6832

$$SS_{xy} = \sum xy = \frac{\sum x \sum y}{n} = 1080 - \frac{(31.25)(164)}{5} = 55$$

$$\overline{x} = \frac{\sum x}{n} = \frac{31.25}{5} = 6.25$$

$$\overline{y} = \frac{\sum y}{n} = \frac{164}{5} = 32.8$$

$$SS_x = \sum x^2 - \frac{(\sum x)^2}{n} = 197.8125 - \frac{(31.25)^2}{5} = 2.5$$

$$b = \frac{SS_{xy}}{SS_x} = \frac{55}{2.5} = 22$$

$$a = \overline{y} - b\overline{x} = 32.8 - 22(6.25) = -104.7$$

$$y = -104.7 + 22x$$

(c) $$SS_y = \sum y^2 - \frac{(\sum y)^2}{n} = 6832 - \frac{(164)^2}{5} = 1452.8$$

$$S_e = \sqrt{\frac{SS_y - b \cdot SS_{xy}}{n-2}} = \sqrt{\frac{1452.8 - 22(55)}{5-2}} = \sqrt{\frac{242.8}{3}} = 8.9963$$

(d) Let $x = 6.5$. Then $y = -105.7 + 22(6.5) = 38.3\%$.

(e) For a 75% confidence interval with $n - 2 = 5 - 2 = 3$ degrees of freedom, $t_c = 1.423$.

$$E = t_c S_e \sqrt{1 + \frac{1}{n} + \frac{(x-\overline{x})^2}{SS_x}}$$

$$= 1.423(8.9963)\sqrt{1 + \frac{1}{5} + \frac{(6.5-6.25)^2}{2.5}}$$

$$= 12.8017\sqrt{1.225}$$

$$= 14.2$$

$y + E = 38.3 + 14.2 = 52.5$

$y - E = 38.3 - 14.2 = 24.1$

A 75% confident forecast for the percentage of unidentified artifacts at a given elevation is between 24.1 and 52.5.

8. (a)

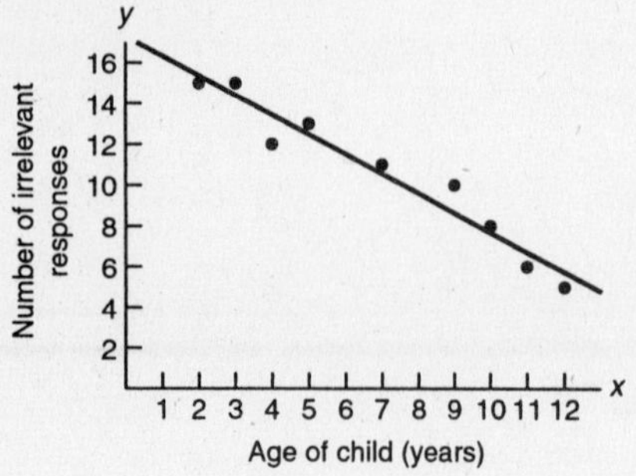

(b)

x	y	xy	x^2	y^2
2	15	30	4	225
3	15	45	9	225
4	12	48	16	144
5	13	65	25	169
7	11	77	49	121
9	10	90	81	100
10	8	80	100	64
11	6	66	121	36
12	5	60	144	25
63	95	561	549	1109

$$SS_{xy} = \sum xy - \frac{\sum x \sum y}{n} = 561 - \frac{(63)(95)}{9} = -104$$
$$\overline{x} = \frac{\sum x}{n} = \frac{63}{9} = 7$$
$$\overline{y} = \frac{\sum y}{n} = \frac{95}{9} = 10.556$$
$$SS_x = \sum x^2 - \frac{(\sum x)^2 n}{n} = 549 - \frac{(63)^2}{9} = 108$$
$$b = \frac{SS_{xy}}{SS_x} = \frac{-104}{108} = -0.963$$
$$a = \overline{y} - b\overline{x} = 10.556 - (-0.963)(7) = 17.297$$
$$y = 17.297 - 0.963x$$

(c) $$SS_y = \sum y^2 - \frac{(\sum y)^2}{n} = 1109 - \frac{(95)^2}{9} = 106.222$$
$$S_e = \sqrt{\frac{SS_y - b \cdot SS_{xy}}{n-2}} = \sqrt{\frac{106.222 - (-0.963)(-104)}{9-2}} = \sqrt{\frac{6.07}{7}}$$
$$= 0.9312$$

(d) Let $x = 9.5$. Then $y = 17.297 - 0.963(9.5) = 81.5 \approx 8$

(e) For a 99% confidence interval with $n - 2 = 9 - 2 = 7$ degrees of freedom, $t_c = 3.499$.

$$E = t_c S_e \sqrt{1 + \frac{1}{n} + \frac{(x - \overline{x})^2}{SS_x}}$$
$$= 3.499(0.9312)\sqrt{1 + \frac{1}{9} + \frac{(9.5 - 7)^2}{108}}$$
$$= 3.2583\sqrt{1.1690}$$
$$= 3.5$$
$$y + E = 8 + 3.5 = 11.5$$
$$y - E = 8 - 3.5 = 4.5$$

A 99% confident forecast for the number of irrelevant responses for a child who is 9.5 years old is between 4.5 and 11.5.

9. (a) See the graph in your text. Yes. The pattern of residuals appears randomly scattered around the horizontal line at zero.

(b) No, there does not appear to be any outliers on the residual graph.

10. (a)

x	y	yp	$y-yp$
27	30	27.1	2.9
44	19	16.9	2.1
32	24	24.1	–0.1
47	13	15.1	–2.1
23	29	29.5	–0.5
40	17	19.3	–2.3
34	21	22.9	–1.9
52	14	12.1	1.9

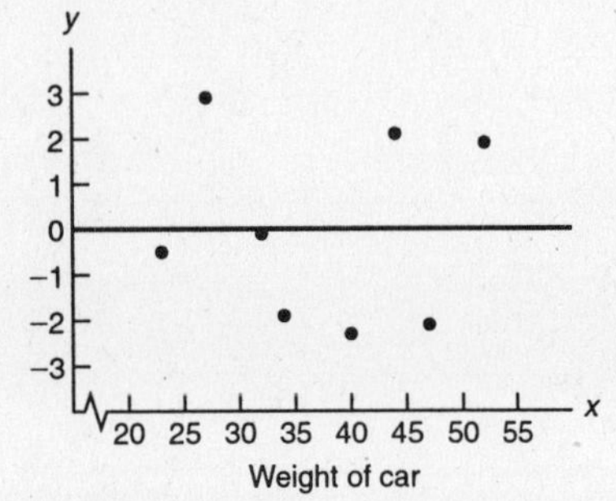

(b) The least-squares model is appropriate for these data because the points in the residual plot from part (a) are random and unstructured about the horizontal line at 0.

Section 12.5

1. (a) Results check.

(b) $H_0: \rho = 0$
$H_1: \rho < 0$

$$t = \frac{r\sqrt{n-2}}{\sqrt{1-r^2}} = \frac{-0.377\sqrt{8-2}}{\sqrt{1-(-0.377)^2}} = -0.997$$

$d.f. = n-2 = 8-2 = 6$

At 5% level of significance, $t_0 = -1.943$.

P value > 0.125

Since –0.997 > –1.943 and P value > 0.125, we fail to reject H_0. The sample evidence does not support a negative correlation.

(c) $H_0: \beta = 0$
$H_1: \beta \neq 0$

$$t = \frac{b-\beta}{\frac{S_e}{\sqrt{SS_x}}} = \frac{-0.468-0}{\frac{7.443}{\sqrt{252.40}}} = -0.999$$

$d.f. = n-2 = 8-2 = 6$

At 5% level of significance, $t_0 = \pm 2.447$.

P value > 0.250

Since –2.447 < –0.999 < 2.447, we fail to reject H_0. The sample evidence does not support a nonzero slope.

2. (a) Results check.

(b) $H_0: \rho = 0$
$H_1: \rho \neq 0$
$$t = \frac{r\sqrt{n-2}}{\sqrt{1-r^2}} = \frac{0.187\sqrt{9-2}}{\sqrt{1-(0.187)^2}} = 0.504$$
$d.f. = n - 2 = 9 - 2 = 7$

At 5% level of significance, $t_0 = \pm 2.365$.
P value > 0.250
Since $-2.365 < 0.504 < 2.365$ and P value > 0.250, we fail to reject H_0. The sample evidence does not support a nonzero correlation.

(c) $H_0: \beta = 0$
$H_1: \beta \neq 0$
$$t = \frac{b-\beta}{\frac{S_e}{\sqrt{SS_x}}} = \frac{0.172-0}{\frac{6.369}{\sqrt{350.70}}} = 0.506$$
$d.f. = n - 2 = 9 - 2 = 7$
At 5% level of significance, $t_0 = \pm 2.365$.
P value > 0.250
Since $-2.365 < 0.506 < 2.365$ and P value > 0.250, we fail to reject H_0. The sample evidence does not support a nonzero slope. (The t–values in (b) and (c) differ due to roundoff error.)

3. (a) Results check.

(b) $H_0: \rho = 0$
$H_1: \rho < 0$
$$t = \frac{r\sqrt{n-2}}{\sqrt{1-r^2}} = \frac{-0.976\sqrt{7-2}}{\sqrt{1-(-0.976)^2}} = -10.02$$
$d.f. = n - 2 = 7 - 2 = 5$

At 1% level of significance, $t_0 = -3.365$.
P value < 0.005
Since $-10.02 < -3.365$ and P value < 0.005, we reject H_0. The sample evidence supports a negative correlation.

(c) $H_0: \beta = 0$
$H_1: \beta < 0$
$$t = \frac{b-\beta}{\frac{S_e}{\sqrt{SS_x}}} = \frac{-0.054-0}{\frac{0.166}{\sqrt{940.89}}} = -9.98$$
$d.f. = n - 2 = 7 - 2 = 5$
At 1% level of significance, $t_0 = -3.365$.
P value < 0.005
Since $-9.98 < -3.365$ and P value < 0.005, we reject H_0. The sample evidence supports a negative slope.

4. **(a)** Results check.

(b) $H_0: \rho = 0$
$H_1: \rho > 0$

$$t = \frac{r\sqrt{n-2}}{\sqrt{1-r^2}} = \frac{0.984\sqrt{5-2}}{\sqrt{1-(0.984)^2}} = 9.57$$

$d.f. = n-2 = 5-2 = 3$

At 1% level of significance, $t_0 = 4.541$.
P value < 0.005
Since $9.57 > 4.541$ and P value < 0.005, we reject H_0. The sample evidence supports a positive correlation.

(c) $H_0: \beta = 0$
$H_1: \beta > 0$

$$t = \frac{b-\beta}{\frac{S_e}{\sqrt{SS_x}}} = \frac{6.876-0}{\frac{2.532}{\sqrt{12.25}}} = 9.505$$

$d.f. = n-2 = 5-2 = 3$

At 1% level of significance, $t_0 = 4.541$.
P value < 0.005
Since $9.505 > 4.541$ and P value < 0.005, we reject H_0. The sample evidence supports a positive slope.
Note: carrying more decimal places than shown in (b) gives $t = 9.505$, the same as in (c).

5. **(a)** Results check.

(b) $H_0: \rho = 0$
$H_1: \rho > 0$

$$t = \frac{r\sqrt{n-2}}{\sqrt{1-r^2}} = \frac{0.956\sqrt{6-2}}{\sqrt{1-(0.956)^2}} = 6.517$$

$d.f. = n-2 = 6-2 = 4$

At 1% level of significance, $t_0 = 3.747$.
P value < 0.005
Since $6.517 > 3.747$ and P value < 0.005, we reject H_0. The sample evidence supports a positive correlation.

(c) $H_0: \beta = 0$
$H_1: \beta > 0$

$$t = \frac{b-\beta}{\frac{S_e}{\sqrt{SS_x}}} = \frac{0.758-0}{\frac{0.1527}{\sqrt{1.733}}} = 6.535$$

$d.f. = n-2 = 6-2 = 4$

At 1% level of significance, $t_0 = 3.747$.
P value < 0.005
Since $6.535 > 3.747$ and P value < 0.005, we reject H_0. The sample evidence supports a positive slope.

6. (a) Results check.

(b) $H_0: \rho = 0$

$H_1: \rho > 0$

$$t = \frac{r\sqrt{n-2}}{\sqrt{1-r^2}} = \frac{0.977\sqrt{5-2}}{\sqrt{1-(0.977)^2}} = 7.936$$

$d.f. = n-2 = 5-2 = 3$

At 1% level of significance, $t_0 = 4.541$.

P value < 0.005

Since 7.936 > 4.541 and P value < 0.005, we reject H_0. The sample evidence supports a positive correlation.

(c) $H_0: \beta = 0$

$H_1: \beta > 0$

$$t = \frac{b-\beta}{\frac{S_e}{\sqrt{SS_x}}} = \frac{0.879-0}{\frac{0.1522}{\sqrt{188.26}}} = 7.924$$

$d.f. = n-2 = 5-2 = 3$

At 1% level of significance, $t_0 = 4.541$.

P value < 0.005

Since 7.924 > 4.541 and P value < 0.005, we reject H_0. The sample evidence supports a positive slope.

7. (a) $H_0: \rho = 0$

$H_1: \rho \neq 0$

$$t = \frac{r\sqrt{n-2}}{\sqrt{1-r^2}} = \frac{0.90\sqrt{6-2}}{\sqrt{1-(0.90)^2}} = 4.129$$

$d.f. = n-2 = 6-2 = 4$

At 1% level of significance, $t_0 = \pm 4.604$.

Since −4.604 < 4.129 < 4.604 and 0.01 < P value, we do not reject H_0. The correlation coefficient ρ is not significantly different from zero at the 0.01 level of significance.

(b) $H_0: \rho = 0$

$H_1: \rho \neq 0$

$$t = \frac{r\sqrt{n-2}}{\sqrt{1-r^2}} = \frac{0.90\sqrt{10-2}}{\sqrt{1-(0.90)^2}} = 5.840$$

$d.f. = n-2 = 10-2 = 8$

At 1% level of significance, $t_0 = \pm 3.355$.

Since 5.840 > 3.355 and P value < 0.01, we reject H_0. The correlation coefficient ρ is significantly different from zero at the 0.01 level of significance.

(c) From part (a) to part (b), n increased from 6 to 10, the test statistic t increased from 4.12 to 5.840, and the critical values t_0 decreased (in absolute value) from 4.604 to 3.355. For the same $r = 0.90$ and the same level of significance $\alpha = 0.01$, we rejected H_0 for the larger n but not for the smaller n.

In general, as n increases, the degrees of freedom $(n-2)$ increase and the critical value(s) become(s) closer to zero. Also, as n increases, the test statistic $\left(t = \frac{r\sqrt{n-2}}{\sqrt{1-r^2}}\right)$ moves farther from zero. The combination of the critical value(s) approaching zero while the test statistic moves farther out into the tail of the t–distribution means we are more likely to reject H_0 for larger n (using the same r and α).

Chapter 12 Review

1. H_0 : Test time and test results are independent.

H_1 : Test time and test results are not independent.

$$x^2 = \sum \frac{(0-E)^2}{E}$$

$$= \frac{(23-18.93)^2}{18.93} + \frac{(42-42.60)^2}{42.60} + \frac{(65-71)^2}{71}$$

$$+ \frac{(12-9.47)^2}{9.47} + \frac{(17-21.07)^2}{21.07} + \frac{(48-47.40)^2}{47.40}$$

$$+ \frac{(85-79.00)^2}{79.00} + \frac{(8-10.53)^2}{10.53}$$

$$= 3.924$$

Since there are 2 rows and 4 columns, *d.f.* = (2 – 1)(4 – 1) = 3. For $\alpha = 0.01$, the critical value is $\chi^2_{0.01} = 11.34$.

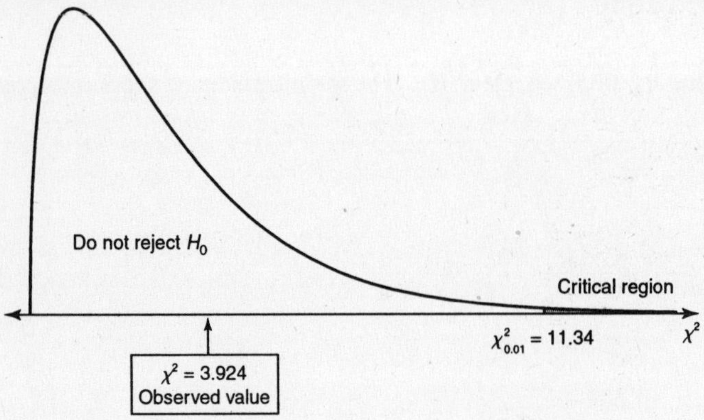

Since the sample statistic falls outside the critical region, do not reject H_0. There is insufficient evidence to conclude that time to do a test and test results are not independent.

2. H_0 : Teacher ratings and student grades are independent.

H_1 : Teacher ratings and student grades are not independent.

$$x^2 = \sum \frac{(0-E)^2}{E}$$

$$= \frac{(14-10)^2}{10} + \frac{(18-13.33)^2}{13.33} + \frac{(15-21.67)^2}{21.67} + \frac{(3-5)^2}{5}$$

$$+ \frac{(25-30)^2}{30} + \frac{(35-40)^2}{40} + \frac{(75-65)^2}{65} + \frac{(15-15)^2}{15}$$

$$+ \frac{(21-21)^2}{20} + \frac{(27-26.67)^2}{26.67} + \frac{(40-43.33)^2}{43.33}$$

$$+ \frac{(12-10)^2}{10}$$

$$= 9.792$$

Since there are 3 rows and 4 columns, *d.f.* = (3 – 1)(4 – 1) = 6. For $\alpha = 0.01$, the critical value is $\chi^2_{0.01} = 16.81$.

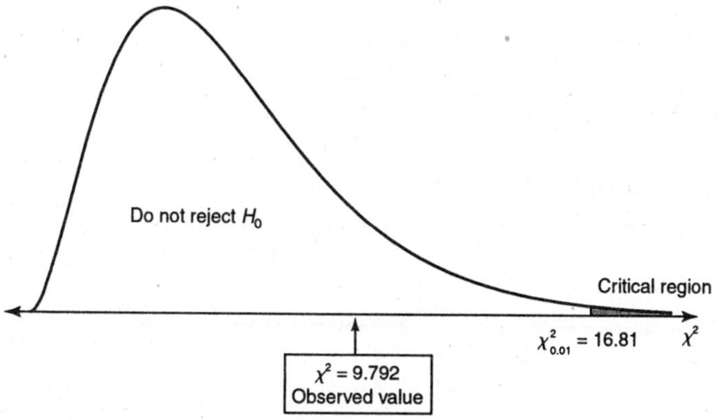

Since the sample statistic falls outside the critical region, do not reject H_0. There is insufficient evidence to conclude that teacher ratings and student grades are not independent.

3. H_0 : The distributions are the same.

H_1 : The distributions are different.

$$x^2 = \sum \frac{(0-E)^2}{E}$$

$$= \frac{(15-42)^2}{42} = \frac{(25-31.5)^2}{31.5} + \frac{(70-63)^2}{63} + \frac{(80-52.5)^2}{52.5}$$

$$+ \frac{(20-21)^2}{21}$$

$$= 33.93$$

$d.f.$ = (number of E entries) – 1 = 5 – 1 = 4. The critical value is $\chi^2_{0.01} = 15.09$.

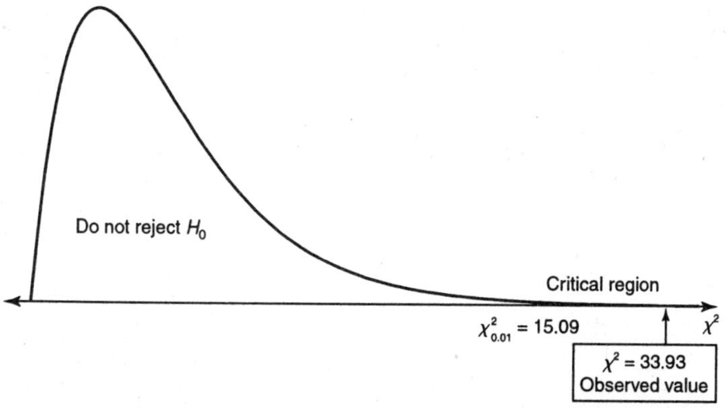

Since the sample statistic falls inside the critical region, we reject H_0. We conclude that the distributions are different.

4. H_0 : The distributions are the same.

H_1 : The distributions are different.

$$x^2 = \sum \frac{(0-E)^2}{E}$$
$$= \frac{(48-40)^2}{40} + \frac{(75-60)^2}{60} + \frac{(77-100)^2}{100}$$
$$= 10.64$$

d.f. = (number of *E* entries) – 1 = 3 – 1 = 2. The critical value is $\chi^2_{0.01} = 9.21$.

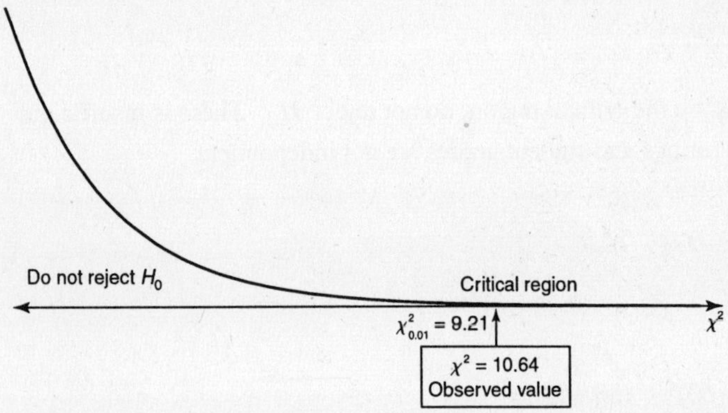

Since the sample statistic falls inside the critical region, we reject H_0. We conclude the distributions are different.

5. (a) H_0: $\sigma^2 = 15$

H_1: $\sigma^2 \neq 15$

Since ≠ is in H_1, a two-tailed test is used.

For *d.f.* = 22 – 1 = 21, the critical values are $\chi^2_{0.025} = 35.48$ and $\chi^2_{0.975} = 10.28$.

$$\chi^2 = \frac{(n-1)s^2}{\sigma^2} = \frac{(22-1)(14.3)}{15} = 20.02$$

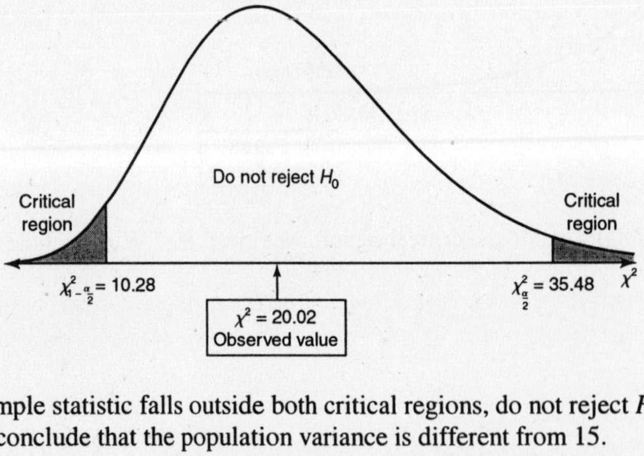

Since the sample statistic falls outside both critical regions, do not reject H_0. There is insufficient evidence to conclude that the population variance is different from 15.

6. $H_0 : \sigma^2 = 0.0625$

$H_1 : \sigma^2 > 0.0625$

Since > is in H_1, a right-tailed test is used.

For *d.f.* = 12 – 1 = 11, the critical value is $\chi^2_{0.05} = 19.68$.

$$x^2 = \frac{(n-1)s^2}{\sigma^2} = \frac{(12-1)(0.38)^2}{(0.25)^2} = 25.41$$

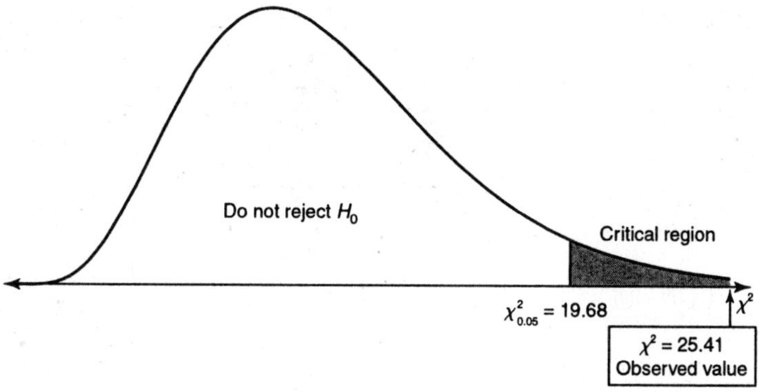

Since the sample statistic falls inside the critical region, we reject H_0. We conclude that the variance has increased.

7. **(a)** Let $x = 9.5$. Then $y = -3.682 + 0.966(9.5) = 5.495$.

(b) $$SS_y = \sum y^2 - \frac{\left(\sum y\right)^2}{n} = 116.03 - \frac{(23.7)^2}{5} = 3.692$$

$$S_e = \sqrt{\frac{SS_y - b \cdot SS_{xy}}{n-2}} = \sqrt{\frac{3.692 - (0.966)(2.926)}{5-2}} = 0.537$$

For an 80% confidence interval with $n - 2 = 5 - 2 = 3$ degrees of freedom, $t_c = 1.638$.

$$E = t_c S_e \sqrt{1 + \frac{1}{n} + \frac{(x-\overline{x})^2}{SS_x}}$$

$$= 1.638(0.537)\sqrt{1 + \frac{1}{5} + \frac{(9.5-8.72)^2}{3.028}}$$

$$= (0.8796)(1.1836)$$

$$= 1.0411$$

$y + E = 5.495 + 1.0411 = 6.54$

$y - E = 5.495 - 1.0411 = 4.45$

An 80% confident forecast for the per capita retail sales (thousands of dollars) is between 4.45 and 6.54 thousand dollars.

(c) $H_0 : p = 0$

$H_1 : p > 0$

$$t = \frac{r\sqrt{n-2}}{\sqrt{1-r^2}} = \frac{0.875\sqrt{5-2}}{\sqrt{1-(0.875)^2}} = 3.13$$

$d.f. = n - 2 = 5 - 2 = 3$

At 5% level of significance, $t_0 = 2.353$.

$0.025 < P$ value < 0.05

Since $3.13 > 2.353$ and $p < 0.05$, we reject H_0. The sample evidence supports a positive correlation.

(d) $b = \frac{SS_{xy}}{SS_x} = \frac{2.926}{3.028} = 0.966$

$H_0: \beta = 0$

$H_1: \beta > 0$

$t = \frac{b-\beta}{\frac{Se}{\sqrt{SS_x}}} = \frac{0.966-0}{\frac{0.537}{\sqrt{3.028}}} = 3.13$

$d.f. = n-2 = 5-2 = 3$

At the 5% level of significance, $t_0 = 2.353$.

$0.025 < P \text{ value} < 0.05$

Since $3.13 > 2.353$ and $P < 0.05$, we reject H_0. The sample evidence supports a positive slope.

8. (a) Let $x = 5$. Then $y = 0.633 + 0.694(5) = 4.103$.

(b) $SS_y = \sum y^2 - \frac{(\sum y)^2}{n} = 72.41 - \frac{(21.3)^2}{7} = 7.597$

$S_e = \sqrt{\frac{SS_y - b \cdot SS_{xy}}{n-2}} = \sqrt{\frac{7.597-(0.694)(8.759)}{7-2}} = 0.551$

For an 80% confidence interval with $n-2=7-2-5$ degrees of freedom, $t_c - 1.476$.

$n-2=7-2-5$ degrees of freedom, $t_c = 1.476$.

$E = t_c S_e \sqrt{1+\frac{1}{n}+\frac{(x-\bar{x})^2}{SS_x}}$

$= (1.476)(0.551)\sqrt{1+\frac{1}{7}+\frac{(5-3.47)^2}{12.614}}$

$= (0.8133)(1.1526)$

$= 0.9374$

$y + E = 4.103 + 0.9374 = 5.04$

$y - E = 4.103 - 0.9374 = 3.17$

An 80% confident forecast for the percentage change in consumer prices for the past year in Australia, Canada, France, Italy, Spain, and the United States is between 3.17% and 5.04%.

(c) $H_0: p = 0$

$H_1: p \neq 0$

$t = \frac{r\sqrt{n-2}}{\sqrt{1-r^2}} = \frac{0.895\sqrt{7-2}}{\sqrt{1-(0.895)^2}} = 4.487$

$d.f. = n-2 = 7-2 = 5$

At 1% level of significance, $t_0 = 4.032$.

$P \text{ value} < 0.01$

Since $4.487 > 4.032$ and $p < 0.01$, we reject H_0. The sample evidence supports a correlation different from 0.

(d) $b = \dfrac{SS_{xy}}{SS_x} = \dfrac{8.759}{12.614} = 0.694$

$H_0 : \beta = 0$

$H_1 : \beta \neq 0$

$$t = \frac{b - \beta}{\frac{S_e}{\sqrt{SS_x}}} = \frac{0.694 - 0}{\frac{0.551}{\sqrt{12.614}}} = 4.473$$

$d.f. = n - 2 = 7 - 2 = 5$

At 1% level of significance, $t_0 = 4.032$.

P value < 0.01

Since $4.473 > 4.032$ and $P < 0.01$, we reject H_0. The sample evidence supports a slope different from 0.

Part V

Test Item File Sample Questions And Answers

Chapter 1
Getting Started

Section A: Static Items

1. Write a brief but complete essay in which you answer the following questions: What is statistics? What is descriptive statistics? What is inferential statistics? Where is statistics used in everyday life?

2. Write a brief but complete essay in which you discuss three examples of modern life in which you think that statistics might be used. For each example describe the population under study and what a sample might be. In this context what do we mean by the term "descriptive statistics"? What do we mean by the term "inferential statistics"?

3. To estimate the amount of time State College students who live off campus spend to commute to classes each week, a random sample of 45 such students was surveyed.
(a) What is the population?
(b) What is the sample?
(c) Would this sample necessarily be representative of the time part-time off-campus students spend commuting to class each week? Why or why not?

4. Data may be classified by one of the four levels of measurement. Which is the lowest level?
A. interval B. population C. nominal D. ordinal E. ratio

5. An insurance company wants to know the proportion of medical doctors in New York involved in at least one malpractice suit in the last three years. They survey a random sample of 200 medical doctors in New York.
(a) What is the population?
(b) What is the sample?
(c) Could the insurance company generalize and say that this sample is representative of all medical doctors in rural areas of the country? Why or why not?

6. A company making radar detection devices maintains quality control by testing a random sample of 20 such devices produced each day.
(a) What is the population?
(b) What is the sample?
(c) If the manager decided to test a random sample of 100 devices on Monday and did no further sampling for the rest of the week, could she draw any conclusions about the quality of production for the entire week? Explain your answer.

7. Data may be classified by one of the four levels of measurement. Which is the highest level?

A. interval B. nominal C. ratio D. sample E. ordinal

8. Write a brief but complete essay in which you describe four different ways to produce data. Include an example illustrating each method.

9. Which level of measurement is suitable for putting data into categories?

A. inferential B. nominal C. ordinal D. interval E. ratio

10. Choose the word that does *not* complete the following statement:
Statistics is the study of how to ________, ________, ________, and ________ numerical information from data.

A. randomize B. analyze C. collect D. interpret E. organize

11. A professor of social science at a certain university questioned 200 students out of a total enrollment of 12,243 concerning their opinions of the quality of various university services. Identify the variable.

A. the professor B. social science C. 200 students
D. the students' opinions E. the total enrollment

12. English 105: Great Writers of America has 447 students. To estimate the amount of time students in this section spent studying for the midterm exam a random sample of 40 students were surveyed.
(a) What is the population?
(b) What is the sample?
(c) Would this sample necessarily be representative of the time students enrolled in English 105 spent studying for the comprehensive final exam? Why or why not?

13. The cook for a summer camp wants to know the eating preferences of the campers. Twenty-five of the 113 campers are surveyed concerning their eating preferences.
(a) What is the population?
(b) What is the sample?
(c) Would this sample necessarily be representative of the eating preferences of the camp counselors? Why or why not?

14. Choose the word that does *not* complete the following statement:
Descriptive statistics involves methods of ________, ________, and ________ information from ________ or populations.

A. samples B. summarizing C. organizing D. calculating E. picturing

15. A neighbor's dog is described by the following information:
(a) Name: Rocky
(b) Age: 6
(c) Weight: 60 lb
(d) Temperament: Friendly
(e) Color: Brown
For the information (a) to (e) list the highest level of measurement as ratio, interval, ordinal, or nominal.

16. Lake Lulu in Rocky Mountain National Park has several common species of trout along with the rare greenback trout. The National Park Service wants to estimate the proportion of greenback trout in Lake Lulu. Using a large net at random locations, the rangers caught (and released) a total of 587 trout of which 91 were the rare greenback trout.
(a) What is the population?
(b) What is the sample?
(c) Would this sample necessarily be representative of the trout in a different lake at higher elevation? Why or why not?

17. A simple random sample of *n* measurements from a population is a subset of the population selected in a manner such that
(a) every sample of size *n* from the population has an equal chance of being __________ and
(b) every member of the population has an equal chance of being __________ in the sample.

18. A random-number table is to be used to pick a random sample of 40 sheep out of a population of 300 sheep. If the blocks of five digits need to be regrouped, how many digits should be included in each block?

A. 3 B. 5 C. 1 D. 4 E. 2

19. The county sheriff wants to know the proportion of drivers who make an illegal left turn as they leave the post office parking lot. A sample of 200 cars is observed.
(a) What is the implied population?
(b) What is the sample?
(c) Would this sample necessarily reflect the proportion of cars making illegal left turns at all intersections? Explain your answer.

20. At a hospital nursing station the following information is available about a patient.

(a) Name:	Jim Wood
(b) Age:	17
(c) Weight:	165 lb
(d) Height:	6'1"
(e) Blood type:	A
(f) Temperature:	96.8° F
(g) Condition:	Fair
(h) Date of admission:	January 21, 1998
(i) Response to treatment:	Excellent

For the information (a) to (i) list the highest level of measurement as ratio, interval, ordinal, or nominal.

21. A construction supervisor needs to purchase a ladder that will be strong enough to support the heaviest of his employees. Explain why or why not taking a simple random sample of his employees will determine the appropriate ladder to purchase.

22. A hikers log of favorite hikes contains the following information:

(a) Destination:	Mirror Lake
(b) Distance:	6.5 miles
(c) Difficulty level:	Moderate
(d) Change in elevation:	3,000 ft
(e) Time to hike (one way):	3 hours
(f) Last year hiked:	1997

For the information in parts (a) through (f) list the highest level of measurement as ratio, interval, ordinal or nominal and explain your choices.

23. At a used car lot the following information is obtained about one of the cars on the lot.

(a) Make and Model:	Ford Escort
(b) Model year:	1990
(c) Color:	blue
(d) Number of cylinders:	4
(e) Gas consumption:	35 miles per gallon
(f) Condition:	Clean, in good repair
(g) Sticker price:	$8,817.00
(h) Body registration number:	TA1592763C51
(i) Number of miles on odometer:	36,719
(j) Sales Person Claim:	This is the best car on the lot for you.

For the information (a) to (j) list the highest level of measurement that applies: ratio, interval, ordinal, or nominal.

24. A book inventory record contains the following information:
(a) Title: More Mysteries
(b) Author: Roger Mortimer
(c) Date of publication: 1998
(d) List price: $25.00
(e) Number in stock: 6
For the information in parts (a) through (e) list the highest level of measurement as ratio, interval, ordinal or nominal and explain your choices.

25. Write a brief but complete essay in which you describe each of the four levels of measurement discussed in this chapter. Give an example of data at each of the four levels of measurement.

26. Write a brief essay in which you discuss some of the aspects of surveys. Give specific examples to illustrate your main points.

27. Herman is a movie critic who keeps a record of the films he sees. His film records include the following information:
(a) Title: Adventures with Aunt Agatha
(b) Genre: Mystery
(c) Length: 150 minutes
(d) Date released: 1928
(e) Rating: 3 stars (on a scale from one star to five stars)
For the information in parts (a) through (e) list the highest level of measurement as ratio, interval, ordinal, or nominal and explain your choices.

28. Myra compiles information about various possible vacation sites. One of her records included the following information:
(a) Location: Glacier Point, Alaska
(b) Distance: 1,525 miles
(c) Number of motels: 2
(d) Mean summer temperature: 42°F
(e) Outdoor activities rating: 4 (on a scale from 1, poor, to 5, excellent)
For the information in parts (a) through (e) list the highest level of measurement as ratio, interval, ordinal, or nominal and explain your choices.

29. What technique for gathering data (sampling, experiment, simulation or census) do you think was used in each of the following studies?
(a) The manager of an automobile repair shop selects a random sample of service records and records the total amount of time each vehicle was in the facility.
(b) The same manager tests a computerized diagnostic machine by comparing its performance on a random sample of 20 vehicles with the evaluation of a professional mechanic for the same 20 vehicles.
(c) The same manager surveys every customer who has had a car serviced to determine the quality of the customer's service and the customer's level of satisfaction with the service.
(d) An automobile manufacturer uses a computer simulation program to test the aerodynamic properties of a proposed new automobile body design.
(e) A service manager uses computer software to simulate a new arrangement of automotive workstations to see if the arrangement will provide more efficient service.

30. What techniques for gathering data (sampling, experiment, simulation or census) do you think were used in the following studies?
(a) 100 customers at a grocery store sampled a new food product and gave their opinion about the quality of the product. The grocery store has about 2,000 customers per day.
(b) An exit poll of 1,000 voters in a mayoral election for a large city showed 656 voters supported the incumbent.
(c) A computer was used to simulate two different assembly processes for car stereos to see which process produced more stereos per hour.
(d) All the owners in a subdivision were polled to determine their level of satisfaction with the exterior maintenance service.

31. What technique for gathering data (sampling, experiment, simulation or census) do you think was used in the following studies?
(a) 100 members of a large labor union were polled regarding a proposed new contract.
(b) A computer program was used to simulate movements of the earth's crust near an earthquake fault in an attempt to predict earthquakes.
(c) To test the safety of refrigerated packaged meat products 6 packages were chosen and their bacteria counts determined.
(d) All residents of a town were surveyed to determine the demand for home dairy delivery service.
(e) A volunteer group of 20 elderly residents of a care facility were given a program of nutrition and regular exercise. Their level of mental alertness was measured before and after they completed the program.

32. What technique for gathering data (sampling, experiment, simulation or census) do you think was used in each of the following studies?
(a) A computer program was used to model global weather patterns and to produce long-range weather forecasts for a rural agricultural region.
(b) A random sample of 1,000 residents of a major metropolitan area was surveyed to determine the level of support for a new sports complex among all residents of the area.
(c) To determine the effect of a new fertilizer on productivity of tomato plants one group of plants is treated with the new fertilizer while a second group is grown without such treatment. The number of ripe tomatoes produced by each group is counted.
(d) A study was done regarding the number of home runs scored by major league baseball teams playing at altitudes over 5,000 feet. Data for all major league baseball games played at this altitude was used in the study.

33. (a) Describe how you would use a random number table to get a random sample of 50 people who had listed phone numbers in a phone book that had 7500 people with listed numbers.
(b) Use the line from the random number table given below to identify the first three people in your sample.
94456 48396 73780 06436 86641 69239

34. (a) Describe how you would use a random number table to get a random sample of 12 men's dress shirts taken from a shipment of 475 men's dress shirts.
(b) Use the line from the random number table given below to identify the first five shirts in your sample.
08811 82711 57392 25252 30333

35. (a) How would you use a random number table to get a random sample of 8 families from a subdivision? The subdivision contains 92 families.
(b) Use the line from the random number table below to identify the first four families in your sample.
66330 33393 95261 95349

36. (a) Describe how you could use a random number table to simulate the experiment of tossing one die 275 times. The results of tossing a die once can be any of the digits 1, 2, 3, 4, 5, or 6.
(b) Use the line from the random number table below to identify the first ten outcomes in the simulation.
29553 18432 13630

37. (a) Describe how you could use a random number table to choose 18 workers from a work force of 6225 workers.
(b) Use the line from the random number table below to identify the first three workers chosen.
10072 55980 64688 68239 20461

38. (a) Describe how you could use a random number table to simulate the experiment of throwing two distinct dice.
(b) Use the line from the random number table below to identify the first five outcomes.
25464 59098 27436 89421 80754

39. (a) Describe how you could use random walk to simulate the arrival and departure of birds at a bird feeder. Assume that the birds arrive and depart one at a time and that they are as likely to depart as to arrive.
(b) Use the line from the random number table below to determine the number of birds at the feeder for the first 10 arrivals and/or departures. Use even for arrival and odd for departure. Initially there are two birds at the feeder.
25138 26810 07093

40. (a) A business employs 736 people. Describe how you could get a simple random sample of size 30 for a survey regarding professional training opportunities.
(b) Use the line from the random number table below to identify the first five employees to be included in the sample.
71886 56450 36507 09395 96951

41. Write a brief but complete essay in which you describe the topics: stratified sampling, systematic sampling, cluster sampling, and convenience sampling. List some advantages and disadvantages of each sampling method. How do these methods compare to "simple" random sampling?

42. Suppose you want to survey a sample of the registered voters of Chicago regarding their opinions about a proposed increase in sales tax. Categorize the following sampling methods using the categories: stratified sample, systematic sample, cluster sample, convenience sample, simple random sample.
(a) Divide the registered voters according to age and sample from each of the age categories.
(b) Select a random sample of area codes in the city and then sample from the registered voters in the selected area codes.
(c) Number all of the registered voters and use a random number table to identify the voters to include in the sample.
(d) Survey the registered voters who work in the voter registration office.

43. Identify the type of sampling (cluster, convenience, random, stratified, systematic) which would be used to get each of the following samples:
(a) For a period of two days measure the length of time each fifth person coming into a bank waits in line for teller service.
(b) Take a random sample of five zip codes from the Chicago metropolitan region and count the number of students enrolled in the first grade for every elementary school in each of the zip code areas.
(c) Divide the users of the Internet into different age groups and then select a random sample from each age group to survey about the amount of time they spend on the Internet each month.
(d) Survey five friends regarding their opinion about the quality of food in the student cafeteria.
(e) Pick a random sample of students enrolled at your college and determine the number of credit hours they have each accumulated toward their degree program.

44. To determine monthly rental prices of apartment units in the San Francisco Bay area, samples were constructed in the following ways. Identify the technique used to produce each sample. (cluster, convenience, simple random, stratified, systematic)
(a) Number all the units in the area and use a random number table to select the apartments to include in the sample.
(b) Classify the apartment units according to the number of bedrooms and then take a random sample from each of the classes.
(c) Classify the apartments according to zip code and take a random sample from each of the zip code regions.
(d) Look in the newspaper and choose the first apartments you find that list rents.

45. The owner of a specialty store in a shopping mall would like to estimate the demand for a proposed new inventory item. She would like to draw a sample of potential customers and survey them regarding their interest or lack of interest in the product. Identify the sampling technique (cluster, convenience, simple random, stratified, systematic) for each sampling technique described below.
(a) Choose every tenth person who comes into the mall at the entrance nearest her store.
(b) Select every person who makes a purchase at her store on some convenient day.
(c) Obtain a list of all residents of the region in which the mall is located. Count them and use a random number table to select the sample.
(d) Obtain a list of all residents of the region in which the mall is located. Divide them into groups by income level and choose a random sample from residents in each income level.

46. Use the line from the random number table below to simulate drawing a seven card hand from a standard deck of cards with no jokers. Specify which of the cards correspond to each of the 52 numbers 1 through 52. List both the numbers you have chosen and the corresponding cards.
44940 60430 22834 14130 96593 23298 56203 92671

47. The manager of a large department store would like to sample the people who have credit cards issued by the store about their shopping preferences. Describe a way the manager could draw each of the following:
(a) A simple random sample
(b) A stratified sample
(c) A systematic sample
(d) A cluster sample

48. What technique (observational study or experiment) for gathering data do you think was used in the following study? In a national forest, 87 deer were caught, tagged, and then released back into the wild. Two weeks later, 62 deer were caught and 43 were found to have tags. From this, it was possible to estimate how many deer live in the forest.

49. What technique (observational study or experiment) for gathering data do you think was used in the following study? Texas Parks and Wildlife administered medicine to 50 sheep that were all suffering from a form of mange. Two months later, 42 showed noticeable signs of improvement.

50. What technique (observational study or experiment) for gathering data do you think was used in the following study? In a national park, a park ranger noticed 37 bears. Of these, 16 were brown bears and 21 were American black bears.

Section B: Algorithms

Objective 1.1: What is Statistics?

51. Is the data set "the number of leaves on a branch" quantitative or qualitative?

52. Is the data set "dollars of sales per week" quantitative or qualitative?

53. Is the data set "average weight of dog breeds" quantitative or qualitative?

54. Does the following represent a sample or a population?
calories created while burning one ounce of coal taken from a railcar full of coal

55. Does the following represent a sample or a population?
amount of rain on your porch that falls into a glass used to record precipitation

56. Does the following represent a sample or a population?
the colors of the flowers on a branch of a randomly chosen tree in a forest

57. Identify the population and suggest a sample that could be used to answer the question: What subject do ninth-grade students in your school consider the most difficult?

58. Categorize the following measurement according to level: nominal, ordinal, interval, or ratio:
score on final exam (based on 100 possible points)

59. Categorize the following measurement according to level: nominal, ordinal, interval, or ratio:
gross income for past 10 years

60. Categorize the following measurement according to level: nominal, ordinal, interval, or ratio:
guidebook rating bird watching locations: poor, fair, good

61. Categorize the following measurement according to level: nominal, ordinal, interval, or ratio:
time of road trip

62. Is the data set "zip codes" quantitative or qualitative?

63. Classify the following data set as quantitative or qualitative:
Sponsors of local baseball teams.

Objective 1.2: Random Samples

64. Do a phone survey of casinos in your state. Is this a random sample for determining the odds of getting a certain number throwing a pair of dice?

65. Do a phone survey of casinos in your state. Is this a random sample for determining the odds of getting a certain number throwing a pair of dice?

66. Which is the best random sample for predicting whether a state ballot measure will pass?

A. Do a telephone survey of 1000 randomly selected homes across the state.

B. Do a person-to-person poll of your neighborhood.

C. Listen to radio talk shows.

D. Interview the people who wrote the ballot measure.

67. Anastasia is the president of the student council. She wants to find out whether people in her school would prefer carpet or hardwood floors in the new school library. She asks an equal number of students from each grade. What kind of sampling method is she using?

68. What type of sampling technique is described by the following?
questioning the parents waiting for their children at a school

69. Questioning equal percentages of men and women describes what type of sampling technique?

70. At Rosa's summer job with a research company, she must get a random sample of people from her town to answer a question about health habits. Which of the following methods could be used to get a random sample?

A. gather responses from her apartment building

B. selecting people to respond randomly by computer

C. ask teachers and students

D. have citizens mail in responses

Objective 1.3: Introduction to Experimental Design

71. Decide whether a sample or a census better determines the following:
the most popular television show among students at your school last night

72. Decide whether a sample or a census better determines the following:
the number of students in the drama club

73. Is the following survey question biased or unbiased?
Should the school's parking lot be repaved to increase our safety?

74. Is the following survey question biased or unbiased?
Do you think Johnson is a good police chief in spite of his recent personal problems?

75. For which of the following surveys would you use a sample?

A. how many athletes there are on the hockey team

B. the most popular television show among students at your school last night

C. the number of roses in a vase

D. the type of shoes preferred by students in your math class

76. For which of the following survey would you use a sample?

A. the number of houses on your block.

B. the number of teenagers in your community who perform voluntary community service

C. the number of players on a football team

D. how many people in the school band play wind instruments

77. Which type of bias is most likely to result from using the following survey?
A high school psychology teacher asks his students to survey their friends to determine the percentage of high school students that live more than 5 miles from school.

78. Decide whether you would use a sample to learn about the group. Explain
the number of spectators at a professional baseball game who are wearing hats

79. Identify the population and suggest a sample that could be used to answer the question:
How many teenagers in your community do volunteer work at least once a month?

80. Decide whether the group is a sample or a population.
the number of leaves on a branch of a randomly chosen tree in a forest

Answer Key for Chapter 1
Getting Started

Section A: Static Items

[1] See text.

[2] See text.

(a) Population: The time spent commuting by each off-campus student of State College.
(b) Sample: The time spent commuting to class by each of the 45 students.
[3] (c) Not necessarily; part-time students may come to campus less frequently.

[4] C.

(a) Population: The response (has or has not been involved in a malpractice suit in the last three years) from all medical doctors in New York.
(b) Sample: The responses to the same question from the 200 medical doctors in the survey.
(c) No. The sample was not restricted to medical doctors from rural areas in New York.
[5] Many of those in the sample could have come from New York City.

(a) Population: The test results of all radar detection devices produced on a given day by the company.
(b) Sample: The test results of the sample of 20 devices selected from the day's production.
(c) By testing only the devices made on Monday, the manager has no information about
[6] those produced during the rest of the week.

[7] C.

[8] See text.

[9] B.

[10] A.

[11] D.

[12] (a) Population: The times spent studying for the midterm exam by the 447 students in English 105.
(b) Sample: The times spent studying for the midterm exam by the 40 students surveyed.
(c) No. Students would probably spend more time studying for a comprehensive final exam than they would for a midterm.

[13] (a) Population: The eating preferences of the 113 campers.
(b) Sample: The eating preferences of the 25 campers surveyed.
(c) No. The camp counselors eating preferences may differ from the campers due to age, religion, region of origin, etc.

[14] D.

[15] (a) Nominal; (b) Ratio; (c) Ratio; (d) Ordinal; (e) Nominal

[16] (a) Population: The species of each trout in Lake Lulu.
(b) Sample: The species of the 587 trout in the sample.
(c) No. Other lakes may have other trout species and the species may occur in different proportions than species of trout in Lake Lulu.

[17] (a) selected; (b) included

[18] A.

[19] (a) Population: Observation results (legal or illegal) of all drivers who exit the post office parking lot under the present law enforcement patterns.
(b) Sample: Observation results for the drivers of the 200 cars in the sample.
(c) No. These observations apply only to the cars exiting the post office parking lot under study.

[20] (a) Nominal; (b) Ratio; (c) Ratio; (d) Ratio; (e) Nominal; (f) Interval; (g) Ordinal; (h) Interval; (i) Ordinal.

[21] There is no guarantee that the heaviest employee will be included in the sample. So, taking a simple random sample of employees is not an appropriate way to determine which ladder to purchase.

[22] (a) Nominal; (b) Ratio; (c) Ordinal; (d) Ratio; (e) Ratio; (f) Interval.

[23] (a) Nominal; (b) Interval; (c) Nominal; (d) Ratio; (e) Ratio; (f) Ordinal; (g) Ratio; (h) Nominal; (i) Ratio; (j) Ordinal.

[24] (a) Nominal; (b) Nominal; (c) Interval; (d) Ratio; (e) Ratio.

[25] Essay. See text.

[26] Essay. See text.

[27] (a) Nominal; (b) Nominal; (c) Ratio; (d) Interval; (e) Ordinal.

[28] (a) Nominal; (b) Ratio; (c) Ratio; (d) Interval; (e) Ordinal.

[29] (a) Sampling; (b) Experiment; (c) Census; (d) Simulation; (e) Simulation.

[30] (a) Sampling; (b) Sampling; (c) Simulation (of an experiment); (d) Census.

[31] (a) Sampling; (b) Simulation; (c) Experiment; (d) Census; (e) Experiment.

[32] (a) Simulation; (b) Sampling; (c) Experiment; (d) Census.

[33] (a) Number the people from 0001 to 7500. Read consecutive groups of four digits until you have fifty distinct numbers from 1 through 7500.
(b) 6483, 4368, 6641

[34] (a) Number the shirts from 001 through 475. Read consecutive groups of three digits until you have twelve distinct numbers from 1 to 475.
(b) 088, 118, 271, 157, 392

[35] (a) Number the families from 01 to 92. Read consecutive groups of two digits until you have eight distinct numbers from 1 to 92.
(b) 66, 33, 03, 93

[36] (a) Associate the numbers 1 through 6 with the number showing on the top face of the die. Read consecutive single digits until you have 275 digits from 1 through 6.
(b) 1, 5, 5, 3, 1, 4, 3, 2, 1, 3

(a) Number the workers from 0001 to 6225. Read consecutive groups of four digits until you have 18 distinct numbers from 1 to 6225.
[37] (b) 1007, 2559, 2046

(a) Associate pairs of digits, each in the range 1 through 6, with the numbers showing on the top faces of the two dice. Read consecutive pairs of digits discarding any pair for which either digit is outside the range 1 through 6.
[38] (b) (2, 5), (4, 6), (4, 5), (4, 3), (2, 1)

(a) Let digits of one parity (even or odd) represent arrivals and let those of the opposite parity represent departures. Read single digits, adding 1 to the total number of birds for each arrival and subtracting 1 from the total number of birds for each departure.
(b)

Time	0	1	2	3	4	5	6	7	8	9	10
Birds	2	3	2	1	0	1	2	3	4	3	4

[39]

(a) Number the employees from 001 to 736. Read consecutive groups of three digits until you have 30 distinct numbers from 1 to 736.
[40] (b) 718, 645, 036, 507, 093

[41] Essay; see text.

(a) Stratified sample
(b) Cluster sample
(c) Simple random sample
[42] (d) Convenience sample

(a) Systematic sampling
(b) Cluster sampling
(c) Stratified sampling
(d) Convenience sampling
[43] (e) Simple random sampling

(a) Simple random sampling
(b) Stratified sampling; the strata are the number of bedrooms.
(c) Cluster sampling
[44] (d) Convenience sampling

(a) Systematic sampling
(b) Convenience sampling
(c) Simple random sampling
[45] (d) Stratified sampling

[46] Choose some ordering for the cards. For example, order the suits: hearts, diamonds, spades and clubs. Then order the cards within each suit from, say, ace through king. Read pairs from the random number table discarding those greater than 52. The list 44, 06, 04, 30, 22, 41, 13, corresponds to the hand: 5 of clubs, 6 of hearts, 4 of hearts, 4 of spades, 9 of diamonds, 2 of clubs, king of hearts.

[47] (a) Simple random sample: Number all the credit card holders consecutively from 1 to N. Read consecutive groups which have the same number of digits as N.
(b) Stratified sample: Categorize the credit card holders by some criterion such as age. Choose a random sample of credit card holders from each category.
(c) Systematic sample: Choose every 50th credit card holder from a list of all credit card holders.
(d) Categorize credit card holders by ZIP code. Choose a random sample of ZIP codes and survey every credit card holder in each of the ZIP codes chosen.

[48] observational study

[49] experiment

[50] observational study

Section B: Algorithms

Objective 1.1: What is Statistics?

[51] Quantitative

[52] Quantitative

[53] Quantitative

[54] Sample

[55] Sample

[56] Sample

[57] The population includes all the ninth-grade students in your school. A sample might include all the students in your class.

[58] Ratio

[59] Ratio

[60] Ordinal

[61] Interval

[62] qualitative

[63] qualitative

Objective 1.2: Random Samples

[64] No, this is not a random sample

[65] No, this is not a random sample

[66] A.

[67] Stratified

[68] Convenience

[69] Stratified

[70] C.

Objective 1.3: Introduction to Experimental Design

[71] Sample

[72] Census

[73] Biased

[74] Biased

[75] B.

[76] B.

[77] Nonrepresentative sample

[78] Yes. Explanations may vary.

[79] Answers may vary. Sample answer: The population is all the teenagers in your community. A sample could be all the students in your school.

[80] sample

Chapter 2
Organizing Data

Section A: Static Items

1. Write a brief but complete essay in which you answer the following questions: What is statistics? What is descriptive statistics? What is inferential statistics? Where is statistics used in everyday life?

2. Write a brief but complete essay in which you discuss three examples of modern life in which you think that statistics might be used. For each example describe the population under study and what a sample might be. In this context what do we mean by the term "descriptive statistics"? What do we mean by the term "inferential statistics"?

3. To estimate the amount of time State College students who live off campus spend to commute to classes each week, a random sample of 45 such students was surveyed.
   (a) What is the population?
   (b) What is the sample?
   (c) Would this sample necessarily be representative of the time part-time off-campus students spend commuting to class each week? Why or why not?

4. Data may be classified by one of the four levels of measurement. Which is the lowest level?

   A. ratio B. nominal C. population D. interval E. ordinal

5. An insurance company wants to know the proportion of medical doctors in New York involved in at least one malpractice suit in the last three years. They survey a random sample of 200 medical doctors in New York.
   (a) What is the population?
   (b) What is the sample?
   (c) Could the insurance company generalize and say that this sample is representative of all medical doctors in rural areas of the country? Why or why not?

6. A company making radar detection devices maintains quality control by testing a random sample of 20 such devices produced each day.
   (a) What is the population?
   (b) What is the sample?
   (c) If the manager decided to test a random sample of 100 devices on Monday and did no further sampling for the rest of the week, could she draw any conclusions about the quality of production for the entire week? Explain your answer.

7. Data may be classified by one of the four levels of measurement. Which is the highest level?

A. nominal B. ordinal C. interval D. ratio E. sample

8. Write a brief but complete essay in which you describe four different ways to produce data. Include an example illustrating each method.

9. Which level of measurement is suitable for putting data into categories?

A. interval B. ordinal C. ratio D. nominal E. inferential

10. Choose the word that does *not* complete the following statement:
Statistics is the study of how to ________, ________, ________, and ________ numerical information from data.

A. analyze B. organize C. interpret D. collect E. randomize

11. A professor of social science at a certain university questioned 200 students out of a total enrollment of 12,243 concerning their opinions of the quality of various university services. Identify the variable.

A. social science B. the total enrollment C. 200 students

D. the professor E. the students' opinions

12. English 105: Great Writers of America has 447 students. To estimate the amount of time students in this section spent studying for the midterm exam a random sample of 40 students were surveyed.
(a) What is the population?
(b) What is the sample?
(c) Would this sample necessarily be representative of the time students enrolled in English 105 spent studying for the comprehensive final exam? Why or why not?

13. The cook for a summer camp wants to know the eating preferences of the campers. Twenty-five of the 113 campers are surveyed concerning their eating preferences.
(a) What is the population?
(b) What is the sample?
(c) Would this sample necessarily be representative of the eating preferences of the camp counselors? Why or why not?

14. Choose the word that does *not* complete the following statement:
Descriptive statistics involves methods of ________, ________, and ________ information from ________ or populations.

A. samples B. organizing C. picturing D. summarizing E. calculating

15. A neighbor's dog is described by the following information:

(a) Name: Rocky
(b) Age: 6
(c) Weight: 60 lb
(d) Temperament: Friendly
(e) Color: Brown

For the information (a) to (e) list the highest level of measurement as ratio, interval, ordinal, or nominal.

16. Lake Lulu in Rocky Mountain National Park has several common species of trout along with the rare greenback trout. The National Park Service wants to estimate the proportion of greenback trout in Lake Lulu. Using a large net at random locations, the rangers caught (and released) a total of 587 trout of which 91 were the rare greenback trout.

(a) What is the population?
(b) What is the sample?
(c) Would this sample necessarily be representative of the trout in a different lake at higher elevation? Why or why not?

17. A simple random sample of *n* measurements from a population is a subset of the population selected in a manner such that

(a) every sample of size *n* from the population has an equal chance of being __________ and
(b) every member of the population has an equal chance of being __________ in the sample.

18. A random-number table is to be used to pick a random sample of 40 sheep out of a population of 300 sheep. If the blocks of five digits need to be regrouped, how many digits should be included in each block?

A. 5 B. 4 C. 2 D. 1 E. 3

19. The county sheriff wants to know the proportion of drivers who make an illegal left turn as they leave the post office parking lot. A sample of 200 cars is observed.

(a) What is the implied population?
(b) What is the sample?
(c) Would this sample necessarily reflect the proportion of cars making illegal left turns at all intersections? Explain your answer.

20. At a hospital nursing station the following information is available about a patient.

(a) Name:	Jim Wood
(b) Age:	17
(c) Weight:	165 lb
(d) Height:	6'1"
(e) Blood type:	A
(f) Temperature:	96.8° F
(g) Condition:	Fair
(h) Date of admission:	January 21, 1998
(i) Response to treatment:	Excellent

For the information (a) to (i) list the highest level of measurement as ratio, interval, ordinal, or nominal.

21. A construction supervisor needs to purchase a ladder that will be strong enough to support the heaviest of his employees. Explain why or why not taking a simple random sample of his employees will determine the appropriate ladder to purchase.

22. A hikers log of favorite hikes contains the following information:

(a) Destination:	Mirror Lake
(b) Distance:	6.5 miles
(c) Difficulty level:	Moderate
(d) Change in elevation:	3,000 ft
(e) Time to hike (one way):	3 hours
(f) Last year hiked:	1997

For the information in parts (a) through (f) list the highest level of measurement as ratio, interval, ordinal or nominal and explain your choices.

23. At a used car lot the following information is obtained about one of the cars on the lot.

(a) Make and Model:	Ford Escort
(b) Model year:	1990
(c) Color:	blue
(d) Number of cylinders:	4
(e) Gas consumption:	35 miles per gallon
(f) Condition:	Clean, in good repair
(g) Sticker price:	$8,817.00
(h) Body registration number:	TA1592763C51
(i) Number of miles on odometer:	36,719
(j) Sales Person Claim:	This is the best car on the lot for you.

For the information (a) to (j) list the highest level of measurement that applies: ratio, interval, ordinal, or nominal.

24. A book inventory record contains the following information:

(a) Title:	More Mysteries
(b) Author:	Roger Mortimer
(c) Date of publication:	1998
(d) List price:	$25.00
(e) Number in stock:	6

For the information in parts (a) through (e) list the highest level of measurement as ratio, interval, ordinal or nominal and explain your choices.

25. Write a brief but complete essay in which you describe each of the four levels of measurement discussed in this chapter. Give an example of data at each of the four levels of measurement.

26. Write a brief essay in which you discuss some of the aspects of surveys. Give specific examples to illustrate your main points.

27. Herman is a movie critic who keeps a record of the films he sees. His film records include the following information:

(a) Title:	Adventures with Aunt Agatha
(b) Genre:	Mystery
(c) Length:	150 minutes
(d) Date released:	1928
(e) Rating:	3 stars (on a scale from one star to five stars)

For the information in parts (a) through (e) list the highest level of measurement as ratio, interval, ordinal, or nominal and explain your choices.

28. Myra compiles information about various possible vacation sites. One of her records included the following information:

(a) Location:	Glacier Point, Alaska
(b) Distance:	1,525 miles
(c) Number of motels:	2
(d) Mean summer temperature:	42°F
(e) Outdoor activities rating:	4 (on a scale from 1, poor, to 5, excellent)

For the information in parts (a) through (e) list the highest level of measurement as ratio, interval, ordinal, or nominal and explain your choices.

29. What technique for gathering data (sampling, experiment, simulation or census) do you think was used in each of the following studies?
(a) The manager of an automobile repair shop selects a random sample of service records and records the total amount of time each vehicle was in the facility.
(b) The same manager tests a computerized diagnostic machine by comparing its performance on a random sample of 20 vehicles with the evaluation of a professional mechanic for the same 20 vehicles.
(c) The same manager surveys every customer who has had a car serviced to determine the quality of the customer's service and the customer's level of satisfaction with the service.
(d) An automobile manufacturer uses a computer simulation program to test the aerodynamic properties of a proposed new automobile body design.
(e) A service manager uses computer software to simulate a new arrangement of automotive workstations to see if the arrangement will provide more efficient service.

30. What techniques for gathering data (sampling, experiment, simulation or census) do you think were used in the following studies?
(a) 100 customers at a grocery store sampled a new food product and gave their opinion about the quality of the product. The grocery store has about 2,000 customers per day.
(b) An exit poll of 1,000 voters in a mayoral election for a large city showed 656 voters supported the incumbent.
(c) A computer was used to simulate two different assembly processes for car stereos to see which process produced more stereos per hour.
(d) All the owners in a subdivision were polled to determine their level of satisfaction with the exterior maintenance service.

31. What technique for gathering data (sampling, experiment, simulation or census) do you think was used in the following studies?
(a) 100 members of a large labor union were polled regarding a proposed new contract.
(b) A computer program was used to simulate movements of the earth's crust near an earthquake fault in an attempt to predict earthquakes.
(c) To test the safety of refrigerated packaged meat products 6 packages were chosen and their bacteria counts determined.
(d) All residents of a town were surveyed to determine the demand for home dairy delivery service.
(e) A volunteer group of 20 elderly residents of a care facility were given a program of nutrition and regular exercise. Their level of mental alertness was measured before and after they completed the program.

32. What technique for gathering data (sampling, experiment, simulation or census) do you think was used in each of the following studies?
(a) A computer program was used to model global weather patterns and to produce long-range weather forecasts for a rural agricultural region.
(b) A random sample of 1,000 residents of a major metropolitan area was surveyed to determine the level of support for a new sports complex among all residents of the area.
(c) To determine the effect of a new fertilizer on productivity of tomato plants one group of plants is treated with the new fertilizer while a second group is grown without such treatment. The number of ripe tomatoes produced by each group is counted.
(d) A study was done regarding the number of home runs scored by major league baseball teams playing at altitudes over 5,000 feet. Data for all major league baseball games played at this altitude was used in the study.

33. (a) Describe how you would use a random number table to get a random sample of 50 people who had listed phone numbers in a phone book that had 7500 people with listed numbers.
(b) Use the line from the random number table given below to identify the first three people in your sample.
94456 48396 73780 06436 86641 69239

34. (a) Describe how you would use a random number table to get a random sample of 12 men's dress shirts taken from a shipment of 475 men's dress shirts.
(b) Use the line from the random number table given below to identify the first five shirts in your sample.
08811 82711 57392 25252 30333

35. (a) How would you use a random number table to get a random sample of 8 families from a subdivision? The subdivision contains 92 families.
(b) Use the line from the random number table below to identify the first four families in your sample.
66330 33393 95261 95349

36. (a) Describe how you could use a random number table to simulate the experiment of tossing one die 275 times. The results of tossing a die once can be any of the digits 1, 2, 3, 4, 5, or 6.
(b) Use the line from the random number table below to identify the first ten outcomes in the simulation.
29553 18432 13630

37. (a) Describe how you could use a random number table to choose 18 workers from a work force of 6225 workers.
(b) Use the line from the random number table below to identify the first three workers chosen.
10072 55980 64688 68239 20461

38. (a) Describe how you could use a random number table to simulate the experiment of throwing two distinct dice.
(b) Use the line from the random number table below to identify the first five outcomes.
25464 59098 27436 89421 80754

39. (a) Describe how you could use random walk to simulate the arrival and departure of birds at a bird feeder. Assume that the birds arrive and depart one at a time and that they are as likely to depart as to arrive.
(b) Use the line from the random number table below to determine the number of birds at the feeder for the first 10 arrivals and/or departures. Use even for arrival and odd for departure. Initially there are two birds at the feeder.
25138 26810 07093

40. (a) A business employs 736 people. Describe how you could get a simple random sample of size 30 for a survey regarding professional training opportunities.
(b) Use the line from the random number table below to identify the first five employees to be included in the sample.
71886 56450 36507 09395 96951

41. Write a brief but complete essay in which you describe the topics: stratified sampling, systematic sampling, cluster sampling, and convenience sampling. List some advantages and disadvantages of each sampling method. How do these methods compare to "simple" random sampling?

42. Suppose you want to survey a sample of the registered voters of Chicago regarding their opinions about a proposed increase in sales tax. Categorize the following sampling methods using the categories: stratified sample, systematic sample, cluster sample, convenience sample, simple random sample.
(a) Divide the registered voters according to age and sample from each of the age categories.
(b) Select a random sample of area codes in the city and then sample from the registered voters in the selected area codes.
(c) Number all of the registered voters and use a random number table to identify the voters to include in the sample.
(d) Survey the registered voters who work in the voter registration office.

43. Identify the type of sampling (cluster, convenience, random, stratified, systematic) which would be used to get each of the following samples:
(a) For a period of two days measure the length of time each fifth person coming into a bank waits in line for teller service.
(b) Take a random sample of five zip codes from the Chicago metropolitan region and count the number of students enrolled in the first grade for every elementary school in each of the zip code areas.
(c) Divide the users of the Internet into different age groups and then select a random sample from each age group to survey about the amount of time they spend on the Internet each month.
(d) Survey five friends regarding their opinion about the quality of food in the student cafeteria.
(e) Pick a random sample of students enrolled at your college and determine the number of credit hours they have each accumulated toward their degree program.

44. To determine monthly rental prices of apartment units in the San Francisco Bay area, samples were constructed in the following ways. Identify the technique used to produce each sample. (cluster, convenience, simple random, stratified, systematic)
(a) Number all the units in the area and use a random number table to select the apartments to include in the sample.
(b) Classify the apartment units according to the number of bedrooms and then take a random sample from each of the classes.
(c) Classify the apartments according to zip code and take a random sample from each of the zip code regions.
(d) Look in the newspaper and choose the first apartments you find that list rents.

45. The owner of a specialty store in a shopping mall would like to estimate the demand for a proposed new inventory item. She would like to draw a sample of potential customers and survey them regarding their interest or lack of interest in the product. Identify the sampling technique (cluster, convenience, simple random, stratified, systematic) for each sampling technique described below.
(a) Choose every tenth person who comes into the mall at the entrance nearest her store.
(b) Select every person who makes a purchase at her store on some convenient day.
(c) Obtain a list of all residents of the region in which the mall is located. Count them and use a random number table to select the sample.
(d) Obtain a list of all residents of the region in which the mall is located. Divide them into groups by income level and choose a random sample from residents in each income level.

46. Use the line from the random number table below to simulate drawing a seven card hand from a standard deck of cards with no jokers. Specify which of the cards correspond to each of the 52 numbers 1 through 52. List both the numbers you have chosen and the corresponding cards.
44940 60430 22834 14130 96593 23298 56203 92671

47. The manager of a large department store would like to sample the people who have credit cards issued by the store about their shopping preferences. Describe a way the manager could draw each of the following:
(a) A simple random sample
(b) A stratified sample
(c) A systematic sample
(d) A cluster sample

48. What technique (observational study or experiment) for gathering data do you think was used in the following study? In a national forest, 87 deer were caught, tagged, and then released back into the wild. Two weeks later, 62 deer were caught and 43 were found to have tags. From this, it was possible to estimate how many deer live in the forest.

49. What technique (observational study or experiment) for gathering data do you think was used in the following study? Texas Parks and Wildlife administered medicine to 50 sheep that were all suffering from a form of mange. Two months later, 42 showed noticeable signs of improvement.

50. What technique (observational study or experiment) for gathering data do you think was used in the following study? In a national park, a park ranger noticed 37 bears. Of these, 16 were brown bears and 21 were American black bears.

Section B: Algorithms

Objective 2.1: Bar Graphs, Circle Graphs, and Time Plots

51. Curious to find out why some students failed to turn in assignments, Frank interviewed a college campus and made a list of possible causes. Below is a table showing his results. Make a Pareto chart showing the data.

Cause	Frequency
Had to work overtime at job	12
Didn't understand assignment	9
Had too much work in other classes	6
Had social engagement	13
Too tired	8
Other	3

52. Tenth-grade students were surveyed to determine how many hours per day they spent on various activities. The results are represented in the circle graph below. About how many hours altogether was spent on socializing and watching TV?

How Students Spend Their Time

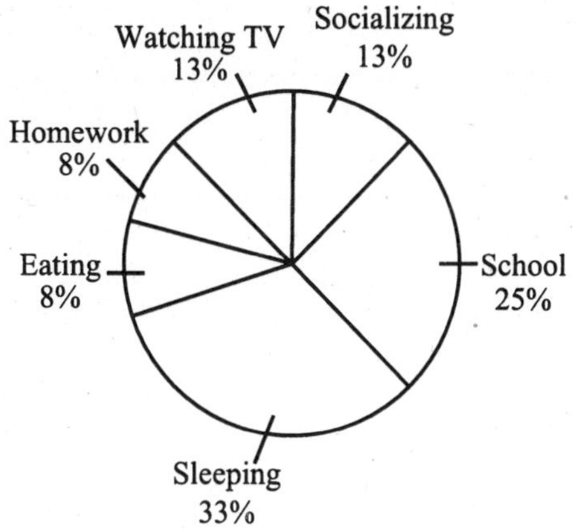

53. Shilain interviewed his extended family to find out which sports were most favored. Below is a table showing his results. Make a circle graph to represent the data.

Percentage of Sports Preferred by Family	
Baseball	4
Basketball	8
Football	5
Soccer	3

54. Suppose a garden plot contains 3 zinnias, 7 gladiolas, 16 marigolds, 20 hollyhocks, and 23 lilies. If you were to make a circle graph of the flowers in the plot, what angles (to the nearest degree) should be assigned to each flower?

55. A survey about the art program at a school found that 250 students wanted an after-school art program, 150 students wanted art classes during the day, and 100 students thought that the school should have no art program. If a circle graph were made from this data, what would be the measure of the central angle for the "after-school" group?

56. Jamie interviewed her college's female basketball team to find out why they participated in school sports. Below is a table showing her results. Construct a Pareto chart for the following data.

Reason	Frequency
Love for the sport	22
Competitive drive	17
Sense of team work	19
Personal challenge	31
Physical fitness	30
Other	13

57. The number of hours required in each discipline of a college core curriculum is represented by the circle graph below. What percent of these hours is in History and English combined?

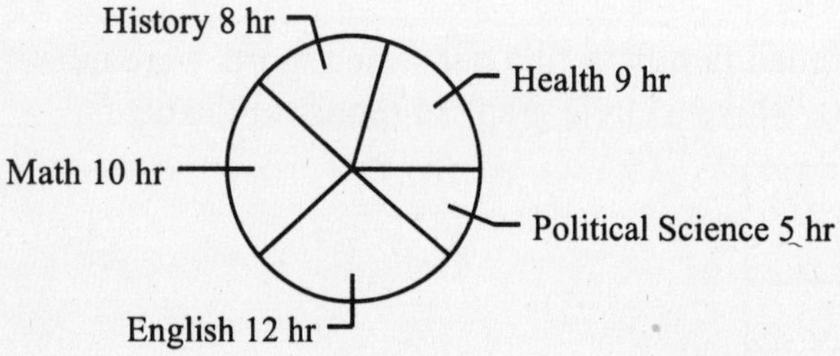

58. In a young single person's monthly budget, \$100 is spent on food, \$140 is spent on housing, and \$160 is spent on other items. Which represents these data with a circle graph?

Objective 2.2: Frequency Distributions and Histograms

59. On a certain day in July, the high temperatures in 18 European cities were compiled into this table.

Temperature interval F°	61 – 70	71 – 80	81 – 90	91 – 100	101 – 110
Frequency	2	6	3	1	6

a. Construct a frequency histogram from the given data.
b. Draw the frequency line on this histogram.

60. For the following set of numbers, draw a frequency histogram using the intervals 1–3, 4–6, 7–9, and 10–12:
8, 12, 4, 8, 1, 10, 1, 9, 1, 3, 5, 1, 9, 4, 10, 7, 8, 9, 3, 4

61. Use the histogram below to find the total frequency for the interval from 1.5 to 2.5.

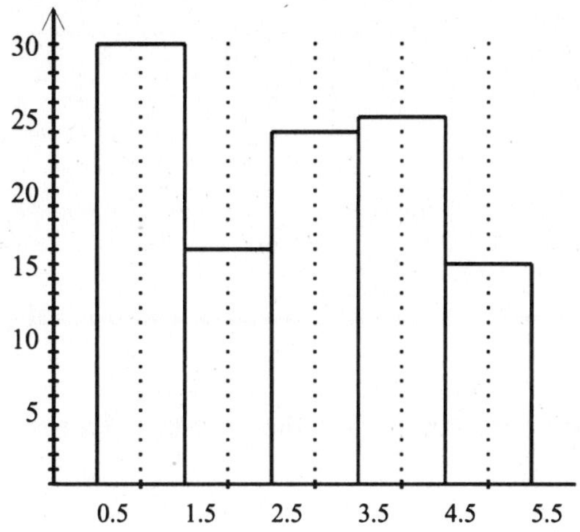

62. Draw a relative frequency histogram for the intervals 10–12, 13–15, 16–18, and 19–21 using the following data:
14, 17, 12, 16, 18, 10, 15, 15, 11, 16, 11, 16, 11, 20, 19, 16, 15, 15, 14, 16

63. The golf scores for the 20 members of a country club were as follows:
81, 76, 107, 95, 119, 92, 83, 74, 108, 88, 95, 74, 83, 76, 97, 82, 79, 91, 93, 89
Create a relative frequency histogram using ten-point intervals to show the distribution of the scores.

64. Test scores for the 25 members of the general math class were as follows:
72, 59, 91, 64, 86, 53, 98, 77, 85, 62, 54, 71, 85, 93, 66, 78, 87, 69, 68, 71, 75, 66, 63, 74, 79
Create a relative frequency histogram using ten-point intervals to show the distribution of the scores.

65. The retirement ages of 30 police detectives are recorded in the following frequency table.

Age	Frequency	Relative Frequency
51–53	4	0.08
54–56	12	0.24
57–59	11	0.22
60–62	10	0.20
63–65	8	0.16
66–68	5	0.10

(a) Construct a frequency histogram.
(b) Construct a relative frequency histogram.

66. The table below shows the number of days that received rain over a three-month period in Carlsville.

Rainfall in inches	0.0 – 0.2	0.2 – 0.4	0.4 - 0.6	0.6 - 0.8	0.8 - 1.0	1.0 - 1.2	1.2 - 1.4	1.4 - 1.6
Number of Days	37	11	11	6	7	6	6	4

Draw a histogram that represents this data.

67. Draw a histogram with bin width 1.0 for this data:
0.1, 5.6, 5.6, 2.8, 5.6, 1.8, 6.5, 2.5, 2.2, 4.5, 3.2, 3.8, 1.7, 1.7, 0.8, 4.5, 1.6, 0.1, 4.1, 0.6

68. The histogram below shows the ages of the people attending the showing of a new movie. Describe the distribution of these ages.

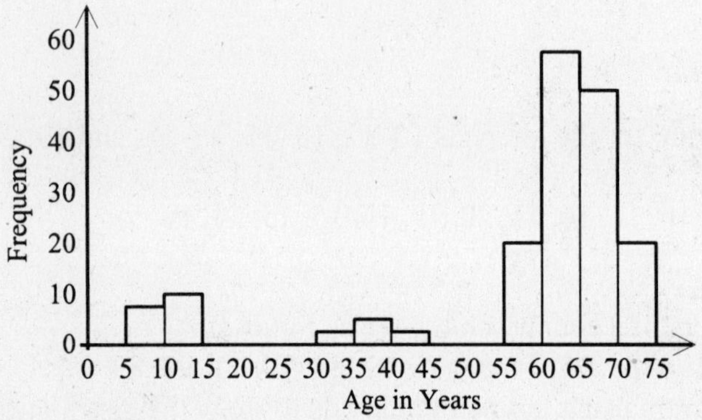

69. Determine whether the data in the table appear to be skewed to the right, skewed to the left, bimodal, uniform, or symmetrical:

Value	Frequency
10 – 19	19
20 – 29	17
30 – 39	35
40 – 49	62
50 – 59	117
60 – 69	133
70 – 80	113

70. On a certain day in July, the high temperatures in 28 European cities were compiled into this table.

Temperature interval F°	61–70	71–80	81–90	91–100	101–110
Frequency	5	5	8	2	8

a. Construct a frequency histogram from the given data.
b. Draw the frequency line on this histogram.

71. The retirement ages of a group of police detectives are recorded in the following frequency table. Construct the relative frequency histogram for the data shown in the table below.

Age	Frequency	Relative Frequency
50–52	1	0.045
53–55	3	0.136
56–58	5	0.227
59–61	6	0.273
62–64	7	0.318

72. Determine whether the data in the table appear to be skewed to the right, skewed to the left, bimodal, uniform, or symmetrical.

Score	Students
71–75	25
76–80	30
81–85	22
86–90	14
91–95	8
96–100	6

73. Construct a histogram with intervals 33-35, 36-38, 39-41, and 42-44 for the following data.
43, 35, 44, 41, 39, 34, 44, 35, 36, 35, 41, 40, 41, 44, 35, 36, 37, 43, 34, 34

74. The test scores for the 25 members of the General Math class are represented in the histogram below. What percentage of the students had scores from 90 to 100?

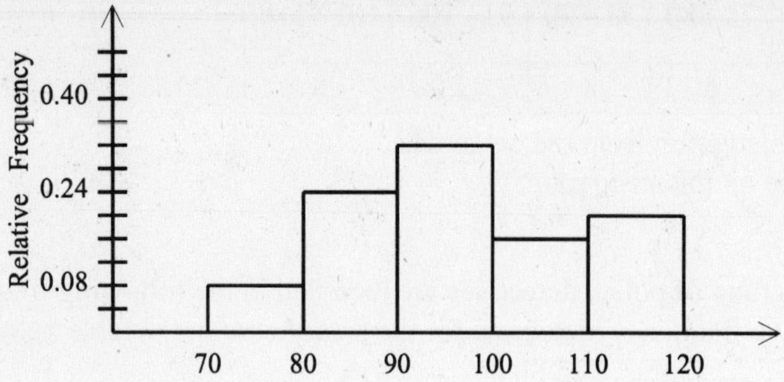

75. Retirement Ages of Police Detectives

Age	Tally	Frequency
50-52	|	1
53-55	||||	4
56-58	||||	4
59-61	~~||||~~ |	6
62-64	~~||||~~ |||	8
65-67	~~||||~~ ||	7

Display the set of data in a histogram.

Objective 2.3: Stem-and-Leaf Displays

76. What are the stems for the data set 54, 43, 58, 25, 67, 36, 61?

77. Make a stem–and–leaf diagram for the following data:
89, 98, 60, 73, 68, 86, 63, 66, 65, 94, 83, 85, 76, 62, 75

78. The data below show the average daily high temperature for Chicago, Illinois, for twelve recent spring and summer months. Construct a stem-and-leaf diagram for the data: 64, 61, 57, 68, 78, 75, 50, 83, 71, 58, 62, 80

79. Yvette made the following stem–and–leaf diagram of the high and low temperatures for 10 days.

Lows	Stem	Highs
5 3 3 2	3	6 6
9 2 0	4	
2 5 9	5	1 2 5 9
	6	3 4 4 7

$0|3 = 30°\text{F}$ $3|8 = 38°\text{F}$

How much greater was the highest temperature than the lowest temperature?

80. Construct a stem–and–leaf diagram for the following data.
31, 50, 53, 71, 37, 51, 30, 32, 35, 38, 45, 47, 44, 26, 40

Answer Key for Chapter 2
Organizing Data

Section A: Static Items

[1] See text.

[2] See text.

(a) Population: The time spent commuting by each off-campus student of State College.
(b) Sample: The time spent commuting to class by each of the 45 students.
[3] (c) Not necessarily; part-time students may come to campus less frequently.

[4] B.

(a) Population: The response (has or has not been involved in a malpractice suit in the last three years) from all medical doctors in New York.
(b) Sample: The responses to the same question from the 200 medical doctors in the survey.
(c) No. The sample was not restricted to medical doctors from rural areas in New York.
[5] Many of those in the sample could have come from New York City.

(a) Population: The test results of all radar detection devices produced on a given day by the company.
(b) Sample: The test results of the sample of 20 devices selected from the day's production.
(c) By testing only the devices made on Monday, the manager has no information about
[6] those produced during the rest of the week.

[7] D.

[8] See text.

[9] D.

[10] E.

[11] E.

[12] (a) Population: The times spent studying for the midterm exam by the 447 students in English 105.
(b) Sample: The times spent studying for the midterm exam by the 40 students surveyed.
(c) No. Students would probably spend more time studying for a comprehensive final exam than they would for a midterm.

[13] (a) Population: The eating preferences of the 113 campers.
(b) Sample: The eating preferences of the 25 campers surveyed.
(c) No. The camp counselors eating preferences may differ from the campers due to age, religion, region of origin, etc.

[14] E.

[15] (a) Nominal; (b) Ratio; (c) Ratio; (d) Ordinal; (e) Nominal

[16] (a) Population: The species of each trout in Lake Lulu.
(b) Sample: The species of the 587 trout in the sample.
(c) No. Other lakes may have other trout species and the species may occur in different proportions than species of trout in Lake Lulu.

[17] (a) selected; (b) included

[18] E.

[19] (a) Population: Observation results (legal or illegal) of all drivers who exit the post office parking lot under the present law enforcement patterns.
(b) Sample: Observation results for the drivers of the 200 cars in the sample.
(c) No. These observations apply only to the cars exiting the post office parking lot under study.

[20] (a) Nominal; (b) Ratio; (c) Ratio; (d) Ratio; (e) Nominal; (f) Interval; (g) Ordinal; (h) Interval; (i) Ordinal.

[21] There is no guarantee that the heaviest employee will be included in the sample. So, taking a simple random sample of employees is not an appropriate way to determine which ladder to purchase.

[22] (a) Nominal; (b) Ratio; (c) Ordinal; (d) Ratio; (e) Ratio; (f) Interval.

[23] (a) Nominal; (b) Interval; (c) Nominal; (d) Ratio; (e) Ratio; (f) Ordinal; (g) Ratio; (h) Nominal; (i) Ratio; (j) Ordinal.

[24] (a) Nominal; (b) Nominal; (c) Interval; (d) Ratio; (e) Ratio.

[25] Essay. See text.

[26] Essay. See text.

[27] (a) Nominal; (b) Nominal; (c) Ratio; (d) Interval; (e) Ordinal.

[28] (a) Nominal; (b) Ratio; (c) Ratio; (d) Interval; (e) Ordinal.

[29] (a) Sampling; (b) Experiment; (c) Census; (d) Simulation; (e) Simulation.

[30] (a) Sampling; (b) Sampling; (c) Simulation (of an experiment); (d) Census.

[31] (a) Sampling; (b) Simulation; (c) Experiment; (d) Census; (e) Experiment.

[32] (a) Simulation; (b) Sampling; (c) Experiment; (d) Census.

[33] (a) Number the people from 0001 to 7500. Read consecutive groups of four digits until you have fifty distinct numbers from 1 through 7500.
(b) 6483, 4368, 6641

[34] (a) Number the shirts from 001 through 475. Read consecutive groups of three digits until you have twelve distinct numbers from 1 to 475.
(b) 088, 118, 271, 157, 392

[35] (a) Number the families from 01 to 92. Read consecutive groups of two digits until you have eight distinct numbers from 1 to 92.
(b) 66, 33, 03, 93

[36] (a) Associate the numbers 1 through 6 with the number showing on the top face of the die. Read consecutive single digits until you have 275 digits from 1 through 6.
(b) 1, 5, 5, 3, 1, 4, 3, 2, 1, 3

(a) Number the workers from 0001 to 6225. Read consecutive groups of four digits until you have 18 distinct numbers from 1 to 6225.
[37] (b) 1007, 2559, 2046

(a) Associate pairs of digits, each in the range 1 through 6, with the numbers showing on the top faces of the two dice. Read consecutive pairs of digits discarding any pair for which either digit is outside the range 1 through 6.
[38] (b) (2, 5), (4, 6), (4, 5), (4, 3), (2, 1)

(a) Let digits of one parity (even or odd) represent arrivals and let those of the opposite parity represent departures. Read single digits, adding 1 to the total number of birds for each arrival and subtracting 1 from the total number of birds for each departure.
(b)

Time	0	1	2	3	4	5	6	7	8	9	10
Birds	2	3	2	1	0	1	2	3	4	3	4

[39]

(a) Number the employees from 001 to 736. Read consecutive groups of three digits until you have 30 distinct numbers from 1 to 736.
[40] (b) 718, 645, 036, 507, 093

[41] Essay; see text.

(a) Stratified sample
(b) Cluster sample
(c) Simple random sample
[42] (d) Convenience sample

(a) Systematic sampling
(b) Cluster sampling
(c) Stratified sampling
(d) Convenience sampling
[43] (e) Simple random sampling

(a) Simple random sampling
(b) Stratified sampling; the strata are the number of bedrooms.
(c) Cluster sampling
[44] (d) Convenience sampling

(a) Systematic sampling
(b) Convenience sampling
(c) Simple random sampling
[45] (d) Stratified sampling

Choose some ordering for the cards. For example, order the suits: hearts, diamonds, spades and clubs. Then order the cards within each suit from, say, ace through king. Read pairs from the random number table discarding those greater than 52. The list 44, 06, 04, 30, 22, 41, 13, corresponds to the hand: 5 of clubs, 6 of hearts, 4 of hearts, 4 of spades, 9 of
[46] diamonds, 2 of clubs, king of hearts.

(a) Simple random sample: Number all the credit card holders consecutively from 1 to N. Read consecutive groups which have the same number of digits as N.
(b) Stratified sample: Categorize the credit card holders by some criterion such as age. Choose a random sample of credit card holders from each category.
(c) Systematic sample: Choose every 50th credit card holder from a list of all credit card holders.
(d) Categorize credit card holders by ZIP code. Choose a random sample of ZIP codes and
[47] survey every credit card holder in each of the ZIP codes chosen.

[48] observational study

[49] experiment

[50] observational study

Section B: Algorithms

Objective 2.1: Bar Graphs, Circle Graphs, and Time Plots

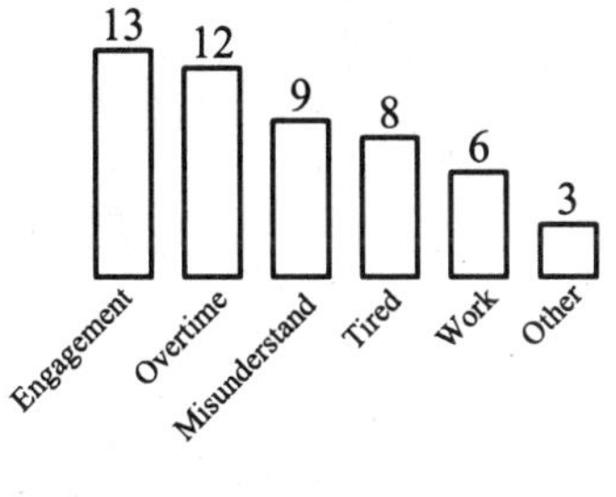

[51]

[52] 6 hr

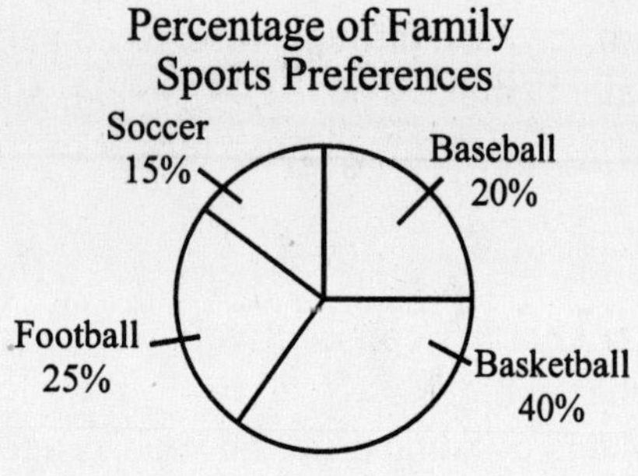

[53] ______________________

[54] 16°, zinnias; 37°, gladiolas; 83°, marigolds; 104°, hollyhocks; 120°, lilies

[55] 180°

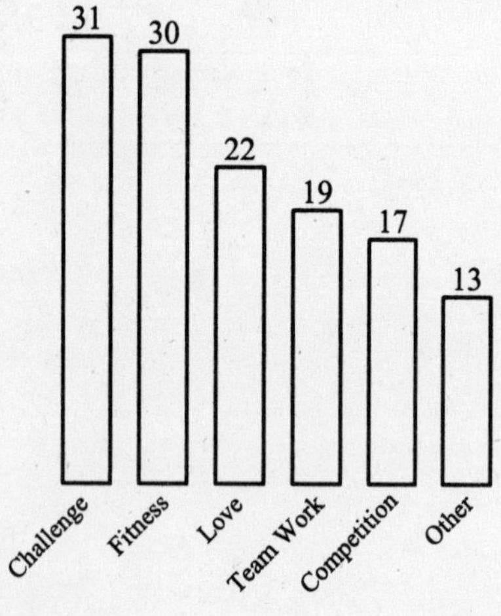

[56] ______________________

[57] 45.5%

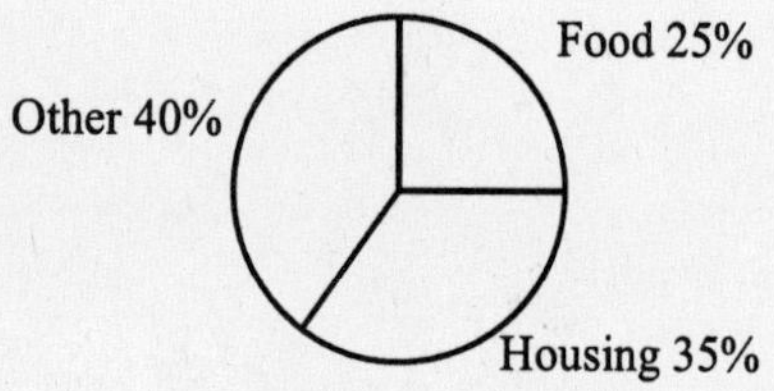

[58] ______________________

Objective 2.2: Frequency Distributions and Histograms

a. and b.

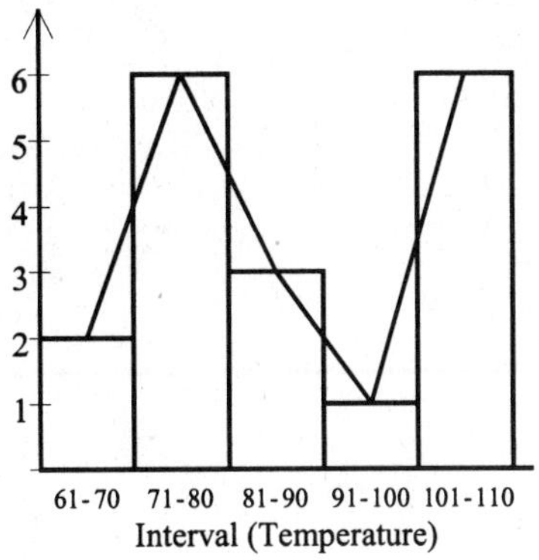

[59]

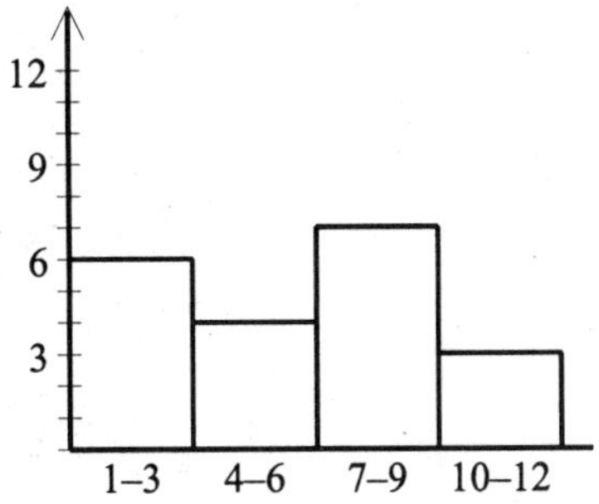

[60]

[61] 16

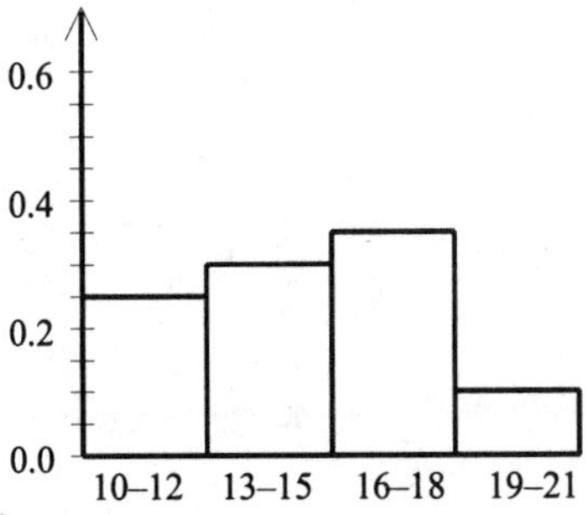

[62]

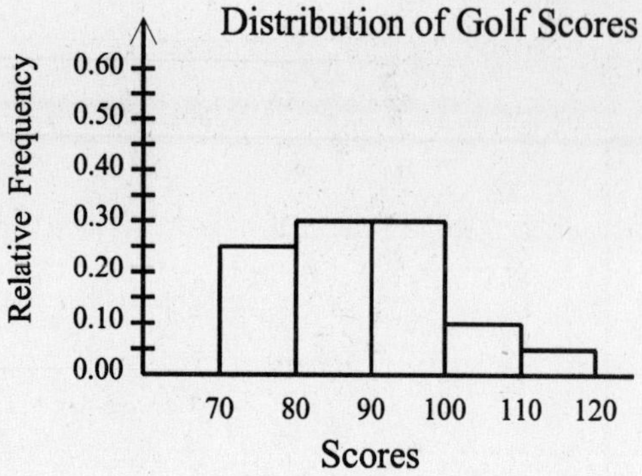
Distribution of Golf Scores
Relative Frequency
0.60
0.50
0.40
0.30
0.20
0.10
0.00
70
80
90
100
110
120
Scores

[63] ____________________

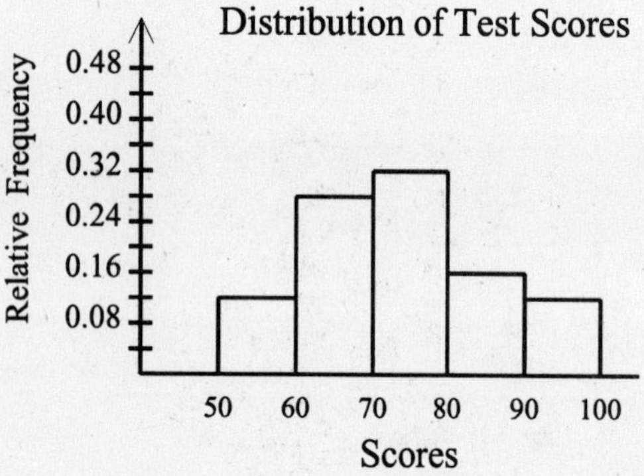
Distribution of Test Scores
Relative Frequency
0.48
0.40
0.32
0.24
0.16
0.08
50
60
70
80
90
100
Scores

[64] ____________________

(a)

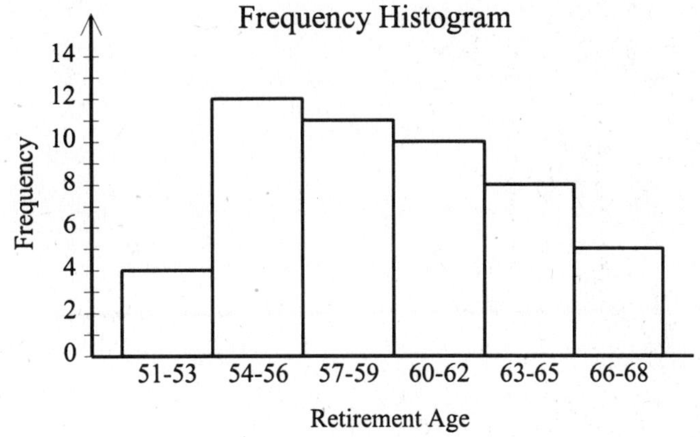
Frequency Histogram
Frequency
14
12
10
8
6
4
2
0
51-53 54-56 57-59 60-62 63-65 66-68
Retirement Age

(b)

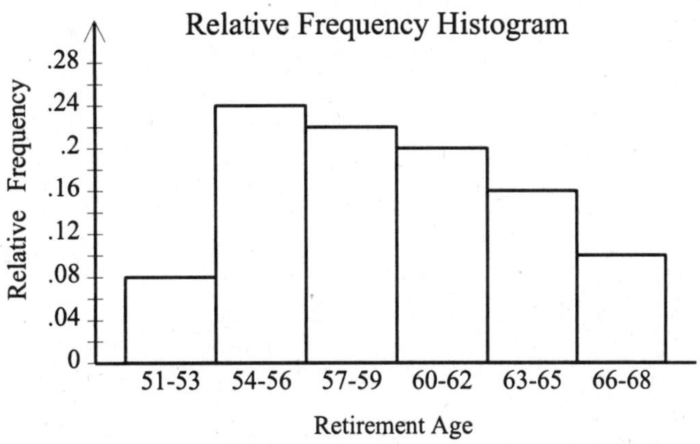
Relative Frequency Histogram
Relative Frequency
.28
.24
.2
.16
.12
.08
.04
0
51-53 54-56 57-59 60-62 63-65 66-68
Retirement Age

[65] ______________________________

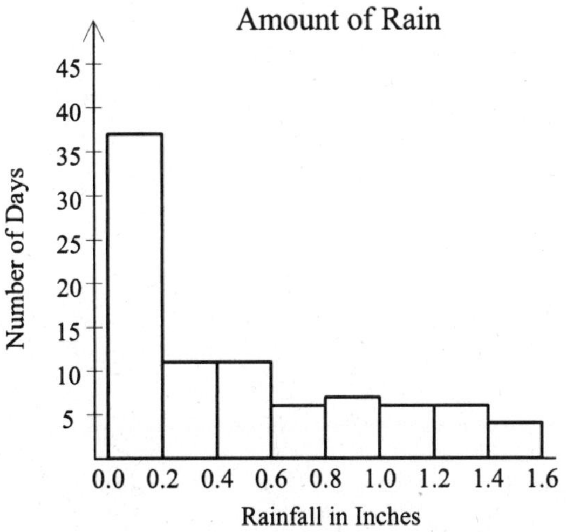
Amount of Rain
Number of Days
45
40
35
30
25
20
15
10
5
0.0 0.2 0.4 0.6 0.8 1.0 1.2 1.4 1.6
Rainfall in Inches

[66] ______________________________

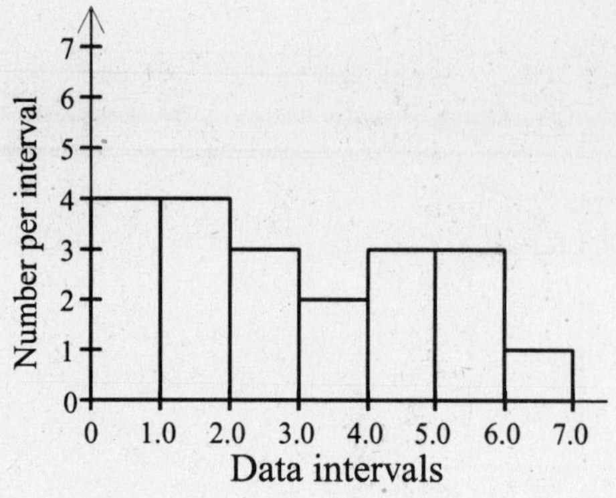

[67]

[68] Skewed to the left

[69] Skewed to the left

a. and b.

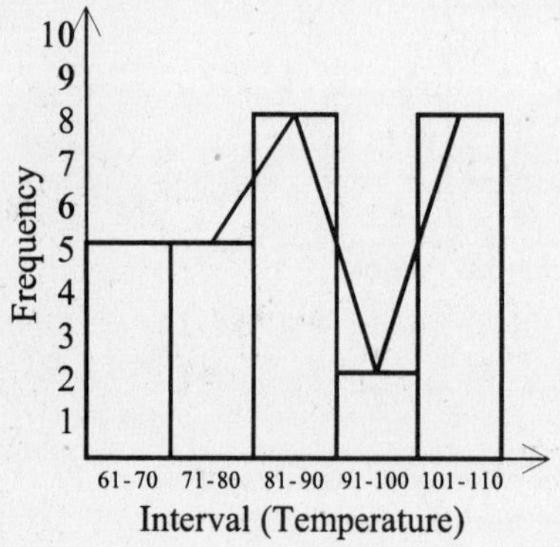

[70]

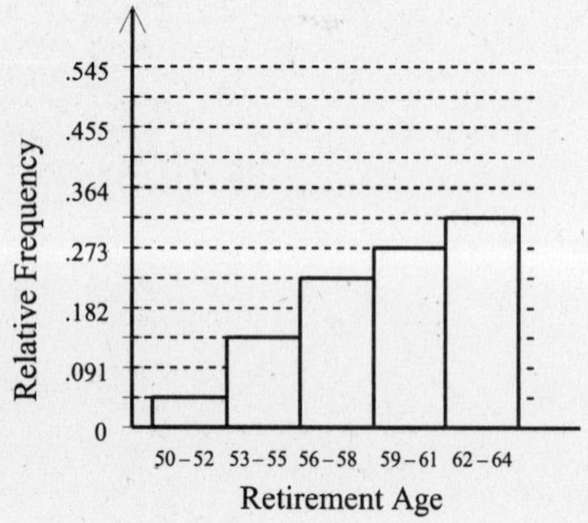

[71]

[72] Skewed right

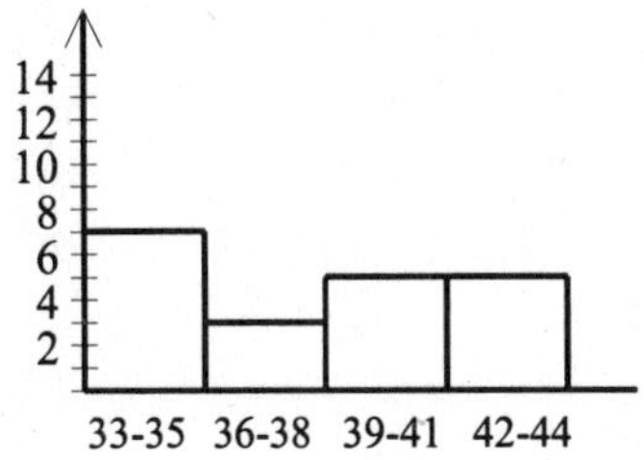

[73]

[74] 32.0%

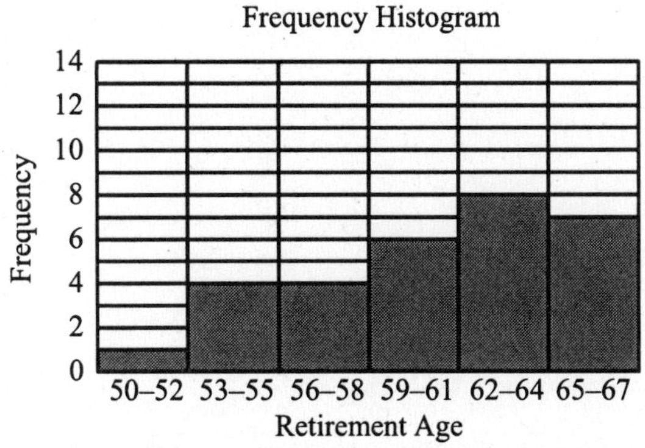

[75]

Objective 2.3: Stem-and-Leaf Displays

[76] 2, 3, 4, 5, 6

Stem	Leaf
6	0 2 3 5 6 8
7	3 5 6
8	3 5 6 9
9	4 8

[77] 6|4 = 64

Stem	Leaf
5	0, 7, 8
6	1, 2, 4, 8
7	1, 5, 8
8	0, 3

[78] $6|4 = 64$

[79] 35°

Stem	Leaf
2	6
3	0 1 2 5 7 8
4	0 4 5 7
5	0 1 3
6	
7	1

[80] $6|4 = 64$

Chapter 3
Averages and Variation

Section A: Static Items

1. The heights of ten varsity basketball players at Mountain View College were measured in inches and found to be:

   74 82 78 72 78 73 78 72 78 81

   Find the mean, the median and the mode of the heights.

2. Statistical Abstracts (117th edition) reports gasoline excise taxes, in cents per gallon, in the west (mountain region) as follows:

   28 26 9 22 19 18 19 24

   Find the mean, the median, and the mode of these taxes.

3. The prices per ounce in cents of ten shampoos sold at a market are:

   11 54 27 31 69 15 57 63 31 72

   Find the mean, median, and mode of the prices per ounce.

4. In an effort to estimate the size of elk herds wintering in Rocky Mountain National Park, the rangers used a small airplane to spot count groups of elk in the park. They found 15 groups of elk and recorded the size of each group as:

   21 15 19 16 18 17 17 20
   25 7 16 10 16 18 12

   (a) Compute the mean, the median and the mode of the data.
   (b) Describe what each number tells you about the sizes of groups of elk in Rocky Mountain National Park.

5. Statistical Abstracts, (117th edition), reports average travel time to work for commuters in the New England states to be (to the nearest minute):

   19 22 18 23 19 21

   Find the mean, the median and the mode of these commuting times.

6. Find the mean of the following ages of retirees:

   79 75 83 91 66 68 72 81 64
   68 71 66 86 64 80 69 66 65

   A. 68.5 B. 72 C. 66 D. 73 E. 70

7. Petroleum pollution in oceans is known to increase the growth of a certain bacteria. Brian did a project for his ecology class in which he made a bacteria count (per 100 milliliters) in nine random samples of sea water. His counts gave the following readings: 17 23 18 19 21 16 12 15 18
(a) Find the range.
(b) Find the sample mean.
(c) Find the sample standard deviation.

8. In the process of tuna fishing, porpoises are sometimes accidentally caught and killed. A U.S. oceanographic institute wants to study the number of porpoises killed in this way. Records from eight commercial tuna fishing fleets gave the following information about the number of porpoises killed in a three month period:
2 6 18 9 0 15 3 10
(a) Find the range.
(b) Find the sample mean.
(c) Find the sample standard deviation.

9. A swimmer was timed while swimming across a small lake six times. Find the median of her six times (min:sec):
3:20 3:08 2:51 3:31 2:49 2:56

A. 3:08 B. 3:02 C. 3:06 D. 3:00 E. 2:58

10. Black Hole Pizza Parlor instructs its cooks to put a "handful" of cheese on each large pizza. A random sample of six such handfuls were weighed. The weights to the nearest ounce were:
3 2 3 4 3 5
(a) Find the mode, the median and the mean weight of the handfuls of cheese.
(b) Find the range and the standard deviation of the weights.
(c) A new cook used to play football. He has large hands. His handful of cheese weighs 6 ounces. Replace the 2 ounce data value by 6 ounces. Recalculate the mode, median and mean. Which average changed the most? Comment on the changes.

11. Response time for eight emergency phone calls in a large metropolitan area were measured to the nearest minute and found to be:
7 10 8 5 8 6 8 9
Find the mean, the median and the mode of the response times.

12. Find the mode of the following set of numbers:
1 4 3 2 2 1 3 3 4 1 2 4 4 1 2 3

A. 2 B. 4 C. 2.5 D. 3 E. no mode

13. City Hospital has a temporary shortage of nurses, so nurses have been working overtime. A random sample of six nurses reported that the number of overtime hours they worked last week were:

7 2 4 5 4 3

(a) Compute the mode, the median and the mean for this data set.
(b) Compute the range and standard deviation.
(c) Suppose that a recording error occurred and the data value of 7 was supposed to be a 2. Replace 7 by 2 in the data. Recalculate the mode, the median and the mean and comment on the changes produced in the averages by the change in the data.

14. Myra did a study of the number of weeks 10 novels were on a national best-seller list. The results were as follows:

7 15 48 2 12 20 10 15 102 24

(a) Compute the mean, the median and the mode of this data set.
(b) Compute a 10% trimmed mean for the data set.
(c) How did the extreme values affect each average?

15. Complete a 5% trimmed mean for the following set of ages (in years) of giant tortoises:

5 12 16 24 33 7 9 13 42 62 21 41
29 20 36 3 17 81 50 22

16. Edgar and Ann are travel agents. They recorded the number of international airline tickets each sold for the last five months.

Ann: 10 4 12 0 18
Edgar: 8 9 12 7 8

(a) Compute the sample mean and the sample standard deviation for Ann and for Edgar.
(b) Who had the more consistent sales over the five month period? Give a reason for your answer.

17. For the past ten years the daily high temperature on January 1 in a mountain town (in degrees Fahrenheit) was:

3 25 −8 17 −2 10 −12 21 4 6

(a) Find the range.
(b) Find the sample mean.
(c) Find the sample standard deviation.
(d) Find the coefficient of variation.

18. Find the mean of the following credit card interest rates (%):

7.9 11.9 24.5 21.1 15.9 9.9

A. 15.2% B. 15.4% C. 15.8% D. 15.0% E. 16.0%

19. According to data provided by the Statistical Abstract of the United States (117th edition), the number of daily newspapers in the five states in the midwest has a mean $\mu_1 = 30.8$ with standard deviation $\sigma_1 = 38.3$. For five states in the Pacific region $\mu_2 = 61.4$ with standard deviation $\sigma_2 = 19.07$.
(a) Compute the coefficient of variation for each region.
(b) Which region has the greater variation in the number of newspapers as a fraction of the mean?

20. From information found in the Statistical Abstract of the United States (117th edition) it was determined that the median income for households in each of the fifty states, in thousands of dollars, has mean $\mu = 33.81$ and standard deviation $\sigma = 5.10$.
(a) Use Chebyshev's Theorem to find an interval in which at least 75% of the states' median incomes will lie.
(b) Use Chebyshev's Theorem to find an interval in which at least 93.8% of the states' median incomes lie.

21. The heights in feet of six mountains in a certain mountain range are the following:
11,290 12,560 10,740 14,110 15,670 12,970
Find the mean and median heights.

A. Mean: 12,885 ft; median: 11,925 ft
B. Mean: 12,885 ft; median: 12,765 ft
C. Mean: 12,890 ft; median: 12,130 ft
D. Mean: 12,890 ft; median: 11,925 ft
E. Mean: 12,890 ft; median: 12,765 ft

22. From years of experience fishing for trout in the Yellowstone River you know that the mean length of trout you catch is 14.7 inches with standard deviation 1.5 inches.
(a) Use Chebyshev's Theorem to find an interval for the lengths of trout which will contain the lengths of at least 75% of the fish you catch.
(b) Use Chebyshev's Theorem to find an interval for the lengths of trout which will contain the lengths of at least 93.8% of the fish you catch.

23. The Statistical Abstract (117th edition) reports the number of hours per year individuals spent watching television for 1990 through 1995. Information is given for independent TV stations and for basic cable.

Independent Stations:	340	227	159	162	172	183
Basic Cable:	260	340	359	375	388	468

(a) Compute the median, the population mean and the population standard deviation for independent stations.
(b) Compute the median, the population mean and the population standard deviation for basic cable.
(c) Compute the coefficient of variation for both data sets.
(d) What do these numbers tell you about the two distributions?

24. Choose the number to add to the data set below that will make the mean, median, and mode equal.

1 4 2 5 9 8 6

A. 6 B. 5 C. 3 D. 4 E. 2

25. Retail prices of five name brand 50 inch television sets and five brand name 25 inch television sets were advertised as follows:

50 inch sets	1,700	1,900	2,500	3,200	1,500
25 inch sets	444	425	499	599	375

(a) Compute the range, sample mean, sample standard deviation and coefficient of variation for the 50 inch TV sets.
(b) Compute the range, sample mean, sample standard deviation and coefficient of variation for the 25 inch TV sets.
(c) Compare the answers for parts (a) and (b) and comment on the differences in the two sets of data.

26. The Jackson Dormitory complex has a computer lab to give students easy access to computers. However there is a complaint that a few students seem to monopolize the equipment. To study usage patterns, a random sample of 160 users was selected. The time each student spent on the computer during one session was measured. The times (to the nearest minute) were:

Time	1 – 60	61 – 120	121 – 180	181 – 240
Frequency	70	50	30	10

(a) Estimate the sample mean time for student sessions on the computer.
(b) Estimate the sample standard deviation.

27. The depth in feet of a harbor was measured at 10 different locations. Find, to the nearest hundredth, the sample standard deviation and the population standard deviation of the following measurements:

23 31 18 22 19 25 29 17 30 26

A. 5.06 ft; 4.80 ft B. 5.06 ft; 4.79 ft C. 5.08 ft; 4.83 ft
D. 5.09 ft; 4.83 ft E. 5.05 ft; 4.80 ft

28. A random sample of 316 students were asked how many credit hours they were taking this semester. The results follow:

Credit Hours	1 – 5	6 – 10	11 – 15	16 – 20
Frequency	42	65	116	93

(a) Estimate the sample mean number of credit hours taken by the students.
(b) Estimate the sample variance and the sample standard deviation.

29. By sampling different locations along Dry Creek, the number of tree frogs per willow tree was determined. The results were the following (frogs per tree):

20 59 32 47 101 92 11 76 36 26

Compute the range and sample standard deviation. Round to the nearest hundredth if necessary.

30. A study was done showing the age distribution of people doing volunteer work for a random sample of 545 volunteers.

Age	14 – 17	18 – 24	25 – 44	45 – 64	65 – 80
Frequency	142	125	72	124	82

(a) Estimate the sample mean age of volunteers.
(b) Estimate the sample standard deviation.

31. The resting heart-rates for 10 members of a school's soccer team were measured by the school nurse. The results were the following (beats per minute):

55 58 57 62 64 61 56 60 63 65

Compute the sample mean, sample standard deviation, and the coefficient of variation. Round to the nearest hundredth if necessary.

A. $\bar{x} = 60.0$; $s = 3.48$; $CV = 5.78\%$
B. $\bar{x} = 60.0$; $s = 3.47$; $CV = 5.78\%$
C. $\bar{x} = 60.1$; $s = 3.47$; $CV = 5.79\%$
D. $\bar{x} = 60.1$; $s = 3.48$; $CV = 5.79\%$
E. $\bar{x} = 60.0$; $s = 3.48$; $CV = 5.78\%$

32. A random sample of households in an upscale community was surveyed about their yearly monetary charitable donations. The mean number of hours was found to be $\bar{x} = \$2,709.26$, with a standard deviation of $s = \$1,115.52$. Find an interval A to B for the monetary charitable donations into which 75% of the households would fit.

A. \$478.22 to \$4,940.30
B. \$480 to \$4,940.30
C. \$468.22 to \$4,940.30
D. \$478.22 to \$4,840.30
E. \$478.22 to \$4,939.30

33. A corporation manufactures tires at six factories. The costs to produce a type of tire are the following (dollars per tire):

13.47 14.21 13.63 14.72 14.01 12.98

Compute the range, mean, standard deviation, and coefficient of variation. Round to the nearest hundredth if necessary.

34. An illegal pain-killing drug is sometimes given to horses to get them to run faster than normal. A veterinarian working for the racing commission is studying the drug to determine how long the drug stays in the horse's blood stream. A random sample of horses were given the drug. In the table below, x = time (in hours) for the drug level in the blood stream to go back to zero. f is the frequency, i.e. the number of horses.

x	2	4	6	8	10	12	14	16
f	1	3	8	11	9	6	4	2

(a) Compute the sample mean recovery time for this drug.
(b) Compute the sample standard deviation.

35. Allied Airlines is doing a study of late arrival times for the New York to Atlanta run. A random sample of late flights gave the following information. In the table below, x = number of minutes the flight is late and f = number of flights.

x	10	20	30	40	50	60	70	80	90
f	15	19	10	7	3	2	2	1	1

(a) Compute the sample mean, $\bar{x}$ of the late arrival times.
(b) Compute the sample standard deviation of the late arrival times.

36. A supermarket sells eight brands of 8-oz cans of baked beans, with the following prices (dollars per can):

0.98 1.29 1.59 1.69 1.09 1.49 1.19 1.38

Compute the population mean and population standard deviation.

A. $\mu = \$1.34$; $\sigma = \$0.23$ B. $\mu = \$1.32$; $\sigma = \$0.24$ C. $\mu = \$1.33$; $\sigma = \$0.22$

D. $\mu = \$1.33$; $\sigma = \$0.23$ E. $\mu = \$1.34$; $\sigma = \$0.25$

37. The average points per game scored by eight of the ten high school basketball teams of the Northeast Division in 2001 are as follows:

48.6 39.4 62.1 58.7 44.6 52.3 49.8 60.0

Compute the sample mean and sample standard deviation. Round to the nearest tenth if necessary.

A. $\bar{x} = 51.9$; $s = 7.4$ B. $\bar{x} = 51.9$; $s = 7.9$ C. $\bar{x} = 51.9$; $s = 7.8$

D. $\bar{x} = 51.8$; $s = 7.9$ E. $\bar{x} = 50.9$; $s = 7.9$

38. A traveling salesperson sells ten types of soap. The prices per bar are as follows (dollars per bar):

0.48 0.79 0.56 1.09 1.12 2.24 0.98 0.87 1.69 1.98

Compute the range, population mean, and population standard deviation.

A. Range = \$1.75; μ = \$1.18; σ = \$0.57
B. Range = \$1.76; μ = \$1.17; σ = \$0.57
C. Range = \$1.76; μ = \$1.18; σ = \$0.57
D. Range = \$1.76; μ = \$1.18; σ = \$0.60
E. Range = \$1.77; μ = \$1.18; σ = \$0.60

39. The average hours worked per week in 2001 by the eight employees of Johnson Electric are as follows.

36.0 44.3 40.2 39.6 41.8 40.6 37.8 41.5

Compute the population mean, population standard deviation, and find an interval *A* and *B* for the number of hours worked per week into which at least 75% of the employees fit. Round to the nearest hundredth if necessary.

40. Statistical Abstracts, (117th edition), lists the number of hazardous waste sites on the National Priority List by state. For the states in the Mountain West region the numbers are:

9 10 3 18 11 10 16 1

Find the median, the population mean, and the population standard deviation.

41. The annual salaries earned by members of an accounting firm are:

150,000 90,000 30,000 30,000 30,000 30,000 30,000 20,000

(a) Calculate the sample median and the sample mean.
(b) If you were considering taking an entry level job with the firm, which of these numbers would give you a better estimate of your potential starting salary?

42. The Richter scale is used to measure the magnitude of earthquakes. The scale readings are such that 1 represents an earthquake that is observed only on instruments, while 3 indicates a quake felt only by people near its epicenter. A reading of 7 represents a major quake, usually accompanied by substantial destruction. A sample of 15 earthquakes occurring in the United States had readings on the Richter scale of

1.0 2.1 1.3 5.1 3.4 2.1 1.4 1.5
2.0 3.1 4.6 1.2 1.8 2.7 3.5

(a) Compute the five-number summary and the interquartile range.
(b) Make a box-and-whisker plot.

43. In one personality assessment test, a group of questions relate to self-acceptance. A random sample of 15 scores on the self-acceptance portion are

5 20 22 27 30 17 12 15
16 9 18 13 12 28 19

(a) Compute the five-number summary and the interquartile range.
(b) Make a box-and-whisker plot.

44. The Softex Plastic Company tested a random sample of 12 pieces of new plastic for breaking strength. They obtained the following measurements in units of 1,000 pounds per square inch:

2.0 2.3 4.1 3.3 1.9 3.5 3.4 1.8 4.0 2.7 2.6 3.9

(a) Compute the five-number summary and the interquartile range.
(b) Make a box-and-whisker plot.
(c) Find the range.
(d) Find the mean.
(e) Find the sample standard deviation.

45. Use the table of data below and the computation formula for grouped data to find the sample mean and sample standard deviation.

x	f	xf	x^2f
1	2	2	2
2	4	8	16
3	5	15	45
4	8	32	128
5	7	35	175
6	3	18	108

A. $\bar{x} \approx 4.31; s \approx 1.42$ B. $\bar{x} \approx 3.79; s \approx 56.76$ C. $\bar{x} \approx 3.79; s \approx 1.42$

D. $\bar{x} \approx 3.76; s \approx 1.42$ E. $\bar{x} \approx 1.38; s \approx 3.94$

46. The cost of one serving of peanut butter (in cents) for a random sample of 19 jars of peanut butter was found to be:

22 27 32 26 26 19 26 14 21 20 21 22
12 32 17 9 16 17 16

(a) Compute the five-number summary and the interquartile range.
(b) Make a box-and-whisker plot.

47. A random sample of 100 newspaper subscribers gave the following information about the number of days per week they read (any part of) the sports section, where x = number of days per week the sports section was read and f = number of subscribers.

x	0	1	2	3	4	5	6	7
f	21	10	7	8	10	11	14	19

Use the computation formula for grouped data to find the mean and standard deviation of the number of days per week the sports section was read. Round to the nearest hundredth if necessary.

48. Hotel occupancy rates often dictate how easy it might be to reserve a room at the last minute and determine the average cost of a room. Rooms are often discounted in areas with low occupancy. Occupancy rates across the United States for major cities are given below:

56	89	79	71	70	60	60	61	62	63	64	65
81	73	68	72	56	59	60	62	73	64	72	67

(a) Compute the five-number summary and the interquartile range.
(b) Make a box-and-whisker plot.

49. The number of hours of volunteer service given last month by nurses at the Red Cross emergency station were as follows:

10	8	1	3	5	6	8	15	8	7	3	4
6	3	8	9	8	2	4	16	11	12	7	

Find the mean, the median and the mode of the hours of volunteer service.

50. A municipal transit manager, trying to determine where to locate a new bus stop, surveyed a random sample of riders of the bus line in question concerning the number of city blocks they walked to a particular bus stop. Their responses are shown in the table below.

Blocks walked, x	0 – 2	3 – 5	6 – 8	9 – 11
Number of riders, f	6	7	5	2

Using the midpoints of the blocks walked intervals, estimate the mean blocks walked and the standard deviation.

A. $\bar{x} = 4.45; s \approx 2.16$ B. $\bar{x} = 4.05; s \approx 2.96$ C. $\bar{x} = 1.1; s \approx 2.96$

D. $\bar{x} = 4.45; s \approx 1.61$ E. $\bar{x} = 4.45; s \approx 2.96$

51. Ten students took an aptitude test in technical electronics. There were 1,000 possible points. The scores were as follows:

351 988 348 450 290 965 360 346 332 318

(a) Compute the mean of all ten scores.
(b) Compute the median of all ten scores and compare your answer to that of part (a).
(c) If all ten scores were to be used, which average (mean or median) would best describe most of the students?
(d) Compute a 10% trimmed mean.
(e) Which of the three averages best describes most of the students?

52. At an athletic event, members of the college weightlifting team lifted the following weights (in pounds):

190 173 200 188 190 175
190 180 200 177 179 190

(a) Find the mean, the median and the mode of these numbers.
(b) In your own words define the terms mean, median, and mode. Discuss how your definitions fit into the context of this problem. How do the mean, the median and the mode each represent in a different way an estimate of the team's weightlifting capacity?

53. Use the computation formula for grouped data and the table below to find the mean salary and standard deviation for a certain small business.

Wage, x	7.85	9.25	10.51	16.98	25.60
Employees with wage x, f	4	6	3	2	1

A. $\bar{x} \approx 11.12$; $s \approx 2.54$ B. $\bar{x} \approx 4.39$; $s \approx 4.78$ C. $\bar{x} \approx 11.12$; $s \approx 4.78$

D. $\bar{x} \approx 4.39$; $s \approx 2.54$ E. $\bar{x} \approx 11.12$; $s \approx 13.05$

54. A random sample of six credit card accounts gave the following information about the payment due on each card:

\$53.18 \$71.12 \$115.10 \$27.30 \$36.19 \$66.48

(a) Find the range.
(b) Find the sample mean.
(c) Find the sample standard deviation.

55. Sir Charles Bradley wrote six novels before he died. The page length of these novels are:
280 318 279 356 410 305
Since these numbers represent all novels written by Sir Charles, the data is population data.
(a) Find the range.
(b) Find the population mean.
(c) Find the population standard deviation.

56. A psychology test to measure memory skills was given to a random sample of 43 students. In the results which follow, x is the student score and f is the frequency with which students obtained the score.

x	0 – 10	11 – 21	22 – 32	33 – 43	44 – 54
f	1	12	18	9	3

(a) Estimate the mean score.
(b) Estimate the sample standard deviation of the scores.

57. The members of a random sample of 61 Pittsburgh policemen were asked how many years (rounded to the nearest year) they had served on the city's police force. The results follow. x represents the number of years of service and f is the number of policemen.

x	1 – 5	6 – 10	11 – 15	16 – 20	21 – 25
f	20	12	14	10	5

(a) Estimate the mean number of years of service.
(b) Estimate the standard deviation for the number of years of service.

58. Find the interquartile range of the following set of data:

1	3	8	4	10	11	15	2
12	6	7	8	9	18	19	5

A. 8 B. 7.5 C. 7 D. 9 E. 6

59. Jan is a political science student who is studying the incidence of state senators in her legislature who are marked absent when a bill is voted on in the senate. Jan used state government documents to obtain a random sample of 60 bills that were voted on in the senate. For each bill Jan obtained the following information. x represents the number of senators absent when the vote on a bill was taken and f is the number of bills for which this level of absenteeism occurred.

x	0 – 4	5 – 9	10 – 14	15 – 19	20 – 24
f	7	18	23	8	4

(a) Estimate the mean number of senators absent for a vote.
(b) Estimate the sample standard deviation for the number of senators absent for a vote.

60. Find Q_1, Q_2, Q_3 and the interquartile range of the following set of data:

1.2	3.6	5.1	4.8	10.2
6.1	7.3	9.5	2.4	11.1
0.7	1.9	4.6	2.1	9.8

61. A random sample of readings at a radioactive spill site gave the following information where x is the radiation level in microcuries and f is the number of readings at that level.

x	5	10	15	20	25	30	35
f	6	18	11	5	3	1	1

(a) Compute the sample mean of the radiation levels.
(b) Compute the sample standard deviation.

62. The Dow Jones Industrial Average is a measure used by investors to track the general stock market. From the first business day of January to the last business day of December the number of consecutive days the Dow Jones Industrial Average went up were:

1 1 1 1 1 1 1 1 1 1 1 1 1 1 1 1 1 1 1 1
2 2 2 2 2 2 2 2 2 2 2 2 3 3 3 3 3 3 3 3
3 3 4 4 4 4 4 4 4 5 6 6 6 6 9

(a) Give the five-number summary.
(b) Make a box-and-whisker plot.

63. Give the five-number summary for the following set of ages of children attending a birthday party:

6 7 5 8 7 9 11 10
4 5 6 8 7 9 9 7
5 5 6 4 3 8 7 7

A. low value = 4; $Q_1 = 5$; median = 8; $Q_3 = 6$; high value = 9

B. low value = 3; $Q_1 = 8$; median = 6; $Q_3 = 7$; high value = 10

C. low value = 3; $Q_1 = 5$; median = 6; $Q_3 = 8$; high value = 11

D. low value = 3; $Q_1 = 5$; median = 7; $Q_3 = 8$; high value = 10

E. low value = 3; $Q_1 = 5$; median = 7; $Q_3 = 8$; high value = 11

64. Counts of white cells in blood were done for 30 blood samples. The number of white cells (in thousands) per cubic millimeter of blood for each sample was:

3.8 4.9 5.0 7.3 6.1 7.7 5.8 8.1 8.2 7.6
5.7 5.2 9.5 10.1 10.0 9.6 8.7 7.5 4.9 8.9
6.3 7.2 8.1 9.2 5.7 8.4 6.9 4.2 7.1 4.9

(a) Find the interquartile range.
(b) Give the five-number summary.
(c) Make a box-and-whisker plot.

65. Twenty-four people were asked to rate the flavor of Jane's Jams 'n' Jellies brand grape jelly, on a scale of 1 to 10. The results follow:

8	7	6	9	9	5	3	8
10	9	8	8	6	5	6	7
8	9	1	9	7	8	6	2

(a) Give the five-number summary.
(b) Make a box-and-whisker plot.

66. Advantage VCR's carry a one-year warranty. However the terms of the warranty require that a registration card be sent in. One of the questions on the registration card asks the owner's age. For a random sample of 50 registration cards, the ages given were:

17	19	21	22	23	24	25	26	27	27
27	27	27	28	31	31	31	32	32	32
32	33	33	34	35	35	36	37	37	37
38	38	39	40	41	42	42	43	43	45
45	46	47	47	47	51	51	52	55	55

(a) Find the interquartile range.
(b) Give a five-number summary for the data.
(c) Make a box-and-whisker plot of the data.

67. McElroy Discount Fashions claims to have a fairly low mark-up percent on the items they sell. A random sample of 50 items showed the mark-up percent over cost to be (in percents):

15	21	22	25	27	27	27	28	29	31
33	35	36	36	37	37	37	42	42	42
45	46	47	48	49	49	51	51	51	52
53	55	55	58	59	61	65	67	68	71
72	72	72	72	72	75	75	77	78	85

(a) Find the interquartile range.
(b) Give a five-number summary of the data.
(c) Make a box-and-whisker plot for the data.

68.

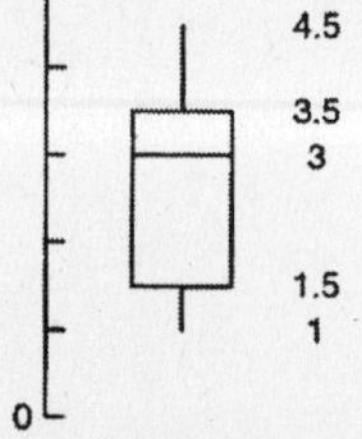

From the box-and-whisker plot above, select the correct five-number summary.

A. 4.5, 3.5, 3, 1.5, 1
B. 1, 2, 3, 4, 5
C. 1, 1.5, 3, 3.5, 4.5
D. 1, 1.5, 3, 3.5, 5
E. 1, 1.5, 3.5, 3.5, 4.5

69. The median score on a national engineering aptitude test is 450 and the 90th percentile is 680.
(a) What fraction of those who took the test had scores at or over 680?
(b) What fraction of those who took the test had scores between 450 and 680 inclusive?

70. Compute the *IQR* for the following set of data:

1.1	5.8	4.9	16.2	13.8
3.0	7.4	6.2	11.1	15.2
9.6	5.7	2.8	8.3	12.9

A. 8.9 B. 6.2 C. 7.2 D. 8 E. 9.9

71. Marjorie is doing a study of her monthly living expenses. Monthly totals for the last year were (in dollars):

1,300	1,160	1,380	1,440	1,000	3,500
900	1,050	1,220	1,120	1,300	880

(a) Find the mean of this data.
(b) Find a 10% trimmed mean.
(c) Which mean would be a better estimate of her living expenses for a typical month?
(d) Which mean would be more useful if she wanted to allow for possible emergencies?

72. Home Security Systems is studying the time utilization of its sales force. A random sample of 40 sales calls showed that the representatives spend an average (mean) time of 42 minutes on the road for each sales call. The standard deviation of the call times is 5 minutes.
(a) Compute the coefficient of variation (*CV*) for this data.
(b) Use Chebyshev's Theorem to find the smallest time interval in which we can expect to find at least 75% of the data.

73. Job Finders did a study of mid-management jobs in large cities. A sample of such jobs showed the average (mean) number of applicants for each position to be 37 with standard deviation 6.
(a) Compute the coefficient of variation (*CV*) for this data.
(b) Use Chebyshev's Theorem to find the smallest time range in which we can expect at least 75% of the data to fall.

74. Which data set results in the following box-and-whisker plot?

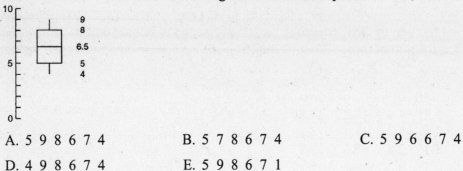

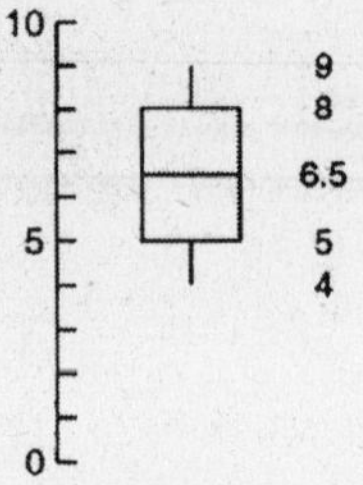

A. 5 9 8 6 7 4 B. 5 7 8 6 7 4 C. 5 9 6 6 7 4

D. 4 9 8 6 7 4 E. 5 9 8 6 7 1

75. A study was done of the number of years new car buyers keep their cars before trading them in. The data was rank ordered and the following information was obtained.

Percentile	25th	50th	75th	90th
Value	3.6 years	8.2 years	10.5 years	15.1 years

(a) What percentage of the sample keep their cars fewer than 3.6 years?
(b) What percentage of the sample keep their cars from 3.6 years to 10.5 years?
(c) What percentage of the sample keep their cars more than 15.1 years?

76. Many travelers make airline reservations and then do not show up for their scheduled flights. Air Connect did a study comparing the number of no-shows before and after it implemented a policy penalizing ticket-holders who do not show up. Box-and-whisker plots of the two data sets are shown below. Compare the two plots. Discuss the locations of the medians, the location of the middle section of the data, and the amount of variation in each set of data. What effect, if any, does the policy appear to have had?

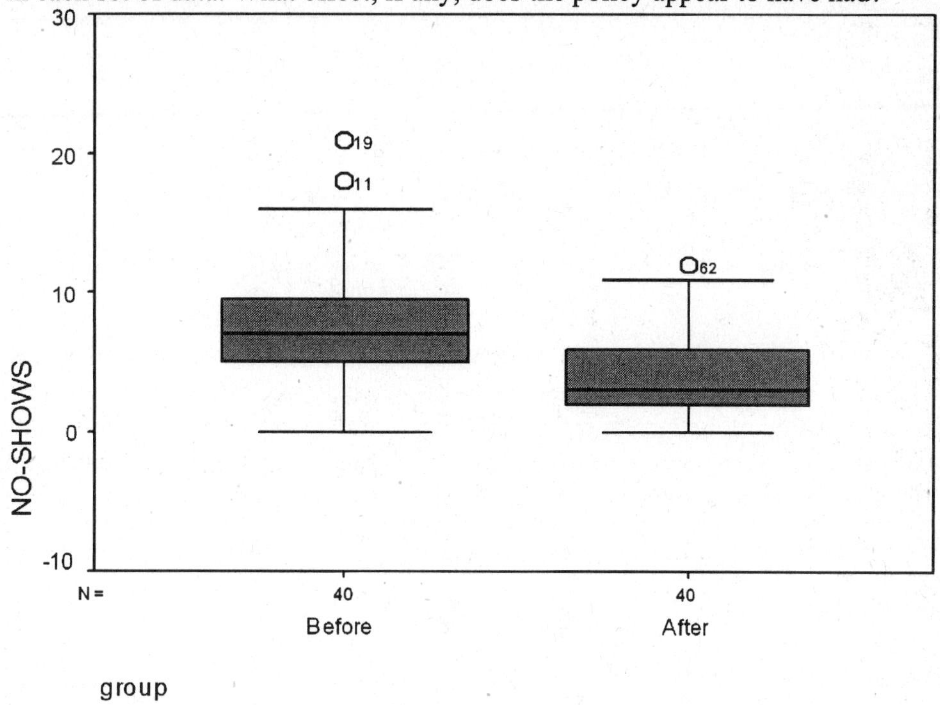

77. Find the median, mean, and standard deviation of the following complete set of a football team's numbers:

00	90	08	81	12	18
21	24	27	77	34	52
42	43	62	50	40	56
58	44	66	70	71	32
80	10	84	88	01	99

A. median = 51; $\mu = 47$; $\sigma \approx 28.2$

B. median = 46; $\mu = 48$; $\sigma \approx 28.3$

C. median = 44; $\mu = 48$; $\sigma \approx 28.3$

D. median = 46; $\mu = 48$; $\sigma \approx 28.2$

E. median = 47; $\mu = 48$; $\sigma \approx 28.2$

78. Roger scored at the 65th percentile on a standardized reading test.
(a) What fraction of the population had scores at or above his score?
(b) What fraction of the population scored below Roger's score?

79. Martha got a score of 89 out of 100 on a teacher competency test. Her score was at the 70th percentile for that test. Sharon scored 80 out of 100 on a different test of comparable difficulty level. Her score was at the 75th percentile?
(a) If the tests are equivalent in scope and level of difficulty, which person did better in terms of knowledge of material?
(b) Who did better compared to the rest of the people who took the same test?

80. Compute the 10% trimmed mean for the following set of data:

1.1	5.6	8.9	18.2	19.1
7.2	8.2	7.4	6.9	15.5
3.6	4.7	11.8	12.4	17.7
2.5	10.3	8.7	7.1	6.4

A. 8.9 B. 9.2 C. 8.8 D. 9.0 E. 9.1

81. A study was done comparing annual insurance premiums for suburban drivers and rural drivers. Box-and-whisker plots for the data are shown below. Compare the two plots. Discuss the location of the medians, the middle half of the data set and the variation within each data set. Does there appear to be a difference between premiums for suburban drivers and premiums for rural drivers?

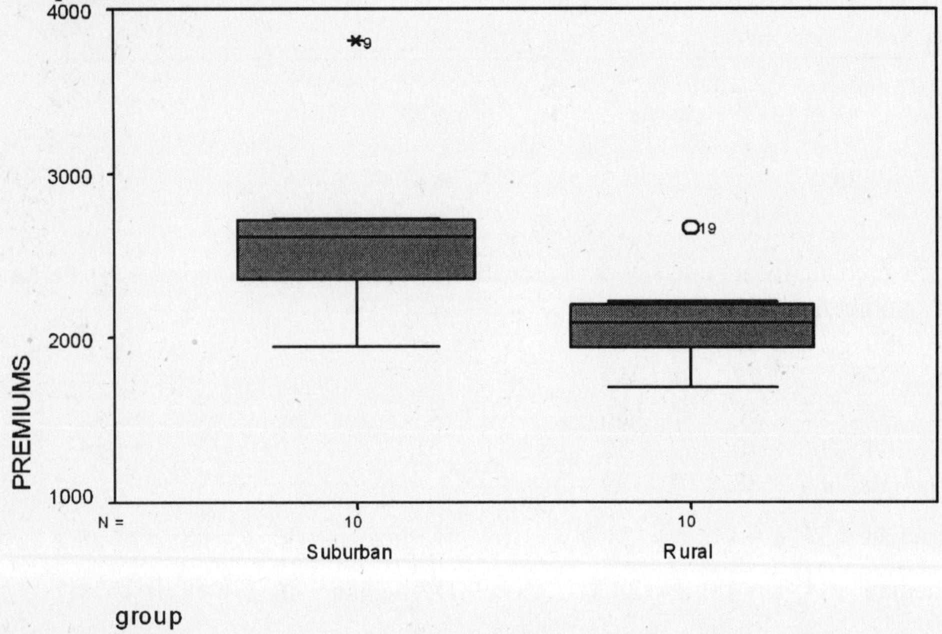

Section B: Algorithms

Objective 3.1: Measures of Central Tendency: Mode, Median, and Mean

82. Find the mode of 95, 64, 57, 67, and 57.

83. In a mathematics class, half of the students scored 81 on an achievement test. Most of the remaining students scored 79 except for a few students who scored 48. What is the relationship between the mean and the median?

84. If Mr. Sanders reports student progress using the test score occurring most frequently, which measure of central tendency does he report?

85. In five compartments, there were 9 bundles, 12 bundles, 10 bundles, 15 bundles, and 9 bundles, respectively. If the bundles were rearranged so that each compartment held the same number of bundles, how many bundles would be in each compartment?

86. Mr. De Angelo recorded the number of sick days taken last year by each employee, as shown in the table.

Employees	Sick Days
Dylan	2
Lloyd	5
Sarita	3
Will	6
Kurt	0
Isabel	4
Lucio	5
Lynn	3
Ryan	3

What is the mean number of days employees were sick?

87. Find the mode for each set of data:
15, 12, 16, 14, 12, 16, 15, 14, 16, 18

88. The manager of the customer service department of a telephone company wants to hire more employees to answer customer calls. To justify her decision, she chose a random sample of 60 callers and timed how long each was on hold before a customer service employee could attend to their call. She then graphed her results in the histogram shown below.

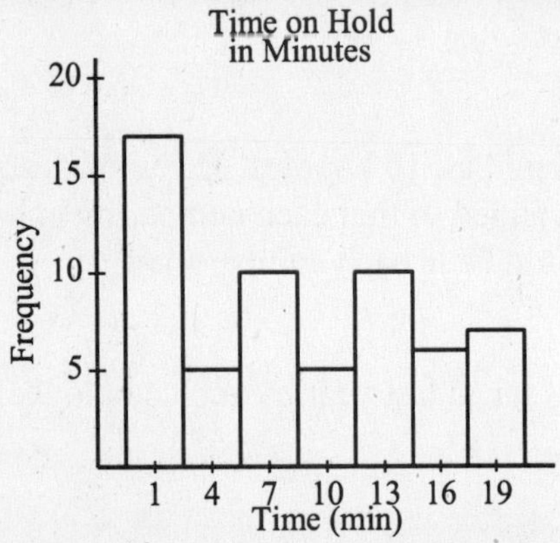

Approximate the mean of the distribution and compute an approximate 15% trimmed mean for the data.

89. Find the outliers in the data. Then find the mean when outliers are included and again when they are not included. If there are no outliers, find the mean of the data.
76, 69, 79, 23, 74, 66, 77, 73, 115, 73

Objective 3.2: Measures of Variation

90. Find the range of each set of data:
3, 15, 22, 3, 23, 25, 27, 28, 15, 15, 2, 22, 3, 27, 2, 27

91. Find the sample variance s^2:
7, 15, 18, 4, 23, 17, 20, 15, 16

92. The following are the ages of the employees at a small law firm. The mean of the ages of people working in this field is 41. Find the sample variance s^2 for the following sample.

25	38	36	34	33	56	22
45	57	54	49	26	48	46
44	42	41	43	51	39	28

93. Find the population standard deviation σ:
14, 16, 18, 22, 5, 12, 4, 20, 6

94. Find the sample standard deviation s:
22, 5, 28, 18, 27, 24, 20, 14, 4

95. The following are the ages of the non-faculty staff at a small college. The mean of the ages of people working in this field is 38. Find the sample standard deviation s for the following sample.

47	33	30	26	55	39	25
35	46	56	48	23	40	52
42	28	37	41	24	36	45

96. For a random sample of five computer companies, the profits per employees in the same units were:
22.0, 27.6, 32.9, 52.8, 49.7
Compute the sample standard deviation and the coefficient of variation.

97. Mrs. Lee is the director of a student volunteer program that donates work time to various community projects. For a random sample of students in the program, Mrs. Lee found the mean was $\bar{x} = 26.8$ hours each semester, with a standard deviation $s = 2$ hours each semester. Find an interval A to B for the number of hours volunteered into which at least 93.8% of the students in this program would fit.

98. The owner of a magazine company is curious to find out how many readers respond to polls published in her magazine. For three years she records how many responses she receives. The mean number of responses per magazine poll is $\bar{x} = 505$ with a standard deviation $s = 29$. Determine a Chebyshev interval about the mean in which at least 75% of the data fall.

99. The numbers below represent the number of gallons of water used daily by ten people. Find the range of the data.
25, 24, 30, 27, 25, 29, 25, 21, 23, 28

100. Mrs. Bell is the director of a student volunteer program that donates work time to various community projects. For a random sample of students in the program, Mrs. Bell found the mean was $\bar{x} = 34$ hours each semester, with a standard deviation $s = 2.4$ hours each semester. Find an interval A to B for the number of hours volunteered into which at least 88.9% of the students in this program would fit.

101. The hatching success for a particular game bird was recorded by a ornithologist. The percentage of successful nests for five different locations are 24.4, 41.4, 41.5, 43.6, 59.1. Find the sample standard deviation and the coefficient of variation.

102. The manager of the customer service department of a telephone company wants to hire more employees to answer customer calls. To justify her decision, she chose a random sample of 55 callers and timed how long each was on hold before a customer service employee could attend to their call. She then graphed her results in the histogram shown below.

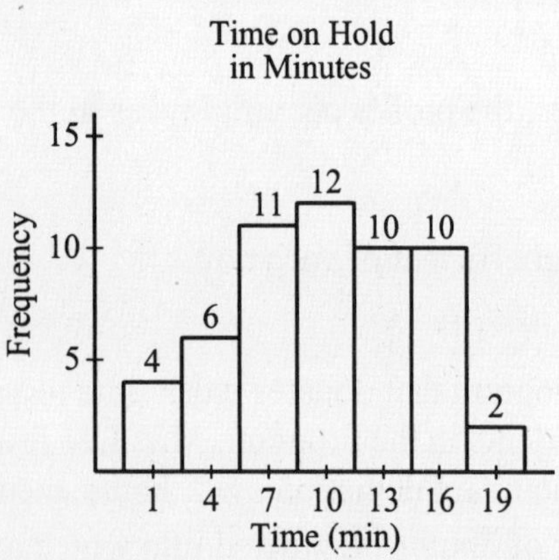

Approximate the standard deviation of the distribution.

Objective 3.3: Percentiles and Box-and-Whisker Plots

103. Suppose you score 88% on your midterm and score 72% on your final. Using the weights of 38% for the midterm and 62% for the final exam, find the weighted average of your scores.

104. The following are the scores of 49 students on a recent algebra test. If Jane scored 84 on the test, what percentile would that score represent?

49	64	79	48	92	57	59
93	51	66	95	80	48	65
48	53	81	63	67	59	88
45	56	80	56	90	79	96
87	62	54	65	84	84	51
58	46	90	89	48	48	81
53	46	51	57	47	82	46

105. The table below shows Dana's test scores and percentiles in each of her four subjects.

Subject	Raw Score	Percentile
English	87	93
Math	69	64
Science	71	62
History	84	72

What is Dana's percentile for her science test score? What does this percentile show?

106. Ms. Harbison drew a box-and-whisker plot to represent her students' scores on a midterm test.

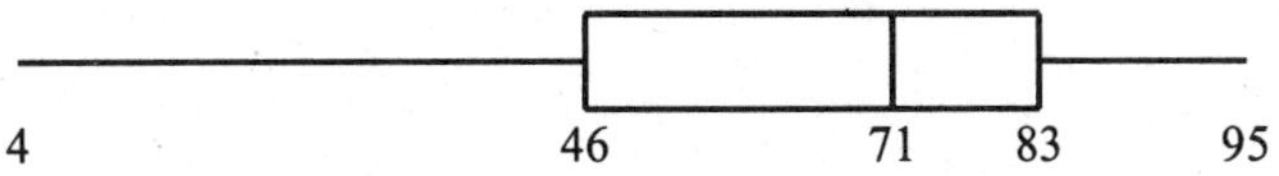

Steve earned 46 on the test. Describe how the scores of Steve's classmates compared to Steve's score.

107. Is the statement below true based on the data shown by the box-and-whisker plot?
The interquartile range is 21.

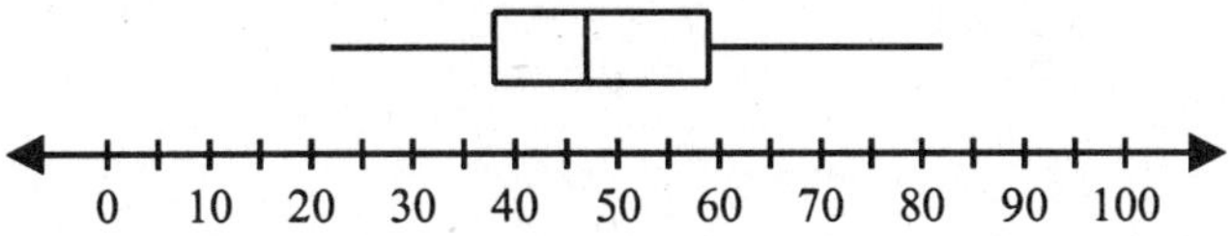

108. Find the upper quartile, lower quartile, and interquartile range for the set of data.
23, 37, 31, 18, 26, 14, 33, 13

109. Find the first, second, and third quartiles of the data. Then draw a box-and-whisker plot of the data.
32, 34, 35, 28, 36, 32, 30, 29, 30, 38, 40, 31, 29, 31, 36

110. A college professor discovered that 126 of the 155 students that took a history test scored lower than 75. What percentile does the score of 75 represent?

111. Find the lower extreme.

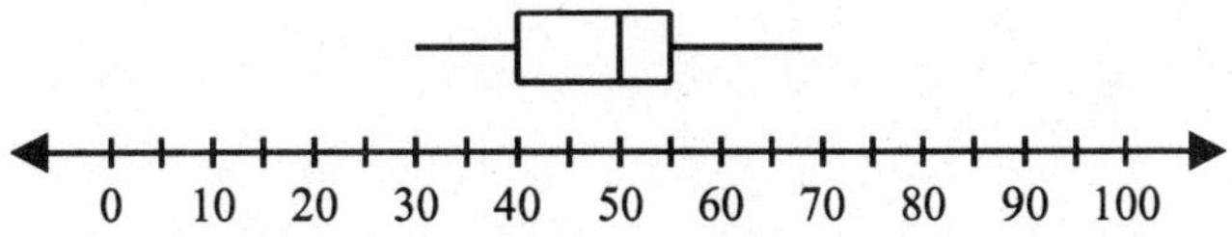

Answer Key for Chapter 3
Averages and Variation

Section A: Static Items

[1] Mean = 76.6; median = 78; mode = 78

[2] Mean = 20.6; median = 20.5; mode = 19

[3] Mean = 43; median = 42.5; mode = 31

[4] Mean = 16.5; median = 17; mode = 16
For discussion, apply definitions in text to the content of this problem.

[5] Mean = 20.3; median = 20; mode = 19

[6] D.

[7] Range = 11; $\bar{x} = 17.67$; $s = 3.24$

[8] Range = 18; $\bar{x} = 7.88$; $s = 6.36$

[9] B.

[10]
(a) Mode = 3; median = 3; mean = 3.3
(b) Range = 3; standard deviation = 1.03
(c) Mode = 3; median = 3.5; mean = 4.0
The mean changed the most. It is the average most sensitive to extreme values.

[11] Mean = 7.63 minutes; median = 8 minutes; mode = 8 minutes

[12] E.

(a) Mode = 4; median = 4; mean = 4.17
(b) Range = 5; sample standard deviation = 1.72
(c) Mode: none; median = 3.5; mean = 3.33
The mode no longer exists since both 4 and 2 occur the maximum number of times. The
[13] mean changed quite a bit because the largest data value changed.

(a) Mean = 25.5; median = 15; mode = 15
(b) Trimmed mean = 18.9
(c) The extreme values increased the mean noticeably. They did not affect the median or
[14] the mode.

[15] 5% trimmed mean: 25.5 years

(a) Ann: mean = 8.8; standard deviation = 7.01
Edgar: mean = 8.8; standard deviation = 1.92
(b) Edgar had the more consistent sales since his sales showed less variability over the
[16] period.

(a) Range = 37 degrees
(b) Sample mean = 6.4 degrees
(c) Sample standard deviation = 12.1 degrees
[17] (d) $CV = 189\%$

[18] A.

(a) Midwest: $CV = 124.4\%$; Pacific Region: $CV = 31.06\%$
[19] (b) The midwest region has greater variability.

(a) 23.61 thousand to 44.01 thousand
[20] (b) 13.41 thousand to 54.21 thousand

[21] E.

(a) 11.7 inches to 17.7 inches
[22] (b) 8.7 inches to 20.7 inches

(a) Median = 177.5; $\mu = 207.2$; $\sigma = 63.53$
(b) Median = 367.0; $\mu = 365.0$; $\sigma = 61.81$
(c) Independent stations: $CV = 30.7\%$; Basic cable: $CV = 16.9\%$
(d) The number of people watching basic cable TV is generally greater than the number
[23] watching independent TV stations and there is less variation in the data for basic cable TV.

[24] B.

(a) Range = 1,700; $\bar{x}$ = 2,160; s = 691.38; CV = 32.0%
(b) Range = 224; $\bar{x}$ = 468.4; s = 85.45; CV = 18.2%
(c) There is greater variability in prices for the 50 inch TV sets than there is for the 25
[25] inch TV sets.

(a) The mean is approximately 83 minutes.
[26] (b) The sample standard deviation is approximately 55.8 minutes.

[27] A.

(a) $\bar{x}$ = 12.1 credit hours

[28] (b) s^2 = 25.1; s = 5.01 credit hours

[29] Range = 90; s = 31.00

(a) $\bar{x} \approx$ 36.72 years
[30] (b) $s \approx$ 20.98 years

[31] D.

[32] A.

[33] Range = \$1.74; μ = \$13.84; σ = \$0.56; CV = 4.02%

(a) $\bar{x}$ = 9.09 hours
[34] (b) s = 3.30 hours

(a) $\bar{x}$ = 28.17 minutes
[35] (b) s = 18.73 minutes

[36] A.

[37] B.

[38] C.

[39] $\mu = 40.23$ hr; $\sigma = 2.37$ hr; 35.49 hr to 49.97 hr

[40] (a) Median = 10; $\mu = 9.75$; $\sigma = 5.38$

[41] (a) Median = 30,000; mean = \$51,250
(b) The median

[42] (a) Five-number summary: (1.0, 1.4, 2.1, 3.4, 5.1); $IQR = 2.0$
(b)

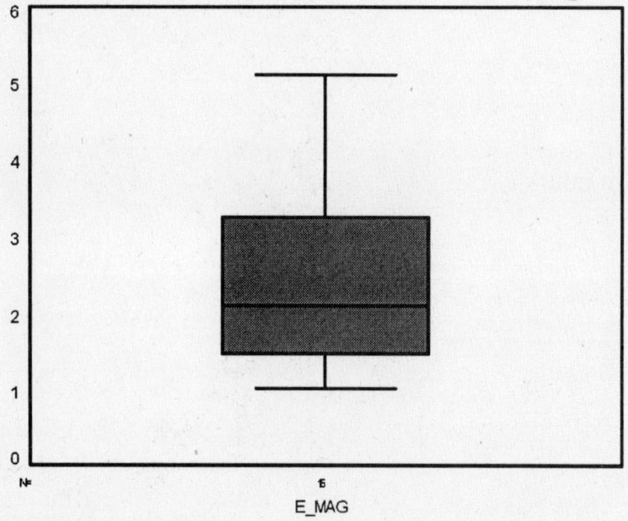

[43] (a) Five-number summary: (5, 12, 17, 22, 30); $IQR = 10$
(b)

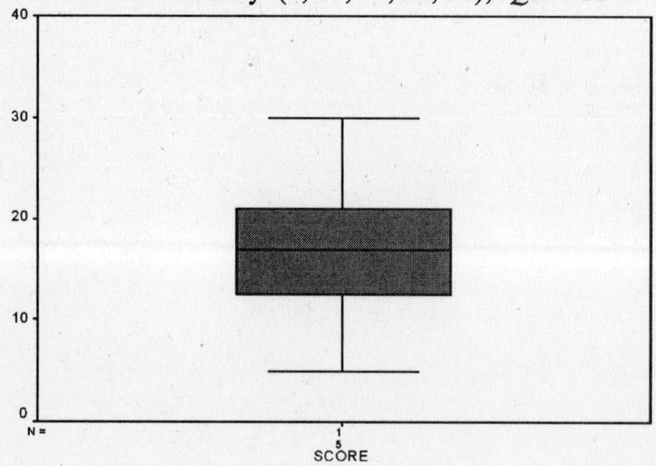

(a) Five-number summary: (1.8, 2.2, 3.0, 3.7, 4.1); *IQR* = 1.5

(b)

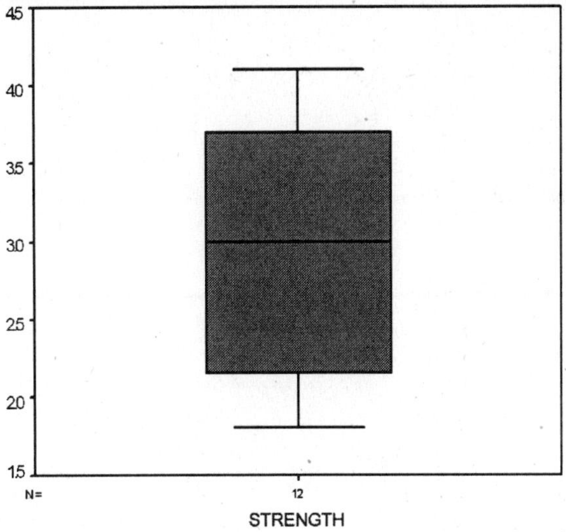

(c) Range = 2.3

(d) Mean = 2.96

[44] (e) Standard deviation = 0.85

[45] C.

(a) Five-number summary: (9, 16, 21, 26, 32); *IQR* = 10

(b)

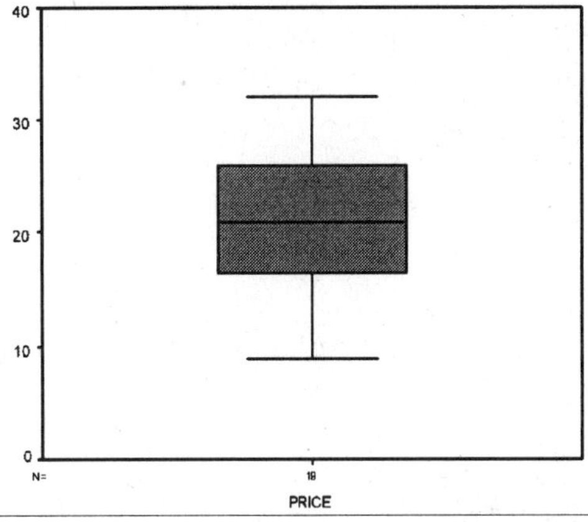

[46]

[47] $\bar{x}$ = 3.6 days per week; s = 2.63

(a) Five-number summary: (56, 60.5, 64.5, 72, 89); $IQR = 11.5$

(b)

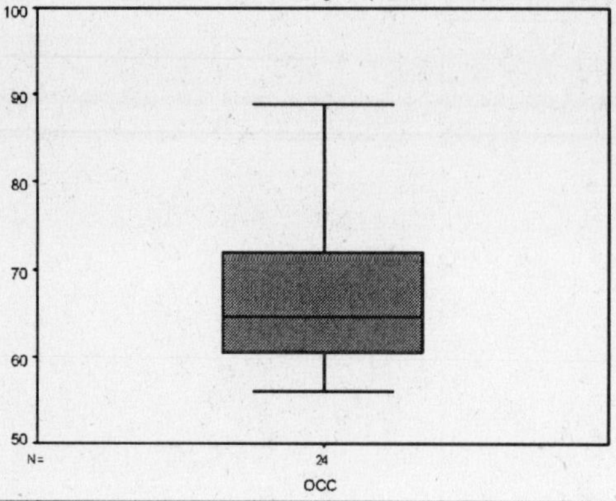

[48]

[49] (a) Mean = 7.13 hr.; median = 7 hr.; mode = 8 hr

[50] E.

(a) Mean = 474.8 points
(b) Median = 349.5 points; The median is less than the mean.
(c) The median
(d) Trimmed mean = 433.75 points
[51] (e) The median; 349.5 points

(a) Mean = 186 lb; median = 189 lb; mode = 190 lb
[52] (b) Describe highlights of the three averages. See text.

[53] C.

(a) Range = \$87.80
(b) Mean = \$61.56
[54] (c) Standard deviation = \$31.21

(a) Range = 131 pages
(b) $\mu = 324.7$ pages
[55] (c) $\sigma = 46.1$ pages

(a) Mean $\approx$ 27.26 points
[56] (b) Standard deviation $\approx$ 10.32 points

(a) Mean $\approx$ 10.38 years.
[57] (b) Standard deviation $\approx$ 6.62 years

[58] C.

(a) Mean ≈ 10.67 senators
[59] (b) Standard deviation ≈ 5.28 senators

[60] $Q_1 = 2.1$; $Q_2 = 4.8$; $Q_3 = 9.5$; $IRQ = 7.4$

(a) Mean ≈ 13.67 microcuries
[61] (b) Standard deviation ≈ 6.78 microcuries

(a) Five-number summary: (1, 1, 2, 3, 9)
(b)

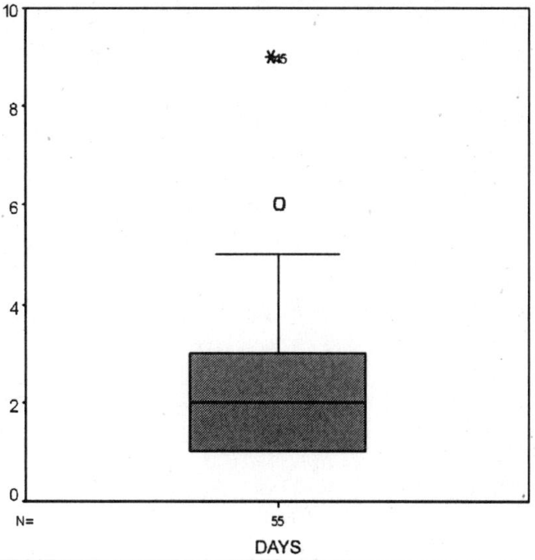

[62]

[63] E.

(a) $IQR = 2.7$
(b) Five-number summary = (3.8, 5.7, 7.25, 8.4, 10.1).
(c)

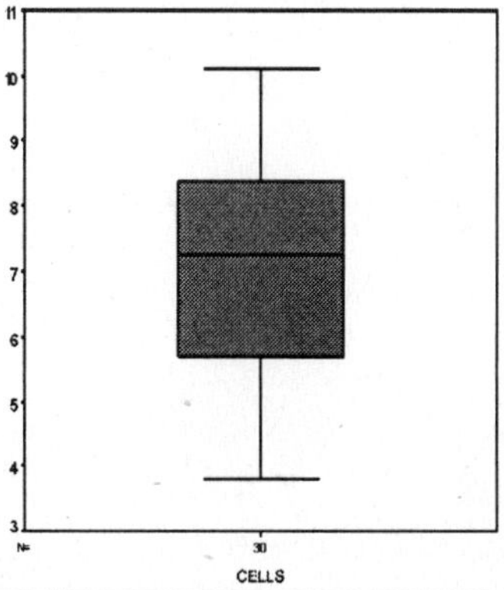

[64]

(a) Five-number summary = (1, 6, 7.5, 8.5, 10)

(b)

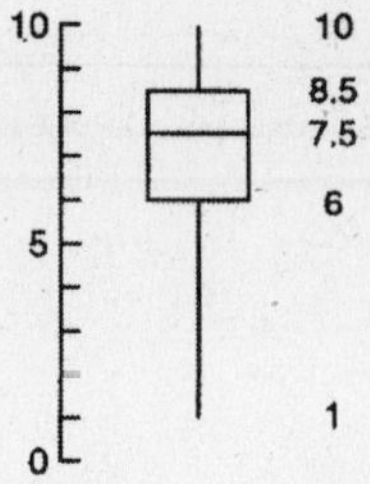

[65]

(a) *IQR* = 16

(b) Five-number summary: (17, 27, 35, 43, 55)

(c)

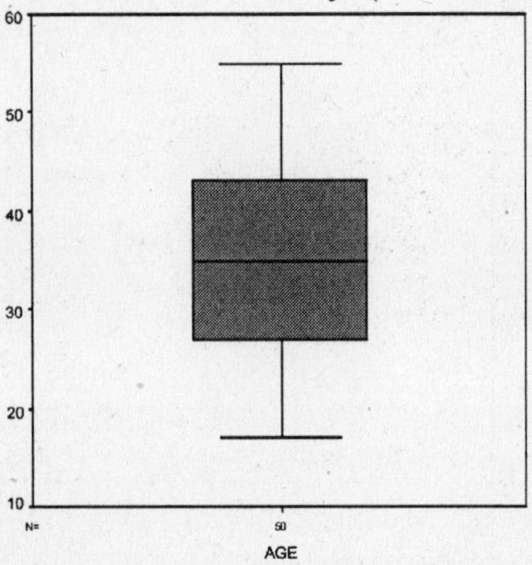

[66]

(a) *IQR* = 31

(b) Five-number summary: (15, 36, 49, 67, 85)

(c)

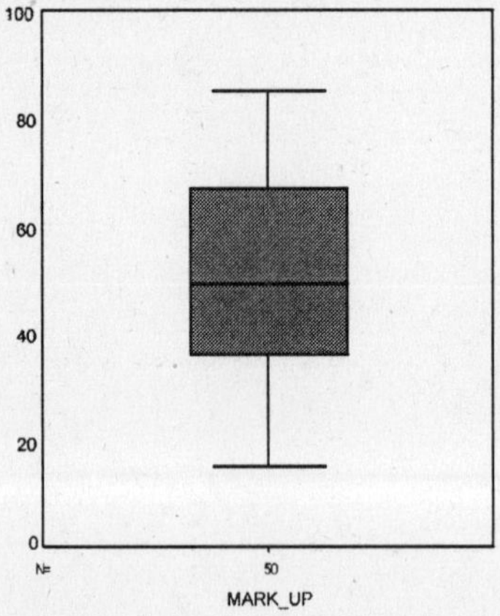

[67]

[68] C.

(a) 10% of those who took the test had scores over 680.

[69] (b) 40% of those who took the test had scores between 450 and 680.

[70] D.

(a) Mean = $1,354
(b) Trimmed mean = $1,187
(c) The trimmed mean would be better.
[71] (d) The usual mean would be better.

(a) $CV = 11.9\%$
[72] (b) Interval from 32 minutes to 52 minutes

(a) $CV = 16.22\%$
[73] (b) Interval from 25 applicants to 49 applicants

[74] A.

(a) 25% of the people kept their cars fewer than 3.6 years.
(b) 50% of the people kept their cars from 3.6 years to 10.5 years.
[75] (c) 10% of the people kept their cars over 15.1 years.

[76] The median number of travelers who do not show up was noticeably fewer after the policy was put into effect. It appears that the policy discourages no-shows.

[77] B.

(a) 35% of the population scored above Roger's score.
[78] (b) 65% of the population scored below Roger's score.

(a) Martha did better on content.
[79] (b) Sharon scored better compared to the other people who took the same test.

[80] A.

[81] Premiums are less on the whole for rural drivers. There is also less variation in the premiums charged to rural drivers.

Section B: Algorithms

Objective 3.1: Measures of Central Tendency: Mode, Median, and Mean

[82] 57

[83] The mean is less than the median.

[84] Mode

[85] 11

[86] 3.44

[87] 16

[88] $\bar{x} = 8.6,\ \bar{t} = 8.1$

[89] 115 and 23 are outliers; mean with outliers = 72.5; mean without outliers = 73.375

Objective 3.2: Measures of Variation

[90] 26

[91] 36.00

[92] 104.0

[93] 6.32

[94] 8.79

[95] 10.2

[96] $s = 12.00,\ CV = 0.32$

[97] At least 93.8% of the students volunteered from 18.8 to 34.8 hours each semester.

[98] 447 to 563 responses per ad

[99] 9

[100] At least 88.9% of the students volunteered from 26.8 to 41.2 hours each semester.

[101] $s = 11.00,\ CV = 0.26$

[102] $s = 4.8$

Objective 3.3: Percentiles and Box-and-Whisker Plots

[103] 78.08%

[104] 78

[105] 62; She scored better than 62% of the students.

[106] Answers may vary. Sample answer: about $\frac{3}{4}$ scored higher; about $\frac{1}{4}$ scored lower

[107] Yes

[108] 32, 16, 16

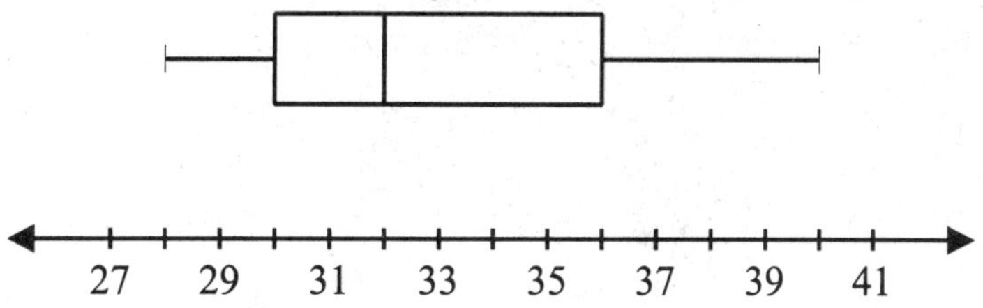

[109] first quartile: 30, second quartile: 32, third quartile: 36

[110] 81

[111] 30

Chapter 4
Regression and Correlation

Section A: Static Items

1. In a random sample of eight military contracts involving cost overruns, the following information was obtained. x = bid price of the contract (in millions of dollars) and y = cost of overrun (expressed as a percent of the bid price).

x	6	10	3	5	9	18	16	21
y	31	25	39	35	29	12	17	8

(a) Draw a scatter diagram for the data.
(b) Find the slope, b, and the intercept, a, for the least-squares line. Write the equation of the least-squares line.
(c) Graph the least-squares line on your scatter diagram.
(d) Find the standard error of estimate, S_e.
(e) If an overrun contract was bid at 12 million dollars, what does the least-squares line predict for the cost of overrun (as a percent of bid price)?

2. The central office of a large manufacturing company manages ten similar plants in different locations. The general manager is interested in the relation between profits and percent of operating capacity being used at each plant. In the data below,

x = percent of operating capacity in use and y = profits (in hundreds of thousands of dollars).

x	50	61	77	80	82	85	88	81	95	99
y	24	26	35	31	35	34	39	40	38	42

(a) Draw a scatter plot for the data.
(b) Find the slope, b, and the intercept, a, for the least-squares line. Write the equation of the least-squares line.
(c) Graph the least-squares line on your scatter diagram.
(d) Find the standard error of estimate, S_e.
(e) If the plant is working at 75% of its operating capacity, what does the least-squares line predict for the profits from that plant?

3. A government economist obtained the following data from a simple random sample of ten private companies in the natural gas exploration business. In the table below, x is the capital investment the company puts into exploration and development (in millions of dollars) and y is the volume of natural gas discovered (adjusted to sea level pressure and in billions of cubic feet).

x	5.8	8.3	1.7	7.7	5.6	2.4	15.8	10.1	12.4	7.9
y	9.1	17.4	4.7	12.3	3.8	4.6	18.1	19.3	16.9	12.6

(a) Draw a scatter diagram for this data.
(b) Find the slope, b, and the intercept, a, for the least-squares line. Write the equation of the least-squares line.
(c) Graph the least-squares line on your scatter diagram.
(d) Find the standard error of estimate, S_e.
(e) If a company had invested 6.5 million dollars, what would the least-squares line predict for the volume of natural gas discovered?

4. The United States Navy did a study of its recruiting efforts one year. They wanted to examine the relationship between the number of recruiting offices located in a particular city and the number of people recruited in that city. Recruiting data from 11 cities is given in the table below. x = number of recruiting offices. y = number of recruits.

x	2	3	1	4	9	8	5	7	12	15	10
y	60	100	20	80	120	180	100	140	220	280	240

(a) Draw a scatter diagram for the data.
(b) Find the slope, b, and the intercept, a, for the least-squares line. Write the equation of the least-squares line.
(c) Graph the least-squares line on your scatter diagram.
(d) Find the standard error of estimate, S_e.
(e) If a city has six recruiting offices, what would you predict to be the number of recruits enlisted?

5. A college counselor claims that there is a negative correlation between x = the number of days a student is absent from class and y = the student's score on a midterm exam. A random sample of 30 French students provided data pairs (x, y). The sample correlation coefficient was $r = -0.67$. Assuming that x and y are normally distributed, is the counselors claim justified at the 1% significance level?

6. A certain psychologist counsels people who are getting divorced. A random sample of six of her clients provided the following data. In the table below, x = number of years of courtship before marriage and y = number of years of marriage before divorce.

x	3	0.5	2	1.5	5	1
y	9	6	14	10	20	6.5

(a) Draw a scatter diagram for the data.
(b) From the scatter diagram would you say that r is closer to 1, 0, or –1?
(c) Find r.

7. Some children experience "sugar highs," a form of hyperactivity caused by a large influx of sugar into the blood stream. In an effort to study this phenomenon, Dr. Adams gave different amounts of sugar to a random sample of fourth-grade children. He then measured their brain wave activity. In the table below, x is the blood sugar level in milligrams per milliliter of blood and y is the brain wave activity in microvolts.

x	5.2	6.5	9.5	11.0	5.0	5.8	10.7	8.2	7.7
y	9.1	15.1	32.3	38.7	7.8	11.7	36.5	24.4	22.0

(a) Draw a scatter diagram for the data.
(b) From the scatter diagram would you say that r is closer to 1, 0, or –1?
(c) Find r.

8. One factor in determining the shelf life of a dry cereal is the moisture content. Since standard packaging techniques do not completely prevent cereal from absorbing moisture, the longer it stays on the shelf, the more moisture it absorbs. A new cereal called Tops is being tested. In the data that follow, x is the time on the shelf in days and y is the moisture content measured on a scale from 1 to 10.

x	0	6	10	16	24	30	37
y	2.9	3.2	3.5	3.6	3.9	4.2	4.6

(a) Draw a scatter diagram.
(b) Find the equation of the least-squares line.
(c) Graph the least-squares line on your scatter diagram.
(d) At 30 days on the shelf, what is the moisture content of Tops predicted by the least-squares line?
(e) Find the standard error of estimate, S_e.

9. One factor in determining the shelf life of a dry cereal is the moisture content. Since standard packaging techniques do not completely prevent cereal from absorbing moisture, the longer it stays on the shelf, the more moisture it absorbs. A new cereal called Tops is being tested. In the data that follow, x is the time on the shelf in days and y is the moisture content measured on a scale from 1 to 10.

x	0	6	10	16	24	30	37
y	2.9	3.2	3.5	3.6	3.9	4.2	4.6

(a) Find the value of the correlation coefficient, r.
(b) Use the 1% significance level to test for a positive population correlation between the time on the shelf and the moisture content.
(c) Find the coefficient of determination, r^2.
(d) What percent of the variation in moisture content is explained by the least squares line and the variation in time on the shelf?

10. Merchants are always concerned when temperatures are extremely cold because they claim that they have fewer shoppers the colder it gets. At Paradise Mall the merchant association did a study. The results are contained in the table below. x is the noonday temperature in °F and y is the number of customers visiting the mall (in hundreds) that day.

x	0	10	20	30	40
y	2	3	10	15	17

(a) Draw a scatter diagram.
(b) Find the equation of the least-squares line.
(c) Graph the least-squares line on your scatter diagram.
(d) Find the standard error of estimate, S_e.
(e) Use the least-squares line to predict the number of shoppers if the noonday temperature is 15°F.

11. Dr. Young is an economist. She claims that the prime interest rate and the unemployment rate are positively correlated. To support her claim, she took a random sample of six data pairs, (x, y) where x = prime interest rate and y = unemployment rate. The sample correlation coefficient turned out to be $r = 0.73$. Assume that both x and y are normally distributed. Determine if r is significant at the 1% significance level.

12. Merchants are always concerned when temperatures are extremely cold because they claim that they have fewer shoppers the colder it gets. At Paradise Mall, the merchant association did a study. The results are contained in the table below. x is the noonday temperature in °F and y is the number of customers visiting the mall (in hundreds) that day.

x	0	10	20	30	40
y	2	3	10	15	17

(a) Find the correlation coefficient, r.
(b) Use a 5% significance level to test the claim that there is a positive population correlation between the noonday temperature and the number of customers.
(c) Find the coefficient of determination, r^2.
(d) What percent of the variation in the number of customers is explained by the variation in temperature using the least-squares line?

13. Joanne is the general manager of a new large office complex with eight new photocopying machines. Over the past year she obtained the following data. x = the number of copies made by a machine (rounded to the nearest thousand) and y = total maintenance cost of this machine (rounded to the nearest dollar).

x	22	18	27	29	32	25	20	10
y	38	32	45	49	54	42	35	19

(a) Draw a scatter diagram.
(b) Find the equation of the least-squares line.
(c) Graph the least-squares line on your scatter diagram.
(d) Find the standard error of estimate, S_e.
(e) Predict the maintenance cost of a machine that has made 28 (thousand) copies.

14. Joanne is the general manager of a new large office complex with eight new photocopying machines. Over the past year she obtained the following data. x = the number of copies made by a machine (rounded to the nearest thousand) and y = total maintenance cost of this machine (rounded to the nearest dollar).

x	22	18	27	29	32	25	20	10
y	38	32	45	49	54	42	35	19

(a) Find the value of the correlation coefficient, r.
(b) Use a 1% significance level to test for a positive population correlation between the number of copies made and the maintenance cost.
(c) Find the coefficient of determination, r^2.
(d) What percent of the variation in cost is explained by the variation in machine use using the least-squares line?

15. John is the manager of the local Wall Red Variety store. He has taken a random sample of seven business days from the last several months to get the data in the table below. x = total number of people in the store (in hundreds) and y = total days receipts (in thousands of dollars).

x	5.8	6.1	3.2	2.4	7.5	8.1	4.4
y	12.6	13.4	7.6	5.8	15.9	17.0	9.7

(a) Draw a scatter diagram.
(b) Find the equation of the least-squares line.
(c) Graph the least-squares line on your scatter diagram.
(d) Find the standard error of estimate, S_e.
(e) On a day when Wall Red has 5 (hundred) customers in the store, what does the least-squares line predict for the total sales?

16. John is the manager of the local Wall Red Variety store. He has taken a random sample of seven business days from the last several months to get the data in the table below. x = total number of people in the store (in hundreds) and y = total day's receipts (in thousands of dollars).

x	5.8	6.1	3.2	2.4	7.5	8.1	4.4
y	12.6	13.4	7.6	5.8	15.9	17.0	9.7

(a) Find the value of the correlation coefficient, r.
(b) Use the 5% significance level to test the claim that there is a positive population correlation between the number of shoppers and the day's receipts.
(c) Find the coefficient of determination, r^2.
(d) What percent of the total variation in receipts is due to the least-squares line and the variation in the number of shoppers?

17. Urban travel times and distances are important factors in the analysis of traffic flow patterns. A traffic engineer in Los Angeles obtained the following data from area freeways. x = miles traveled and y = time in minutes to travel x miles in an automobile.

x	5	9	3	11	20	15	12	25
y	9	13	6	16	28	21	16	31

(a) Draw a scatter diagram.
(b) Find the equation of the least-squares line.
(c) Use the least-squares line to predict the time it will take to travel 22 miles.

18. Urban travel times and distances are important factors in the analysis of traffic flow patterns. A traffic engineer in Los Angeles obtained the following data from area freeways. x = miles traveled and y = time in minutes to travel x miles in an automobile.

x	5	9	3	11	20	15	12	25
y	9	13	6	16	28	21	16	31

(a) Find the value of the correlation coefficient, r.
(b) Use the 5% significance level to test for a positive population correlation between the distance traveled and the travel time.
(c) Find the coefficient of determination.
(d) What percent of the variation in travel time is explained by the least-squares line and the variation in the distance traveled?

19. For a sociology class, Margaret is studying the cost of basic necessities for families of various sizes. She determined market baskets of food for families of various sizes, then priced the market baskets at a randomly chosen list of supermarkets. The results are given in the table below. x represents the size of the family. y represents the cost of the market basket in dollars.

x	1	2	3	4	5	6	8	10
y	43	65	94	117	152	179	210	265

(a) Draw a scatter diagram.
(b) Find the equation of the least-squares line.
(c) Graph the least-squares line on your scatter diagram.
(d) Find the standard error of estimate, S_e.
(e) Predict the cost of food for a family of 7 people.

20. For a sociology class, Margaret is studying the cost of basic necessities for families of various sizes. She determined market baskets of food for families of various sizes, then priced the market baskets at a randomly chosen list of supermarkets. The results are given in the table below. x represents the size of the family. y represents the cost of the market basket in dollars.

x	1	2	3	4	5	6	8	10
y	43	65	94	117	152	179	210	265

(a) Find the sample correlation coefficient, r.
(b) Use the 5% significance level to test for a positive correlation between the size of a family and the cost for food.

21. An ecology class did an experiment to determine the distance a car will travel on one gallon of gasoline at different speeds. A random sample of 12 cars of about the same age, model and condition were filled with gasoline and driven around an oval track at constant speed for a fixed distance. The gasoline mileage was then calculated. The results are summarized below. x represents the speed of the car in miles per hour. y represents the gasoline mileage in miles per gallon.

x	30	35	40	45	50	55	60	65	70	75	80	85
y	39	36	37	35	35	31	33	30	26	27	25	24

(a) Draw a scatter diagram for this data.
(b) Find the equation of the least-squares line.
(c) Graph the least-squares line on your scatter diagram.
(d) Predict the gasoline mileage when the speed is 58 mph.
(e) Find the standard error of estimate, S_e.

22. An ecology class did an experiment to determine the distance a car will travel on one gallon of gasoline at different speeds. A random sample of 12 cars of about the same age, model and condition were filled with gasoline and driven around an oval track at constant speed for a fixed distance. The gasoline mileage was then calculated. The results are summarized below. x represents the speed of the car in miles per hour. y represents the gasoline mileage in miles per gallon.

x	30	35	40	45	50	55	60	65	70	75	80	85
y	39	36	37	35	35	31	33	30	26	27	25	24

(a) Find the correlation coefficient, r.
(b) Use the 5% significance level to test for a negative population correlation between gasoline mileage and vehicle speed.

23. Statistical Abstracts (117th edition) lists the yield and the price of a bushel of soybeans in leading soybean producing states. A random sample of ten states produced the data below. x = yield (Bu/acre). y = price (dollars/Bu.).

x	44	41	45	32	31	38	35	37	37	38
y	6.80	6.95	6.75	7.10	7.10	6.90	7.00	6.65	6.70	6.75

(a) Draw a scatter diagram for this data.
(b) Find the correlation coefficient, r.
(c) Calculate the coefficient of determination, r^2.
(d) What does r^2 tell you about the variation in yield and the variation in price?

24. Statistical Abstracts (117th edition) lists yields of corn (Bu./acre) and price of corn (dollars/Bu.) in the 21 top corn producing states. A random sample of eight of these states produced the data in the table below. x represents corn yield. y represents price.

x	114	111	119	113	124	114	115	111
y	3.24	3.22	3.14	3.38	3.24	3.11	3.20	3.33

(a) Draw a scatter diagram for this data.
(b) Calculate the correlation coefficient, r.
(c) Use the 5% significance level to determine whether the population correlation coefficient is non-zero.

25. Some students claim that they can tell the cost of a textbook by looking at its thickness. To test this claim, they picked a random sample of eight hardbound textbooks of the same height and width. They measured the thickness and recorded the price. In the table below, x represents the thickness of the book and y represents the price.

x	1	1	2	0.5	1.3	2.7	1.5	1.5
y	32	24	40	23	35	65	42	38

(a) Draw a scatter diagram for this data.
(b) Find the equation of the least-squares line.
(c) Graph the least-squares line on your scatter plot.
(d) Predict the price of a textbook that is 2.5 inches thick.

26. Some students claim that they can tell the cost of a textbook by looking at its thickness. To test this claim they picked a random sample of eight hardbound textbooks of the same height and width. They measured the thickness and recorded the price. In the table below, x represents the thickness of the book and y represents the price.

x	1	1	2	0.5	1.3	2.7	1.5	1.5
y	32	24	40	23	35	65	42	38

(a) Find the correlation coefficient, r.
(b) Find the coefficient of determination, r^2.
(c) What fraction of the variation in the cost of a textbook is explained by the variation in thickness using the least-squares line?

27. If you have bought or sold real estate you know that there is a difference between the asking price and the selling price for real estate. A random sample of eight real estate transactions in the same suburb produced the data below. x is the asking price and y is the selling price. Units are thousands of dollars.

x	699.0	239.9	192.0	189.0	159.0	120.0	173.9	192.0
y	615.5	228.0	185.6	159.2	145.5	115.0	165.4	175.7

(a) Draw a scatter diagram for this data.
(b) Find the equation of the least-squares line.
(c) Predict the selling price of a house in this neighborhood for which the asking price is 145.0 thousand dollars.

28. If you have bought or sold real estate you know that there is usually a difference between the asking price and the selling price for real estate. A random sample of eight real estate transactions in the same neighborhood produced the data below. x is the asking price and y is the selling price. Units are thousands of dollars.

x	699.0	239.9	192.0	189.0	159.0	120.0	173.9	192.0
y	615.5	228.0	185.6	159.2	145.5	115.0	165.4	175.7

(a) Calculate the sample correlation coefficient, r, for this data.
(b) Calculate the coefficient of determination, r^2 for this data.
(c) What percent of the variation in selling price is explained by the least-squares line and the variation in asking price.
(d) What would you say about a possible linear relation between the selling price and the asking price for homes in this neighborhood?

29. A political scientist is interested in the relation between the number of registered voters in a congressional district and the number who actually voted in the last general election. A survey of 10 congressional districts chosen at random produced the data below. x = % of the residents of the district who are registered voters.
y = % of the residents of the district who voted in the last general election.

x	81.6	68.4	58.9	73.7	93.3	60.0	55.5	72.6	70.6	59.4
y	58.2	51.3	42.7	52.2	61.1	45.7	42.3	44.3	34.2	44.1

(a) Find the least-squares line for this data.
(b) If a district has 75% of its residents registered to vote and conditions are about the same as they were at the last general election, predict the percentage of residents who will actually vote in the next general election.

30. Is there a connection between the amount states spend on highways and the amount they collect in tax on motor fuels? Statistical Abstracts (117th edition) reports data for the fifty states. A random sample of ten states produced the data in the table below. x represents the amount collected in tax on motor fuels and y represents the amount spent on highways. The units are hundreds of millions of dollars.

x	4.45	1.96	0.77	2.26	5.81	0.90	11.91	3.30	1.26	5.02
y	7.80	4.15	2.82	6.98	11.60	2.58	26.51	6.13	2.66	7.98

(a) Draw a scatter diagram for this data.
(b) Find the sample correlation coefficient, r.
(c) Find the equation of the least-squares line.
(d) Graph the least-squares line on your scatter plot.
(e) Predict the highway expenditure for a state which collects 4.5 (hundred million) in motor fuels tax.

31. Is there a correlation between the average annual precipitation for a city and the average number of rainy days in the city? Statistical abstracts (117th edition) gives the information below for a random sample of ten cities. x = average annual precipitation in inches and y = mean number of rainy days.

x	33.36	36.25	44.14	35.82	32.93	55.37	14.40	33.12	61.90	63.96
y	83	114	127	126	125	109	101	108	114	122

(a) Calculate the sample correlation coefficient, r.
(b) Use the 5% significance level to test for a positive population correlation between the average annual precipitation and the mean number of rainy days.

32. Is there a correlation between the average annual precipitation for a city and the average number of rainy days in the city? Statistical abstracts (117th edition) gives the information below for a random sample of ten cities.

x	33.36	36.25	44.14	35.82	32.93	55.37	14.40	33.12	61.90	63.96
y	83	114	127	126	125	109	101	108	114	122

(a) Find the equation of the least-squares line.
(b) Predict the average number of rainy days for a city that gets an average of 40.0 inches of rainfall per year.
(c) Calculate the standard error of estimate S_e.

33. Bill is investigating the connection between average inventory and sales revenue in small specialty shops in shopping malls. A random sample of eight such stores produced the data below. x represents the inventory and y represents the sales. Units are thousands of dollars.

x	10.3	8.6	12.7	15.3	6.3	7.5	9.3	7.3
y	6.9	4.8	7.6	9.3	3.6	4.2	5.6	3.8

(a) Draw a scatter diagram of the data.
(b) Calculate the sample correlation coefficient.
(c) Calculate the coefficient of determination.
(d) What does the coefficient of determination tell you about the variation in sales revenue and the variation in average inventory?

34. Mary Sue wants to know if there is a connection between attendance at craft fairs and the number of exhibitors who have booths at the fair. For a random sample of seven local craft fairs, she chose a random day of the fair and recorded that attendance and the number of exhibitors. In the data below, x represents the number of exhibitors and y represents the attendance in hundreds of people.

x	35	55	75	95	100	135	150
y	1.2	2.1	4.2	5.4	5.8	6.2	9.5

(a) Draw a scatter diagram for this data.
(b) Find the equation of the least-squares line.
(c) Predict the attendance when there are 80 exhibitors.

35. Mary Sue wants to know if there is a connection between attendance at craft fairs and the number of exhibitors who have booths at the fair. For a random sample of seven local craft fairs, she chose a random day of the fair and recorded that attendance and the number of exhibitors. In the data below, x represents the number of exhibitors and y represents the attendance in hundreds of people.

x	35	55	75	95	100	135	150
y	1.2	2.1	4.2	5.4	5.8	6.2	9.5

(a) Calculate the sample correlation coefficient, r.
(b) Calculate the coefficient of determination, r^2.
(c) What does the coefficient of determination tell you about the variation in attendance and the variation in the number of exhibitors?

36. Statistical Abstracts (117th edition) reports the total dollar value of automobile loans and the interest rates on new car loans for a ten-year period. In the table below, x = interest rate and y = dollar value of automobile loans (in billions of dollars).

x	10.45	10.86	12.08	11.78	11.13	9.28	8.08	8.13	9.57	9.05
y	266.1	285.5	291.0	282.4	259.3	257.1	279.8	317.2	350.8	377.3

(a) Find the equation of the least-squares line.
(b) Predict the total dollar value of automobile loans for a year when the interest rate is 8.5%.
(c) Calculate the sample correlation coefficient, r.
(d) Calculate the coefficient of determination, r^2.
(e) What percent of the variation in the dollar value of automobile loans can be explained by the variation in auto loan interest rates using the least-squares line?

37. Statistical Abstracts (117th edition) reports the total dollar value of credit card debt (in billions of dollars) and credit card interest rates for a ten-year period. In the table below, x = interest rate expressed as a percent and y = total dollar value of credit card debt.

x	17.9	17.8	18.0	18.2	18.2	17.8	16.8	15.7	16.0	15.6
y	153.3	174.5	198.6	223.3	245.3	257.8	287.0	339.3	413.9	462.4

(a) Calculate the sample correlation coefficient, r.
(b) What does the sign of r tell you about the relation between credit card interest rates and credit card debt?
(c) Find the equation of the least-squares line.
(d) Predict the amount of credit card debt in a year when the credit card interest rate is 16.5%.

38. Max is a sales manager for a company that sells its products over the phone. For a random sample of 16 phone calls he records the length of time spent on the phone in minutes and the dollar value of the sale. The result is a set of data pairs (x, y) where x = length of the call and y = amount of the sale. The data had a sample correlation coefficient of 0.65. Assuming that x and y are normally distributed, determine if there is a population correlation (either positive or negative) between the length of the call and the dollar value of the sale. Use a 5% significance level.

39. Robert wants to know if there is any connection between a person's cumulative college grade point average and the person's starting salary in his or her first full time professional job. A random sample of 9 college graduates produced the data below. x represents the cumulative college GPA and y the starting salary (in thousands of dollars).

x	3.5	3.8	3.2	2.7	3.5	2.5	2.0	2.8	3.0
y	29	36	28	23	21	24	18	26	29

(a) Draw a scatter diagram for this data.
(b) Calculate the correlation coefficient, r.
(c) Would you say that the linear relationship is strong, moderate, or weak?
(d) Find the equation of the least-squares line.
(e) Predict the starting salary for a graduate with a 3.6 GPA.

40. As director of personnel for a prosperous company, you have just hired a new public relations person. The final salary arrangements are negotiated depending on the number of years experience the new public relations person brings to the company. After checking with several other companies in your area, you obtain the following data. In the table below, x = number of years of experience and y = annual salary in thousands of dollars.

x	1	2	15	11	9	6
y	33	37	61	57	45	42

(a) Draw a scatter diagram for your data.
(b) Find the equation of the least-squares line.
(c) Graph the least-squares line on your scatter diagram.
(d) Find the standard error of estimate, S_e.
(e) If your candidate for the public relations position has 8 years of experience, what would the least-squares line suggest for the salary?

41. Big Mike claims that there is a negative correlation between the number of speeding tickets and the number of CB radios being sold. Mike randomly selected 20 towns of about the same size in his general locality. For each town, he found the number of CB radios sold that year and the number of speeding tickets given that year. The result was a set of data pairs (x, y) where x = number of CB radios sold and y = number of speeding tickets issued. The sample correlation coefficient is $r = -0.40$. Assuming that x and y are normally distributed, determine if r is significant at the 1% significance level.

42. A personal financial advisor claims that a person's income and level of life insurance are positively correlated. A random sample of 24 of his clients produced a set of data pairs (x, y) where x = personal income level and y = total dollar value of life insurance for that individual. After going through the calculations, he found the sample correlation coefficient to be $r = 0.38$. Assume x and y are normally distributed. Use a 1% significance level to test for a population correlation between a person's income and level of life insurance.

43. Seven children in the third grade were given a verbal test to study the relationship between the number of words used in a response and the silence interval before the response. A psychologist asked each child a number of questions, and then recorded the total number of words used and the time that elapsed before the child began answering the questions. The following data were obtained. In the table below,
x = total silence time in seconds and y = total number of words in the response.

x	25	16	19	27	38	42	31
y	60	55	50	64	73	70	65

(a) Draw a scatter diagram for the data.
(b) Compute the correlation coefficient, r.
(c) Compute the coefficient of determination, r^2.
(d) What percent of the variation in the number of words used can be explained by the corresponding variation in silence time and the least-squares line?

44. Ms. Pins owns a garment factory which makes tailored blouses and shirts. The supervisor of the factory is afraid to let the employees work overtime as he thinks that workers make more errors toward the end of a work day when they are tired. The employees would like to work overtime and collect overtime wages. They claim that the longer they work the better they get at it. To settle the dispute, Ms. Pins randomly selected 16 20-minute periods throughout the day. For each period she calculated the percentage, y, of defective items produced and the number of hours, x, the employees had been working at the start of the time period. The sample correlation coefficient between x and y was $r = 0.48$. Ms. Pins says she will take the manager's advice if a test indicates that $\rho > 0$ at the 1% significance level. Otherwise she will let the employees work overtime. What should Ms. Pins do?

45. A government economist is studying a possible relationship between a substantial income tax increase and the possible effect on the Dow Jones stock market index. He identifies eight periods in the past decade in which large tax increases have been imposed and records x = amount of tax increase and y = Dow Jones index. This data has sample correlation coefficient –0.58. Use the 5% significance level to determine whether there is a negative population correlation.

46. A random sample of 15 married couples had their heights measured with the following results. x = man's height, y = woman's height.

x	63	73	67	69	65	75	72	77	66	69	80	79	85	74	68
y	63	68	66	67	66	73	67	68	69	66	74	72	77	62	66

(a) Draw a scatter diagram.
(b) Find the equation of the least-squares line.
(c) Use the least-squares line to predict the height of a married woman if her husband is 69 inches tall.

47. A random sample of 15 married couples had their heights measured with the following results. x = man's height, y = woman's height.

x	63	73	67	69	65	75	72	77	66	69	80	79	85	74	68
y	63	68	66	67	66	73	67	68	69	66	74	72	77	62	66

(a) Find the correlation coefficient, r.
(b) Use the 5% significance level to test for a population correlation (either positive or negative).
(c) Calculate the coefficient of determination, r^2.
(d) Discuss the meaning of r^2 in this context.

48. Rob is studying hospital administration. He found forecasts by eight analysts predicting percentages of hospital occupancy for five and ten years from now. In the table below x = percent occupancy five years from now and y = percent of occupancy ten years from now.

x	62	64	58	68	72	63	65	63
y	67	66	65	74	78	69	65	66

(a) Draw a scatter diagram for this data.
(b) Find the equation of the least-squares line.
(c) Graph the least-squares line on your scatter diagram.
(d) Predict the percentage of beds occupied in hospitals ten years from now if 70% are occupied five years from now.

49. Professor Stanislaw is directing an oceanography study in which ocean water is tested for oxygen content at different depths. A field study produced 30 data pairs (x, y) where x = sample depth and y = oxygen content. The sample correlation coefficient is $r = -0.63$. Use a 1% significance level to test for a population negative correlation between depth of sample and oxygen content.

50. You are the general manager of a car-rental service at the airport. In an effort to estimate the maintenance costs you obtain the following data from a random sample of 11 cars. x = number of miles the car has been driven (in thousands) and y = total maintenance cost in dollars.

x	3	5	12	8	3	16	12	9	2	4	10
y	90	150	320	208	98	390	275	240	75	115	265

(a) Draw a scatter diagram for your data.
(b) Find the equation of the least-squares line.
(c) Graph the least-squares line on your scatter diagram.
(d) If you have a car which has been driven 7 thousand miles, what does the least-squares line predict for the maintenance cost for the car?

51. You are the general manager of a car-rental service at the airport. In an effort to estimate the maintenance costs you obtain the following data from a random sample of 11 cars. x = number of miles the car has been driven (in thousands) and y = total maintenance cost in dollars.

x	3	5	12	8	3	16	12	9	2	4	10
y	90	150	320	208	98	390	275	240	75	115	265

(a) Calculate the sample correlation coefficient, r.
(b) Use the 1% significance level to test for a positive population correlation between the number of miles driven and maintenance cost.

52. Nancy and Pete run the Wall Red Discount Store. Nancy claims that there is a positive correlation between the length of time a customer shops in the store and the amount of money spent by the customer. Behind her one-way mirrors Nancy observed a random sample of 12 customers. For each customer, she determined x = length of time in the store and y = amount of money spent in the store. The correlation coefficient was 0.75. If x and y are normally distributed, does the data support Nancy's claim at the 5% significance level?

53. Betty is an industrial psychologist who developed a test to measure aggressiveness in salespeople. A random sample of 14 salespeople took the test. For each salesperson, the test score and last year's sales were recorded. The sample correlation coefficient was $r = -0.60$. For the type of sales work involved (funeral insurance), she does not know whether to expect a positive or negative correlation. Use a 5% significance level to test for a population correlation either way.

54. For Sherlock Holmes the "science of deduction" was the ultimate key to solving every mystery. Distance between footprints (length of stride) told the master detective the height of the criminal he was pursuing. What kind of relation is there between a person's height and length of stride. A random sample of 12 people provided the data below. x = length of stride in inches. y = height in inches.

x	15	17	19	11	16	22	17	25	12	15	25	23
y	66	67	76	62	72	78	75	87	64	68	81	85

(a) Draw a scatter diagram for the data.
(b) Find the equation of the least-squares line.
(c) Graph the least-squares line on your scatter diagram.
(d) If footprints indicate that the suspect had a stride of 20 inches, what would the least squares line predict for the height?

55. For Sherlock Holmes the "science of deduction" was the ultimate key to solving every mystery. Distance between footprints (length of stride) told the master detective the height of the criminal he was pursuing. What kind of relation is there between a person's height and length of stride. A random sample of 12 people provided the data below. x = length of stride in inches. y = height in inches.

x	15	17	19	11	16	22	17	25	12	15	25	23
y	66	67	76	62	72	78	75	87	64	68	81	85

(a) Find the sample correlation coefficient, r.
(b) Use the 1% significance level to test for a positive population correlation between a person's height and length of stride.

56. The following data is taken from the Statistical Abstract of the United States (117th edition). In this data, x = percent interest on U.S. 10-year treasury notes and y = price of a barrel of crude oil. Data for a seven-year period is given below.

x	8.6	7.9	7.0	5.9	7.7	6.6	6.4
y	21.4	17.0	16.0	13.9	12.6	13.6	16.8

(a) Draw a scatter diagram for the data.
(b) Find the equation of the least-squares line for this data.
(c) Estimate the price of a barrel of crude oil when the interest on 10-year treasury notes is 7.5%.

57. The following data is taken from the Statistical Abstract of the United States (117th edition). In this data, x = percent interest on U.S. 10-year treasury notes and y = price of a barrel of crude oil. Data for a seven-year period is given below.

x	8.6	7.9	7.0	5.9	7.7	6.6	6.4
y	21.4	17.0	16.0	13.9	12.6	13.6	16.8

(a) Find the sample correlation coefficient, r.
(b) Use the 5% significance level to test for a population correlation between x and y (either positive or negative).
(c) Find the coefficient of determination, r^2.
(d) What does r^2 tell you about the variation in the price of a barrel of crude oil and the variation in the percent interest of U.S. 10-year treasury notes?

58. A study of the aerospace industry listed the number of employees and the profits of leading aerospace companies. A random sample of seven companies produced the data below. x = number of employees (in ten thousands) and y = profits (hundreds of millions of dollars).

x	12.5	9.3	8.7	7.2	3.8	4.5	3.7
y	10.4	7.3	6.5	4.6	1.2	0.8	0.6

(a) Draw a scatter diagram for the data.
(b) Find the equation of the least-squares line.
(c) Graph the least-squares line on your scatter diagram.
(d) Estimate the annual profits of an aerospace company with 8.2 (ten thousands) employees.

59. A study of the aerospace industry listed the number of employees and the profits of leading aerospace companies. A random sample of seven companies produced the data below. x = number of employees (in ten thousands) and y = profits (hundreds of millions of dollars).

x	12.5	9.3	8.7	7.2	3.8	4.5	3.7
y	10.4	7.3	6.5	4.6	1.2	0.8	0.6

(a) Calculate the sample correlation coefficient, r.
(b) Calculate the coefficient of determination, r^2.
(c) What fraction of the variation in profits can be explained by the variation in number of employees using the least-squares line?

60. The Food and Drug Administration is examining the effect of different doses of a new drug on the pulse rate of human subjects. The results of a study on six people is given in the accompanying table. x = drug dose (mg/kg body weight) and y = drop in pulse rate/min.

x	2.50	3.00	3.50	4.50	5.50	6.00
y	8	11	9	16	19	20

(a) Draw a scatter diagram for your data.
(b) Find the equation of the least-squares line.
(c) Graph the least-squares line on your scatter diagram.
(d) Estimate the predicted drop in pulse rate for a dose of 2.75 mg/kg body weight.

61. The Food and Drug Administration is examining the effect of different doses of a new drug on the pulse rate of human subjects. The results of a study on six people is given in the accompanying table. x = drug dose (mg/kg body weight) and y = drop in pulse rate/min.

x	2.50	3.00	3.50	4.50	5.50	6.00
y	8	11	9	16	19	20

(a) Calculate the sample correlation coefficient, r.
(b) Use the 5% significance level to test for a positive population correlation between the drug dose and the drop in pulse rate.
(c) Calculate the coefficient of determination, r^2.
(d) What fraction of the variation in drop of pulse rate is explained by the variation in drug dose using the least-squares line?

62. The medical profession tells us that we should be concerned about not only the calorie content of food, but also the amount of cholesterol. In the data that follows, x = total calories and y = milligrams of cholesterol in 3.5 oz. servings of different kinds of fish.

x	184	203	216	211	151	236
y	49	77	87	64	73	161

(a) Calculate the sample correlation coefficient, r.
(b) Use the 5% significance level to test for a positive population correlation between the amount of cholesterol and the total calories.
(c) Calculate the coefficient of determination, r^2.
(d) What fraction of the variation in the amount of cholesterol can be explained by the variation in total calories using the least-squares line.

63. A medical doctor specializing in gerontology (study of aging) suspects that older people have trouble getting necessary vitamins from the food they eat and may require vitamin supplements. A random sample of eight people over 50 gave the following information. In the table below, x = age in years and y = measure of vitamin B12 in the bloodstream expressed as a percent of the amount of B12 that normally is present. No patient in the sample took vitamin pills.

x	68	57	92	63	77	86	62	81
y	60	80	55	92	64	55	84	60

(a) Compute the sample correlation coefficient, r.
(b) Compute the coefficient of determination, r^2.
(c) What percent of the variation in y can be explained by the variation in x using the least-squares line?

64. Pilots who eject from a high-velocity fighter jet receive a tremendous jerk as their parachutes open. A new type of parachute has been developed that automatically increases deceleration time, according to the air speed of the opening chute. A random sample of 10 pilots tested the new parachute by ejecting at randomly selected speeds over 300 miles per hour. Their ejection speed and deceleration time were measured. In the table below, x = ejection speed in miles per hour. y = deceleration time in seconds.

x	450	510	545	345	395	600	315	410	360	480
y	9	15	17	8	10	19	7	9	5	16

(a) Calculate the sample correlation coefficient, r.
(b) Calculate the coefficient of determination, r^2.
(c) What percent of the variation in y can be explained by the corresponding variation in x using the least-squares line?
(d) What percent of the variation is unexplained?

65. Write a brief but complete essay in which you answer the following questions:
(a) What is the coefficient of determination? How can we interpret the meaning of the coefficient of determination?
(b) What is the least-squares criterion? Why do we say that the least-squares line is the "best fitting" line to a given set of (x, y) data values?
(c) What is a response variable? What are explanatory variables? In the case of multiple regression, what do we mean by a regression model?
(d) What is the standard error of estimate? How is it used to construct confidence intervals for the least-squares prediction? Do you think that a large standard error of estimate would result in a more "accurate" prediction than a smaller standard error of estimate would? Explain.
(e) Where is the student's t distribution used in simple regression? Where is it used in multiple regression?
(f) List at least three areas of modern everyday life where you think regression could be used. Describe how it would be used and list the variables in each example.

66. A college professor wants to see if there is a correlation between the variables x = the number of days a student is absent from class and y = the student's percentage score on a midterm exam. The records of seven students provided the following values for x and y.

x	2	3	0	1	2	1	4
y	83	74	89	62	79	95	65

(a) Draw a scatter diagram for this data.

(b) Find $\bar{x}$, $\bar{y}$, and b. Then find the equation of the least-squares line.

(c) Graph the least-squares line on your scatter diagram. Be sure to use the point $(\bar{x}, \bar{y})$ as one of the points on the line.

(d) Find the standard error of estimate S_e.

(e) What would the least-squares line give as prediction for the score of a student absent for 3 days prior to this midterm exam to the nearest percentage point?

67. Some students think that you can guess the price of a textbook by its weight. To test to see if a correlation exists, ten textbooks were chosen at random. The table below shows the corresponding weights and prices. The variables are x = weight of a textbook in pounds, and y = the price of the textbook to the nearest dollar.

x	10	5	4	15	9	8	8	5	6.5	4
y	65	37	26	73	68	62	59	43	39	33

(a) Draw a scatter diagram for this data.

(b) Find $\bar{x}$, $\bar{y}$, and b. Then find the equation of the least-squares line.

(c) Graph the least-squares line on your scatter diagram. Be sure to use the point $(\bar{x}, \bar{y})$ as one of the points on the line.

(d) Find the standard error of estimate S_e.

(e) What would the least-squares line give as prediction for the price of a textbook that weighs 7 pounds to the nearest cent?

68. Choose the best answer from among choices A.–E. below. For a set of $n = 8$ data points, assume it is known that $\bar{x} = 11.875$, $\bar{y} = 33.75$, $SS_x = 58.875$, $SS_y = 35.5$, and $SS_{xy} = 42.75$. Find the least-squares line for this data.

A. $y = 0.6030x + 26.5897$ B. $y = 0.8304x + 23.8889$ C. $y = 0.7261x + 25.1274$

D. $y = 1.3772x + 17.3958$ E. $y = 1.2042x + 19.4498$

69. Choose the best answer from among choices A.–E. below. For a set of $n = 8$ data points, assume it is known that $\sum x = 96$, $\sum y = 268$, $\sum x^2 = 1{,}212$, $\sum y^2 = 9{,}024$, and $\sum xy = 3{,}267$. Find the least-squares line for this data.

A. $y = 1.18x + 19.34$ B. $y = 1.11x + 20.18$ C. $y = 0.9x + 22.7$

D. $y = 0.85x + 23.3$ E. $y = 0.77x + 24.26$

70. Choose the best answer from among choices A.–E. below. For a set of $n = 8$ data points, assume it is known that $\sum x = 97$, $\sum y = 270$, $\sum x^2 = 1{,}231$, $\sum y^2 = 9{,}168$, and $\sum xy = 3{,}328$. Find the standard error of estimate S_e.

A. 0.5103 B. 0.5580 C. 0.4832 D. 0.5834 E. 0.4534

71. Choose the best answer from among choices A.–E. below. For a set of $n = 8$ data points, assume it is known that $\bar{x} = 12.875$, $\bar{y} = 34.25$, $SS_x = 64.875$, $SS_y = 55.5$, and $SS_{xy} = 55.25$. Find the standard error of estimate S_e.

A. 1.7748 B. 1.4532 C. 1.0276 D. 1.1865 E. 1.0180

72. A college professor wants to see if there is a correlation between the variables x = the number of days a student is absent from class and y = the student's percentage score on a midterm exam. The records of seven students provided the following values for x and y.

x	2	3	0	1	2	1	4
y	83	74	89	62	79	95	65

(a) Compute coefficient of the correlation r.
(b) Compute the coefficient of determination r^2.
(c) What percent of the variation in y can be explained by the corresponding variation in x using the least-squares line? What percentage of the variation is unexplained?

73. Some students think that you can guess the price of a textbook by its weight. To see if a correlation exists, ten textbooks were chosen at random. The table below shows the corresponding weights and prices. The variables are x = weight of a textbook in pounds, and y = the price of the textbook to the nearest dollar.

x	10	5	4	15	9	8	8	5	6.5	4
y	65	37	26	73	68	62	59	43	39	33

(a) Compute coefficient of the correlation r.
(b) Compute the coefficient of determination r^2.
(c) What percent of the variation in y can be explained by the corresponding variation in x using the least-squares line? What percentage of the variation is unexplained?

74. Choose the best answer from among choices A.–E. below. A police department wants to find out if a correlation exists between x = number of muggings per night during warmer months of the year in a particular park not far from downtown and y = degrees Fahrenheit. This is to help decide how police duties should be allocated on these nights. Some data points for some nights from the spring and summer of the past year are listed below.

x	5	7	8	4	11	12	8	7
y	79	82	83	81	86	89	91	84

Compute the coefficient of correlation r. Compute the coefficient of determination r^2. What percentage of the variation in y can be explained by the corresponding variation in x using the least-squares line? What percentage of the variation is unexplained?

A. $r = 0.7132$; $r^2 = 0.5087$; 51% explained; 49% unexplained

B. $r = 0.7217$; $r^2 = 0.5208$; 52% explained; 48% unexplained

C. $r = 0.7877$; $r^2 = 0.6204$; 62% explained; 38% unexplained

D. $r = 0.6911$; $r^2 = 0.4776$; 48% explained; 52% unexplained

E. $r = 0.5111$; $r^2 = 0.2612$; 26% explained; 74% unexplained

75. Choose the best answer from among choices A.–E. below. Let x be the list price (in thousands of dollars) for a random selection of different models of automobiles on a new car lot. Let y be the dealer invoice (in thousands of dollars) for the given vehicle.

x	15.8	16.5	19.2	20.7	21.7	22.0	23.2	24.7
y	14.6	15.1	16.7	18.9	19.8	21.1	22.1	22.5

Compute the coefficient of correlation r. Compute the coefficient of determination r^2. What percentage the variation in y can be explained by the corresponding variation in x using the least-squares line? What percentage of the variation is unexplained?

A. $r = 0.9662$; $r^2 = 0.9336$; 93% explained; 7% unexplained

B. $r = 0.9835$; $r^2 = 0.9674$; 97% explained; 3% unexplained

C. $r = 0.9686$; $r^2 = 0.9381$; 94% explained; 6% unexplained

D. $r = 0.9596$; $r^2 = 0.9208$; 92% explained; 8% unexplained

E. $r = 0.9804$; $r^2 = 0.9612$; 96% explained; 4% unexplained

76. Choose the best answer from among choices A.–E. below. Let x be the age of a Shetland pony (in months) and let y be the average weight of the pony (in kilograms). Consider the data points in the table below.

x	3	6	9	12	15	18	21	24
y	60	95	120	140	160	170	177	185

Compute the coefficient of correlation r. Compute the coefficient of determination r^2.What percentage of the variation in y can be explained by the corresponding variation in x using the least-squares line? What percentage of the variation is unexplained?

A. $r = 0.9738$; $r^2 = 0.9483$; 95% explained; 5% unexplained

B. $r = 0.9554$; $r^2 = 0.9128$; 91% explained; 9% unexplained

C. $r = 0.9835$; $r^2 = 0.9673$; 97% explained; 3% unexplained

D. $r = 0.9604$; $r^2 = 0.9224$; 92% explained; 8% unexplained

E. $r = 0.9662$; $r^2 = 0.9336$; 93% explained; 7% unexplained

77. Choose the best answer from among choices A.–E. below. Let x be the magnitude of an earthquake (on the Richter scale), and let y be the depth (in kilometers) of the quake below the surface at the epicenter. The following is based on information taken from the National Earthquake Information Service of the U.S. Geological survey.

x	2.9	4.2	3.3	4.5	2.6	3.2	3.4
y	5.0	10.0	11.2	10.0	7.9	3.9	5.5

Compute the coefficient of correlation r. Compute the coefficient of determination r^2. What percentage of the variation in y can be explained by the corresponding variation in x using the least-squares line? What percentage of the variation is unexplained?

A. $r = 0.4713$; $r^2 = 0.2221$; 22% explained; 78% unexplained

B. $r = 0.5111$; $r^2 = 0.2612$; 26% explained; 74% unexplained

C. $r = 0.6061$; $r^2 = 0.3674$; 37% explained; 63% unexplained

D. $r = 0.4782$; $r^2 = 0.2287$; 23% explained; 77% unexplained

E. $r = 0.4886$; $r^2 = 0.2387$; 24% explained; 76% unexplained

78. Choose the best answer from among choices A.–E. below. Let x be the body weight of a child (in kilograms), and let y be the metabolic rate of the child (100 kcal/24 h).

x	3	5	9	11	15	17	19	21
y	1.4	2.7	5.0	6.0	7.1	7.8	8.3	8.8

Compute the coefficient of correlation r. Compute the coefficient of determination r^2. What percentage of the variation in y can be explained by the corresponding variation in x using the least-squares line? What percentage of the variation is unexplained?

A. $r = 0.9840$; $r^2 = 0.9682$; 97% explained; 3% unexplained

B. $r = 0.0961$; $r^2 = 0.9237$; 92% explained; 8% unexplained

C. $r = 0.9785$; $r^2 = 0.9574$; 96% explained; 4% unexplained

D. $r = 0.9662$; $r^2 = 0.9336$; 93% explained; 7% unexplained

E. $r = 0.9268$; $r^2 = 0.8589$; 86% explained; 14% unexplained

79. Choose the best answer from among choices A.–E. below. Let x be the average number of employees in a group health insurance plan, and let y be the average administrative cost as a percentage of claims.

x	3	7	15	35	40	50	75	85
y	40	35	30	25	23	21	18	15

Compute the coefficient of correlation r. Compute the coefficient of determination r^2. What percentage of the variation in y can be explained by the corresponding variation in x using the least-squares line? What percentage of the variation is unexplained?

A. $r = -0.9182$; $r^2 = 0.8431$; 84% explained; 16% unexplained

B. $r = -0.9549$; $r^2 = 0.9119$; 91% explained; 9% unexplained

C. $r = -0.9711$; $r^2 = 0.9430$; 94% explained; 6% unexplained

D. $r = -0.9874$; $r^2 = 0.9750$; 98% explained; 2% unexplained

E. $r = -0.9262$; $r^2 = 0.8578$; 86% explained; 14% unexplained

Section B: Algorithms

Objective 4.1: Introduction to Paired Data and Scattered Diagrams

80. The scatter plot shows how the length of the average frost-free period in 15 U.S. cities varies by latitude.

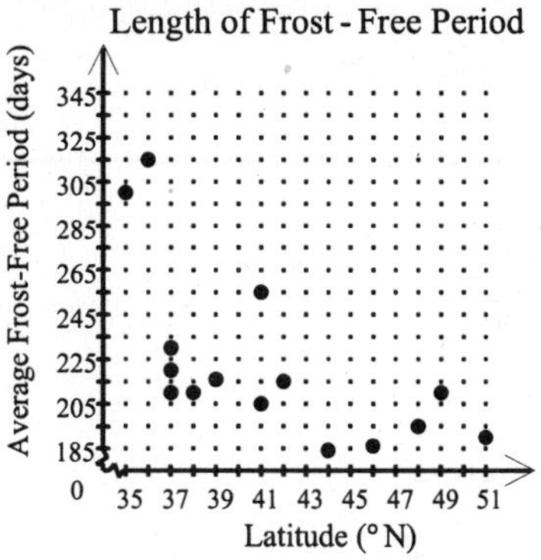

The latitude of a particular city is 48° N. About how many frost-free days does this city have in a year?

81. Describe the relationship of size to time in the scatter plot below.

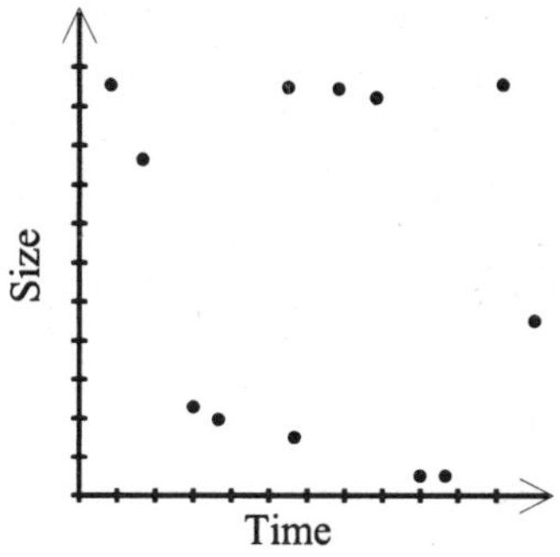

82. Sketch a size versus time scatter plot that shows a moderate positive relationship.

83. Use the table to tell whether there is a *positive correlation*, a *negative correlation*, or *no correlation* between girth and weight.

Fish	Weight	Length (in.)	Girth (in.)
Redear Sunfish	1 lb 15.5 oz	13	12.75
Small Mouth Bass	7 lb 9 oz	22	18.25
Warmouth	1 lb 14.2 oz	11	12.75
Green Sunfish	11 oz	10	9.25
Walleye	19 lb 15.3 oz	37	21.5

84. Examine the scatterplots below. The axes all have the same scale.

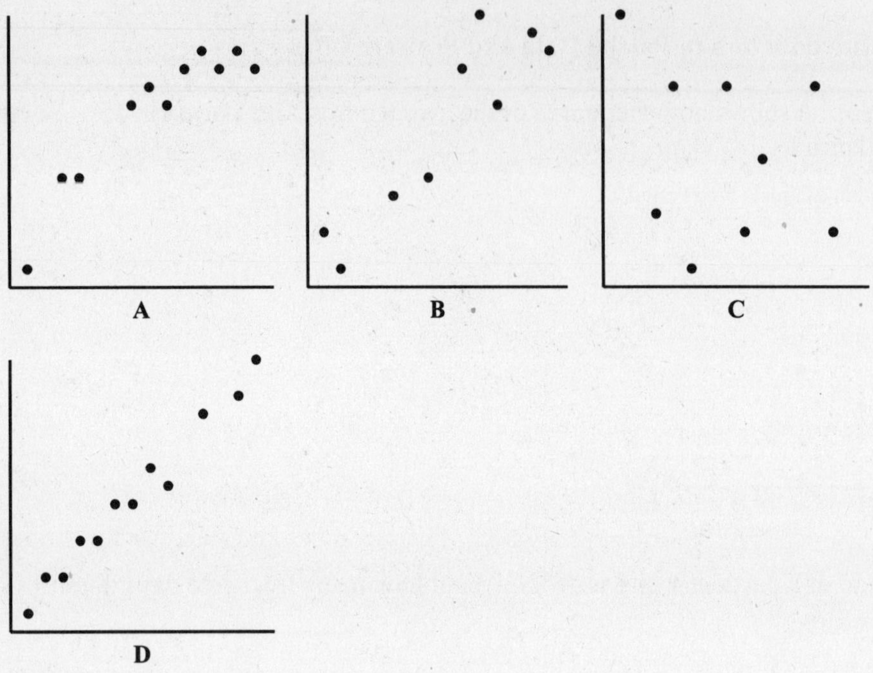

a. Which plot shows the weakest correlation between its two variables? Explain.

Plot: ______

b. Do any of the plots illustrate a negative correlation? If so, which plot? Explain.

Yes ______ *No* ______ *Plot:* ______

c. Which plot shows the strongest correlation? Explain.

Plot: ______

85. A technician at the local oil change shop kept track of the number of cars that received an oil change during the day, and the time of day the cars were serviced. The data is displayed below.

oil changes	3	3	4	4	7	6	7
time of day	10 a.m.	11 a.m.	Noon	1 p.m.	2 p.m.	3 p.m.	4 p.m.

Display the data on a scatter plot, and determine whether the correlation of the data can be represented by a linear model. If there is a linear correlation, is there a positive or negative correlation?

Objective 4.2: Linear Regression and Confidence Bounds for Prediction

86. The number of cooks at a large restaurant varies due to absenteeism on any given night. In a random sample of meals prepared over several nights, the following data was obtained, where x = number of cooks absent and y = number of incorrect meals prepared.

x	1	3	7	0	6
y	1	4	9	1	7

a. Draw a scatter plot for the data.
b. Find $\bar{x}$, $\bar{y}$, and b. Then find the equation of least-squares line.

87. Mr. Smiley, a used car dealer, uses the "Red Book" to help him determine the value of used cars that his customers trade in. Although the Red Book provides values of each car model according to it's features and current condition, it does not indicate the value determined by the odometer reading. Mr. Smiley knows that the odometer reading is a critical factor for used-car buyers. To examine this issue further, he randomly selects 10 three-year old cars of the same make and model that were sold during the past three months. He wants to find the regression line based on the price and the number of miles on the odometer. Using the information below, determine SS_x, SS_{xy}, SS_y, and b.

Car	Odometer Reading	Selling Price
1	44,400	4970
2	42,400	5630
3	40,400	5680
4	40,100	5390
5	43,400	5180
6	42,200	5550
7	38,500	5490
8	40,700	5640
9	40,300	5580
10	41,500	5200

88. Find the least squares regression equation for the data in the table below:

x	14	17	19	23	27	29	33	37	42	45	48	50	53	58	61
y	46	46	47	48	48	49	50	50	51	52	53	53	54	55	55

89. It is thought that the more time a student spends watching TV the night before an exam, the less likely he or she is to receive a good grade. Let x be the number of hours of TV watched before the day of an exam by six different students. Let y be the percentage they received on the exam. Graph the least-squares line. Plot $(\bar{x}, \bar{y})$ as one of the points on the line.

Student	Hours (x)	Score (y)
1	0	93
2	5	53
3	4	54
4	6	40
5	1	82
6	2	74

90. Using the $b = 0.0674003$, the tabulated data, and the equation $a = \bar{y} - b\bar{x}$, find a.

x	11	12	17	21	22	25	30	35	40	42
y	42	17	32	20	25	12	35	37	41	16

91. Using $b = 0.665052$, the tabulated data, and the equation $a = \bar{y} - b\bar{x}$, find a.

x	7	11	12	13	14	16	17	19	20	21	24	25
y	15	28	20	36	18	16	30	32	42	35	40	13

92. The table below shows the population growth (in thousands) for a country over a given period of time. From the data below, find the least squares line and forecast the population for the year 2010.

x	1937	1941	1946	1951	1952	1954	1955	1958	1963	1965
y	12	20	33	42	51	63	76	86	93	106

93. A real estate broker takes a random sample of eight homes in the same residential area as a house that she is listing. Use the data she collected to obtain a point estimate for the mean price of all 2600-square-foot homes in that area.

Size (100 sq ft)	Price ($1000)
38	365
23	248
22	227
25	238
21	210
31	341
20	176
30	270

94. An admissions counselor at a college is screening candidates that are applying for early admission as juniors based on PSAT scores. He randomly selects nine students attending the college and compares their PSAT and SAT scores. Obtain a point estimate for the mean SAT score of juniors with a score of 119 on their PSAT.

PSAT	SAT
112	1142
126	1241
117	1171
110	1113
132	1252
116	1121
121	1161
128	1147
129	1227

95. Given the linear regression equation $x_1 = 2.4 + 2.6x_2 + 2.7x_3 + 2.0x_4$, list the coefficients with their corresponding explanatory variables.

96. The number of cooks at a large restaurant varies due to absenteeism on any given night. In a random sample of meals prepared over several nights, the following data was obtained, where x = number of cooks absent and y = number of incorrect meals prepared.

x	8	15	11	13	14
y	10	16	13	15	14

a. Find a scatter plot for the data.
b. Find $\bar{x}$, $\bar{y}$, and b. Then find the equation of least-squares line.

97. Using $b = 0.0111398$, the tabulated data, and the equation $a = \bar{y} - b\bar{x}$, find a.

x	13	17	21	25	27	30	35	40	43	46
y	52	73	53	42	60	47	45	43	62	68

98. Using $b = -0.693169$, the tabulated data, and the equation $a = \bar{y} - b\bar{x}$, find a.

x	9	13	15	19	22	23
y	31	42	25	24	20	31

99. An admissions counselor at a college is screening candidates that are applying for early admission as juniors based on PSAT scores. He randomly selects ten students attending the college and compares their PSAT and SAT scores. Obtain a point estimate for the mean SAT score of juniors with a score of 124 on their PSAT.

PSAT	SAT
153	1148
131	1078
136	1063
139	1056
143	1150
154	1120
135	1115
150	1065
161	1217
164	1094

100. Given the linear regression equation $x_1 = 7.0 + 4.6x_2 + 5.2x_3 + 7.8x_4$, list the coefficients with their corresponding explanatory variables. If $x_2 = x_3 = x_4$, interpret the meaning of the coefficient of the new independent variable.

101. Some students decided to take a college-entrance exam after first using some practice software for the test. The table below shows the time spent practicing and the percent of gain compared to the first test score.

Practice Time (hr)	5	6	8	10	12	15	20
Gain (%)	8.0	9.3	10.1	11.6	13.2	15.0	16.9

Find an equation of the line of best fit for the data. Use x to represent practice time and y to represent percent of gain. Round coefficients to the nearest hundredth.

102. The table shows the relationship between the time a student spends working out each week and his percentage improvement on speed times. On graph paper, make a scatter plot of percentage improvement vs. work-out hours and then draw the line of best fit. Using the line of best fit, estimate how many work-out hours would be expected if the percentage gain is 18%.

Work-out Hours	6	8	10	12	14	16	18
Percentage Improvement	20	17	21	31	34	35	42

103. The table shows the number of times a cricket chirped in 1 minute and the temperature at the time of the chirps.

Temperature (°C)	18	8	15	14	10	7	14	6	14	9	14	13	15
Number of chirps	20	17	18	18	16	15	17	15	16	15	17	16	17

A linear model that approximates this data is
$N = 0.313t + 12.9$
where N represents the number of chirps and t represents the temperature. Plot the actual data and the model on the same graph. How closely does the model represent the data?

Objective 4.3: The Linear Correlation Coefficient

104. What conclusion can be made when r, the correlation coefficient between two variables, is 0?

105. A real estate agent believes that the selling price of a house in a particular town is directly related to the number of square feet the house has. In order to justify his assumption, he took a sample of 8 homes that had been recently sold, and recorded the price and size. Using the data collected, compute r, the correlation coefficient.

House Size $(100\ \text{ft}^2)$	Selling Price (in thousands)
12.2	129
12.8	149
13.5	156
14.5	166
15.7	193
17.3	209
18.9	234
19.5	255

106. It has been said that the older a person becomes, the slower his or her reaction time is when driving. The following data was collected during a driving simulation study. Let $x =$ the age of the person and let $y =$ their reaction time, in seconds, to a particular part of the test.

x	16	20	21	35	43	55	67
y	0.6	0.9	1.0	1.2	1.3	1.5	1.7

Use this data to find SS_x, SS_y, SS_{xy}, r, and b.

107. Although the negative effects of sleep deprivation are still widely debated, most believe that a person is unable to function properly when having had little or no sleep the night prior. Six people participated in a sleep study. Each participant received different amounts of sleep, in hours, per night. Upon waking, each participant completed a series of puzzlers. Let $x =$ the number of hours each participant slept and $y =$ their score on the puzzler exam.

Participant	1	2	3	4	5	6
x	1	2	3	5	7	9
y	43	44	50	53	60	68

Find SS_x, SS_y, SS_{xy}, r, and b.

108. It has been said that the older a person becomes, the slower their reaction time is when driving is. The following data was collected during a driving simulation study. Let x = the age of the person and let y = their reaction time, in seconds, to a particular part of the test.

x	16	20	21	35	43	55	67
y	1.4	2.3	3.0	3.7	4.6	5.6	8.7

Use this data to find SS_x, SS_y, SS_{xy}, r, b, and r^2.

109. A real-estate agent believes that the selling price of a house for a particular town is directly related to the number of square feet the house has. In order to justify his assumption, he took a sample of 8 homes that had been recently sold, and recorded the price and size. Using the data collected, compute r^2, the coefficient of determination.

House Size $(100\ \text{ft}^2)$	Selling Price (in thousands)
12.9	145
13.1	158
14.5	172
15.3	187
17.1	197
17.8	210
18.8	221
19.9	255

Answer Key for Chapter 4
Regression and Correlation

Section A: Static Items

(a) and (c)

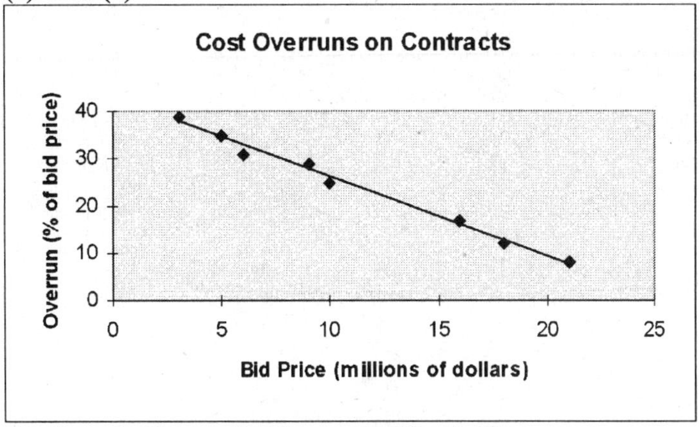

(b) $b = -1.68$; $a = 42.99$; $y = 42.99 - 1.68x$.
(d) $S_e = 1.228$.
[1] (e) If $x = 12$ million, the predicted value of y is 22.82%.

(a) and (c)

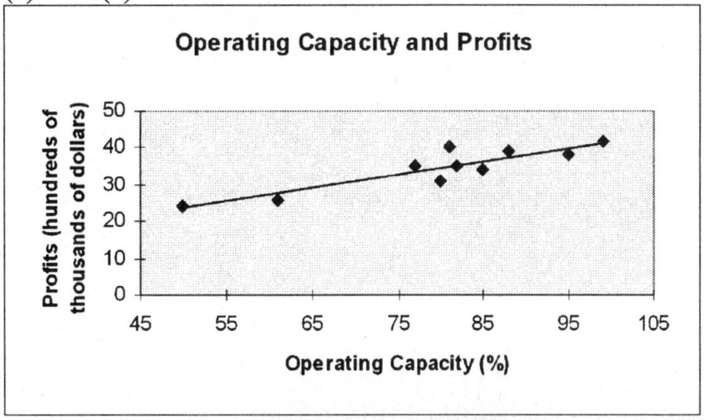

(b) $b = 0.364$; $a = 5.347$; $y = 5.347 + 0.364x$;
(d) $S_e = 2.645$;
[2] (e) If $x = 75\%$, the predicted value of y is 32.65 thousand dollars.

(a) and (c)

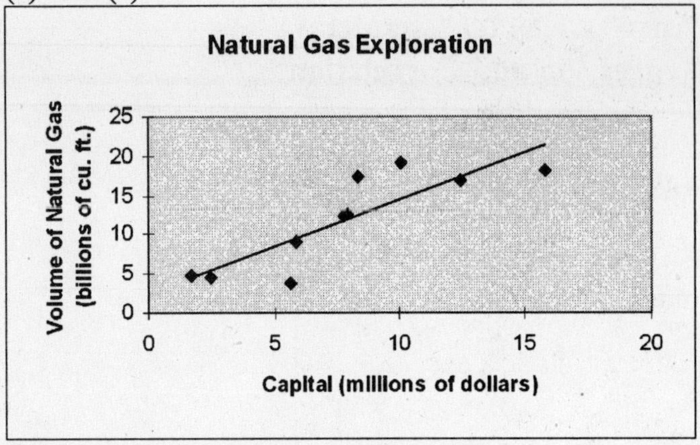

(b) $b = 1.20$; $a = 2.55$; $y = 2.55 + 1.20x$.
(d) $S_e = 3.33$.
[3] (e) If $x = 6.5$ million, the predicted value of y is 10.36 billion cubic feet.

(a) and (c)

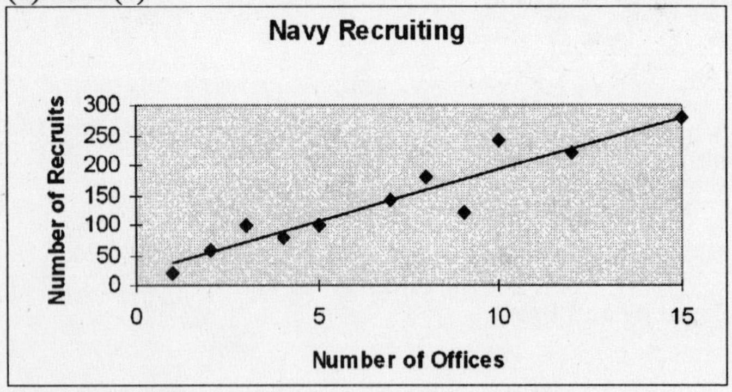

(b) $b = 17.42$; $a = 19.66$; $y = 19.66 + 17.42x$.
(d) $S_e = 28.04$.
[4] (e) If $x = 6$, the predicted value of y is 124.17 recruits.

[5] H_0: $\rho = 0$; H_1: $\rho > 0$; Critical r value is –0.42; $r = –0.67$; Reject H_0. The counselor's claim is justified.

(a)

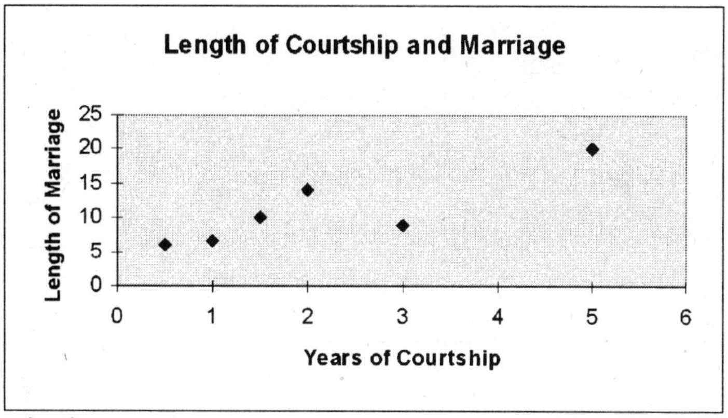

(b) r is closer to 1.

[6] (c) $r = 0.869$.

(a)

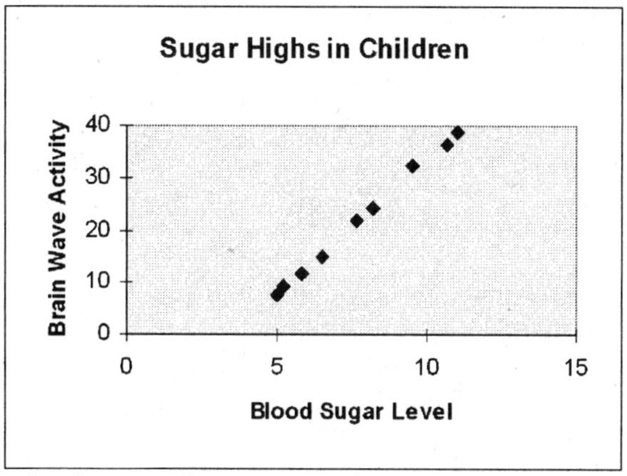

(b) r is closer to 1.

[7] (c) $r = 0.9989$.

(a) and (c)

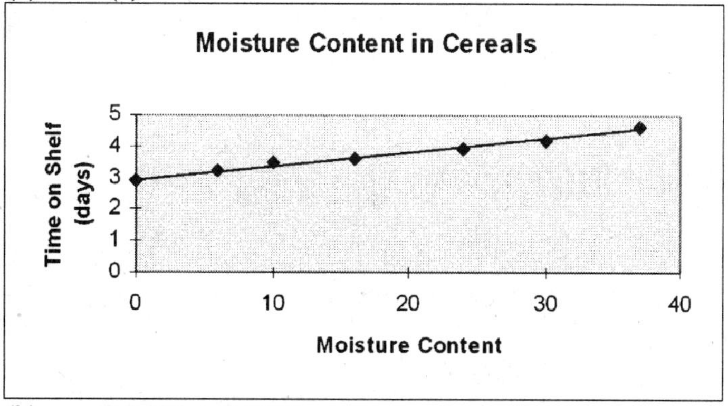

(b) $y = 2.9404 + 0.0432x$.

(d) If $x = 30$ days the predicted value of y is 4.24.

[8] (e) $S_e = 0.0774$.

(a) $r = 0.9926$.
(b) H_0: $\rho = 0$; H_1: $\rho > 0$; Critical r value is 0.83; $r = 0.9926$; Reject H_0. There is a positive correlation between number of days on the shelf and moisture content.
(c) $r^2 = 0.9853$.
(d) About 98.5% of the variation in moisture content can be explained by the variation in
[9] time on the shelf using the least-squares line.

(a) and (c)

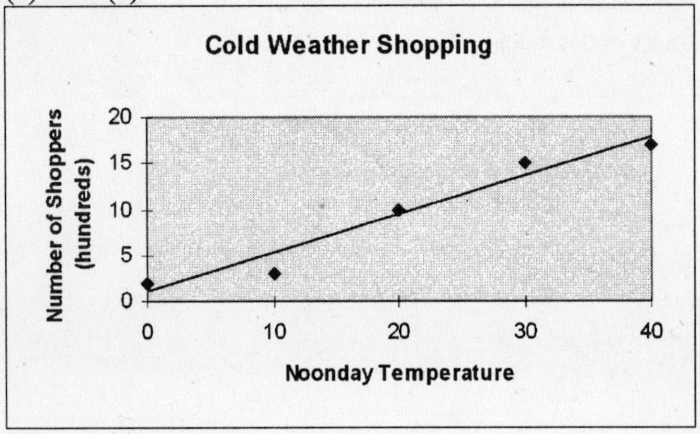

(b) $y = 1.0 + 0.42x$.
(d) $S_e = 1.71$.
[10] (e) If $x = 15°$ F, the predicted value of y is 7.3 hundred shoppers.

H_0: $\rho = 0$; H_1: $\rho > 0$; Critical r value is 0.88; $r = 0.73$; Do not reject H_0. We cannot conclude, at the 1% significance level, that there is a positive correlation between the
[11] prime interest rate and the unemployment rate.

(a) $r = 0.976$.
(b) $H_0 : \rho = 0$; $H_1 : \rho > 0$; Critical r value is 0.81; $r = 0.976$; Reject H_0. There appears to be a positive correlation between the noonday temperature and the number of shoppers.
(c) $r^2 = 0.9525$.
(d) About 95% of the variation in the number of shoppers is explained by the variation in
[12] temperature using the least-squares line.

(a) and (c)

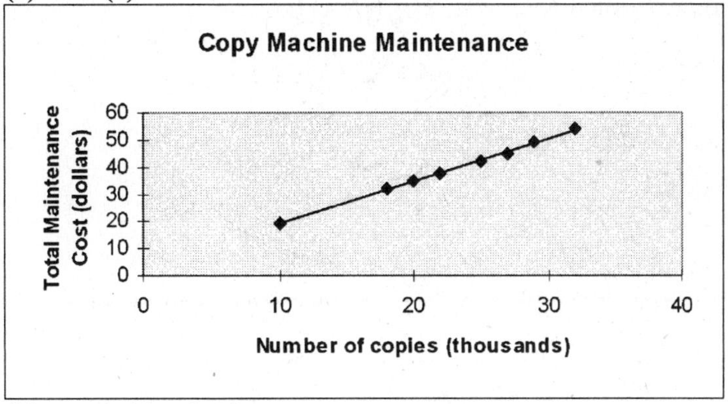

(b) $y = 3.47 + 1.56x$.
(d) $S_e = 0.4625$.
[13] (e) If $x = 28$ (thousand), $y = \$47.15$.

(a) $r = 0.999$.
(b) $H_0: \rho = 0$; $H_1 > 0$; Critical r value is 0.79; $r = 0.999$; Reject H_0. There seems to be a positive correlation between the number of copies and the maintenance cost.
(c) $r^2 = 0.998$;
(d) About 99.8% of the variation in maintenance cost is explained by the variation in the
[14] number of copies using the least-squares line.

(a) and (c)

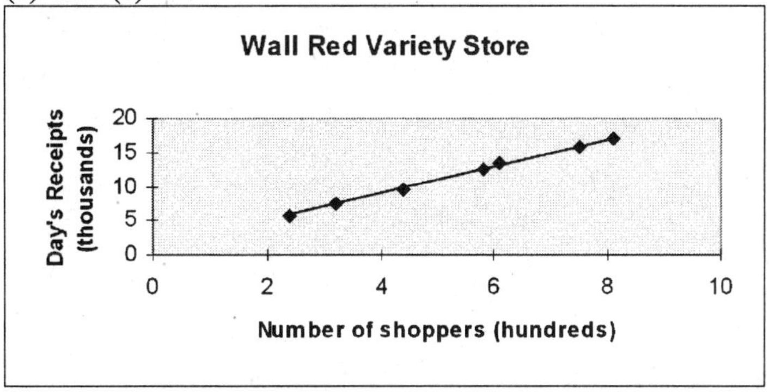

(b) $y = 1.187 + 1.965x$.
(d) $S_e = 0.146$.
[15] (e) If $x = 5$ (hundred), the predicted value of y is 11.012 thousand dollars.

(a) $r = 0.9994$.
(b) H_0: $\rho = 0$; $H_1 : \rho > 0$; Critical r value is 0.67; $r = 0.9997$; Reject H_0. There appears to be a positive correlation between the number of shoppers and the day's receipts.
(c) $r^2 = 0.9990$.
(d) About 99.9% of the variation in the days receipts can be explained by the variation in
[16] the number of shoppers using the least-squares line.

(a)

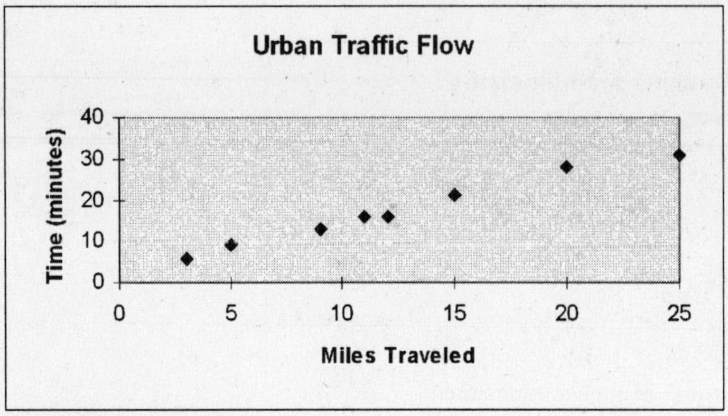

(b) $y = 2.763 + 1.179x$.

[17] (c) If $x = 22$ miles, the predicted value of y is 28.7 minutes.

(a) $r = 0.9945$.

(b) H_0: $\rho = 0$; H_1: $\rho > 0$;Critical r value is 0.62; $r = 0.9945$; Reject H_0. There appears to be a positive correlation between the distance traveled and the travel time.

(c) $r^2 = 0.9891$.

(d) About 98.9% of the variation in travel time is explained by the variation in distance

[18] traveled using the least-squares line.

(a) and (c)

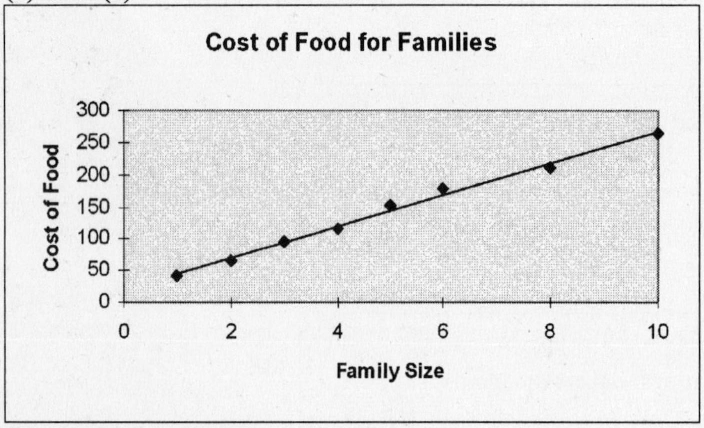

(b) $y = 20.2 + 24.7x$.

(d) $S_e = 6.779$.

[19] (e) If $x = 7$, the predicted value for y is \$193.10.

(a) $r = 0.996$.

(b) H_0: $\rho = 0$; H_1: $\rho > 0$;Critical value of r is 0.62; $r = 0.996$; Reject H_0. There is a

[20] positive correlation between the size of a family and the cost of its food.

(a) and (c)

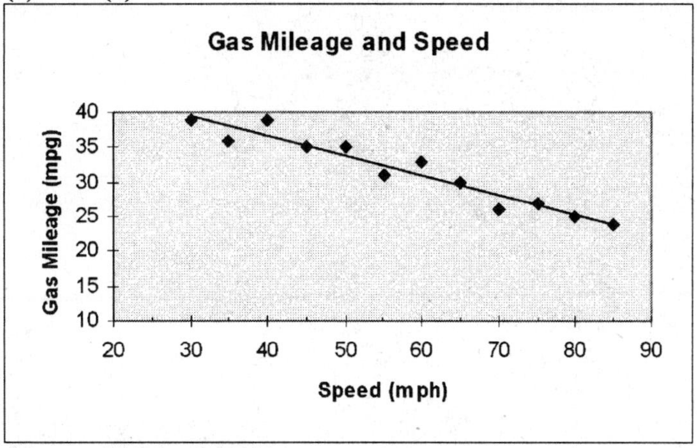

(b) $y = 47.26 - 0.2741x$

(d) If $x = 58$ mph, the predicted value of y is 31.36 mpg.

[21] (e) $S_e = 1.2799$.

(a) $r = -0.9709$.

(b) H_0: $\rho = 0$; H_1: $\rho < 0$;Critical value of r is –0.66; $r = -0.9709$; Reject H_0. There appears

[22] to be a negative correlation between gasoline mileage and speed.

(a)

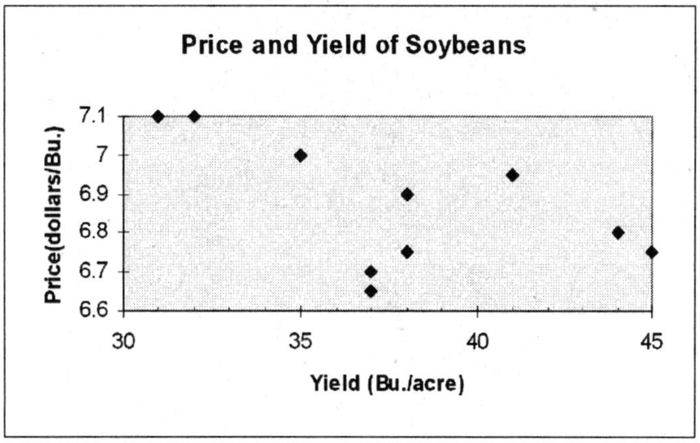

(b) $r = -0.5932$.

(c) $r^2 = 0.3519$.

(d) About 35% of the variation in the price of soybeans can be explained by the variation

[23] in yield using the least-squares line.

(a)

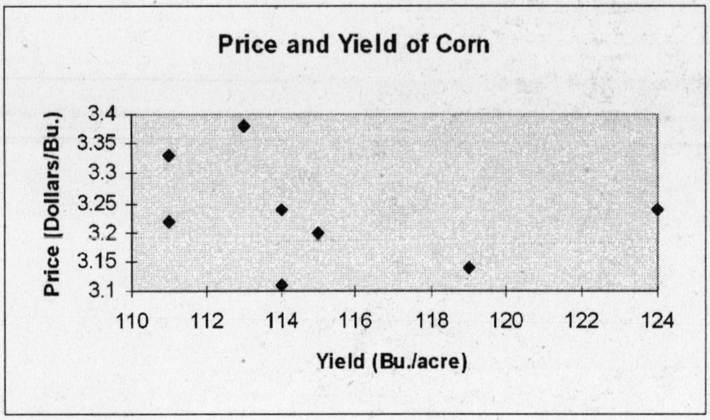

(b) $r = -0.2989$.

(c) H_0: $\rho = 0$; H_1: $\rho \neq 0$;Critical r values are ± 0.71; $r = -0.2989$; Do not reject H_0. We

[24] cannot conclude that there is any significant correlation.

(a) and (c)

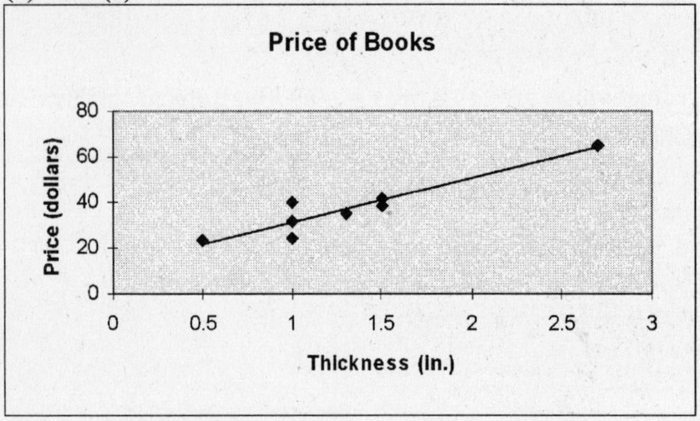

(b) $y = 11.00 + 18.35x$

[25] (d) If $x = 2.5$ inches the predicted price is \$56.88.

(a) $r = 0.9426$.

(b) $r^2 = 0.8885$.

(c) About 89% of the variation in the cost of a textbook is explained by the variation in

[26] thickness using the least-squares line.

(a)

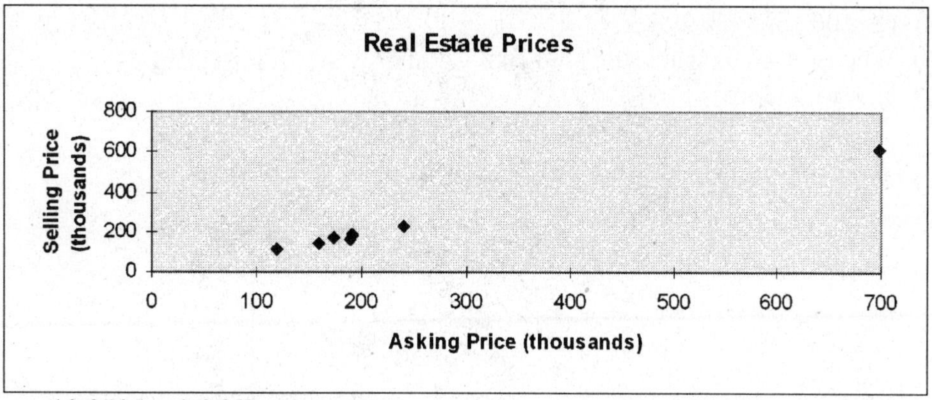

(b) $y = 10.9705 + 0.8663x$.

[27] (c) If $x = 145.0$ thousand, the predicted value of y is 136.6 thousand.

(a) $r = 0.9988$.

(b) $r^2 = 0.9976$.

(c) About 99.8% of the variation in selling price is explained by the variation in asking price using the least-squares line.

(d) It appears that there may be a linear correlation between selling price and asking price

[28] for homes in the same neighborhood.

(a) $y = 13.0294 + 0.4983x$

[29] (b) If $x = 75\%$, the predicted value of y is 50.4%.

(a) and (d)

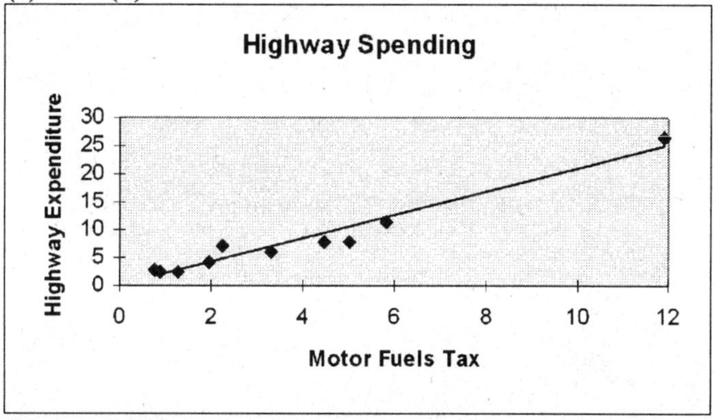

(b) $r = 0.979$.

(c) $y = 0.0892 + 2.0807x$.

[30] (e) If $x = 4.5$, the predicted value for y is 9.45.

(a) $r = 0.3355$.

(b) H_0: $\rho = 0$; H_1: $\rho > 0$; Critical value of r is 0.54. $r = 0.3355$; Do not reject H_0. We cannot conclude that there is a correlation between the average annual precipitation and

[31] the mean number of rainy days per year.

(a) $y = 100.6365 + 0.2982x$.
(b) When $x = 40.0$ inches, the predicted value of y is 112.5645 days.
[32] (c) $S_e = 13.6388$.

(a)

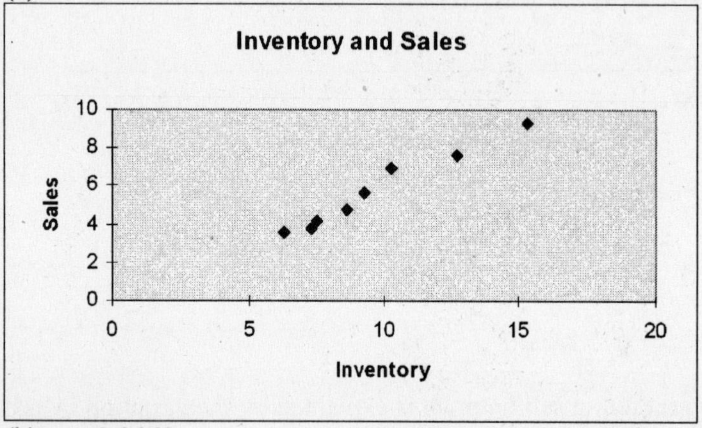

(b) $r = 0.9469$.
(c) $r^2 = 0.8966$.
(d) About 90% of the variation in sales can be explained by variation in inventory using
[33] the least-squares line.

(a)

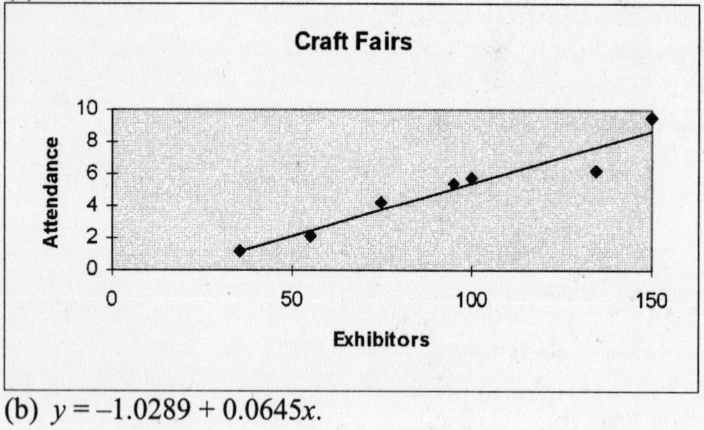

(b) $y = -1.0289 + 0.0645x$.
[34] (c) If $x = 80$, the predicted value for y is 4.13 or 413 people.

(a) $r = 0.9614$.
(b) $r^2 = 0.9243$.
(c) About 92% of the variation in attendance is explained by the variation in the number
[35] of exhibitors using the least-squares line.

(a) $y = 394.5074 - 9.7458x$.
(b) If $x = 8.5\%$ the predicted value of y is 311.7 billion dollars.
(c) $r = -0.3493$.
(d) $r^2 = 0.1220$.
(e) Only 12.2% of the variation in dollar value of automobile loans can be explained by
[36] the variation in interest rates on new car loans using the least-squares line.

(a) $r = -0.8853$.
(b) Credit card debt tends to decrease when credit card interest rates rise.
(c) $y = 1{,}727.4814 - 84.4152x$.
[37] (d) If $x = 16.5\%$, the predicted value of y is 334.6 billion dollars.

H_0: $\rho = 0$; H_1: $\rho \neq 0$; Critical r value is 0.50; $r = 0.65$; Reject H_0. There appears to be a
[38] correlation between the length of the call and the dollar value of the sale.

(a)

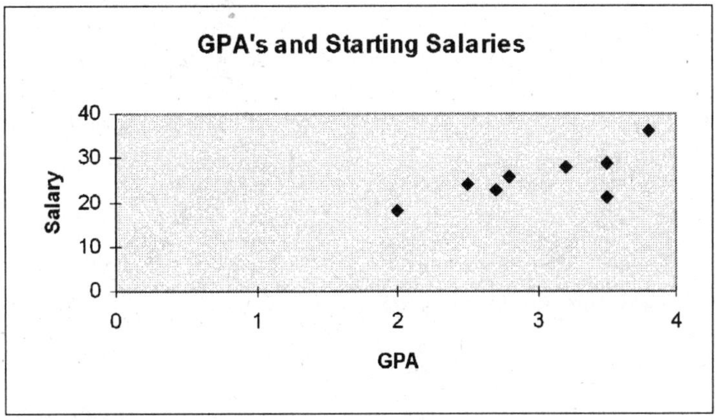

(b) $r = 0.722$.
(c) The linear correlation is moderate.
(d) $y = 5.7266 + 6.7578x$.
[39] (e) If $x = 3.6$, the predicted value of y is 30.05 thousand dollars.

(a) and (c)

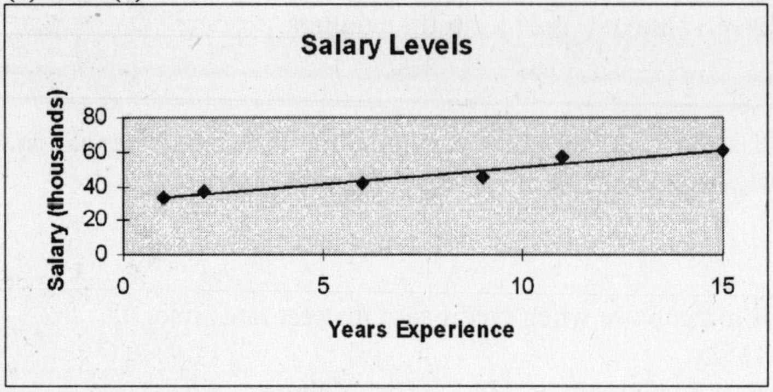

(b) $y = 31.2342 + 1.9908x$.
(d) $S_e = 3.0343$.
[40] (e) If $x = 8$ years, $y = 47.16$ thousand dollars.

[41] H_0: $\rho = 0$; H_1: $\rho < 0$ Critical r value is –0.52; $r = -0.40$; Do not reject H_0. r is not significant at the 1% significance level.

[42] H_0: $\rho = 0$; H_1: $\rho \neq 0$; Critical r value is 0.52; $r = 0.38$. Do not reject H_0. There does not appear to be a significant correlation between peoples' income and their level of life insurance.

(a)

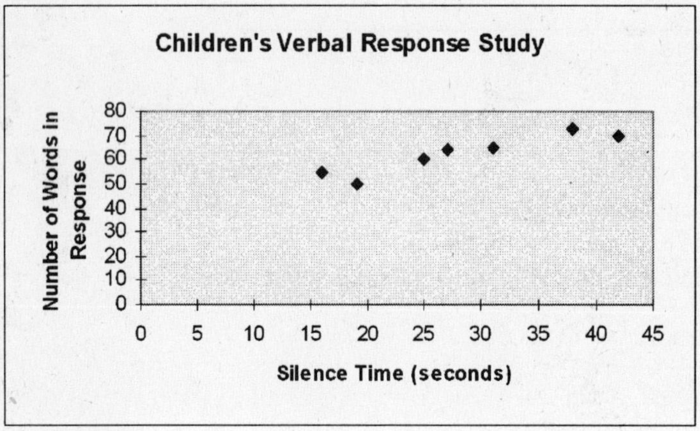

(b) $r = 0.9247$.
(c) $r^2 = 0.8551$.
[43] (d) About 85.5% of the variation in the number of words in the response is explained by the variation in silence time using the least-squares line.

[44] H_0: $\rho = 0$; H_1: $\rho > 0$; Critical r value is 0.57; $r = 0.48$; Do not reject H_0. She should let her employees work overtime.

H_0: $\rho = 0$; H_1: $\rho < 0$; Critical r value is –0.62; r = –0.58; Do not reject H_0. We cannot conclude that there is a negative correlation between substantial tax increases and the Dow
[45] Jones index.

(a)

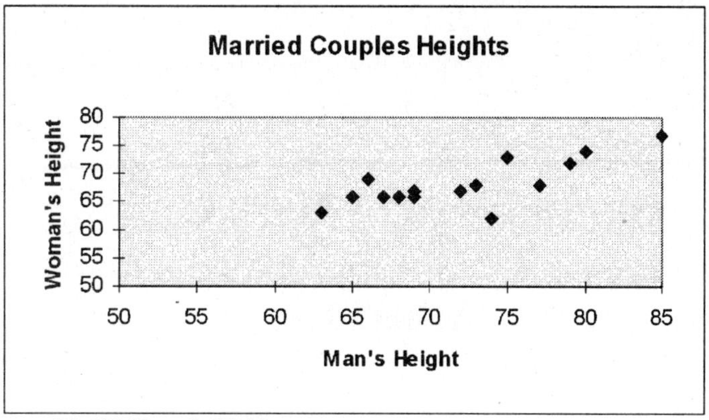

(b) $y = 31.8563 + 0.5048x$.
[46] (c) If x = 69 inches, the predicted value for y is 66.69 inches.

(a) $r = 0.7661$.
(b) H_0: $\rho = 0$; H_1: $\rho \neq 0$; Critical r value is ±0.51; $r = 0.7661$; Reject H_0. It appears that there is a correlation between the man's height and the woman's height in married couples.
(c) $r^2 = 0.5869$.
(d) For married couples, about 58.7% of the variation in women's height is explained by
[47] the variation in men's height using the least-squares line.

(a) and (c)

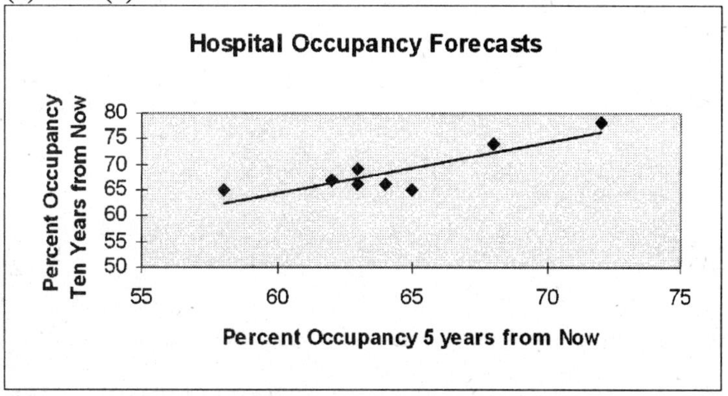

(b) $y = 5.4951 + 0.9826x$.
[48] (d) If x = 70%, the predicted value of y is 74.28%.

H_0: $\rho = 0$; H_1: $\rho < 0$; Critical r value is –0.42; r = –0.63; Reject H_0. At the 1% significance level, the data support the claim that there is a negative correlation between oxygen
[49] content and depth in ocean water.

(a) and (c)

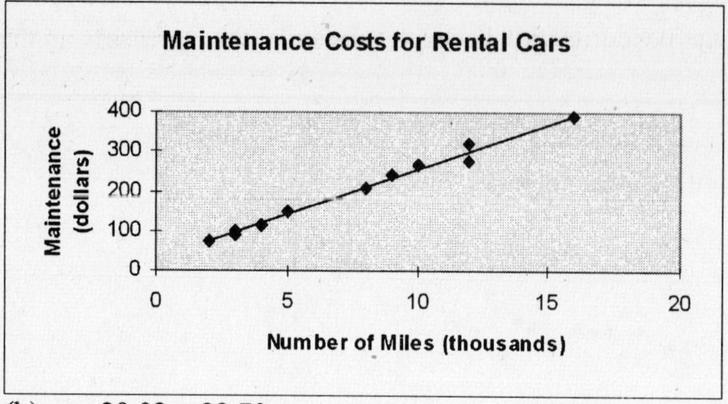

(b) $y = 29.02 + 22.70x$.

[50] (d) If $x = 7$ (thousand) miles the predicted value of y is \$187.92.

(a) $r = 0.9940$.

(b) H_0: $\rho = 0$; H_1: $\rho > 0$; Critical r value is 0.69; $r = 0.9940$; Reject H_0. There seems to be

[51] a positive correlation between mileage and maintenance cost for rental cars at this agency.

H_0: $\rho = 0$; H_1: $\rho > 0$; Critical r value is 0.50; $r = 0.75$; Reject H_0. There appears to be a positive correlation between the time a customer spends in the store and the amount of

[52] money the customer spends.

H_0: $\rho = 0$; H_1: $\rho \neq 0$; Critical r value is ± 0.53; $r = -0.60$; There appears to be a correlation

[53] between aggressiveness and amount of funeral insurance sold.

(a) and (c)

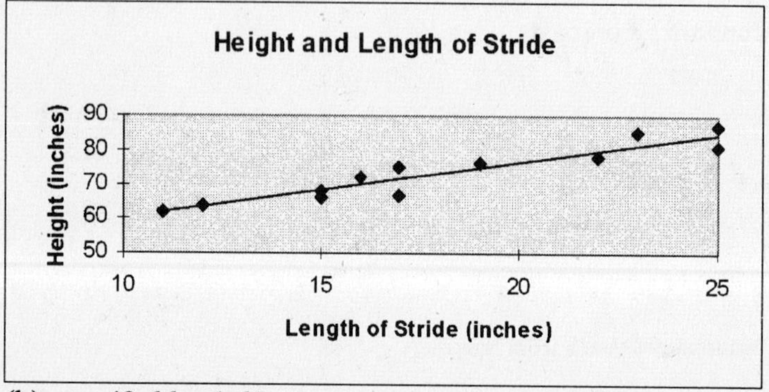

(b) $y = 43.66 + 1.65x$.

[54] (d) If $x = 20$ inches, the predicted value of y is 76.6 inches.

(a) $r = 0.9461$.

(b) H_0: $\rho = 0$; H_1: $\rho > 0$; Critical r value is 0.66; $r = 0.9461$; Reject H_0. There appears to

[55] be a positive correlation between length of stride and height.

(a)

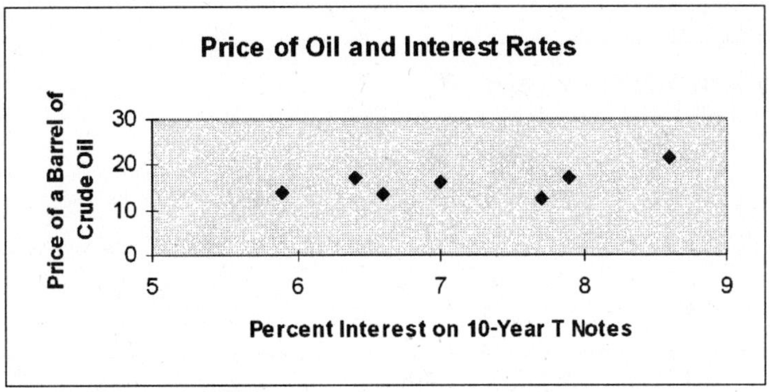

(b) $y = 2.6085 + 1.8571x$.

[56] (c) If $x = 7.5\%$ the predicted value of y is \$16.54 per barrel.

(a) $r = 0.5968$;

(b) H_0: $\rho = 0$; H_1: $\rho \neq 0$; Critical r value is ± 0.75; $r = 0.5968$; Do not reject H_0. We cannot conclude that there is a correlation between the percent interest on U.S. 10-year treasury notes and the price of a barrel of crude oil.

(c) $r^2 = 0.3562$.

(d) About 35.6% of the variation in the price of a barrel of crude oil can be explained by the variation in the percent interest on U.S. 10-year treasury notes using the least-squares

[57] line.

(a) and (c)

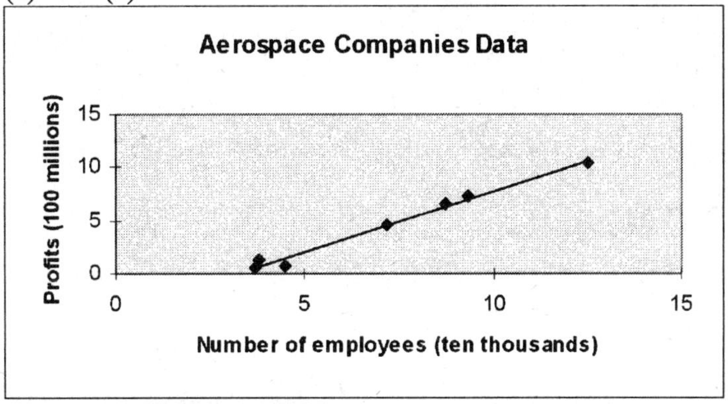

(b) $y = -3.6097 + 1.1402x$.

[58] (d) If $x = 8.2$, the predicted value of y is 5.7.

(a) $r = 0.9946$.

(b) $r^2 = 0.9872$.

(c) About 98.7% of the variation in profits can be explained by the variation in number of

[59] employees using the least-squares line.

(a) and (c)

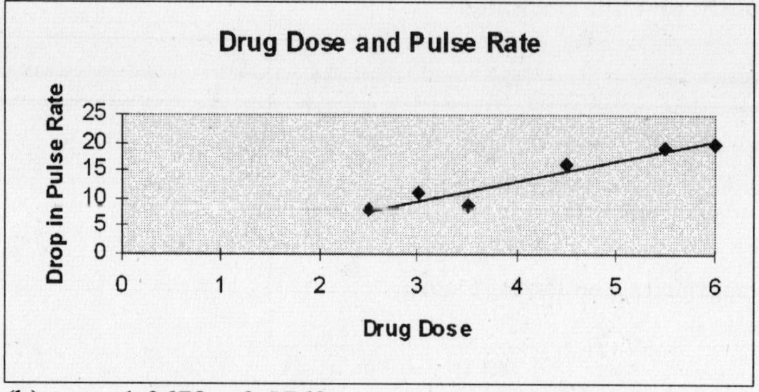

(b) $y = -1.0678 + 3.5763x$.

[60] (d) If $x = 2.75$, the predicted value of y is 8.767.

(a) $r = 0.9658$.
(b) H_0: $\rho = 0$; H_1: $\rho > 0$; Critical r value is 0.73; $r = 0.9658$; Reject H_0. There seems to be a positive correlation between the dose size and the drop in pulse rate.
(c) $r^2 = 0.9328$.
(d) About 93.3% of the variation in drop of pulse rate is explained by the variation in drug

[61] dose, using the least-squares line.

(a) $r = 0.6351$.
(b) H_0: $\rho = 0$; H_1: $\rho > 0$; Critical r value is 0.73; $r = 0.6351$; Do not reject H_0. We cannot conclude that there is a positive correlation between the total calories in food and the amount of cholesterol in it.
(c) $r^2 = 0.4034$.
(d) About 40.3% of the variation in the amount of cholesterol in a sample is explained by

[62] the total calories in the sample using the least-squares line.

(a) $r = -0.8383$.
(b) $r^2 = 0.7027$.
(c) About 70.3% of the variation in the vitamin B12 level is explained by variation in age

[63] using the least-squares line.

(a) $r = 0.9310$.
(b) $r^2 = 0.8668$.
(c) About 86.7% of the variation in deceleration time is explained by variation in ejection speed using the least-squares line.

[64] (d) 13.3% of the variation is unexplained.

[65] Essay. See text.

(a) and (c)

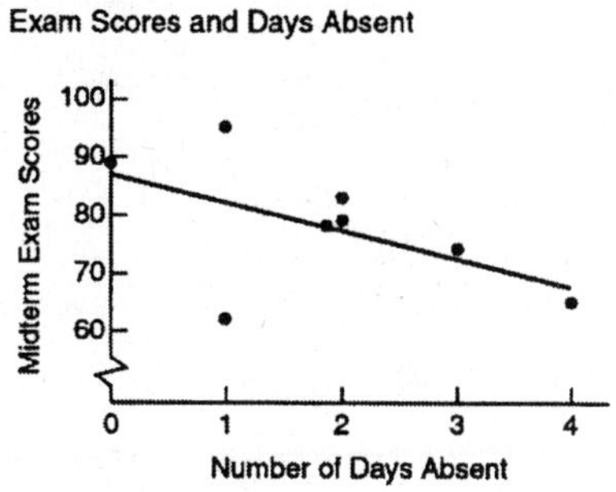

(b) $\bar{x} = 1.86$; $\bar{y} = 78.14$; $b = -4.868$; $y = -4.868x + 87.184$

(d) $S_e = 11.13$

[66] (e) 72.6%

(a) and (c)

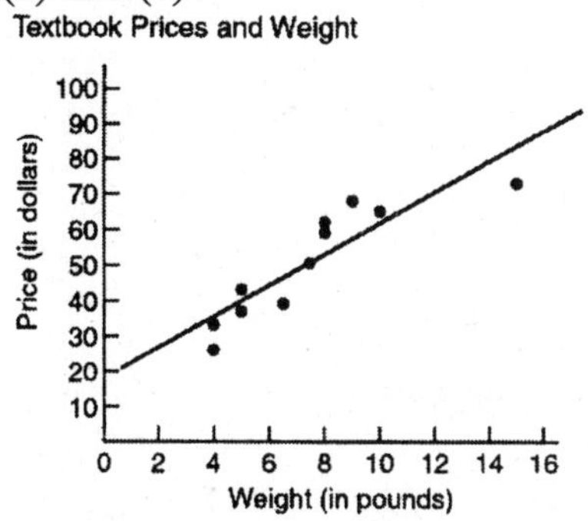

(b) $\bar{x} = 7.45$; $\bar{y} = 50.5$; $b = 4.3812$; $y = 4.3812 + 17.86$

(d) $S_e = 8.0863$

[67] (e) $48.53

[68] C.

[69] D.

[70] B.

[71] D.

(a) $r = -0.5417$

(b) $r^2 = 0.2935$

[72] (c) 29% explained; 71% unexplained

(a) $r = 0.8895$
(b) $r^2 = 0.7911$
[73] (c) 79% explained; 21% unexplained

[74] B.

[75] B.

[76] E.

[77] B.

[78] A.

[79] B.

Section B: Algorithms

Objective 4.1: Introduction to Paired Data and Scattered Diagrams

[80] About 195

[81] There is no correlation.

Answers may vary. Sample answer:

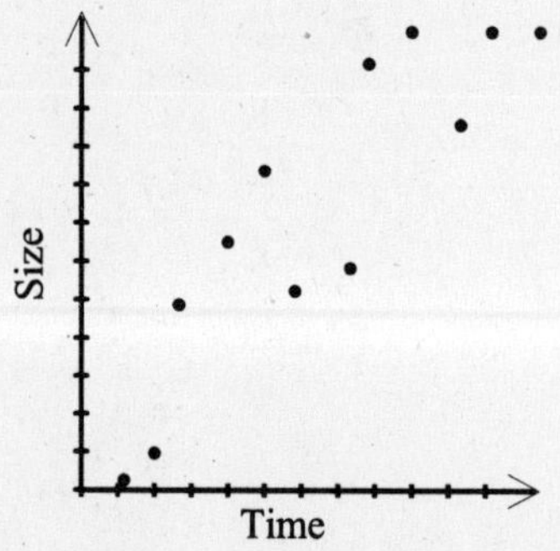

[82]

[83] positive correlation

From the *Patterns of Association* unit, this task requires students to determine which of four given scatterplots illustrates the weakest and strongest correlations and a negative correlation.

a. The weakest correlation is shown in plot **C**, since the points tend to lie farther from a line.

b. Plot **C** shows a negative correlation since the points tend to go down from left to right; that is, as the value of the variable on the horizontal axis increases, the value of the variable on the vertical axis decreases.

[84] c. Plot **D** shows the strongest correlation since the points are most nearly linear.

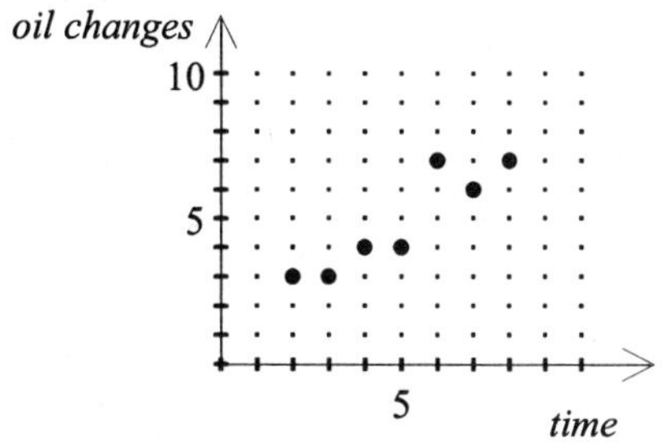

[85] It is a linear model with a positive correlation.

Objective 4.2: Linear Regression and Confidence Bounds for Prediction

a.

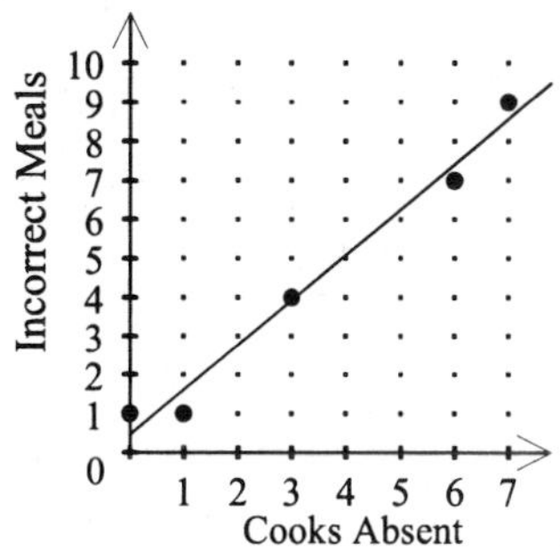

[86] b. $\bar{x} = 3.4$, $\bar{y} = 4.4$, $b = 1.16$, and least-squares line = $y = 0.456 + x$

$SS_x = 27{,}449{,}000$, $SS_{xy} = -2{,}290{,}900$

[87] $SS_y = 515{,}690$, $b = -0.08$

[88] $\hat{y} = 0.2022x + 42.9703$

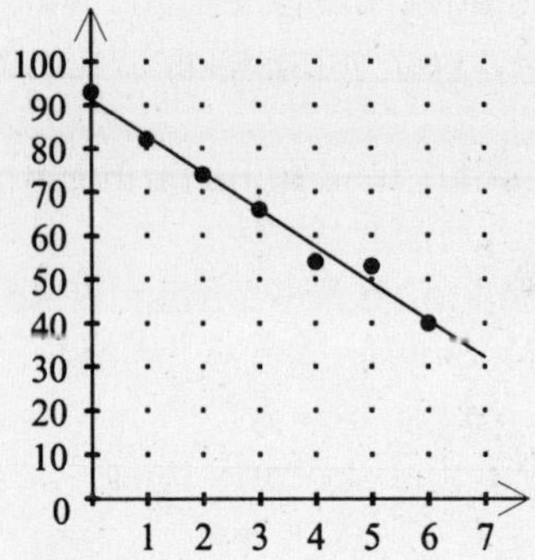

[89]

[90] 25.9813

[91] 16.0546

[92] $y = 3.48476x - 6744.74$; 260

[93] $257,000

[94] 1163

[95] Constant term is 2.4; 2.6 with x_2; 2.7 with x_3; and 2 with x_4

a.

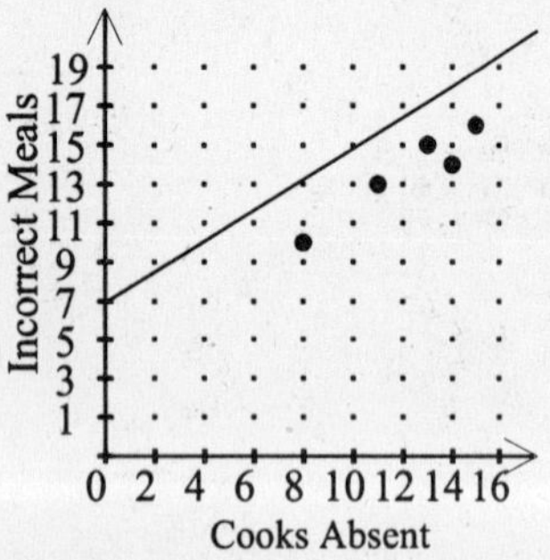

[96] b. $\bar{x} = 12.2$, $\bar{y} = 13.6$, $b = 0.79$, and least-squares line $= y = 3.962 + 0.79x$

[97] 54.1691

[98] 40.5017

[99] 1060

[100] Constant term is 7.0; 4.6 with x_2; 5.2 with x_3; and 7.8 with x_4; If $x_2 = x_3 = x_4$, then the coefficient of the resulting variable will become the slope of the equation.

[101] $y = 0.60x + 5.53$

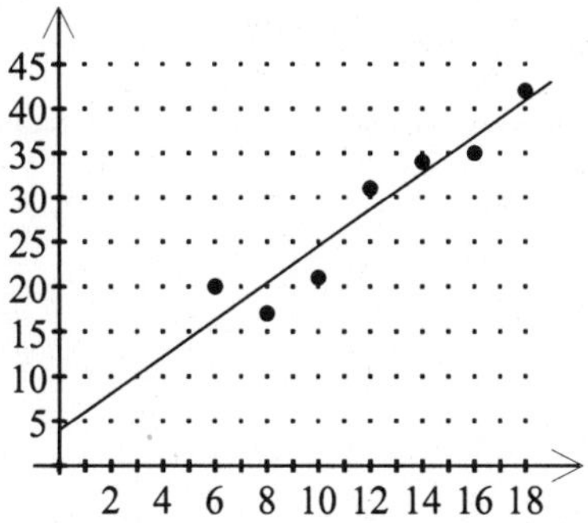

[102] Answers may vary. Sample answer: 9 hr

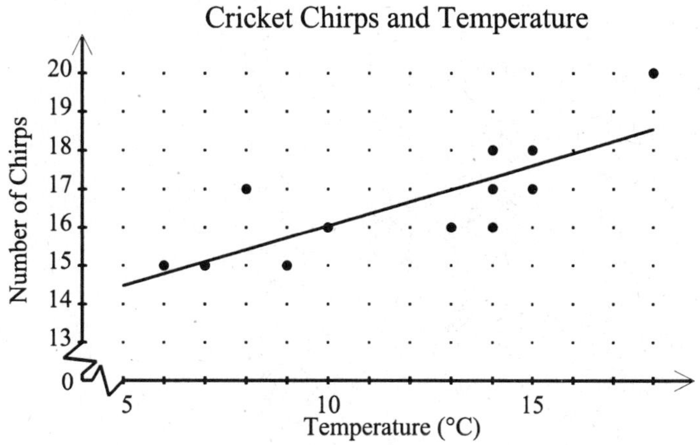

[103] The model is a "good fit" for the actual data.

Objective 4.3: The Linear Correlation Coefficient

[104] Answers may vary. Sample answer: There is no linear relation for the points of the scatter diagram.

[105] 0.99

[106] $SS_x = 2249.429$, $SS_y = 0.834$, $SS_{xy} = 41.843$

$r = 0.966$, and $b = 0.019$.

$SS_x = 47.500,\ SS_y = 464.000,\ SS_{xy} = 147.000$

[107] $r = 0.990$, and $b = 3.095$

$SS_x = 2249.429,\ SS_y = 35.509,\ SS_{xy} = 273.871$

[108] $r = 0.969$, $b - 0.122$, and $r^2 = 0.939$

[109] 0.98

Chapter 5
Elementary Probability Theory

Section A: Static Items

1. Answer each question and give a brief explanation for each answer.
   (a) If $P(A) = 0.48$, what is the value of $P(not\ A)$?
   (b) If $P(not\ A) = 0.75$, what is the value of $P(not\ A)$?
   (c) Is it possible that $P(A) = 17/3$?
   (d) Is it possible that $P(A) = 3/17$?
   (e) Does $P(A) + P(not\ A)$ always equal 1?

2. If event A is certain to occur, what is $P(A)$?

   A. 0.25 B. 0.5 C. 0.75 D. 0 E. 1

3. Alice, Carol, Bob and Dick all volunteer to be subjects for a psychology experiment. One person is to be chosen at random from this group. Each person is equally likely to be chosen.
   (a) Describe the sample space for this problem.
   (b) What is the probability that Bob will be chosen?
   (c) What is the probability that Bob or Dick will be chosen?
   (d) What is the probability that a woman will be chosen?
   (e) What is the probability that Alice will not be chosen?

4. A coin is to be tossed 1,000 times. What is the probability that the 785th toss is heads?

   A. 1 B. 0 C. $\frac{1}{2}$ D. $\frac{1}{4}$ E. $\frac{3}{4}$

5. The Quick Turn Taxi Company wants to estimate the probability that a taxi will be in a wreck during a one-month period. Last month 2 out of their 200 taxis were involved in a wreck. What would you estimate the probability to be? What method did you use to get your estimate?

6. The student Executive Cabinet consists of six members including Sarah and Michael. A reporter for the student newspaper wants to contact any member of the cabinet and calls one at random.
   (a) What is the probability that Sarah is called?
   (b) What is the probability that Michael or Sarah is called?
   (c) What is the probability that neither Michael nor Sarah is called?

7. Andrea did a survey in her statistics class and found that 18 students out of 30 use the library at least twice a week.
(a) What is the probability that a student chosen at random from the class uses the library at least twice a week?
(b) What is the probability that a student chosen at random from the class uses the library less often than twice a week?
(c) If two students are chosen at random from the class what is the probability that both of them use the library at least twice a week?
(d) What is the probability that neither of them uses the library at least twice a week.

8. A football team's coach predicts that his team will win 9 out of 10, or 90%, of its games. Which method did he most likely use to make this prediction?

A. probability
B. law of large numbers
C. intuition
D. equally likely outcomes
E. relative frequency

9. Suppose that you roll a single die.
(a) What is the probability that the number on the face is a 3 or a 6?
(b) What is the probability that the number on the face is not a 2?
(c) What is the probability that the number on the face is a 2 or a 4 or a 6?

10. Out of 1000 car-door handles tested, 23 were found to be defective.
(a) Use the probability formula for relative frequency to determine the probability that a car-door handle is defective.
(b) Using the result from part (a), how many car-door handles out of 5,000 do you expect to be defective?

11. You draw one card from a standard deck of 52 cards.
(a) What is the probability that the card is the ace of spades?
(b) What is the probability that the card is a spade?
(c) What is the probability that the card is an ace or a spade?

12. You draw two cards from a standard deck of 52 cards and do not replace the first card before you draw the second.
(a) What is the probability that the first card is the king of diamonds and the second is the king of hearts?
(b) What is the probability that the first card is the king of diamonds and the second card is a diamond?
(c) What is the probability that the first card is not a king and the second card is also not a king?

13. The enrollment of an elementary school is 1,250. Sixty children had their teeth checked by a dentist, and 12 children were found to have one or more cavities.
(a) Use the probability formula for relative frequency to determine the probability that a child at this school has at least one cavity.
(b) Using the result from part (a), how many children in the entire school do you expect to have at least one cavity?

14. You roll two fair dice, one red and one green.
(a) What is the probability of getting a number less than 5 on both?
(b) What is the probability of getting a sum of 9 on the two dice?
(c) What is the probability of getting a five on both?

15. In the long run, as the sample size increases and increases, the relative frequencies of ____________ get closer and closer to the theoretical (or actual) probability value. Which choice best fills in the blank?

A. guesses B. outcomes C. choices D. probabilities E. numbers

16. You roll two fair dice, a red one and a green one.
(a) What is the probability of getting a sum of 8 on the two dice?
(b) What is the probability of getting a 3 on the red die and a 5 on the green one?
(c) What is the probability of getting a 3 on one die and a 5 on the other?

17. An urn contains 8 balls identical in every respect except color. There are 4 blue balls, 3 red ones and 1 white one.
(a) If you draw one ball from the urn what is the probability that it is blue or white?
(b) If you draw two balls without replacing the first one, what is the probability that the first is red and the second is white?
(c) If you draw two balls without replacing the first one, what is the probability that one ball is red and the other is white?

18. A large apartment complex has 750 tenants. Seventy-five tenants were called at random and asked if they would like a swimming pool installed for the tenants' use. Thirty-five answered that they would like a pool. Choose which number of tenants best represents the likely number of tenants that would like to have a pool.

A. 400 B. 350 C. 750 D. 300 E. 250

19. Ten equally qualified people apply for a job. Three will be selected to interview. How many different combinations of three people can be interviewed?

20. A single six person jury must be selected from a pool of 15 jurors. How many ways can a group of six people be selected?

21. A jar is filled with blue, green, red, yellow, and purple jelly beans. Out of 30 jelly beans drawn from the jar at random, three were blue, five were green, six were red, nine were yellow, and seven were purple.
(a) When determining the probability of selecting at random a particular color of jelly bean, what is the sample space?
(b) What is the probability of selecting a yellow jelly bean?
(c) What is the probability of selecting a red jelly bean?

22. Two student officers can go on an all-expense-paid trip to a conference in Hawaii. If there are five officers and all want to go, how many ways can the two officers be chosen?

23. A coin is tossed three times. Complete this sample space with a choice below (T = tails, H = heads):

HHH, HHT, HTH, HTT, THT, TTH, TTT.

A. TTH B. THH C. HTT D. TTT E. THT

24. In a market survey a random sample of 100 people were asked two questions. Did they buy Sparkle toothpaste last month, and did they see an ad for Sparkle on TV last month? The responses are in the accompanying table.

	Bought Sparkle	**Did not buy Sparkle**	**Row total**
Saw ad	32	33	65
Did not see ad	17	18	35
Column total	49	51	100

For a person selected at random from the sample:
(a) Find the probability that a person saw the ad.
(b) Find the probability that a person bought Sparkle.
(c) Find the probability that the person bought Sparkle, given that she or he saw the ad.
(d) Find the probability that a person saw the ad and bought Sparkle.

25. If the probability that an event will happen is 0.27, what is the probability that the event won't happen?

A. 0.23 B. 0.73 C. 0.63 D. 1.73 E. 0.83

26. In purchasing a sound system you have 4 choices for speakers, another 2 choices for receivers and 5 choices for CD players. How many different systems can you construct consisting of 2 identical speakers, one receiver and 1 CD player?

27. A company is considering five sites for plants. One plant will design products, the other will manufacture the product. How many choices are there for placing the two plants on the sites under consideration? Only one plant may be placed on a single site.

28. If a coin is tossed four times, what is the probability that heads will appear exactly three times?

A. $\frac{1}{8}$ B. $\frac{1}{4}$ C. $\frac{1}{2}$ D. $\frac{1}{16}$ E. $\frac{1}{6}$

29. In an experiment you toss a coin three times. Use a tree diagram to record all possible outcomes for the three tosses.

30. The Aim n' Shoot Camera Company wants to estimate the probability that one of their cameras is defective. A random sample of 400 new cameras showed that 24 were defective.
(a) How would you estimate the probability that a new Aim n' Shoot camera is defective?
(b) What is your estimate that a new Aim n' Shoot camera is not defective?
(c) Either a camera is defective or it is not. What is the sample space for this problem?
(d) Are the events in this sample space equally likely?

31. Choose the statement below that is *not* a fact about probabilities.

A. $P(not\ A) = 1 + P(A)$

B. $P(A) + P(not\ A) = 1$

C. The probability of an event *A* is denoted by *P*(h).

D. The sum of the probabilities of outcomes in a sample space is 1.

E. The probability event *A* does not occur is denoted by *P*(*not A*).

32. A random sample of 100 adults were surveyed. They were asked if they regularly watch NFL football games. They were also asked if their favorite team had ever won the Super Bowl. The results follow.

	Team won	**Team did not win**	**Row total**
Watch games	22	53	75
Do not watch games	13	12	25
Column total	35	65	100

For a person selected at random from the sample:
(a) Find *P*(favorite team won the Super Bowl).
(b) Find *P*(watch NFL games).
(c) Find *P*(watch NFL games, given favorite team won Super Bowl).
(d) Find *P*(favorite team won and watch NFL games).

33. A coin is tossed three times. What is the probability that the result is heads for all three tosses?

A. $\frac{1}{6}$ B. $\frac{1}{4}$ C. $\frac{1}{2}$ D. $\frac{1}{3}$ E. $\frac{1}{8}$

34. Suppose that you throw a die and toss a coin. Use a tree diagram to list all outcomes of the experiment.

35. There are five true-false questions on a test. How many different arrangements of answers are possible?

36. Five black and five white marbles are in a box. Without looking, three marbles are taken from the box without replacement. What is the probability that all three are black?

A. $\frac{1}{12}$ B. $\frac{25}{144}$ C. $\frac{1}{10}$ D. $\frac{1}{8}$ E. $\frac{3}{50}$

37. Ten horses are entered in a horse race. How many permutations of 1st, 2nd and 3rd place winners are there?

38. (a) Seven equally qualified students apply for a scholarship, but only five scholarships, equal in value, can be given. How many ways can the five winners be selected?
(b) If the five scholarships are all different in value how many ways can the five winners be chosen?

39. A twelve-sided die is rolled three times. What is the probability that the result is twelve each time it is rolled?

A. $\frac{1}{12}$ B. $\frac{1}{2{,}184}$ C. $\frac{1}{1{,}728}$ D. $\frac{1}{144}$ E. $\frac{1}{1{,}320}$

40. (a) Eight students wish to select a subcommittee of four to talk to the president of the college about tuition costs. In how many ways can the subcommittee be formed?
(b) There are twelve members of the campus botany club. How many ways can you form a slate of candidates for the four offices of the organization?

41. The city council has three liberal members (one of whom is the mayor) and two conservative members. One member is selected at random to testify in Washington DC.
(a) What is the probability that this member is a liberal?
(b) What is the probability that this member is a conservative?
(c) What is the probability that the mayor is chosen?
(d) What is the probability that the mayor is not chosen?

42. June wants to become a policewoman. She must take a physical exam and a written exam. Records of past candidates indicate that the probability of passing the physical exam is 0.82. The probability that the candidate passes the written exam given that he or she has passed the physical exam is 0.58. What is the probability that June passes both exams?

43. While shopping for a new shirt, John counts 48 shirts on a sale rack. There are 6 extra-large, 18 large, 13 medium, and 11 small sized shirts. John wears size large.
(a) If John grabs one shirt from the rack without looking for the size, what is the probability it will be the right size?
(b) What is the probability it will be too large or too small?
(c) If he takes four shirts off the rack, one at a time, what is the probability that the first is a small, the second is a medium, the third is a large, and the fourth is an extra-large shirt?

44. The dean of women at Brookfield College found that 20% of the female students are majoring in engineering. If 53% of the students at Brookfield are women what is the probability that a Brookfield student chosen at random will be a woman engineering major?

45. Compute the probability of rolling seven with two six-sided dice.

A. $\frac{1}{6}$ B. $\frac{5}{36}$ C. $\frac{1}{9}$ D. $\frac{1}{36}$ E. $\frac{1}{12}$

46. Airlines have many different fares and service categories. Records of the last several years show that 12% of the tickets Columbia Airways sells are first class, 28% are excursion fares, 35% are executive commuter fares and 25% are coach. If a person boarding a Columbia flight is selected at random, what is the probability that the person is holding a first class or executive commuter ticket? What is the probability that the person is holding a coach or excursion fare ticket?

47. An urn contains 8 balls identical in every way except for color. There are 2 red balls, 5 green balls and 1 blue ball.
(a) You draw two balls from the urn but replace the first before drawing the second. Find the probability that the first ball is red and the second ball is green.
(b) You draw two balls from the urn but do not replace the first ball before drawing the second. Find the probability that the first ball is red and the second ball is green.

48. One make of cellular telephone comes in 3 models. Each model comes in 2 colors (dark green and white). If the store wants to display each model in each color, how many cellular telephones must be displayed? Make a tree diagram showing the outcomes for selecting a model and a color.

49. What is the probability of rolling eleven twice in a row with two six-sided dice?

A. $\frac{1}{9}$ B. $\frac{1}{324}$ C. $\frac{1}{18}$ D. $\frac{1}{11}$ E. $\frac{2}{35}$

50. An urn contains 9 balls identical in every way except in color. There are 3 red balls, 4 blue balls and 2 white balls.
(a) You draw 2 balls from the urn but replace the first before drawing the second. Find the probability that the first ball is white and the second ball is blue.
(b) You draw 2 balls from the urn but do not replace the first before drawing the second. Find the probability that the first ball is white and the second ball is blue.

51. There are 24 students in the dorm who are enrolled in a science course. How many ways can they be assigned to two laboratory sections of 12 students each? The labs have different instructors.

52. Use the following table of 100 people (55 male, 45 female) surveyed to answer the questions below.

	Likes Video Games	Dislikes Video Games	No Opinion
Male	49	5	1
Female	20	19	6
Total	69	24	7

(a) Of all 100 people in the survey, what is the probability that someone is male and dislikes video games?
(b) Of all 100 people in the survey, what is the probability that someone is female and has no opinion of video games?

53. Student Life did a survey of students in which they asked if the person is a part-time student or a full-time student. They also asked if the person had voted in the most recent student election. The results follow:

	Part-time Student	Full-time Student	Row Total
Voted	15	20	35
Did not vote	25	30	55
Column Total	40	50	90

If a student is selected at random from this group of 90 students, find the probability that:
(a) The student voted in the most recent election.
(b) The student voted in the most recent election, given that the student is a part-time student.
(c) The student voted in the most recent election and the student is a part-time student.
(d) Are the events "student voted" and "student is a part-time student" independent or not? Explain.

54. Which statement below is *not* a basic probability rule for events A and B?

A. For any event A: $0 \le P(A) \le 1$

B. Events A and B are mutually exclusive if $P(A \text{ and } B) \ne 0$

C. Complement of A: $P(\text{not } A) = 1 - P(A)$

D. Events A and B are independent events if $P(A) = P(A, \text{given } B)$

E. $P(\text{entire sample space}) = 1$

55. The college placement office did a study to determine how graduates got their first full-time jobs after graduation. Interviews of 1000 students produced the following information:

Source	Number
On-campus Interviews	350
Networking	400
Employment Agency	200
Job Fair	50

(a) Find the probability the job came from each of the above sources.
(b) Do the probabilities add up to 1? Should they?
(c) Find the probability that the job came from on-campus interviews or from networking.
(d) Find the probability that the job did not come from an employment agency.

56. If you draw three cards from a standard deck of 52 cards without replacement, what is the probability that the first is a diamond, the second is a club, and the third is a heart?

57. An urn contains 10 balls, all identical except for color. There are 4 red balls, 2 blue balls, 3 yellow balls and 1 green ball.
(a) You draw one ball at random from the urn. Find the probability that you get a blue ball or a yellow ball.
(b) Are the outcomes of part (a) mutually exclusive? Give a reason for your answer.
(c) Find the probability that the ball is not red.
(d) Find the probability that the ball is blue, yellow or green.

58. The age distribution of senators in a recent congress was as follows:

Age	Under 40	40 – 49	50 – 59	60 – 69	70 – 79	80 & Over
Number	5	30	36	22	5	2

For a senator selected at random from this group find the probability that:
(a) The senator is under 40 years old.
(b) The senator is in her or his 40's or 50's.
(c) The senator is at least 60 years old.
(d) The senator is under 50 years old.

59. Sixty children were asked if their favorite color was blue, pink, or neither. Below are the results.

	Blue	**Pink**	**Neither**
Boys	12	0	16
Girls	6	17	9
Total	18	17	25

What is the probability that pink is *not* a girl's favorite color?

A. $\frac{15}{32}$ B. $\frac{17}{32}$ C. $\frac{17}{28}$ D. $\frac{43}{60}$ E. $\frac{15}{28}$

60. A new grading policy has been proposed by the dean of the College of Education for all education majors. All faculty and students in the college were asked to give their opinions about the new policy. The results are given below:

	Favor	Neutral	Opposed	Row Total
Students	353	75	191	619
Faculty	11	5	18	34
Column Total	364	80	209	653

Suppose that someone is selected at random from the College of Education (either student or faculty). Let us use the following notation for events: S = Student; F = faculty; Fa = in favor; N = neutral; O = opposed. In this notation, *P*(O, *given* S) represents the probability that a person selected opposes the grading policy given that the person is a student.

(a) Compute *P*(Fa), *P*(Fa, *given* F), and *P*(Fa, *given* S).
(b) Compute *P*(F *and* Fa).
(c) Are the events "F" and "Fa" independent? Explain.
(d) Compute *P*(S, *given* O) and *P*(S, *given* N).
(e) Compute *P*(S *and* Fa) and *P*(S *and* O).
(f) Are the events "S" and "Fa" independent? Explain.
(g) Compute *P*(Fa *or* O). Are these events mutually exclusive?

61. A box contains 15 marbles: 5 blue, 5 red, and 5 green. If three marbles are removed, one at a time, without replacement, what is the probability that first a red is removed, then two greens?

A. $\frac{10}{273}$ B. $\frac{2}{91}$ C. $\frac{1}{27}$ D. $\frac{4}{135}$ E. $\frac{25}{546}$

62. The following table shows how 558 people applying for a credit card were classified according to home ownership and length of time in present job.

	Less than 2 years	2 or more years	Row Total
Owner	73	194	267
Renter	210	81	291
Column Total	283	275	558

Suppose that a person is chosen at random from the 558 applicants. We will use the following notation for events: O = person owns home; R = person rents home; L = person has had present job less than 2 years; M = person has had present job 2 years or more.

(a) Compute *P*(O), *P*(O, *given* L), *P*(O, *given* M).
(b) Compute *P*(O *and* M) and *P*(O *or* M).
(c) Compute *P*(R), *P*(R, *given* L), and *P*(R, *given* M).
(d) Compute *P*(R *and* L) and *P*(R *or* L).
(e) Are the events O and M independent? Explain.
(f) Are the events O and M mutually exclusive? Explain.

63. Two cards are drawn from a standard deck of 52 without replacement. What is the probability that a king is drawn first and a queen of hearts next?

A. $\frac{1}{2,652}$ B. $\frac{1}{169}$ C. $\frac{1}{663}$ D. $\frac{1}{676}$ E. $\frac{4}{663}$

64. Brookridge College did a survey of former students who were incoming freshmen 8 years ago. Of these former students some graduated from college and some dropped out and did not graduate. Some were earning salaries at or above the upper middle income level and others were not. The data is tabulated below:

	Below upper middle income	**At or above upper middle income**	**Row Total**
College graduate	105	406	511
College dropout	351	291	642
Column total	456	697	1,153

Suppose a person is selected from the people classified above. We will use the following notation for events: G = person graduated from college; D = person dropped out of college; B = salary is below the upper middle income level; A = salary is at or above upper middle income level.

(a) Compute *P*(A), *P*(A, *given* G), *P*(A, *given* D).
(b) Compute *P*(B), *P*(B, *given* G), *P*(B, *given* D).
(c) Compute *P*(G), *P*(G, *given* A), *P*(G, *given* B).
(d) Compute *P*(D), *P*(D, *given* A), *P*(D, *given* B).
(e) Compute *P*(G *and* A) and *P*(D *or* B).
(f) Are the events A and G independent? Explain.
(g) Are the events B and D mutually exclusive? Explain.

65. A hockey team plays two games. Use a tree diagram to list the possible win, loss, and the sequences the team can experience for the set of three games.

66. Cut-Rate Sam's Appliance Store has received a shipment of 12 toasters. Sam does not know that 8 are defective.

(a) Pete is the first person to buy one of these toasters. What is the probability that it is defective?
(b) Iris is the next person after Pete to buy one of these toasters. What is the probability that her toaster is defective if Pete's was defective?
(c) What is the probability that both Pete's and Iris's toasters are defective?
(d) What is the probability that at least one of the two toasters is defective?
(e) What is the probability that neither toaster is defective?

67. What is the probability at 1 in the tree diagram below if the balls are drawn from an urn without replacement and there are six blue and four green balls in the urn?

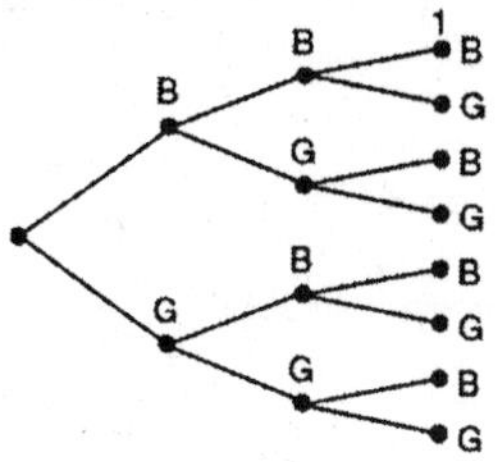

A. $\frac{1}{6}$ B. $\frac{1}{30}$ C. $\frac{3}{10}$ D. $\frac{3}{25}$ E. $\frac{3}{125}$

68. (a) How many ways can you choose a set of 8 beads from 12 distinct beads?
(b) How many ways can you arrange the 8 beads you have chosen on a string?

69. (a) How many ways can you choose a panel of 5 judges from a pool of 10 different judges?
(b) How many ways can you seat the five judges in a row?

70. How many ways can you arrange 8 different brands of cereal in a row on a display shelf?

71. What is the probability at 1 in the tree diagram below if two marbles are drawn from a box without replacement and there are five yellow, four orange, and three red balls in the box?

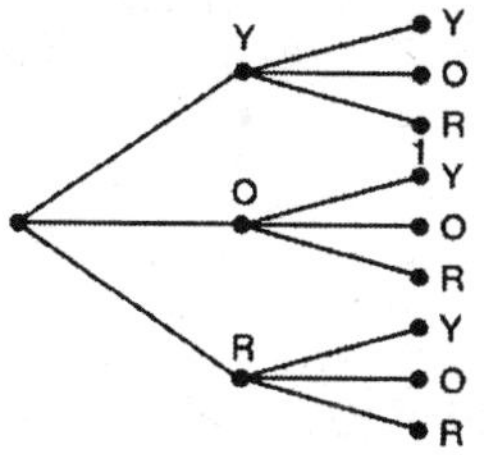

A. $\frac{5}{44}$ B. $\frac{5}{33}$ C. $\frac{1}{6}$ D. $\frac{5}{36}$ E. $\frac{1}{11}$

72. Betty belongs to a 24 member hiking club. The club received two free annual passes to the national parks and decided to give them to two members chosen by lottery. Each member was eligible to win at most one pass. All 24 members took part in the lottery.
(a) Find the probability that Betty did not win the first pass.
(b) Find the probability that Betty won the second pass given that she did not win the first pass.
(c) Find the probability that Betty won the second pass.
(d) Find the probability that Betty won either the first or the second pass.
(e) Find the probability that Betty did not win either pass.

73. An urn contains six balls numbered 1 through 6.
(a) How many ways can you select exactly 4 of these six balls?
(b) How many ways can you arrange the six letters {S, S, S, S, F, F}?

74. A silk shirt comes in a choice of five colors, four sizes, and with short or long sleeves. How many distinct styles does the shirt come in?

A. 10 B. 8 C. 36 D. 40 E. 20

75. (a) How many ways can you choose 4 people to play bridge from 9 bridge players? Assume that everyone is willing to play everyone else.
(b) How many ways can you seat the four people you have chosen around the bridge table?

76. In some skating competitions the order in which a group of five skaters skate is determined by lottery. Skating last (fifth position) is often considered to be advantageous. In a competition Andrew, Bill, Charles, David and Edward are the last group to skate.
(a) Andrew draws first. What is the probability that Andrew draws a position other than 5?
(b) Bill in the second skater to draw. What is the probability that Bill will draw a position other than 5 given that Andrew did not draw a five?
(c) What is the probability that the position 5 will not be drawn by either Andrew or Bill?

77. A boat is sold with a variety of options. It has six different colors, four different engines, three different seat layouts, and with a convertible roof or without. In how many different styles is this boat available?

A. 24 B. 120 C. 72 D. 180 E. 144

78. Mark is searching for a friend who has just moved into a new condo. Since he has forgotten the building number and the condo number he has decided to search door to door. The complex has three buildings. Each building has two floors and each floor has two condos.
(a) Draw a tree diagram showing the buildings, floors and condos to be checked.
(b) What is the maximum number of condos Mark will have to search to find his friend's condo?

79. Compute the number of ordered seating arrangements for seven people in a car that seats five.

A. 60 B. 840 C. 42 D. 5,040 E. 2,520

80. Fanciful Flights airline has direct flights from Denver, Colorado to San Francisco, Chicago and New York City. From San Francisco it has direct flights to Singapore, Hong Kong and Sidney. From Chicago it has direct flights to Dublin and London and from New York City it has direct flights to Paris, Rome and Tel Aviv.
(a) Draw a tree diagram to illustrate all flights the airline offers to international destinations.
(b) How many such routes does the airline offer?

81. Laura is training for a week-long mountain cycling tour. She has 12 short hilly routes from which to choose mid-week rides.
(a) How many ways can she choose four different rides from the list for the first week's training if the order matters?
(b) How many ways can she choose four different rides if the order does not matter?
(c) If she has chosen the first weeks rides, how many ways can she choose four more different rides for the second week? Assume that order does not matter.

82. A caterer has given a party organizer 12 choices for a five-course meal. Compute the number of ordered possibilities for the catered dinner.

A. 2,020 B. 3,991,680 C. 95,040 D. 5,040 E. 120

83. How many ways can you choose a project management team consisting of a project manager, a systems analyst and three programmers? There are two project managers, 4 systems analysts and 6 programmers to choose from.

84. The manager of a water-sports store has found that 5% of her customers purchase sailboats and 15% of her customers purchase boat trailers. However the probability that a customer will purchase a boat trailer given that she or he has purchased a sailboat is 80%.
(a) Find the probability that a customer will purchase a sailboat and trailer.
(b) Find the probability that a customer will purchase a sailboat or a trailer.

85. Nine aspiring athletes will be selected out of 17 trying out for their school basketball team. How many different groups of nine are there?

86. How many ways can you have 4 sunny days and 3 cloudy days (with no sun) in one week? (Hint: Number the days and choose the numbers of the sunny days.)

87. How many ways can you have 5 work days and 2 days off in one week. (Hint: Number the days and choose the numbers of the days off.)

88. Six hopeful musicians will be selected from 12 auditioning for a symphony. How many different groups of six are there?

A. 924 B. 13,366,080 C. 110,880 D. 5,544 E. 665,280

89. The manager of a restaurant has found that 40% of the evening dinner customers order appetizers and 60% of them order desserts. He has also observed that 70% of these customers who have ordered appetizers also order desserts.
(a) Calculate the probability that an evening dinner customer will order both an appetizer and a dessert.
(b) Calculate the probability that an evening dinner customer will order an appetizer or a dessert.
(c) Calculate the probability that an evening dinner customer will not order either an appetizer or a dessert.

90. How many sequences of three letters can you make from the letters {A, B, C, D, E} if
(a) The letters must all be different.
(b) Repetitions are allowed.

Section B: Algorithms

Objective 5.1: What is Probability?

91. Determine the probability of the event:
You will roll the number 3 on a die.

92. A tetrahedron is a solid figure with 4 faces, each of which is an equilateral triangle numbered 1 through 4.

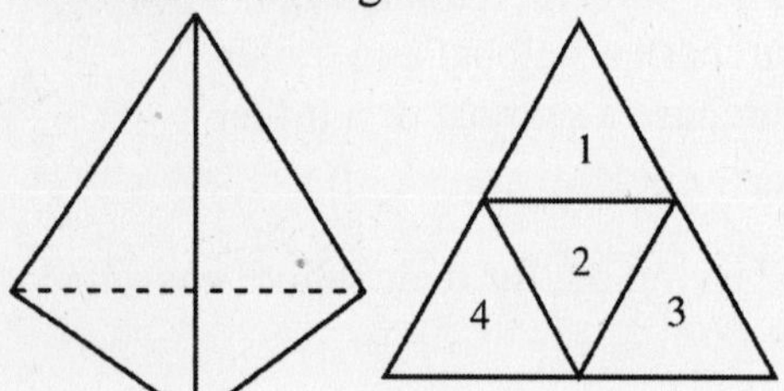

Ben rolled a tetrahedron, noted which face landed down, and recorded the results in the table below:

Number	Frequency
1	8
2	10
3	7
4	11

Using the result in the table, what is the probability that Ben will get an even number on the next roll of the tetrahedron?

93. Hiro is in the spirit club. There are 29 students in the club. Three of them will be picked at random to go on a retreat. What is the probability that Hiro will *not* be picked to attend the retreat?

94. The probability of drawing a red ball from an urn is $\frac{5}{33}$. What is the probability that a red ball will *not* be drawn from the urn?

95. A spinner is numbered from 1 through 9 with each number equally likely to occur. What is the probability of obtaining a number less than 4 or greater than 7 in a single spin?

96. You work at a T-shirt printing business. Of 2400 T-shirts shipped, 168 are printed improperly. If you choose a T-shirt out of the shipment, what is the chance that it is printed correctly?

97. The probability of drawing a red ball from an urn is $\frac{9}{44}$. What is the probability that a red ball will *not* be drawn from the urn?

Objective 5.2: Some Probability Rules-Compound Events

98. From a committee of 6 women and 5 men, two names are drawn to lead the committee. What is the probability that both people drawn will be women?

99. Are the following events independent?
Draw a red card from a deck with an equal number of red and blue cards, and then draw another red card.

100. A bag contains 2 purple marbles and 4 green marbles. One marble is drawn at random and not replaced. Then a second marble is drawn. Find $P(\text{green, purple})$.

101. Suppose you roll one standard six-sided die. Are the 2 events below disjoint?
Event 1: The die shows an even number.
Event 2: The die shows a power of 3.

102. A bag contains 6 white marbles and 4 yellow marbles. One marble is drawn at random and not replaced. Then a second marble is drawn. Find $P(\text{yellow, white})$.

103. Identify the disjoint events.
a. picking a 5 from a deck of cards, rolling a 3 with a standard six-sided die
b. with one roll of a standard six-sided die, rolling a 5, rolling a 3
c. with two rolls of a standard six-sided die, rolling a 5, rolling a 3
d. rolling a 5 with a standard six-sided die, drawing a 5 from a deck of cards

104. Mary has a bag that contains 9 green marbles, 7 red marbles, and 5 blue marbles. Without looking, Mary draws a marble from the bag. What is the probability that Mary will draw a red or a green marble?

105. After the introduction of a new soft drink, a taste test is conducted to see how it is being received. Of those who participated, 40 said they preferred the new soft drink, 136 preferred the old soft drink, and 24 could not tell any difference. What is the probability that a person in this survey preferred the new soft drink?

106. Highest Degree Earned in 1993 (Number in thousands)

	Total	Doctorate	Professional	Master's	Bachelor's	Associate
Total	188,683	1295	2193	8355	24,943	8301
Male	90,555	996	1580	4222	12,554	3598
Female	98,128	299	614	4134	12,390	4703

What is the probability that a woman chosen at random has a professional degree? Round to the nearest thousandth.

Objective 5.3: Trees and Counting Techniques

107. At a pizza parlor, Paola has a choice of pizza toppings and sizes. The topping choices are mushrooms, ham, and black olives. The size choices are medium and large. Draw a tree diagram that shows the number of possible pizza combinations that Paola can order.

108. Ernest went to the mall to buy a shirt for a friend. His style choices for the shirt are striped and v-neck. Both of these choices come in orange, green, and blue. Draw a tree diagram that shows the number of shirt choices that are available to Ernest.

109. Suppose you can choose from a white, yellow, or tan shirt and blue, beige, or black jeans. Make a tree diagram and list the outcomes. How many total outcomes are there?

110. Suppose you are choosing a 4-digit personal access code. This code is made up of 2 numbers (1-9 which can be repeated), followed by 2 letters (A-Z which also can be repeated). Find the total number of possibilities.

111. Mary and Trent designed the game *Lucky Draw* for their school carnival. To play, draw a card from a standard deck of cards. Replace the card after each draw. If you draw a red card you lose. If you draw a black card, you get to draw again. If you draw a black card on the second draw, you win the prize. If you draw a red card on the second draw, you lose. Make a tree diagram to help you answer the question below.

a. What is the probability of winning Lucky Draw?
b. If 128 people play Lucky Draw, how many would you expect to win?
c. Explain why this is not a fair game.
d. How would you change the rules to make Lucky Draw a fair game?

Evaluate.

112. 10!

113. $P_{9,3}$

114. $C_{6,2}$

115. Evaluate:
7!

116. There are 11 students participating in a spelling bee. How many ways can the students who go first, second, third and fourth be chosen?

117. A committee is to consist of two members. If there are seven men and six women available to serve on the committee, how many different committees can be formed?

118. Suppose you are choosing a wall color from among 5 different paint colors and also choosing an accent color from among 2 different paint colors and 3 different levels of shine (flat, semi-gloss, and glossy). How many choices do you have?

119. In how many different ways can three diamonds be drawn from a standard deck of cards if the order of selection is important?

120. Barry, Lupe, Julio, Jim, and Maria are in the math club. The club advisor will assign students to 3-person teams at the next Math Team competition. How many 3-person teams can be formed from these five students?

Answer Key for Chapter 5
Elementary Probability Theory

Section A: Static Items

(a) $P(not\ A) = 0.52$
(b) $P(A) = 0.25$
(c) No
(d) Yes
[1] (e) Yes

[2] E.

(a) {Alice, Bob, Carol, Dick}
(b) P(Bob) = 0.25
(c) P(Bob *or* Dick) = 0.50
(d) P(woman) = 0.5
[3] (e) P(*not* Alice) = 0.75

[4] C.

[5] $P = 0.01$; Relative frequency

(a) P(Sarah) = 1/6 = 0.1667
(b) P(Michael *or* Sarah) = 1/3 = 0.3333
[6] (c) P(*not* Michael *or* Sarah) = 2/3 = 0.6667

(a) P(library user) = 0.6
(b) P(*not* library user) = 0.4
(c) P(2 library users) 0.3517
[7] (d) P(2 library non-users) 0.1517

[8] C.

(a) P(3 *or* 6) = 1/3 = 0.3333
(b) P(*not* 2) = 5/6 = 0.8333
[9] (c) P(2 *or* 4 *or* 6) = 3/6 = 0.5

[10] (a) 0.023; (b) 115 defective car-door handles

(a) P(ace of spades) = 0.0192
(b) P(spade) = 0.25
[11] (c) P(ace *or* spade) = 0.3077

(a) P((king of diamonds, king of hearts)) = 0.0004
(b) P((king of diamonds, diamond)) = 0.0045
[12] (c) P(neither card is a king) = 0.8507

[13] (a) 0.2; (b) 250 children

(a) P(number less than 5 on both dice) = 0.4444
(b) P(sum = 8) = 0.1389
[14] (c) P(both dice show 5) = 0.0278

[15] B.

(a) P(sum = 8) = 0.1389
(b) P(red die shows 3 and green die shows 5) = 0.0278
[16] (c) P(dice show 3 and 5 in either order) = 0.0556

(a) P(blue or white) = 0.625
(b) P((red, white)) = 0.0536
[17] (c) P(red and white in either order) = 0.10712

[18] B.

[19] $C(10,3) = 120$

[20] $C(15,6) = 5{,}005$

(a) The sample space is the colors of jelly beans: blue, green, red, yellow, and purple.;
[21] (b) 0.3; (c) 0.2

[22] $C(5,2) = 10$

[23] B.

(a) *P*(saw ad) = 0.65
(b) *P*(bought Sparkle) = 0.49
(c) *P*(bought Sparkle, *given* ad) = 0.4923
[24] (d) *P*(saw ad *and* bought Sparkle) = 0.3200

[25] B.

[26] 40 choices

[27] *P*(5,2) = 20

[28] B.

Tree Diagram for Coin Toss

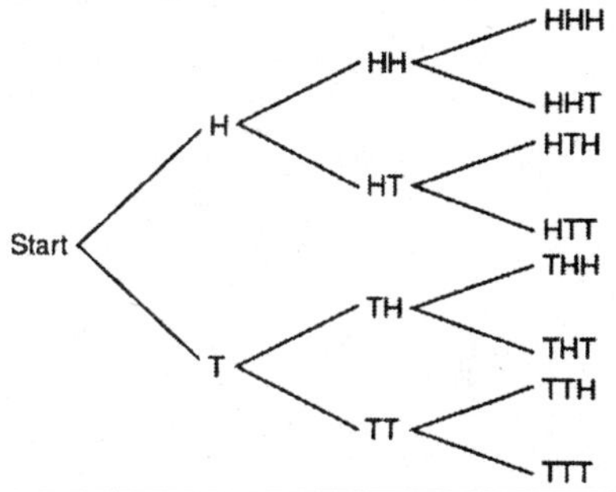

[29]

(a) *P*(defective) = 0.06
(b) *P*(*not* defective) = 0.94
(c) Sample space: {defective, not defective}
[30] (d) No, the events are not equally likely.

[31] A.

(a) *P*(team won) = 0.35
(b) *P*(watch games) = 0.75
(c) *P*(watch games, *given* team won) = 0.6286
[32] (d) *P*(team won *and* watch games) = 0.2200

[33] E.

Tree Diagram for die and coin

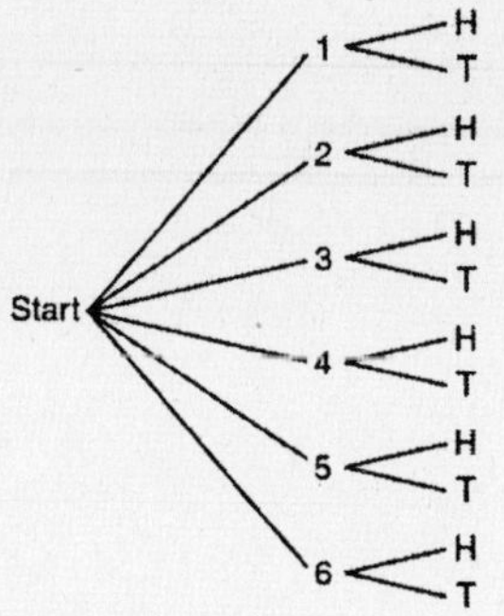

[34]

[35] Number of arrangements of answers =

[36] A.

[37] Number of arrangements of winners = $P(10,3) = 720$

(a) Number of winners = $C(7,5) = 21$
[38] (b) Number of winners = $P(7,5) = 2,520$

[39] C.

(a) Number of subcommittees = $C(8,4) = 70$
[40] (b) Number of slates = $P(12,4) = 11,880$

(a) $P(\text{liberal}) = 0.60$
(b) $P(\text{conservative}) = 0.40$
(c) $P(\text{mayor}) = 0.20$
[41] (d) $P(\textit{not}\ \text{mayor}) = 0.80$

[42] $P(\text{June passes both exams}) = 0.4756$

[43] (a) $\frac{3}{8}$ (b) $\frac{5}{8}$ (c)

[44] $P(\text{woman}\ \textit{and}\ \text{engineering major}) = 0.106$

[45] A.

P(1st class *or* executive commuter) = 0.47
[46] P(coach *or* excursion) = 0.53

(a) P((R,G)) = 0.1563
[47] (b) P((R,G)) = 0.1786

Number of telephones = 6
Tree Diagram for Phone Types

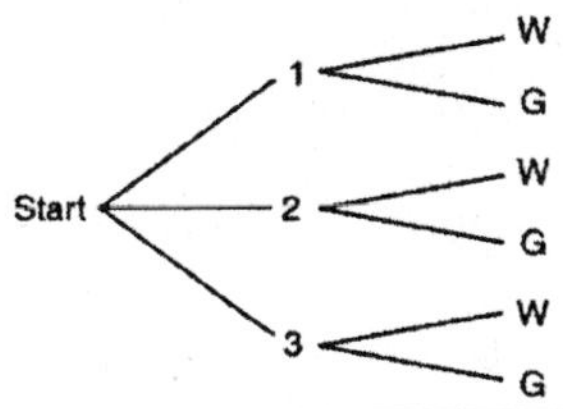

[48]

[49] B.

(a) P((white, blue)) = 0.0988
[50] (b) P((white, blue)) = 0.1111

[51] Number of assignments = $C(24,12)$ = 2,704,156

[52] (a) $\frac{1}{20}$; (b) $\frac{3}{50}$

(a) P(student voted) = 0.3889
(b) P(student voted, *given* part-time student) = 0.375
(c) P(part-time student *and* student voted) = 0.1667
[53] (d) No, the events are not independent.

[54] B.

(a)

Job Source	Probability
On-campus interviews	0.350
Networking	0.400
Employment Agency	0.200
Job fair	0.050

(b) Yes, the probabilities add up to one.
(c) *P*(on-campus interview *or* networking) = 0.750
[55] (d) *P*(*not* employment agency) = 0.800

[56] $P = \dfrac{169}{10{,}200}$

(a) *P*(blue *or* yellow) = 0.50
(b) Yes, they are mutually exclusive.
(c) *P*(*not* red) = 0.60
[57] (d) *P*(blue *or* yellow *or* green) = 0.60

(a) *P*(under 40) = 0.05
(b) *P*(40's *or* 50's) = 0.66
(c) *P*(at least 60) = 0.29
[58] (d) *P*(under 50) = 0.35

[59] A.

(a) *P*(Fa) = 0.5574; *P*(Fa, *given* F) = 0.3235; *P*(Fa, *given* S) = 0.5703
(b) *P*(F *and* Fa) = 0.0168
(c) No, they are not independent. *P*(Fa) ≠ *P*(Fa, *given* F)
(d) *P*(S, *given* O) = 0.9139; *P*(S, *given* N) = 0.9375
(e) *P*(S *and* Fa) = 0.5406; *P*(S *and* O) = 0.2925
(f) No. They are not independent.
[60] (g) *P*(Fa or O) = 0.8775; Yes, Fa and O are mutually exclusive.

[61] A.

(a) P(O) = 0.4785; P(O, *given* L) = 0.2580; P(O, *given* M) = 0.7055
(b) P(O *and* M) = 0.3477; P(O *or* M) = 0.6237
(c) P(R) = 0.5215; P(R, *given* L) = 0.7420; P(R, *given* M) = 0.2945
(d) P(R *and* L) = 0.3763; P(R *or* L) = 0.6523
(e) No, they are not independent. $P(\text{O}) \neq P(\text{O}, \textit{given}\ \text{M})$
(f) No, they are not mutually exclusive. Some owners have been at the present job two or

[62] more years.

[63] C.

(a) P(A) = 0.6045; P(A, *given* G) = 0.7945; P(A, *given* D) = 0.4533
(b) P(B) = 0.3955; P(B, *given* G) = 0.2055; P(B, *given* D) = 0.5467
(c) P(G) = 0.4432; P(G, *given* A) = 0.5825; P(G, *given* B) = 0.2303
(d) P(D) = 0.5568; P(D, *given* A) = 0.4175; P(D, *given* B) = 0.7697
(e) P(G *and* A) = 0.3521; P(D *or* B) = 0.6479
(f) No, A and G are not independent. Graduates are more likely to be at or above middle income in this population than members of the population as a whole are.
(g) No, B and D are not mutually exclusive. Some dropouts are below the upper middle

[64] income level.

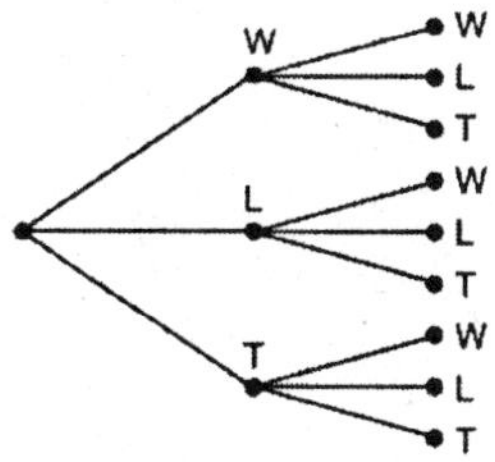

[65]

(a) P(Pete's toaster defective) = 0.6667
(b) P(Iris's toaster defective, *given* that Pete's is) = 0.6364
(c) P(both toasters defective) = 0.4242
(d) P(Pete's *or* Iris's toaster defective) = 0.9091

[66] (e) P(neither toaster defective) = 0.0909

[67] A.

(a) Number of ways = $C(12,8)$ = 495

[68] (b) Number of ways = $P(8,8)$ = 8! = 40,320

(a) Number of ways = $C(10,5)$ = 252

[69] (b) Number of ways = $P(5,5)$ = 5! = 120

[70] Number of ways = 8! = 40,320

[71] B.

(a) *P*(Betty did not win first pass) = 0.9583
(b) *P*(Betty won second pass, *given* that she did not win the first) = 0.0435
(c) *P*(Betty won the second pass) = 0.0417
(d) *P*(Betty won the first pass *or* the second pass) = 0.0833
[72] (e) *P*(Betty did not win either pass) = 0.9167

(a) Number of ways = *C*(6,4) = 15
[73] (b) Number of ways = *C*(6,4) = 15

[74] D.

(a) Number of ways = *C*(9,4) = 126
[75] (b) Number of ways = 4!/4 = 3! = 6

(a) *P*(*not* 5) = 0.80
(b) *P*(*not* 5 on second draw, *given not* 5 on first) = 0.75
[76] (c) *P*(*not* 5 on first draw *and not* 5 on second draw) = 0.60

[77] E.

(a) Tree diagram for search

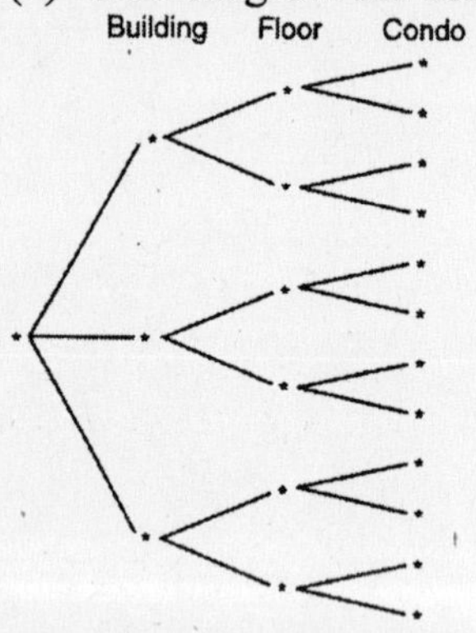

[78] (b) Number of condos = 12

[79] E.

(a) Tree diagram for flights

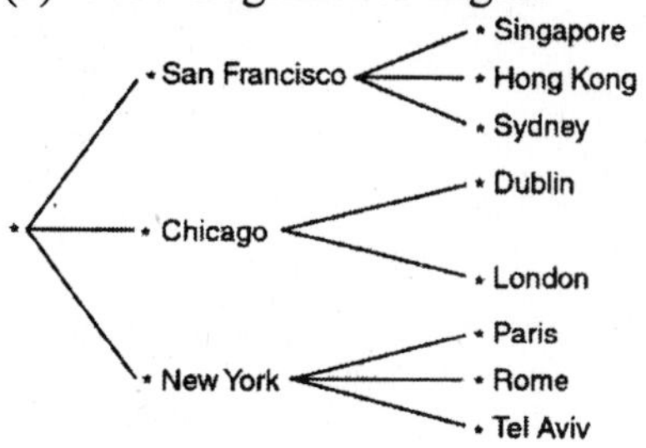

[80] (b) Number of routes = 8

(a) Number of ways = $P(12,4) = 9{,}720$
(b) Number of ways = $C(12,4) = 495$
[81] (c) Number of ways = $C(8,4) = 70$

[82] C.

[83] Number of ways = $2 \times 4 \times C(6, 3) = 160$

(a) P(boat *and* trailer) = 0.04
[84] (b) P(boat *or* trailer) = 0.16

[85] Combinations: 24,310

[86] Number of ways = $C(7,4) = 35$

[87] Number of ways = $C(7,2) = 21$

[88] A.

(a) P(appetizer and dessert) = 0.28
(b) P(appetizer or dessert) = 0.72
[89] (c) P(not(appetizer or dessert)) = 0.28

(a) Number of sequences = $P(5,3) = 60$
[90] (b) Number of sequences = $5^3 = 125$

Section B: Algorithms

Objective 5.1: What is Probability?

[91] $\frac{1}{6}$

[92] $\frac{21}{36}$

[93] $\frac{26}{29}$

[94] $\frac{28}{33}$

[95] $\frac{5}{9}$

[96] 93%

[97] $\frac{35}{44}$

Objective 5.2: Some Probability Rules-Compound Events

[98] $\frac{3}{11}$

[99] No

[100] $\frac{4}{15}$

[101] Yes

[102] $\frac{4}{15}$

[103] with one roll of a standard six-sided die, rolling a 5, rolling a 3

[104] $\frac{16}{21}$

[105] $\frac{1}{5}$

[106] 0.006

Objective 5.3: Trees and Counting Techniques

[107] 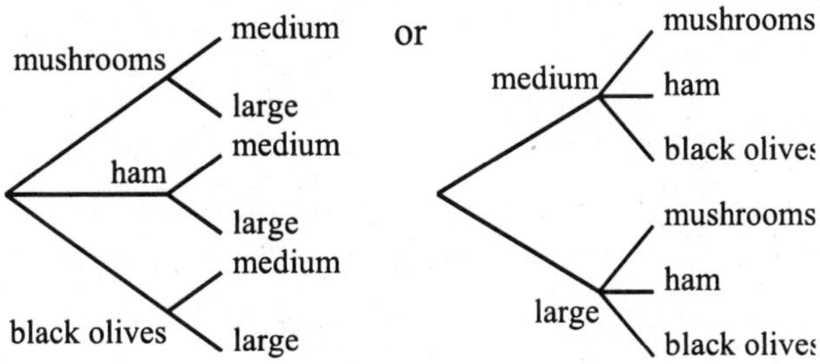

[108] 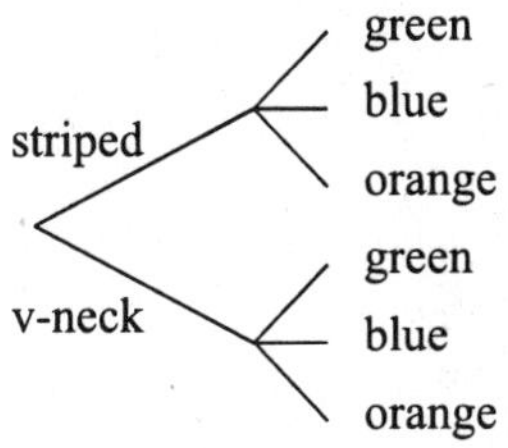

[109] 9

[110] 54,756

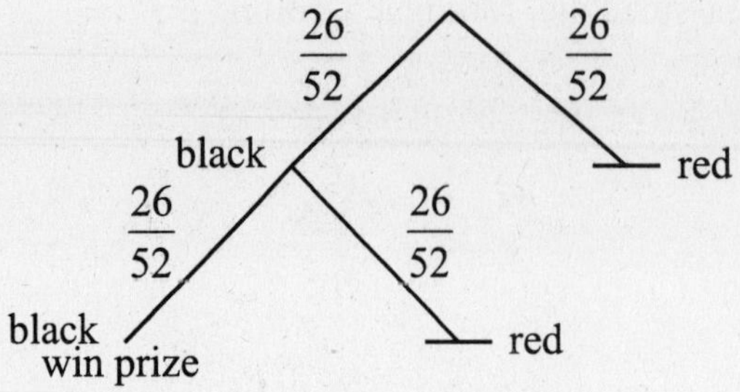

a. $\frac{1}{4}$

b. 32 people

c. Sample answer: The chances of winning and losing are not equal.

[111] d. Sample answer: If you draw a red card, you win.

[112] 3,628,800

[113] 504

[114] 15

[115] 5040

[116] 7920

[117] 78

[118] 30

[119] 1716

[120] 10

Chapter 6
The Binomial Distribution and Related Topics

Section A: Static Items

1. Identify each of the random variables as continuous or discrete.
   (a) The number of cows in a pasture
   (b) The number of electrons in a molecule
   (c) The voltage on a power line
   (d) The volume of milk given by a cow per milking
   (e) The distance from Cape Canaveral to a point chosen at random in the Sea of Tranquility on the moon
   (f) The number of words in a book chosen at random

2. Identify each of the random variables as continuous or discrete.
   (a) The time in hours that you sleep on a random weekday night
   (b) The home team score in a basketball game
   (c) The number of ducks on a pond
   (d) A score on the reading comprehension portion of the SAT exam
   (e) The volume of water in Lake Powell
   (f) The number of fish in Lake Lulu

3. At Cape College the business students run an investment club. Each fall they create investment portfolios in multiples of $1,000 each. Records from the past several years show the following probabilities of profits (rounded to the nearest $50). In the table below, x = profit per $1,000 and $P(x)$ is the probability of earning that profit.

x	0	50	100	150	200
$P(x)$	0.15	0.35	0.25	0.20	0.05

   (a) Find the expected value of the profit in a $1,000 portfolio.
   (b) Find the standard deviation of the profit.
   (c) What is the probability of a profit of $150 or more in a $1,000 portfolio?

4. A franchise chain of small grocery stores has kept records of the number of bad checks passed in its stores. They used the data to get a probability distribution for the number of bad checks passed in a store each week. In the table below x = number of bad checks and $P(x)$ is the probability that x bad checks will be passed in a week.

x	0	1	2	3	4
$P(x)$	0.3	0.3	0.2	0.1	0.1

(a) Calculate the expected number of bad checks the chain will get in one week.
(b) Calculate the standard deviation for the number of bad checks.
(c) What is the probability that two or more bad checks will be passed in a week?
(d) What is the probability that no bad checks will be passed in a week?

5. George is a computer salesman who usually visits 6 customers each day. Over the years, George has recorded the number of sales per day. He used this data to estimate the probability of 0 sales, 1 sale, 2 sales and so on up to 6 sales per day. In the table below, x = number of sales per day and $P(x)$ is the probability of making x sales.

x	0	1	2	3	4	5	6
$P(x)$	0.047	0.187	0.311	0.276	0.138	0.037	0.004

(a) The number of sales can be thought of as a random variable. Is it discrete or continuous? Explain.
(b) Compute the expected value for the number of sales on a typical day.
(c) Compute the standard deviation for the number of sales.

6. Carol is a dental assistant. She usually has five clients per day. Over the years, she has recorded the number of cancellations and used the data to estimate the probability of 0 through 5 cancellations per day. In the table below x = number of cancellations per day and $P(x)$ = the probability of x cancellations.

x	0	1	2	3	4	5
$P(x)$	0.328	0.410	0.205	0.051	0.005	0.001

(a) The number of cancellations is a random variable. Is it discrete or continuous? Explain.
(b) Compute the expected number of cancellations per day.
(c) Compute the standard deviation for the number of cancellations.

7. A local cab company is interested in the number of pieces of luggage a cab carries on a taxi run. A random sample of 260 taxi runs gave the following information. x = number of pieces of luggage and f is the frequency with which taxi runs carried x pieces of luggage.

x	0	1	2	3	4	5	6	7	8	9	10
f	42	51	63	38	19	16	12	10	6	2	1

(a) Find the probability distribution for x.
(b) Make a histogram of the probability distribution.
(c) Estimate the probability that a taxi run will have from 0 to 4 pieces of luggage (including 0 and 4).
(d) Compute the expected value of x.
(e) Compute the standard deviation for x.

8. The Army gives a battery of exams to all new recruits. One exam measures a person's ability to work with technical machinery. This exam was given to a random sample of 360 new recruits. In the table below x = score on the test and f is the frequency of new recruits with score x.

x	1	2	3	4	5	6	7	8	9	10
f	28	42	79	83	51	36	18	12	7	4

(a) Find the probability distribution for these scores.
(b) Draw a histogram of the probability distribution.
(c) The ground-to-air missile battalion needs people with a score of 7 or higher on the exam. What is the probability that a new recruit will meet this criterion?
(d) The kitchen battalion can use people with scores of three or less. What is the probability that a new recruit is not overqualified for this work?
(e) Compute the expected value of these scores.
(f) Compute the standard deviation of these scores.

9. Reggie Richman has a poor driving record and must take out a special insurance policy. He wants to insure his $58,000 sports car. Based on previous driving records, AnyState has estimated the probability of various levels of loss per year as indicated in the table below. x = dollar amount of loss. The company will pay no benefits for any other partial losses.

% of loss	100	50	25	10	0
x	58,000	29,000	14,500	5,800	0
$P(x)$	0.05	0.12	0.28	0.42	0.13

(a) Calculate the expected loss on Reggie's car.
(b) How much should AnyState charge Reggie for insurance if it wants to make a profit of $1,000 on his policy.

10. Mr. Dithers wants to insure his yacht for $280,000. The Big Rock insurance company estimates that, over a period of one year, a total loss may occur with probability 0.005, a 50% loss with probability 0.01, a 25% loss with probability 0.05, a 10% loss with probability 0.08. Big Rock will pay for no other partial losses.
(a) What is the expected loss on this yacht?
(b) How much should the company charge Mr. Dithers for insurance if it wants to make a profit of $1000 per year on his policy?

11. The student senate is sponsoring a car raffle to raise money to buy playground equipment for disadvantaged children. The senate received a donated car worth $7,000 and sold 3,750 raffle tickets at $5 each.
(a) If you buy 10 tickets, what is the probability that you will win the car? What is the probability that you will not win the car?
(b) Compute your expected earnings. (Your expected earnings is the product of the value of the car and the probability that you will win it.)
(c) How much did you contribute to the playground fund? (The amount of the contribution is the difference between what you paid for the tickets and your expected earnings.)

12. The Old West Historical Museum is trying to raise money for building restoration. A travel agent donated an all-expense paid trip to Hawaii for two, valued at $3,587.00. Friends of the Museum sold 1950 tickets at $5.00 per ticket.
(a) If you buy 8 tickets, what is the probability that you will win the trip? What is the probability that you will not win the trip?
(b) What are your expected earnings from this raffle?
(c) How much do you contribute to the Museum through your ticket purchase? (The amount contributed is the difference between the cost of the tickets and your expected earnings.)

13. To estimate the number of books required for a course, the student council at Lees Field College took a random sample of 100 courses and obtained the information below. x = number of books and f = number of courses requiring x books.

x	0	1	2	3	4	5	6
f	5	48	22	11	9	3	2

(a) Make a table showing the probability distribution for the number of required books.
(b) Make a histogram of the probability distribution.
(c) Find the expected number of books per course and the standard deviation.
(d) If the average book costs $42 and you take 5 courses, how much can you expect to spend for books?

14. Identify which one of the following random variables is continuous.

A. The number of fish caught by a fishing boat

B. The number of tables sold at a furniture store

C. The number of traffic accidents in a city

D. The number of gallons of water in a reservoir

E. The number of coins contained in a slot machine

15. Identify which one of the following random variables is discrete.

A. The number of gallons of concrete used at a construction site

B. The number of beverages sold at a lemonade stand

C. The temperature of a pot roast cooking in an oven

D. The time required for a runner to finish a marathon

E. The number of inches of rainfall in a county

16. Identify each random variable as continuous or discrete.
(a) Speed of an automobile
(b) The number of doughnuts left in the pantry
(c) The air temperature of a public park
(d) The weight of a professional wrestler
(e) The number of restaurant patrons

17. Choose the answer below that identifies a value for y that results in a valid probability distribution.

x	0	1	2
$P(x)$	0.35	0.5	y

A. $y = 0.6$

B. $y = 0.15$

C. $y = -1.85$

D. $y = 1.85$

E. $y = 0.06$

18. A police commissioner is interested in the number of phone calls into the police dispatcher that result in criminal charges actually being filed. For a random sample of 190 nights (7 pm to 7 am), the following information was obtained where x = number of calls resulting in criminal charges and f = frequency with which this many calls occurred (i.e., number of nights).

x	25	26	27	28	29	30	31	32	33	34
f	11	16	17	22	24	17	23	26	21	13

Assuming these 190 nights represent the population of all nights in this police precinct, what do you estimate the probability is that on a randomly selected night, there will be from 27 to 31 (including 27 and 31) calls to the dispatcher that result in criminal charges being filed?

A. 0.46

B. 0.67

C. 0.33

D. 0.21

E. 0.54

19. A police commissioner is interested in the number of phone calls into the police dispatcher that result in criminal charges actually being filed. For a random sample of 190 nights (7 pm to 7 am), the following information was obtained where x = number of calls resulting in criminal charges and f = frequency with which this many calls occurred (i.e., number of nights).

x	25	26	27	28	29	30	31	32	33	34
f	11	16	17	22	24	17	23	26	21	13

Assuming these 190 nights represent the population of all nights in this police precinct, what do you estimate the probability is that on a randomly selected night, there will be from 25 to 29 (including 25 and 29) calls to the dispatcher that result in criminal charges being filed?

A. 0.29

B. 0.28

C. 0.47

D. 0.71

E. 0.18

20. A police commissioner is interested in the number of phone calls into the police dispatcher that result in criminal charges actually being filed. For a random sample of 190 nights (7 pm to 7 am), the following information was obtained where x = number of calls resulting in criminal charges and f = frequency with which this many calls occurred (i.e., number of nights).

x	25	26	27	28	29	30	31	32	33	34
f	11	16	17	22	24	17	23	26	21	13

Assuming these 190 nights represent the population of all nights in this police precinct, use relative frequencies to find $P(x)$ when x = 25, 26, 27, 28, 29, 30, 31, 32, 33, and 34 rounded to two decimal places. Use these probabilities to find the expected number of calls to the dispatcher that result in criminal charges being filed.

A. 32.13

B. 20.16

C. 23.12

D. 29.75

E. 27.09

21. For a particular lake in the U.S., the following table shows the percentage of fishermen who caught x number of fish during a 6-hour period while fishing from shore.

x	0	1	2	3 or more
%	52	27	14	7

(a) Find the probability that a fisherman selected at random fishing from shore catches one or more fish during a 6-hour period.
(b) Find the probability that a fisherman selected at random fishing from shore catches two or more fish during a 6-hour period.
(c) Compute μ, the expected value of the number of fish caught per fisherman in a 6-hour period (round 3 or more to 3). Round to two decimal places.
(d) Compute σ the standard deviation of the number of fish caught per fisherman in a 6-hour period (round 3 or more to 3). Round to two decimal places.

22. Eric teaches ceramics in his studio. He estimates that one out of every five people who call for information about a class will sign up for the class. Last week he received nine calls. Find the probability that four or fewer of the people who called will sign up for a class.

23. Dorothy leads wilderness expeditions. She has found that 15% of those who attend a promotional meeting will sign up for an expedition at the meeting. If 20 people attend a meeting:
(a) What is the probability that none of them will sign up?
(b) What is the probability that five or more of them will sign up?

24. In Summit County 65% of the voter population are Republicans. What is the probability that a random sample of ten Summit County voters will contain:
(a) Exactly eight Republicans?
(b) Exactly two Republicans?
(c) No Republicans?
(d) All Republicans?

25. The college health center did a campus-wide survey of students and found that 15% of the students smoke cigarettes. A group of nine students randomly come together and sit at the same table on the plaza in front of the library. Find the probability that:
(a) No student at the table smokes.
(b) At least one student at the table smokes.
(c) More than two students smoke.
(d) From one to five smoke (including one and five).

26. The mayor said that local fire stations are so well equipped and staffed that they are able to get to a fire in five minutes or less about 80% of the time. If this is so, find the probability that out of the next ten fire calls:
(a) The fire department arrives in five minutes or less for every call.
(b) It takes five minutes or less for five or more of the calls.
(c) It takes the fire department more than five minutes to arrive for all ten calls.
(d) For at least three calls it takes more than five minutes for the fire department to arrive.

27. A TV sports commentator claims that 45% of all football injuries are knee injuries. Assuming that this claim is true, what is the probability that in a game with 5 reported injuries:
(a) All are knee injuries?
(b) None are knee injuries?
(c) At least three are knee injuries?
(d) No more than 2 are knee injuries?

28. A safety engineer claims that 18% of all automobile accidents are due to mechanical failure. Assume that this is correct and use the formula for binomial probabilities to find the probability that exactly three out of eight automobile accidents is due to mechanical failure.

29. Martha estimates the probability that she will receive at least one marketing telephone call at home during the hours of 5 pm to 7 pm on a weekday night to be 2/3. Use the formulas for computing binomial probabilities to answer the following questions:
(a) What is the probability that she will receive at least one call on all five of the next five weekday nights?
(b) What is the probability that she will not receive a call on any of the next five weekday nights?
(c) What is the probability that she will receive a call on at least four of the next five weekday nights?

30. It is estimated that four out of five patients who have an artery bypass heart operation survive eight years without needing another bypass operation. Of seven patients who recently had such an operation, what is the probability that:
(a) All will survive eight years without needing another bypass operation?
(b) At least four will survive eight years without needing another bypass operation?
(c) Exactly four will survive eight years without needing another bypass operation?

31. The probability of an adverse reaction to a pneumonia shot is 0.15. Pneumonia shots are given to a group of 12 people.
(a) What is the probability that none of them will have an adverse reaction to the pneumonia shot?
(b) What is the probability that more than half of them will have an adverse reaction to the pneumonia shot?
(c) What is the probability that 1 or fewer of the 12 people have an adverse reaction to the pneumonia shot?
(d) What is the probability that two or more of them will have an adverse reaction to the pneumonia shot?

32. The probability that a restaurant patron will request seating in the outdoor patio is 0.45. A random sample of 7 people call to make reservations.
(a) Find the probability that fewer than three of them request outdoor seating.
(b) Find the probability that three or more of them request outdoor seating.
(c) Find the probability that none of them request outdoor seating.

33. It is claimed that 70% of the cars on the Valley Highway are going faster than 65 miles per hour. A random sample of 12 cars was observed under normal driving conditions with no police car in sight.
(a) What is the probability that all of them were going faster than 65 miles per hour?
(b) What is the probability that fewer than half of them were going over 65 miles per hour?

34. A fair quarter is flipped three times.
(a) Find the probability of getting exactly three heads.
(b) Find the probability of getting two or more heads.
(c) Find the probability of getting fewer than two heads.
(d) Find the probability of getting exactly two heads.

35. A fair quarter is flipped 11 times.
(a) Find the probability of getting exactly 6 heads.
(b) Find the probability of getting more than 6 heads.
(c) Find the probability of getting fewer than 6 heads.
(d) Find the probability of getting four to nine heads (including four and nine).

36. Richard has just been given a ten-question multiple choice test in his history class. Each question has five answers only one of which is correct. Since Richard has not attended class recently, he does not know any of the answers. Assume that Richard guesses randomly on all ten questions.
(a) Find the probability that he will answer all ten questions correctly.
(b) Find the probability that he will answer five or more questions correctly.
(c) Find the probability that he will answer none of the questions correctly.
(d) Find the probability that he will answer at least three questions correctly.

37. A tourist bureau for a western state conducted a study which showed that 65% of the people who seek information about the state actually come for a visit. The office receives 15 requests for information on the state.
(a) Find the probability that all 15 of the people visit the state.
(b) Find the probability that at least 9 of the people visit the state.
(c) Find the probability that no more than 8 of the people visit the state.
(d) Find the probability that from 4 to 10 people visit the state (including 4 and 10).

38. A certain type of penicillin will cause a skin rash in 10% of the patients receiving it.
(a) If the penicillin is given to a random sample of 15 patients what is the probability that no more than 2 will have a skin rash?
(b) If it is given to a random sample of 9 patients, what is the probability that at least one will have a skin rash?
(c) If it is given to a random sample of 12 patients what is the probability that more than two will have a skin rash?

39. A recent report stated that 35% of teenagers of driving age owned a car. Consider a random sample of eight teenagers of driving age.
(a) Find the probability that five or more of them do not own a car.
(b) Find the probability that three or fewer of them do own a car.

40. A recent medical survey reported that 45% of the respondents to a poll on patient care felt that doctors usually explain things well to their patients. Assuming that the poll reflects the feelings of all patients toward their doctors, find the probability that for 12 patients selected at random:

(a) Four of more agree with the statement.

(b) No more than 8 do not agree with the statement.

41. Sam is a computer salesman who has a history of making successful calls one-fourth of the time.

(a) What is the probability that he will be successful on at least three of the next five calls?

(b) What is the probability that he will be successful on none of the next five calls?

42. In rolling two fair dice simultaneously, the probability of rolling a 7 or 11 is $\frac{2}{9}$. What is the approximate probability of rolling a 7 or 11 two or more times in ten rolls of two dice?

A. 0.30

B. 0.69

C. 0.77

D. 0.23

E. 0.08

43. In rolling two fair dice simultaneously, the probability of rolling a 7 or 11 is $\frac{2}{9}$. What is the approximate probability of rolling a 7 or 11 at least once in 5 rolls of two dice?

A. 0.28

B. 0.41

C. 0.63

D. 0.72

E. 0.69

44. In rolling two fair dice simultaneously, the probability of rolling "snake eyes," that is, a 1 on each die for a sum of 2, is $\frac{1}{36}$. What is the probability of rolling either 1 or 2 snake eyes in 10 rolls of two dice?

A. 0.25

B. 0.22

C. 0.32

D. 0.24

E. 0.31

45. Assume that 67% of the cars on a particular freeway are traveling faster than 70 miles per hour. A random sample of 15 cars was observed under normal driving conditions with no police car in sight. What is the probability that 8 or more of them were going faster than 70 miles per hour?

A. 0.70

B. 0.83

C. 0.67

D. 0.92

E. 0.96

46. Willard has just been given a ten-question multiple choice test in one of his classes. Each question has five answers only one of which is correct. Since Willard has not attended class recently, he does not know any of the answers. Assume that Willard guesses randomly on all ten questions. Find the probability that he will answer, at most, 3 questions correctly.

A. 0.88

B. 0.77

C. 0.80

D. 0.70

E. 0.50

47. Assume that 65% of the cars on a freeway are traveling faster than 75 miles per hour. A random sample of 12 cars was observed under normal driving conditions with no police car in sight. What is the probability that more than 6 of them were going faster than 75 miles per hour?

A. 98%

B. 65%

C. 92%

D. 88%

E. 79%

48. Willard has just been given a 20-question multiple choice test in one of his classes. Each question has five answers only one of which is correct. Since Willard has not attended class recently, he does not know any of the answers. Assume that Willard guesses randomly on all 20 questions. Find the probability that he will answer, at most, 5 questions correctly.

A. 0.97

B. 0.41

C. 0.91

D. 0.80

E. 0.63

49. In the past year, Valentino's Pizza Restaurant grossed more than $1,500 a day for about 70% of its business days. During the next 8 business days, what is the probability (rounded to two decimal places) that Valentino's will gross more than $1,500 a day for
(a) 7 or more days?
(b) 5 or more days?
(c) less than 4 days?

50. In rolling two fair dice simultaneously, the probability of rolling a double (that is, the same number face up on each die), a 7, or an 11 is $\frac{7}{18}$. If two dice are rolled simultaneously 5 times, what is the probability (to two decimal places) of this occurring
(a) 0 times?
(b) 5 times?
(c) at most, 3 times?
(d) more than 3 times?

51. The probability that a theater patron will request seating on the main floor is 0.35. A random sample of 6 patrons call for tickets. Let r be the number who request main-floor seating.
(a) Find $P(r)$ for $r = 0, 1, 2, 3, 4, 5$, and 6.
(b) Make a histogram for the r probability distribution.
(c) What is the expected number out of 6 who will request main-floor seating?
(d) Find the standard deviation of r.

52. Long-term history has shown that 65% of all elected offices in a rural county have been won by Republican candidates. This year there are 5 offices up for public election in the county. Let r be the number of public offices won by Republicans.
(a) Find $P(r)$ for $r = 0, 1, 2, 3, 4$, and 5.
(b) Make a histogram for the r probability distribution.
(c) What is the expected number of Republicans who will win office in the coming election?
(d) What is the standard deviation of r?

53. Long-term history has shown that 75% of all elected offices in an urban district have been won by Democrats. This year there are 8 public offices up for election.
(a) Find the probability that five or more offices are won by Democrats.
(b) Find the probability that all of the offices are won by Democrats.
(c) What is the expected number of Democrats who will win office in this election?

54. A student has an option of using the American Heritage Dictionary, Webster's or Random House in English 103. 1/3 of the students use Webster's. 2/3 of the students use one of the other dictionaries.
(a) Find the probability that out of 4 English 103 students, 3 or more use Webster's.
(b) If 375 students are registered for English 103 next term and each student buys a dictionary at the bookstore, what is the expected number of Webster's dictionaries the bookstore will need?

55. Jim is an automobile salesman at Courtesy Cars, Incorporated. He has a history of making a sale for about 15% of all prospective customers that he takes for a test drive. On most days he takes about 8 prospects for test drives. Let r be the number of sales on the days when Jim has eight customers.
(a) Find $P(r)$ for $r = 0, 1, 2, 3, 4, 5, 6, 7$, and 8.
(b) Make a histogram for the r probability distribution.
(c) What is the expected number of cars Jim will sell on the days when he has eight prospective customers?
(d) What is the standard deviation of the r probability distribution?

56. Jim is an automobile salesman at Courtesy Cars, Incorporated. He has a history of making a sale for about 15% of all prospective customers that he takes for a test drive. How many prospective customers must Jim have if he wants to be 80% sure of making at least one sale?

57. The owners of a motel in Florida have noticed that in the long run about 40% of the people who stop to inquire about a room for the night actually rent a room.
(a) How many inquiries must the owners answer to be 99% sure of renting at least one room?
(b) If 25 separate inquiries are made about rooms, what is the expected number of rentals coming from these inquiries?

58. The athletic director at Smoky Hills College is recruiting freshman basketball players. From past experience he knows that the probability that an athlete whom he contacts will come to Smoky Hills is 75%.
(a) What is the minimum number of players he should contact so that the probability of at least 6 successful recruits is 90% or higher?
(b) If he contacts eight players, what is the expected number of the eight who will come to Smoky Hills?

59. Sharon makes telephone appeals for donations of used clothing and household goods for a charitable organization. She knows from experience that about 30% of the calls result in donations.
(a) How many calls must she make to be 90% sure of getting at least two donations.
(b) What is the expected number of donations from 15 calls.

60. A statewide survey of adults found that only 40% linked the greenhouse effect with the phenomenon of global warming. Suppose 20 adults of the region were asked about the greenhouse effect.
(a) What is the probability that fewer than 5 of them linked the greenhouse effect to global warming?
(b) What is the probability that 16 or more of them linked the greenhouse effect to global warming?
(c) What is the expected number of people in this group who would see the connection between the greenhouse effect and global warming?

61. An insurance company says that 15% of all fires are caused by arson. A random sample of five fire insurance claims are under study. Let r be the number of claims in this sample from fires that were started by arson.
(a) Make a histogram for the probability distribution of r.
(b) What is the expected number of arson fires among five fires?
(c) What is the standard deviation of r?

62. Safety engineers say that an automobile tire with less than 1/32 of an inch of rubber tread is not safe. It has been estimated that 10% of all cars in a western state have unsafe tires. The highway patrol recently stopped 7 cars and did a safety check including a check of tire tread. Let r be the number of cars in the sample that had unsafe tires.
(a) Make a histogram for the probability distribution of r.
(b) What is the expected number of cars in the sample with unsafe tires?
(c) What is the standard deviation of r?

63. Garfield College has an art appreciation course in the humanities program that is taken on a pass/fail basis. Over a long period of time the art department has observed that about 80% of the students enrolled in the course pass the course. Sixteen students are enrolled in a typical section of the course this term.
(a) Find the probability that all of the students in a 16-student section pass the course.
(b) Find the probability that 10 or fewer of the students in a 16-student section pass the course.
(c) Find the expected number of students in the section who will pass the course.

64. A new serum is claimed to be 70% effective in preventing the common cold. A random sample of 11 people are injected with the serum. The serum is considered successful if someone who has taken the serum survives the winter without a cold.
(a) Find the probability that the serum is effective for all 11 people.
(b) Find the expected number of people in the sample who survive the winter without a cold.
(c) Find the standard deviation of this probability distribution.

65. A biologist has found that 40% of all brown bears are infected with trichinosis.
(a) What is the expected number of infected brown bears in a random sample of 27 brown bears?
(b) What is the standard deviation of the probability distribution?

66. Salvage Products is a mail-order firm that sells merchandise at a substantial discount because some of the items are damaged or have missing parts. Customers may not return the merchandise. The catalogue lists rubber raincoats at 80% off the regular price. However, many of the raincoats leak. The catalogue states that 40% of the raincoats in stock leak.
(a) How many of these raincoats should you order to be 97% sure that at least one does not leak?
(b) If you order 10 raincoats what is the expected number which do not leak?

67. The quality-control inspector of a production plant will reject a batch of automobile batteries if three or more defectives are found in a random sample of eleven batteries taken from the batch. Suppose the batch contains 7% defective batteries. Estimate the mean and standard deviation, μ and σ of this binomial distribution. What is the expected number of defective batteries the inspector will find?

A. $\mu = 0.77$, $\sigma = 0.85$; the inspector will expect to find 1 defective battery.

B. $\mu = 0.72$, $\sigma = 0.77$; the inspector will expect to find 0 defective battery.

C. $\mu = 0.85$, $\sigma = 0.77$; the inspector will expect to find 1 defective battery.

D. $\mu = 0.77$, $\sigma = 0.85$; the inspector will expect to find 0 defective battery.

E. $\mu = 0.77$, $\sigma = 0.72$; the inspector will expect to find 1 defective battery.

68. The quality-control inspector of a production plant will reject a batch of automobile batteries if three or more defectives are found in a random sample of eleven batteries taken from the batch. Suppose the batch contains 7% defective batteries. What is the probability that the batch will be accepted?

A. 0.96

B. 0.95

C. 0.97

D. 0.99

E. 0.04

69. The quality-control inspector of a production plant will reject a batch of lamps if two or more defectives are found in a random sample of six lamps taken from the batch. Suppose the batch contains 18% defective lamps.
(a) Make a histogram showing the probability of $r = 0, 1, 2, 3, 4, 5, 6$ defective lamps in a random sample of six lamps.
(b) Find μ. What is the expected number of defective lamps the inspector will find?
(c) What is the probability (to two decimal places) that the batch will be accepted?
(d) Find σ (to two decimal places).

70. In a binomial distribution with $n = 9$ trials, the probability of success is $p = 0.52$. Find μ and σ to two decimal places.

A. $\mu = 4.32$, $\sigma = 2.25$

B. $\mu = 2.16$, $\sigma = 2.25$

C. $\mu = 4.32$, $\sigma = 1.50$

D. $\mu = 4.68$, $\sigma = 1.50$

E. $\mu = 1.50$, $\sigma = 4.32$

Section B: Algorithms

Objective 6.1: Introduction to Random Variables and Probability Distributions

71. Classify the data as discrete or continuous.
How many seconds it takes to run a lap around the track.

72. A doctor developed a test to measure anxiety levels. The possible scores are 0, 1, 2, 3, 4, 5, and 6, with 6 indicating the highest level of anxiety. She administered the test to 2000 volunteers between the ages of 21 and 35. The table below shows her results. Make a probability distribution graph of this data.

Score	Number of Subjects
0	107
1	235
2	421
3	589
4	382
5	188
6	78

73. Find the population standard deviation of the data set.

x	5	6	5	5	3	5	4	3
$P(x)$	0.32	0.41	0.45	0.35	0.37	0.39	0.46	0.31

74. Two dice are rolled. If the total is even, then player A receives 2 points. If the total is 5, then player B receives 9 points. Find the expected value for each player, and determine if the game is fair.

75. A coin is flipped twice. If heads comes up either time, add 5 points. If tails comes up twice, subtract 4 points. What is the expected value of this game?

76. Following is a probability distribution table for a scratch-off game for a local charity.

Prize, x	Probability, $P(x)$
\$7.00	$\frac{7}{16}$
\$9.00	$\frac{5}{16}$
\$11.00	$\frac{3}{16}$
Win nothing	$\frac{1}{16}$

(a) Find $\sum x$, $\sum P(x)$ and $\sum x \cdot P(x)$.

(b) These scratch-off cards are sold for \$8.50 each. How much money does the local charity expect to make if 1000 scratch-off cards are sold? Explain.

77. Is the time to run 100 meters a continuous or a discrete quantity?

78. Find the mean and the standard deviation for the following data.

x	5	8	7	9	2	3	4	5
$P(x)$	0.25	0.29	0.21	0.22	0.3	0.28	0.27	0.35

79. The possible scores on an experiment conducted by the university lab were 0, 1, 2, 3, 4, 5, and 6, with 6 indicating the highest level. The test was administered to 2000 volunteers between the ages of 21 and 35. The table below shows her results. Find the probability distribution graph of this data.

Score	Number of Subjects
0	107
1	189
2	378
3	618
4	409
5	224
6	75

80. Find the expected value of the random variable with the given probability distribution.

Outcome	91	24	87	64	40
Probability	0.02	0.27	0.29	0.31	0.11

81. A number cube is rolled and a coin is flipped. If the number cube shows an odd number, or the coin shows tails, or both, we get 6 points. If neither event occurs, we get 4 points. What is the expected value of this game?

Objective 6.2: Binomial Probabilities

82. A company guarantees customer satisfaction on the purchase of a product, or the company will refund the purchase price of the product. Previous experience has shown that 6% of all purchases are returned. What is the probability that no more than 1 of the next 8 purchases will be returned? Round answer to the nearest thousandth.

83. There are twelve cars in a parking lot on a very cold day. Suppose the probability of any one of them not starting is 0.13. What is the probability that exactly three of the cars do not start? Round answer to the nearest thousandth.

84. A biologist is studying a new hybrid squash. She knows that the probability of this hybrid germinating is 0.70. In a controlled environment she plants 12 seeds. What is the probability that exactly 8 seeds will germinate? Round answer to the nearest thousandth.

85. A microchip manufacturer knows that, when the process goes wrong, the percentage of defective chips produced is 8%. If a defective batch of chips is being inspected, find the probability that the fourth chip tested is the first to be defective. Round answer to nearest thousandth.

86. The chance that a new type of rocket launches successfully is 0.77. What is the probability that the fourth launch will be the first successful launch? Round answer to nearest thousandth.

87. Ben and Chris both want to use the family car tonight. Chris proposes a challenge. He will throw a ball at a target. Chris gets the car if he hits the target two times. Ben gets the car if Chris misses two times. Ben agrees to the contest. The probability that Chris hits the target each time is 0.7. What is the probability that it will take exactly three throws for one of them to get the car?

88. A technician is designing a new portable CD player that requires six batteries for its power. The probability that any one of these batteries will fail is 0.05. If one battery fails, the CD player will not run. The technician wants to be sure that the CD player will receive enough power from the batteries. What is the probability that of six batteries three will not fail?

89. A professional baseball player's batting average is .304, meaning that the probability of getting a hit on any given at bat is .304. What is the probability that his next hit will occur on his third time at bat?

90. A number cube is tossed 12 times. What is the probability of obtaining exactly 4 threes? Give the answer as a four-place decimal.

91. A fair coin is tossed 8 times. What is the probability of obtaining exactly 6 heads? Express the answer both in terms of ${}_nC_k$ and as a four-place decimal.

92. Kathy is playing a new version of the game Twenty Questions called Nine Questions. She flips a coin. If it comes up heads she gets asked a question and wins a prize. If it comes up tails, she is asked a question and a point is scored against her. The game is over when she has 9 prizes or has 9 points against her. What is the probability that the game will end in exactly eleven flips of the coin?

Objective 6.3: Additional Properties of the Binomial Distribution

93. The probability of a successful outcome in a scientific experiment is 0.29. Suppose the experiment is performed 4 times.
(a) Construct a histogram for this binomial distribution.
(b) Find the mean and standard deviation for the distribution.

94. At Yoshi's Seafood Palace, the probability that a lone diner will leave a tip is 0.25. During one lunch hour, six people dine at Yoshi's by themselves. Make a graph of the binomial probability distribution that shows the probabilities that 0, 1, 2, 3, 4, 5, or all 6 lone diners leave tips.

95. Let $n = 18$, $p = 0.7$, and $q = 0.3$. Find the mean and standard deviation for the Binomial distribution.

96. At Stanley's Steak House, the probability that a lone diner will leave a tip is 0.7. During one lunch hour, six people dine at Stanley's by themselves. Find the graph of the binomial probability distribution that shows the probabilities that 0, 1, 2, 3, 4, 5, or all 6 lone diners leave tips.

97. Let $n = 16$, $p = 0.3$, and $q = 0.7$. Find the mean and standard deviation for the distribution.

98. The probability of the success of a new scientific experiment is 0.4. Suppose the experiment is tested 4 times.
a. Construct a histogram to represent the binomial distribution.
b. Find the mean and standard deviation of the distribution.

99. For the binomial distribution with $n = 100$ and $p = 0.25$, find $P(43)$, the mean, and the standard deviation.

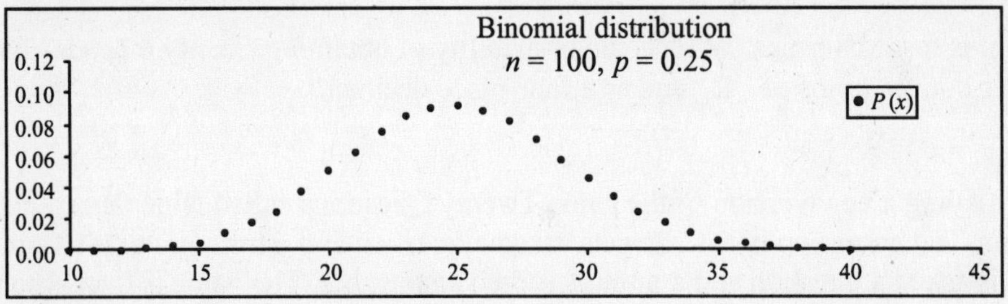

100. For the binomial distribution with $n = 100$ and $p = 0.5$, find $P(64)$, the mean, and the standard deviation.

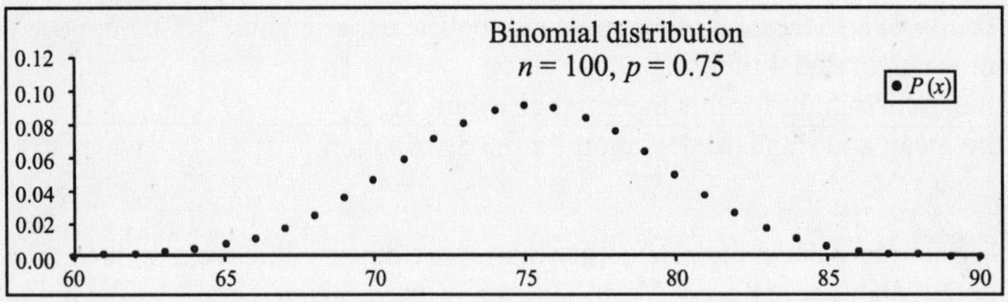

Answer Key for Chapter 6
The Binomial Distribution and Related Topics

Section A: Static Items

(a) Discrete
(b) Discrete
(c) Continuous
(d) Continuous
(e) Continuous
[1] (f) Discrete

(a) Continuous
(b) Discrete
(c) Discrete
(d) Discrete
(e) Continuous
[2] (f) Discrete

(a) Expected profit = \$82.50
(b) Standard deviation = \$55.40
[3] (c) $P(\text{profit} \geq \$150) = 0.25$

(a) Expected number of bad checks = 1.40
(b) Standard deviation = 1.28
(c) P(2 or more bad checks) = 0.4
[4] (d) P(no bad checks) = 0.3

(a) Discrete
(b) Expected number of sales = 2.398
[5] (c) Standard deviation = 1.201

(a) Discrete
(b) Expected number of cancellations = 0.998
[6] (c) Standard deviation = 0.893

(a)

x	0	1	2	3	4	5	6	7	8	9	10
$P(x)$	0.162	0.196	0.242	0.146	0.073	0.062	0.046	0.038	0.023	0.008	0.004

(b) See graph below.
(c) P(0 to 4 pieces of luggage) = 0.819
(d) Expected number of pieces of luggage = 2.558
(e) Standard deviation = 2.163

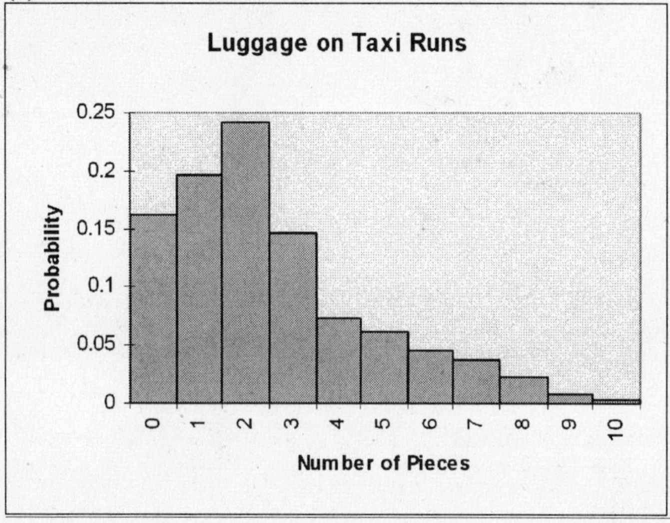

[7]

(a)

x	1	2	3	4	5	6	7	8	9	10
$P(x)$	0.078	0.117	0.219	0.231	0.142	0.100	0.050	0.033	0.019	0.011

(b) See graph below.
(c) P(score is 7 or greater) = 0.113
(d) P(score is 3 or less) = 0.414
(e) Expected score = 4.098
(f) Standard deviation = 1.942

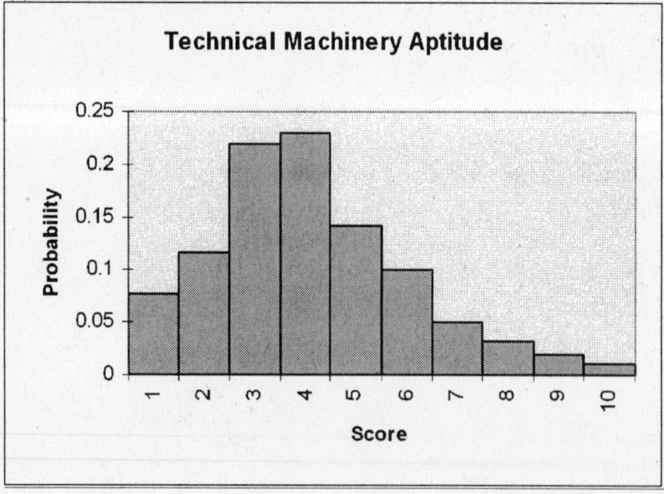

[8]

(a) Expected loss = $12,876
[9] (b) Premium = $13,876

(a) Expected loss = $8,540
[10] (b) Premium = $9,540

(a) P(win) = 0.0027; P(not win) = 0.9973
(b) Expected earnings = $18.90
[11] (c) Contribution = $31.10

(a) P(win) = 0.0041; P(not win) = 0.9959
(b) Expected earnings = $14.71
[12] (c) Contribution = $25.29

(a)

x	0	1	2	3	4	5	6
$P(x)$	0.05	0.48	0.22	0.11	0.09	0.03	0.02

(b) See graph below.
(c) Expected number of books per course = 1.88
(d) Estimated book cost for 5 courses = $394.80

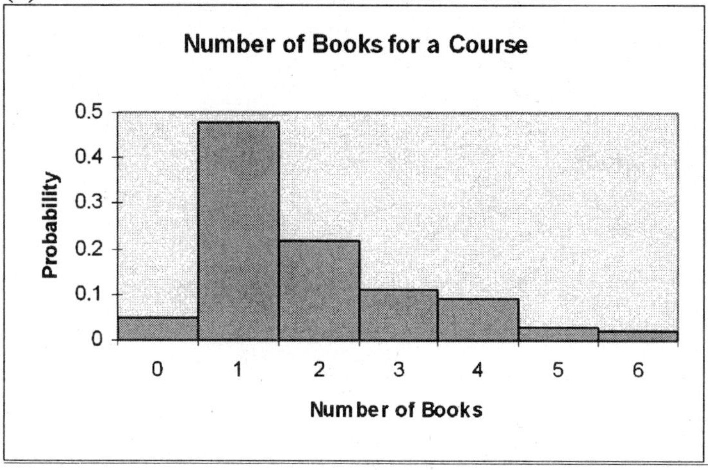

[13]

[14] D.

[15] B.

(a) Continuous
(b) Discrete
(c) Continuous
(d) Continuous
[16] (e) Discrete

[17] B.

[18] E.

[19] C.

[20] D.

(a) 0.48
(b) 0.21
(c) 0.76
[21] (d) 0.94

$n = 9, p = 0.2$
[22] P(four or fewer sign up) = 0.98

$n = 20, p = 0.15$
(a) P(no one signs up) = 0.039
[23] (b) P(5 or more sign up) = 0.170

$n = 10, p = 0.65$
(a) P(exactly eight Republicans) = 0.176
(b) P(exactly two Republicans) = 0.004
(c) P(no Republicans) = 0.000
[24] (d) P(all Republicans) = 0.014

$n = 9, p = 0.15$
(a) P(no one smokes) = 0.232
(b) P(at least one smokes) = 0.768
(c) P(more than two smoke) = 0.141
[25] (d) P(from one to five smoke) = 0.768

$n = 10, p = 0.80$
(a) P(prompt response for all calls) = 0.107
(b) P(prompt response for five or more calls) = 0.992
(c) P(prompt response for none of the calls) = 0.000
[26] (d) P(slow response for at least three calls) = 0.322

$n = 5, p = 0.45$
(a) P(all are knee injuries) = 0.019
(b) P(none are knee injuries) = 0.050
(c) P(at least three are knee injuries) = 0.408
[27] (d) P(no more than two are knee injuries) = 0.593

$n = 8, p = 0.18$
[28] P(three accidents are due to mechanical failure) = $C(8,3)(0.18)^3(0.82)^5 = 0.121$

$n = 5, p = 2/3$
(a) P(called on all 5 nights) = $C(5,5)\ (2/3)^5(1/3)^0 = 0.132$
(b) P(called on none of 5 nights) = $C(5,0)(2/3)^0(1/3)^5 = 0.004$
[29] (c) P(called on at least 4 nights) = $P(4) + P(5) = 0.461$

$n = 7, p = 0.8$
(a) P(all seven survive) = 0.210
(b) P(at least four survive) = 0.967
[30] (c) P(exactly four survive) = 0.115

$n = 12, p = 0.15$
(a) P(no adverse reactions) = 0.142
(b) P(more than half have adverse reactions) = 0.001
(c) P(one or fewer have adverse reactions) = 0.443
[31] (d) P(two or more have adverse reactions) = 0.556

$n = 7, p = 0.45$
(a) P(fewer than three request outdoor seating) = 0.316
(b) P(three or more request outdoor seating) = 0.684
[32] (c) P(none request outdoor seating) = 0.015

$n = 12, p = 0.70$
(a) P(all 12 are speeding) = 0.014
[33] (b) P(fewer than 6 are speeding) = 0.039

$n = 3, p = 0.50$
(a) P(exactly three heads) = 0.125
(b) P(two or more heads) = 0.50
(c) P(fewer than two heads) = 0.50
[34] (d) P(exactly two heads) = 0.375

$n = 11, p = 0.50$
(a) P(exactly six heads) = 0.226
(b) P(more than six heads) = 0.274
(c) P(fewer than six heads) = 0.50
[35] (d) P(from four to nine heads) = 0.882

$n = 10, p = 0.20$
(a) P(all ten correct) = 0.000
(b) P(five or more correct) = 0.033
(c) P(none correct) = 0.107
[36] (d) P(at least three correct) = 0.322

$n = 15, p = 0.65$
(a) P(all 15 people visit the state) = 0.002
(b) P(at least 9 people visit the state) = 0.756
(c) P(no more than 8 people visit the state) = 0.245
[37] (d) P(from 4 to 10 people visit the state) = 0.648

$p = 0.10$
(a) For $n = 15$, P(no more than two get a skin rash) = 0.816
(b) For $n = 9$, P(at least one gets a skin rash) = 0.612
[38] (c) For $n = 12$, P(more than 2 get a skin rash) = 0.110

$n = 8, p = 0.35$
(a) P(five or more do not own a car) = 0.707
[39] (b) P(three or fewer do own a car) = 0.707

$n = 12, p = 0.45$
(a) P(four or more agree) = 0.866
[40] (b) P(8 or fewer do not agree) = 0.866

$n = 5, p = 0.25$
(a) P(three or more successes) = 0.104
[41] (b) P(no successes) = 0.237

[42] B.

[43] D.

[44] D.

[45] D.

[46] A.

[47] E.

[48] D.

(a) 0.26
(b) 0.81
[49] (c) 0.06

(a) 0.09
(b) 0.01
(c) 0.92
[50] (d) 0.08

(a)

x	0	1	2	3	4	5	6
$P(x)$	0.075	0.244	0.328	0.236	0.095	0.020	0.002

(b) See graph below
(c) Expected number to request main-floor seating = 2.1
(d) Standard deviation = 1.17

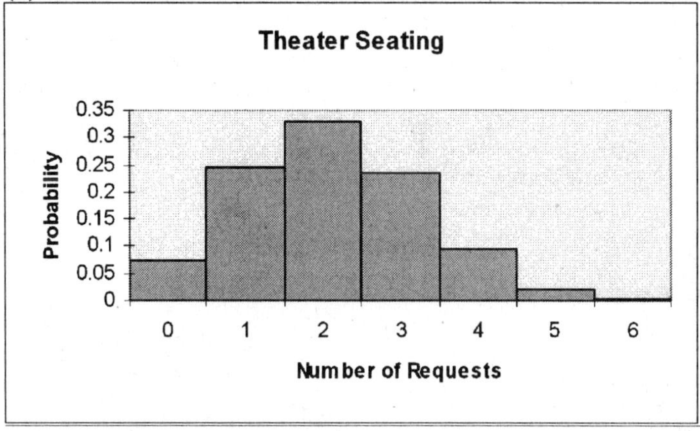

[51]

(a)

x	0	1	2	3	4	5
$P(x)$	0.005	0.049	0.181	0.336	0.312	0.116

(b) See graph below.
(c) Expected number of Republicans who win = 3.25
(d) Standard deviation = 1.067

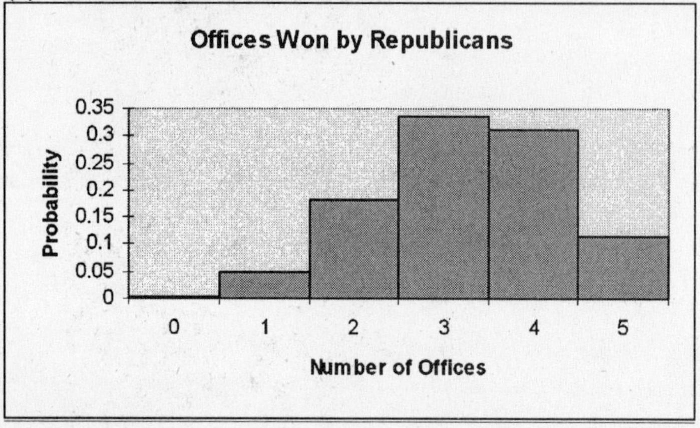

[52]

$n = 8, p = 0.75$
(a) P(Democrats win five or more) = 0.886
(b) P(Democrats win eight) = 0.100
[53] (c) Expected number of Democrats = 6

$n = 4, p = 1/3$
(a) P(3 or more use Webster's) = $P(3) + P(4) = 0.111$
[54] (b) For $n = 375$, expected number of Webster's dictionaries = 125

(a)

x	0	1	2	3	4	5	6	7	8
$P(x)$	0.272	0.385	0.238	0.084	0.018	0.003	0.000	0.000	0.000

(b) See graph below.
(c) Expected number of cars sold = 1.2
(d) Standard deviation = 1.01

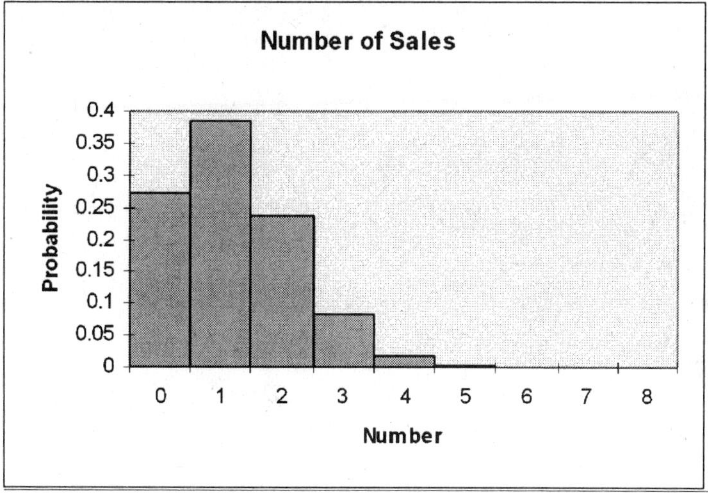

[55]

$p = 0.15$

[56] For $n = 10$, $P(r \geq 1) \geq 0.8$ and $P(r = 0) \leq 0.2$. Contact 10 prospects.

$p = 0.4$

(a) For $n = 9$, $P(r \geq 1) = 0.99$. Answer 9 inquiries.

[57] (b) For $n = 25$, the expected number of rooms rented is 10.

$p = 0.75$

(a) For $n = 10$, $P(r \geq 6) = 0.923$. Contact 10 players.

[58] (b) For $n = 8$, the expected number who will come is 6.

$p = 0.30$

(a) For $n = 12$, $P(r \geq 2) = 0.915$. Make 12 calls.

[59] (b) For $n = 15$, the expected number of donations is 4.5.

$n = 20, p = 0.40$

(a) P(fewer than 5 link the greenhouse effect with global warming) = 0.05

(b) P(16 or more link the greenhouse effect with global warming) = 0.0003

[60] (c) Expected number who link the greenhouse effect with global warming = 8

(a)

x	0	1	2	3	4	5
$P(x)$	0.444	0.392	0.138	0.024	0.002	0.000

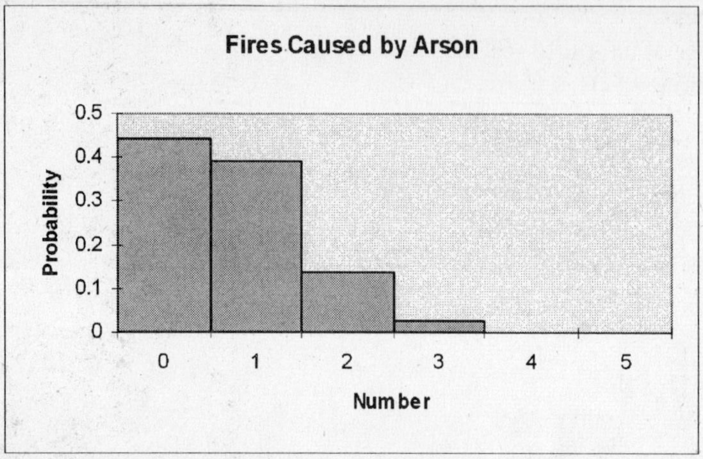

(b) Expected number of fires caused by arson = 0.75

[61] (c) Standard deviation = 0.798

(a)

x	0	1	2	3	4	5	6	7
$P(x)$	0.478	0.372	0.124	0.023	0.003	0.000	0.000	0.000

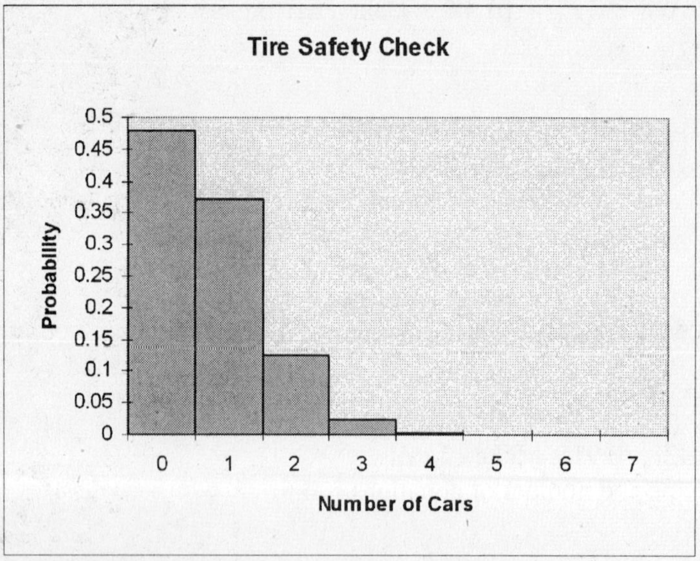

(b) Expected number of cars with unsafe tires = 0.70

[62] (c) Standard deviation = 0.794

$n = 16, p = 0.80$

(a) P(all pass) = 0.028

(b) P(ten or fewer pass) = 0.082

[63] (c) Expected number who pass = 12.8

$n = 11, p = 0.70$
(a) P(all protected) = 0.020
(b) Expected number who are protected = 7.7
[64] (c) Standard deviation = 1.52

$n = 27, p = 0.40$
(a) Expected number of infected bears = 10.8
[65] (b) Standard deviation = 2.55

Success: raincoat does not leak. $p = 0.60$
(a) For $n = 4$, P(at least one does not leak) = 0.974.
Order 4 raincoats.
[66] (b) For $n = 10$, the expected number which do not leak = 6.

[67] A.

[68] A.

(a)

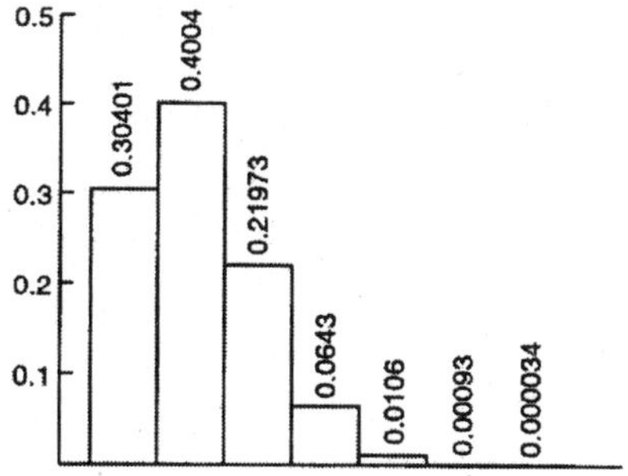

(b) $\mu = 1.08$; the inspector will expect to find 1 defective lamp.
(c) 0.70
[69] (d) 0.94

[70] D.

Section B: Algorithms

Objective 6.1: Introduction to Random Variables and Probability Distributions

[71] Discrete

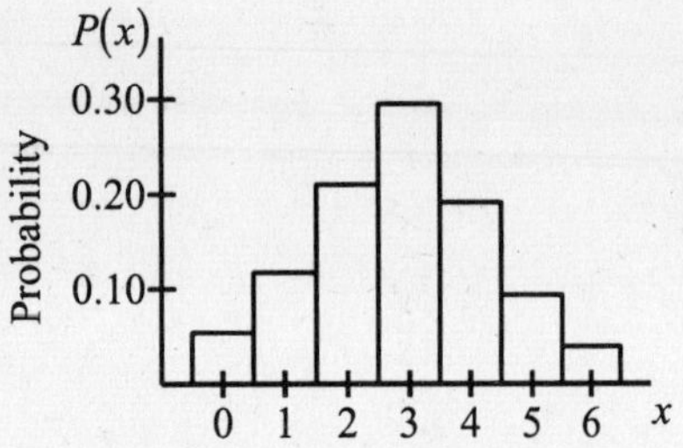

[72]

[73] $\sigma = 16.45$

[74] A: 1 point; B: 1 point; The game is fair.

[75] 2.75 points

[76] (a) $\sum x = \$27.00$; $\sum P(x) = 1$; $\sum x \cdot P(x) = \$7.94$
(b) The local charity would expect to make an average of \$8.50 – \$7.94 or \$0.56 per scratch-off card. For 1000 scratch-off cards, they would expect to make (1000)(\$0.56) or \$560.00.

[77] continuous

[78] Mean = 11.29; standard deviation = 9.56

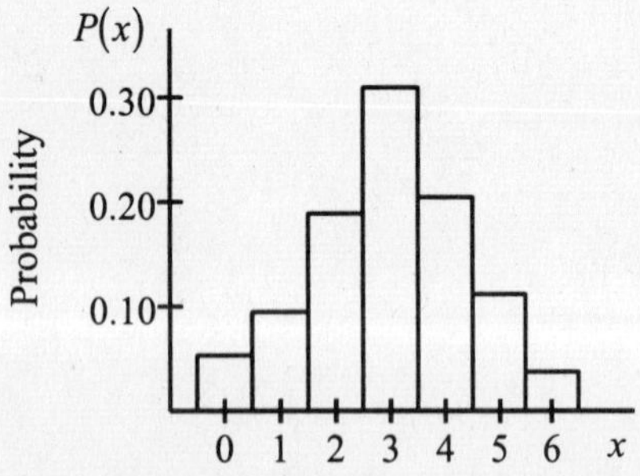

[79]

[80] 57.77

[81] 5.5

Objective 6.2: Binomial Probabilities

[82] 0.921

[83] 0.138

[84] 0.231

[85] 0.062

[86] 0.009

[87] 0.420

[88] 0.21%

[89] 0.147

[90] 0.0888

[91] ${}_8C_2(0.5)^8 \approx 0.1094$

[92] 0.044

Objective 6.3: Additional Properties of the Binomial Distribution

(a)

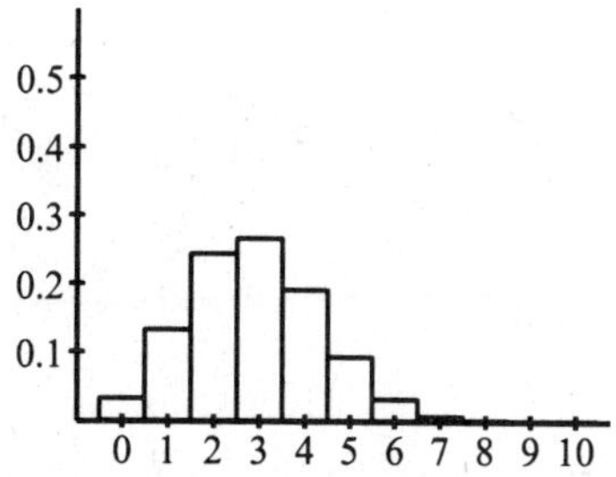

[93] (b) Mean = 1.2, standard deviation = 0.908

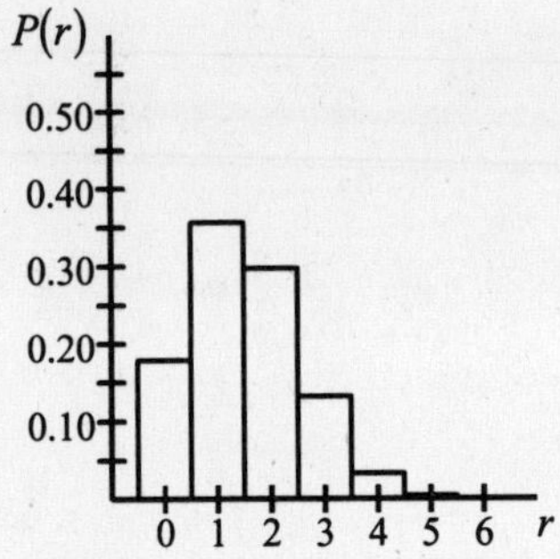

[94]

[95] $\mu = 12.6,\ \sigma = 1.94$

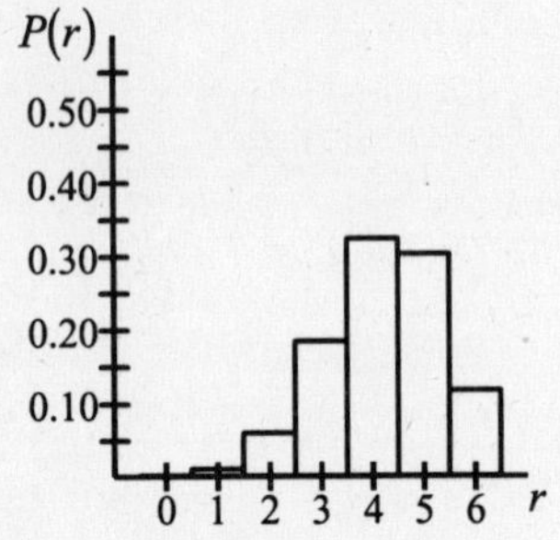

[96]

[97] $\mu = 4.8,\ \sigma = 1.83$

a.

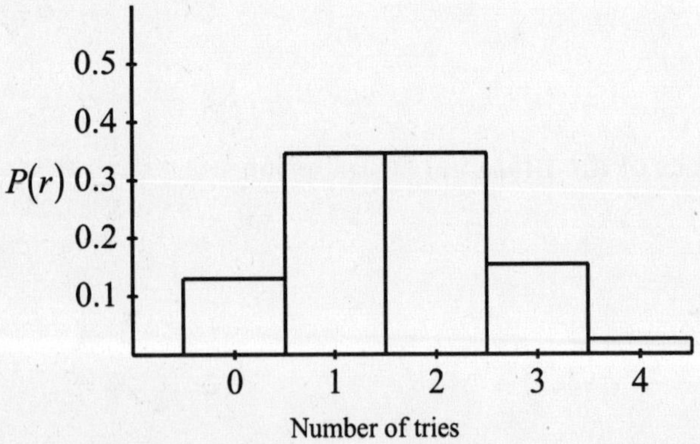

[98] b. Mean = 1.6; Standard deviation = 0.98

[99] $P(43) = 0.00004,\ \mu = 25,\ \sigma = 4.33$

[100] $P(64) = 0.00156,\ \mu = 50,\ \sigma = 5$

Chapter 7
Normal Distributions

Section A: Static Items

1. Statistical Abstracts (117th edition) provides information about state per capita income taxes. The distribution is mound shaped with mean $\mu = \$1,701$ and standard deviation $\sigma = \$672$.
(a) Use the empirical rule to find an interval centered about the mean in which about 68% of the data will fall.
(b) Estimate a range of values centered about the mean in which about 95% of the state per capita income taxes will fall.

2. Which statement is *not* an important property of a normal curve?

   A. Approximately 99.8% of the data values will be within four standard deviations on each side of the mean.

   B. The transition points between cupping upward and downward occur above $\mu + \sigma$ and $\mu - \sigma$.

   C. It is symmetrical about a vertical line through μ.

   D. The curve approaches the horizontal axis but never touches or crosses it.

   E. The curve is bell-shaped with the highest point over the mean μ.

3. The running times of feature movies have a mound shaped distribution with mean $\mu = 110$ minutes and standard deviation $\sigma = 30$ minutes.
(a) Use the empirical rule to find an interval centered about the mean in which 68% of the data will fall.
(b) Estimate a range of values centered about the mean in which about 95% of the data will fall.

4. The percentage growth rate of stocks of companies that specialize in construction and residential home building has a mound shaped distribution with mean approximately 9.8% and standard deviation 3.8%.
(a) Estimate a range of values centered about the mean in which about 68% of the data will fall.
(b) Estimate a range of values centered about the mean in which about 95% of the data will fall.

5. In a normal distribution, if $\mu = 6$ and $\sigma = 3$, where are the transition points located?

   A. 3, 9

   B. 0, 12

   C. 4, 8

   D. –3, 15

   E. 5, 7

6. The percentage growth rate of stocks of companies that specialize in recreation equipment and appliances has a mound shaped distribution with mean estimated to be 13.1% and standard deviation 6.0%.
(a) Estimate a range of values centered about the mean in which 68% of the data will fall.
(b) Estimate a range of values centered about the mean in which 95% of the data will fall.

7. A certain brand of light bulb lasts according to a normal distribution with μ = 1,000 hr and σ = 50 hr.
(a) What is the probability that a light bulb selected at random will last between 1,000 and 1,050 hr?
(b) What is the probability that a light bulb selected at random will last between 850 hr and 900 hr?

8. Eggs Inc. is a restaurant that specializes in breakfasts. The company has established a target mean of 175 customers per morning with standard deviation 25. For 12 consecutive days the number of breakfasts was tabulated. In the table below t = time in days and x = number of breakfasts.
(a) Make a control chart for the above data.
(b) Determine whether the process is in statistical control. If it is not, specify which out-of-control signals are present.

t	1	2	3	4	5	6	7	8	9	10	11	12
x	165	115	110	130	180	192	241	199	183	212	188	235

9. The Highway Patrol has a target of 28 traffic tickets per week with standard deviation 5 tickets per week for a stretch of mountain highway. The number of tickets issued for 15 consecutive weeks is given below. t = number of the week, x = number of tickets.

t	1	2	3	4	5	6	7	8	9	10	11	12	13	14	15
x	22	20	15	16	25	30	30	34	30	32	36	40	33	35	30

(a) Make a control chart for the above data.
(b) Determine whether the process is in statistical control. If it is not specify which out-of-control signals are present.

10. The lifetime of a disposable water filter is normally distributed with mean μ = 30 gal and standard deviation σ = 6 gal. What is the probability that a filter selected at random will last from 36 gal to 42 gal?

A. 0.27

B. 0.34

C. 0.135

D. 0.475

E. 0.1585

11. Lewis earned 85 on his biology midterm and 81 on his history midterm. In the biology class the mean score was 79 with standard deviation 5. In the history class the mean score was 76 with standard deviation 3.

(a) Convert each score to a standard z score.
(b) On which test did he do better compared to the rest of the class?

12. Last fiscal year, the average number of defective televisions produced by a factory each day was 17.6, with a standard deviation of 5.2. For the first ten days of this fiscal year, the number of defective televisions produced is shown in the following table:

Day	1	2	3	4	5	6	7	8	9	10
Defective Televisions	19	15	13	16	24	29	34	25	20	10

(a) Make a control chart for the daily number of defective televisions produced.
(b) Do the data indicate that the production is "in control"? Explain your answer.

13. Statistical Abstracts (117th edition) lists the number of federal officials indicted for the years 1985 to 1995. The mean number of officials indicted in this period is 574 and the standard deviation is 184. In the table below t = number of year (with year 1 = 1985) and x = number of indictments.

t	1	2	3	4	5	6	7	8	9	10	11
x	563	596	651	629	695	615	803	624	627	571	527

(a) Make a control chart for the above data.
(b) Determine if the process is in statistical control. If it is not, specify which out-of-control signals are present.

14. Jan earned 86 on her political science midterm and 82 on her chemistry midterm. In the political science class the mean score was 80 with standard deviation 4. In the chemistry class the mean score was 70 with standard deviation 6.

(a) Convert each midterm score to a standard z score.
(b) On which test did she do better compared to the rest of the class?

15. A control chart contains the data points:

11	15	16	17	16	18	20	19	16	17	13	10	12	15
1	2	3	4	5	6	7	8	9	10	11	12	13	14

$\mu = 14.3$ and $\sigma = 3.1$. Which "out-of-control" indicator is evident from this data?

A. A run of nine consecutive points on one side of the center line.

B. One point falls beyond the 2σ level.

C. At least two of three consecutive points lie beyond the 2σ level on the same side of the center line.

D. One point falls beyond the 3σ level.

E. At least four points beyond the σ line.

16. Bob earned 116 on his accounting midterm and 82 on his biology midterm. For the accounting exam the mean score was 102 with standard deviation 14. For the biology exam the mean score was 70 with standard deviation 5.
(a) Convert each midterm score to a standard score.
(b) On which test did he do better compared to the rest of the class?

17. Troy took a standardized test to try to get credit for first-year Spanish by examination. His standardized score was 1.9. The mean score for the exam was 100 with standard deviation 12. The language department requires a raw score of 120 to get credit by exam for first-year Spanish.
(a) Compute Troy's raw score.
(b) Will Troy get credit for first-year Spanish based on this exam?

18. The graduation rate for seniors at a certain university in any particular year is, on average, 72.1% and is normally distributed. The population standard deviation is 3.5%. What is the probability, in any given year, that between 79.1% and 82.6% of the seniors will graduate?

A. 0.4985

B. 0.047

C. 0.27

D. 0.0235

E. 0.135

19. Leo has a pizza business. The times to prepare and deliver pizzas are approximately normally distributed with mean $\mu = 20$ minutes and standard deviation $\sigma = 10$ minutes.
(a) Suppose that Leo advertises "If it takes more than 30 minutes to get your pizza, you get it free." What fraction of his take-out pizzas will he have to give away free?
(b) How long a time should Leo allow if he wants to give away no more than 5% of the pizzas?

20. Weights of Pacific yellowfin tuna follow a normal distribution with mean weight 68 lb and standard deviation 12 lb. A Pacific yellowfin tuna is caught at random.
(a) Find the probability that its weight is less than 48 lb.
(b) Find the probability that its weight is more than 100 lb.
(c) Find the probability that its weight is between 48 lb and 100 lb.

21. A normal distribution has $\mu = 5$ and $\sigma = 1.5$. Find z for $x = 2.5$.

A. $z = -1.66$

B. $z = -1.67$

C. $z = 1.66$

D. $z = 1.4$

E. $z = 1.67$

22. In order to make a little money for the holidays, Marla is raising turkeys. The male of the species she has selected has a mean weight of 17 lb at maturity and a standard deviation of 1.8 lb. The weights are known to be normally distributed. What is the probability of raising a male turkey that weighs at least 21 lb at maturity?

23. The Snack Pack of potato chips is advertised to weigh 3.5 oz. The weights are normally distributed with mean 3.5 oz and standard deviation 0.2 oz. A Snack Pack of potato chips is selected at random.
(a) Find the probability that it weighs less than 3.0 oz.
(b) Find the probability that it weighs more than 3.7 oz.

24. Jose received an 82 on his math exam. $\mu = 73$ and $\sigma = 8$ for the entire class. Compute the number of standard deviations Jose's score is from the mean.

A. $z = 0.805$

B. $z = 1.125$

C. $z = -1.014$

D. $z = -1.125$

E. $z = 1.014$

25. The snow pack on the summit of Wolf Creek Pass, Colorado on March 1 has been measured for many years. It is normally distributed with mean $\mu = 78.1$ inches and standard deviation $\sigma =$ 10.4 inches. A year is selected at random.
(a) Find the probability that the snowpack is less than 60 inches.
(b) Find the probability that the snowpack is more than 85 inches.
(c) Find the probability that the snowpack is between 60 inches and 85 inches.

26. In Manoa Valley on the island of Oahu (Hawaii) the annual rainfall averages 43.6 inches with standard deviation 7.5 inches. A year is chosen at random.
(a) Find the probability that the annual rainfall is more than 56 inches.
(b) Find the probability that the annual rainfall is less than 32 inches.
(c) Find the probability that the annual rainfall is between 32 inches and 56 inches.

27. A normal distribution has $\mu = 13$ and $\sigma = 2.1$. To transform this into standard normal distribution,
(a) how many units, and in which direction along the *x*-axis, must the distribution be shifted?
(b) will the standard normal distribution be wider or narrower than the original?

28. Find the z value so that
(a) 8% of the area under the standard normal curve lies to the left of $-z$.
(b) 93% of the area under the standard normal curve lies between $-z$ and z.

29. The life of the Corn Delight popcorn maker is normally distributed with mean 20 months and standard deviation 2 months. The manufacturer will replace a Corn Delight popper if it breaks during the guarantee period.
(a) If the manufacturer guarantees the popper for 18 months what fraction of the poppers will he probably have to replace?
(b) How long should the guarantee period be if the manufacturer does not want to replace more than 3% of the poppers? (Round to the nearest month.)

30. A normal distribution was transformed to a standard normal distribution. The original μ and σ were 17 and 3.4, respectively. For $z = 2$, what is the corresponding x value?

A. 11.9

B. 7

C. 37.4

D. 23.8

E. 18.7

31. The operating life of a Sports Master fishing reel is normally distributed with mean 48 months and standard deviation 6 months. The manufacturer will replace any Sports Master reel that malfunctions during the guarantee period.
(a) If the guarantee period is 45 months, what fraction of the reels will Sports Master have to replace?
(b) How long should the guarantee period be if the manufacturer does not want to replace more than 5% of the reels?

32. Use a standard normal distribution table to find the area under the curve to the left of $z = -1.17$.

A. 0.8790

B. 0.8577

C. 0.1190

D. 0.1210

E. 0.1423

33. The operating life of an Arctic Air portable cooler is normally distributed with mean 75 months and standard deviation 9 months. The manufacturer will replace any cooler that breaks during the guarantee period.
(a) If the guarantee period is for 64 months, what fraction of the coolers will the manufacturer have to replace?
(b) How long should the guarantee run if the manufacturer does not want to replace more than 2.5% of the coolers?

34. Recently an investor tabulated the percent growth for a random sample of 25 large mutual funds. The distribution was mound shaped and symmetric. It had a mean $\mu = 25.7\%$ and standard deviation $\sigma = 7.1\%$.
(a) Estimate an interval about the mean in which about 68% of these percent gains would fall.
(b) Estimate an interval centered about the mean in which about 95% of these percent gains would fall.

35. Use a standard normal distribution table to find the area under the curve to the left of $z = 1.57$.

A. 0.0582

B. 0.9292

C. 0.0708

D. 0.9418

E. 0.9525

36. George is a senior citizen who takes medication to stabilize his (systolic) blood pressure. He takes his blood pressure every day and, from long experience, he knows that the mean pressure is 137 with standard deviation 6. Assume the blood pressure follows a normal distribution. For the past ten days his blood pressure has been recorded in the following table. t = number of the day. x = blood pressure.

t	1	2	3	4	5	6	7	8	9	10
x	128	133	130	134	139	140	139	142	158	161

(a) Make a control chart for the systolic blood pressure readings.
(b) Determine whether the blood pressure distribution is in statistical control. If it is not, indicate which out-of-control signals are present.
(c) If George's blood pressure changes too much it is recommended that his medication be changed. Looking at the control chart, do you think it might be time to consider changing his medication? Explain.

37. Let x represent the life of a AA battery in a portable radio. The x distribution has $\mu = 44$ hours and $\sigma = 3.6$ hours. Convert each of the following x intervals into z intervals. Round to the nearest hundredth if necessary.
(a) $40 \le x \le 46$
(b) $37 \le x \le 51$
(c) $41 \le x \le 45$

38. From long experience the store manager at the local W-Mart Store knows that the number of customers in the store each day follows a normal distribution with mean 611 customers and standard deviation 36 customers. The table below gives the number of customers in the store each day for the last ten days. t = number of the day, x = number of customers.

t	1	2	3	4	5	6	7	8	9	10
x	541	592	620	605	548	535	510	489	500	490

(a) Make a control chart for the daily number of customers at W-Mart.
(b) Determine if the distribution of customers is in statistical control. If it is not, indicate which out-of-control signals are present.
(c) The manager of W-Mart usually does more advertising and has more sales if the number of customers drops off. Looking at the control chart what would you recommend at this time?

39. Use a standard normal distribution table to find the area under the curve between $z = -1.08$ and $z = 2.13$.

A. 0.8234

B. 0.8433

C. 0.1766

D. 0.8339

E. 0.1567

40. Let x represent the life of a 60-watt light bulb. The x distribution has a mean $\mu = 1{,}000$ hours with standard deviation $\sigma = 75$ hours. Convert each of the following z intervals into x intervals.

(a) $0 \le z \le 1.25$

(b) $-1.5 \le z \le 2.4$

(c) $1.25 \le z \le 2.25$

41. A high school counselor was given the following z intervals in information about a vocational training aptitude test. The test scores had a mean $\mu = 450$ points and standard deviation $\sigma = 35$ points. Convert each interval into a x interval. x = test score.

(a) $-1.14 \le z \le 2.27$

(b) $z \le -2.88$

(c) $z \ge 1.85$

42. The following are six grade-point averages for six students.

Joseph: 2.01	Sabine: 2.64
Xena: 3.67	Ron: 2.38
Maria: 3.22	Ali: 3.59

If these grade-point averages were used to create a standard normal distribution, which student's grade-point average would have a z-value of 0.488?

A. Xena

B. Sabine

C. Maria

D. Ali

E. Ron

43. David and Laura are both applying for a position on the ski patrol. David took the advanced first aid class at his college. His score on the comprehensive final exam was 173 points. The final exam scores followed a normal distribution with mean 150 points and standard deviation 25 points. Laura took an advanced first aid class at the plant where she works. For the method of testing there her cumulative score on all exams was 88. The cumulative scores followed a normal distribution with mean 65 and standard deviation 10. Both courses are comparable in content and level of difficulty. There is only one position available on the ski patrol and David and Laura are equally qualified as skiers.

(a) Both David and Laura scored 23 points above the mean for their respective tests. Does this mean that they both gave the same performance on the first aid course? Explain your answer.

(b) Calculate the z scores for David and Laura. On the basis of test results which of them should get the job? Explain your answer.

44. Jim scored 630 on at national bankers' examination in which the mean is 600 and the standard deviation is 70. June scored 530 on the Hoople College bankers' examination for which the mean is 500 and the standard deviation is 25.
(a) Calculate the standard z score for Jim and for June.
(b) If Jim and June both apply for a job at Hoople State Bank and each examination has equal weight in the hiring criteria, who has the better chance based on test results? Explain your answer.

45. Let x have a normal distribution with $\mu = 13$ and $\sigma = 3$. Find the probability that an x value selected at random from this distribution is between 8 and 16.

A. 0.0062

B. 0.7938

C. 0.1649

D. 0.8413

E. 0.9938

46. Researchers at a pharmaceutical company have found that the effective time duration of a safe dosage of a pain relief drug is normally distributed with mean 2 hours and standard deviation 0.3 hour. For a patient selected at random:
(a) What is the probability that the drug will be effective for 2 hours or less?
(b) What is the probability that the drug will be effective for 1 hour or less?
(c) What is the probability that the drug will be effective for 3 hours or more?

47. Quality control for Comfort-Ease Computer Peripherals, Inc. has done studies showing that its voice recognition device has a mean life of 7.5 years with standard deviation 2.2 years. The manufacturer will replace any such device that wears out during the guarantee period.
(a) If the guarantee period is 5 years, what fraction of the voice recognition devices will the manufacturer have to replace?
(b) How long should the guarantee run if the manufacturer wishes to replace no more than 10% of the devices?

48. Let x have a normal distribution with $\mu = 20$ and $\sigma = 4$. Find the probability that an x value selected at random from this distribution is between 19 and 21.

A. 1.0000

B. 0.4013

C. 0.5987

D. 0.1974

E. 0.8026

49. The life of a Freeze Breeze electric fan is normally distributed with mean 4 years and standard deviation 1.2 years. The manufacturer will replace any fan that wears out during the guarantee period.
(a) What fraction of the fans will have to be replaced if the fans are guaranteed for three years?
(b) How long should the fans be guaranteed if the manufacturer does not want to replace more than 5% of them? (Give the answer to the nearest month.)

50. The Zinger is a modern sports car with an epoxy body. The manufacturer constructs Zinger bodies under pressure in a mold, and must wait a certain length of time to be sure that enough resin bonding and hardening have occurred before removing the body from the mold. Chemical engineers have found that the hardening time is normally distributed with mean 28.7 hours and standard deviation 6.4 hours. What is the minimal length of time the body should stay in the mold if the manufacturer wants to be 85% sure that the bonds have hardened?

51. Which of the following ranges of z values corresponds to a probability of 0.8213?

A. $-1.13 \le z \le 1.91$

B. $-1 \le z \le 1$

C. $-1.13 \le z \le 1.65$

D. $-1 \le z \le 1.65$

E. $-1.65 \le z \le 2$

52. As the manager of a large manufacturing plant, you are concerned about employee absenteeism. The plant is staffed so that operations are still efficient when the average number of employees absent each shift has mean $\mu = 10.5$ with standard deviation $\sigma = 2$. For the most recent 12 shifts the number of absent employees is given in the table below. t = number of shift, x = number absent.

t	1	2	3	4	5	6	7	8	9	10	11	12
x	8	4	7	10	9	11	9	15	12	16	9	11

(a) Make a control chart for the above data.
(b) Determine whether the number of workers absent is in statistical control. If it is not specify which out-of-control signals are present.

53. A certain portable CD player is known to function properly for a mean 3.6 years with a standard deviation of 0.6 year. The player is sold with a 3-year warranty. What is the probability that the player will break during the warranty period?

54. Peter is a door-to-door sales representative who makes a sale at only 8% of his house calls. If Peter makes 219 house calls today, what is the probability that:
(a) he makes a sale at 12 or fewer houses?
(b) he makes a sale at 20 or fewer houses?
(c) he makes a sale at 15 to 25 houses?

55. A toll-free computer software support line for the Magnet Computer System has established a target length of time for each customer call. The calls are targeted to have a mean duration of 7 minutes with standard deviation 2 minutes. For one help technician the most recent ten calls had duration given in the table below.
t = number of call, x = duration.

t	1	2	3	4	5	6	7	8	9	10
x	8	4	7	10	9	11	9	15	12	16

(a) Make a control chart showing the lengths of calls.
(b) Determine whether the length of calls is in statistical control. If it is not specify which out-of-control signals are present.

56. An air conditioner repairman is facing significant competition in his local area, so he wants to give a guarantee to his customers when he services their air conditioners. According to the data he has collected over the years, the mean time between necessary servicing of an air conditioner is 26 months, with a standard deviation of 6 months. How long should his guarantee period be (in months) if he does not want to perform a free service for more than 10% of his customers?

57. In Summit County 35% of registered voters are Democrats. What is the probability that in a random sample of 300 voters:
(a) 100 or more are Democrats?
(b) between 100 and 120 are Democrats?
(c) 90 or fewer are Democrats?

58. Quality control reports that are not available to consumers indicate that 20% of Hot Shot ovens have faulty control panels. The owner of an apartment complex has just purchased 120 of these ovens. Assume that the ovens purchased constitute a random sample.
(a) Can you use the normal approximation to the binomial distribution to estimate the probability that between 22 and 35 of them have faulty control panels? Explain.
(b) Estimate the probability that between 22 and 35 of them (including 22 and 35) have faulty control panels.

59. Find z such that 81% of the area under the standard normal curve lies between $-z$ and z.

A. 1.30

B. 1.32

C. 1.31

D. 1.33

E. 1.29

60. Coast Guard inspectors have found that 16% of all lifeboats on large cruise ships have a faulty latch release. A large cruise ship has 73 lifeboats. What is the probability that 9 or more of the lifeboats have a faulty latch release?

61. Find z such that 62% of the area under the standard normal curve lies between $-z$ and z.

A. 0.19

B. 0.38

C. 0.86

D. 0.88

E. 0.87

62. Records at the College of Engineering show that 62% of all freshmen who declare EE (electrical engineering) as their intended major field of study eventually graduate with an EE major. This fall, the College of Engineering has 316 freshmen who have declared an EE major.
(a) What is the probability that between 200 and 225 of them (including 200 and 225) will graduate with an EE major?
(b) Can you use a normal approximation to the binomial distribution to estimate this probability? Explain.

63. One model of an imported automobile is known to have defective seat belts. An extensive study has found that 17% of the seat belts were incorrectly installed at the factory. A car dealer has just received a shipment of 196 cars of this model. What is the probability that in this sample:
(a) 22 or fewer have defective seat belts?
(b) 30 or more have defective seat belts?
(c) between 25 and 50 have defective seat belts?

64. Find z so that 3.5% of the standard normal curve lies to the right of z.

A. 1.82

B. –0.04

C. 1.81

D. –0.05

E. 1.80

65. The probability that a person with a telephone has an unlisted number is 0.15. The district manager of a political action group is phoning people urging them to vote. In a district with 416 households, all households have phones. What is the probability that the district manager will find that
(a) 50 or fewer households in the district have unlisted phone numbers?
(b) 345 or more have listed phone numbers?
(c) between 40 and 80 have unlisted numbers?

66. Find z so that 42% of the standard normal curve lies to the left of z.

A. –0.17

B. –0.21

C. –0.19

D. –0.20

E. –0.18

67. The probability of an adverse reaction to a flu shot is 0.02. If the shot is given to 1000 people selected at random, what is the probability that:
(a) 15 or fewer people will have an adverse reaction?
(b) 25 or more people will have an adverse reaction?
(c) between 20 and 30 people will have an adverse reaction?

68. For a binomial distribution with $n = 100$ and $r = 83$, using a normal probability distribution to approximate this binomial distribution, compute μ and σ.

A. $\mu = 17$; $\sigma = 3.76$

B. $\mu = 83$; $\sigma = 9.11$

C. $\mu = 83$; $\sigma = 4.12$

D. $\mu = 17$; $\sigma = 4.14$

E. $\mu = 83$; $\sigma = 3.76$

69. Canter Political Polling Service has found that the probability that a registered voter selected at random will answer and return a short questionnaire on a well-publicized issue is 0.31. If 1,000 questionnaires are sent to registered voters selected at random, what is the probability that:
(a) 350 or more will be returned?
(b) 320 or fewer will be returned?
(c) between 315 and 355 will be returned?

70. Bob received a z score of 0.9 on a college entrance exam. If the raw scores have a mean of 540 and a standard deviation of 80 points, what is his raw score?

A. 406

B. 468

C. 134

D. 612

E. 566

71. The state tourism board reports that 52% of the residents of the state plan to take summer vacations in the state. An independent company did a survey of 200 people.
(a) Find the probability that fewer than 100 are planning in-state vacations.
(b) Find the probability that 120 or more are planning in-state vacations.

72. Determine $P(z \geq 2.09)$ with a standard normal distribution table.

A. 0.9817

B. 0.9634

C. 0.183

D. 0.0366

E. 0.0183

73. A postal worker has observed that 72% of the customers who buy stamps request particular commemorative stamps. For a random sample of 80 customers find the probability that:
(a) 50 or more ask for the special stamps.
(b) From 50 to 65 people ask for the commemorative stamps. (Include 50 and 65.)

74. On Professor Grindstone's final exam the mean score was 78 and the standard deviation 10. He wants to "curve" the exam so that the middle 50% of the students get C's.
(a) Find a z score for which the area under the standard normal curve from $-z$ to z is 50%.
(b) Find the range of scores which will be assigned a grade of C.

75. The daily attendance of an economics class follows a normal distribution with a mean of 26 students and a standard deviation of four students.
(a) Find the probability that the attendance will be less than 22 students.
(b) Find the probability that the attendance will be 28 or more students.

76. The daily attendance at a weekly flea market follows a normal distribution with mean 875 people and standard deviation 150 people. For a day chosen at random:
(a) Find the probability that the attendance will be less than 500 people.
(b) Find the probability that the attendance will be 700 people or more.

77. The summer duck population at a lake in a state park is approximately normally distributed with a mean of 9,640 ducks and a standard deviation of 820 ducks. Let x be the random variable that represents the size of the duck population in the summer of a given year.
(a) Convert $7{,}918 < x < 10{,}255$ to a z interval.
(b) Convert $-1.25 < z < 0.90$ to an x interval.
(c) If 7,590 represents an unusually low population, and 11,690 represents an unusually high number, give z numbers ranges that represent unusually low and high summer duck populations.

78. The attendance at home games of the college football team follows a normal distribution with mean 9,500 and standard deviation 600. For a game chosen at random:
(a) Find the probability that the attendance will be between 8,000 and 11,000.
(b) Find the probability that the attendance will be over 11,000.

79. The life of a certain brand of car stereo is normally distributed with a mean of 5.2 years and a standard deviation of 0.9 year. What is the probability that a car stereo of this type chosen at random will last seven years or more?

A. 0.281

B. 0.7190

C. 0.0228

D. 0.228

E. 0.9772

80. Martha manages a gift shop at a national park. Her daily revenues during the summer months follow a normal distribution with mean \$2,400 and standard deviation \$500. For a day chosen at random:
(a) Find the probability that her revenue will be between \$1,500 and \$3,000.
(b) Find the probability that her revenue will be under \$3,600.

81. The ages of applicants to an engineering school are normally distributed with a mean of 20.4 years and a standard deviation of 1.2 years. What is the probability that an applicant is younger than 18 years old or older than 22 years old?

A. 0.0228

B. 0.5573

C. 0.4427

D. 0.1146

E. 0.098

82. The nightly attendance at the university summer outdoor theater series follows a normal distribution with mean 475 people and standard deviation 150 people.
(a) Find the probability that on a night chosen at random the attendance will be greater than 175 people.
(b) Find the probability that on a night chosen at random the attendance will be between 400 and 900 people.

83. For a certain psychiatric evaluation, z scores above 2.18 and below –2.33 are considered abnormal. What is the probability that a person evaluated by this will be considered normal?

A. 0.5122

B. 0.9714

C. 0.0245

D. 0.9755

E. 0.4878

84. Bill runs a bicycle repair shop. During the summer months the number of jobs per week follows a normal distribution with mean 45 and standard deviation 9. For a week chosen at random:
(a) Find the probability that he will have 24 or more jobs.
(b) Find the probability that he will have between 30 and 60 jobs.

85. The number of visitors per day at a national park during the summer months follows a normal distribution with mean 10,500 and standard deviation 2,750. For a summer day chosen at random:
(a) Find the probability that the number of visitors will be between 6,000 and 15,000.
(b) Find the probability that the number of visitors will be less than 5,000.

Section B: Algorithms

Objective 7.1: Graphs of Normal Probability Distributions

86. The heights of 750 students at a local university were recorded and found to be approximated by the normal curve below. Find the mean and standard deviation for these data.

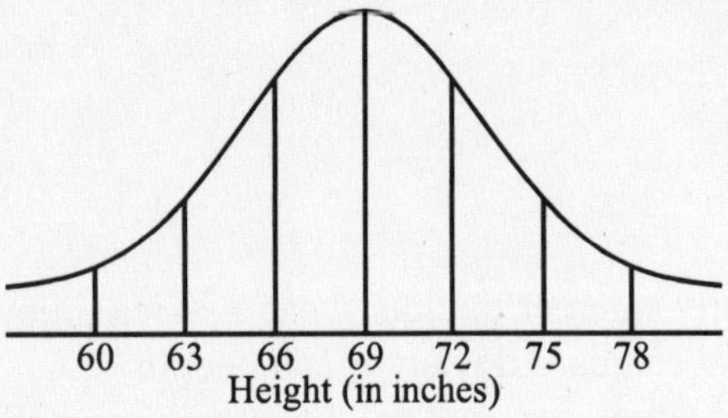

87. The heights of 750 employees at a software company were recorded and found to be approximated by the normal curve below. Find the mean and standard deviation for these data.

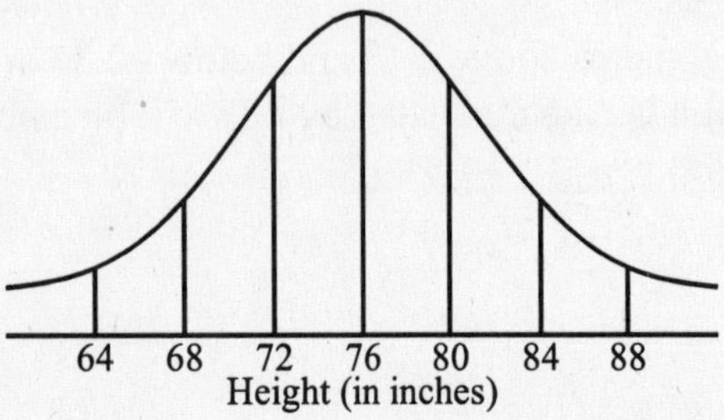

88. The height of all the 11-year-old boys in the United States is distributed as shown:

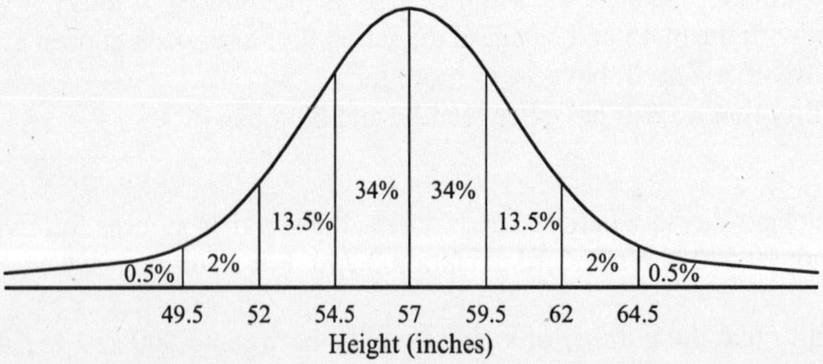

What is the probability that a boy chosen at random from that age group will have a height more than 62 inches?

89. The height of all the 11-year-old girls in Canada is distributed as shown.

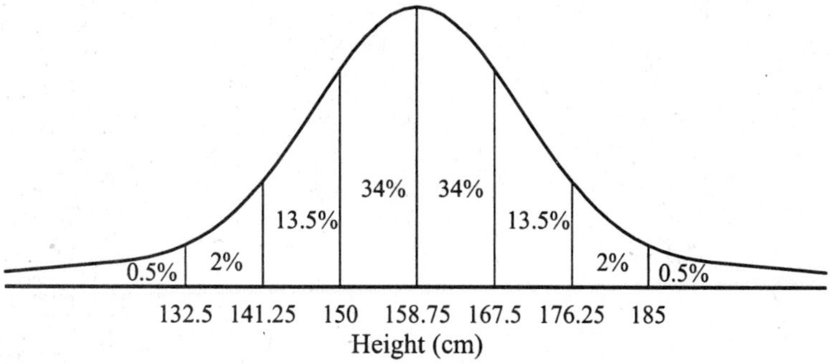

What is the probability that a girl chosen at random from that age group will have a height less than 167.5 centimeters?

90. Sally got a summer job as the manager of a large pizza and pasta restaurant. Part of her responsibility is to make sure that the wait staff properly stocks their tables at the end of the night, including the hot pepper and Parmesan shakers. Although the wait staff must have their stations stocked before they leave, Sally knows that there will always be a few tables that are not properly stocked. Every 15 days, Sally has a general staff meeting. Before the meeting she examines several control charts of the restaurant. For the past month she has found the following number of tables not properly stocked with hot pepper and Parmesan. Make a control chart for these data.

Day	1	2	3	4	5	6	7	8
x = number of tables	34	23	27	33	21	29	29	22
Day	9	10	11	12	13	14	15	
x = number of tables	27	29	24	27	35	34	18	

91. Below is a graph of the distribution of standard scores of 15,000 students on an eleventh grade Algebra Readiness Test.

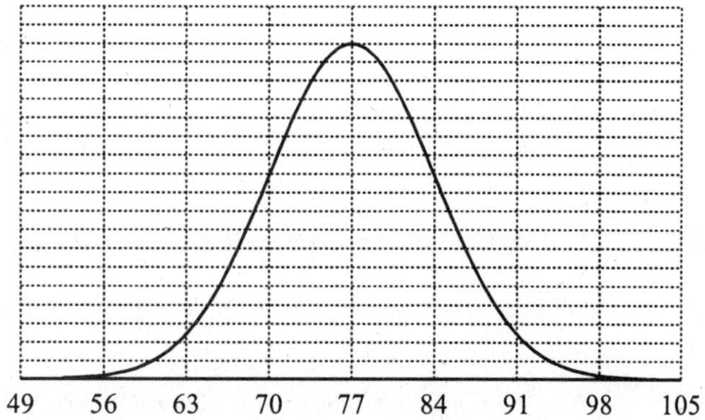

a. This distribution is approximately normal. Estimate the mean and standard deviation of this distribution.
b. About how many of the 15,000 students had scores between 77 and 98? Explain.

92. The heights of 720 employees at a software company were recorded and found to be approximated by the normal curve below. Find the mean and standard deviation for the data.

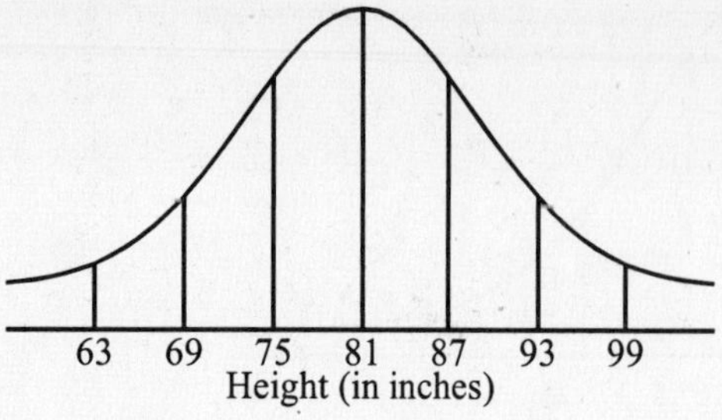

93. The height of all the 12-year-old girls in the United States is distributed as shown.

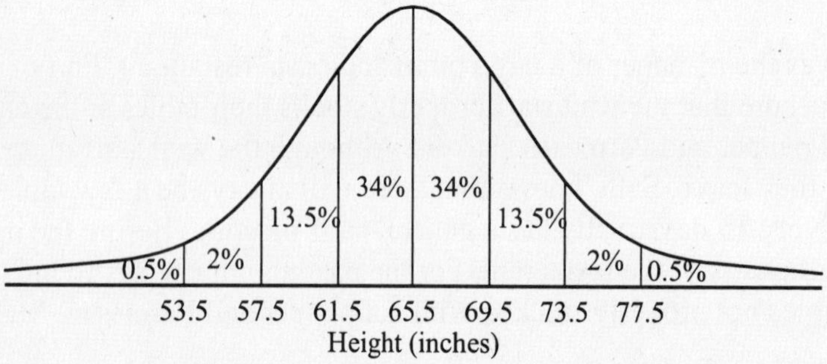

What is the probability that a girl chosen at random from that age group will have a height more than 53.5 inches?

94. Alex got a job working at his university's library. Part of his job involves making sure that all returned materials are properly shelved. While shelving material, Alex often comes across books that were incorrectly placed. For the past fifteen days he has been keeping track of the number of books he finds in the wrong location. Make a control chart for the data.

Day	1	2	3	4	5	6	7	8
x = number of books	51	31	31	46	51	30	47	53
Day	9	10	11	12	13	14	15	
x = number of books	32	51	30	29	27	52	48	

Objective 7.2: Standard Units and Areas Under the Standard Normal Distribution

95. The Jackson triplets, Jenny, John, and James are in different math classes at City High. On their final exams, Jenny scored 80 on a test with a mean of 70 and a standard deviation of 10; John scored 83 on a test with a mean of 71 and a standard deviation of 9; and James scored 81 on a test with a mean of 72 and a standard deviation of 8.5. Who had the best score?

96. Angus's z score on a college entrance exam is 0.6. If the raw scores have a mean of 488 and standard deviation of 53 points, what is her raw score? Round to nearest tenth.

97. Lian's z score on a biology final is 1.5. If the raw scores have a mean of 423 and standard deviation of 60 points, what is his raw score? Round to the nearest tenth.

98. What is the area under the standard normal curve to the left of $z = -1.25$?

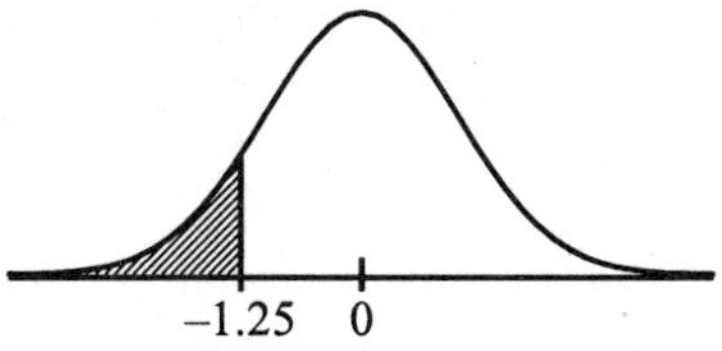

99. What is the area under the standard normal curve between $z = -1.68$ and $z = 1.68$?

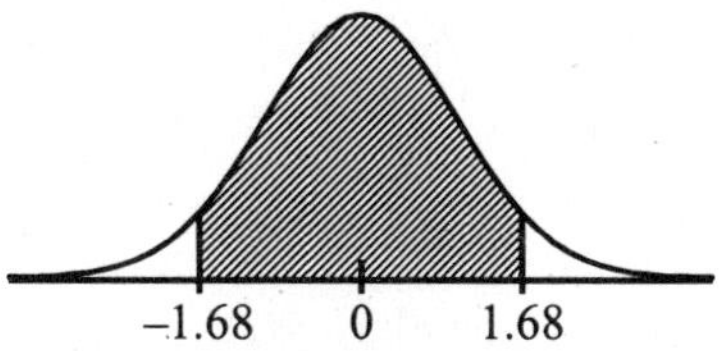

100. What is the area under the standard normal curve between $z = -1.13$ and $z = 1.44$?

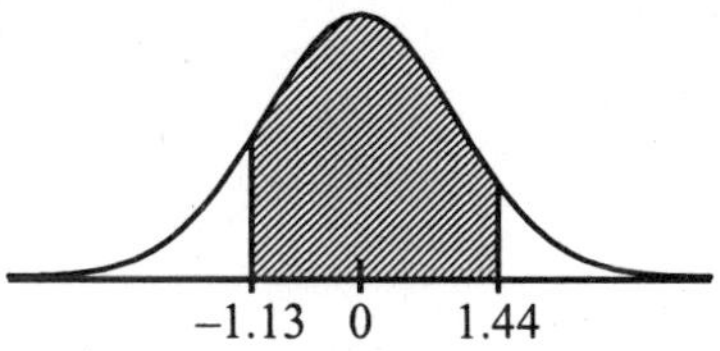

101. What is the area under the standard normal curve in the two tails, one to the left of $z = 1.12$ and the other to the right of $z = 1.4$?

102. Ramsis's z score on a organic chemistry midterm is 1.1. If the raw scores have a mean of 502 and standard deviation of 67 points, which is his raw score?

103. The Jackson triplets, Bethany, Belinda, and Bionka are in different math classes at City Community College. On their final exams, Bethany scored 71 on a test with a mean of 62 and a standard deviation of 6.5; Belinda scored 66 on a test with a mean of 58 and a standard deviation of 8; and Bionka scored 74 on a test with a mean of 68 and a standard deviation of 6.5. Who had the best score?

Objective 7.3: Areas Under Any Normal Curve

104. The scores for a standardized reading test are found to be normally distributed with a mean of 500 and a standard deviation of 70. If the test is given to 700 students, how many are expected to have scores between 500 and 570?

105. Last year, the personal best high jumps of track athletes in a nearby state were normally distributed with a mean of 205 centimeters and a standard deviation of 16 centimeters. What is the probability that a randomly selected high jumper has a personal best between 221 and 253 centimeters? Round answer to nearest hundredth.

106. Music Madness makes and sells CD players. Their guarantee policy will refund full purchase price of the CD players if they fail during the guarantee period. The research department has done tests and found that the CD player life is normally distributed with a mean life of 29 months and a standard deviation of 6 months. How long can the guarantee period be if management does not want to refund the purchase price on more than 10% of the Music Madness CD players?

107. Find the z value so that 68% of the area under the standard normal curve lies between $-z$ and z.

108. Find the z value so that 3% of the area under the standard normal curve lies to the right of z.

109. The personal savings of the Young Saver Club were normally distributed with a mean of \$650 and a standard deviation of \$72. What is the probability that a randomly selected saver has an account total between \$650 and \$794?

110. Find the z value so that 12% of the area under the standard normal curve lies to the left of z.

Objective 7.4: Normal Approximation to the Binomial Distribution

111. Find a binomial distribution, including a value for n, p, and q, that is approximated by a normal distribution.

112. One study found that the probability of a certain species of slug reaching its second year of survival is 0.71. To test the validity of this study, a random sample of 500 slugs were collected for a new study, and observed over a 5-year period. Use the normal approximation to the binomial to find the probability that between 340 and 370 of the slugs survive their first two years of life. Round answer to nearest hundredth.

113. Which is a binomial distribution that is approximated by a normal distribution?

A. $n = 7$, $p = 0.41$, and $q = 0.59$

B. $n = 7$, $p = 0.11$, and $q = 0.89$

C. $n = 38$, $p = 0.41$, and $q = 0.59$

D. $n = 24$, $p = 0.11$, and $q = 0.89$

114. One study claims that the probability of a person living beyond the age of 55 years is 0.73. To test the validity of this study, a random sample of 500 people was considered for a new study. Use the normal approximation to the binomial to find the probability that between 350 and 380 of the people survive beyond the age of 55 years. Round to the nearest hundredth.

115. A study found that the probability of a high school student passing his or her final high school exams is 0.60. To test the validity of this study, a random sample of 400 high school students were collected for a new study, and observed over a 5-year period. Use the normal approximation to the binomial to find the probability that between 225 and 255 of the slugs survive their first two years of life. Round to the nearest hundredth.

Answer Key for Chapter 7
Normal Distributions

Section A: Static Items

(a) 68% interval: ($1,029, $2,373)
[1] (b) 95% range: $357 to $3,045

[2] A.

(a) 68% interval: (80 min., 140 min.)
[3] (b) 95% range: 50 min. to 170 min

(a) 68% range: 6.0% to 13.6%
[4] (b) 95% range: 2.2% to 17.4%

[5] A.

(a) 68% range: 7.1% to 19.1%
[6] (b) 95% range: 1.1% to 25.1%

(a) 0.34
[7] (b) 0.0235

(a)

$\mu - 3\sigma$	$\mu - 2\sigma$	μ	$\mu + 2\sigma$	$\mu + 3\sigma$
100	125	175	225	250

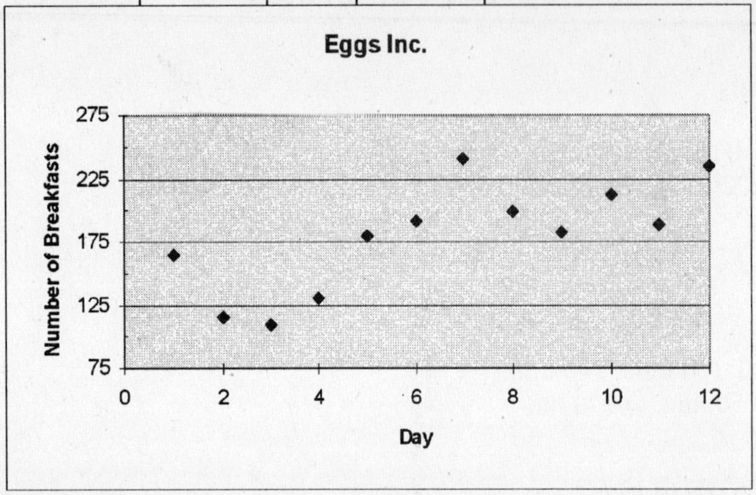

(b) The process is not in statistical control. Out-of-control signal III is present for days 1, 2 and
[8] 3.

(a)

$\mu - 3\sigma$	$\mu - 2\sigma$	μ	$\mu + 2\sigma$	$\mu + 3\sigma$
13	18	28	38	43

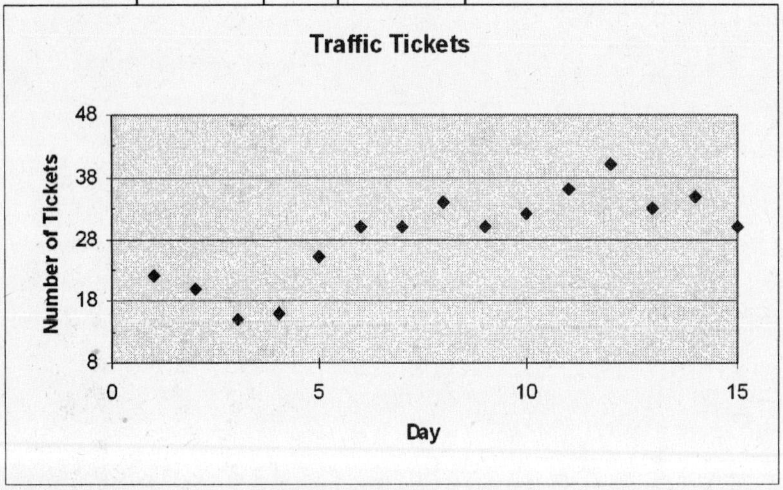

(b) The process is out of statistical control. Out-of-control signal II is present at days 6 through
[9] 15 and out-of-control signal III is present at days 2, 3, and 4.

[10] C.

(a) Biology: $z = 1.20$; History: $z = 1.67$
[11] (b) He did better, compared to the rest of the class, on the history test.

(a)

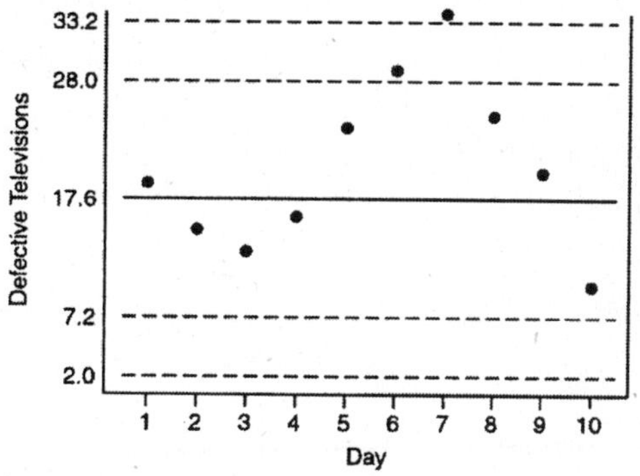

(b) No. There are two indications that production is "out of control": Signal I: one point beyond the 3σ level (on day 7), and Signal III: Two of three consecutive points beyond the 2σ level on
[12] the same side of the center line (days 6 and 7).

(a)

$\mu-3\sigma$	$\mu-2\sigma$	μ	$\mu+2\sigma$	$\mu+3\sigma$
22	206	574	942	1,126

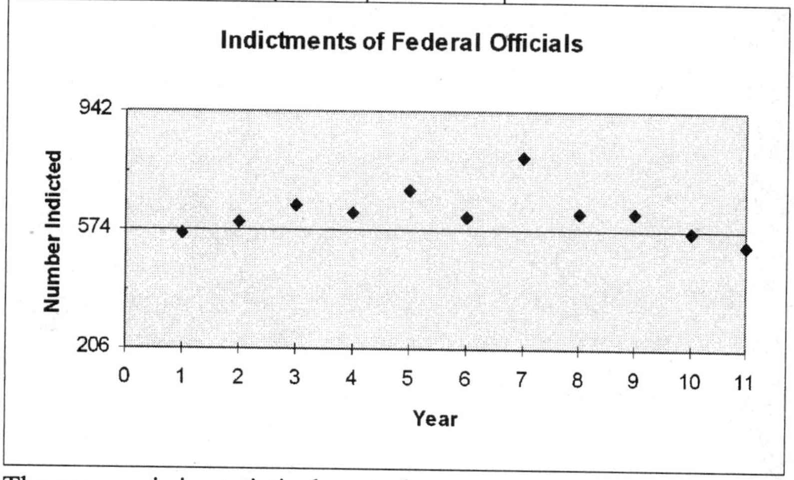

[13] (b) The process is in statistical control.

(a) Political science: $z = 1.50$; Chemistry: $z = 2.00$
[14] (b) She did better, compared to the rest of the class, on the chemistry test.

[15] A.

(a) Accounting: $z = 1.00$; Biology: $z = 2.40$
[16] (b) He did better, comparedto the rest of the class, on the biology test.

(a) Troy's raw score: $x = 122.8$
[17] (b) Yes, Troy's score is high enough to meet the standard.

[18] D.

(a) $P(x \geq 30) = 0.1587$. He will give away about 16% of his pizzas.
[19] (b) $z = 1.645$; Guarantee delivery within 36 minutes.

(a) $P(x < 48 \text{ lb}) = 0.0475$
(b) $P(x > 100 \text{ lb}) = 0.0038$
[20] (c) $P(48 \text{ lb} \leq x \leq 100 \text{ lb}) = 0.9487$

[21] B.

[22] $P(x \geq 21 \text{ lb}) = 0.0132$

(a) $P(x < 3.0 \text{ oz}) = 0.0062$
[23] (b) $P(x > 3.7 \text{ oz}) = 0.1587$

[24] B.

(a) $P(x < 60 \text{ in.}) = 0.0409$
(b) $P(x > 85 \text{ in.}) = 0.2546$
[25] (c) $P(60 \text{ in.} \leq x \leq 85 \text{ in.}) = 0.7045$

(a) $P(x > 56 \text{ in.}) = 0.0495$
(b) $P(x < 32 \text{ in.}) = 0.0606$
[26] (c) $P(32 \text{ in.} \leq x \leq 56 \text{ in.}) = 0.8899$

(a) 13 units to the left
[27] (b) narrower

(a) 8% of the area lies to the left of $-z$ for $z = 1.41$
[28] (b) 93% of the area lies between $-z$ and z for $z = 1.81$

(a) $P(x \leq 18 \text{ months}) = 0.1587$; He will have to replace about 16% of them.
[29] (b) $z = -1.88$; Guarantee them for 16 months.

[30] D.

(a) $P(x \leq 45 \text{ months}) = 0.3085$; 31% of the reels will have to be replaced.

[31] (b) $z = -1.645$; Guarantee them for 38 months.

[32] D.

(a) $P(x \leq 64) = 0.1112$; 11% of them will have to be replaced.

[33] (b) $z = -1.96$; Guarantee them for 57 months.

(a) 68% interval: (18.6%, 32.8%)

[34] (b) 95% interval: (11.5%, 39.9%)

[35] D.

(a)

$\mu - 3\sigma$	$\mu - 2\sigma$	μ	$\mu + 2\sigma$	$\mu + 3\sigma$
119	125	137	149	155

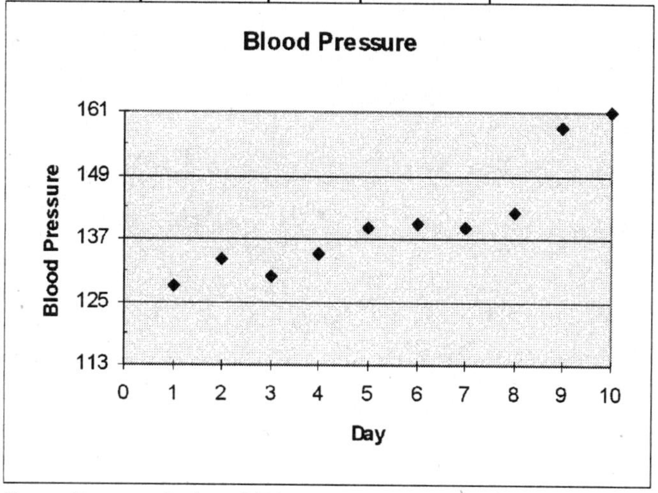

(b) Out-of-control signal I is present at days 9 and 10. Out-of-control signal III is present at days 8, 9, and 10.

[36] (c) Yes. George should see his doctor.

(a) $-1.11 \leq z \leq .56$

(b) $-1.94 \leq z \leq 1.94$

[37] (c) $-0.83 \leq z \leq 0.28$

(a)

$\mu - 3\sigma$	$\mu - 2\sigma$	μ	$\mu + 2\sigma$	$\mu + 3\sigma$
503	539	611	683	719

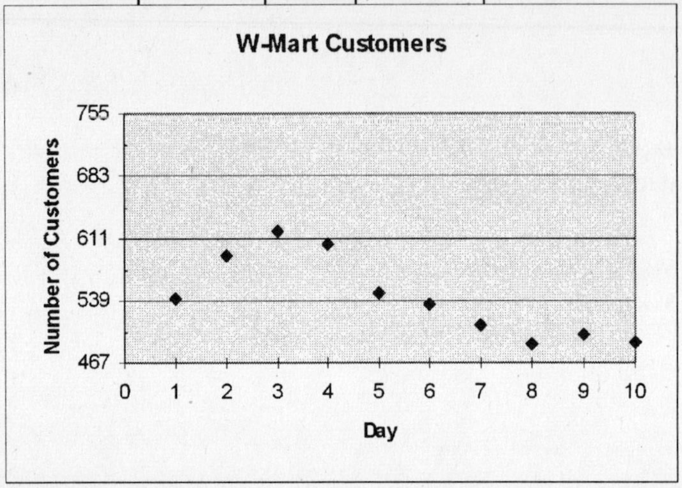

(b) Out-of-control signals I and III are present at days 8, 9 and 10.

[38] (c) Recommend more advertising and sales.

[39] B.

(a) $1{,}000 \le x \le 1{,}093.75$

(b) $887.5 \le x \le 1{,}180$

[40] (c) $1{,}093.75 \le x \le 1{,}168.75$

(a) $410.1 \le x \le 529.45$

(b) $x \le 349.2$

[41] (c) $x \ge 514.75$

[42] C.

(a) No. The standard deviation on David's test was greater than the standard deviation on Laura's test. We should convert to z scores to compare their performances.

(b) For David, $z = 0.92$. For Laura, $z = 2.39$. Laura did better compared to the rest of her group than David did compared to his. If the populations who took the two tests were

[43] comparable, choose Laura.

(a) For Jim, $z = 0.43$. For June, $z = 1.2$

[44] (b) June has the higher z score and the better chance on the basis of the exam.

[45] B.

(a) $P(x \le 2) = 0.50$
(b) $P(x \le 1) = 0.0004$
[46] (c) $P(x \ge 3) = 0.0004$

(a) $P(x \le 5) = 0.1271$; About 13% of them will have to be replaced.
[47] (b) $z = -1.645$; Guarantee them for about 3.8 years or 46 months.

[48] D.

(a) $P(x \le 3) = 0.2033$; About 20% of them will be replaced.
[49] (b) $z = -1.645$; Guarantee the fans for 24 months.

[50] $z = 1.05$; The body should stay in the mold about 35.4 hours.

[51] C.

(a)

$\mu - 3\sigma$	$\mu - 2\sigma$	μ	$\mu + 2\sigma$	$\mu + 3\sigma$
4.5	6.5	10.5	14.5	16.5

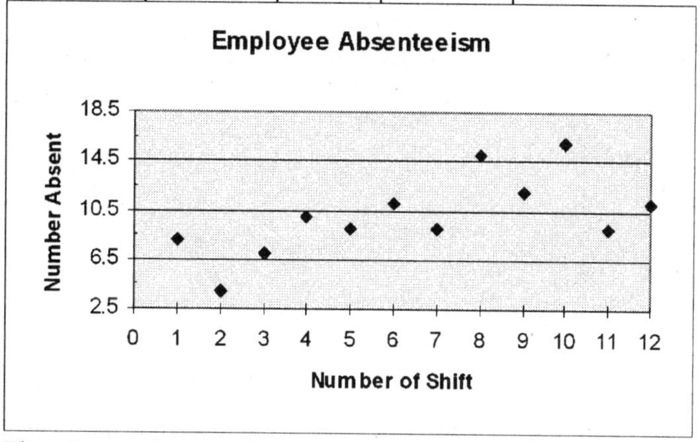

(b) The process is out of control. Out-of-control signal I is present at shift 2. Out-of-control
[52] signal III is present at shifts 8, 9, and 10.

[53] $P(x \le 3) = P(z \le -1) = 0.1587$

(a) $P(r \le 12) = P(x \le 12.5) = P(z \le -1.25) = 0.1056$
(b) $P(r \le 20) = P(x \le 20.5) = P(z \le 0.74) = 0.7704$
[54] (c) $P(15 \le r \le 25) = P(14.5 \le x \le 25.5) = P(-0.75 \le z \le 1.99) = 0.7501$

(a)

$\mu-3\sigma$	$\mu-2\sigma$	μ	$\mu+2\sigma$	$\mu+3\sigma$
1	3	7	11	13

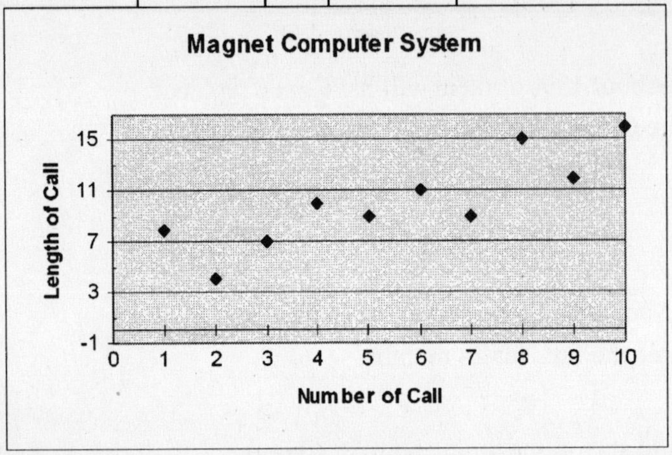

(b) The process is out of statistical control. Out-of-control signals I and III are present at calls

[55] 7, 8, 9, and 10.

[56] 18 months

(a) $P(r \geq 100) = P(x \geq 99.5) = P(z \geq -0.67) = 0.7486$
(b) $P(100 \leq r \leq 120) = P(99.5 \leq x \leq 120.5) = P(-0.67 \leq z \leq 1.88) = 0.7185$

[57] (c) $P(r \leq 90) = P(x \leq 90.5) = P(z \leq -1.76) = 0.0392$

(a) Since *np* and *nq* are greater than 5 the normal approximation to the binomial distribution is suitable.

[58] (b) $P(22 \leq r \leq 35) = P(21.5 \leq x \leq 35.5) = P(-0.57 \leq z \leq 2.62) = 0.7113$

[59] C.

[60] $P(r \geq 9) = P(x \geq 8.5) = P(z \geq -1.02) = 0.8461$

[61] D.

(a) $P(200 \leq r \leq 225) = P(199.5 \leq x \leq 225.5) = P(0.41 \leq z \leq 3.43) = 0.3406$
(b) Since both *np* and *nq* are greater than 5, the normal approximation to the binomial is

[62] suitable.

(a) $P(r \le 22) = P(x \le 22.5) = P(z \le -2.06) = 0.0197$

(b) $P(r \ge 30) = P(x \ge 29.5) = P(z \ge -0.73) = 0.7673$

[63] (c) $P(25 \le r \le 50) = P(24.5 \le x \le 50.5) = P(-1.68 \le z \le 3.27) = 0.9530$

[64] C.

(a) $P(r \le 50) = P(x \le 50.5) = P(z \le -1.63) = 0.0516$

(b) $P(r \le 71) = P(x \le 71.5) = P(z \le 1.25) = 0.8944$

[65] (c) $P(40 \le r \le 80) = P(39.5 \le x \le 80.5) = P(-3.14 \le z \le 2.49) = 0.9928$

[66] D.

(a) $P(r \le 15) = P(x \le 15.5) = P(z \le -1.02) = 0.1539$

(b) $P(r \ge 25) = P(x \ge 24.5) = P(z \ge 1.02) = 0.1539$

[67] (c) $P(20 \le r \le 30) = P(19.5 \le x \le 30.5) = P(-0.11 \le z \le 2.37) = 0.5349$

[68] E.

(a) $P(r \ge 350) = P(x \ge 349.5) = P(z \ge 2.70) = 0.0035$

(b) $P(r \le 320) = P(x \le 320.5) = P(z \le 0.72) = 0.7642$

[69] (c) $P(315 \le r \le 355) = P(314.5 \le x \le 355.5) = P(0.31 \le z \le 3.11) = 0.3774$

[70] D.

(a) $P(r < 100) = P(x \le 99.5) = P(z \le -0.64) = 0.2611$

[71] (b) $P(r \ge 120) = P(x \ge 119.5) = P(z \ge 2.19) = 0.0143$

[72] E.

(a) $P(r \ge 50) = P(x \ge 49.5) = P(z \ge -2.02) = 0.9783$

[73] (b) $P(50 \le r \le 65) = P(49.5 \le x \le 65.5) = P(-2.02 \le z \le 1.97) = 0.9539$

(a) 50% of the area will be between $-z$ and z for $z = 0.675$.

[74] (b) C range is from 71.25 to 84.75.

(a) $P(x < 22) = 0.1587$

[75] (b) $P(x \ge 28) = 0.6915$

[76] (a) $P(x < 500) = 0.0062$
(b) $P(x \geq 700) = 0.8790$

[77] (a) $-2.1 < z < 0.75$
(b) $8{,}615 < x < 10{,}378$
(c) $z \leq -2.5$ represents an unusually low population. $z \geq 2.5$ represents an unusually high population.

[78] (a) $P(8{,}000 \leq x \leq 11{,}000) = 0.9876$
(b) $P(x > 11{,}000) = 0.0062$

[79] C.

[80] (a) $P(1{,}500 \leq x \leq 3{,}000) = 0.8490$
(b) $P(x < 3{,}600) = 0.9918$

[81] D.

[82] (a) $P(x > 175) = 0.9772$
(b) $P(400 < x < 900) = 0.6892$

[83] D.

[84] (a) $P(x \geq 24) = 0.9901$
(b) $P(30 \leq x \leq 60) = 0.9050$

[85] (a) $P(6{,}000 \leq x \leq 15{,}000) = 0.8990$
(b) $P(x < 5{,}000) = 0.0228$

Section B: Algorithms

Objective 7.1: Graphs of Normal Probability Distributions

[86] Mean: 69, standard deviation: 3

[87] Mean: 76, standard deviation: 4

[88] 2.5%

[89] 84%

[90]

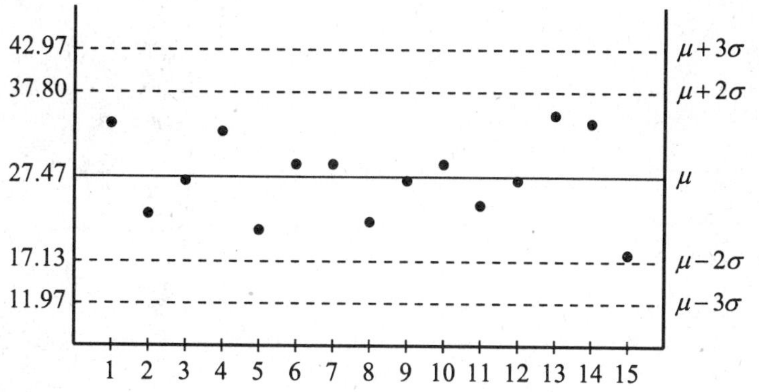

[91]
a. Mean ≈ 77; standard deviation ≈ 7
b. Since the distribution is normal and 77 and 98 are at the mean and 3 deviations above the mean, respectively, about 50% of the scores should be between these two values. This translates to 50% of 15,000 or 7500 students.

[92] Mean: 81; standard deviation: 6

[93] 99.5%

[94]

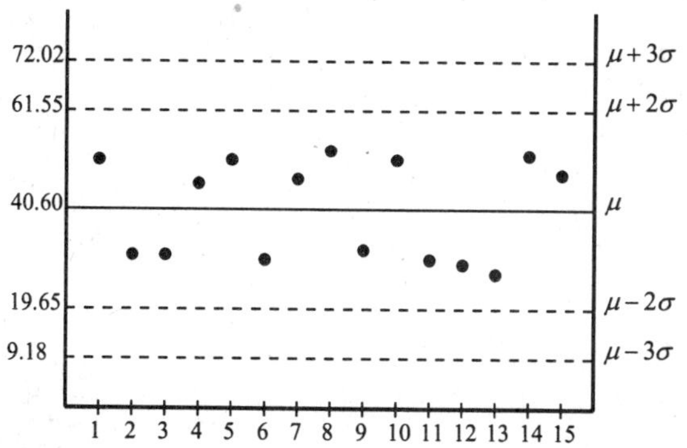

Objective 7.2: Standard Units and Areas Under the Standard Normal Distribution

[95] John

[96] 519.8

[97] 513.0

[98] 0.1056

[99] 0.9070

[100] 0.7958

[101] 0.2121

[102] 575.7

[103] Bethany

Objective 7.3: Areas Under Any Normal Curve

[104] 239

[105] 0.16

[106] 21 months

[107] 0.99

[108] 1.88

[109] 0.475

[110] −1.175

Objective 7.4: Normal Approximation to the Binomial Distribution

[111] Answers may vary. Sample answer: $n = 23$, $p = 0.43$, and $q = 0.57$

[112] 0.87

[113] C.

[114] 0.87

[115] 0.83

Chapter 8
Introduction to Sampling Distributions

Section A: Static Items

1. Carol wants to study the volume of winter wheat production for last year for all rural counties of eastern Colorado, Nebraska, and Kansas.
   (a) Describe the population of interest to her.
   (b) Identify two population parameters that she would be likely to estimate for her study.
   (c) What statistics could she use to estimate these parameters.
   (d) How could she get values for these statistics?

2. Alex wants to study the number of people who visit Colorado National Monument each day in a given year. The monument is open and accessible year-round.
   (a) Describe the population for this study.
   (b) Identify two population parameters that he would be likely to estimate for his study.
   (c) What statistics could he use to estimate these parameters?
   (d) How could he get values for these statistics?

3. Karen would like to find out what fraction of children in her city who will enter kindergarten next year have had all of the necessary immunizations.
   (a) Describe the population of interest to Karen.
   (b) What population parameter would she be likely to want to estimate?
   (c) What statistic could she use to estimate this parameter?
   (d) How could she get a value for this statistic?

4. Mark would like to know what fraction of people who live in his state plan to purchase a digital TV within the next year.
   (a) Describe the population of interest to Mark.
   (b) What population parameter would he be likely to want to estimate?
   (c) What statistic could he use to estimate this parameter?
   (d) How could he get a value for this statistic?

5. Identify the definition of *sampling distribution.*

   A. A set of measurements (or counts), either existing or conceptual

   B. A numerical descriptive measure of a sample

   C. A numerical descriptive measure of a population

   D. A conclusion about the value of a population parameter based on information about the corresponding sample statistic and probability

   E. A probability distribution for a sample statistic

6. Identify the definition of *population parameter*.

A. A set of measurements (or counts), either existing or conceptual

B. A conclusion about the value of a population parameter based on information about the corresponding sample statistic and probability

C. A numerical descriptive measure of a sample

D. A numerical descriptive measure of a population

E. A probability distribution for a sample statistic

7. Identify the definition of *statistical inference*.

A. A probability distribution for a sample statistic

B. A numerical descriptive measure of a population

C. A numerical descriptive measure of a sample

D. A conclusion about the value of a population parameter based on information about the corresponding sample statistic and probability

E. A set of measurements (or counts), either existing or conceptual

8. Identify the definition of *sample statistic*.

A. A conclusion about the value of a population parameter based on information about the corresponding sample statistic and probability

B. A probability distribution for a sample statistic

C. A numerical descriptive measure of a sample

D. A numerical descriptive measure of a population

E. A set of measurements (or counts), either existing or conceptual

9. Identify an example of a *population*.

A. Seven cards chosen at random from a 52-card deck

B. The lengths of all trout in a lake

C. The automobiles bought by Americans polled in a telephone survey

D. A week of television shows watched by Americans as reported in a survey

E. Registered Oklahoma voters who voted in a U.S. presidential election

10. Identify an example of a random sample of a population.

A. The television shows watched by all Americans

B. The automobiles bought by Americans in a single year

C. The lengths of trout caught by fisherman surveyed at a lake on a weekend

D. All of the cards in a 52-card deck

E. Registered Oklahoma voters

11. What are the two principal ways a sample statistic is used to make inferences about corresponding population parameters?

12. Where are sampling distributions used in the study of statistics?

13. Which of the following variables does NOT signify a population parameter?

A. σ B. σ^2 C. $\bar{x}$ D. p E. μ

14. Which of the following variables does NOT signify a sample statistic?

A. μ B. $\bar{x}$ C. s D. s^2 E. $\hat{p}$

15. (a) Give three examples of a parameter and the corresponding sample statistic.
(b) If the sample size n is sufficiently large, what does the Central Limit theorem tell us about the relationship between the mean of the sample means, $\bar{x}$, for samples of size n and the mean of the original population of x values?

16. (a) When we use methods of statistics to estimate a population parameter using a sample statistic, are we using methods of descriptive statistics or of inferential statistics?
(b) What is the name of the theorem that gives us information about the distribution of sample means, $\bar{x}$ computed from random samples of size n? State the theorem.

17. Sandra wants to study the allocation of government money to universities to be used for scientific research. She has taken a random sample of 60 universities and recorded the amount each university received in federal government grants for scientific research. The data has a mound-shaped histogram but does not appear to follow a normal distribution.
(a) Is it reasonable for her to use the normal distribution to approximate the distribution of individual grant amounts? Give a reason for your answer.
(b) Could she use the normal distribution to study the distribution of sample means for random samples of 50 university grant amounts? Explain.

18. The number of boxes of girl scout cookies sold by each of the girl scouts in a midwestern city has a distribution which is approximately normal with mean $\mu = 75$ boxes and standard deviation $\sigma = 30$ boxes.
(a) Find the probability that a scout chosen at random sold between 60 and 90 boxes of cookies.
(b) Find the probability that the sample mean number of boxes of cookies sold by a random sample of 36 scouts is between 60 and 90 boxes.
(c) What difference do you observe in the two answers? Explain.

19. Roger has read a report that the weights of adult male Siberian tigers have a distribution which is approximately normal with mean $\mu = 390$ lb. and standard deviation $\sigma = 65$ lb.
(a) Find the probability that an individual male Siberian tiger will weigh more than 450 lb.
(b) Find the probability that a random sample of 4 male Siberian tigers will have sample mean weight more than 450 lb.

20. Statistical Abstracts (117th edition) reports sale price of unleaded gasoline (in cents per gallon) at the refinery. The distribution is mound-shaped with mean $\mu = 80.04$ cents per gallon and standard deviation, $\sigma = 4.74$ cents per gallon.
(a) Are we likely to get good results if we use the normal distribution to approximate the distribution of sample means for samples of size 9? Explain.
(b) Find the probability that for a random sample of size 36, the sample mean price will be between 79 and 82 cents per gallon.

21. Statistical Abstracts (117th edition) provides information about the number of registered automobiles, including taxis, by state. The number of registered automobiles per 1,000 residents follows an approximately normal distribution with mean $\mu = 514.6$ and standard deviation $\sigma = 84.6$. In the questions below, x = number of registered automobiles per 1,000 residents.

(a) Find the probability that for a random sample of 4 states the sample mean, $\bar{x}$ is between 475 and 525.

(b) Find the probability that for a random sample of 16 states the sample mean, $\bar{x}$ is between 475 and 525.
(c) Compare your answers for parts (a) and (b) and give a reason for the difference.

22. Statistical Abstracts (117th edition) tabulates the number of licensed drivers in each state. The distribution of the numbers of licensed drivers per 1,000 residents has a mound-shaped distribution with mean $\mu = 679.2$ and standard deviation $\sigma = 50.02$. In the questions below, x = number of licensed drivers per 1,000 residents.
(a) Is it appropriate to use the normal distribution to approximate the distribution of x? Explain.

(b) Find the probability that for a random sample of 30 states the sample mean, $\bar{x}$ is less than 650.

23. Statistical Abstracts (117th edition) reports lengths of 78 shuttle flights. The distribution is somewhat asymmetric with mean $\mu = 8.56$ days and standard deviation $\sigma = 3.33$ days.
(a) Is the normal distribution a good approximation to the distribution of sample mean length of flights for random samples of size 4? Explain.
(b) Find the probability that the sample mean for a random sample of 40 flights will be greater than 9 days.

24. Statistical Abstracts (117th edition) reports state per-capita expenditures for education. These expenditures have a distribution which is mound-shaped and has mean $\mu = \$1,106.96$ and standard deviation $\sigma = \$202.10$. In the questions below, x represents the state per-capita expenditure for education.
(a) Find the probability that the sample mean expenditure for random samples of size 30 is greater than $1,000.
(b) Find the probability that the sample mean expenditure for random samples of size 36 is between $1,000 and $1,200.

25. Life spans for red foxes follow an approximately normal distribution with mean 7 years and standard deviation 3.5 years.
(a) Find the probability that a red fox chosen at random will live at least 10 years.
(b) Find the probability that a random sample of 4 red foxes will have a sample mean life span of over 10 years.

26. Carl is doing a project for his ecology class. He read that the average person uses about 100 gallons of water per day. He estimates the standard deviation to be about 30 gallons per day.
(a) Assuming that Carl's figures are correct, find the probability that a random sample of 49 people will have a sample mean water use of 90 gallons per day or less.
(b) Find the probability that a random sample of 64 people will have a sample mean water use over 110 gallons per day.

27. Wheat production on an acre of land varies depending on factors such as soil, climate, fertilizer program, insect control and rainfall. Near Dodge City, Kansas the mean yield over the past five years has been 22 bushels per acre with standard deviation 6 bushels per acre. Assuming that this year's conditions are similar to those of the last five years, what is the probability that a random sample of 36 one-acre plots have a sample mean yield of 24 bushels or more?

28. The Internal Revenue Service has studied millions of tax returns. They have determined that the mean deduction taken for charitable contributions by people with incomes between twenty-five and thirty thousand dollars per year is about $800. The standard deviation is estimated to be about $170. What is the probability that the sample mean deduction for charitable contributions based on a random sample of 100 returns is more than $830?

29. Read-a-Lot Bookstores, Inc., claims that the annual gross revenue for Read-a-Lot stores has mean $\mu = \$300,000$ with standard deviation $\sigma = \$72,000$. A random sample of 36 stores in the franchise is selected.

(a) Find the probability that the mean annual gross revenue, $\bar{x}$ for these stores is more than \$305,000.
(b) Find the probability that the mean annual gross revenue is less than \$295,000.
(c) Find the probability that the mean annual gross revenue is between \$295,000 and \$305,000.

30. The general manager of an amusement park wants to estimate the length of time people spend in the park. From data collected over several months it was found that the average time an individual spends in the park is $\mu = 3.4$ hr. with standard deviation $\sigma = 1.1$ hr. It was not known whether the distribution is normal but it appears to be mound-shaped.
(a) Thirty-five people just got off the city bus and went into the amusement park. Find the probability that the sample mean time these people spend in the park is more than 3 hr.
(b) Find the probability that the sample mean time these people spend in the park is between 3 hr. and 3.5 hr.

31. The calorie content of Speedy Supper Chunky Beef Dinners is normally distributed with mean $\mu = 580$ and standard deviation $\sigma = 35$. The calorie count varies because the portions vary slightly and the fat content of the beef varies.
(a) You are on a strict dict and have allowed only 600 calories for dinner. Find the probability that the Speedy Supper Chunky Beef Dinner you happen to eat has 600 or fewer calories.
(b) If you had a random sample of 10 Speedy Supper Chunky Beef Dinners, what is the probability that the sample mean calorie count, $\bar{x}$ for this random sample of 10 dinners is 600 calories or less?
(c) Can we assume that the $\bar{x}$ values follow a normal distribution? Explain.

32. The life of one species of poplar tree is normally distributed with mean 20 years and standard deviation 2.3 years.
(a) If one tree is selected at random, what is the probability that it will live 21 years or more?
(b) If nine trees are selected at random, what is the probability that the sample mean life of the nine trees will be 21 years or more?
(c) Can we assume that the sample means follow a normal distribution? Explain.

33. At one of the Colorado ski resorts a lift attendant has been measuring the time a skier waits in the lift line. Over a long period of time the lift attendant has found the mean waiting time to be 6.3 minutes with standard deviation 1.8 minutes.
(a) What is the probability that a random sample of 49 skiers will have a sample mean waiting time of 7 or more minutes?
(b) What is the probability that a random sample of 64 skiers will have a sample mean waiting time of less than 6 minutes?

34. The body temperature of healthy children is normally distributed with mean 98.6° F and standard deviation 0.9° F.
(a) If one healthy child is chosen at random, what is the probability that the body temperature is more than 99.6° F?
(b) If six healthy children are chosen at random what is the probability that the sample mean body temperature $\bar{x}$ is more than 99.6° F?

35. It has been estimated that the average tax refund for individuals filing income tax returns during the past year was $1,250. The standard deviation was estimated to be $300. A random sample of 36 income tax returns was selected.
(a) Find the probability that the sample mean tax refund is less than $1,125.
(b) Find the probability that the sample mean tax refund is more than $1,400.
(c) Find the probability that the sample mean tax refund is between $1,125 and $1,400.

36. The electric company has estimated that the mean number of kilowatts of electricity used by a household per month is 365 kW with standard deviation 30 kW. To check their estimates a random sample of 40 households was chosen.
(a) If the power company's estimates are correct, find the probability that the sample mean amount of electricity will be between 358 and 372 kW.
(b) If the actual sample mean was 385 kW would you question the power company's figures? Explain.

37. The manager of Mammon Savings and Loan has computed that the mean number of dollars borrowed for the purchase of an economy automobile is $9,373.18 with standard deviation $1,016.12. During the next few weeks 50 customers came in for a loan to purchase an economy car.
(a) Find the probability that $\bar{x}$ the sample mean for the 50 customers, is more than $9,500.
(b) Find the probability that the sample mean is less than $9,000.

38. The diameters of grapefruit in a certain citrus grove are normally distributed with mean 4.6 inches and standard deviation 1.3 inches. A random sample of 10 of these grapefruit are put in a bag and sold.

(a) Find the probability that the sample mean $\bar{x}$ for this sample is greater than 5 inches.
(b) Find the probability that the sample mean is between 4 inches and 5 inches.
(c) Find the probability that the sample mean is less than 4 inches.

39. True Sound cassette tapes have playing times that are normally distributed with mean 120 minutes and standard deviation 5.4 minutes.

(a) Find the probability that a tape selected at random will have a playing time between 115 and 125 minutes.
(b) A random sample of 4 tapes is chosen. Find the probability that the sample mean playing time for this sample of 4 tapes is between 115 and 125 minutes.
(c) Compare your answers to parts (a) and (b). Explain the reason for the difference.

40. The time for first-time applicants for a driver's license to complete the driving portion of the test has mean $\mu = 20$ min. and standard deviation $\sigma = 5$ minutes. One morning 40 first-time applicants show up to take the driving test. Find the probability that the sample mean time for this sample of 40 people will be between 18 and 21 minutes.

41. Margot has found that the mean time to run a job at the college copy center is $\mu = 12.6$ min. with standard deviation $\sigma = 10$ minutes. She selects a random sample of 64 jobs.

(a) What is the probability that the sample mean copying time for this sample is between 14 and 15 minutes?
(b) What is the probability that the sample mean copying time for this sample is between 10 and 12 minutes?

42. Nature Kiss Granola comes in a medium-size box. The weight of granola in the box is normally distributed with mean $\mu = 14$ oz. and standard deviation $\sigma = 0.4$ oz.

(a) What is the probability that the weight of granola in a box of this cereal will be between 13.9 oz. and 14.2 oz.?
(b) What is the probability that sample mean weight of granola in a random sample of 12 boxes of this cereal will be between 13.9 and 14.2 oz.?
(c) Compare your answers for parts (a) and (b). Explain the difference.

43. Hank has a hot dog stand across the street from a little league ball field. He has observed that the daily demand for hot dogs on a weekend follows a normal distribution with mean $\mu = 400$ and standard deviation $\sigma = 50$.
(a) Find the probability that the demand for hot dogs on a weekend day chosen at random will be less than 300.
(b) Find the probability that the sample mean demand for a random sample of 4 weekend days will be less that 300.
(c) Compare your answers to parts (a) and (b). Explain the difference.

44. Sheila has a part-time job assembling computer work stations. She has found that the mean time it takes her to assemble a station is normally distributed with mean $\mu = 90$ min. and standard deviation $\sigma = 15$ minutes.
(a) Find the probability that it will take her more than 105 minutes to assemble one work station.
(b) Over the past few weeks she has assembled six work stations. Find the probability that the sample mean time for this sample of size six is more than 105 minutes.
(c) Discuss the difference between your answers to parts (a) and (b).

45. Joe's Auto Repair offers customers a free 15 point inspection when they bring their cars in for service. The inspection often reveals the need for additional repairs. The cost for such additional repairs is approximately normally distributed with mean $\mu = \$250$ and standard deviation $\sigma = \$100$. Find the probability that a random sample of 16 cars which receive the inspection have sample mean cost of additional repairs less than \$175.

46. At Mountains and Plains College, the time it takes a new student to complete advising and registration has a mound-shaped distribution with mean $\mu = 55$ min. and standard deviation $\sigma = 20$ min.
(a) A random sample of 36 students comes in to register. Find the probability that the sample mean registration time for these students is less than 45 minutes.
(b) Find the probability that the sample mean registration time for these students is greater than 50 minutes.

47. Mary Jo uses llamas as pack animals for her wilderness expeditions. The weights of supplies carried by the llamas follows a normal distribution with mean $\mu = 62.5$ lb. and standard deviation $\sigma = 6.25$ lb. For a recent expedition Mary Jo used 3 llamas. Find the probability that the sample mean weight of supplies for this sample is over 65 lb.

48. Stuart has observed that the daily revenues from his coffee house follow an approximately normal distribution with mean $\mu = \$550$ and standard deviation $\sigma = \$175$.
(a) Find the probability that the revenues for a day chosen at random will be between \$500 and \$600.
(b) Find the probability that the sample mean revenues for a random sample of 7 days will be between \$500 and \$600.

49. Suppose that x has a distribution with $\mu = 20$ and $\sigma = 15$.
(a) If a random sample of size $n = 40$ is drawn, find $\mu_{\bar{x}}$, $\sigma_{\bar{x}}$, and $P(20 \le \bar{x} \le 23)$ (round each value, including any z-values, to two decimal places, and then find probability to four decimal places).
(b) If a random sample of size $n = 80$ is drawn, find $\mu_{\bar{x}}$, $\sigma_{\bar{x}}$, and $P(20 \le \bar{x} \le 23)$ (Use rounding convention discussed in part (a).)
(c) Why should you expect the probability of part (b) to be higher than that of part (a)? (*Hint*: consider the standard deviations in parts (a) and (b).)

50. Suppose that x has a distribution with $\mu = 17$ and $\sigma = 8$.
(a) If random samples of size $n = 25$ are selected, can we say anything about the $\bar{x}$ distribution of sample means?
(b) If the original x distribution is normal, can we say anything about the $\bar{x}$ distribution from samples of size $n = 25$? Find $P(15 \le \bar{x} \le 20)$ to three decimal places.

51. Suppose that x has a distribution with $\mu = 65$ and $\sigma = 40$.
(a) If random samples of size $n = 20$ are selected, can we say anything about the $\bar{x}$ distribution of sample means?
(b) If random samples of size $n = 80$ are selected, can we say anything about the $\bar{x}$ distribution of sample means? Find $P(\bar{x} \ge 70)$ to four decimal places. (Round $\sigma_{\bar{x}}$ and any z-values to two decimal places.)

52. Coal is carried from a mine in West Virginia to a power plant in New York in hopper cars on a long train. The automatic hopper car loader is set to put 85 tons of coal into each car. The actual weights of coal loaded into each car is *normally distributed* with mean $\mu = 85$ tons and standard deviation $\sigma = 2$ tons.
(a) What is the probability that one car chosen at random will have less than 83 tons of coal?
(b) What is the probability that five cars chosen at random will have a mean load weight $\bar{x}$ of less than 83 tons of coal?

53. Coal is carried from a mine in West Virginia to a power plant in New York in hopper cars on a long train. The automatic hopper car loader is set to put 90 tons of coal into each car. The actual weights of coal loaded into each car is *normally distributed* with mean $\mu = 90$ tons and standard deviation $\sigma = 3.3$ tons. What is the probability that ten cars chosen at random will have a mean load weight $\bar{x}$ of more than 92 tons of coal?

A. 0.4726 B. 0.0274 C. 0.2291 D. 0.0537 E. 0.2709

Refer to this data for the following questions.

Let x be a random variable that represents white blood cell count per cubic milliliter of whole blood. Assume that x has a distribution that is approximately normal with mean $\mu = 7{,}500$ and estimated standard deviation $\sigma = 1{,}750$.

54. Refer to the data above regarding white blood cell count.
Suppose that a doctor uses the average $\bar{x}$ for two tests taken about a week apart. What is the probability of $\bar{x} < 5{,}000$?

A. 0.0764 B. 0.4783 C. 0.0021 D. 0.4979 E. 0.0217

55. Refer to the data above regarding white blood cell count.
Suppose that a doctor uses the average $\bar{x}$ for three tests taken about a week apart. What is the probability of $\bar{x} < 5{,}000$?

A. 0.0035 B. 0.0764 C. 0.0068 D. 0.4236 E. 0.4932

Refer to this data for the following questions.

Sam is a night watchman in a large warehouse. During one complete round of the warehouse, he must check in at 40 different checkpoints, all of which are about the same distance apart. At each checkpoint he inserts a key that tells a central computer the time he checked in. Let x be a random variable representing the length of time from one check-in to the next. While monitoring Sam's check-in times for the past year, the computer found the mean of the x distribution to be $\mu = 6.2$ minutes with standard deviation $\sigma = 1.7$ minutes.

56. Refer to data above regarding check-in time intervals.
In one complete round of the warehouse, there are 40 check-in time intervals. What is the probability that the average check-in time interval will be from 6 to 7 minutes?

A. 0.2704 B. 0.9837 C. 0.7689 D. 0.8531 E. 0.4985

57. Refer to the data above regarding check-in time intervals.
In two complete rounds of the warehouse, there are 80 check-in time intervals. What is the probability that the average check-in time interval will be from 6 to 7 minutes?

A. 0.8531 B. 0.9837 C. 0.7689 D. 0.2704 E. 0.4985

Refer to this data for the following questions.

The taxi and takeoff time for commercial jets is a random variable x with a mean of 8 minutes and a standard deviation of 3 minutes. Assume that the distribution of taxi and takeoff times is approximately normal. You may assume that the jets are lined up on a runway so that one taxies and takes off immediately after the other, and they take off one at a time on a given runway.

58. Refer to the data above regarding taxi and takeoff times.
What is the probability that for 36 jets on a given runway total taxi and takeoff time will be less than 300 minutes? That is, the average will be less than $\frac{300}{36} = 8.33$ minutes per plane?

A. 0.1587 B. 0.6331 C. 0.7454 D. 0.8413 E. 0.3669

59. Refer to the data above regarding taxi and takeoff times.
What is the probability that for 36 jets on a given runway total taxi and takeoff time will be more than 270 minutes, that is, on average more than $\frac{270}{36} = 7.5$ minutes per plane?

A. 0.8413 B. 0.6331 C. 0.7454 D. 0.1587 E. 0.3669

Refer to this data for the following questions.

It is important that each operation on an assembly line be completed in a predictable amount of time. In the assembly-line production of one particular vehicle, the head light installation process is designed to take an average of $\mu = 5.7$ minutes with standard deviation $\sigma = 1.3$ minutes for each vehicle. Assume that the assembly times follow a *normal* distribution and that the vehicles are lined up so that head light units are installed on one vehicle right after another.

60. Refer to the data above regarding assembly times.
What is the probability that the assembly time to install head light units on 9 vehicles is less than 50 minutes, that is, on average less than $\frac{50}{9} = 5.56$ minutes per vehicle?

A. 0.1711 B. 0.1255 C. 0.8289 D. 0.4562 E. 0.3745

61. Refer to the data above regarding assembly times.
What is the probability that the assembly time to install head light units on 9 vehicles is more than 55 minutes, that is, on average less than $\frac{55}{9} = 6.11$ minutes per vehicle?

A. 0.8289 B. 0.1711 C. 0.4562 D. 0.1255 E. 0.3745

Section B: Algorithms

Objective 8.1: Sampling Distributions

62. Is the following a parameter in statistics?
σ

63. Is the following an example of statistics?
i

64. Which of the following is *not* an example of a parameter?
$\hat{p}$, ρ, π, s

65. Name a parameter in statistics.

66. Name an example of statistics.

67. Which of the following is not an example of statistics?
e, $\bar{x}$, s^2, $\hat{p}$

Objective 8.2: The Central Limit Theorem

68. Let x be a random variable representing the amount of water, in cups, that each adult consumed in Springfield Mississippi yesterday. Consider a sampling distribution of sample means $\bar{x}$. As the sample size becomes increasingly large, what distribution does the $\bar{x}$ distribution approach?

69. If x has a normal distribution with mean $\mu = 24$ and standard deviation $\sigma = 5$, describe the distribution of $\bar{x}$ values for sample size n, where $n = 9$, $n = 49$, $n = 121$. How do the $\bar{x}$ distributions compare to the various sample sizes?

70. Given $\sigma = 0.22$ and $n = 5$, find the standard error of the mean. Round answer to thousandths place.

71. Given $\sigma = 0.14$ and $n = 14$, find the standard error of the mean. Round answer to thousandths place.

72. Given $\sigma = 0.43$ and $n = 10$, find the standard error of the mean. Round answer to thousandths place.

73. Given $\sigma = 0.8$ and $n = 14$, find the standard error of the mean. Round answer to thousandths place.

74. Suppose x has a normal distribution with mean $\mu = 23$ and a standard deviation of $\sigma = 1$. If you draw random samples of size 9 from the x distribution, and $\bar{x}$ represents the sample mean, how could you standardize the $\bar{x}$ distribution?

75. Suppose you know that the x distribution has a mean of $\mu = 73$ and a standard deviation of $\sigma = 10$, but you do not know if the distribution is normal. You draw samples of size 26 from the x distribution. Let $\bar{x}$ represent the sample mean. Is the sample size large enough? If so, how could you standardize the $\bar{x}$ distribution?

76. The manager of a cereal factory has observed that the amount of cereal in each "14-ounce" box is actually a normally distributed random variable, with a mean of 14.4 ounces and a standard deviation of 0.4 ounce. Find the probability that if a customer buys one box of cereal, that the cereal will contain more than 14 ounces. Round to the nearest ten-thousandth.

77. Mr. Lee is the foreman of Juicy Joo, a bottling plant in Chicago. He has observed that the amount of juice in each "18-ounce" bottle is normally distributed, with a mean of 18.8 ounces and a standard deviation of 0.2 ounces. Find the probability that if a customer buys a carton of four bottles of juice that the mean of the four will be greater than 18 ounces. Round answer to nearest ten-thousandth.

78. The dean of a liberal arts school claims that the average weekly income of graduates of her school one year after graduation is \$560. Assume that the dean's claim is correct. Given that the distribution of weekly incomes has a standard deviation of 95, what is the probability that 50 randomly selected graduates have an average weekly income of less than \$540? Round answer to nearest ten-thousandth.

79. Given $\sigma = 0.62$ and $n = 8$, find the standard error of the mean.

80. Suppose you know that the x distribution has a mean of $\mu = 84$ and a standard deviation of $\sigma = 15$, but you do not know if the distribution is normal. You draw samples of size 42 from the x distribution. Let $\bar{x}$ represent the sample mean. Is the sample size large enough? If so, how could you standardize the $\bar{x}$ distribution?

81. The manager of a chocolate factory has observed that the number of chocolates in each "38-ounce" box is actually a normally distributed random variable, with a mean of 38.7 ounces and a standard deviation of 0.7 ounces. Find the probability that if a customer buys one box of chocolates the box will contain more than 38 ounces.

82. The dean of a business school claims that the average weekly income of graduates of his school one year after graduation is \$610. Assume that the dean's claim is correct. Given that the distribution of weekly incomes has a standard deviation of 90, what is the probability that 40 randomly selected graduates have an average weekly income of less than \$570?

83. Given that $n = 38$, $p = 0.07$, and $q = 0.93$, determine if the normal approximation for the distribution of $\hat{p}$ is appropriate. Support your answer.

Objective 8.3: Class Project Illustrating the Central Limit Theorem

84. The possible outcomes of rolling a six-sided die are 1, 2, 3, 4, 5, and 6. Create a random sample for the experiment for $n = 10$ times. Find the sample means for each sample, the sample mean of the sample means, and standard deviation of the sample means. Create the following table of the distribution and construct a histogram. Also determine if the distribution relates to the Central Limit Theorem.

Interval	Frequency	Percent	Hypothetical Normal Distribution
$\bar{x} - 3s$ to $\bar{x} - 2s$			2 or 3%
$\bar{x} - 2s$ to $\bar{x} - s$			13 or 14%
$\bar{x} - s$ to $\bar{x}$			About 34%
$\bar{x}$ to $\bar{x} + s$			About 34%
$\bar{x} + s$ to $\bar{x} + 2s$			13 or 14%
$\bar{x} + 2s$ to $\bar{x} + 3s$			2 or 3%

85. The possible outcomes of rolling an eight-sided die are 1, 2, 3, 4, 5, 6, 7, and 8. Create a random sample for the experiment for $n = 16$ times. Find the sample means for each sample, the sample mean of the sample means, and standard deviation of the sample means. Create the following table of the distribution and construct a histogram. Also determine if the distribution relates to the Central Limit Theorem.

Interval	Frequency	Percent	Hypothetical Normal Distribution
$\bar{x} - 3s$ to $\bar{x} - 2s$			2 or 3%
$\bar{x} - 2s$ to $\bar{x} - s$			13 or 14%
$\bar{x} - s$ to $\bar{x}$			About 34%
$\bar{x}$ to $\bar{x} + s$			About 34%
$\bar{x} + s$ to $\bar{x} + 2s$			13 or 14%
$\bar{x} + 2s$ to $\bar{x} + 3s$			2 or 3%

86. The possible outcomes of spinning a spinner with six spaces each marked with a digit from 1 to 6 are 1, 2, 3, 4, 5, and 6. Create a random sample for the experiment for $n = 13$ times. Find the sample means for each sample, the sample mean of the sample means, and standard deviation of the sample means. Create the following table of the distribution and construct a histogram. Also determine if the distribution relates to the Central Limit Theorem.

Interval	Frequency	Percent	Hypothetical Normal Distribution
$\bar{x} - 3s$ to $\bar{x} - 2s$			2 or 3%
$\bar{x} - 2s$ to $\bar{x} - s$			13 or 14%
$\bar{x} - s$ to $\bar{x}$			About 34%
$\bar{x}$ to $\bar{x} + s$			About 34%
$\bar{x} + s$ to $\bar{x} + 2s$			13 or 14%
$\bar{x} + 2s$ to $\bar{x} + 3s$			2 or 3%

87. The possible outcomes of spinning a spinner with seven spaces each marked with a digit between 0 and 7 are 0, 1, 2, 4, 5, 6, and 7. Create a random sample for the experiment for $n = 13$ times. Find the sample means for each sample, the sample mean of the sample means, and standard deviation of the sample means. Create the following table of the distribution and construct a histogram. Also determine if the distribution relates to the Central Limit Theorem.

Interval	Frequency	Percent	Hypothetical Normal Distribution
$\bar{x} - 3s$ to $\bar{x} - 2s$			2 or 3%
$\bar{x} - 2s$ to $\bar{x} - s$			13 or 14%
$\bar{x} - s$ to $\bar{x}$			About 34%
$\bar{x}$ to $\bar{x} + s$			About 34%
$\bar{x} + s$ to $\bar{x} + 2s$			13 or 14%
$\bar{x} + 2s$ to $\bar{x} + 3s$			2 or 3%

88. The possible outcomes of pulling out a number in spades from a deck of cards are 1, 2, 3, 4, 5, 6, 7, 8, 9, and 10. Create a random sample for the experiment for $n = 12$ times. Find the sample means for each sample, the sample mean of the sample means, and standard deviation of the sample means. Create the following table of the distribution and construct a histogram. Also determine if the distribution relates to the Central Limit Theorem.

Interval	Frequency	Percent	Hypothetical Normal Distribution
$\bar{x} - 3s$ to $\bar{x} - 2s$			2 or 3%
$\bar{x} - 2s$ to $\bar{x} - s$			13 or 14%
$\bar{x} - s$ to $\bar{x}$			About 34%
$\bar{x}$ to $\bar{x} + s$			About 34%
$\bar{x} + s$ to $\bar{x} + 2s$			13 or 14%
$\bar{x} + 2s$ to $\bar{x} + 3s$			2 or 3%

89. The possible outcomes of pulling out a number in clubs from a deck of cards are 1, 2, 3, 4, 5, 6, 7, 8, 9, and 10. Create a random sample for the experiment for $n = 14$ times. Find the sample means for each sample, the sample mean of the sample means, and standard deviation of the sample means. Create the following table of the distribution and construct a histogram. Also determine if the distribution relates to the Central Limit Theorem.

Interval	Frequency	Percent	Hypothetical Normal Distribution
$\bar{x} - 3s$ to $\bar{x} - 2s$			2 or 3%
$\bar{x} - 2s$ to $\bar{x} - s$			13 or 14%
$\bar{x} - s$ to $\bar{x}$			About 34%
$\bar{x}$ to $\bar{x} + s$			About 34%
$\bar{x} + s$ to $\bar{x} + 2s$			13 or 14%
$\bar{x} + 2s$ to $\bar{x} + 3s$			2 or 3%

90. The possible outcomes of spinning a spinner with eight spaces each marked with a digit from between 0 and 9 are 0, 2, 3, 4, 5, 6, 7 and 9. Create a random sample for the experiment for $n = 14$ times. Find the sample means for each sample, the sample mean of the sample means, and standard deviation of the sample means. Create the following table of the distribution and construct a histogram. Also determine if the distribution relates to the Central Limit Theorem.

Interval	Frequency	Percent	Hypothetical Normal Distribution
$\bar{x} - 3s$ to $\bar{x} - 2s$			2 or 3%
$\bar{x} - 2s$ to $\bar{x} - s$			13 or 14%
$\bar{x} - s$ to $\bar{x}$			About 34%
$\bar{x}$ to $\bar{x} + s$			About 34%
$\bar{x} + s$ to $\bar{x} + 2s$			13 or 14%
$\bar{x} + 2s$ to $\bar{x} + 3s$			2 or 3%

91. Create a random sample of 9 numbers using digits from 0 to 9. Find the sample means for each sample, the sample mean of the sample means, and standard deviation of the sample means. Create the following table of the distribution and construct a histogram. Also determine if the distribution relates to the Central Limit Theorem.

Interval	Frequency	Percent	Hypothetical Normal Distribution
$\bar{x}-3s$ to $\bar{x}-2s$			2 or 3%
$\bar{x}-2s$ to $\bar{x}-s$			13 or 14%
$\bar{x}-s$ to $\bar{x}$			About 34%
$\bar{x}$ to $\bar{x}+s$			About 34%
$\bar{x}+s$ to $\bar{x}+2s$			13 or 14%
$\bar{x}+2s$ to $\bar{x}+3s$			2 or 3%

Answer Key for Chapter 8
Introduction to Sampling Distributions

Section A: Static Items

(a) The amounts of winter wheat produced in each rural county of eastern Colorado, Nebraska and Kansas
(b) The population mean, μ, and the standard deviation, σ
(c) Use $\bar{x}$ to estimate μ and s to estimate σ.
(d) Choose a large random sample of rural counties in eastern Colorado, Nebraska and Kansas. Find out the winter wheat production in these counties and calculate the sample
[1] mean and the sample standard deviation.

(a) The number of visitors per day to Colorado National Monument
(b) The population mean, μ, and the population standard deviation, σ
(c) The sample mean, $\bar{x}$ and the sample standard deviation, s
(d) Choose a large random sample of days. Look up the number of visitors to the Colorado National Monument for each day. Calculate the sample mean and the sample standard
[2] deviation.

(a) The response to the question "Has your child had all her or his vaccinations?" from all parents of children in the city preparing to enter kindergarten next year
(b) The proportion, p, of all children who have had their vaccinations
(c) Use r/n.
(d) Choose a large random sample of children. Determine their vaccination status and
[3] count the number who have had all their vaccinations.

(a) The responses to the question "Are you planning to buy a digital TV within the next year?" from all adults in the state
(b) The proportion, p, of them who plan to buy a digital TV during the next year
(c) Use r/n.
(d) Take a large random sample of adults in the state. Ask them whether they are planning
[4] to buy a digital TV during the next year. Count the number of positive responses.

[5] E.

[6] D.

[7] D.

[8] C.

[9] B.

[10] C.

[11] The two principal types of inferences that are made are to estimate the value of a population parameter, and to formulate a decision about the value of a population parameter.

[12] Sampling distributions are used for statistical inference.

[13] C.

[14] A.

[15] (a) Parameters: μ, σ, p; Corresponding statistics: $\bar{x}$ s, r/n.
(b) They are the same.

[16] (a) Inferential statistics
(b) Central Limit Theorem; see text.

[17] (a) No, the distribution of individual grant amounts may not be approximately normal.
(b) Yes, for a variable with a mound-shaped distribution, sample means for samples of size 50 will have a distribution which is approximately normal.

[18] (a) $P(60 \le x \le 90) = 0.3830$
(b) $P(60 \le \bar{x} \le 90) = 0.9974$
(c) The probability is higher in (b) because the standard deviation for $\bar{x}$ is smaller.

[19] (a) $P(x > 450) = 0.1788$
(b) $P(\bar{x} > 450) = 0.0322$

[20] (a) No, the sample size is too small.
(b) $P(79 \le \bar{x} \le 82) = 0.90$

[21] (a) For $n = 4$, $P(475 \le \bar{x} \le 525) = 0.4251$
(b) For $n = 16$, $P(475 \le \bar{x} \le 525) = 0.6572$
(c) The probability in part (b) is larger because the standard deviation is smaller.

(a) No, we do not know whether the distribution of x values is normal.
[22] (b) $P(\bar{x} < 650) = 0.0007$

(a) No, the sample size is too small.
[23] (b) $P(\bar{x} > 9) = 0.2005$

(a) For $n = 30$, $P(\bar{x} > 1{,}000) = 0.9981$
[24] (b) For $n = 36$, $P(1{,}000 \le \bar{x} \le 1{,}200) = 0.9964$

(a) $P(x \ge 10) = 0.1949$
[25] (b) For $n = 4$, $P(\bar{x} > 10) = 0.0436$

(a) For $n = 49$, $P(\bar{x} \le 90) = 0.0099$
[26] (b) For $n = 64$, $P(\bar{x} > 110) = 0.0038$

[27] $P(\bar{x} \ge 24) = 0.0228$

[28] $P(\bar{x} > 830) = 0.0392$

(a) $P(\bar{x} > 305{,}000) = 0.3372$
(b) $P(\bar{x} < 295{,}000) = 0.3372$
[29] (c) $P(295{,}000 \le \bar{x} \le 305{,}000) = 0.3256$

(a) $P(\bar{x} > 3.0) = 0.9842$
[30] (b) $P(3.0 \le \bar{x} \le 3.5) = 0.6896$

(a) $P(x \le 600) = 0.7157$
(b) $P(\bar{x} \le 600) = 0.9649$
[31] (c) Yes, since x has a normal distribution, $\bar{x}$ has a normal distribution for any n.

(a) $P(x \ge 21) = 0.3336$
(b) $P(\bar{x} \ge 21) = 0.0968$
[32] (c) Yes, since x has a normal distribution $\bar{x}$ has a normal distribution for all n.

(a) For $n = 49$, $P(\bar{x} \ge 7) = 0.0033$
[33] (b) For $n = 64$, $P(\bar{x} < 6) = 0.0918$

(a) $P(x > 99.6) = 0.1335$
[34] (b) $P(\bar{x} > 99.6) = 0.0033$

(a) $P(\bar{x} < 1{,}125) = 0.0062$
(b) $P(\bar{x} > 1{,}400) = 0.0013$
[35] (c) $P(1{,}125 \le \bar{x} \le 1{,}400) = 0.9925$

(a) $P(358 \le \bar{x} \le 372) = 0.8612$
(b) $P(\bar{x} \ge 385) = 0.0000$ to four decimal places. The sample mean is unlikely to be as high
[36] as 385kW if the electric company's figures are correct.

(a) $P(\bar{x} > 9{,}500) = 0.1849$
[37] (b) $P(\bar{x} < 9{,}000) = 0.0047$

(a) $P(\bar{x} > 5) = 0.1660$
(b) $P(4 \le \bar{x} \le 5) = 0.7619$
[38] (c) $P(\bar{x} < 4) = 0.0721$

(a) $P(115 \le x \le 125) = 0.6476$
(b) $P(115 \le \bar{x} \le 125) = 0.9356$
[39] (c) The probability in part (b) is greater because the standard deviation is smaller.

[40] $P(18 \le \bar{x} \le 21) = 0.8935$

(a) $P(14 \le \bar{x} \le 15) = 0.1040$
[41] (b) $P(10 \le \bar{x} \le 12) = 0.2968$

(a) $P(13.9 \le x \le 14.2) = 0.2902$
(b) $P(13.9 \le \bar{x} \le 14.2) = 0.7660$
[42] (c) The probability in part (b) is larger because the standard deviation is smaller.

(a) $P(x < 300) = 0.0228$
(b) $P(\bar{x} < 300) = 0.0001$
(c) The probability for part (b) is smaller because the standard deviation is smaller and the
[43] values of $\bar{x}$ tend to be closer to the mean.

(a) $P(x > 105) = 0.1587$
(b) $P(\bar{x} > 105) = 0.0071$
(c) The probability for part (b) is smaller because the standard deviation is smaller and the
[44] values of $\bar{x}$ tend to be closer to the mean.

[45] $P(\bar{x} < 175) = 0.0013$

(a) $P(\bar{x} < 45) = 0.0013$
[46] (b) $P(\bar{x} > 50) = 0.9332$

[47] $P(\bar{x} > 65) = 0.2451$

(a) $P(500 \le x \le 600) = 0.2282$
[48] (b) $P(500 \le \bar{x} \le 600) = 0.5528$

(a) $\mu_{\bar{x}} = 20$; $\sigma_{\bar{x}} = 2.37$; 0.3980
(b) $\mu_{\bar{x}} = 20$; $\sigma_{\bar{x}} = 1.68$; 0.4633
[49] (c) The standard deviation of part (b) is smaller, resulting in a narrower distribution.

(a) No, the sample size is too small.
[50] (b) Normal with mean 17 and standard deviation $\frac{8}{5}$; 0.864

(a) No, the sample size is too small.
[51] (b) Normal with mean 65 and standard deviation 4.47; 0.1314

(a) 0.1587
[52] (b) 0.0122

[53] B.

[54] E.

[55] C.

[56] C.

[57] A.

[58] C.

[59] A.

[60] E.

[61] B.

Section B: Algorithms

Objective 8.1: Sampling Distributions

[62] Yes

[63] No

[64] π

[65] Answers may vary. Sample answer: σ^2

[66] Answers may vary. Sample answer: s

[67] e

Objective 8.2: The Central Limit Theorem

[68] A normal distribution.

[69] Answers may vary. Sample answer: All the $\bar{x}$ distributions will be normal with mean 24. The standard deviations will be $\frac{5}{3}$, $\frac{5}{7}$, and $\frac{5}{11}$, respectively.

[70] 0.098

[71] 0.037

[72] 0.136

[73] 0.214

[74] $z = \dfrac{\overline{x} - 23}{0.3}$

[75] No

[76] 0.8413

[77] 1.0000

[78] 0.5832

[79] 0.219

[80] Yes; $z = \dfrac{\overline{x} - 84}{2.3}$

[81] 0.8413

[82] 0.3300

[83] No, because $(38)(0.07) \leq 5$

Objective 8.3: Class Project Illustrating the Central Limit Theorem

Step 1: The student will create a random sample of 10 possible outcomes. There should be at least 50 such samples in all in the class.

Step 2: The student will compute the sample mean $\overline{x}$ for his or her sample of size 10.

Step 3: Collect all the values of the sample mean of each sample of size 10 and calculate the sample mean of the sample means, $\overline{x_{\overline{x}}}$. Also calculate the standard deviation, *s*, of the list of the sample means.

Step 4: Using the values of $\overline{x_{\overline{x}}}$ and *s* complete the first column in the table given below. Determine the frequency of sample means in each interval and use the tally to complete the second column. Then determine the percent of the data of each interval and complete the third column. The percentages in the fourth column are the hypothetical percentages from a normal distribution. Compare the percentages in the third and fourth columns.

Interval	Frequency	Percent	Hypothetical Normal Distribution
$\overline{x}-3s$ to $\overline{x}-2s$			2 or 3%
$\overline{x}-2s$ to $\overline{x}-s$			13 or 14%
$\overline{x}-s$ to $\overline{x}$			About 34%
$\overline{x}$ to $\overline{x}+s$			About 34%
$\overline{x}+s$ to $\overline{x}+2s$			13 or 14%
$\overline{x}+2s$ to $\overline{x}+3s$			2 or 3%

Step 5: Using the frequencies in the second column of the above table construct a histogram of the sample means with six bars.

Step 6: Now ask students to create sample sizes of sizes $n = 20,\ 30,$ and $40.$ For each sample size repeat steps 2 to 5 and determine if the results tend to demonstrate the Central
[84] Limit Theorem.

Step 1: The student will create a random sample of 16 possible outcomes. There should be at least 50 such samples in all in the class.

Step 2: The student will compute the sample mean $\overline{x}$ for his or her sample of size 16.

Step 3: Collect all the values of the sample mean of each sample of size 16 and calculate the sample mean of the sample means, $\overline{x}_{\overline{x}}$. Also calculate the standard deviation, s, of the list of the sample means.

Step 4: Using the values of $\overline{x}_{\overline{x}}$ and s complete the first column in the table given below. Determine the frequency of sample means in each interval and use the tally to complete the second column. Then determine the percent of the data of each interval and complete the third column. The percentages in the fourth column are the hypothetical percentages from a normal distribution. Compare the percentages in the third and fourth columns.

Interval	Frequency	Percent	Hypothetical Normal Distribution
$\overline{x}-3s$ to $\overline{x}-2s$			2 or 3%
$\overline{x}-2s$ to $\overline{x}-s$			13 or 14%
$\overline{x}-s$ to $\overline{x}$			About 34%
$\overline{x}$ to $\overline{x}+s$			About 34%
$\overline{x}+s$ to $\overline{x}+2s$			13 or 14%
$\overline{x}+2s$ to $\overline{x}+3s$			2 or 3%

Step 5: Using the frequencies in the second column of the above table construct a histogram of the sample means with six bars.

Step 6: Now ask students to create sample sizes of sizes $n = 32$, 48, and 64. For each sample size repeat steps 2 to 5 and determine if the results tend to demonstrate the Central
[85] Limit Theorem.

Step 1: The student will create a random sample of 13 possible outcomes. There should be at least 50 such samples in all in the class.

Step 2: The student will compute the sample mean $\overline{x}$ for his or her sample of size 13.

Step 3: Collect all the values of the sample mean of each sample of size 13 and calculate the sample mean of the sample means, $\overline{x}_{\overline{x}}$. Also calculate the standard deviation, s, of the list of the sample means.

Step 4: Using the values of $\overline{x}_{\overline{x}}$ and s complete the first column in the table given below. Determine the frequency of sample means in each interval and use the tally to complete the second column. Then determine the percent of the data of each interval and complete the third column. The percentages in the fourth column are the hypothetical percentages from a normal distribution. Compare the percentages in the third and fourth columns.

Interval	Frequency	Percent	Hypothetical Normal Distribution
$\overline{x}-3s$ to $\overline{x}-2s$			2 or 3%
$\overline{x}-2s$ to $\overline{x}-s$			13 or 14%
$\overline{x}-s$ to $\overline{x}$			About 34%
$\overline{x}$ to $\overline{x}+s$			About 34%
$\overline{x}+s$ to $\overline{x}+2s$			13 or 14%
$\overline{x}+2s$ to $\overline{x}+3s$			2 or 3%

Step 5: Using the frequencies in the second column of the above table construct a histogram of the sample means with six bars.

Step 6: Now ask students to create sample sizes of sizes $n = 26,\ 39,$ and 52. For each sample size repeat steps 2 to 5 and determine if the results tend to demonstrate the Central
[86] Limit Theorem.

Step 1: The student will create a random sample of 13 possible outcomes. There should be at least 50 such samples in all in the class.

Step 2: The student will compute the sample mean $\overline{x}$ for his or her sample of size 13.

Step 3: Collect all the values of the sample mean of each sample of size 13 and calculate the sample mean of the sample means, $\overline{x}_{\overline{x}}$. Also calculate the standard deviation, s, of the list of the sample means.

Step 4: Using the values of $\overline{x}_{\overline{x}}$ and s complete the first column in the table given below. Determine the frequency of sample means in each interval and use the tally to complete the second column. Then determine the percent of the data of each interval and complete the third column. The percentages in the fourth column are the hypothetical percentages from a normal distribution. Compare the percentages in the third and fourth columns.

Interval	Frequency	Percent	Hypothetical Normal Distribution
$\overline{x}-3s$ to $\overline{x}-2s$			2 or 3%
$\overline{x}-2s$ to $\overline{x}-s$			13 or 14%
$\overline{x}-s$ to $\overline{x}$			About 34%
$\overline{x}$ to $\overline{x}+s$			About 34%
$\overline{x}+s$ to $\overline{x}+2s$			13 or 14%
$\overline{x}+2s$ to $\overline{x}+3s$			2 or 3%

Step 5: Using the frequencies in the second column of the above table construct a histogram of the sample means with six bars.

Step 6: Now ask students to create sample sizes of sizes $n=26,\ 39,$ and 52. For each sample size repeat steps 2 to 5 and determine if the results tend to demonstrate the Central
[87] Limit Theorem.

Step 1: The student will create a random sample of 12 possible outcomes. There should be at least 50 such samples in all in the class.

Step 2: The student will compute the sample mean $\overline{x}$ for his or her sample of size 12.

Step 3: Collect all the values of the sample mean of each sample of size 12 and calculate the sample mean of the sample means, $\overline{x}_{\overline{x}}$. Also calculate the standard deviation, *s*, of the list of the sample means.

Step 4: Using the values of $\overline{x}_{\overline{x}}$ and *s* complete the first column in the table given below. Determine the frequency of sample means in each interval and use the tally to complete the second column. Then determine the percent of the data of each interval and complete the third column. The percentages in the fourth column are the hypothetical percentages from a normal distribution. Compare the percentages in the third and fourth columns.

Interval	Frequency	Percent	Hypothetical Normal Distribution
$\overline{x}-3s$ to $\overline{x}-2s$			2 or 3%
$\overline{x}-2s$ to $\overline{x}-s$			13 or 14%
$\overline{x}-s$ to $\overline{x}$			About 34%
$\overline{x}$ to $\overline{x}+s$			About 34%
$\overline{x}+s$ to $\overline{x}+2s$			13 or 14%
$\overline{x}+2s$ to $\overline{x}+3s$			2 or 3%

Step 5: Using the frequencies in the second column of the above table construct a histogram of the sample means with six bars.

Step 6: Now ask students to create sample sizes of sizes $n = 24,\ 36,$ and 48. For each sample size repeat steps 2 to 5 and determine if the results tend to demonstrate the Central
[88] Limit Theorem.

Step 1: The student will create a random sample of 14 possible outcomes. There should be at least 50 such samples in all in the class.

Step 2: The student will compute the sample mean $\bar{x}$ for his or her sample of size 14.

Step 3: Collect all the values of the sample mean of each sample of size 14 and calculate the sample mean of the sample means, $\overline{x}_{\bar{x}}$. Also calculate the standard deviation, s, of the list of the sample means.

Step 4: Using the values of $\overline{x}_{\bar{x}}$ and s complete the first column in the table given below. Determine the frequency of sample means in each interval and use the tally to complete the second column. Then determine the percent of the data of each interval and complete the third column. The percentages in the fourth column are the hypothetical percentages from a normal distribution. Compare the percentages in the third and fourth columns.

Interval	Frequency	Percent	Hypothetical Normal Distribution
$\bar{x}-3s$ to $\bar{x}-2s$			2 or 3%
$\bar{x}-2s$ to $\bar{x}-s$			13 or 14%
$\bar{x}-s$ to $\bar{x}$			About 34%
$\bar{x}$ to $\bar{x}+s$			About 34%
$\bar{x}+s$ to $\bar{x}+2s$			13 or 14%
$\bar{x}+2s$ to $\bar{x}+3s$			2 or 3%

Step 5: Using the frequencies in the second column of the above table construct a histogram of the sample means with six bars.

Step 6: Now ask students to create sample sizes of sizes $n=28$, 42, and 56. For each sample size repeat steps 2 to 5 and determine if the results tend to demonstrate the Central
[89] Limit Theorem.

Step 1: The student will create a random sample of 14 possible outcomes. There should be at least 50 such samples in all in the class.

Step 2: The student will compute the sample mean $\overline{x}$ for his or her sample of size 14.

Step 3: Collect all the values of the sample mean of each sample of size 14 and calculate the sample mean of the sample means, $\overline{x}_{\overline{x}}$. Also calculate the standard deviation, s, of the list of the sample means.

Step 4: Using the values of $\overline{x}_{\overline{x}}$ and s complete the first column in the table given below. Determine the frequency of sample means in each interval and use the tally to complete the second column. Then determine the percent of the data of each interval and complete the third column. The percentages in the fourth column are the hypothetical percentages from a normal distribution. Compare the percentages in the third and fourth columns.

Interval	Frequency	Percent	Hypothetical Normal Distribution
$\overline{x}-3s$ to $\overline{x}-2s$			2 or 3%
$\overline{x}-2s$ to $\overline{x}-s$			13 or 14%
$\overline{x}-s$ to $\overline{x}$			About 34%
$\overline{x}$ to $\overline{x}+s$			About 34%
$\overline{x}+s$ to $\overline{x}+2s$			13 or 14%
$\overline{x}+2s$ to $\overline{x}+3s$			2 or 3%

Step 5: Using the frequencies in the second column of the above table construct a histogram of the sample means with six bars.

Step 6: Now ask students to create sample sizes of sizes $n = 28$, 42, and 56. For each sample size repeat steps 2 to 5 and determine if the results tend to demonstrate the Central

[90] Limit Theorem.

Step 1: The student will create a random sample of 9 possible outcomes. There should be at least 50 such samples in all in the class.

Step 2: The student will compute the sample mean $\bar{x}$ for his or her sample of size 9.

Step 3: Collect all the values of the sample mean of each sample of size 9 and calculate the sample mean of the sample means, $\bar{x}_{\bar{x}}$. Also calculate the standard deviation, s, of the list of the sample means.

Step 4: Using the values of $\bar{x}_{\bar{x}}$ and s complete the first column in the table given below. Determine the frequency of sample means in each interval and use the tally to complete the second column. Then determine the percent of the data of each interval and complete the third column. The percentages in the fourth column are the hypothetical percentages from a normal distribution. Compare the percentages in the third and fourth columns.

Interval	Frequency	Percent	Hypothetical Normal Distribution
$\bar{x} - 3s$ to $\bar{x} - 2s$			2 or 3%
$\bar{x} - 2s$ to $\bar{x} - s$			13 or 14%
$\bar{x} - s$ to $\bar{x}$			About 34%
$\bar{x}$ to $\bar{x} + s$			About 34%
$\bar{x} + s$ to $\bar{x} + 2s$			13 or 14%
$\bar{x} + 2s$ to $\bar{x} + 3s$			2 or 3%

Step 5: Using the frequencies in the second column of the above table construct a histogram of the sample means with six bars.

Step 6: Now ask students to create sample sizes of sizes $n = 18$, 27, and 36. For each sample size repeat steps 2 to 5 and determine if the results tend to demonstrate the Central
[91] Limit Theorem.

Chapter 9
Estimation

Section A: Static Items

1. In September a biological research team caught, weighed, and released a random sample of 54 chipmunks in Rocky Mountain National Park. The mean of the sample weights was $\bar{x} = 8.7$ oz with sample standard deviation $s = 1.4$ oz. Find a 90% confidence interval for the mean September weight of all chipmunks in Rocky Mountain National Park.

2. A sociologist would like to estimate the total amount of federal, state and city taxes paid by a family of four with income level \$50,000 per year. A random sample of 100 families produced sample mean $\bar{x} = \$4{,}258$ with sample standard deviation $s = \$1{,}098$. Find a 95% confidence interval for the population mean.

3. Using a standard normal distribution table, find the critical value, z_c, for $P(-z_{0.97} < z < z_{0.97}) = 0.97$.

4. A random sample of 60 employees of a large corporation had a sample mean $\bar{x} = 32.5$ accumulated vacation days with sample standard deviation $s = 18.5$ days. Find a 90% confidence level for the population mean number of accumulated vacation days.

5. A college career center conducted a study of skill levels required for jobs. A random sample of 200 job listings showed 120 required a high level of technical skill. Find a 90% confidence interval for the population proportion of jobs requiring a high level of technical skill.

6. For $s = 1.1$, $n = 120$, and $z_{0.95} = 1.96$, find the maximal error tolerance, E.

A. 5.08 B. 0.279 C. 0.018 D. 0.197 E. 0.026

7. An air pollution index (from 1 to 25 with 25 being the most serious pollution) takes into account the amount of carbon dioxide in the air. A random sample of 42 readings taken one hour before sunset had a mean index reading of 9.7 with standard deviation 3.2. Find a 90% confidence interval for the population mean index reading.

8. A random sample of 200 businesses showed that 120 of them ban smoking on their premises. Find a 95% confidence interval for the population proportion of all businesses than ban smoking on their premises.

9. Given that $P(-0.227 < \bar{x} - \mu < 0.227) = 0.90$ and $n = 150$, find s.

A. 1.69 B. 03 C. 1.54 D. 4.57 E. 1.66

10. The national program planner for the Girl Scouts wants to estimate the average number of girls per troop. A random sample of 81 troops shows sample mean $\bar{x} = 14.8$ girls with standard deviation 9.0 girls. Find a 99% confidence interval for the population mean number of girls per troop.

11. A random sample of 50 calls initiated on cellular car phones had a mean duration of 3.5 minutes with standard deviation 1.2 minutes. Find a 99% confidence interval for the population mean duration of telephone calls initiated on cellular car phones.

12. Given $s = 2.4$, $c = 0.85$, $n = 110$, and $\bar{x} = 9.65$, compute the confidence interval for μ.

13. The director of the city summer recreation program would like to estimate the proportion of children in the summer learn to swim program who do not know how to swim. A random sample of 80 registration records from last year showed 36 children who did not know how to swim. It is reasonable to assume that the population for this year has the same characteristics as the population in last year's program. Find a 90% confidence interval for the proportion of children entering the program who do not know how to swim.

14. Karl would like to estimate the proportion of eligible voters in a large metropolitan district who are registered to vote. A random sample of 175 eligible voters showed 110 of them were registered.
(a) Find a 95% confidence interval for the population proportion of registered voters.
(b) Find the margin of error for this estimate.

15. A candy store sells packages of saltwater taffy. Because the size and weight of the individually wrapped taffy pieces varies, the number of pieces per pound is not consistent. The storekeeper counted a sample of 50 one-pound packages and found that $\bar{x} = 13.5$ pieces and $s = 1.8$ pieces. The storekeeper wants to label the packages with a range that represents a 95% confidence interval. Compute this range.

16. Roger would like to estimate the proportion of registered voters in his city who voted in the last general election. A random sample of 225 registered voters showed 98 actually voted.
(a) Find a 90% confidence interval for the population proportion of registered voters who voted in the last general election.
(b) Find the margin of error for this estimate.

17. The postmaster in a large city found that 56 packages out of a random sample of 400 packages had insufficient postage. Find a 99% confidence interval for the population proportion of packages with insufficient postage.

18. Given the confidence interval, $6.71 < \mu < 7.23$, and that $s = 1.15$ and $n = 75$, find the confidence level, c.

A. 0.85 B. 0.90 C. 0.99 D. 0.95 E. 0.80

19. A random sample of 48 fish taken from a popular fishing lake showed 15 to be yellow perch. Let p represent the population proportion of yellow perch in the lake.
(a) Find a point estimate for p.
(b) Find an 85% confidence interval for p.

20. A biologist has found the average weight of 12 randomly selected mud turtles to be 8.7 lb. with standard deviation 3.6 lb. Find a 90% confidence interval for the population mean weight of all such turtles.

21. Given the confidence interval, $11.29 < \mu < 12.01$, and that $n = 90$ and $c = 0.90$, find the sample standard deviation, s.

A. 1.44 B. 2.08 C. 4.15 D. 0.48 E. 0.72

22. The Aloha Taxi Company of Honolulu, Hawaii wants to estimate the mean life of tires on its cabs. A random sample of 15 tires from the cabs had a mean life of 18,280 mi. with standard deviation 1,310 mi. Find a 90% confidence interval for the population mean life of the tires.

23. A park ranger has timed the mating calls of 22 bull elk. The mean duration of these calls is 14.6 sec. with standard deviation 2.8 sec. Find a 95% confidence interval for the population mean duration of mating calls of the bull elk.

24. Use a student's t distribution table to find the critical value t_c for a 0.95 confidence level for a t distribution with sample size $n = 9$.

A. 2.365 B. 2.896 C. 2.306 D. 1.860 E. 3.355

25. The weights of grapefruit follow a normal distribution. A random sample of 12 new hybrid grapefruit has a mean weight of 1.7 lb. with sample standard deviation 0.24 lb. Find a 95% confidence interval for the population mean weight of the hybrid grapefruit.

26. What is the average cost of a 30-second ad on prime time TV? A random sample of 12 such ads had an average cost of $122,000 with sample standard deviation $15,000. Find a 95% confidence interval for the cost of the ads.

27. For $t_{0.98} = 3.747$ we use a student's t distribution table to find n.

A. 6 B. 4 C. 5 D. 7 E. 3

28. A random sample of 15 novels making a national best seller list was selected. The average length of time these novels stayed on the best seller list was 5.2 weeks with standard deviation 2.3 weeks. Find a 99% confidence interval for the average number of weeks a book is on this best seller list.

29. The Tender Turkey Farm hatches turkey eggs. A random sample of 100 eggs hatched 84 chicks. Let p be the proportion of eggs from Tender Turkey Farm that hatch.
(a) Find a point estimate for p.
(b) Find a 90% confidence interval for p.

30. A random sample of ten jumbo burgers from a local diner had the following weights (in ounces):

12.1	12.4	11.8	12.6	11.9
11.5	12.2	11.3	11.7	12.3

If μ is the mean weight of all jumbo burgers served at the diner, find a 99% confidence interval for μ.

31. Scott wants to estimate the cost of a night at a Los Angeles hotel. He called a random sample of 8 hotels and got a sample mean cost for one night of $145 with sample standard deviation $40.
(a) Find an 85% confidence interval for the population mean cost.
(b) What size sample does he need to say with 90% confidence that the sample mean is within $10 of the population mean?

32. A biology student is doing a study of the nesting habits of the American Robin. She checks a random sample of 6 nests and finds a sample mean $\bar{x} = 4.9$ eggs with sample standard deviation $s = 1.4$. Find a 90% confidence interval for the population mean number of eggs.

33. The 95% confidence interval for a certain set of random sample data is $4.36 < \mu < 4.68$. The sample size is 14. Find s.

A. 0.31 B. 0.27 C. 0.28 D. 0.55 E. 0.54

34. Country Boy seed Company wants to estimate the yield from their corn seed on prime Iowa farm land. A random sample of 20 one acre plots gave an average yield of 170 bushels per acre with standard deviation 6.4 bushels per acre. Find a 99% confidence interval for the true mean yield per acre.

35. The heights of a random sample of six fifth-graders are the following (in inches):

49.4	59.5	50.8
55.0	60.1	53.3

Compute a 90% confidence interval for all fifth-graders from this data, letting μ represent the average height of all fifth-graders.

36. An official of the National Parks Service would like to determine what proportion of the visitors to Yellowstone National Park would be willing to pay an additional \$25 for a national park pass. A random sample of 275 visitors were surveyed. 90 of them said they would be willing to pay the additional amount. Let p be the population proportion of visitors who would be willing to pay \$25 more.
(a) Find a point estimate for p.
(b) Find a 95% confidence interval for p.
(c) Express the results of (a) and (b) using the margin of error.

37. A preliminary study of the fall lobster catch showed that for 35 lobsters selected at random the mean weight was 2.9 lb with standard deviation 0.6 lb.
(a) Find a 90% confidence interval for the population mean weight of the fall lobsters.
(b) How many more lobsters should be included in the sample if we want to say with 90% confidence that the population mean weight of the fall lobster catch is within 0.1 lb of the sample mean?

38. A set of random sample data has $s = 3.4$ and $\bar{x} = 21.8$. Compute the maximal error of estimate if the degrees of freedom are 11 in number and the confidence level is 85%.

A. 1.13 B. 1.51 C. 1.59 D. 1.52 E. 1.58

39. A member of the House of Representatives wants to determine the proportion of voters in her district who favor a flat income tax. A random sample of 200 voters in her district showed 89 in favor. Let p represent the proportion of voters who favor a flat income tax.
(a) Find a point estimate for p.
(b) Find a 95% confidence interval for p.
(c) Does the data indicate a majority of the voters oppose the tax? Explain.

40. A member of the state legislature wants to determine what fraction of the voters in his district favor a proposal refunding excess state sales tax revenues to the tax-payers. A random sample of 145 voters in his district show 97 in favor. Let p be the population proportion in favor of the proposal.
(a) Find a 95% confidence interval for p.
(b) What is the margin of error for this estimate?
(c) What size sample does he need in order to state with 95% confidence that the sample estimate is within 0.05 of p?

41. The weights of eight two-year-olds were selected at random from a hospital's records. They are the following (in pounds):

29.0	27.4	35.1	31.2
30.6	33.8	26.3	29.7

Let μ be the mean weights of all the two-year-olds' weights listed in the hospital's records, and find a 98% confidence interval for μ.

A. $27.2 < \mu < 33.6$ B. $27.0 < \mu < 33.8$ C. $28.3 < \mu < 32.5$

D. $27.9 < \mu < 32.9$ E. $27.4 < \mu < 33.8$

42. The Director of a Museum would like to know what fraction of the museum associates make purchases through the gift shop catalogue.
(a) If no preliminary study is done, how large a sample must be taken if the director is to say with 90% confidence that the sample estimate is within 2% of the population proportion?
(b) A preliminary study showed that out of 60 associates, 12 have used the gift shop catalogue. What size sample does the director need in order to say with 90% confidence that the sample estimate is within 2% of the population proportion?

43. A random sample of 80 mature coast redwoods was selected, and the heights were measured. The mean height was $\bar{x} = 281$ feet with $s = 24$ feet. Compute a 99% confidence interval if μ represents the mean height of all mature coast redwoods.

A. $279 < \mu < 283$ B. $274 < \mu < 288$ C. $277 < \mu < 285$

D. $275 < \mu < 287$ E. $280 < \mu < 282$

44. The marketing director for a cereal company would like to know what proportion of households that receive free samples of the cereal with their newspapers later purchase the cereal. A random sample of 175 households showed that 45 purchased the cereal after receiving the free sample.
(a) Find a point estimate for p.
(b) Find a 95% confidence interval for p.
(c) Express the results of (a) and (b) using the margin of error.

45. The time it takes broken bones of a given size and type of break to heal is normally distributed. A random sample of 17 simple arm breaks occurring in people in their twenties took an average of 53 days to heal with standard deviation 5 days. Find a 90% confidence interval for the population mean healing time for all broken bones of this type.

46. Five hundred members of a health club out of a total of 10,000 are selected at random and asked if they are satisfied or unsatisfied with the club's facilities. Out of the 500, 375 said they were satisfied. Compute the point estimate for the probability that any of the 10,000 members selected at random will answer that they are satisfied with the club's facilities.

A. $\hat{p} = 0.25$ B. $\hat{p} = 0.75$ C. $\hat{p} = 1.33$ D. $\hat{p} = 0.0875$ E. $\hat{p} = 0.0375$

47. The past few years a nursery has kept records of blue spruce trees they have replanted. Over this period it was found that 421 out of 518 blue spruce trees survived replanting. Let p be the population proportion of blue spruce trees that survive replanting.
(a) Find a point estimate for p.
(b) Find a 99% confidence interval for p. Give a brief interpretation of the meaning of your confidence interval.
(c) Assuming that the number of trees, r, which survive replanting is a binomial random variable, what additional conditions must be satisfied to ensure the accuracy of your confidence interval?

48. Why do people enroll in fitness programs? A random sample of 500 people were surveyed. 335 of them reported that an important reason for enrolling was to prevent heart disease.
(a) Find a point estimate for p.
(b) Find a 90% confidence interval for p.

49. Compute $\sigma = \sqrt{\hat{p}\hat{q}/n}$ given $\hat{p} = 0.6$ and $n = 35$.

A. 0.993 B. 0.007 C. 0.917 D. 0.041 E. 0.083

50. In a study regarding public opinion about American involvement in foreign affairs, a random sample of 40 people showed that 25 people favored increasing U.S. involvement in foreign affairs. How many more people should be included in the sample for the study to state with 90% confidence that the sample proportion is within 0.05 of the population proportion?

51. A seed company advertises that the mean time from planting to harvest for a new variety of zucchini seeds is 50 days. The standard deviation is estimated to be 10.3 days. If we wanted to verify this claim how large a sample would we need in order to state with 95% confidence that our sample mean differs from the population mean by no more than 3 days?

52. Compute the maximal error tolerance of the error of estimate $|\hat{p} - p|$ for a confidence level of 99%, given $\hat{p} = 0.7$ and $n = 60$.

A. 0.85 B. 0.009 C. 0.07 D. 0.991 E. 0.15

53. A legislative aide is doing research on the duration of calls initiated on cellular car phones. A preliminary study showed that the standard deviation of these calls is about 1.8 minutes. How large a sample will be needed for the aide to be able to state with 99% confidence that the sample mean duration time for the calls differs from the population mean by no more than 0.25 minute?

54. A doctor is testing a new sinus medication. A random sample of subjects will be given the medication and the number who are cured will be counted. How large a sample does he need to be able to state with 95% confidence that the sample proportion cured is within 0.03 of the actual population proportion?

55. If $n = 70$ and $r = 21$, compute a 98% confidence interval for p.

56. A random sample of 40 cups of coffee from a vending machine had a sample mean volume of coffee dispensed equal to 7.1 oz with sample standard deviation $s = 0.3$ oz. Find a 90% confidence interval for the mean amount of coffee dispensed per cup.

57. The Junetag Company makes washing machines. In one phase of production it is necessary to estimate the drying time of enamel paint on metal sheeting. Analysis of a random sample of 420 sheets shows that the average drying time is 56 hr with standard deviation 12 hr. Find a 99% confidence interval for the population mean drying time.

58. For the confidence interval $0.64 < p < 0.76$, given that $n = 100$, determine the confidence level, c.

A. 0.84 B. 0.83 C. 0.85 D. 0.81 E. 0.82

59. A random sample of 100 felony trials in a large city in the Midwest shows the mean waiting time between arrest and trial is 173 days with standard deviation 28 days. Find a 99% confidence interval for the mean time interval between arrest and trial.

60. A local C. J. Nickel department store took a random sample of 49 charge accounts and found that the average balance due was $63.19 with standard deviation $13.50. Find a 95% confidence interval for the average balance due at this store.

61. For the confidence interval $0.83 < p < 0.91$, given that the confidence level is 0.95, determine n.

A. 272 B. 11 C. 50 D. 6 E. 68

62. Pro Computer Services has computer monitoring of all outgoing and incoming telephone calls to record the duration of technical consulting calls. A random sample of 40 calls has sample mean $\bar{x} = 15.95$ min with standard deviation $s = 5.43$ min. Find a 90% confidence interval for the population mean duration of technical consulting calls at Pro Computer Services.

63. From a survey of 18 patients, symptoms of a new flu virus have been determined to have a mean duration of 9.7 days with standard deviation 4.8 days. Find a 95% confidence interval for the population mean duration of these flu symptoms.

64. A random sample of 212 El Cheapo brand mechanical pencils showed that 53 were defective. Let p represent the proportion of mechanical pencils that are defective to all mechanical pencils produced by El Cheapo.
(a) What is the point estimate for p?
(b) Find a 95% confidence interval for p.
(c) The confidence interval was computed using a normal approximation. Was this justified?

65. A random sample of four boxes of Purr Lots cat food had weights (in ounces) of 12.7, 11.6, 11.2, and 12.4. Find a 99% confidence interval for the mean weight of all such boxes of Purr Lots cat food.

66. The average cholesterol level for a random sample of 26 adult women from Dairyhaven, Wisconsin is 263 units (mg. per dl. of blood). The sample standard deviation is 43 units. Find a 99% confidence interval for the mean cholesterol level of all adult women in Dairyhaven, Wisconsin.

67. Compute an 85% confidence interval for the following situation: Out of 160 basketballs purchased, 16 have leaks. Let p represent the proportion of basketballs with leaks to all basketballs produced by the manufacturer.

A. $0.07 < p < 0.13$ B. $0.09 < p < 0.11$ C. $0.05 < p < 0.15$

D. $0.08 < p < 0.12$ E. $0.06 < p < 0.14$

68. In wine making, acidity of the grape is a crucial factor. A pH range from 3.1 to 3.6 is considered very acceptable. A random sample of 12 bunches of grapes was taken from a California vineyard. The sample mean acidity was 3.38 with sample standard deviation 0.20. Find a 99% confidence interval for the mean acidity of the entire harvest of grapes from this vineyard.

69. Certain rivers in Labrador are known to have phenomenally large brook trout. A biologist using catch-and-release tagging of trout obtained the following information about the annual weight gain of seven trout (in pounds). The results are: 0.73 1.05 0.81 1.12 0.92 1.17 0.98. Find a 90% confidence interval for the mean annual weight gain of all brook trout from these rivers.

70. For a particular research project, the maximum error of estimate must be no larger than 0.10 unit, with a confidence level of 99%. A preliminary sample study produced a sample standard deviation of $s = 2.5$ units. How large must the sample taken be to ensure the above requirements for the research project?

A. 4160 B. 65 C. 2081 D. 64 E. 4161

71. Shoplifting has been a problem in a large men's clothing store. Using special security measures to monitor shoplifting, it was found that there were attempts to shoplift the following dollar values of items of merchandise during the last nine weeks: \$356 \$285 \$310 \$375 \$290 \$325 \$331 \$342 \$335.
(a) Find a 90% confidence interval for the population mean shoplifting loss per week.
(b) Estimate the mean shoplifting losses for a four-week month.

72. Air Connect airlines found that 88 out of a random sample of 121 passengers purchased round-trip tickets. Let p be the proportion of all Air Connect passengers who purchase round-trip tickets.
(a) Find a point estimate for p.
(b) Find a 95% confidence interval for p.
(c) Express the results of parts (a) and (b) using the margin of error.

73. A company with 2,200 employees is planning to alter their health insurance plan. A random sample of 330 employees, when asked whether or not they are satisfied with their current coverage, resulted in 132 replying with "satisfied" and 198 replying with "unsatisfied." Let p represent the proportion of employees of the entire company that are unsatisfied with their coverage.
(a) What is the point estimate for p?
(b) Find a 99% confidence interval for p.
(c) The confidence interval was computed using a normal approximation. Was this justified?

74. An automobile manufacturer used a random sample of 50 cars of a certain model to estimate the mileage this model car gets in highway driving. The sample standard deviation was found to be 5.7 mpg. How large a sample do we need to state with 90% confidence that the sample mean mileage is within 1 mpg of the population mean for all cars of this model.

75. Jim is doing a research project in political science to determine the proportion of registered voters, *p*, in his district who favor capital punishment.
(a) If no preliminary sample is taken to estimate *p*, how large a sample is necessary if he wants to state with 95% confidence that the margin of error is 2%?
(b) Jim did a preliminary study in which he found that in a random sample of 932 registered voters 136 favored capital punishment. How many more voters should be included in the sample if the margin of error is to be 2%?

76. A biologist is studying varieties of fungi in a state park. A preliminary study found that *p* was approximately 0.42, where *p* represents the proportion of the dominant variety of fungi to all varieties occurring in the park. For a 95% confidence level, how large a sample should be taken to ensure that the point estimate for *p* will be in error either way by less than 0.02?

A. 570 B. 2,340 C. 47 D. 1,194 E. 24

77. Union officials want to estimate the percent, *p*, of workers in the Big Bend Metalworks who favor a strike. The union wants to have 90% confidence that its estimate for *p* is within 2.5% of the true proportion of workers who favor a strike.
(a) If no accurate estimate for *p* is available, how large a sample of workers is necessary?
(b) A preliminary sample of 100 workers gave $r/n = 0.35$. How many more workers should be included in this sample?

78. A random sample of 40 Salt Lake City teachers showed the standard deviation of teaching experience to be 5.3 years. How many more teachers should be included in the sample to have 95% confidence that the sample mean number of years teaching experience is within six months of the population mean?

79. A car company is investigating whether or not to issue a recall of a model of car that has shown signs of having a defective emissions system. If no preliminary study is made to estimate *p*, the probability that the emissions system is defective, how large a sample should the investigator use to obtain a 98% confidence level that the point estimate for *p* will be in error either way by less than 0.03?

A. 20 B. 1,509 C. 1,508 D. 6,032 E. 6,033

80. Aerodynamic engineers want to know the mean shear force on the wings of an FX5000 fighter jet as the jet comes out of a 3,000 ft. vertical dive. A random sample of FX5000 jets flown by Air Force fighter pilots was used for 37 test flights. The sample standard deviation of wing shear force was $s = 948$ foot-pounds. The engineers want to have 95% confidence that the sample mean is within 250 foot-pounds of the population mean. How many more test flights should be made?

81. Rocky mows lawns for a living. For a random sample of 40 6,000 sq. ft suburban lawns the sample mean time to mow the lawn and trim is 75 minutes with standard deviation 10 minutes. Find an 80% confidence interval for the mean time it takes Rocky to mow all such lawns.

82. A random sample of 100 art books showed a sample average price $40.52 with sample standard deviation $15.30. Find a 90% confidence interval for the population mean price of art books.

83. Use a standard normal distribution table to compute the critical value for an 89% level of confidence.

A. 1.63 B. 1.62 C. 1.59 D. 1.61 E. 1.60

Section B: Algorithms

84. Using an Areas of a Standard Normal Distribution table, find a number $z_{0.70}$ so that 70% of the area under the standard normal curve lies between $-z_{0.70}$ and $z_{0.70}$.

85. Using an Areas of a Standard Normal Distribution table, find a number $z_{0.90}$ so that 90% of the area under the standard normal curve lies between $-z_{0.90}$ and $z_{0.90}$.

86. Given that $n = 53$ and $s = 6$, find E, the maximal error of tolerance, using a 99% level of confidence.

87. Given that $n = 53$ and $s = 5$, find E, the maximal error of tolerance, using a 98% level of confidence.

88. Mrs. Warrack, a retired archeologist, enjoys hiking in Badger Creek every other day. Usually she travels 3 miles per hike. During the past year Mr. Warrack has recorded the times of her hikes. For a sample of 70 times, the mean was $\bar{x} = 19.93$ minutes and the standard deviation was $s = 1.30$ minutes. Let μ be the mean hiking time for the entire distribution of Mrs. Warrack's 3-mile hike times (taken over the past year). Find a 0.95 confidence interval μ.

89. Find the standard error of the mean for a sample with 100 values, a mean of 35, and a standard deviation of 2.7. Then find the range for a 5% level of confidence.

90. Suppose the weights of 132 5-year-old kindergarten boys in a school district are measured and used to estimate the mean weight of all 5-year-old boys in the U.S. If the mean weight of the boys in the sample is 40.6 pounds, with a standard deviation of 10.9 pounds, approximate a 95% confidence interval for the mean weight of all U.S. 5-year-old boys.

91. The compact florescent light bulbs made by a company are guaranteed to last 10,000 hours The quality control lab tests samples of 48 bulbs from each production batch and has rated the most recent batch at 10,240 hours, with a standard deviation of 800 hours. Use a 99% confidence interval to see if fewer than half the bulbs can be expected to fail before 10,000 hours of use. Round hour values to the nearest integer.

92. A group of 34 asthma patients who all used the same medication were treated with a new medication. Their pulmonary output test scores improved by an average of 5.7%, with a standard deviation of 6.5%. If all asthma patients using the same original medication were given the same new treatment, estimate that the population's mean improvement falls within two standard errors of the mean of the trial group and write a prediction interval for the population's mean improvement.

93. Using an Areas of a Standard Normal Distribution table, find a number $z_{0.98}$ so that 98% of the area under the standard normal curve lies between $-z_{0.98}$ and $z_{0.98}$.

Objective 9.1: Estimating with Large Samples

Objective 9.2: Estimating with Small Samples

94. Given that $n = 18$, find the degrees of freedom, *d.f.*

95. Find t_c for a 0.95 confidence level of a t distribution with sample size $n = 6$.

96. Find t_c for a 0.90 confidence level of a t distribution with sample size $n = 14$.

97. Find t_c for a 0.90 confidence level of a t distribution with sample size $n = 23$.

98. Thome works at Demetre's, a popular Greek restaurant in Seattle. The IRS (Internal Revenue Service) is auditing Thome this year on her taxes. They believe that Thome made more in tips than she actually claimed on her W-2. In order to justify her claims, Thome took the following random sample of six credit card receipts, each of which indicated her tip: \$2.80, \$9.95, \$2.10, \$5.80, \$1.76, \$1.90.
Find a 0.99 confidence interval for the population mean of tips received by Thome.

99. Find t_c for a 0.90 confidence level of a t distribution with sample size $n = 15$.

100. Thome works at Demetre's, a popular Greek restaurant in Seattle. The IRS (Internal Revenue Service) is auditing Thome this year on her taxes. They believe that Thome made more in tips than she actually claimed on her W-2. In order to justify her claims, Thome took the following random sample of six credit card receipts, each of which indicated her tip: \$3.80, \$6.70, \$3.04, \$11.00, \$9.35, \$4.05. Find a 0.90 confidence interval for the population mean of tips received by Thome.

101. Suppose the IRS (Internal Revenue Service) is auditing a waiter. The agent believes that the waiter made more in tips than he actually claimed on his W-2 form. In order to justify his claims, the waiter took the following random sample of six credit card receipts, each of which indicated his tip, to the audit session: \$3.50, \$7.65, \$4.95, \$4.75, \$4.80, \$6.30. Find an appropriate 95% confidence interval for the population mean of tips received by the waiter. Show the calculation.

Objective 9.3: Estimating in the Binomial Distribution

102. A random sample of 146 gumballs were taken from a giant gumball machine at an amusement park. Of the 146 gumballs, 55 were grape flavored. Let p represent the proportion of gumballs that were grape. What is a point estimate for p?

103. A random sample of 171 CDs purchased from a popular music store showed that 77 of the CDs were country. Let p represent the proportion of CDs sold by this store that are country. What is a point estimate for p?

104. Of a random sample of 385 people, 59 said they trust their mechanic to give them a fair and reasonable quote on car repairs. Let p represent the proportion of the people satisfied with their mechanic's quotes. Find a 90% confidence interval for p.

105. The NBC News/Wall Street Journal poll of September 9-12, 1999, found that 13% of 1010 adults nationwide thought Babe Ruth was the greatest American male athlete of the 20th century. What is the 95% confidence interval for this result?
(source: PollingReport.com, www.pollingreport.com)

106. When 376 voters were polled, 31% said they were voting "Yes" on an initiative measure. Find the margin of error and an interval which is likely to contain the true population proportion.

107. When 526 voters were polled, 79% said they were voting "Yes" on an initiative measure. Find the margin of error and an interval which is likely to contain the true population proportion.

108. Jaelle read in the newspaper that 23% of voters in her city were voting "no" on a local initiative measure. The poll claimed a margin of error of ±4%. Jaelle wanted to know how many voters were polled and wrote an equation to solve. What should she answer be?

Objective 9.4: Choosing the Sample Size

109. Imagine you are responsible for estimating the percentage of TV households that are tuned to the late evening news on a particular night in your county. You want to randomly sample n households and determine r, the number of households tuned into the late evening news. You want to have 75% confidence that your sample percentage has a margin of error of not more than 0.06. Assuming that nothing is known about the proportion of households tuned in to television after 10 P.M., how many TV households must be surveyed?

110. At many gasoline stations, customers have the option to have their car washed for a few extra dollars over the price of their gasoline. It's your job to determine the proportion of customers in your neighborhood that take advantage of this option and have their car washed while getting gasoline. Assuming you know nothing about the estimate p, how large a sample of customers is necessary to be 95% sure that the point of estimate $\hat{p}$ will be within a distance of 0.07 of p?

111. A large aircraft parts manufacturing company in Los Angeles has over 1900 employees. The president of the company wants to estimate the average engineering experience (in years) of her employees. A preliminary random sample of 70 employees yields a sample standard deviation of $s = 3.0$ years. The president wants to be 98% confident that the sample mean $\bar{x}$ does not differ from the population mean by more than half a year. How large a sample should be used?

112. Determine the sample size (to the nearest whole number) required to ensure that $\bar{x}$ will be within 0.99 of μ with 95% confidence, given that $\sigma = 4.1$.

113. Determine the sample size (to the nearest whole number) required to ensure that $\bar{x}$ will be within 0.7 of μ with 90% confidence, given that $\sigma = 5$.

Answer Key for Chapter 9
Estimation

Section A: Static Items

[1] Interval from 8.3 oz to 9.1 oz

[2] Interval from $4,043 to $4,473

[3] $z_{0.97} = 2.17$

[4] Interval from 28.6 days to 36.4 days

[5] Interval from 0.54 to 0.66

[6] D.

[7] Interval from 8.9 to 10.5

[8] Interval from 0.53 to 0.67

[9] A.

[10] Interval from 12.2 girls to 17.4 girls

[11] Interval from 3.1 min to 3.9 min

[12] $9.32 < \mu < 9.98$

[13] Interval from 0.36 to 0.54

[14] (a) Interval from 0.557 to 0.700
(b) Margin of error: 7.2%

[15] The storekeeper can conclude with 95% confidence that each package contains 13 to 14 pieces of taffy.

[16] (a) Interval from 0.381 to 0.490
(b) Margin of error: 5.4%

[17] Interval from 0.123 to 0.157

[18] D.

[19] (a) $\hat{p} = 0.3125$
(b) Interval from 0.2162 to 0.4088

[20] Interval from 6.83 lb to 10.57 lb

[21] B.

[22] Interval from 17,684 mi to 18,876 mi

[23] Interval from 13.57 sec to 15.63 sec

[24] C.

[25] Interval from 1.55 lb to 1.85 lb

[26] Interval from 112.47 thousand dollars to 131.53 thousand dollars

[27] C.

[28] Interval from 3.43 weeks to 6.97 weeks

[29] (a) $\hat{p} = 0.84$
(b) Interval from 0.7797 to 0.9003

[30] $11.56 < \mu < 12.41$

[31] (a) Interval from \$122.13 to \$167.87
(b) Sample size: 34 hotels

[32] Interval from 3.75 to 6.05 robin eggs

[33] C.

[34] Interval from 165.7 to 174.3 bushels per acre

[35] $51.0 < \mu < 58.3$

(a) $\hat{p} = 0.3273$
(b) Interval from 0.2718 to 0.3828
(c) About 33% of those surveyed said that they would pay an additional $25 for a national
[36] park pass. The margin of error is 5.5%.

(a) Interval from 2.73 lb to 3.07 lb
[37] (b) You need 63 more. (Sample size: 98)

[38] D.

(a) $\hat{p} = 0.445$
(b) Interval from 0.410 to 0.4801
(c) We can state with 95% confidence that between 41% and 48% favor the tax. It appears
[39] that a majority oppose the tax.

(a) Interval from 0.592 to 0.746
(b) Margin of error: 7.7%
[40] (c) Sample size: 341

[41] A.

(a) Sample size without estimate: 1,692
[42] (b) Sample size with estimate: 1,083

[43] B.

(a) $\hat{p} = 0.2571$
(b) Interval from 0.1924 to 0.3219
(c) About 25.7% of the people who receive cereal samples buy the cereal. The margin of
[44] error is 6.5%.

[45] Interval from 50.9 days to 55.1 days

[46] B.

(a) $\hat{p} = 0.8127$
(b) Interval from 0.7685 to 0.8569
We can say with 99% confidence that the mean number of all blue spruce trees from the nursery which survive replanting is between 76.9% and 85.7%.
[47] (c) *np* and *nq* must both be greater than 5.

(a) $\hat{p} = 0.670$
[48] (b) Interval from 0.635 to 0.705

[49] E.

[50] 214 more people are needed. (Sample size = 254)

[51] Sample size: 46 seeds

[52] E.

[53] Sample size: 346 calls

[54] Sample size: 1,067 subjects

[55] $0.17 < p < 0.43$

[56] Interval from 7.02 oz to 7.18 oz

[57] Interval from 54.49 hours to 57.51 hours

[58] D.

[59] Interval from 165.78 days to 180.22 days

[60] Interval from \$59.41 to \$66.97

[61] A.

[62] Interval from 14.54 min to 17.36 min

[63] Interval from 7.31 days to 12.09 days

[64]
(a) $\hat{p} = 0.25$
(b) $0.19 < p < 0.31$
(c) $np = 53 > 5$ and $nq = 159 > 5$, so the approximation was justified.

[65] Interval from 9.95 oz to 14.00 oz

[66] Interval from 239.50 units to 286.50 units

[67] A.

[68] Interval from 3.2007 to 3.5593

[69] Interval from 0.8507 lb to 1.0865 lb

[70] E.

[71]
(a) Interval from \$309.50 to \$354.84
(b) Estimated shoplifting loss for 4 week month = \$1,310.67

[72]
(a) $\hat{p} = 0.7273$
(b) Interval from 0.6479 to 0.8067
(c) It is estimated that about 73% of the customers buy round trip tickets. The margin of error is 7.9%.

[73]
(a) $\hat{p} = 0.6$
(b) $0.53 < p < 0.67$
(c) $np = 198 > 5$ and $nq = 132 > 5$, so the approximation was justified.

[74] Sample size: 88 cars

[75] (a) Sample size without estimate: 2,401 registered voters
(b) Sample size with estimate: 265 more voters are needed. (Sample size = 1,197)

[76] B.

[77] (a) Sample size without estimate: 1,083 workers
(b) 885 more workers are needed. (Sample size: 985)

[78] 392 more teachers are needed. (Sample size: 432)

[79] B.

[80] 19 more test flights are needed. (Sample size: 56)

[81] Interval from 73.0 min to 77.0 min

[82] Interval from $38.00 to $43.04

[83] E.

Section B: Algorithms

[84] $z_{0.70} = 1.04$, so that $P(-1.04 < z < 1.04) = 0.70$.

[85] $z_{0.90} = 1.645$, so that $P(-1.645 < z < 1.645) = 0.90$.

[86] $E = 2.13$

[87] $E = 1.60$

[88] $19.63 < \mu < 20.23$

[89] Standard error of the mean = 0.27
Range at 5% level of confidences is approximately 34.47 to 35.53

The standard error of the sample can be estimated as $\frac{10.9}{\sqrt{132}} \approx 0.9$. A 95% confidence interval for the population mean is within two standard errors of the sample mean. The

[90] interval is from $40.6 - 2(0.9)$ to $40.6 + 2(0.9)$, or between 38.8 pounds and 42.4 pounds.

The 99% confidence interval for the mean life of this batch of bulbs is found within 2.5 standard errors of the sample mean. The standard error can be estimated by dividing the standard deviation by the square root of 48: $\frac{800}{\sqrt{48}} \approx 115$ hours. The confidence interval for the mean life of all the bulbs in the production batch has a lowest value of $10{,}240 - 2.5(115) \approx 10{,}010$ hours. That means that it is possible that fewer than 50% of the

[91] bulbs will fail before 10,000 hours of use.

The standard error is $\frac{6.5}{\sqrt{34}} \approx 1.1\%$. The population's mean improvement should fall

[92] between $5.7 - 2(1.1) = 3.5\%$ and $5.7 + 2(1.1) = 7.9\%$.

[93] $P(-2.33 < z < 2.33) = 98$

Objective 9.1: Estimating with Large Samples

Objective 9.2: Estimating with Small Samples

[94] $d.f. = 17$

[95] $t_{0.95} = 2.571$

[96] $t_{0.90} = 1.771$

[97] $t_{0.90} = 1.717$

[98] $-1.31 < \mu < 9.42$

[99] $t_{0.90} = 1.761$

[100] $3.63 < \mu < 9.01$

The sample is small, so the interval should be determined using the t distribution. A 95% confidence interval will fall within 2.6 standard errors of the sample mean.

Tip Amount ($)	Sample Mean	Deviations	Squared Deviations
3.50	5.33	−1.83	3.3489
7.65	5.33	2.32	5.3824
4.95	5.33	−0.38	0.1444
4.75	5.33	−0.58	0.3364
4.80	5.33	−0.53	0.2809
6.30	5.33	0.97	0.9409

The sum of squares is 10.4339. The standard deviation is $\sqrt{\frac{10.4339}{5}} \approx 1.44457$. The standard error is $\frac{1.44457}{\sqrt{6}} \approx 0.59$. The 95% confidence interval for the population mean is $5.33 \pm 2.6(0.59)$, so the population mean should fall between \$3.80 and \$6.86.

[101]

Objective 9.3: Estimating in the Binomial Distribution

[102] $\hat{p} = 0.38$

[103] $\hat{p} = 0.45$

[104] $0.12 < p < 0.18$

The maximum error in the sample's proportion is $2 \times \sqrt{\frac{(0.13)(1-0.13)}{1010}} \approx 2.2\%$. The

[105] 95% confidence interval is 10.8% to 15.2%

The margin of error is $2 \times \sqrt{\frac{(0.31)(1-0.31)}{376}} \approx 4.8\%$. The 95% confidence interval is

[106] 0.31 ± 4.8, or between 26.2% and 35.8%.

[107] $\pm 4.4\%$; between 74.6% and 83.4%

[108] 625

Objective 9.4: Choosing the Sample Size

[109] 92

[110] 196

[111] 196

[112] 66

[113] 138

Chapter 10
Hypothesis Testing

Section A: Static Items

For each hypothesis test please provide the following information:

(a) State the null and the alternate hypotheses.
(b) Identify the sampling distribution to be used: the standard normal distribution or the student's t distribution. Find the critical values.
(c) Sketch the critical region and show the critical value(s) on the sketch.
(d) Compute the z or t value of the sample test statistic and show its location on the sketch of part (c).
(e) Find the P value or an interval containing the P value for the sample test statistic.
(f) Based on your answers to parts (a) through (e) decide whether to reject or not reject the null hypothesis at the given significance level. Explain your conclusion in simple non-technical terms.

1. The Smoky Bear Trucking Company claims that the average weight of a fully loaded moving van is 12,000 lb. The highway patrol decides to check this claim. A random sample of 30 Smoky Bear moving vans shows that the average weight is 12,100 lb. with a standard deviation of 800 lb. Construct a hypothesis test to determine whether the average weight of a Smoky Bear moving van is more than 12,000 lb. Use a 5% level of significance.

2. The Magic Dragon Cigarette Company claims that their cigarettes contain an average of only 10 mg of tar. A random sample of 100 Magic Dragon cigarettes shows the average tar content to be 11.5 mg with standard deviation 4.5 mg. Construct a hypothesis test to determine whether the average tar content of Magic Dragon cigarettes exceeds 10 mg. Use a 5% level of significance.

3. Jerry is doing a project for his sociology class in which he tests the claim that the Pleasant View housing project contains family units of average size 3.3 people (the national average). A random sample of 64 families from Pleasant View project shows a sample mean of 4.3 people per family unit with sample standard deviation 1.3. Construct a hypothesis test to determine whether the average size of a family unit in Pleasant View is different from the national average of 3.3. Use a 5% level of significance.

4. The Mammon Savings and Loan Company claims that the average amount of money on deposit in a savings account in their bank is \$7,500. Suppose a random sample of 49 accounts shows the average amount on deposit to be \$6,850 with sample standard deviation \$1,200. Construct a hypothesis test to determine whether the average amount on deposit per account is different from \$7,500. Use a 1% level of significance.

For each hypothesis test please provide the following information:

(a) State the null and the alternate hypotheses.
(b) Identify the sampling distribution to be used: the standard normal distribution or the student's t distribution. Find the critical values.
(c) Sketch the critical region and show the critical value(s) on the sketch.
(d) Compute the z or t value of the sample test statistic and show its location on the sketch of part (c).
(e) Find the P value or an interval containing the P value for the sample test statistic.
(f) Based on your answers to parts (a) through (e) decide whether to reject or not reject the null hypothesis at the given significance level. Explain your conclusion in simple non-technical terms.

5. Some professional football players seem to earn tremendous amounts of money. However, their careers as professional players are short. One sports magazine reported that the average career length is 4.3 years. A random sample of 40 retired players showed a sample mean career length of 5.2 years with standard deviation 2.3 years. Construct a hypothesis test to determine whether the average career in professional football is longer than 4.3 years. Use a 5% level of significance.

6. Killer bees have migrated into this country. There is fear that they will spread across the nation. However, they cannot survive in cold climates. It is thought that they cannot tolerate temperatures below 36° F. To test this claim a random sample of 9 killer bee hives were subjected to colder and colder temperatures until they died. The temperatures at which the hives died were recorded. The mean temperature was 37° F with standard deviation 4° F. Assuming that the killing temperature level is normally distributed, test the claim that the mean killing temperature is different from 36° F. Use a 1% level of significance.

7. Statistical Abstracts (117th edition) reports that the average amount spent annually for food by householders under 25 years of age is \$2,690. A random sample of 16 people under 25 years of age who live in a university neighborhood were surveyed. The survey showed that they spent a sample mean \$3,220 with sample standard deviation \$750. Test the claim that the mean for this neighborhood is greater than the national average. Use a 5% significance level.

8. Statistical Abstracts (117th edition) reports that the national average amount a single person spends annually for housing is \$10,465. A random sample of 20 householders living in the San Francisco Bay area had a sample mean housing cost \$14,575 with standard deviation \$4,580. Test to see if the mean housing cost in the San Francisco Bay area is higher than the national average. Use a 1% significance level.

For each hypothesis test please provide the following information:

(a) State the null and the alternate hypotheses.
(b) Identify the sampling distribution to be used: the standard normal distribution or the student's t distribution. Find the critical values.
(c) Sketch the critical region and show the critical value(s) on the sketch.
(d) Compute the z or t value of the sample test statistic and show its location on the sketch of part (c).
(e) Find the P value or an interval containing the P value for the sample test statistic.
(f) Based on your answers to parts (a) through (e) decide whether to reject or not reject the null hypothesis at the given significance level. Explain your conclusion in simple non-technical terms.

9. Statistical Abstracts (117th edition) reports that the average annual expenditure for health care by individuals 25 to 34 years old is \$1,096. A random sample of 24 athletes between the ages of 25 and 34 had a sample mean expenditure of \$950 with sample standard deviation \$425. Test to see if the mean expenditure for health care for athletes between the ages of 25 and 34 is different from the national average. Use a 1% significance level.

10. Last year the average daily change in the Dow Jones Industrial Index was 7.3 points. A random sample of 15 trading days this year showed the average change to be 5.2 points with standard deviation 10.2. Test the claim that the average change in the index is less this year than it was last year. Use a 5% significance level.

11. Statistical Abstracts (117th edition) reports that the average number of days per year that it rains more than 0.01 inches in Albuquerque, New Mexico is 61. A random sample of 6 years of weather records from a mountain community west of Albuquerque had sample mean 75 days per year with sample standard deviation 12.4 days per year. Test to see if the mean number of rainy days in the mountain community is different from that in Albuquerque. Use a 5% significance level.

12. Ruth is concerned about the spending habits of teens. She read a report that the national weekly spending average for teens in the age group 12 to 15 years is \$42. She took a random sample of 60 teens who live in a rural area and found that they spent an average of \$36 per week with sample standard deviation \$7.50. Test the claim that rural teens from this area spend less than the national average. Use a 1% significance level.

13. The U.S. census reported that it takes workers an average of 28 minutes to get home from work. A random sample of 100 workers in a large metropolitan area showed a sample mean time to get home from work that was 32 minutes with standard deviation 10 minutes. Is this data statistically significant evidence that workers in a large city take longer than 28 minutes to get home from work? Use a 5% significance level.

For each hypothesis test please provide the following information:

(a) State the null and the alternate hypotheses.
(b) Identify the sampling distribution to be used: the standard normal distribution or the student's t distribution. Find the critical values.
(c) Sketch the critical region and show the critical value(s) on the sketch.
(d) Compute the z or t value of the sample test statistic and show its location on the sketch of part (c).
(e) Find the P value or an interval containing the P value for the sample test statistic.
(f) Based on your answers to parts (a) through (e) decide whether to reject or not reject the null hypothesis at the given significance level. Explain your conclusion in simple non-technical terms.

14. According to Statistical Abstracts (117th edition) 27% of the adults in the United States visited an art museum at least once last year. A random sample of 200 residents of a large city showed that 80 of them had visited an art museum during the past year. Test to see if the proportion of people in this area who visit art museums is higher than the national average. Find the P value for your test statistic and use it to determine if the data is statistically significant. Use a 5% significance level.

15. The Dog Days Lawn Service advertises that it will completely maintain your lawn at an average cost per customer of $55 per month. A random sample of 18 Dog Days customers had a sample mean cost of $59.50 with sample standard deviation $10.50. Do the data support the claim that the average cost is more than $55? Use a 5% significance level.

16. An overnight package delivery service has a promotional discount rate in effect this week only. For several years the mean weight of a package delivered by this company has been 10.7 oz. A random sample of 12 packages mailed this week has sample mean weight 11.81 oz with standard deviation 2.24 oz. Test the claim that the mean weight of all packages mailed this week is greater than 10.7 oz. Use a 1% significance level.

17. The Food and Drug Administration has approved a prescription medication that contains 4 mg of codeine per pill. To determine if the pills actually contain 4 mg of codeine, a random sample of 16 pills were analyzed. The mean codeine content was 4.3 mg with sample standard deviation 0.08 mg. Test the hypothesis that the actual codeine content of the pills is different from 4 mg. Use a 1% level of significance. Assume that the production of the pills is such that the amount of codeine per pill follows a normal distribution.

18. The board of real estate developers claims that 55% of all voters will vote for a bond issue to construct a massive new water project. A random sample of 215 voters was taken and 96 said that they would vote for the new water project. Test to see if this data indicates that less than 55% of all voters favor the project. Use a 1% significance level.

For each hypothesis test please provide the following information:

(a) State the null and the alternate hypotheses.
(b) Identify the sampling distribution to be used: the standard normal distribution or the student's t distribution. Find the critical values.
(c) Sketch the critical region and show the critical value(s) on the sketch.
(d) Compute the z or t value of the sample test statistic and show its location on the sketch of part (c).
(e) Find the P value or an interval containing the P value for the sample test statistic.
(f) Based on your answers to parts (a) through (e) decide whether to reject or not reject the null hypothesis at the given significance level. Explain your conclusion in simple non-technical terms.

19. A city council member gave a speech in which she said that 18% of all private homes in the city had been undervalued by the county tax assessor's office. In a follow-up story the local newspaper reported that it had taken a random sample of 91 private homes. Using professional Realtors to evaluate the property and checking against county tax records it found that 14 of the homes had been undervalued. Does this data indicate that the proportion of private homes that are undervalued by the county tax assessor is different from 18%? Use a 5% significance level.

20. A washing machine manufacturer says that 85% of its washers last five years before repairs are necessary. A random sample of 100 washers showed that 73 of them lasted five years before they needed repair. Does this data indicate that the manufacturer's claim is too high? Use a 1% significance level.

21. A survey done one year ago showed that 45% of the population participated in recycling programs. In a recent poll a random sample of 1,250 people showed that 588 participate in recycling programs. Test the hypothesis that the proportion of the population who participate in recycling programs is greater than it was one year ago. Use a 5% significance level.

22. The manager of Let's Go Hamburger Restaurant wants to study customer demand for V-Burgers, new meatless hamburgers. The salesperson for V-Burgers claims that 75% of people who normally eat hamburgers also like V-Burgers. The manager surveyed a random sample of 92 customers and found that 54 of them said they would like to try V-Burgers. Does this data indicate that the salesperson's claim is too high? Use a 5% significance level.

For each hypothesis test please provide the following information:

(a) State the null and the alternate hypotheses.
(b) Identify the sampling distribution to be used: the standard normal distribution or the student's t distribution. Find the critical values.
(c) Sketch the critical region and show the critical value(s) on the sketch.
(d) Compute the z or t value of the sample test statistic and show its location on the sketch of part (c).
(e) Find the P value or an interval containing the P value for the sample test statistic.
(f) Based on your answers to parts (a) through (e) decide whether to reject or not reject the null hypothesis at the given significance level. Explain your conclusion in simple non-technical terms.

23. According to one newspaper article 47% of the households receiving magazine sweepstakes solicitations in the mail do not respond to them. An intensive TV campaign was aired for two weeks prior to the mailing of the Fantastic Fantasy sweepstakes offering. In a random sample of 300 households it was found that 153 responded to the sweepstakes mailing. Is this evidence that the response to the Fantastic Fantasy sweepstakes was higher than 47%? Use a 5% significance level.

24. A large waterfront warehouse installed a new security system a few years ago. Under the old security system the warehouse managers estimated that they were losing an average of $678 worth of merchandise to thieves each week. A random sample of 48 weekly records under the new system showed that they were still losing an average of $650 of merchandise a week. The sample standard deviation was $93. Does this indicate that the loss each week under the new system is different (either more or less) than it was under the old system? Use a 5% significance level.

25. Dental associations recommend that the time lapse between routine dental check-ups should average six months. A random sample of 36 patient records at one dental clinic showed the average time between routine check-ups to be 7.2 months with standard deviation 2 months. Do the sample data indicate that patients wait longer than recommended between dental checkups? In your test, calculate the P value for the sample test statistic and determine if the data is statistically significant. Use a 1% significance level.

26. A recent study showed that in California the average single-family home owner lived at one address 8.0 years before moving. In Sonoma County, a random sample of 116 single-family home owners lived at one address an average of 8.7 years with standard deviation 2.8 years. Test the claim that single-family home owners in Sonoma County live in one place longer than the state average. Calculate the P value for your sample test statistic and determine if the data is statistically significant. Use a 1% significance level.

For each hypothesis test please provide the following information:

(a) State the null and the alternate hypotheses.
(b) Identify the sampling distribution to be used: the standard normal distribution or the student's t distribution. Find the critical values.
(c) Sketch the critical region and show the critical value(s) on the sketch.
(d) Compute the z or t value of the sample test statistic and show its location on the sketch of part (c).
(e) Find the P value or an interval containing the P value for the sample test statistic.
(f) Based on your answers to parts (a) through (e) decide whether to reject or not reject the null hypothesis at the given significance level. Explain your conclusion in simple non-technical terms.

27. Statistical Abstracts (117th edition) reports that the average monthly rate for basic cable TV is \$24.41. In the southeast region of the country, a random sample of 63 cable users paid an average amount of \$25.98 per month with standard deviation \$4.50. Use this data to test the claim that the average cost for basic cable TV service in the southeast is different from the national average. Calculate the P value for your test statistic and determine if the data is statistically significant. Use a 1% significance level.

28. A pharmaceutical company makes tranquilizers that are claimed to have an effective period of 2.8 hr. Researchers in a hospital used the drug on a random sample of 81 patients and found the mean effective period to be 2.63 hr with sample standard deviation 0.6 hr. Does this indicate that the company's claim is too high? Use a 1% level of significance.

29. Robert Jeffries is an industrial psychologist working for an auto manufacturing company in Detroit. His job is to invent ways to speed up production without increasing cost. After examining the production line he thinks that he can reorder the assembly sequence and reduce the assembly time. Under the old routine, the mean time for door assembly was 8.50 min. Under Bob's new method a random sample of 47 door jobs showed that the doors were done in a mean time of 8.21 min. with sample standard deviation 0.86 min. Does this sample indicate that Bob's new method has reduced the mean assembly time? Calculate the P value for your test statistic and determine if the data is statistically significant. Use a 5% significance level.

For each hypothesis test please provide the following information:

(a) State the null and the alternate hypotheses.
(b) Identify the sampling distribution to be used: the standard normal distribution or the student's t distribution. Find the critical values.
(c) Sketch the critical region and show the critical value(s) on the sketch.
(d) Compute the z or t value of the sample test statistic and show its location on the sketch of part (c).
(e) Find the P value or an interval containing the P value for the sample test statistic.
(f) Based on your answers to parts (a) through (e) decide whether to reject or not reject the null hypothesis at the given significance level. Explain your conclusion in simple non-technical terms.

30. Walter Gleason is an astronomer who has been studying radio signals from the planet Jupiter. Over a long period of time, the planet's electromagnetic field has been sending low-frequency signals with a mean frequency of 14.2 megahertz. A recent space module that went by Jupiter recorded a major volcanic eruption that may have covered a large surface on the planet. For the past several months Walter has been measuring Jupiter's radio signals of a random time schedule. A group of 75 measurements gave a mean 14.49 megahertz with sample standard deviation 0.9 megahertz. Test to see if the mean radio frequency of these signals has changed. Calculate the P value for your test statistic and determine if this data is statistically significant. Use a 1% significance level.

31. Julia Ching is a botanist hired by the city of Tulsa to study the problem of finding a better summer watering plan for the many acres of lawn in city parks. Julia believes that less frequent but longer watering is better, because it gives the grass a better root system. The water board says that it has been using an average of 137,000 gallons of water a day for city parks, and Julia cannot go above that average. After Julia's new watering system was put into effect, a random sample of 40 watering days showed the mean amount of water being used was 132,119 gallons with standard deviation 6,410 gallons. Test the claim that Julia's system uses less water on the average than the amount used by the old system. Use $\alpha = 1\%$.

32. The owner of a private fishing pond says that the average length of fish in the pond is 15 inches. A sample of 23 fish are caught from various locations in the pond. The average length of these fish is 10.7 in. with standard deviation 4.8 in. Does the data support the claim that the average fish length is less than 15 in.? Use a 1% significance level.

33. A random sample of 400 families in the city of Minneapolis showed that 192 of them owned pets. The city council claims that 53% of the families in the city own pets. Does the data indicate that the actual percentage of families owning pets is different from 53%? Use a 5% significance level.

For each hypothesis test please provide the following information:

(a) State the null and the alternate hypotheses.
(b) Identify the sampling distribution to be used: the standard normal distribution or the student's t distribution. Find the critical values.
(c) Sketch the critical region and show the critical value(s) on the sketch.
(d) Compute the z or t value of the sample test statistic and show its location on the sketch of part (c).
(e) Find the P value or an interval containing the P value for the sample test statistic.
(f) Based on your answers to parts (a) through (e) decide whether to reject or not reject the null hypothesis at the given significance level. Explain your conclusion in simple non-technical terms.

34. Statistical Abstracts (117th edition) reports that the average number of miles driven per year by drivers in the U.S. is 11,372. A random sample of 8 undergraduate women from Wyoming drove the following number of miles last year:
10,545 11,630 12,250 11,540 11,115 11,745 11,950 11,140
(a) Find the sample mean and the sample standard deviation.
(b) Test the claim that the annual number of miles driven by undergraduate women from Wyoming is greater than the national average. Use a 5% significance level.

35. A recent study has shown that the average outstanding balance on credit cards is $3,012.50. Twelve faculty at Eastern State College gave information about their outstanding balances. The sample mean was $2,625.50 with sample standard deviation $543.15. Test the claim that the faculty at Eastern State College have an average outstanding balance which is less than the national average. Use a 1% significance level.

36. Sometimes people who are required to take a medication develop a chemical response that makes the medication ineffective. A random sample of 200 people taking insulin showed that 14 had developed blood chemicals acting against the insulin. The pharmaceutical company manufacturing this brand of insulin says that only 3% of the people who use insulin will develop such chemical reactions. Use a 1% level of significance to test the claim that the actual proportion of people who develop such inhibiting chemicals is higher than 3%.

37. A survey of small businesses across the United States reported that 34% of all small businesses are based in the home. In a mountainous Colorado county where the towns are small, a local survey showed that 30 out of 65 small businesses selected at random are located in the home rather than in an office outside of the home. Is the proportion of small businesses located in the home different in this county than it is in the U.S. in general? Use a 5% level of significance.

For each hypothesis test please provide the following information:

(a) State the null and the alternate hypotheses.
(b) Identify the sampling distribution to be used: the standard normal distribution or the student's t distribution. Find the critical values.
(c) Sketch the critical region and show the critical value(s) on the sketch.
(d) Compute the z or t value of the sample test statistic and show its location on the sketch of part (c).
(e) Find the P value or an interval containing the P value for the sample test statistic.
(f) Based on your answers to parts (a) through (e) decide whether to reject or not reject the null hypothesis at the given significance level. Explain your conclusion in simple non-technical terms.

38. Internal Revenue Service employees who process tax returns say that 10% of all tax returns contain arithmetic errors in excess of $1,200. A random sample of 817 tax returns showed that 86 of them contain such errors. Does this data indicate that the proportion of income tax returns with errors in excess of $1,200 is different from 10%? Use a 1% significance level.

39. At a national convention for travel agents, Ellen heard a report that 27% of the people who reserve a trip more than 10 months in advance cancel the trip. After the convention, Ellen did a study for her travel agency. From a random sample of 78 reservations made more than 10 months in advance, 23 were canceled. Does this data indicate that the proportion of cancellations made at her agency is different from the percentage reported at the convention? Use a 1% significance level.

40. Shortage of landfills for dumps and a growing awareness of the environmental impact of our lifestyle is leading to increased recycling. The Environmental Protection Agency says that 14.7% of general household trash consists of plastic and glass. Responsible Citizens ran a neighborhood-by-neighborhood campaign with telephone calls and door-to-door visits to encourage recycling of plastic and glass. Then they conducted a study of the household trash being put out for general trash pickup. They studied a random sample of 200 containers of trash. The trash was sorted and it was found that it contained 24 containers of glass and plastic. Was the percentage of trash that consisted of plastic and glass less than the national average in the neighborhoods in which Responsible Citizens had been active? Use a 5% significance level.

41. A real estate broker says that 68% of all retired couples prefer a condominium to a single-family house. A random sample of 75 retired couples showed that 61 preferred a condominium to single-family homes. Use a 5% significance level to test the claim that the realtor's estimate is too low.

For each hypothesis test please provide the following information:

(a) State the null and the alternate hypotheses.
(b) Identify the sampling distribution to be used: the standard normal distribution or the student's t distribution. Find the critical values.
(c) Sketch the critical region and show the critical value(s) on the sketch.
(d) Compute the z or t value of the sample test statistic and show its location on the sketch of part (c).
(e) Find the P value or an interval containing the P value for the sample test statistic.
(f) Based on your answers to parts (a) through (e) decide whether to reject or not reject the null hypothesis at the given significance level. Explain your conclusion in simple non-technical terms.

42. A dentist claims that 15% of all appointments are canceled. In one month, 15 out of a random sample of 136 appointments were canceled. Does this data indicate that the percentage of cancellations is different than 15%? Use a 5% significance level.

43. A hair stylist read a report that 12% of all people who make appointments are late for their appointments, creating scheduling problems and delays for other customers. A random sample of 300 of the hair stylists appointments in the last three months showed 24 late arrivals. Does this data indicate that the proportion of late appointments for this hair stylist is different from the reported 12%? Use a 5% significance level.

44. The Police Department in Los Angeles claims that the incidence of violent crime is up compared to last year, when 19.7% of all crimes were classified as violent crimes. A member of the city council took a random sample of 877 crime reports filed in the LA police records this year and found that 178 were classified by the police as violent crimes. Using a 5% significance level, test the claim that the incidence of violent crimes has increased.

45. You are using a computer package to conduct a statistical test of a hypothesis. The computer gives you a P value of 0.0328.
(a) Do you reject the null hypothesis at the 5% significance level?
(b) Do you reject the null hypothesis at the 1% significance level?

46. A random sample of 530 union bricklayers showed that 40 had been laid off at least once in the last 2 years. An independent random sample of 640 non-union bricklayers showed that 60 had been laid off at least once in the last 2 years. Does this data indicate that the proportion of bricklayers who have experienced recent layoffs is greater for the non-union people than for the union members? Use a 5% significance level.

For each hypothesis test please provide the following information:

(a) State the null and the alternate hypotheses.
(b) Identify the sampling distribution to be used: the standard normal distribution or the student's t distribution. Find the critical values.
(c) Sketch the critical region and show the critical value(s) on the sketch.
(d) Compute the z or t value of the sample test statistic and show its location on the sketch of part (c).
(e) Find the P value or an interval containing the P value for the sample test statistic.
(f) Based on your answers to parts (a) through (e) decide whether to reject or not reject the null hypothesis at the given significance level. Explain your conclusion in simple non-technical terms.

47. The Big Break Moving Company claims that a typical family moves an average of once every 5.2 years. In a random sample of 100 families, the average length of time between moves was 5.7 years with standard deviation 1.8 years. Does this data indicate that the average moving time is different from 5.2 years? Use a 5% level of significance.

48. You are using a computer to conduct a test of a hypothesis at the 5% significance level. The reported P value is 0.0612. Should you reject the null hypothesis or not?

49. A statistical study was conducted to see if depression levels are different for people subjected to 5 hours of music in a minor key as compared to people who were subjected to 5 hours of music in a major key. The null hypothesis is that there is no difference. The alternate hypothesis is that there is a difference. The P value of the sample test statistic is 0.023.
(a) What conclusion do you reach at the 1% significance level?
(b) What conclusion do you reach at the 5% significance level?

50. On the computer output of a hypothesis test, which is a right-tail test, the P value of the sample test statistic is 0.007. Can we reject the null hypothesis at the 1% significance level?

51. Rhonda had read a report that only 25% of the welfare recipients in her city are either employed or enrolled in a job training program. She believes that it is higher. She collected a large random sample of data from welfare recipients and ran a computer analysis of the data. For a right-tail test for one proportion the computer output showed a P value of 0.075.
(a) If she uses a 5% significance level, what should her conclusion be?
(b) If she uses a 1% significance level, what should her conclusion be?
(c) What is the smallest significance level at which she can reject H_0?

For each hypothesis test please provide the following information:

(a) State the null and the alternate hypotheses.
(b) Identify the sampling distribution to be used: the standard normal distribution or the student's t distribution. Find the critical values.
(c) Sketch the critical region and show the critical value(s) on the sketch.
(d) Compute the z or t value of the sample test statistic and show its location on the sketch of part (c).
(e) Find the P value or an interval containing the P value for the sample test statistic.
(f) Based on your answers to parts (a) through (e) decide whether to reject or not reject the null hypothesis at the given significance level. Explain your conclusion in simple non-technical terms.

52. Donald wants to see if the proportion of people who perceive themselves as overweight is different for men than for women. He collects a large random sample of data and runs a computer analysis. The two-tail test for two proportions gives a P value 0.027. Is the data statistically significant at the 1% significance level? Explain.

53. Big Shot Pizza Company makes the advertising claim of being able to deliver an order in 30 minutes or less to anywhere in its delivery area. An independent consumer agency is doing a study to test the truth of this claim. What should be used as the null hypothesis? What should be used as the alternate hypothesis?

54. A car manufacturer advertises that its new subcompact models get 36 miles per gallon (mpg). Assume then that the null hypothesis is H_0: $\mu = 36$ and the alternate hypothesis is H_1: $\mu < 36$. If based on sample evidence, we reject H_0 in favor of H_1 when in fact the average number of miles per gallon achieved by the subcompact car is 36 mpg or higher, what type of error has been committed: a type I error or a type II error?

55. A car manufacturer advertises that its new subcompact models get 36 miles per gallon (mpg). Assume then that the null hypothesis is H_0: $\mu = 36$ and the alternate hypothesis is H_1: $\mu < 36$. If based on sample evidence, we fail to reject H_0 in favor of H_1 when in fact H_1 is true, what type of error has been committed: a type I error or a type II error?

56. Michael is a fulltime bartender. The IRS is auditing his tax return this year. Michael claims that on an average evening, he made \$40 in tips last year. To support this claim, he sent the IRS a random sample of records for 43 of these evenings. Using these records, the IRS computed the sample average and found it to be $\bar{x} = 50$ with sample standard deviation of $s = 32$. Find the associated sample z value. Do these receipts indicate that Michael received more than an average of \$40 worth of tips last year to a 1% level of significance?

For each hypothesis test please provide the following information:

(a) State the null and the alternate hypotheses.
(b) Identify the sampling distribution to be used: the standard normal distribution or the student's t distribution. Find the critical values.
(c) Sketch the critical region and show the critical value(s) on the sketch.
(d) Compute the z or t value of the sample test statistic and show its location on the sketch of part (c).
(e) Find the P value or an interval containing the P value for the sample test statistic.
(f) Based on your answers to parts (a) through (e) decide whether to reject or not reject the null hypothesis at the given significance level. Explain your conclusion in simple non-technical terms.

57. Michael is a fulltime bartender. The IRS is auditing his tax return this year. Michael claims that on an average evening, he made $40 in tips last year. To support this claim, he sent the IRS a random sample of records for 43 of these evenings. Using these records, the IRS computed the sample average and found it to be $\bar{x} = 50$ with sample standard deviation of $s = 32$. Find the associated sample z value. Do these receipts indicate that Michael received more than an average of $40 worth of tips last year to a 5% level of significance?

58. Choose the best answer from among choices A.–E. below. Last year the average age of students attending Fisher College was 20.6 years. In order to meet the demands of older students, steps were taken that included adding evening and weekend classes. A sample of 80 students enrolled this semester was studied. The average age of these students was $\bar{x} = 21.3$ with sample standard deviation $s = 2.8$. Find the associated z value and P value of the sample statistic $\bar{x}$. Does this confirm the alternate hypothesis H_1: $\mu > 20.6$, that the average age of students this semester is higher than it was last year using the 1% significance level? That is, do we reject the null hypothesis H_0: $\mu = 20.6$?

A. $z = 2.89$; P value = 0.0019; reject H_0.

B. $z = 2.24$; P value = 0.0125; do not reject H_0.

C. $z = 2.43$; P value = 0.0075; reject H_0.

D. $z = 1.85$; P value = 0.0322; do not reject H_0.

E. $z = 1.79$; P value = 0.0367; do not reject H_0.

For each hypothesis test please provide the following information:

(a) State the null and the alternate hypotheses.
(b) Identify the sampling distribution to be used: the standard normal distribution or the student's t distribution. Find the critical values.
(c) Sketch the critical region and show the critical value(s) on the sketch.
(d) Compute the z or t value of the sample test statistic and show its location on the sketch of part (c).
(e) Find the P value or an interval containing the P value for the sample test statistic.
(f) Based on your answers to parts (a) through (e) decide whether to reject or not reject the null hypothesis at the given significance level. Explain your conclusion in simple non-technical terms.

59. Choose the best answer from among choices A.–E. below. Last year the average age of students attending Fisher College was 20.6 years. In order to meet the demands of older students, steps were taken that included adding evening and weekend classes. A sample of 70 students enrolled this semester was studied. The average age of these students was $\bar{x} = 21.5$ with sample standard deviation $s = 3.1$. Find the associated z value and P value of the sample statistic $\bar{x}$. Does this confirm the alternate hypothesis H_1: $\mu > 20.6$, that the average age of students this semester is higher than it was last year using the 1% significance level? That is, do we reject the null hypothesis H_0: $\mu = 20.6$?

A. $z = 2.43$; P value = 0.075; reject H_0.

B. $z = 1.79$; P value = 0.0367; do not reject H_0.

C. $z = 2.24$; P value = 0.0125; do not reject H_0.

D. $z = 1.85$; P value = 0.0322; do not reject H_0.

E. $z = 2.89$; P value = 0.0019; reject H_0.

For each hypothesis test please provide the following information:

(a) State the null and the alternate hypotheses.
(b) Identify the sampling distribution to be used: the standard normal distribution or the student's t distribution. Find the critical values.
(c) Sketch the critical region and show the critical value(s) on the sketch.
(d) Compute the z or t value of the sample test statistic and show its location on the sketch of part (c).
(e) Find the P value or an interval containing the P value for the sample test statistic.
(f) Based on your answers to parts (a) through (e) decide whether to reject or not reject the null hypothesis at the given significance level. Explain your conclusion in simple non-technical terms.

60. Choose the best answer from among choices A.–E. below. Last year the average age of students attending Fisher College was 20.6 years. In order to meet the demands of older students, steps were taken that included adding evening and weekend classes. A sample of 80 students enrolled this semester was studied. The average age of these students was $\bar{x} = 21.2$ with sample standard deviation $s = 2.9$. Find the associated z value and P value of the sample statistic $\bar{x}$. Does this confirm the alternate hypothesis H_1: $\mu > 20.6$, that the average age of students this semester is higher than it was last year using the 1% significance level? That is, do we reject the null hypothesis H_0: $\mu = 20.6$?

A. $z = 2.89$; P value = 0.0019; reject H_0.

B. $z = 1.79$; P value = 0.0367; do not reject H_0.

C. $z = 1.85$; P value = 0.0322; do not reject H_0.

D. $z = 2.24$; P value = 0.0125; do not reject H_0.

E. $z = 2.43$; P value = 0.0075; reject H_0.

For each hypothesis test please provide the following information:

(a) State the null and the alternate hypotheses.
(b) Identify the sampling distribution to be used: the standard normal distribution or the student's t distribution. Find the critical values.
(c) Sketch the critical region and show the critical value(s) on the sketch.
(d) Compute the z or t value of the sample test statistic and show its location on the sketch of part (c).
(e) Find the P value or an interval containing the P value for the sample test statistic.
(f) Based on your answers to parts (a) through (e) decide whether to reject or not reject the null hypothesis at the given significance level. Explain your conclusion in simple non-technical terms.

61. Choose the best answer from among choices A.–E. below. Last year the average age of students attending Fisher College was 20.6 years. In order to meet the demands of older students, steps were taken that included adding evening and weekend classes. A sample of 70 students enrolled this semester was studied. The average age of these students was $\bar{x} = 21.6$ with sample standard deviation $s = 2.9$. Find the associated z value and P value of the sample statistic $\bar{x}$. Does this confirm the alternate hypothesis H_1: $\mu > 20.6$, that the average age of students this semester is higher than it was last year using the 1% significance level? That is, do we reject the null hypothesis H_0: $\mu = 20.6$?

A. $z = 2.43$; P value = 0.0075; reject H_0.

B. $z = 2.24$; P value = 0.0125; do not reject H_0.

C. $z = 2.89$; P value = 0.0019; reject H_0.

D. $z = 1.79$; P value = 0.0367; do not reject H_0.

E. $z = 1.85$; P value = 0.0322; do not reject H_0.

For each hypothesis test please provide the following information:

(a) State the null and the alternate hypotheses.
(b) Identify the sampling distribution to be used: the standard normal distribution or the student's t distribution. Find the critical values.
(c) Sketch the critical region and show the critical value(s) on the sketch.
(d) Compute the z or t value of the sample test statistic and show its location on the sketch of part (c).
(e) Find the P value or an interval containing the P value for the sample test statistic.
(f) Based on your answers to parts (a) through (e) decide whether to reject or not reject the null hypothesis at the given significance level. Explain your conclusion in simple non-technical terms.

62. Choose the best answer from among choices A.–E. below. Last year the average age of students attending Fisher College was 20.6 years. In order to meet the demands of older students, steps were taken that included adding evening and weekend classes. A sample of 80 students enrolled this semester was studied. The average age of these students was $\bar{x} = 21.1$ with sample standard deviation $s = 2.5$. Find the associated z value and P value of the sample statistic $\bar{x}$. Does this confirm the alternate hypothesis H_1: $\mu > 20.6$, that the average age of students this semester is higher than it was last year using the 1% significance level? That is, do we reject the null hypothesis H_0: $\mu = 20.6$?

A. $z = 2.89$; P value = 0.0019; reject H_0.

B. $z = 2.24$; P value = 0.0125; do not reject H_0.

C. $z = 1.79$; P value = 0.0367; do not reject H_0.

D. $z = 1.85$; P value = 0.0322; do not reject H_0.

E. $z = 2.43$; P value = 0.0075; reject H_0.

63. A particular type of avalanche studied in Canada had an average thickness of $\mu = 67$ cm. This type of avalanche was studied in a region of the southwest United States. A random sample of 16 such avalanche thicknesses had a mean of $\bar{x} = 61.7$ with a standard deviation of $s = 11.3$. Assume this thickness has an approximately normal distribution. Use a 1% level of significance to test the claim that this mean avalanche thickness in this U.S. region is less than that in Canada. State the critical value t_0, the sample t value, and the conclusion that is reached.

For each hypothesis test please provide the following information:

(a) State the null and the alternate hypotheses.
(b) Identify the sampling distribution to be used: the standard normal distribution or the student's t distribution. Find the critical values.
(c) Sketch the critical region and show the critical value(s) on the sketch.
(d) Compute the z or t value of the sample test statistic and show its location on the sketch of part (c).
(e) Find the P value or an interval containing the P value for the sample test statistic.
(f) Based on your answers to parts (a) through (e) decide whether to reject or not reject the null hypothesis at the given significance level. Explain your conclusion in simple non-technical terms.

64. A particular type of avalanche studied in Canada had an average thickness of $\mu = 67$ cm. This type of avalanche was studied in a region of the southwest United States. A random sample of 16 such avalanche thicknesses had a mean of $\bar{x} = 61.7$ with a standard deviation of $s = 11.3$. Assume this thickness has an approximately normal distribution. Use a 5% level of significance to test the claim that this mean avalanche thickness in this U.S. region is less than that in Canada. State the critical value t_0, the sample t value, and the conclusion that is reached.

65. About 82% of all U.S. health professionals responded to a poll that if they had the chance to start their careers all over again, they would. A random sample of 62 health professionals from a single state showed that 43 would choose a career in the health professions again. Does this indicate that the proportion of health professionals in this state who would choose their professions again is less than the national rate of 82%? Use a 1% level of significance. State the critical value z_0, the proportion $\hat{p}$, the associated z value, and the conclusion that is reached.

Section B: Algorithms

Objective 10.1: Introduction to Hypothesis Testing

66. Identify the alternative hypothesis and determine whether the test is a one- or two-tailed test.

H_0: $\sigma^2 = 7.5$

H_a: $\sigma^2 < 7.5$

67. Use the information given to decide which type of error, if any, was made.
H_0: $\mu = 104$ gm
H_a: $\mu < 104$ gm
Given the sample, you do not reject the null hypothesis. Classify that decision if, in fact, $\mu < 104$ gm.

68. Use the information given to decide if the situation below describes a Type II error.
H_0: $\mu = \$104$
H_a: $\mu > \$104$
The null hypothesis is rejected and $\mu = \$104$.

69. Based on the null hypothesis below, identify whether the result is a Type I error, a Type II error, or not an error.
H_0: The answer on question one of the midterm is True.
Result: You answer True and get the answer correct.

70. Identify the null hypothesis and determine whether the test is a one- or two-tailed test.
H_0: $\mu = 7.7$
H_a: $\mu > 7.7$

71. Use the information given to decide which situation describes a Type II error.
H_0: $\mu = 60$ years
H_a: $\mu < 60$ years

72. Use the information given to decide which type of error, if any, was made.
H_0: $\mu = \$102$
H_a: $\mu \neq \$102$
Given the sample, you do not reject the null hypothesis. Classify that decision if, in fact, $\mu = \$102$.

Objective 10.2: Test Involving the Mean (Large Samples)

73. What would be the test statistic, z, to test the claim $\mu = 132.3$, given a sample of size $n = 169$ for which $\bar{x} = 131$? Assume that $\sigma = 32.19$.

74. For the data below: $\sum x = 772,\ \sum x^2 = 17200,\ \mu = 21.$

25 16 15 27 16 30 20 23 17 20 19 24 26 15 20 18 24 24

16 23 23 17 24 23 25 24 18 23 20 15 28 17 21 20 26 30

Calculate the test statistic, z, for the data.

75. The manager of a large home and gardening hardware store is thinking of establishing a new billing system for the store's credit card customers. After some analysis, she determines that the new system will be cost-effective only if the mean monthly account is $83 or greater. The manager randomly selects 200 accounts, and finds the mean to be $78. She knows that the accounts are normally distributed with a standard deviation of $33. Is there enough evidence at the 1% significance level to conclude that the new system will be cost-effective?

76. The owner of a book store is thinking of establishing a new payment system for the store's purchase of books from the various wholesale outlets. After some analysis, she determines that the new system will be cost-effective only if the amount of money spent each month on an average was greater than $93. She randomly selects the account payments of 150 months, and finds the mean to be $100. The accounts are normally distributed with a standard deviation of $31. Define H_0 and H_1 and the sample test statistic, z. Is there enough evidence at the 1% significance level to conclude that the new system will be cost-effective?

77. Test the claim that $\mu \le 461$, given a sample with $n = 68$ for which $\bar{x} = 456.4$. Assume that $\sigma = 22.96$ and test at a level of significance $\alpha = 0.05$.

Objective 10.3: The P Value in Hypothesis Testing

78. A nursery claims that the trees they sell will grow 18 inches during the first year after being planted. A consumer advocate group believes that the actual growth is much less. To test this conjecture, the group bought 30 of the trees at random and collected the following data:
Inches Grown During the First Year After Planting

10	23	19	23	9	22
21	15	17	16	10	10
17	21	11	11	22	14
17	20	14	16	13	21
12	16	19	23	18	17

(a) What is the p value you would get from the data?
(b) At a 5% significance level, do the data support the consumer group's conjecture?

Calculate the p value from the following information.

79. H_0: $\mu \leq 75$; H_1: $\mu > 75$; $z = 1.45$

80. Right-tailed test; $z = 2.45$

81. Left-tailed test; $z = -1.64$

82. Would the following be significant at a 5% significance level?
H_0: $\mu = 340$, H_a: $\mu \neq 340$, $\bar{x} = 336.6$, $s = 18$, $n = 96$

83. What is the greatest level of significance that can be claimed for the following?
H_0: $\mu = 461$, H_a: $\mu \neq 461$, $\bar{x} = 458.7$, $s = 6.62$, $n = 55$

Calculate the p value from the following information.

84. H_0: $\mu \geq 75$; H_1: $\mu < 75$; $z = 1.58$

85. Two-tailed test; $z = 3.32$

Objective 10.4: Tests Involving the Mean (Small Samples)

86. Test the claim that $\mu \geq 338$, given a sample with $n = 16$ for which $\bar{x} = 344.8$, and test at $\alpha = 0.01$. Assume that the population is normally distributed with $s = 10.88$.

87. What is the greatest level of significance $(0.01,\ 0.05,\ 0.10)$ that can be claimed for the following? Assume that the population is normally distributed.
H_0: $\mu = 464$, H_a: $\mu \neq 464$, $\bar{x} = 466.3$, $s = 4.31$, $n = 16$

88. Ms. Gristeller manages a factory that produces sulfuric acid. She has observed that the output is normally distributed with a mean of 8251 liters per hour. She has been recently informed that the government is thinking of imposing a new law that would require her to alter the way in which the acid is manufactured. Ms. Gristeller is uncertain if the new law will reduce output, however she informs the faculty to comply with the proposed law. She records the output for 16 consecutive hours in the table below. With a 5% significance level, do these data indicate that the new legislation will reduce output?
Hourly Output

7553	8688	8418	8374
8617	8145	8055	7979
8558	7566	7970	7591
7457	7796	8499	7541

89. According to a national survey, the average annual family expenditure on soft drinks is $1301. The expenditure on soft drinks of 20 lower-income families is displayed in the table below. Assume that the population is normally distributed.

$1380	$1263	$1272	$1637	$1494	$1145	$1108	$1839	$1268	$1613
$1392	$1188	$1512	$1679	$1424	$1258	$1448	$1127	$1211	$1135

What is the test statistic needed to determine if the amount spent by the sample families is different than the national average?

90. What would be the test statistic to test the claim that $\mu \le 102.8$, given a sample with $n = 14$ for which $\bar{x} = 104.3$? Assume that $\sigma = 20.23$.

Objective 10.5: Tests Involving a Proportion

91. Farmer Wiley has produced a new hybrid tomato. He knows that for the parent tomato plants, the proportion of seeds germinating is 80%. Even though the proportion of seeds germination for the new hybrid tomato is unknown, he claims that it is also 80%. To test this claim, farmer Wiley's wife tests 490 hybrid seeds and found that 465 germinated. Using a 5% level of significance, test the claim that the proportion germinating for the hybrid is 80%.

92. The proportion of males getting arrested for a crime is 50% in a city. It is also known that of 45 people arrested last month 37 were males. Using a 5% level of significance, test the claim that the proportion of arrests is 50%.

93. What would be the test statistic, z, to test the claim $\rho \leq 0.6$ at a 1% level of significance given a sample of size $n = 67$, with 40 number of successes and sample proportion of $p = 0.6$?

94. What would be the test statistic, z, to test the claim $\rho \geq 0.5$ at a 1% level of significance given a sample of size $n = 69$, with 63 number of successes and sample proportion of $p = 0.5$?

95. What is the P-value of the test used to claim $\rho \leq 0.8$ at a 1% level of significance given a sample of size $n = 350$, with 260 number of successes and sample proportion of 80%?

Answer Key for Chapter 10
Hypothesis Testing

Section A: Static Items

[1] H_0: μ = 12,000; H_1: μ > 12,000; z_0 = 1.645; z = 0.685; P value: 0.2467; Do not reject H_0. We cannot conclude that the average weight of a Smoky Bear moving van is greater than 12,000 lb.

[2] H_0: μ = 10 mg; H_1: μ > 10 mg; z_0 = 1.645; z = 3.33; P value: 0.0004; Reject H_0. The average amount of tar is greater than 10 mg.

[3] H_0: μ = 3.3; H_1: $\mu \neq$ 3.3; z_0 = ±1.96; z = 6.15; P value: 0.0000 to four decimal places. Reject H_0. The average size of a family in Pleasant View is different from the national average.

[4] H_0: μ = \$7,500; H_1: $\mu \neq$ \$7,500; z_0 = ±2.58; z = −3.79; P value: 0.0002; Reject H_0. The average amount on deposit is different from \$7,500.

[5] H_0: μ = 4.3 yr; H_1: μ > 4.3 yr; z_0 = 1.645; z = 2.47; P value: 0.0068; Reject H_0. The average career in professional football is longer than 4.3 years.

[6] H_0: μ = 36° F; H_1: $\mu \neq$ 36° F; t_0 = 3.355; t = 0.75; P value is greater than 0.25. Do not reject H_0. We cannot conclude that the mean killing temperature for the bees is different from 36°F.

[7] H_0: μ = \$2,690; H_1: μ > \$2,690; t_0 = 1.753; t = 2.8267; P value is between 0.005 and 0.01. Reject H_0. The mean amount spent for food by people under 25 years of age in this neighborhood is greater than the national average.

[8] H_0 : μ = \$10,465; H_1 : μ > \$10,465; t_0 = 2.539; t = 4.0132; P value is less than 0.005. Reject H_0. The mean cost of housing in the San Francisco Bay area is greater than the national average.

[9] H_0 : μ = \$1,096; H_1 : $\mu \neq$ \$1,096; t_0 = ±2.807; t = −1.6829; P value is between 0.10 and 0.15. Do not reject H_0. The mean expenditure for health care by athletes between the ages of 25 and 34 is not different from the national average expenditure for all persons in this age range.

[10] $H_0: \mu = 7.3$; $H_1: \mu < 7.3$; $t_0 = -1.761$; $t = -0.7974$; *P* value is greater than 0.125. Do not reject H_0. The index is not less this year than it was last year.

[11] $H_0: \mu = 61$; $H_1: \mu \neq 61$; $t_0 = \pm 2.571$; $t = 2.7656$; *P* value is between 0.02 and 0.05. Reject H_0. The mean number of rainy days is different in the mountain community than it is in Albuquerque.

[12] $H_0: \mu = \$42$; $H_1: \mu < \$42$; $z_0 = -2.33$; $z = -6.20$; *P* value: 0.0000 to four decimal places. Reject H_0. Rural teens from the area spend less than the national average.

[13] $H_0: \mu = 28$ min; $H_1: \mu > 28$ min; $z_0 = 1.645$; $z = 4.00$; *P* value: 0.0000 to four decimal places; Reject H_0. It takes workers in the city longer than 28 minutes to get to work.

[14] $H_0: p = 0.27$; $H_1: p > 0.27$; $z_0 = 1.645$; $z = 4.14$; *P* value: 0.0000 to four decimal places. Reject H_0. The proportion of people who visit art museums in the city is greater than the national average. The data is statistically significant at the 5% significance level.

[15] H_0: $\mu = \$55$; H_1: $\mu > \$55$; $t_0 = 1.74$; $t = 1.81$; *P* value is between 0.025 and 0.05. Reject H_0. The mean cost for the lawn service is greater than \$55.

[16] H_0: $\mu = 10.7$ oz; H_1: $\mu > 10.7$ oz; $t_0 = 2.718$; $t = 1.7166$; *P* value is between 0.05 and 0.075. Do not reject H_0. We cannot conclude that the mean weight of all packages mailed this week is greater than 10.7 oz.

[17] $H_0: \mu = 4$ mg; $H_1: \mu \neq 4$ mg; $t_0 = \pm 2.947$; $t = 14.52$; *P* value is less than 0.01. Reject H_0. The actual codeine content of the pills is not 4 mg.

[18] $H_0: p = 0.55$; $H_1: p < 0.55$; $z_0 = -2.33$; $z = -3.08$; *P* value: 0.001; Reject H_0. Fewer than 55% of all voters favor the project.

[19] $H_0: p = 0.18$; $H_1: p \neq 0.18$; $z_0 = \pm 1.96$; $z = -0.6494$; *P* value: 0.5156; Do not reject H_0. We cannot conclude that the proportion of undervalued homes is different from 18%.

[20] $H_0: p = 0.85$; $H_1: p < 0.85$; $z_0 = -2.33$; $z = -3.36$; *P* value: 0.0004; Do not reject H_0. It appears that the manufacturer's claim is too high.

[21] $H_0: p = 0.45$; $H_1: p > 0.45$; $z_0 = 1.645$; $z = 1.45$; P value: 0.0735; Do not reject H_0. We cannot conclude that the proportion of the population who recycle has increased since last year.

[22] $H_0: p = 0.75$; $H_1: p < 0.75$; $z_0 = -1.645$; $z = -3.61$; P value: 0.0002; Reject H_0. The salesperson's claim is probably too high.

[23] $H_0: p = 0.47$; $H_1: p > 0.47$; $z_0 = 1.645$; $z = 1.39$; P value: 0.0823; Do not reject H_0. We cannot conclude that the response to the Fantastic Fantasy sweepstakes was greater than 47%.

[24] $H_0: \mu = \$678$; $H_1: \mu \neq \$678$; $z_0 = \pm 1.96$; $z = -2.08$; P value: 0.0376; Reject H_0. The mean loss each week under the new system is different from the weekly loss under the old system.

[25] $H_0: \mu = 6$ mo; $H_1: \mu > 6$ mo; $z_0 = 2.33$; $z = 3.60$; P value: 0.0002; Reject H_0. The mean time between checkups is greater than 6 months.

[26] $H_0: \mu = 8.0$ yr; $H_1: \mu > 8.0$ yr; $z_0 = 2.33$; $z = 2.69$; P value: 0.0036; Reject H_0. Home owners in Sonoma County move less frequently. The data is statistically significant at the 1% significance level.

[27] $H_0: \mu = \$24.41$; $H_1: \mu \neq \$24.41$; $z_0 = \pm 2.58$; $z = 2.77$; P value: 0.0056; Reject H_0. The mean amount people in the southeast pay for basic cable TV is different from the national average. The data is statistically significant at the 1% significance level.

[28] $H_0: \mu = 2.8$ hr; $H_1: \mu < 2.8$ hr; $z_0 = -2.33$; $z = -2.55$; P value: 0.0054; Reject H_0. The company's claim is too high.

[29] $H_0: \mu = 8.50$; $H_1: \mu < 8.50$; $z_0 = -1.645$; $z = -2.31$; P value: 0.0104; Reject H_0. The new method appears to be faster. The data is statistically significant at the 5% significance level.

[30] $H_0: \mu = 14.2$; $H_1: \mu \neq 14.2$; $z_0 = \pm 2.58$; $z = 2.79$; P value: 0.0052; Reject H_0. The mean radio frequency has changed. The data is statistically significant at the 1% significance level.

[31] $H_0: \mu = 137{,}000$; $H_1: \mu < 137{,}000$; $z_0 = -2.33$; $z = -4.815$; P value: 0.0000 to four decimal places; Reject H_0. Julia's system uses less water on the average.

[32] $H_0: \mu = 15$ in.; $H_1: \mu < 15$ in.; $t_0 = -2.508$; $t = -4.29$; P value is less than 0.005; Reject H_0. The average length of the fish is less than 15 in.

[33] $H_0: p = 0.53$; $H_1: p \neq 0.53$; $z_0 = \pm 1.96$; $z = -2.00$; P value: 0.0456; Reject H_0. The proportion of families owning pets is different from 53%.

[34] (a) $\bar{x} = 11{,}489.375$ mi; $s = 538.8442$ mi
(b) $H_0: \mu = 11{,}372$ mi; $H_1: \mu > 11{,}372$ mi; $t_0 = 1.895$; $t = 0.6161$; P value is greater than 0.125; Do not reject H_0. We cannot conclude that the mean number of miles driven by undergraduate women in the state of Wyoming is greater than the national average.

[35] $H_0: \mu = \$3{,}012.50$; $H_1: \mu < \$3{,}012.50$; $t_0 = -2.718$; $t = -2.4682$; P value is between 0.01 and 0.025; Do not reject H_0. We cannot conclude that the faculty at Eastern State College have a mean credit card balance which is less than the national average.

[36] $H_0: p = 0.03$; $H_1: p > 0.03$; $z_0 = 2.33$; $z = 3.32$; P value: 0.0005; Reject H_0. The proportion of people who develop inhibiting chemicals is higher than 3%.

[37] $H_0: p = 0.34$; $H_1: p \neq 0.34$; $z_0 = \pm 1.96$; $z = 2.07$; P value: 0.0384; Reject H_0. The proportion of businesses which are located in the home is different from 34% in this county.

[38] $H_0: p = 0.10$; $H_1: p \neq 0.10$; $z_0 = \pm 2.58$; $z = 0.50$; P value: 0.617; Do not reject H_0. We cannot conclude that the proportion of tax returns with arithmetic errors in excess of $1,200 is different from 10%.

[39] $H_0: p = 0.27$; $H_1: p \neq 0.27$; $z_0 = \pm 2.58$; $z = 0.50$; P value: 0.617; Do not reject H_0. We cannot conclude that the proportion of cancellations at Ellen's travel agency is different from 27%.

[40] $H_0: p = 0.147$; $H_1: p < 0.147$; $z_0 = -1.645$; $z = -1.08$; P value: 0.1401; Do not reject H_0. We cannot conclude that the percentage of trash consisting of plastic and glass from neighborhoods in which Responsible Citizens have been active is less than the percentage for the country as a whole.

[41] $H_0: p = 0.68$; $H_1: p > 0.68$; $z_0 = 1.645$; $z = 2.48$; P value: 0.0066; Reject H_0. The data indicates that the realtor's claim is too low.

[42] $H_0: p = 0.15$; $H_1: p \neq 0.15$; $z_0 = \pm 1.96$; $z = -1.30$; P value: 0.1936; Do not reject H_0. We cannot conclude that the proportion of cancellations is different from 15%.

[43] $H_0: p = 0.12$; $H_1: p \neq 0.12$; $z_0 = \pm 1.96$; $z = -2.13$; P value: 0.0332; Reject H_0. The proportion of late appointments for this hair stylist is different from 0.12.

[44] $H_0: p = 0.197$; $H_1: p > 0.197$; $z_0 = 1.645$; $z = 0.45$; P value: 0.3264; Do not reject H_0. We cannot conclude that the incidence of violent crime has increased.

[45] (a) If P value = 0.0328 reject the null hypothesis at the 5% significance level.
(b) If P value = 0.0328 do not reject the null hypothesis at the 1% significance level.

[46] H_0: $p_1 = p_2$; H_1: $p_1 < p_2$; $z_0 = -1.645$; $z = -1.11$; P value: 0.1335; Do not reject H_0. We cannot conclude that the proportion of non-union bricklayers who have experienced layoffs is greater than that of union bricklayers.

[47] $H_0: \mu = 5.2$; $H_1: \mu \neq 5.2$; $z_0 = \pm 1.96$; $z = 2.78$; P value: 0.0054; Reject H_0. The average moving time is different from 5.2 years.

[48] With P value 0.0612, Do not reject the null hypothesis at the 5% significance level.

[49] (a) With P value 0.023, do not reject H_0 at the 1% significance level.
(b) Reject H_0 at the 5% significance level.

[50] Yes, if the P value is 0.007, we reject H_0 at the 1% significance level.

[51] (a) If the P value is 0.075 she cannot conclude that the proportion is higher at the 5% significance level.
(b) She cannot conclude that the proportion is higher at the 1% significance level.
(c) The smallest significance level at which she can conclude that the proportion is higher is 7.5%.

[52] No, if the P value is 0.027 the data is not statistically significant at the 1% significance level.

[53] The claim $\mu = 30$ minutes is in question, so the null hypothesis is H_0: $\mu = 30$. The alternate hypothesis is H_1: $\mu > 30$

[54] A type I error has occurred.

[55] A type II error has occurred.

[56] $z = 2.05$; No.

[57] $z = 2.05$; Yes.

[58] B.

[59] A.

[60] C.

[61] C.

[62] C.

[63] H_0: $\mu = 67$; $H_1 < 67$; left-tailed; $t_0 = -2.602$; sample $t = -1.876$; $0.025 < P$ value $<$ 0.050; Do not reject at the 1% level. Evidence does not support a claim that the average avalanche thickness in this U.S. region is less than that in Canada.

[64] H_0: $\mu = 67$; H_1: $\mu < 67$; left-tailed; $t_0 = -1.753$; sample $t = -1.876$; $0.025 < P$ value $<$ 0.050; Reject at the 5% level. Evidence does support a claim that the average avalanche thickness in this U.S. region is less than that in Canada.

[65] H_0: $p = 0.82$; H_1: $p < 0.82$; left-tailed; $z_0 = -2.33$; $\hat{p} = 0.6935$ with $z = -2.5926$. Reject H_0 at the 1% level. Evidence does support the claim that the average proportion of health professionals in this state who would choose their careers again is less than the national rate.

Section B: Algorithms

Objective 10.1: Introduction to Hypothesis Testing

[66] $\sigma^2 < 7.5$, One-tailed

[67] Type II error

[68] No

[69] Not an error

[70] $\mu = 7.7$, One-tailed

[71] The null hypothesis is not rejected and $\mu < 60$ years.

[72] Type I

Objective 10.2: Test Involving the Mean (Large Samples)

[73] -0.5250

[74] 0.6212

[75] Yes

H_0: $\mu = 93$, H_1: $\mu > 100$
$z = 2.77$
[76] Yes

[77] $z = -1.65$; it is significant at a $\alpha = 0.05$ level.

Objective 10.3: The P Value in Hypothesis Testing

(a) 0.0384
[78] (b) Yes

[79] 0.0735

[80] 0.0071

[81] 0.0505

[82] No

[83] $\alpha = 0.01$

[84] 0.0571

[85] 0.0010

Objective 10.4: Tests Involving the Mean (Small Samples)

[86] $t = 2.5$; it is not significant at a $\alpha = 0.01$ level.

[87] $\alpha = 0.05$

[88] Yes

[89] 1.47

[90] 0.2774

Objective 10.5: Tests Involving a Proportion

[91] $z = 8.24$; Reject H_0

[92] $z = 4.32$; Reject H_0

[93] $z = -0.05$

[94] $z = 6.86$

[95] $P = 0.0038$

Chapter 11
Organizing Data

Section A: Static Items

Section B: Algorithms

Objective 11.1: Tests Involving Paired Differences(Dependant Samples)

1. In order to test how the environment effects the ability of a child to understand concepts better, an experiment was conducted on six pairs of identical twins. In the given table, row B represents the writing age in months of the randomly selected member of a twin pair in an experimental group whereas row A represents the writing age in months of the other member of the twin pair who was not part of the experimental group. Using a 5% level of significance, test if there is a difference in the writing ages of the two groups.

	Twin 1	Twin 2	Twin 3	Twin 4	Twin 5	Twin 6
Row A	59	61	65	60	66	58
Row B	70	72	68	61	65	69

2. In the given table let the first row represent the scores of 5 cricketers in their first match of cricket and let the second row represent their scores in their fifth match of cricket. Determine the P-value of the t-test of the scores given below at the 5% level of significance if it has been claimed that the average score for the fifth match was more than that of the first match.

	Player 1	Player 2	Player 3	Player 4	Player 5
First match	34	36	63	78	64
Fifth match	78	77	76	68	62

3. A survey was conducted to study the number of wolves in two mountain regions. The survey was conducted in 7 different parts of the two mountains and the results were recorded in the table shown below. Using a 2.5% level of significance determine if the number of wolves in the second mountain was higher than the first mountain.

	Part 1	Part 2	Part 3	Part 4	Part 5	Part 6	Part 7
First mountain	28	32	29	30	31	28	32
Second mountain	37	32	39	35	41	38	35

4. Tests were conducted on a patient of low blood pressure to determine if an exercise of 14 minutes on the exercising machine will have an effect on his blood pressure and the results were recorded in the table given below. Use a 2% level of significance to test if the blood pressure will be different before and after a 14-minute exercise on the exercising machine used for the experiment.

Test	1	2	3	4	5	6	7
Before	136	132	131	133	130	134	129
After	149	138	137	132	134	146	143

5. The following table gives the results of election polls in 6 towns taken last year and those taken two years ago when a more extensive election campaign had been run by the competing parties. Use a 1% level of significance to determine if there is a decrease in the average number of votes acquired last year as compared to the year before.

	Town 1	Town 2	Town 3	Town 4	Town 5	Town 6
Poll results two years ago	43	75	62	52	38	73
Poll results for last year	44	38	29	55	39	32

6. After an export organization receives a new shipment of goods, the shipment is checked to assess the number of damaged and undamaged goods. The following table shows the number of goods that were damaged and the number of goods that remain undamaged after shipment. Use a 5% level of significance to determine if the number of goods that remained undamaged after shipment were greater than the number of goods that were damaged.

	1st Shipment	2nd Shipment	3rd Shipment
Undamaged goods	3333	3456	3510
Damaged goods	3346	2980	2887
	4th Shipment	5th Shipment	6th Shipment
Undamaged goods	3498	3500	3245
Damaged goods	3500	2380	2470

7. Two construction companies were given the job of constructing two identical buildings that are 20-stories high for a housing organization. Each company took exactly 7 months to finish the job. Row A shows the the number of hours of work utilized by the first company each month. Row B shows the the number of hours of work utilized by the second company each month. Using a 5% level of significance determine if the first company would do a job slower than the second company.

	Job 1	Job 2	Job 3	Job 4	Job 5	Job 6	Job 7
Row A	518	520	519	518	514	517	519
Row B	525	524	529	514	526	516	522

Objective 11.2: Inferences About the Differences of Two Means(Large, Independant Samples)

8. A teacher wants to examine how long it takes a class of sixth grade students to read a book 600 pages long. She divides her class into two groups. The first group consists of 44 students with a mean of 43 hours and standard deviation of 16 hours. The second group consists of 45 students with a mean of 49 hours and standard deviation of 14 hours. The teacher makes a claim that the first group is able to read the book faster than the second group. Is the claim justified at a 5% level of significance ?

9. A scientist designs two methods of experimentation to test the effectiveness of a new medication for people having recurrent headaches. He calculated the time in minutes that each subsequent trial takes for the medication to start showing an effect on the patient. For the first experiment he had taken 51 trials which gave an average of 19 minutes and a standard deviation of 15 minutes. For the second experiment he had taken 52 trials which gave an average of 14 minutes and a standard deviation of 13 minutes. He then claims that the second experiment gave a faster method of using the tested medication than the first experiment. Using a 5% level of significance, determine if his claim is justified.

10. A random sample of 57 days was taken for a mountain in the Alps and it was found that the sample mean for the number of days that it remained snow covered was 37 days and the corresponding sample standard deviation was 14 days. Another random sample of 58 days was taken for another mountain in the Alps and it was found that the sample mean for the number of days that it remained snow covered was 43 days and the corresponding sample standard deviation was 16 days. Would this indicate that the second mountain remained snow covered for a greater number of days than the first mountain? Use a 5% level of significance.

11. A club that offered various camping and fishing expeditions ran a study on people liking each of the outdoor activities in which each answer marked as "very important" was awarded 6 points and each answer marked as "important" was awarded 2 points. A random sample of 183 people that liked camping had a mean of 4.5 points and standard deviation of 1.3 points. A random sample of 187 people that liked fishing had a mean of 4.9 points and standard deviation of 1.4 points. Use a 5% level of significance to determine the *P*-value of the *z*-test if it is claimed that there is more a preference for fishing than for camping.

12. Two random samples of 160 football players and 159 basketball players were used to determine if there is a difference in the average weights in kilograms of football and basketball players. The sample mean of the 160 football players was 81 kilograms and the standard deviation was 23 kilograms. The sample mean of the 159 basketball players was 80 kilograms and the standard deviation was 27 kilograms. Using a 1% level of significance determine if there is a difference between the population means of the averages weight of football and basketball players.

13. A newspaper firm wants to assess the responses to half-page advertisements in the newspaper as compared to full-page advertisements in the newspaper on a scale of 1 to 10 in order to determine the cost effectiveness of each. Assume that the given data is a representative of all their half-page and full-page advertisements.

Half-page advertisements: $n_1 = 154$, $\overline{x_1} = 6.01$, $s_1 = 1.6$

Full-page advertisements: $n_2 = 155$, $\overline{x_2} = 7.05$, $s_2 = 1.4$

Find a 75% confidence interval for the distribution. (Round off to the nearest tenth.)

14. It is believed by an insurance company that the life expectancy of men is more than that of women. Random samples of 254 men and 250 women were taken to compare the life expectancies of the two on an average. It was discovered that the sample mean for the life expectancy of men was 53 years with a sample standard deviation of 36 years whereas the sample mean for the life expectancy of women was 57 years with a sample standard deviation of 32 years. Would this justify the claim made by the insurance company? Explain. Use a 5% level of significance.

15. A random sample of 47 animal shelters in a northern state was compared with a random sample of 48 animal shelters in a southern state. The mean of the sample of animal shelters in the northern state was 27 animals with a standard deviation of 16 animals. The mean of the sample of animal shelters in the southern state was 33 animals with a standard deviation of 18 animals. Use a 5% level of significance, could we conclude that the population mean of the number of animals in animal shelters in the north is less than the population mean of the number of animals in animal shelters in the south?

Objective 11.3: Inferences About the Differences of Two Means(Small, Independant Samples)

16. A survey was made of the number of thefts occurring in different neighborhoods in a metropolitan city in a month. Then the total number of neighborhoods were divided into two groups. The first group comprised of 19 neighborhoods with a mean of 11 thefts in a month and a standard deviation of 3 thefts in a month. The second group comprised of 18 neighborhoods with a mean of 6 thefts in a month and a standard deviation of 3 thefts in a month. Does the survey show that at a 2.5% level of significance the number of thefts in the first group was greater than the number of thefts in the second group?

17. Two random samples of 52 students were taken from two schools to compare the productivity levels of both the schools. The random sample of 25 students from the first school had a sample mean of 7 hours of productive work in a day and a sample standard deviation of 3.7 hours of productive work in a day. The random sample of 27 students from the second school had a sample mean of 8 hours of productive work in a day and a sample standard deviation of 3.3 hours of productive work in a day. Use a 5% level of significance to determine if the productivity of the second school was more than the productivity of the first school.

18. Two water heaters of different makes had to be tested to compare their effectiveness in heating the water required for an experiment. A random sample of 10 temperature readings of the water heater of the first make had a sample mean of $9.9°$ F and a standard deviation of $3.6°$ F. A random sample of 12 temperature readings of the water heater of the second make had a sample mean of $6.3°$ F and a standard deviation of $3.3°$ F. Use a 5% level of significance to determine if the first water heater was more effective in heating the water than the second water heater.

19. A survey was conducted about the frequency of cases of influenza in children from age groups of ages 3 to 4 and ages 5 to 8 to find the average age at which it can occur. The samples had been chosen from different schools in a certain state. A random sample of 24 children from ages 3 to 4 had a mean of 3.1 years and standard deviation of 1.4 years. A random sample of 26 children from ages 5 to 8 had a mean of 4.1 years and standard deviation of 1.7 years. Use a 1% level of significance to determine if there is a difference in the average ages at which it can occur in both the age groups.

20. The annual mean of profit in millions of 29 randomly chosen bicycle manufacturing companies is \$2.9 per year and the standard deviation is \$1.2 per year. Similarly the annual mean of profit in millions of another set of 23 randomly chosen car manufacturing companies is \$3.5 per year and the standard deviation is \$1.6 per year. Use a 2.5% level of significance to determine the *P*-value if the population mean of the car manufacturing companies is assumed to be more than the population mean of the bicycle manufacturing companies.

21. In order to compare the frequency of the number of cars that would pass two highways, random samples of 15 number of days each were chosen, The mean frequency of the number of cars that passed the first highway in 15 number of days was 74 cars with a standard deviation of 44 cars and the mean frequency of the number of cars that passed the first highway in 15 number of days was 65 cars with a standard deviation of 45 cars. Find a 75% confidence interval for population mean of the first highway being greater than the second highway. Round off to the nearest tenth.

22. Experiments are being conducted to find out the amount of time a certain bacteria can survive outside its host. Each day a new set of bacteria is being tested by two different experimental methods. The first experiment comprised of 28 days with a mean of 104 bacteria having survived at the end of the day and a standard deviation of 36 bacteria at the end of the day. The second experiment comprised of 29 days with a mean of 80 bacteria having survived at the end of the day and a standard deviation of 28 bacteria at the end of the day. Does the survey show that at a 2% level of significance there is a difference in the population means of the two groups?

Objective 11.4: Inferences About the Differences of Two Proprotions

23. A random sample of 118 people was taken from the state of Ohio, of which 31 people took regular exercise. Another random sample of 136 people was taken from the state of Miami, of which 36 people took regular exercise. Does this information show that at a 5% level of significance the population proportion of the number of people that take regular exercise in Ohio is lesser than that in Miami?

24. Suppose that two groups of subjects were randomly chosen for a sleep study. In the first group, the subjects drank a glass of milk before going to sleep and in the second group the subjects did not drink a glass of milk before going to sleep and simply went to sleep. In the first group there are 48 subjects of which 21 people slept without any kind of disturbance. In the second group there are 66 subjects of which 26 people slept without any kind of disturbance. Find a 80% confidence interval for the difference of population proportions. Round to the nearest tenth.

25. In a random sample of 385 people, 242 people had gone through some kind of major or minor eye surgery in the last 10 years. In another random sample of 420 people, 217 people had gone through some kind of major or minor leg surgery in the last 10 years. Using a 1% level of significance find the *P*-value of the *z*-test if it is assumed that the number of people that go through eye surgery is more than the proportion of people that go through leg surgery.

26. Of a random sample of 160 American adults, 68 adults claim that they prefer drinking regular soft drinks to drinking diet soft drinks and of a random sample of 120 American adults 64 adults, claim that they prefer drinking diet soft drinks to drinking regular soft drinks. Would this mean that the proportion of American adults that drink regular soft drinks is less than that of those who drink diet soft drinks? Use a 1% level of significance.

27. A test with only multiple choice questions with only two options each was given to students of a school. Of a random sample of 11 students, 7 students answered more than 50% of the paper correctly and of a random sample of 13 students, 5 students answered less than 50% of the paper correctly. Would this mean that the population proportion of students that would answer more than 50% of the paper correctly was more than that of those that answer less than 50% of the paper correctly? Use a 5% level of significance.

28. A survey was conducted about the number of people that have completed their higher education. The study was conducted in two-ways. A random sample of 141 people were interviewed face-to-face, of which 89 answered that they had completed their higher education. A random sample of 134 people were asked the same question in a questionnaire, of which 68 answered that they had completed their higher education. Let these samples represent the general population. These responses were then checked for accuracy. Use a 5% level of significance to determine if the responses to first method were more accurate than the responses to the second method of survey.

29. Of a random sample of 200 adults, 68 are able to donate blood, and of a random sample of 150 adults, 72 are not able to donate blood due to medical reasons. If these samples represent the entire population, then use a 5% level of significance to determine if the population proportion of adults who are able to donate blood is lesser than those who are not able to donate blood due to medical reasons.

30. Marie wanted to compare the number of vowels in each of the two books she had been asked to read at school. She randomly chose a paragraph from each book to compare the data. There were 120 words in the paragraph she had chosen from the first book with 16 vowels. There were 111 words in the paragraph she had chosen from the first book with 17 vowels. She then used a 5% level of significance to determine if there was a difference between the number of vowels in each book. Use a 5% level of significance to determine the conclusion she must reach while comparing the content of the two books.

Answer Key for Chapter 11
Organizing Data

Section A: Static Items

Section B: Algorithms

Objective 11.1: Tests Involving Paired Differences(Dependant Samples)

[1] Since the t-value corresponding to the sample mean falls within the critical region we can reject the null hypothesis and conclude that at the 5% level of significance there is a difference in the writing ages of the two groups.

[2] $0.025 < P < 0.05$

[3] Since the t-value corresponding to the sample mean falls within the critical region we can reject the null hypothesis and conclude that at the 2.5% level of significance, the number of wolves in the second mountain was higher than the first mountain.

[4] Since the t-value corresponding to the sample mean falls within the critical region we can reject the null hypothesis and conclude that at the 2% level of significance, the blood pressure will be different before and after a 14-minute exercise on the exercising machine used for the experiment.

[5] Since the t-value corresponding to the sample mean does not fall within the critical region we cannot reject the null hypothesis and hence cannot confidently conclude that at the 1% level of significance whether there is any decrease in the average number of votes acquired last year as compared to the year before.

[6] Since the t-value corresponding to the sample mean falls within the critical region we can reject the null hypothesis and conclude that at the 5% level of significance, the number of goods that remained undamaged after shipment was greater than the number of goods that were damaged.

[7] Since the t-value corresponding to the sample test statistic does not fall within the critical region we cannot reject the null hypothesis and hence cannot confidently conclude that at the 5% level of significance whether the first company would do a job faster or slower than the second company.

Objective 11.2: Inferences About the Differences of Two Means(Large, Independant Samples)

[8] Since the z value corresponding to the sample test statistic falls within the critical region we can reject the null hypothesis and conclude that at the 5% level of significance the first group is able to read the book faster than the second group.

[9] Since the z value corresponding to the sample test statistic falls within the critical region, we can reject the null hypothesis and conclude that at the 5% level of significance the second experiment uses a faster method than the first experiment.

[10] Since the z value corresponding to the sample test statistic does not fall within the critical region, we cannot reject the null hypothesis and hence cannot confidently conclude that at the 5% level of significance whether the second mountain remained snow covered for a greater number of days than the first mountain.

[11] $P = 0.0022$

[12] Since the z value corresponding to the sample test statistic does not fall within the critical region, we cannot reject the null hypothesis and cannot confidently conclude that at the 1% level of significance whether the average weight of the football players is lesser than or greater than that of the basketball players.

[13] $-1.2 < \mu_1 - \mu_2 < -0.8$

[14] Since the z value corresponding to the sample test statistic does not fall within the critical region, we cannot reject the null hypothesis and cannot confidently conclude that at the 5% level of significance whether the life expectancy of men is more than that of women.

[15] Since the z value corresponding to the sample test statistic falls within the critical region, we can reject the null hypothesis and conclude that at the 5% level of significance the number of animals in animal shelters in the north is less than the number of animals in animal shelters in the south.

Objective 11.3: Inferences About the Differences of Two Means(Small, Independant Samples)

[16] Since the t-value corresponding to the sample test statistic falls within the critical region, we can reject the null hypothesis and conclude that at the 2.5% level of significance the first group has a greater number of thefts than the second group.

[17] Since the t-value corresponding to the sample test statistic does not fall within the critical region, we cannot reject the null hypothesis and cannot confidently conclude that at the 5% level of significance whether the productivity of the second school was more than the productivity of the first school.

[18] Since the t-value corresponding to the sample test statistic falls within the critical region, we can reject the null hypothesis and conclude that at the 5% level of significance the first water heater was more effective in heating the water then the second water heater.

[19] Since the t-value corresponding to the sample test statistic does not fall within the critical region, we cannot reject the null hypothesis and cannot confidently conclude that at the 1% level of significance whether average age in the age group of ages 3 to 4 has a greater or lesser possibility than the age group of ages 5 to 8.

[20] $0.01 < P < 0.025$

[21] $-10.1 < \mu_1 - \mu_2 < 28.1$

[22] Since the t-value corresponding to the sample test statistic does not fall within the critical region, we cannot reject the null hypothesis and cannot confidently conclude that at the 2% level of significance whether the number of bacteria surviving at the end of the day in the first experiment was greater than or lesser than the number of bacteria surviving at the end of the day in the second experiment.

Objective 11.4: Inferences About the Differences of Two Proprotions

[23] Since the z value corresponding to the sample test statistic does not fall within the critical region, we cannot reject the null hypothesis and cannot confidently conclude that at the 5% level of significance whether population proportion of Ohio is lesser than the population proportion of Miami.

[24] $-5.1 < \mu_1 - \mu_2 < -4.9$

[25] $P = 0.0007$

[26] Since the z value corresponding to the sample test statistic does not fall within the critical region, we cannot reject the null hypothesis and cannot confidently conclude that at the 1% level of significance whether proportion of American adults that drink regular soft drinks is less than that of those who drink diet soft drinks.

[27] Since the z value corresponding to the sample test statistic does not fall within the critical region, we cannot reject the null hypothesis and cannot confidently conclude that at the 5% level of significance whether the proportion of students that would answer more than 50% of the paper correctly was more than that of those that answer less than 50% of the paper correctly.

[28] Since the z value corresponding to the sample test statistic falls within the critical region, we can reject the null hypothesis and conclude that at the 5% level of significance the proportion of responses to the first method was more accurate than the responses to the second method of survey.

[29] Since the z value corresponding to the sample test statistic falls within the critical region, we can reject the null hypothesis and conclude that at the 5% level of significance the population proportion of adults who are able to donate blood is lesser than those who are not able to donate blood due to medical reasons.

[30] Since the z value corresponding to the sample test statistic does not fall within the critical region, we cannot reject the null hypothesis and cannot confidently conclude that at the 5% level of significance whether number of vowels in the first book was less than or more than the number of vowels in the second book.

Chapter 12
Additional Topics Using Inference

Section A: Static Items

Section B: Algorithms

Objective 12.1: Chi Square: Tests of Independance

1. Give the size of the contingency table shown below.

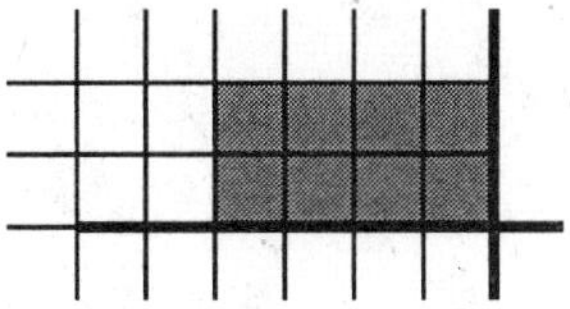

2. Three different headache medicines, A, B, and C were administered to a group of volunteers who suffered from migraines. The amount of time, in hours, that it took for each person to recover from a migraine was recorded in the contingency table below. Find the expected frequency for cell 2 of the contingency table.
   Medicine versus Time to Recover from Migraine.

Medicine	0 – 1 hr	1 – 2 hr	2 – 3 hr	Row Total
A	26	29	27	82
B	28	73	20	121
C	37	47	13	97
Column total	91	149	60	300

3. In a contingency table that has 5 rows and 5 columns, determine the number of degrees of freedom.

4. Eye color is a physical trait that is genetically determined. Each person carries two types of genes for eye color. The gene for brown eyes (B) is dominant over (b), the gene for blue eyes. According to genetics theory, if a man with brown eyes (Bb) and woman with brown eyes (Bb) have a child, the probabilities that the child will have genes (BB), (Bb), or (bb) is given by the table below.

	Father	
Mother	B	b
B	BB	Bb
b	Bb	bb

To test this theory, geneticists took a random sample of 200 children whose mothers and fathers were both (Bb). Their results were as follows: 49 had (BB), 86 had (Bb), and 65 had (bb).

(a) Compute the sample statistic χ^2.

(b) Find the critical value $\chi^2_{0.01}$ for a 0.01 level of significance.

5. For the chi-squared distribution, find $\chi^2_{0.05}$, the mean, and variance for 5 degrees of freedom.

Objective 12.2: Chi Square: Goodness of Fit

6. A game wheel has 36 spaces. There are 9 empty spaces, 16 spaces with 1 point, and 11 spaces with 5 points. The results of 100 spins are listed in the table below. What is the value of the test statistic χ^2 you would use to test if the wheel were out of balance?

Space	Frequency
Empty	30
1 point	49
5 points	21

7. According to recent research, the distribution of the number of children per family in the U.S. is listed in the table below on the left. A random sample of 1000 families with both parents under 40 yielded the data in the table on the right.

National Average		Random Sample	
Number of Children	Percent	Number of Children	Number of Families
More than 3	26.0	More than 3	254
3	20.9	3	211
2	19.5	2	198
1	19.3	1	192
0	14.3	0	145

Calculate χ^2. At a 1% significance level, does it appear that the distribution of the number of children per family for families with both parents under 40 is different from that of families in the U.S. as a whole?

8. A spinner has 25 spaces. There are 7 white spaces, 15 red spaces, and 3 blue spaces. The results of 100 spins are listed in the table below. What is the value of the test statistic χ^2 you would use to test if the spinner was out of balance?

Color	Frequency
White	34
Red	54
Blue	12

9. According to a recent marketing research, the distribution of the ages of people that buy compact discs in the U.S. is listed in the table below. A random sample of 750 people is also presented in the table.

Age in years	Percent of Customers in the Report	Number of Customers in the Report
Less than 20	40.7	190
21-30	21.8	163
31-40	16.1	364
More than 40	21.4	33

Calculate χ^2. At a 5% significance level, does it appear that the distribution of the ages of the customers in the report is different from that of customers in the U.S. as a whole?

10. In the table, let x be the average daily temperature in Fahrenheit in the month of May. A 10-year study produced the given entries in the right column of the table. Use a 1% level of significance to test the claim that the average daily temperature in May has a distribution no different from the sample taken.

Temperature in Fahrenheit	Expected % of temperature	Observed number of days in 10 years
$51 \le x \le 60$	23.0%	29
$61 \le x \le 70$	10.6%	63
$71 \le x \le 80$	25.1%	156
$81 \le x \le 90$	22.3%	136
$91 \le x \le 99$	19.0%	116

Objective 12.3: Testing a Single Variance or Standard Deviation

11. Find the χ^2 value so that 0.5% of the area under the χ^2 curve is to the right of χ^2 when $d.f. = 24$.

12. Find the χ^2 value so that 1% of the area under the χ^2 curve is to the left of χ^2 when $d.f. = 25$.

13. Find the χ^2 value so that 0.5% of the area under the χ^2 curve is to the right of χ^2 when $d.f. = 29$.

14. In an examination a student can obtain a number of 400 possible points and it is generally assumed that the standard deviation is 67 points. A random sample of 15 students gives a standard deviation of 66 points. Use a 10% level of significance to test the claim that the population standard deviation is different from 67 points.

15. Find the χ^2 value so that 97.5% of the area under the χ^2 curve is to the left of χ^2 when $d.f. = 23$.

Objective 12.4: Linear Regression: Confidence Intervals for Predictions

16. Some students decided to take a college entrance exam after first using some practice software for the test. The table below shows the time spent practicing and the percent of gain compared to the first test score.

Practice Time (hr)	4.5	5.5	7	8.5	10	12	14.5
Gain (%)	6.2	8.3	9.1	10.6	11.2	13.8	18.0

a. Find an equation of the line of best fit for the data. Use x to represent practice time and y to represent percent of gain. Round coefficients to the nearest hundredth.
b. Calculate the predicted value of percent of gain for a practice time of 26.1 hours. Round coefficients to the nearest hundredth.

17. The data graphed below came from an experiment that involved the time spent studying and test scores for a number of students.

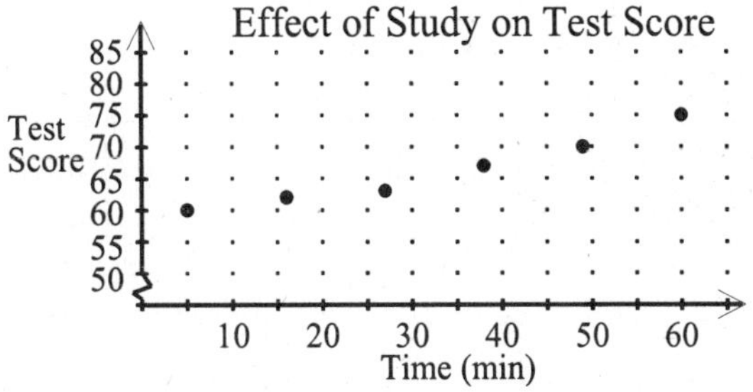

Write the equation that expresses a student's test score as a function of time spent studying. Then find the predicted test score of the student at a given time of 110 minutes. If necessary, round the coefficients to the nearest hundredth.

18. In the following table, x represents the depth in meters of a dive made by a diver and y represents the number of hours it takes the diver to dive down to that depth and then return to the surface. Find the linear regression equation for the given data and then find the standard error of estimate for the data set. Round to the nearest thousandth.

x	134	200	178	187	219	198	147
y	5	7	4	6	7	4	5

19. In the following table, x represents the number of questions David has answered more than Raina in the school exam and y represents the percent of times David gets a higher grade than Raina. Find the linear regression equation for the given data and then find the standard error of estimate for the data set. Round to the nearest thousandth.

x	13	15	20	17	12	18	16
y	21	12	15	23	18	22	14

20. In the following table, x represents the lengths of 8 ball rings in millimeters and y represents their corresponding weights in ounces. Find the linear regression equation for the given data and then find the standard error of estimate for the data set. Round to the nearest thousandth.

x	33	34	27	21	32	20	22	29
y	8	10	12	11	13	9	11	10

21. In the following table, x represents the weight in pounds of a mail order to Oregon and y represents the shipping and handling costs in dollars that 7 different mail-order companies would charge. Find the linear regression equation for the given data. Round coefficients to the nearest thousandth. Then find a 95% confidence interval for a package that has a weight of 34 pounds. Round to the nearest hundredth.

x	25	29	27	21	26	24	30
y	5.7	5.5	6.4	5.9	5.3	6.7	6.15

22. A T-shirt manufacturing company sells its T-shirts to a number of retail outlets in the city. In the following table, x represents the number of T-shirts sold to 7 such retail outlets and y represents the percentage of T-shirts that are sold during the season. Find the linear regression equation for the given data. Round coefficients to the nearest thousandth. Find a 90% confidence interval for 450 number of T-shirts. Round to the nearest hundredth.

x	200	350	100	150	250	300	400
y	84	80	79	78	82	75	77

Objective 12.5: Testing the Correlation Coefficient and the Slope

23. In the following table, x represents the percentage of people who have completed their higher education and y represents their annual salary in thousands of dollars. Use a 5% level of significance to test the claim that $\rho < 0$.

x	8.6	10.1	11.8	9.5	10.3	11.2	11.3
y	54.6	40.4	29.2	41.4	31.2	18	5.8

24. In the following table, x represents the length of a fish in centimeters and y represents its weight in ounces. Use a 5% level of significance to test the claim that $\rho = 0$.

x	7.3	6.7	10.1	9.8	5.6	6.3	7.9
y	5.3	4.1	6.2	5.9	3.3	3.5	5.7

25. In the following table, y represents the cost price of a new model of a car and x represents the corresponding discounted sale price of the car. Use a 5% level of significance to estimate the P-value for the claim that $\rho < 0$.

x	2663	2762	2918	2724	2586	2901	2474
y	2441	2177	1917	2178	2433	2181	2450

26. In the following table, x represents the average percentage of change in prices of consumer goods and y represents the average percentage change in expenditure of a family. Use a 2.5% level of significance to test the claim that $\rho > 0$.

x	3.1	2.2	2.4	4	2.6	2.9	3.9
y	12.2	7	11.2	14.4	11.2	16.4	19.6

27. In the following table, x represents the per capita sales of a wholesale outlet in thousands and y represents the per capita income in thousands of dollars of the outlet. Use a 5% level of significance to estimate the P-value for the claim that $\beta = 0$.

x	15.2	16.4	14.5	15.6	16.1	15	15.2
y	25.2	24.1	25.5	18.8	19.1	24.3	23.5

28. In the following table, x represents the average number of working hours in a week of a part-time faculty member of a school and y represents the average annual salary in thousands of dollars of the respective faculty member. Use a 2% level of significance to test the claim $\beta = 0$.

x	32.5	31	30.5	27.5	25	22.5	20
y	43.5	44	42.5	41.5	39	38.5	39.5

29. A certain amount of money is required to be paid in advance at a club. Every time a person uses the club a small amount of money is deducted from the initial deposited. In the following table x represents the number of times a person visits the club and y represents the amount of money that has been deducted from his or her initial deposit. Use a 5% level of significance to test the claim $\beta < 0$.

x	5	9	7	8	15	11	13
y	59	42	56	38	19	36	22

30. In the following table, x represents the original cost price of manufacturing the new models of a car and y represents the discounted price of the car after it has been in the market for 2 years. Use a 2.5% level of significance to test the claim $\beta > 0$.

x	2090	2060	2110	2100	2080	2070	2120
y	2395	2170	2445	2245	1995	1720	1970

Answer Key for Chapter 12
Additional Topics Using Inference

Section A: Static Items

Section B: Algorithms

Objective 12.1: Chi Square: Tests of Independance

[1] 2×4

[2] 40.73

[3] 16

[4] (a) 6.48
(b) 9.21

[5] $\chi^2_{0.05} = 11.07,\ \mu = 5,\ \sigma^2 = 10$

Objective 12.2: Chi Square: Goodness of Fit

[6] 4.455

[7] 0.236908; No

[8] 1.886

[9] 634.83; Yes

[10] Yes

Objective 12.3: Testing a Single Variance or Standard Deviation

[11] 45.56

[12] 11.52

[13] 52.34

[14] Since the observed value does not lie in the critical region, we cannot reject the null hypothesis and conclude that there is no difference between the population proportion and the sample proportion.

[15] 38.08

Objective 12.4: Linear Regression: Confidence Intervals for Predictions

[16] a. $y = 1.07x + 1.56$
b. 29.49 %

[17] $S = 0.27T + 58.65$; 88.35

[18] $y = 0.020x + 1.896$; 1.233

[19] $y = -0.046x + 18.582$; 4.707

[20] $y = -0.009x + 10.744$; 1.731

[21] $y = -0.020x + 6.461$; $3.68 \le y \le 7.90$

[22] $y = -0.008x + 81.250$; $69.28 \le y \le 86.15$

Objective 12.5: Testing the Correlation Coefficient and the Slope

[23] As the t-value lies within the critical region, we reject the null hypothesis and confidently conclude that at the 5% level of significance there is a negative correlation.

[24] As the t-value lies within the critical region, we reject the null hypothesis and confidently conclude that at the 5% level of significance there is a positive correlation.

[25] $0.005 < P < 0.01$

[26] As the t-value lies within the critical region, we reject the null hypothesis and confidently conclude that at the 2.5% level of significance there is a positive correlation.

[27] $0.2 < P < 0.25$

[28] As the *t*-value lies within the critical region, we reject the null hypothesis and confidently conclude that at the 2% level of significance there is a positive slope.

[29] As the *t*-value lies within the critical region, we reject the null hypothesis and confidently conclude that at the 5% level of significance there is a negative slope

[30] As the *t*-value does not lie within the critical region, we cannot reject the null hypothesis and cannot confidently conclude that at the 2.5% level of significance there is a positive slope